Continuous Distributions

W9-BZX-439

Beta
$0 < \alpha$
$0 < \beta$

$$f(x) = \frac{\Gamma(\alpha + \beta)}{\Gamma(\alpha)\Gamma(\beta)} x^{\alpha-1}(1-x)^{\beta-1}, \qquad 0 < x < 1$$

$$\mu = \frac{\alpha}{\alpha + \beta}, \qquad \sigma^2 = \frac{\alpha\beta}{(\alpha + \beta + 1)(\alpha + \beta)^2}$$

Chi-square
$\chi^2(r)$
$r = 1, 2, \ldots$

$$f(x) = \frac{1}{\Gamma(r/2)2^{r/2}} x^{r/2-1}e^{-x/2}, \qquad 0 \le x < \infty$$

$$M(t) = \frac{1}{(1-2t)^{r/2}}, \qquad t < \frac{1}{2}$$

$$\mu = r, \qquad \sigma^2 = 2r$$

Exponential
$0 < \theta$

$$f(x) = \frac{1}{\theta} e^{-x/\theta}, \qquad 0 \le x < \infty$$

$$M(t) = \frac{1}{1 - \theta t}, \qquad t < \frac{1}{\theta}$$

$$\mu = \theta, \qquad \sigma^2 = \theta^2$$

Gamma
$0 < \alpha$
$0 < \theta$

$$f(x) = \frac{1}{\Gamma(\alpha)\theta^\alpha} x^{\alpha-1}e^{-x/\theta}, \qquad 0 \le x < \infty$$

$$M(t) = \frac{1}{(1-\theta t)^\alpha}, \qquad t < \frac{1}{\theta}$$

$$\mu = \alpha\theta, \qquad \sigma^2 = \alpha\theta^2$$

Normal
$N(\mu, \sigma^2)$
$-\infty < \mu < \infty$
$0 < \sigma$

$$f(x) = \frac{1}{\sigma\sqrt{2\pi}} e^{-(x-\mu)^2/2\sigma^2}, \qquad -\infty < x < \infty$$

$$M(t) = e^{\mu t + \sigma^2 t^2/2}$$

$$E(X) = \mu, \qquad \mathrm{Var}(X) = \sigma^2$$

Uniform
$U(a, b)$
$-\infty < a < b < \infty$

$$f(x) = \frac{1}{b - a}, \qquad a \le x \le b$$

$$M(t) = \frac{e^{tb} - e^{ta}}{t(b - a)}, \qquad t \ne 0; \qquad M(0) = 1$$

$$\mu = \frac{a + b}{2}, \qquad \sigma^2 = \frac{(b - a)^2}{12}$$

80025 75540

RELATED TITLES OF INTEREST FROM PRENTICE HALL

Mathematical Statistics, Vol. I
Peter Bickel and Kjell Doksum
0-13-850363-X

John E. Freund's Mathematical Statistics with Applications 7e
Irwin Miller and Maryless Miller
0-13-142706-7

An Introduction to Mathematical Statistics and Its Applications 4e
Richard Larsen and Morris Marx
0-13-186793-8

Statistics and Data Analysis
Ajit Tamhane and Dorothy Dunlop
0-13-744426-5

Introduction to Mathematical Statistics, 6e
Hogg/Craig/McKean
0-13-008507-3

Miller and Freund's Probability and Statistics for Engineers 7e
Richard Johnson
0-13-143745-3

Probability and Statistics for Engineers and Scientists 7e
Ron Walpole, Ray Myers, Sharon Myers, and Keying Ye
0-13-041529-4

Applied Statistics for Engineers and Scientists
Joe Petruccelli, B. Nandram, and M. Chen
0-13-565953-1

Applied Statistics for Engineers and Scientists: Using Excel and Minitab
David Levine, Patricia Ramsey, and Robert Smidt
0-13-488801-4

Time Series Analysis: Forecasting & Control 3e
George Box, Gwilyn Jenkins, and Greg Reinsel
0-13-060774-6

Contact **www.prenhall.com** to order any of these titles

Applied Multivariate Statistical Analysis 5e
Richard Johnson and Dean Wichern
0-13-092553-5

A Second Course in Statistics: Regression Analysis 6e
William Mendenhall and Terry Sincich
0-13-022323-9

Biostatistics: How it Works
Steve Selvin
0-13-046616-6

Statistics for the Life Sciences 3e
Jeff Witmer and Myra Samuels
0-13-041316-X

Statistics with Applications to the Biological and Health Sciences 3e
Tony Schork and Richard Remington
0-13-022327-1

Biostatistical Analysis 4e
Jerrold Zar
0-13-081542-X

Fundamentals of Probability 3e
Saeed Ghrahramani
0-13-145340-8

A First Course in Probability 7e
Sheldon Ross
0-13-185662-6

Introduction to Stochastic Processes with Biology Applications
Linda Allen
0-13-035218-7

Mathematics of Medical Imaging
Charles Epstein
0-13-067548-2

Contact **www.prenhall.com** to order any of these titles

PROBABILITY
AND
STATISTICAL INFERENCE

SEVENTH EDITION

Robert V. Hogg

University of Iowa

Elliot A. Tanis

Hope College

PEARSON

Prentice
Hall

Upper Saddle River, NJ 07458

Library of Congress Cataloging-in-Publications Data

Hogg, Robert V.
    Probability and statistical inference / Robert V. Hogg,
      Elliot A. Tanis-7th ed.
        p. cm
    Includes bibliographical references and index.
     ISBN 0-13-146413-2
        1. Probabilities. 2. Mathematical statistics. I. Tanis, Elliot A.
          II. Title.
  CIP data available

Executive Acquisitions Editor: *George Lobell*
Editor-in-Chief: *Sally Yagan*
Production Editor: *Jeanne Audino/Bayani Mendoza de Leon*
Assistant Managing Editor: *Bayani Mendoza de Leon*
Senior Managing Editor: *Linda Mihatov Behrens*
Executive Managing Editor: *Kathleen Schiaparelli*
Assistant Manufacturing Manager/Buyer: *Michael Bell*
Marketing Manager: *Halee Dinsey*
Marketing Assistant: *Joon Moon*
Managing Editor, Digital Supplements: *Nicole M. Jackson*
Media Production Editor: *Donna Crilly*
Director of Creative Services: *Paul Belfanti*
Art Director: *Geoffrey Cassar*
Interior Designer: *Kristine Carney*
Cover Designer: *DaSa Design*
Art Editor: *Thomas Benfatti*
Editorial Assistant: *Jennifer Urban*
Manager, Cover Visual Research & Permissions: *Karen Sanatar*
Cover Image: *Lino Tagliapietra, Silea, 2000. Hand-blown colored glass with inserts
of multiple colored filigree zanirico canes, h. 50 × 28 × 18 cm.
Photo: Russell Johnson.*

© 2006, 2001, 1997 by Pearson Education, Inc.
Pearson Prentice Hall
Pearson Education, Inc.
Upper Saddle River, NJ 07458

Previous editions copyright 1993, 1988, 1983, 1977 by Macmillan Publishing
Company, a division of Macmillan, Inc.

All rights reserved. No part of this book may be reproduced, in any form or by any means, without
permission in writing from the publisher.

Pearson Prentice Hall™ is a trademark of Pearson Education, Inc.

Printed in the United States of America

10 9 8 7 6 5 4 3 2 1

ISBN: 0-13-146413-2

Pearson Education, Ltd., *London*
Pearson Education Australia PTY. Limited, *Sydney*
Pearson Education Singapore, Pte., Ltd
Pearson Education North Asia Ltd, *Hong Kong*
Pearson Education Canada, Ltd., *Toronto*
Pearson Education de Mexico, S.A. de C.V.
Pearson Education—Japan, *Tokyo*
Pearson Education Malaysia, Pte. Ltd

CONTENTS

PROLOGUE

The discipline of statistics deals with the collection and the analysis of data. The advances in computing technology, particularly in relationship to changes in science and business, has increased the need for more statistical scientists to examine the huge amount of data being collected. We know that data is not equivalent to information. Once data (hopefully of high quality) are collected, there is a strong need for statisticians to make sense of these data. That is, data must be analyzed, providing information upon which decisions can be made. In light of this great demand, opportunities for the discipline of statistics have never been greater, and there is a special need for more bright young persons going into statistical science.

If we think of fields in which data play a major part, the list is almost endless: accounting, actuarial science, atmospheric science, biological science, economics, educational measurement, environmental science, epidemiology, finance, genetics, manufacturing, marketing, medicine, pharmaceutical industries, psychology, sociology, sports, and on and on. Due to all of these areas in which statistics is useful, it really should be taught as an applied science. Nevertheless, to go very far in such an applied science, it is necessary to understand the importance of creating models for each situation under study. Now no model is ever exactly right, but some are extremely useful as an approximation to the real situation. Most appropriate models in statistics require a certain mathematical background in probability. Accordingly this textbook, while alluding to applications in the examples and the exercises, is really about the mathematics needed for the appreciation of probabilistic models necessary for statistical inferences.

In a sense, statistical techniques are really the heart of the scientific method. Observations are made that suggest conjectures. These conjectures are tested and data collected and analyzed, providing information about the truth of the conjectures. Sometimes these are supported by the data, but often the conjectures need to be modified and more data collected to test the modifications and so on. Clearly, in this iterative process, statistics plays a major role with its emphasis on proper design and analysis of experiments and the resulting inferences upon which decisions can be made. Through statistics, information is provided for taking certain actions: for illustrations, improving manufactured products, providing better services, marketing new products or services, forecasting energy needs, classifying diseases better, and so on.

Statisticians recognize that there are often small errors in their inferences, and they attempt to make the probabilities of those mistakes as small as possible. The reason that these uncertainties even exist is due to the fact that there is variation in the data. Even though experiments are repeated under seemingly the same conditions, the results vary from trial to trial. We try to improve the quality of the data by making them as reliable as possible, but the data simply do not fall on given patterns. There is variation in almost all processes. In light of this uncertainty, the statistician tries to determine the pattern in the best possible way, while always explaining the error structures of the statistical estimates.

This is an important lesson to be learned: Variation is almost everywhere. It is the statistician's job to understand the variation. Often, as in manufacturing, the desire is to reduce variation because the products will be more consistent. In other words, the "car doors will fit better" in the manufacturing of automobiles if the variation is decreased by making each door closer to its target values.

Many statisticians in industry have stressed the need of "statistical thinking" in everyday operations. It is based upon three points, two of which have been mentioned in the preceding paragraph: (1) Variation exists in all processes, (2) Understanding and reducing undesirable variation is a key to success, and (3) All work occurs in a system of interconnected processes. W. Edwards Deming, an esteemed statistician and quality improvement guru, stressed these three points, particularly the third one. He would carefully note that you could not maximize the total operation by maximizing the individual components unless they are independent of each other. However, in most instances, they are highly dependent; and persons in different departments must work together in creating the best products and services. If not, what one unit does to better itself could very well hurt others. He often cited an orchestra as an illustration of the need of the members to work together to create an outcome that is consistent and desirable.

Any student of statistics should understand the nature of variability and the necessity of creating probabilistic models for that variability. We cannot avoid making inferences and decisions in the face of this uncertainty; however, these results are greatly influenced by the probabilistic models selected. Some persons are better model builders than others and accordingly will make better inferences and decisions. The assumptions needed for each statistical model are carefully examined and hopefully the reader will become a better model builder.

Finally, we must mention how dependent modern statistical analyses are upon the computer. The statisticians and computer scientists really should work together in areas of exploratory data analysis and "data mining." Statistical software development is critical today, for the best of it is needed in complicated data analyses. In light of this growing relationship between these two fields, it is good advice for bright students to take substantial offerings in statistics and in computer science.

Students majoring in statistics, computer science, or in a joint program are in great demand in the workplace or in graduate programs. Clearly, they can earn advanced degrees in statistics or computer science or both. But, more important, they are highly desirable candidates for graduate work in other areas: actuarial science, industrial engineering, finance, marketing, accounting, management science, psychology, economics, law, sociology, medicine, health sciences, etc. So many of these fields have been "mathematized" that their programs are begging for majors in statistics and/or computer science. Often these students become "stars" in these other areas. We truly hope that we can interest students enough so that they want to study more statistics. If they do, the opportunities for very successful careers are numerous.

We recognize the importance of the computer in modern statistics by including supplements that usually involve the use of the computer. Some of these illustrate the use of Minitab for calculating probabilities, analyzing data, and applications in statistical inference such as tests of hypotheses and confidence intervals. The power of a computer algebra system (CAS) for theoretical computations is illustrated using *Maple*. The importance of simulation is also demonstrated. Some of these illustrations are printed in the text. There are also many new illustrations on the enclosed CD-ROM and you are encouraged to check these out.

While this book is written primarily as a mathematical introduction to probability and statistics, there are a great many examples and exercises concerned with applications. For illustrations, the reader will find applications in the areas of biology, education, economics, engineering, environmental studies, exercise science, health science, manufacturing, opinion polls, psychology, sociology, and sports. That is, there are many exercises in the text, some illustrating the mathematics of probability and statistics but a great number are concerned with applications.

Different from most textbooks, we have included a prologue, a centerpiece, and an epilogue. The main emphasis in these is that variation occurs in almost every process, and the study of probability and statistics helps us understand this variability. Accordingly, the study of statistics is extremely useful in many fields of endeavor and can lead students to interesting positions in the future.

FEATURES

Throughout the book, figures and real applications will help the student understand probability and statistics and what they can accomplish. For some exercises, it is assumed that calculators or computers are available; thus the solutions will not always involve "nice" numbers. The data sets for all of the exercises are available on the enclosed CD-ROM. The data are provided in different formats so that they should be accessible to most computer programs. Finally, in the first part of the book concerning probability, there are supplementary comments inserted about statistics and computation.

ANCILLARIES

A **Solutions Manual** containing worked out solutions to the even-numbered exercises in the text is available to instructors from the publisher. Many of the numerical exercises were solved using *Maple*. For additional exercises that involve simulations, a separate manual, *Probability & Statistics: Explorations with MAPLE*, second edition, by Zaven Karian and Elliot Tanis, is available for purchase. Several exercises in that manual also make use of the power of *Maple* as a computer algebra system.

The **CD-ROM** contains all of the data sets in various formats. There are also applications of *Minitab* for drawing figures, calculating probabilities, calculating characteristics of a sample, and statistical inference. One folder contains some supplementary *Maple* programs that are useful in probability and statistics. It is these programs that were used for constructing the figures in this text and for other applications. *Maple* was also used to animate some of the figures in the text. All you need is a browser. Simply load in the directory and you can pull up these animations.

If you find any errors in this text, please send them to tanis@hope.edu so that they can be corrected in a future printing. The **errata** will also be posted on http://www.math.hope.edu/tanis/.

ACKNOWLEDGMENTS

We wish to thank our colleagues, students, and friends for many suggestions and for their generosity in supplying data for exercises and examples. In particular we would like to acknowledge the helpful reviews and suggestions of Ching-Yuan Chiang, Richard Dudley, Mark Ghamsary, Paul Joyce, YongHee Kim-Park, Tom Rousseau, David Snyder, and Randall Swift. We also thank the University of Iowa and Hope College for providing office space and encouragement. Finally, our families, through seven editions, have been most understanding during the preparation of all of this material. We would especially like to thank our wives, Ann and Elaine. We truly appreciate their patience and needed their love.

Robert V. Hogg
robert-hogg@uiowa.edu

Elliot A. Tanis
tanis@hope.edu

PREFACE

We are pleased with the reception that was given to the first six editions of *Probability and Statistical Inference*. The seventh edition is still designed for use in a course having from three to six semester hours of credit. No previous study of probability or statistics is assumed, and a standard course in calculus provides an adequate mathematical background. Certain sections have been starred and they are not needed in subsequent sections. This, however, does not mean that these starred sections are unimportant, and we hope many of you will study them.

We still view this book as the basis of a junior or senior level course in the mathematics of probability and statistics that is taught by many departments of mathematics and/or statistics. In particular, the first five chapters provide a substantial one-semester course in probability and probability distributions. While many statisticians are teaching a course at this level by minimizing probability and concentrating on statistics, we have found that those studying statistics, actuarial science, electrical engineering, economics, finance, genetics, and so on need probability as much as statistics. Thus we choose to place about equal emphasis on these two topics. Chapters 6–10 consist of the second semester of a two-semester sequence as they cover topics in statistics and statistical inference. We have discovered that a fairly good four-semester-hour course can be constructed by an instructor by selecting topics from the first five chapters and Chapters 6 and 8.

We have tried to make the seventh edition more "user friendly"; yet we do want to reinforce certain basic concepts of mathematics, particularly calculus. To help the student with methods of algebra of sets and calculus, we include a *Review of Selected Mathematical Techniques* in Appendix A. This review includes a method that makes integration by parts easier. Also, we derive the important *Rule of 72*, which provides an approximation to the number of years necessary for money to double.

ENHANCEMENTS IN THIS EDITION

- There is better and more logical organization, resulting in a major chapter on the normal distribution.
- A new section gives a brief history of probability, indicating how the normal distribution was discovered.
- More real examples and exercises concerning probability were added that will appeal to students of actuarial science, finance, economics, and so on.

- There is a short but excellent Bayesian chapter, including real examples and an indication of how Bayesians prove theorems by establishing "Dutch books."
- In the section on bootstrapping there is an explanation of the origin of this word.
- There are examples of Simpson's paradox.
- Some different statistical techniques, including ordered restricted estimates, have been added.
- There is somewhat more emphasis on the importance of sufficient statistics, noting that such statistics, if they exist, are always used in statistical inferences.
- Tests of hypotheses and confidence intervals are tied together.
- An explanation of the Six Sigma program is in the Epilogue.
- The figures are improved with the use of color.
- The CD-ROM includes not only all of the data sets in different formats, but also many more new applications of Minitab and *Maple*.
- Illustrations of *Maple* as a computer algebra system are given in the text and on the CD-ROM.

IMPORTANT POINTS IN THIS EDITION

Chapter 1 is a basic chapter on probability after considering how discrete data can arise. This is followed by a chapter on discrete probability distributions. Chapter 3 starts with continuous type data, introducing stem-and-leaf diagrams, order statistics, and box plots and includes many standard continuous distribution. Chapter 4 introduces multivariate distributions, which are considered with more emphasis than in past editions, including mixtures of discrete and continuous types. Chapter 5 is the climax of the first semester as it is devoted to the important normal distribution, including a brief history of probability noting how the normal distribution was discovered. In these first five chapters over 100 new "real" probability examples and exercises are provided so that we are less dependent upon problems involving coins, cards, and dice. In particular, we hope the reader recognizes the importance of conditional distributions, correlation, and the use of distributions that are mixtures of the discrete and continuous cases.

We begin the second semester with a strong chapter (Chapter 6) on estimation which includes in a starred section the importance of sufficient statistics as shown by the theorem of Rao and Blackwell. This is followed by a new short but modern chapter (Chapter 7) on Bayesian estimation. George Woodworth and Kate Cowles made a number of suggestions concerning this new Bayesian chapter; in particular, Woodworth taught us about a "Dutch book" used in many Bayesian proofs. Chapter 8 is an excellent chapter on tests of statistical hypotheses, and the connection between tests and confidence intervals is given explicitly. Chapter 9 provides some theory of statistical tests. Chapter 10 is short, but interesting, as it considers some of the methods used to improve the quality of manufactured products, although it can be applied to the service industries too. The Six Sigma program is considered in the Epilogue.

PROBABILITY

1.1 BASIC CONCEPTS

It is usually difficult to explain to the general public what statisticians do. Many think of us as "math nerds" who seem to enjoy dealing with numbers. And there is some truth to that concept. But if we consider the bigger picture, many recognize that statisticians can be extremely helpful in many investigations.

Consider the following:

1. There is some problem or situation that needs to be considered; so statisticians are often asked to work with investigators or research scientists.
2. Suppose that some measure (or measures) are needed to help us understand the situation better. The measurement problem is often extremely difficult, and creating good measures is a valuable skill. For illustration, in higher education, how do we measure good teaching? This is a question to which we have not found a satisfactory answer, although several measures, such as student evaluations, have been used in the past.
3. After the measuring instrument has been developed, we must collect data through observation, possibly the results of a survey or an experiment.
4. Using these data, statisticians summarize the results, often using descriptive statistics and graphical methods.
5. These summaries are then used to analyze the situation. Here it is possible that statisticians make what are called statistical inferences.

6. Finally a report is presented, along with some recommendations that are based upon the data and the analysis of them. Frequently such a recommendation might be to perform the survey or experiment again, possibly changing some of the questions or factors involved. This is how statistics is used in what is referred to as the scientific method, because often the analysis of the data suggests other experiments. Accordingly the scientist must consider different possibilities in his or her search for an answer and thus performs similar experiments over and over again.

The discipline of statistics deals with the *collection* and *analysis of data*. When measurements are taken, even seemingly under the same conditions, the results usually vary. Despite this variability, a statistician tries to find a pattern; yet due to the "noise," not all of the data lie on the pattern. In the face of this variability, he or she must still determine the best way to describe the pattern. Accordingly, statisticians know that mistakes will be made in data analysis, and they try to minimize those errors as much as possible and then give bounds on the possible errors. By considering these bounds, decision makers can decide how much confidence they want to place on these data and the analysis of them. If the bounds are wide, perhaps more data should be collected. If they are small, however, the person involved in the study might want to make a decision and proceed accordingly.

Variability is a fact of life, and proper statistical methods can help us understand data collected under inherent variability. Because of this variability, many decisions have to be made that involve uncertainties. In medical research, interest may center on the effectiveness of a new vaccine for mumps; an agronomist must decide if an increase in yield can be attributed to a new strain of wheat; a meteorologist is interested in predicting the probability of rain; the state legislature must decide whether increasing speed limits will result in more accidents; the admissions officer of a college must predict the college performance of an incoming freshman; a biologist is interested in estimating the clutch size for a particular type of bird; an economist desires to estimate the unemployment rate; an environmentalist tests whether new controls have resulted in a reduction in pollution.

In reviewing the preceding (relatively short) list of possible areas of applications of statistics, the reader should recognize that good statistics is closely associated with careful thinking in many investigations. For illustration, students should appreciate how statistics is used in the endless cycle of the scientific method. We observe Nature and ask questions, we run experiments and collect data that shed light on these questions, we analyze the data and compare the results of the analysis to what we previously thought, we raise new questions, and on and on. Or if you like, statistics is clearly part of the important "plan—do—study—act" cycle: Questions are raised and investigations planned and carried out. The resulting data are studied and analyzed and then acted upon, often raising new questions.

There are many aspects of statistics. Some people get interested in the subject by collecting data and trying to make sense out of these observations. In some cases the answers are obvious and little training in statistical methods is necessary. But if a person goes very far in many investigations, he or she soon realizes that there is a need for some theory to help describe the error structure associated with the various estimates of the patterns. That is, at some point appropriate probability and mathematical models are required to make sense out of complicated data sets. Statistics and the probabilistic foundation on which statistical methods are based can provide the models to help people do this. So in this book, we are more concerned with the mathematical, rather than the applied, aspects of statistics. Still we give enough real examples so that the reader can get a good sense of a number of important applications of statistical methods.

In the study of statistics, we consider experiments for which the outcome cannot be predicted with certainty. Such experiments are called **random experiments**. Each experiment ends in an outcome that cannot be determined with certainty before the experiment is performed. However, the experiment is such that the collection of every possible outcome can be described and perhaps listed. This collection of all outcomes is called the **outcome space**, the **sample space**, or, more simply, the **space** S. The following examples will help illustrate what we mean by random experiments, outcomes, and their associated spaces.

EXAMPLE 1.1-1   Two dice are cast and the total number of spots on the sides that are "up" are counted. The outcome space is $S = \{2, 3, 4, 5, 6, 7, 8, 9, 10, 11, 12\}$. ◄

EXAMPLE 1.1-2   Each of six students selects an integer at random from the first 52 positive integers. We are interested in whether at least 2 of these six integers match (M) or whether all are different (D). Thus $S = \{M, D\}$. ◄

EXAMPLE 1.1-3   A fair coin is flipped successively at random until the first head is observed. If we let x denote the number of flips of the coin that are required, then $S = \{x : x = 1, 2, 3, 4, \ldots\}$, which consists of an infinite, but countable, number of outcomes. ◄

EXAMPLE 1.1-4   A box of breakfast cereal contains one of four different prizes. The purchase of one box of cereal yields one of the prizes as the outcome, and the sample space is the set of four different prizes. ◄

EXAMPLE 1.1-5   In Example 1.1-4, assume that the prizes are put into the boxes randomly. A family continues to buy this cereal until they obtain a complete set of the four different prizes. The number of boxes of cereal that must be purchased is one of the outcomes in $S = \{b : b = 4, 5, 6, \ldots\}$. ◄

EXAMPLE 1.1-6   A fair coin is flipped successively at random until heads is observed on two successive flips. If we let y denote the number of flips of the coin that are required, then $S = \{y : y = 2, 3, 4, \ldots\}$. ◄

EXAMPLE 1.1-7 To determine the percentage of body fat for a person, one measurement that is made is a person's weight under water. If w denotes this weight in kilograms, then the sample space could be $S = \{w : 0 < w < 7\}$, as we know from past experience that this weight does not exceed 7 kilograms. ◀

EXAMPLE 1.1-8 An ornithologist is interested in the clutch size (number of eggs in a nest) for gallinules, a species of bird that lives in a marsh. If we let c equal the clutch size then a possible sample space would be $S = \{c : c = 0, 1, 2, \ldots, 15\}$, as 15 is the largest known clutch size. ◀

Note that the outcomes of a random experiment can be numerical, as in Examples 1.1-3, 1.1-6, 1.1-7, and 1.1-8, but they do not have to be, as shown by Examples 1.1-2, and 1.1-4. Often we "mathematize" those latter outcomes by assigning numbers to them. For instance, in Example 1.1-2, we could denote the outcome $\{D\}$ by the number zero and the outcome $\{M\}$ by the number one. In general, measurements on outcomes associated with random experiments are called **random variables** and these are usually denoted by some capital letter toward the end of the alphabet, like X, Y, and Z.

Note the numbers of outcomes in the sample spaces in these examples. In Examples 1.1-1, 1.1-4, and 1.1-8, the number of outcomes is finite. In Examples 1.1-3 and 1.1-6 the number of possible outcomes is infinite but countable. That is, there are as many outcomes as there are counting numbers (positive integers). The space for Example 1.1-7 is different from the other examples in that the set of possible outcomes is an interval of numbers. Theoretically the weight could be any one of an infinite number of possible weights; here the number of possible outcomes is not countable. However, from a practical point of view, reported weights are selected from a finite number of possibilities because we can read and record the answer to only an accuracy determined by our scale. Many times, however, it is better to conceptualize the space as an interval of outcomes, and Example 1.1-7 is an example of a space of the continuous type.

If we consider a random experiment and its space, we note that under repeated performances of the experiment, some outcomes occur more frequently than others. For illustration, in Example 1.1-3, if this coin-flipping experiment is repeated over and over, the first head is observed on the first flip more often than on the second flip. If we can somehow, by theory or observations, determine the fractions of times a random experiment ends in the respective outcomes, we have described a *distribution of the random variable* (sometimes called a *population*). Often we cannot determine this distribution through theoretical reasoning but must actually perform the random experiment a number of times to obtain guesses or *estimates* of these fractions. The collection of the observations that are obtained from these repeated trials is often called a *sample*. The making of a conjecture about a distribution of a random variable based on the sample is called a *statistical inference*. That is, in statistics, we try to infer from the sample to the population. To understand the background behind statistical inferences that are made from the sample, we need a knowledge of some probability, basic distributions, and sampling distribution theory; these topics are considered in the early part of this book.

Given a sample or set of measurements, we would like to determine methods for describing the data. Suppose that we have some **counting** or **discrete data**. For example, you record the number of children in the family of each of your classmates. Or perhaps your state, on a regular basis, selects five integers out of the first 39 integers for the state lottery; you could count the number of those five integers that are odd.

In general, suppose we repeat a random experiment a number of times, say n. If a certain outcome has occurred f times in these n trials; the number f is called the **frequency** of the outcome. The ratio f/n is called the **relative frequency** of the outcome. A relative frequency is usually very unstable for small values of n, but it tends to stabilize about some number, say p, as n increases. The number p is called the **probability** of the outcome.

To develop an understanding of a particular set of discrete data, we can summarize the data in a frequency table and then construct a histogram of the data. A **frequency table** provides the number of occurrences of each possible outcome. A **histogram** presents graphically the tallied data as illustrated in the following example.

EXAMPLE 1.1-9   In Table 1.1-1, the number of children in each family of 100 students in two statistics classes is recorded.

We shall construct a frequency table in which we tally the number of times that each outcome was observed. We also add a column that gives the relative frequency of each outcome.

In a table of numbers like Table 1.1-1 it is often difficult to detect features of the measurements such as the minimum, the maximum, or the measurement that occurs most often (called the typical value). Often this typical value is called the **mode**. The frequency table, Table 1.1-2, helps to point out these features. A more visual way to present these data is with a **histogram**. To construct a **frequency histogram**, center a rectangle with a base of length one at each observed integer value and make the height equal to the frequency of this integer. To construct a **relative frequency histogram** or a **density histogram**, say $h(x)$, make the height of each rectangle equal to the relative frequency of the outcome. Note, since each base equals one, the area of each rectangle is the respective relative frequency of the midpoint of the base and the total area of a relative frequency histogram is equal to one. This is illustrated in Figure 1.1-1.   ◀

Table 1.1-1: Number of Children per Family																			
2	2	5	3	4	4	3	3	6	4	3	4	4	4	4	2	5	9	2	3
1	3	5	2	4	4	4	3	3	2	2	4	2	2	6	6	1	3	3	3
3	2	3	4	7	3	3	3	2	2	2	2	3	2	3	2	3	2	5	2
3	2	2	2	4	3	3	2	3	2	4	3	3	3	4	2	4	1	2	2
2	4	3	3	3	5	2	3	3	2	2	3	3	4	2	2	2	7	2	3

Table 1.1-2: Frequency Table of Number of Children Per Family			
Number of Children	**Tabulation**	**Frequency**	**Relative Frequency**
1	|||	3	0.03
2	‖‖‖ ‖‖‖ ‖‖‖ ‖‖‖ ‖‖‖ ‖‖‖ ||||	34	0.34
3	‖‖‖ ‖‖‖ ‖‖‖ ‖‖‖ ‖‖‖ ‖‖‖ ||||	34	0.34
4	‖‖‖ ‖‖‖ ‖‖‖ |||	18	0.18
5	‖‖‖	5	0.05
6	|||	3	0.03
7	||	2	0.02
8		0	0.00
9			1	0.01
Total		100	1.00

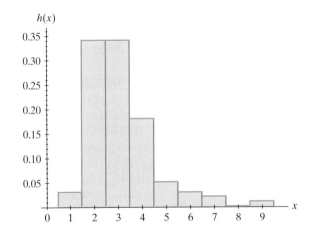

Figure 1.1-1: Relative frequency histogram of number of children per family

One reason for constructing histograms is to picture more easily the data. Also a histogram helps us to use the observed data to say something about the source of the data or the population from which the data were selected, provided that the data were selected randomly. For example, these 100 observations of number of children could perhaps help us say something about the number of children in families having a student at Hope College. If we would like to say something about the number of children in families in the United States, we could select a representative sample of many families and look at the number of children in these families. Note that 0 is a possible outcome for this latter experiment while that is not true of families sending at least one of their children to Hope College.

We have noted that the relative frequency f/n is very unstable for small n but tends to stabilize around some probability p as n increases. Since the relative frequency histogram $h(x)$ represents the relative frequencies of various values of x, we would believe that $h(x)$ approaches, as n increases, some ideal function $f(x)$, which represents the probabilities of the respective values of x. That is, if x_0 is one of the possible x values with frequency f_0, then, as n increases,

$$h(x_0) = \frac{f_0}{n} \to p_0 = f(x_0),$$

where p_0 is that probability about which f_0/n stabilizes. Suppose that we have a certain function $f(x)$ that serves as a model for the probabilities of the outcomes of a random experiment. If we repeat that random experiment a large number of times and construct the corresponding relative frequency histogram $h(x)$, we would anticipate that $h(x)$ would be approximately equal to $f(x)$. We call this function, $f(x)$, the **probability mass function** and abbreviate it **p.m.f.** In order to graphically compare $h(x)$ and $f(x)$, we construct a **probability histogram** of $f(x)$. This is a graphical representation that has a rectangle of height $f(x)$ and a base of length one, centered at each $x \in S$, where $x \in S$ means x belongs to the space S. Since each base is equal to one, the area of each rectangle is equal to the probability $f(x)$.

Let us consider a simple example in which we can intuitively assign a reasonable probability model to the outcomes of a given random experiment.

EXAMPLE 1.1-10   Let us place six chips of the same size in a bowl. The number one is placed on each of three of the chips, the number two on each of two of the other chips, and finally the number three on the last chip. The experiment is to select one chip at random and read the number x on the chip. Here $S = \{1, 2, 3\}$. If this experiment is performed in a "fair" manner, most of us would intuitively assign the probabilities 3/6, 2/6, and 1/6 to the respective x values: $x = 1, x = 2, x = 3$. That is, our model is given by the p.m.f.

$$f(1) = \frac{3}{6}, \qquad f(2) = \frac{2}{6}, \qquad f(3) = \frac{1}{6},$$

or, equivalently, by

$$f(x) = \frac{4 - x}{6}, \qquad x = 1, 2, 3.$$

We simulated this experiment using a computer $n = 600$ times obtaining the respective frequencies 284, 208, 108. The respective relative frequencies are

$$h(1) = \frac{284}{600} = 0.473, \qquad h(2) = \frac{208}{600} = 0.347, \qquad h(3) = \frac{108}{600} = 0.180.$$

In Figure 1.1-2 we plot the probability histogram associated with the p.m.f. $f(x)$. The shaded region represents the relative frequency histogram $h(x)$. Note that $h(x)$ approximates $f(x)$ very well.   ◄

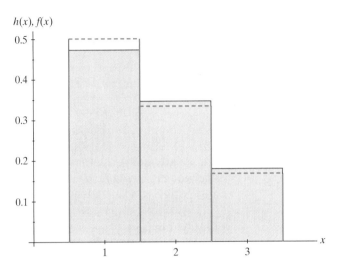

Figure 1.1-2: Relative frequency histogram (shaded) and probability histogram

The value of x that has a greatest probability (or highest relative frequency) is called the **mode** of the distribution (or of the histogram determined by the data). So in the last example, $x = 1$ is the mode.

EXAMPLE 1.1-11 A student of statistics was asked to write down a string of one-digit numbers $(0, 1, 2, \ldots, 9)$ that seemed to her to be random. She recorded the following 40 numbers in this order.

$$8, 2, 6, 0, 3, 9, 1, 6, 5, 8, 7, 4, 9, 5, 0, 5, 2, 7, 5, 2$$
$$4, 8, 1, 3, 6, 5, 2, 8, 4, 1, 0, 8, 1, 2, 7, 6, 1, 9, 4, 0$$

Let us count the number of times she recorded the same number next to itself (like 5, 5), the number of times adjacent numbers were next to each other (like 7, 6 and say 9, 0 are next to each other) and the number of times neither of these occurred. If these numbers were truly selected at random consecutive numbers would be the same about 10% of the times, consecutive numbers would differ by one about 20% of the time (recall 9 and 0 differ by one in this exercise), and consecutive numbers differ by more than one 70% of the time. Her results were as follows.

	Frequency	Relative Frequency
Same	0	$0/39 = 0.00$
Differ by one	6	$6/39 = 0.15$
Differ by more than one	33	$33/39 = 0.85$
Totals	39	$39/39 = 1.00$

It seems as if 0.00 differs too much from 0.10, 0.15 is fairly close to 0.20, and 0.85 is somewhat higher than the expected 0.70. Intuitively, her sequence does not seem to be a very good approximation to a sequence of random digits. Later, in this text, you will study a statistical method to test for the difference.   ◄

In many applied problems, not just those associated with probability, we want to find a model that describes the reality of the situation as well as possible. In truth, no model is exactly right, but often models are close enough to being correct that they are extremely useful in practice. However, before they can be used, models must be tested using data collected from the real situation. In this text, we are interested in constructing probability models for different random experiments. In the next few chapters, we study enough probability that allows us to assign certain reasonable probability models to important cases. Often we do not know everything about these models, that is, they frequently have unknown parameters. Thus we must observe the results of several trials of the experiment to make further inferences about the models; in particular, we must estimate the unknown parameter(s). This process of refining the model using the data is referred to as making a **statistical inference** and is the focus of the latter part of the text.

Statistical Comments (**Simpson's Paradox**) While most of the first five chapters are about probability and probability distributions, from time to time we will mention certain statistical concepts. For illustration, the relative frequency, f/n, is called a **statistic** and it is used to **estimate** a probability p, which is usually unknown. For example, if a major league batter gets $f = 152$ hits in $n = 500$ official at bats during the season, the relative frequency $f/n = 0.304$ is an estimate of his probability of getting a hit and it is often called his batting average for that season.

Once while speaking to a group of coaches, one (Hogg) of us made the comment that it would be possible for batter A to have a higher average than batter B for each season during their careers and yet B could have a better overall average at the end of their careers. While no coach spoke up, you could tell that they were thinking, "And that guy is supposed to know something about math."

Of course, the following simple example convinced them that the statement was true. Suppose they only played two seasons with the following results.

Season	Player A			Player B		
	AB	Hits	Average	AB	Hits	Average
1	500	126	0.252	300	75	0.250
2	300	90	0.300	500	145	0.290
Totals	800	216	0.270	800	220	0.275

Clearly A beats B in the two individual seasons, but B has a better overall average. Note that during their better season (the second) B had more "at bats" than did A. This kind of result is often called **Simpson's paradox** and it happens often in real life. See Exercises 1.1-11, 1.1-12, and 1.1-13.   ○

EXERCISES

1.1-1 Describe the outcome space for each of the following experiments:

    (a) A student is selected at random from a statistics class, and the student's ACT score in mathematics is determined. HINT: ACT test scores in mathematics are integers between 1 and 36, inclusive.

    (b) A candy bar with a 20.4-gram label weight is selected at random from a production line and is weighed.

    (c) A coin is tossed three times, and the sequence of heads and tails is observed.

1.1-2 Describe the outcome space for each of the following experiments:

    (a) Consider families that have three children each and select one such family at random. Describe the sample space S in terms of their children as 3-tuples, agreeing for example, "gbb" would indicate that the youngest is a girl and the two oldest are boys.

    (b) A rat is selected at random from a cage, and its sex is determined.

    (c) A state selects a three-digit integer at random for one of its lottery games.

1.1-3 The *National Geographic* reported the following litter sizes for 20 lions:

$$
\begin{array}{cccccccccc}
2 & 2 & 1 & 3 & 2 & 2 & 3 & 1 & 4 & 3 \\
2 & 1 & 3 & 2 & 2 & 2 & 2 & 1 & 2 & 1
\end{array}
$$

    (a) Construct a frequency table for these data.

    (b) Draw a frequency histogram.

    (c) Draw a relative frequency histogram.

    (d) Is there a typical litter size?

1.1-4 Some ornithologists were interested in the clutch size of the common gallinule. They observed the number of eggs in each of 117 nests, yielding the following data.

$$
\begin{array}{ccccccccccccccc}
7 & 5 & 13 & 7 & 7 & 8 & 9 & 9 & 9 & 8 & 8 & 9 & 9 & 7 & 7 \\
5 & 9 & 7 & 7 & 4 & 9 & 8 & 8 & 10 & 9 & 7 & 8 & 8 & 8 & 7 \\
9 & 7 & 7 & 10 & 8 & 7 & 9 & 7 & 10 & 8 & 9 & 7 & 11 & 10 & 9 \\
9 & 4 & 8 & 6 & 8 & 9 & 9 & 9 & 8 & 8 & 5 & 8 & 8 & 9 & 9 \\
14 & 10 & 8 & 9 & 9 & 9 & 8 & 7 & 9 & 7 & 9 & 10 & 10 & 7 & 6 \\
11 & 7 & 7 & 6 & 9 & 7 & 7 & 6 & 8 & 9 & 4 & 6 & 9 & 8 & 9 \\
7 & 9 & 9 & 9 & 9 & 8 & 8 & 8 & 9 & 9 & 9 & 8 & 10 & 9 & 9 \\
8 & 5 & 7 & 8 & 7 & 6 & 7 & 7 & 7 & 6 & 5 & 9
\end{array}
$$

    (a) Construct a frequency table for these data.

    (b) Draw a histogram.

    (c) What is the mode (the typical clutch size)?

1.1-5 The ornithologists in the last exercise noticed that some of the gallinules had a second brood during the summer. They were interested in comparing the clutch sizes for the second broods with those for the first brood. They were able to collect the following clutch sizes for second broods.

$$
\begin{array}{cccccccccccccccccc}
4 & 4 & 6 & 5 & 5 & 9 & 6 & 5 & 6 & 9 & 7 & 9 & 4 & 8 & 6 & 5 & 9 \\
8 & 8 & 7 & 6 & 8 & 8 & 9 & 10 & 9 & 9 & 7 & 4 & 6 & 8 & 7 & 5
\end{array}
$$

    (a) Construct a frequency table for these data.

    (b) Draw a histogram.

    (c) Is there a typical clutch size for second broods?

1.1-6 Before buying enough cereal to obtain a set of four prizes (Example 1.1-4), a family decided to estimate the number of boxes they would have to buy using simulation. They rolled a four-sided die until they had observed each face at least once. They repeated this 100 times and recorded the numbers of rolls needed as follows:

8	6	6	4	13	11	19	13	4	15	8	13	5	8	9
16	9	5	12	6	4	8	6	6	6	11	6	5	10	5
6	5	8	4	5	14	5	7	5	7	5	8	4	9	8
9	13	8	14	5	24	4	5	6	7	5	4	8	7	6
11	4	5	6	5	10	10	4	5	14	10	15	6	9	8
7	10	14	10	8	8	10	7	9	10	8	10	9	7	5
12	7	16	6	5	11	4	11	5	5					

(a) Construct a frequency table for these data.
(b) Draw a histogram.

1.1-7 For a Texas lottery game, five integers were selected randomly out of the first 39 positive integers. The following table lists the numbers of odd integers out of each set of five integers selected in 100 consecutive drawings.

4	3	2	2	0	3	5	2	1	1	2	4	3	4	1	0	2	4	3	2
3	4	2	1	3	2	1	3	2	2	3	4	3	3	2	3	3	2	3	3
2	3	3	3	4	3	3	2	2	2	3	2	3	0	2	3	2	3	3	2
3	3	2	4	2	2	2	2	3	2	3	3	1	4	3	3	1	2	3	5
3	1	4	1	3	3	2	4	1	2	2	3	2	2	4	1	3	2	1	4

(a) Construct a frequency table for these data.
(b) Draw a histogram.

1.1-8 Consider a bowl containing 10 chips of the same size such that two are marked "one," three are marked "two," three are marked "three," and two are marked "four." Select a chip at random and read the number. Here $S = \{1, 2, 3, 4\}$.

(a) Assign a reasonable p.m.f. $f(x)$ to the outcome space.
(b) Simulate this experiment at least $n = 100$ times and find the relative frequency histogram $h(x)$. HINT: Here you can simulate using a computer; or simply use the table of random numbers (Table IX) and start at a random spot and let an integer in the set $\{0, 1\} = 1$, in $\{2, 3, 4\} = 2$, in $\{5, 6, 7\} = 3$, in $\{8, 9\} = 4$.
(c) Plot $f(x)$ and $h(x)$ on the same graph.

1.1-9 Toss two coins at random and count the number of heads that appear "up." Here $S = \{0, 1, 2\}$. In Chapter 2 we discover that a reasonable probability model is given by the p.m.f. $f(0) = 1/4$, $f(1) = 1/2$, $f(2) = 1/4$. Repeat this experiment at least $n = 100$ times and plot the resulting relative frequency histogram $h(x)$ on the same graph with $f(x)$.

1.1-10 Let the random variable X be the number of tosses of a coin needed to obtain the first head. Here $S = \{1, 2, 3, 4, \ldots\}$. In Chapter 2, we find a reasonable probability model is given by the p.m.f. $f(x) = (1/2)^x$, $x \in S$. Do this experiment a large number of times and compare the resulting relative frequency histogram $h(x)$ to $f(x)$.

1.1-11 In 1985 Al Bumbry of the Baltimore Orioles and Darrell Brown of the Minnesota Twins had the following numbers of hits (H) and official at bats (AB) on grass and artificial turf.

Playing Surface	Bumbry			Brown		
	AB	H	BA	AB	H	BA
Grass	295	77		92	18	
Artificial Turf	49	16		168	53	
Total	344	93		260	71	

(a) Find the batting averages (BA) for each player on grass, namely H/AB.
(b) Find the BA for each player on artificial turf.

(c) Find the season batting averages for the two players.
(d) Interpret your results.

1.1-12 In 1985 Kent Hrbek of the Minnesota Twins and Dion James of the Milwaukee Brewers had the following numbers of hits (H) and official at bats (AB) on grass and artificial turf.

Playing Surface	Hrbek			James		
	AB	H	BA	AB	H	BA
Grass	204	50		329	93	
Artificial Turf	355	124		58	21	
Total	559	174		387	114	

(a) Find the batting averages (BA) for each player on grass, namely H/AB.
(b) Find the BA for each player on artificial turf.
(c) Find the season batting averages for the two players.
(d) Interpret your results.

1.1-13 If we had a choice of two airlines, we would possibly choose the airline with the better "on time performance." So consider Alaska Airlines and America West using data reported by Arnold Barnett, "How Numbers Can Trick You," in *Technology Review*, 1994.

Airline	Alaska Airlines	America West
Destination	Relative Frequency On Time	Relative Frequency On Time
Los Angeles	$\dfrac{497}{559}$	$\dfrac{694}{811}$
Phoenix	$\dfrac{221}{233}$	$\dfrac{4840}{5255}$
San Diego	$\dfrac{212}{232}$	$\dfrac{383}{448}$
San Francisco	$\dfrac{503}{605}$	$\dfrac{320}{449}$
Seattle	$\dfrac{1841}{2146}$	$\dfrac{201}{262}$
Five-City Total	$\dfrac{3274}{3775}$	$\dfrac{6438}{7225}$

(a) For each of the five cities listed, which airline has the better on time performance?
(b) Combining the results, which airline has the better on time performance?
(c) Interpret your results.

1.2 PROPERTIES OF PROBABILITY

In Section 1.1, the collection of all possible outcomes (the *universal set*) of a random experiment is denoted by S and is called the *outcome space*. Given an outcome space S, let A be a part of the collection of outcomes in S, that is, $A \subset S$. Then A is called an **event**. When the random experiment is performed and the outcome of the experiment is in A, we say that **event A has occurred**.

Since, in studying probability, the words *set* and *event* are interchangeable, the reader might want to review **algebra of sets**, found in the Appendix. For convenience, however, here we remind the reader of some terminology:

- $\emptyset$ denotes the **null** or **empty** set;
- $A \subset B$ means A is a **subset** of B;
- $A \cup B$ is the **union** of A and B;
- $A \cap B$ is the **intersection** of A and B;
- A' is the **complement** of A (i.e., all elements in S that are not in A).

Some of these sets are depicted by the shaded region in Figure 1.2-1, in which S is the interior of the rectangles.

Special terminology associated with events that is often used by statisticians includes the following:

1. $A_1, A_2, \ldots, A_k$ are **mutually exclusive events** means that $A_i \cap A_j = \emptyset, i \neq j$, that is, $A_1, A_2, \ldots, A_k$ are disjoint sets;

2. $A_1, A_2, \ldots, A_k$ are **exhaustive events** means $A_1 \cup A_2 \cup \cdots \cup A_k = S$.

So if $A_1, A_2, \ldots, A_k$ are **mutually exclusive and exhaustive** events, we know that $A_i \cap A_j = \emptyset, i \neq j$, and $A_1 \cup A_2 \cup \cdots \cup A_k = S$.

We are interested in defining what is meant by the probability of A, denoted by $P(A)$, and often called the chance of A occurring. To help us understand what is meant by the probability of A, consider repeating the experiment a number of times, say n times. Count the number of times that event A actually occurred throughout these n performances; this number is called the frequency of event A and is denoted by $\mathcal{N}(A)$. The ratio $\mathcal{N}(A)/n$ is called the **relative frequency** of event A in these n repetitions of the experiment. A relative frequency is usually very unstable for small values of n, but it tends to stabilize as n increases. This suggests that we associate with event A a number, say p, that is equal to or approximately equal to the number about which the relative frequency tends to stabilize. This number p can then be taken as the number that the relative frequency of event A will be near in future performances of the experiment. Thus, although we cannot predict the outcome of a random experiment with certainty, we can, for a large value of n, predict fairly accurately the relative frequency associated with event A. The number p assigned to event A is called the **probability** of event A and is denoted by $P(A)$. That is, $P(A)$ represents the proportion of outcomes of a random experiment that terminate in the event A as the number of trials of that experiment increases without bound.

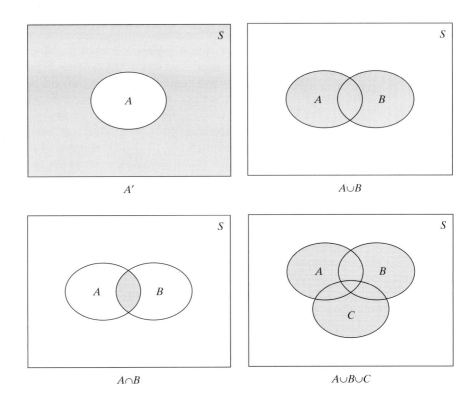

Figure 1.2-1: Algebra of sets

The following example will help to illustrate some of the ideas just presented.

EXAMPLE 1.2-1 A fair six-sided die is rolled six times. If the face numbered k is the outcome on roll k for $k = 1, 2, \ldots, 6$, we say that a match has occurred. The experiment is called a success if at least one match occurs during the six trials. Otherwise, the experiment is called a failure. The sample space is $S = \{$success, failure$\}$. Let $A = \{$success$\}$. We would like to assign a value to $P(A)$. Accordingly, this experiment was simulated 500 times on a computer. Figure 1.2-2 depicts the results of this simulation, and the following table summarizes a few of the results:

n	$N(A)$	$N(A)/n$
50	37	0.740
100	69	0.690
250	172	0.688
500	330	0.660

The probability of event A is not intuitively obvious, but it will be shown in Example 1.5-6 that $P(A) = 1 - (1 - 1/6)^6 = 0.665$. This assignment is certainly supported by the simulation (although not proved by it). ◀

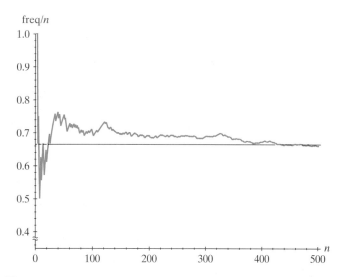

Figure 1.2-2: Fraction of experiments having at least one match

Example 1.2-1 shows that at times intuition cannot be used to assign probabilities although simulation can perhaps help to assign a probability empirically.
The following example gives an illustration where intuition can help in assigning
a probability to an event.

EXAMPLE 1.2-2   A disk 2 inches in diameter is thrown at random on a tiled floor, where each
tile is a square with sides 4 inches in length. Let C be the event that the disk
will land entirely on one tile. In order to assign a value to $P(C)$, consider the
center of the disk. In what region must the center lie to assure that the disk lies
entirely on one tile? If you draw a picture, it should be clear that the center
must lie within a square having sides of length 2 and lying in the center of a
tile. Since the area of this square is 4 and the area of a tile is 16, it makes sense
to let $P(C) = 4/16$.   ◀

Sometimes the nature of an experiment is such that the probability of A
can be assigned easily. For example, when a state lottery randomly selects
a three-digit integer, we would expect each of the 1000 possible three-digit
numbers to have the same chance of being selected, namely 1/1000. If we
let $A = \{233, 323, 332\}$, then it makes sense to let $P(A) = 3/1000$. Or if we
let $B = \{234, 243, 324, 342, 423, 432\}$, then we would let $P(B) = 6/1000$, the
probability of the event B. Probabilities of events associated with many random
experiments are perhaps not quite as obvious and straightforward as was seen
in Example 1.2-2.

So we wish to associate with A a number $P(A)$ about which the relative
frequency $\mathcal{N}(A)/n$ of the event A tends to stabilize with large n. A function
such as $P(A)$, which is evaluated for a set A, is called a **set function**. In this
section we consider the probability set function $P(A)$ and discuss some of its
properties. In succeeding sections we shall describe how the probability set
function is defined for particular experiments.

To help decide what properties the probability set function should satisfy, consider properties possessed by the relative frequency $N(A)/n$. For example, $N(A)/n$ is always nonnegative. If $A = S$, the sample space, then the outcome of the experiment will always belong to S, and thus $N(S)/n = 1$. Also, if A and B are two mutually exclusive events, then $N(A \cup B)/n = N(A)/n + N(B)/n$. Hopefully, these remarks will help to motivate the following definition.

Definition 1.2-1

Probability is a real-valued set function P that assigns to each event A in the sample space S a number $P(A)$, called the probability of the event A, such that the following properties are satisfied:

(a) $P(A) \geq 0$,
(b) $P(S) = 1$,
(c) If $A_1, A_2, A_3, \ldots$ are events and $A_i \cap A_j = \emptyset, i \neq j$, then

$$P(A_1 \cup A_2 \cup \cdots \cup A_k) = P(A_1) + P(A_2) + \cdots + P(A_k)$$

for each positive integer k, and

$$P(A_1 \cup A_2 \cup A_3 \cup \cdots) = P(A_1) + P(A_2) + P(A_3) + \cdots$$

for an infinite, but countable, number of events.

The following theorems give some other important properties of the probability set function. When one considers these theorems, it is important to understand the theoretical concepts and proofs. However, if the reader keeps the relative frequency concept in mind, the theorems should also have some intuitive appeal.

Theorem 1.2-1

For each event A,

$$P(A) = 1 - P(A').$$

Proof: We have

$$S = A \cup A' \qquad \text{and} \qquad A \cap A' = \emptyset.$$

Thus, from properties (b) and (c), it follows that

$$1 = P(A) + P(A').$$

Hence

$$P(A) = 1 - P(A'). \qquad \bigcirc$$

EXAMPLE 1.2-3   A fair coin is flipped successively until the same face is observed on successive flips. Let $A = \{x : x = 3, 4, 5, \ldots\}$; that is, A is the event that it will take three or more flips of the coin to observe the same face on two consecutive flips. To find $P(A)$, we first find the probability of $A' = \{x : x = 2\}$, the complement of A. In two flips of a coin, the possible outcomes are $\{HH, HT, TH, TT\}$, and we assume that each of these four points has the same chance of being observed. Thus

$$P(A') = P(\{HH, TT\}) = \frac{2}{4}.$$

It follows from Theorem 1.2-1 that

$$P(A) = 1 - P(A') = 1 - \frac{2}{4} = \frac{2}{4}.$$ ◄

Theorem 1.2-2 $P(\emptyset) = 0$.

Proof: In Theorem 1.2-1, take $A = \emptyset$ so that $A' = S$. Thus

$$P(\emptyset) = 1 - P(S) = 1 - 1 = 0.$$ ◯

Theorem 1.2-3   If events A and B are such that $A \subset B$, then $P(A) \leq P(B)$.

Proof: Now

$$B = A \cup (B \cap A') \qquad \text{and} \qquad A \cap (B \cap A') = \emptyset.$$

Hence, from property (c),

$$P(B) = P(A) + P(B \cap A') \geq P(A)$$

because from property (a),

$$P(B \cap A') \geq 0.$$ ◯

Theorem 1.2-4   For each event A, $P(A) \leq 1$.

Proof: Since $A \subset S$, we have by Theorem 1.2-3 and property (b) that

$$P(A) \leq P(S) = 1,$$

which gives the desired result.   ◯

Property (a) along with Theorem 1.2-4 shows that, for each event A,

$$0 \le P(A) \le 1.$$

Theorem 1.2-5   If A and B are any two events, then

$$P(A \cup B) = P(A) + P(B) - P(A \cap B).$$

Proof: The event $A \cup B$ can be represented as a union of mutually exclusive events, namely,

$$A \cup B = A \cup (A' \cap B).$$

Hence, by property (c),

$$P(A \cup B) = P(A) + P(A' \cap B). \qquad (1.2\text{-}1)$$

However,

$$B = (A \cap B) \cup (A' \cap B),$$

which is a union of mutually exclusive events. Thus

$$P(B) = P(A \cap B) + P(A' \cap B)$$

and

$$P(A' \cap B) = P(B) - P(A \cap B).$$

If this result is substituted in Equation 1.2-1 we obtain

$$P(A \cup B) = P(A) + P(B) - P(A \cap B),$$

which is the desired result.   ◯

EXAMPLE 1.2-4   A faculty leader was meeting two students in Paris, one arriving by train from Amsterdam and the other arriving by train from Brussels at approximately the same time. Let A and B be the events that the trains are on time, respectively. Suppose from past experience we know that $P(A) = 0.93$, $P(B) = 0.89$, and $P(A \cap B) = 0.87$. Then

$$P(A \cup B) = P(A) + P(B) - P(A \cap B)$$
$$= 0.93 + 0.89 - 0.87 = 0.95$$

is the probability that at least one train is on time.   ◀

Theorem 1.2-6   If A, B, and C are any three events, then

$$P(A \cup B \cup C) = P(A) + P(B) + P(C) - P(A \cap B)$$
$$- P(A \cap C) - P(B \cap C) + P(A \cap B \cap C).$$

Proof: Write

$$A \cup B \cup C = A \cup (B \cup C)$$

and apply Theorem 1.2-5. The details are left as an exercise.   ○

EXAMPLE 1.2-5   Continuing with Example 1.2-4, say that a third student is arriving from Cologne. Let C be the event that this train is on time with $P(C) = 0.91$, $P(B \cap C) = 0.85$, $P(A \cap C) = 0.86$, and $P(A \cap B \cap C) = 0.81$. Then

$$P(A \cup B \cup C) = P(A) + P(B) + P(C) - P(A \cap B) - P(A \cap C)$$
$$- P(B \cap C) + P(A \cap B \cap C)$$
$$= 0.93 + 0.89 + 0.91 - 0.87 - 0.85 - 0.86 + 0.81$$
$$= 0.96$$

is the probability that at least one of these trains is on time.   ◀

Let a probability set function be defined on a sample space S. Let $S = \{e_1, e_2, \ldots, e_m\}$, where each e_i is a possible outcome of the experiment. The integer m is called the total number of ways in which the random experiment can terminate. If each of these outcomes has the same probability of occurring, we say that the m outcomes are **equally likely**. That is,

$$P(\{e_j\}) = \frac{1}{m}, \qquad i = 1, 2, \ldots, m.$$

If the number of outcomes in an event A is h, the integer h is called the number of ways that are favorable to the event A. In this case $P(A)$ is equal to the number of ways favorable to the event A divided by the total number of ways in which the experiment can terminate. That is, under this assumption of equally likely outcomes, we have that

$$P(A) = \frac{h}{m} = \frac{N(A)}{N(S)},$$

where $h = N(A)$ is the number of ways A can occur and $m = N(S)$ is the number of ways S can occur. Exercise 1.2-17 considers this assignment of probability in a more theoretical manner.

It should be emphasized that in order to assign the probability h/m to the event A, we must assume that each of the outcomes $e_1, e_2, \ldots, e_m$ has the same probability $1/m$. This assumption is then an important part of our probability model; if it is not realistic in an application, the probability of the event A cannot be computed this way. Actually we have used this result in the simple case given in Example 1.2-3 because it seemed realistic to assume that each of the possible outcomes in $S = \{HH, HT, TH, TT\}$ had the same chance of being observed.

EXAMPLE 1.2-6   Let a card be drawn at random from an ordinary deck of 52 playing cards. The sample space S is the set of $m = 52$ different cards, and it is reasonable to assume that each of these cards has the same probability for selection, $1/52$. Accordingly, if A is the set of outcomes that are kings, $P(A) = 4/52 = 1/13$ because there are $h = 4$ kings in the deck. That is, $1/13$ is the probability of drawing a card that is a king provided that each of the 52 cards has the same probability of being drawn. ◀

In Example 1.2-6, the computations are very easy because there is no difficulty in the determination of the appropriate values of h and m. However, instead of drawing only one card, suppose that 13 are taken at random and without replacement. We can think of each possible 13-card hand as being an outcome in a sample space, and it is reasonable to assume that each of these outcomes has the same probability. To use the above method to assign the probability of a hand, consisting of seven spades and six hearts, for illustration, we must be able to count the number h of all such hands as well as the number m of possible 13-card hands. In these more complicated situations, we need better methods of determining h and m. We discuss some of these counting techniques in Section 1.3.

EXERCISES

1.2-1 Draw one card at random from a standard deck of cards. The sample space S is the collection of the 52 cards. Assume that the probability set function assigns $1/52$ to each of these 52 outcomes. Let

$$A = \{x : x \text{ is a jack, queen, or king}\},$$

$$B = \{x : x \text{ is a 9, 10, or jack and } x \text{ is red}\},$$

$$C = \{x : x \text{ is a club}\},$$

$$D = \{x : x \text{ is a diamond, a heart, or a spade}\}.$$

Find **(a)** $P(A)$, **(b)** $P(A \cap B)$, **(c)** $P(A \cup B)$, **(d)** $P(C \cup D)$, **(e)** $P(C \cap D)$.

1.2-2 A coin is tossed four times, and the sequence of heads and tails is observed.
   (a) List each of the 16 sequences in the sample space S.
   (b) Let events A, B, C, and D be given by $A = \{$at least 3 heads$\}$, $B = \{$at most 2 heads$\}$, $C = \{$heads on the third toss$\}$, and $D = \{$1 head and 3 tails$\}$. If the probability set function assigns $1/16$ to each outcome in the sample space, find **(i)** $P(A)$, **(ii)** $P(A \cap B)$, **(iii)** $P(B)$, **(iv)** $P(A \cap C)$, **(v)** $P(D)$, **(vi)** $P(A \cup C)$, and **(vii)** $P(B \cap D)$.

1.2-3 A field of beans is planted with three seeds per hill. For each hill of beans, let A_i be the event that i seeds germinate, $i = 0, 1, 2, 3$. Suppose that $P(A_0) = 1/64$, $P(A_1) = 9/64$, and $P(A_2) = 27/64$. Give the value of $P(A_3)$.

1.2-4 Consider the trial on which a 3 is first observed in successive rolls of a four-sided die. Let A be the event that 3 is observed on the first trial. Let B be the event that at least two trials are required to observe a 3. Assuming that each side has probability $1/4$, find **(a)** $P(A)$, **(b)** $P(B)$, and **(c)** $P(A \cup B)$.

1.2-5 A fair eight-sided die is rolled once. Let $A = \{2, 4, 6, 8\}$, $B = \{3, 6\}$, $C = \{2, 5, 7\}$, and $D = \{1, 3, 5, 7\}$. Assume that each face has the same probability.

    (a) Give the values of **(i)** $P(A)$, **(ii)** $P(B)$, **(iii)** $P(C)$, and **(iv)** $P(D)$.
    (b) Give the values of **(i)** $P(A \cap B)$, **(ii)** $P(B \cap C)$, and **(iii)** $P(C \cap D)$.
    (c) Give the values of **(i)** $P(A \cup B)$, **(ii)** $P(B \cup C)$, and **(iii)** $P(C \cup D)$ using Theorem 1.2-5.

1.2-6 If $P(A) = 0.4$, $P(B) = 0.5$, and $P(A \cap B) = 0.3$, find **(a)** $P(A \cup B)$, **(b)** $P(A \cap B')$, and **(c)** $P(A' \cup B')$.

1.2-7 If $S = A \cup B$, $P(A) = 0.7$, and $P(B) = 0.9$, find $P(A \cap B)$.

1.2-8 If $P(A) = 0.4$, $P(B) = 0.5$, and $P(A \cup B) = 0.7$, find

    (a) $P(A \cap B)$.
    (b) $P(A' \cup B')$.

1.2-9 Roll a fair six-sided die three times. Let $A_1 = \{1 \text{ or } 2 \text{ on the first roll}\}$, $A_2 = \{3 \text{ or } 4 \text{ on the second roll}\}$, and $A_3 = \{5 \text{ or } 6 \text{ on the third roll}\}$. It is given that $P(A_i) = 1/3$, $i = 1, 2, 3$; $P(A_i \cap A_j) = (1/3)^2$, $i \neq j$; and $P(A_1 \cap A_2 \cap A_3) = (1/3)^3$.

    (a) Use Theorem 1.2-6 to find $P(A_1 \cup A_2 \cup A_3)$.
    (b) Show that $P(A_1 \cup A_2 \cup A_3) = 1 - (1 - 1/3)^3$.

1.2-10 Prove Theorem 1.2-6.

1.2-11 For each positive integer n, let $P(\{n\}) = (1/2)^n$. Consider the events $A = \{n : 1 \leq n \leq 10\}$, $B = \{n : 1 \leq n \leq 20\}$ and $C = \{n : 11 \leq n \leq 20\}$. Find **(a)** $P(A)$, **(b)** $P(B)$, **(c)** $P(A \cup B)$, **(d)** $P(A \cap B)$, **(e)** $P(C)$, and **(f)** $P(B')$.

1.2-12 Let x equal a number that is selected randomly from the closed interval from zero to one, $[0, 1]$. Use your intuition to assign values to

    (a) $P(\{x : 0 \leq x \leq 1/3\})$.
    (b) $P(\{x : 1/3 \leq x \leq 1\})$.
    (c) $P(\{x : x = 1/3\})$.
    (d) $P(\{x : 1/2 < x < 5\})$.

1.2-13 A typical roulette wheel used in a casino has 38 slots that are numbered $1, 2, 3, \ldots, 36$, $0, 00$, respectively. The 0 and 00 slots are colored green. Half of the remaining slots are red and half are black. Also half of the integers between 1 and 36 inclusive are odd, half are even, and 0 and 00 are defined to be neither odd nor even. A ball is rolled around the wheel and ends up in one of the slots; we assume each slot has equal probability of $1/38$ and we are interested in the number of the slot in which the ball falls.

    (a) Define the sample space S.
    (b) Let $A = \{0, 00\}$. Give the value of $P(A)$.
    (c) Let $B = \{14, 15, 17, 18\}$. Give the value of $P(B)$.
    (d) Let $D = \{x : x \text{ is odd}\}$. Give the value of $P(D)$.

1.2-14 The five numbers 1, 2, 3, 4, and 5 are written respectively on five disks of the same size and placed in a hat. Two disks are drawn without replacement from the hat, and the numbers written on them are observed.

    (a) List the 10 possible outcomes for this experiment as unordered pairs of numbers.
    (b) If each of the 10 outcomes has probability $1/10$, assign a value to the probability that the sum of the two numbers drawn is **(i)** 3; **(ii)** between 6 and 8 inclusive.

1.2-15 Divide a line segment into two parts by selecting a point at random. Use your intuition to assign a probability to the event that the larger segment is at least two times longer than the shorter segment.

1.2-16 Let the interval $[-r, r]$ be the base of a semicircle. If a point is selected at random from this interval, assign a probability to the event that the length of the perpendicular segment from this point to the semicircle is less than $r/2$.

1.2-17 Let $S = A_1 \cup A_2 \cup \cdots \cup A_m$, where events $A_1, A_2, \ldots, A_m$ are mutually exclusive and exhaustive.

(a) If $P(A_1) = P(A_2) = \cdots = P(A_m)$, show that $P(A_i) = 1/m$, $i = 1, 2, \ldots, m$.

(b) If $A = A_1 \cup A_2 \cup \cdots \cup A_h$, where $h < m$, and (a) holds, prove that $P(A) = h/m$.

1.3 METHODS OF ENUMERATION

In this section we develop counting techniques that are useful in determining the number of outcomes associated with the events of certain random experiments. We begin with a consideration of the multiplication principle.

▲ *Multiplication Principle:* Suppose that an experiment (or procedure) E_1 has n_1 outcomes and for each of these possible outcomes an experiment (procedure) E_2 has n_2 possible outcomes. The composite experiment (procedure) $E_1 E_2$ that consists of performing first E_1 and then E_2 has $n_1 n_2$ possible outcomes. ◀

EXAMPLE 1.3-1 Let E_1 denote the selection of a rat from a cage containing one female (F) rat and one male (M) rat. Let E_2 denote the administering of either drug A (A), drug B (B), or a placebo (P) to the selected rat. The outcome for the composite experiment can be denoted by an ordered pair, such as (F, P). In fact, the set of all possible outcomes, namely $(2)(3) = 6$ of them, can be denoted by the following rectangular array:

$$\begin{array}{ccc} (\text{F, A}) & (\text{F, B}) & (\text{F, P}) \\ (\text{M, A}) & (\text{M, B}) & (\text{M, P}) \end{array}$$ ◀

Another way of illustrating the multiplication principle is with a tree diagram like that in Figure 1.3-1. The diagram shows that there are $n_1 = 2$ possibilities

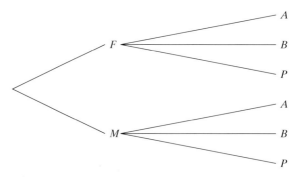

Figure 1.3-1: Tree diagram

(branches) for the sex of the rat and that for each of these outcomes there are $n_2 = 3$ possibilities (branches) for the drug.

Clearly, the multiplication principle can be extended to a sequence of more than two experiments or procedures. Suppose that the experiment E_i has n_i $(i = 1, 2, \ldots, m)$ possible outcomes after previous experiments have been performed. The composite experiment $E_1 E_2 \cdots E_m$, which consists of performing E_1, then $E_2, \ldots$, and finally E_m, has $n_1 n_2 \cdots n_m$ possible outcomes.

EXAMPLE 1.3-2 A certain food service gives the following choices for dinner: E_1, soup or tomato juice; E_2, steak or shrimp; E_3, french fried potatoes, mashed potatoes, or a baked potato; E_4, corn or peas; E_5, Jello, tossed salad, cottage cheese, or cole slaw; E_6, cake, cookies, pudding, brownie, vanilla ice cream, chocolate ice cream, or orange sherbet; E_7, coffee, tea, milk, or punch. How many different dinner selections are possible if one of the listed choices is made for each of $E_1, E_2, \ldots$, and E_7? By the multiplication principle there are

$$(2)(2)(3)(2)(4)(7)(4) = 2688$$

different combinations. ◄

Although the multiplication principle is fairly simple and easy to understand, it will be extremely useful as we now develop various counting techniques.

Suppose that n positions are to be filled with n different objects. There are n choices for filling the first position, $n - 1$ for the second, $\ldots$, and 1 choice for the last position. So by the multiplication principle there are

$$n(n - 1) \cdots (2)(1) = n!$$

possible arrangements. The symbol $n!$ is read "n factorial." We define $0! = 1$; that is, we say that zero positions can be filled with zero objects in one way.

Definition 1.3-1 Each of the $n!$ arrangements (in a row) of n different objects is called a *permutation* of the n objects.

EXAMPLE 1.3-3 The number of permutations of the four letters a, b, c, and d is clearly $4! = 24$. However, the number of possible four-letter code words using the four letters a, b, c, and d if letters may be repeated is $4^4 = 256$, because in this case each selection can be performed in four ways. ◄

If only r positions are to be filled with objects selected from n different objects, $r \leq n$, the number of possible ordered arrangements is

$$_n P_r = n(n - 1)(n - 2) \cdots (n - r + 1).$$

That is, there are n ways to fill the first position, $(n - 1)$ ways to fill the second, and so on until there are $[n - (r - 1)] = (n - r + 1)$ ways to fill the rth position.

In terms of factorials we have that

$$_nP_r = \frac{n(n-1)\cdots(n-r+1)(n-r)\cdots(3)(2)(1)}{(n-r)\cdots(3)(2)(1)} = \frac{n!}{(n-r)!}.$$

Definition 1.3-2   Each of the $_nP_r$ arrangements is called a *permutation of n objects taken r at a time*.

EXAMPLE 1.3-4   The number of possible four-letter code words, selecting from the 26 letters in the alphabet, in which all four letters are different is

$$_{26}P_4 = (26)(25)(24)(23) = \frac{26!}{22!} = 358,800. \qquad \blacktriangleleft$$

EXAMPLE 1.3-5   The number of ways of selecting a president, a vice president, a secretary, and a treasurer in a club consisting of 10 persons is

$$_{10}P_4 = 10 \cdot 9 \cdot 8 \cdot 7 = \frac{10!}{6!} = 5040. \qquad \blacktriangleleft$$

Suppose that a set contains n objects. Consider the problem of drawing r objects from this set. The order in which the objects are drawn may or may not be important. In addition, it is possible that a drawn object is replaced before the next object is drawn. Accordingly, we give some definitions and show how the multiplication principle can be used to count the number of possibilities.

Definition 1.3-3   If r objects are selected from a set of n objects, and if the order of selection is noted, the selected set of r objects is called an *ordered sample of size r*.

Definition 1.3-4   *Sampling with replacement* occurs when an object is selected and then replaced before the next object is selected.

By the multiplication principle, the number of possible ordered samples of size r taken from a set of n objects is n^r when sampling with replacement.

EXAMPLE 1.3-6   A die is rolled five times. The number of possible ordered samples is $6^5 = 7776$. Note that rolling a die is equivalent to sampling with replacement. $\qquad \blacktriangleleft$

EXAMPLE 1.3-7   An urn contains 10 balls numbered 0, 1, 2, ... , 9. If four balls are selected, one at a time and with replacement, the number of possible ordered samples is

$10^4 = 10{,}000$. Note that this is the number of four-digit integers between 0000 and 9999, inclusive.   ◄

Definition 1.3-5   *Sampling without replacement* occurs when an object is not replaced after it has been selected.

By the multiplication principle, the number of possible ordered samples of size r taken from a set of n objects, sampling without replacement, is

$$n(n-1)\cdots(n-r+1) = \frac{n!}{(n-r)!},$$

which is equivalent to $_nP_r$, the number of permutations of n objects taken r at a time.

EXAMPLE 1.3-8   The number of ordered samples of five cards that can be drawn without replacement from a standard deck of 52 playing cards is

$$(52)(51)(50)(49)(48) = \frac{52!}{47!} = 311{,}875{,}200.$$ ◄

Often the order of selection is not important and interest centers only on the selected set of r objects. That is, we are interested in the number of subsets of size r that can be selected from a set of n different objects. In order to find the number of (unordered) subsets of size r, we count, in two different ways, the number of ordered subsets of size r that can be taken from the n distinguishable objects. By equating these two answers, we are able to count the number of (unordered) subsets of size r.

Let C denote the number of (unordered) subsets of size r that can be selected from n different objects. We can obtain each of the $_nP_r$ ordered subsets by first selecting one of the C unordered subsets of r objects and then ordering these r objects. Since the latter can be carried out in $r!$ ways, the multiplication principle yields $(C)(r!)$ ordered subsets; so $(C)(r!)$ must equal $_nP_r$. Thus we have

$$(C)(r!) = \frac{n!}{(n-r)!}$$

or

$$C = \frac{n!}{r!\,(n-r)!}.$$

We denote this answer by either $_nC_r$ or $\binom{n}{r}$; that is,

$$_nC_r = \binom{n}{r} = \frac{n!}{r!\,(n-r)!}.$$

Accordingly, a set of n different objects possesses

$$\binom{n}{r} = \frac{n!}{r!\,(n-r)!}$$

unordered subsets of size $r \leq n$.

We could also say that the number of ways in which r objects can be selected without replacement from n objects, when the order of selection is disregarded, is $\binom{n}{r} = {}_nC_r$, and this can be read as "n choose r." This motivates the following definition.

Definition 1.3-6   Each of the ${}_nC_r$ unordered subsets is called a *combination of n objects taken r at a time,* where,

$$_nC_r = \binom{n}{r} = \frac{n!}{r!\,(n-r)!}.$$

EXAMPLE 1.3-9   The number of possible five-card hands (hands in five-card poker) drawn from a deck of 52 playing cards is

$$_{52}C_5 = \binom{52}{5} = \frac{52!}{5!\,47!} = 2{,}598{,}960.$$ ◀

EXAMPLE 1.3-10   The number of possible 13-card hands (hands in bridge) that can be selected from a deck of 52 playing cards is

$$_{52}C_{13} = \binom{52}{13} = \frac{52!}{13!\,39!} = 635{,}013{,}559{,}600.$$ ◀

The numbers $\binom{n}{r}$ are frequently called **binomial coefficients**, since they arise in the expansion of a binomial. We illustrate this by giving a justification of the binomial expansion

$$(a+b)^n = \sum_{r=0}^{n} \binom{n}{r} b^r a^{n-r}. \tag{1.3-1}$$

In the expansion of

$$(a+b)^n = (a+b)(a+b)\cdots(a+b),$$

either an a or a b is selected from each of the n factors. One possible product is then $b^r a^{n-r}$; this occurs when b is selected from each of r factors and a

from each of the remaining $n - r$ factors. But this latter operation can be completed in $\binom{n}{r}$ ways, which then must be the coefficient of $b^r a^{n-r}$, as shown in Equation 1.3-1.

The binomial coefficients are given in Table I in the Appendix for selected values of n and r. Note that for some combinations of n and r, the table uses the fact that

$$\binom{n}{r} = \frac{n!}{r!\,(n-r)!} = \frac{n!}{(n-r)!\,r!} = \binom{n}{n-r}.$$

That is, the number of ways in which r objects can be selected out of n objects is equal to the number of ways in which $n - r$ objects can be selected out of n objects.

EXAMPLE 1.3-11   Assume that each of the $\binom{52}{5} = 2,598,960$ five-card hands drawn from a deck of 52 playing cards has the same probability for being selected. The number of possible five-card hands that are all spades (event A) is

$$N(A) = \binom{13}{5}\binom{39}{0}$$

because the 5 spades can be selected from the 13 spades in $\binom{13}{5}$ ways after which zero nonspades can be selected in $\binom{39}{0} = 1$ way. We have

$$\binom{13}{5} = \frac{13!}{5!\,8!} = 1287$$

from Table I in the Appendix. Thus the probability of an all-spade five-card hand is

$$P(A) = \frac{N(A)}{N(S)} = \frac{1287}{2,598,960} = 0.000495.$$

Suppose now that the event B is the set of outcomes in which exactly three cards are kings and exactly two cards are queens. We can select the three kings in any one of $\binom{4}{3}$ ways and the two queens in any one of $\binom{4}{2}$ ways. By the multiplication principle, the number of outcomes in B is

$$N(B) = \binom{4}{3}\binom{4}{2}\binom{44}{0},$$

where $\begin{pmatrix} 44 \\ 0 \end{pmatrix}$ gives the number of ways in which 0 cards are selected out of the non-kings and non-queens and of course is equal to one. Thus

$$P(B) = \frac{N(B)}{N(S)} = \frac{\begin{pmatrix} 4 \\ 3 \end{pmatrix}\begin{pmatrix} 4 \\ 2 \end{pmatrix}\begin{pmatrix} 44 \\ 0 \end{pmatrix}}{\begin{pmatrix} 52 \\ 5 \end{pmatrix}} = \frac{24}{2,598,960} = 0.0000092.$$

Finally, let C be the set of outcomes in which there are exactly two kings, two queens, and one jack. Then

$$P(C) = \frac{N(C)}{N(S)} = \frac{\begin{pmatrix} 4 \\ 2 \end{pmatrix}\begin{pmatrix} 4 \\ 2 \end{pmatrix}\begin{pmatrix} 4 \\ 1 \end{pmatrix}\begin{pmatrix} 40 \\ 0 \end{pmatrix}}{\begin{pmatrix} 52 \\ 5 \end{pmatrix}} = \frac{144}{2,598,960} = 0.000055$$

because the numerator of this fraction is the number of outcomes in C. ◄

Now suppose that a set contains n objects of two types, r of one type and $n - r$ of the other type. The number of permutations of n different objects is $n!$. However, in this case, the objects are not all distinguishable. To count the number of distinguishable arrangements, first select r out of the n positions for the objects of the first type. This can be done in $\begin{pmatrix} n \\ r \end{pmatrix}$ ways. Then fill in the remaining positions with the objects of the second type. Thus the number of distinguishable arrangements is

$$_nC_r = \begin{pmatrix} n \\ r \end{pmatrix} = \frac{n!}{r!\,(n-r)!}.$$

Definition 1.3-7   Each of the $_nC_r$ permutations of n objects, r of one type and $n - r$ of another type, is called a *distinguishable permutation*.

EXAMPLE 1.3-12   A coin is flipped 10 times and the sequence of heads and tails is observed. The number of possible 10-tuplets that result in four heads and six tails is

$$\begin{pmatrix} 10 \\ 4 \end{pmatrix} = \frac{10!}{4!\,6!} = \frac{10!}{6!\,4!} = \begin{pmatrix} 10 \\ 6 \end{pmatrix} = 210.$$ ◄

EXAMPLE 1.3-13   Students on a boat send signals back to shore by arranging seven colored flags on a vertical flagpole. If they have four orange and three blue flags, they can send

$$\begin{pmatrix} 7 \\ 4 \end{pmatrix} = \frac{7!}{4!\,3!} = 35$$

different signals. Note that if they had seven flags of different colors, they could send $7! = 5040$ different signals.   ◄

The foregoing results can be extended. Suppose that in a set of n objects, n_1 are similar, n_2 are similar, ... , n_s are similar, where $n_1 + n_2 + \cdots + n_s = n$. The number of distinguishable permutations of the n objects is (see Exercise 1.3-18)

$$\binom{n}{n_1, n_2, \ldots, n_s} = \frac{n!}{n_1! \, n_2! \cdots n_s!}. \tag{1.3-2}$$

EXAMPLE 1.3-14   If the students on the boat have three red flags, four yellow flags, and two blue flags to arrange on a vertical pole, the number of possible signals is

$$\binom{9}{3, 4, 2} = \frac{9!}{3! \, 4! \, 2!} = 1260. \qquad ◄$$

The argument used in determining the binomial coefficients in the expansion of $(a + b)^n$ can be extended to find the expansion of $(a_1 + a_2 + \cdots + a_s)^n$. The coefficient of $a_1^{n_1} a_2^{n_2} \cdots a_s^{n_s}$ is

$$\binom{n}{n_1, n_2, \ldots, n_s} = \frac{n!}{n_1! \, n_2! \cdots n_s!}.$$

This is sometimes called a **multinomial coefficient**.

When r objects are selected out of n objects, we are often interested in the number of possible outcomes. We have seen that for ordered samples, there are n^r possible outcomes when sampling with replacement and $_nP_r$ outcomes when sampling without replacement. For unordered samples, there are $_nC_r$ outcomes when sampling without replacement. Each of the outcomes above is equally likely provided the experiment is performed in a fair manner.

Remark   Although not needed in the study of probability it is interesting to count the number of possible samples of size r that can be selected out of n objects when the order is irrelevant and when sampling with replacement. For example, if a six-sided die is rolled 10 times (or 10 six-sided dice are rolled once), how many possible unordered outcomes are there? To count the number of possible outcomes, think of listing r 0's for the r objects that are to be selected. Then insert $(n - 1)$ |'s to partition the r objects into n sets, the first set giving objects of the first kind, and so on. So if $n = 6$ and $r = 10$ in the die illustration, a possible outcome is

$$0\,0\,|\,|\,0\,0\,0\,|\,0\,|\,0\,0\,0\,|\,0,$$

which says there are two 1's, zero 2's, three 3's, one 4, three 5's, and one 6. In general, each outcome is a permutation of r 0's and $(n - 1)$ |'s. Each distinguishable permutation is equivalent to an unordered sample. The number of distinguishable permutations, and hence the number of unordered samples of size r that can be selected out of n objects when sampling with replacement, is

$$_{n-1+r}C_r = \frac{(n-1+r)!}{r!\,(n-1)!}.$$ ∎

EXERCISES

1.3-1 A boy found a bicycle lock for which the combination was unknown. The correct combination is a four-digit number, $d_1 d_2 d_3 d_4$, where d_i, $i = 1, 2, 3, 4$, is selected from $1, 2, 3, 4, 5, 6, 7, 8$. How many different lock combinations are possible with such a lock?

1.3-2 How many different signals can be made using four flags of different colors on a vertical flagpole if exactly three flags are used for each signal?

1.3-3 How many different license plates are possible if a state uses

    (a) Two letters followed by a four-digit integer (leading zeros permissible and the letters and digits can be repeated)?

    (b) Three letters followed by a three-digit integer?

1.3-4 In designing an experiment, the researcher can often choose many different levels of the various factors in order to try to find the best combination at which to operate. For illustration, suppose the research is studying a certain chemical reaction and can choose four levels of temperature, five different pressures, and two different catalysts.

    (a) To consider all possible combinations, how many experiments would need to be conducted?

    (b) Often in preliminary experimentation, each factor is restricted to two levels. With the three factors noted, how many experiments would need to be run to cover all possible combinations with each of the three factors at two levels? (NOTE: This is often called a 2^3 design.)

1.3-5 How many four-letter code words are possible using the letters in HOPE if

    (a) The letters may not be repeated?

    (b) The letters may be repeated?

1.3-6 The "eating club" is hosting a make-your-own sundae at which the following are provided:

Ice Cream Flavors	Toppings
Chocolate	Caramel
Cookies-n-cream	Hot fudge
Strawberry	Marshmallow
Vanilla	M&M's
	Nuts
	Strawberries

    (a) How many sundaes are possible using one flavor of ice cream and three different toppings?

    (b) How many sundaes are possible using one flavor of ice cream and from zero to six toppings?

(c) How many different combinations of flavors of three scoops of ice cream are possible if it is permissible to make all three scoops the same flavor?

1.3-7 In a state lottery four digits are drawn at random one at a time with replacement from 0 to 9. Suppose that you win if any permutation of your selected integers is drawn. Give the probability of winning if you select

(a) 6, 7, 8, 9.
(b) 6, 7, 8, 8.
(c) 7, 7, 8, 8.
(d) 7, 8, 8, 8.

1.3-8 From a collection of nine paintings, four are to be selected to hang side by side on a gallery wall in positions 1, 2, 3, and 4. In how many ways can this be done?

1.3-9 Some albatrosses return to the world's only mainland colony of royal albatrosses on Otago Peninsula near Dunedin, New Zealand, every two years to nest and raise their young. In order to learn more about the albatross, colored plastic bands are placed on their legs so that they can be identified from a distance. Suppose that three bands are placed on one leg, selecting from the colors red, yellow, green, white, and blue. Find the number of different color codes that are possible for banding an albatross if

(a) The three bands are different colors.
(b) Repeated colors are permissible.

1.3-10 Suppose that Andy Roddick and Roger Federer are playing a tennis match in which the first player to win three sets wins the match. Considering the possible orderings for the winning player, in how many ways could this tennis match end?

1.3-11 The World Series in baseball continues until either the American League team or the National League team wins four games. How many different orders are possible (e.g., *ANNAAA* means the American League team wins in six games) if the series goes

(a) Four games?
(b) Five games?
(c) Six games?
(d) Seven games?

1.3-12 How many different varieties of pizza can be made if you have the following choices: size—small, medium, or large; crust—thin 'n crispy, hand-tossed, or pan; toppings—(cheese is automatic) there are 12 toppings from which you may select from 0 to 12?

1.3-13 A cafe lets you order a deli sandwich your way. There are 6 choices for bread, 4 choices for meat, 4 choices for cheese, and 12 different garnishes (condiments). How many different sandwich possibilities are there if you choose

(a) One bread, one meat, and one cheese?
(b) One bread, one meat, one cheese, and from 0 to 12 garnishes?
(c) One bread; 0, 1, or 2 meats; 0, 1, or 2 cheeses; and from 0 to 12 garnishes?

1.3-14 Pascal's triangle gives a method for calculating the binomial coefficients; it begins as follows:

$$
\begin{array}{ccccccccccc}
 & & & & & 1 & & & & & \\
 & & & & 1 & & 1 & & & & \\
 & & & 1 & & 2 & & 1 & & & \\
 & & 1 & & 3 & & 3 & & 1 & & \\
 & 1 & & 4 & & 6 & & 4 & & 1 & \\
1 & & 5 & & 10 & & 10 & & 5 & & 1 \\
\end{array}
$$

$\vdots \quad \vdots \quad \vdots \quad \vdots \quad \vdots$

The nth row of this triangle gives the coefficients for $(a + b)^{n-1}$. To find an entry in the table other than a 1 on the boundary, add the two nearest numbers in the row directly above. The equation

$$\binom{n}{r} = \binom{n-1}{r} + \binom{n-1}{r-1},$$

called **Pascal's equation**, explains why Pascal's triangle works. Prove that this equation is correct.

1.3-15 Among nine presidential candidates at a debate, three are Republicans and six are Democrats.

    (a) In how many different ways can the nine candidates be lined up?

    (b) How many lineups by party are possible if each candidate is labeled either R or D?

    (c) For each of the nine candidates, you are to decide whether the candidate did a good job or a poor job; that is, give each of the nine candidates a grade of G or P. How many different "score cards" are possible?

1.3-16 Prove:

$$\sum_{r=0}^{n}(-1)^r\binom{n}{r} = 0 \qquad \text{and} \qquad \sum_{r=0}^{n}\binom{n}{r} = 2^n.$$

HINT: Consider $(1-1)^n$ and $(1+1)^n$ or use Pascal's equation and proof by induction.

1.3-17 A poker hand is defined as drawing five cards at random without replacement from a deck of 52 playing cards. Find the probability of each of the following poker hands:

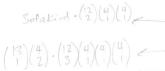

    (a) Four of a kind (four cards of equal face value and one card of a different value).

    (b) Full house (one pair and one triple of cards with equal face value).

    (c) Three of a kind (three equal face values plus two cards of different values).

    (d) Two pairs (two pairs of equal face value plus one card of a different value).

    (e) One pair (one pair of equal face value plus three cards of different values).

1.3-18 Prove Equation 1.3-2. HINT: First select n_1 positions in $\binom{n}{n_1}$ ways, after which select n_2 from the remaining $n - n_1$ positions in $\binom{n-n_1}{n_2}$ ways, and so on. Finally, use the multiplication rule.

1.3-19 There are three teams in a cross country race. Team A has 5 runners, team B has 6 runners, and team C has 7 runners. In how many ways can the runners cross the finish line if we are only interested in the team for which they run? That is, what is the number of distinguishable permutations of 5 A's, 6 B's, and 7 C's? (Note that for scoring purposes, only the scores of the first 5 runners for each team count.)

1.3-20 A box of candy hearts contains 52 hearts of which 19 are white, 10 are tan, 7 are pink, 3 are purple, 5 are yellow, 2 are orange, and 6 are green. If you select 9 pieces of candy randomly from the box, without replacement, give the probability that

    (a) Three of the hearts are white.

    (b) Three are white, two are tan, one is pink, one is yellow, and two are green.

1.3-21 An office furniture manufacturer makes modular storage files. It offers its customers two choices for the base and four choices for the top, and the modular storage files come in five different heights. The customer may choose any combination of the five different-sized modules so that the finished file has a base, a top, and one, two, three, four, five, or six storage modules.

(a) How many choices does the customer have if the completed file has four storage modules, a top, and a base? The order in which the four storage modules are stacked is irrelevant.

(b) The manufacturer would like to use in its advertising the number of different files that are possible—selecting one of the two bases, one of the four tops, and then either one or two or three or four or five or six storage modules, selecting any combination of the five different sizes with the order of stacking irrelevant. What is the number of possibilities?

1.4 CONDITIONAL PROBABILITY

We introduce the idea of conditional probability by means of an example.

EXAMPLE 1.4-1

Suppose that we are given 20 tulip bulbs that are very similar in appearance and told that 8 tulips will bloom early, 12 will bloom late, 13 will be red, and 7 will be yellow, in accordance with the various combinations of Table 1.4-1. If one bulb is selected at random, the probability that it will produce a red tulip (R) is given by $P(R) = 13/20$, under the assumption that each bulb is "equally likely." Suppose, however, that close examination of the bulb will reveal whether it will bloom early (E) or late (L). If we consider an outcome only if it results in a tulip bulb that will bloom early, only eight outcomes in the sample space are now of interest. Thus it is natural to assign, under this limitation, the probability of 5/8 to R; that is, $P(R \mid E) = 5/8$, where $P(R \mid E)$ is read as the probability of R given that E has occurred. Note that

$$P(R \mid E) = \frac{5}{8} = \frac{N(R \cap E)}{N(E)} = \frac{N(R \cap E)/20}{N(E)/20} = \frac{P(R \cap E)}{P(E)},$$

where $N(R \cap E)$ and $N(E)$ are the numbers of outcomes in events $R \cap E$ and E, respectively. ◀

This example is illustrative of a number of common situations. That is, in some random experiments, we are interested only in those outcomes that are elements of a subset B of the sample space S. This means, for our purposes, that the sample space is effectively the subset B. We are now confronted with the problem of defining a probability set function with B as the "new" sample space. That is, for a given event A we want to define $P(A \mid B)$, the probability of A considering only those outcomes of the random experiment that are elements of B. The previous example gives us the clue to that definition. That is, for experiments in which each outcome is equally likely, it makes sense to define $P(A \mid B)$ by

$$P(A \mid B) = \frac{N(A \cap B)}{N(B)},$$

where $N(A \cap B)$ and $N(B)$ are the numbers of outcomes in $A \cap B$ and B, respectively. If we divide the numerator and the denominator of this fraction

Table 1.4-1: Tulip Combinations			
	Early (E)	Late (L)	Totals
Red (R)	5	8	13
Yellow(Y)	3	4	7
Totals	8	12	20

by $N(S)$, the number of outcomes in the sample space, we have

$$P(A \mid B) = \frac{N(A \cap B)/N(S)}{N(B)/N(S)} = \frac{P(A \cap B)}{P(B)}.$$

We are thus led to the following definition.

Definition 1.4-1   The *conditional probability* of an event A, given that event B has occurred, is defined by

$$P(A \mid B) = \frac{P(A \cap B)}{P(B)}$$

provided that $P(B) > 0$.

A formal use of the definition is given in the following example.

EXAMPLE 1.4-2   If $P(A) = 0.4$, $P(B) = 0.5$, and $P(A \cap B) = 0.3$, then $P(A \mid B) = 0.3/0.5 = 0.6$; $P(B \mid A) = P(A \cap B)/P(A) = 0.3/0.4 = 0.75$.   ◄

We can think of the "given B" as specifying the new sample space for which we now want to calculate the probability of that part of A that is contained in B to determine $P(A \mid B)$. The following two examples illustrate this idea.

EXAMPLE 1.4-3   Suppose that $P(A) = 0.7$, $P(B) = 0.3$, and $P(A \cap B) = 0.2$. These probabilities are listed on the Venn diagram in Figure 1.4-1. Given that the outcome of the experiment belongs to B, what then is the probability of A? We are effectively restricting the sample space to B; of the probability $P(B) = 0.3$, 0.2 corresponds to $P(A \cap B)$ and hence to A. That is, 0.2/0.3 = 2/3 of the probability of B corresponds to A. Of course, formally by definition, we also obtain

$$P(A \mid B) = \frac{P(A \cap B)}{P(B)} = \frac{0.2}{0.3} = \frac{2}{3}.$$ ◄

EXAMPLE 1.4-4   A pair of four-sided dice is rolled and the sum is determined. Let A be the event that a sum of 3 is rolled and let B be the event that a sum of 3 or a sum of 5 is rolled. In a sequence of rolls the probability that a sum of 3 is rolled before a sum of 5 is rolled can be thought of as the conditional probability of a sum of

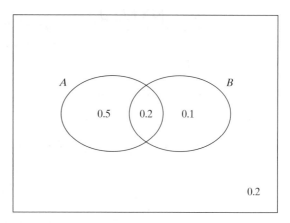

Figure 1.4-1: Conditional probability

3 given that a sum of 3 or 5 has occurred; that is, the conditional probability of A given B

$$P(A \mid B) = \frac{P(A \cap B)}{P(B)} = \frac{P(A)}{P(B)} = \frac{2/16}{6/16} = \frac{2}{6}.$$

Note that for this example, the only outcomes of interest are those having a sum of 3 or a sum of 5, and of these six equally likely outcomes, two have a sum of 3. See Figure 1.4-2. (See also Exercise 1.4-13.)   ◄

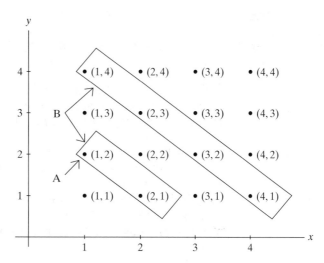

Figure 1.4-2: Dice example

It is interesting to note that conditional probability satisfies the axioms for a probability function, namely, with $P(B) > 0$,

(a) $P(A \mid B) \geq 0$.
(b) $P(B \mid B) = 1$.

(c) If $A_1, A_2, A_3, \ldots$ are mutually exclusive events, then

$$P(A_1 \cup A_2 \cup \cdots \cup A_k \mid B) = P(A_1 \mid B) + P(A_2 \mid B) + \cdots + P(A_k \mid B),$$

for each positive integer k, and

$$P(A_1 \cup A_2 \cup \cdots \mid B) = P(A_1 \mid B) + P(A_2 \mid B) + \cdots$$

for an infinite, but countable, number of events.

Properties (a) and (b) are evident because

$$P(A \mid B) = \frac{P(A \cap B)}{P(B)} \geq 0,$$

since $P(A \cap B) \geq 0$ and $P(B) > 0$, and

$$P(B \mid B) = \frac{P(B \cap B)}{P(B)} = \frac{P(B)}{P(B)} = 1.$$

Property (c) holds because, for the second part of (c),

$$P(A_1 \cup A_2 \cup \cdots \mid B) = \frac{P[(A_1 \cup A_2 \cup \cdots) \cap B]}{P(B)}$$

$$= \frac{P[(A_1 \cap B) \cup (A_2 \cap B) \cup \cdots]}{P(B)}.$$

But $(A_1 \cap B), (A_2 \cap B), \ldots$ are also mutually exclusive events; so

$$P(A_1 \cup A_2 \cup \cdots \mid B) = \frac{P(A_1 \cap B) + P(A_2 \cap B) + \cdots}{P(B)}$$

$$= \frac{P(A_1 \cap B)}{P(B)} + \frac{P(A_2 \cap B)}{P(B)} + \cdots$$

$$= P(A_1 \mid B) + P(A_2 \mid B) + \cdots.$$

The first part of property (c) is proved in a similar manner.

Many times the conditional probability of an event is clear because of the nature of an experiment. The following example illustrates this.

EXAMPLE 1.4-5 At a fair a vendor has 25 helium balloons on strings: 10 balloons are yellow, 8 are red, and 7 are green. A balloon is selected at random and sold. Given that the balloon sold is yellow, what is the probability that the next balloon selected at random is also yellow? Of the 24 remaining balloons, 9 are yellow; so a natural value to assign to this conditional probability is 9/24. ◀

In Example 1.4-5, let A be the event that the first balloon selected is yellow, and let B be the event that the second balloon selected is yellow. Suppose that we are interested in the probability that both selected balloons are yellow. That is, we are interested in finding $P(A \cap B)$. We noted in Example 1.4-5 that

$$P(B \mid A) = \frac{P(A \cap B)}{P(A)} = \frac{9}{24}.$$

Thus, multiplying through by $P(A)$, we have

$$P(A \cap B) = P(A)P(B \mid A) = P(A)\left(\frac{9}{24}\right) \qquad (1.4-1)$$

or

$$P(A \cap B) = \left(\frac{10}{25}\right)\left(\frac{9}{24}\right).$$

That is, Equation 1.4-1 gives us a general rule for the probability of the intersection of two events once we know the conditional probability $P(B \mid A)$.

Definition 1.4-2

The probability that two events, A and B, both occur is given by the *multiplication rule*,

$$P(A \cap B) = P(A)P(B \mid A)$$

or by

$$P(A \cap B) = P(B)P(A \mid B).$$

Sometimes, after considering the nature of the random experiment, one can make reasonable assumptions so that it is easier to assign $P(B)$ and $P(A \mid B)$ rather than $P(A \cap B)$. Then $P(A \cap B)$ can be computed with these assignments. This will be illustrated in Examples 1.4-6 and 1.4-7.

EXAMPLE 1.4-6

A bowl contains seven blue chips and three red chips. Two chips are to be drawn successively at random and without replacement. We want to compute the probability that the first draw results in a red chip (A) and the second draw results in a blue chip (B). It is reasonable to assign the following probabilities:

$$P(A) = \frac{3}{10} \quad \text{and} \quad P(B \mid A) = \frac{7}{9}.$$

The probability of red on the first draw and blue on the second draw is

$$P(A \cap B) = \frac{3}{10} \cdot \frac{7}{9} = \frac{7}{30}.$$

It should be noted that in many instances, it is possible to compute a probability by two seemingly different methods. For illustration, consider Example 1.4-6 but find the probability of drawing a red chip on each of the two draws. Following that example, it is

$$\frac{3}{10} \cdot \frac{2}{9} = \frac{1}{15}.$$

However, we can also find this probability using combinations as follows:

$$\frac{\binom{3}{2}\binom{7}{0}}{\binom{10}{2}} = \frac{\frac{(3)(2)}{(1)(2)}}{\frac{(10)(9)}{(1)(2)}} = \frac{1}{15}.$$

Thus we obtain the same answer, as we should, provided that our reasoning is consistent with the underlying assumptions.

EXAMPLE 1.4-7 From an ordinary deck of playing cards, cards are to be drawn successively at random and without replacement. The probability that the third spade appears on the sixth draw is computed as follows: Let A be the event of two spades in the first five cards drawn, and let B be the event of a spade on the sixth draw. Thus the probability that we wish to compute is $P(A \cap B)$. It is reasonable to take

$$P(A) = \frac{\binom{13}{2}\binom{39}{3}}{\binom{52}{5}} = 0.274 \qquad \text{and} \qquad P(B \mid A) = \frac{11}{47} = 0.234.$$

The desired probability $P(A \cap B)$ is the product of these numbers,

$$P(A \cap B) = (0.274)(0.234) = 0.064. \qquad \blacktriangleleft$$

EXAMPLE 1.4-8 Continuing with Example 1.4-4, in which a pair of four-sided dice is rolled, the probability for rolling a sum of 3 on the first roll and then, continuing the sequence of rolls, rolling a sum of 3 before rolling a sum of 5 is

$$\frac{2}{16} \cdot \frac{2}{6} = \frac{4}{96} = \frac{1}{24}. \qquad \blacktriangleleft$$

The multiplication rule can be extended to three or more events. In the case of three events we have, by using the multiplication rule for two events,

$$P(A \cap B \cap C) = P[(A \cap B) \cap C]$$

$$= P(A \cap B)P(C \mid A \cap B).$$

But

$$P(A \cap B) = P(A)P(B \mid A).$$

Hence

$$P(A \cap B \cap C) = P(A)P(B \mid A)P(C \mid A \cap B).$$

This type of argument can be used to extend the multiplication rule to more than three events, and the general formula for k events can be officially proved by mathematical induction.

EXAMPLE 1.4-9 Four cards are to be dealt successively at random and without replacement from an ordinary deck of playing cards. The probability of receiving in order a spade, a heart, a diamond, and a club is

$$\frac{13}{52} \cdot \frac{13}{51} \cdot \frac{13}{50} \cdot \frac{13}{49},$$

a result that follows from the extension of the multiplication rule and reasonable assignments to the probabilities involved.   ◄

We close this section with a different type of example.

EXAMPLE 1.4-10 A grade school boy has five blue and four white marbles in his left pocket and four blue and five white marbles in his right pocket. If he transfers one marble at random from his left to his right pocket, what is the probability of his then drawing a blue marble from his right pocket? For notation let BL, BR, and WL denote drawing blue from left pocket, blue from right pocket, and white from left pocket, respectively. Then

$$P(BR) = P(BL \cap BR) + P(WL \cap BR)$$
$$= P(BL)P(BR \mid BL) + P(WL)P(BR \mid WL)$$
$$= \frac{5}{9} \cdot \frac{5}{10} + \frac{4}{9} \cdot \frac{4}{10} = \frac{41}{90}$$

is the desired probability.   ◄

EXAMPLE 1.4-11 An insurance company sells a number of different policies; among these are 60% for autos, 40% for homeowners, and 20% are for both of these two. Let A_1 be people with only an auto policy, A_2 with only homeowners, A_3 with both, and A_4 are those with only other types of policies. If a person is selected at random from the policyholders, then $P(A_1) = 0.4$, $P(A_2) = 0.2$, $P(A_3) = 0.2$, and $P(A_4) = 0.2$ as these four events are mutually exclusive and exhaustive. Further let B be the event that a policyholder will renew at least one of the auto or homeowners policies. Say from past experience we can assign the conditional probabilities $P(B \mid A_1) = 0.6$, $P(B \mid A_2) = 0.7$, and $P(B \mid A_3) = 0.8$. Given that the person selected at random has an auto or homeowner policy, what is the conditional probability that the person will renew at least one of these policies? That is,

$$P(B \mid A_1 \cup A_2 \cup A_3) = \frac{P(A_1 \cap B) + P(A_2 \cap B) + P(A_3 \cap B)}{P(A_1) + P(A_2) + P(A_3)}$$

$$= \frac{(0.4)(0.6) + (0.2)(0.7) + (0.2)(0.8)}{0.4 + 0.2 + 0.2}$$

$$= \frac{0.54}{0.80} = \frac{27}{40}$$

$$= 0.675. \quad \blacktriangleleft$$

EXAMPLE 1.4-12 A device has two components, C_1 and C_2, but it will continue to operate with only one active component for a one-year period. The probability that each will fail when both are in operation is 0.01 in that one-year period. However, when one fails, the probability of the other failing is 0.03 in that period due to added strain. Thus the probability that the device fails in one year is

$$P(C_1 \text{ fails})P(C_2 \text{ fails} \mid C_1 \text{ fails}) + P(C_2 \text{ fails})P(C_1 \text{ fails} \mid C_2 \text{ fails}) =$$

$$(0.01)(0.03) + (0.01)(0.03) = 0.0006. \quad \blacktriangleleft$$

EXERCISES

1.4-1 A common test for AIDS is called the ELISA (Enzyme-Linked Immunosorbent Assay) test. Among 1,000,000 people who are given the ELISA test, we can expect results similar to those given in the table.

	B_1: Carry AIDS Virus	B_2: Do Not Carry Aids Virus	Totals
A_1: Test Positive	4,885	73,630	78,515
A_2: Test Negative	115	921,370	921,485
Totals	5,000	995,000	1,000,000

If one of these 1,000,000 people is selected randomly, find the following probabilities: **(a)** $P(B_1)$, **(b)** $P(A_1)$, **(c)** $P(A_1 \mid B_2)$, **(d)** $P(B_1 \mid A_1)$. **(e)** In words, what do parts (c) and (d) say?

1.4-2 The following table classifies 1456 people by their sex and by whether or not they favor a gun law.

	Male (S_1)	Female (S_2)	Totals
Favor (A_1)	392	649	1041
Oppose (A_2)	241	174	415
Totals	633	823	1456

Compute the following probabilities if one of these 1456 persons is selected randomly: **(a)** $P(A_1)$, **(b)** $P(A_1 \mid S_1)$, **(c)** $P(A_1 \mid S_2)$. **(d)** Interpret your answers to parts (b) and (c).

1.4-3 Let A_1 and A_2 be the events that a person is left eye dominant or right eye dominant, respectively. When a person folds their hands, let B_1 and B_2 be the events that their left thumb and right thumb, respectively, are on top. A survey in one statistics class yielded the following table.

	B_1	B_2	Totals
A_1	5	7	12
A_2	14	9	23
Totals	19	16	35

If a student is selected randomly, find the following probabilities: **(a)** $P(A_1 \cap B_1)$, **(b)** $P(A_1 \cup B_1)$, **(c)** $P(A_1 \mid B_1)$, **(d)** $P(B_2 \mid A_2)$. **(e)** If the students had their hands folded and you hoped to select a right eye dominant student, would you select a "right thumb on top" or a "left thumb on top" student? Why?

1.4-4 Two cards are drawn successively and without replacement from an ordinary deck of playing cards. Compute the probability of drawing

(a) Two hearts.
(b) A heart on the first draw, a club on the second draw.
(c) A heart on the first draw, an ace on the second draw.

HINT: In part (c), note that a heart can be drawn by getting the ace of hearts or one of the other 12 hearts.

1.4-5 Suppose that $P(A) = 0.7$, $P(B) = 0.5$, and $P([A \cup B]') = 0.1$.

(a) Find $P(A \cap B)$.
(b) Give $P(A \mid B)$.
(c) Give $P(B \mid A)$.

1.4-6 A hand of 13 cards is to be dealt at random and without replacement from an ordinary deck of playing cards. Find the conditional probability that there are at least three kings in the hand given that the hand contains at least two kings.

1.4-7 Suppose that the genes for eye color for a certain male fruit fly are (R, W) and the genes for eye color for the mating female fruit fly are (R, W), where R and W represent red and white, respectively. Their offspring receive one gene for eye color from each parent.

(a) Define the sample space for the genes for eye color for the offspring.
(b) Assume that each of the four possible outcomes has equal probability. If an offspring ends up with either two red genes or one red and one white gene for eye color, its eyes will look red. Given that an offspring's eyes look red, what is the conditional probability that it has two red genes for eye color?

1.4-8 Suppose that there are 14 songs on a compact disk (CD) and you like 8 of them. When using the random button selector on a CD player, each of the 14 songs is played once in a random order. Find the probability that among the first 2 songs that are played,

(a) You like both of them.
(b) You like neither of them.
(c) You like exactly one of them.

1.4-9 An urn contains four colored balls: two orange and two blue. Two balls are selected at random without replacement, and you are told that at least one of them is orange. What is the probability that the other ball is also orange?

1.4-10 An urn contains 17 balls marked LOSE and 3 balls marked WIN. You and an opponent take turns selecting at random a single ball from the urn without replacement. The person who selects the third WIN ball wins the game. It does not matter who selected the first two WIN balls.

(a) If you draw first, find the probability that you win the game on your second draw.

(b) If you draw first, find the probability that your opponent wins the game on his second draw.

(c) If you draw first, what is the probability that you win? HINT: You could win on your second, third, fourth, ... , or tenth draw, not on your first.

(d) Would you prefer to draw first or second? Why?

1.4-11 In a string of 12 Christmas tree light bulbs, 3 are defective. The bulbs are selected at random and tested, one at a time, until the third defective bulb is found. Compute the probability that the third defective bulb is the

(a) Third bulb tested.

(b) Fifth bulb tested.

(c) Tenth bulb tested.

1.4-12 A small grocery store had l0 cartons of milk, 2 of which were sour. If you are going to buy the sixth carton of milk sold that day at random, compute the probability of selecting a carton of sour milk.

1.4-13 In the gambling game "craps" a pair of dice is rolled and the outcome of the experiment is the sum of the points on the up-sides of the six-sided dice. The bettor wins on the first roll if the sum is 7 or 11. The bettor loses on the first roll if the sum is 2, 3, or 12. If the sum is 4, 5, 6, 8, 9, or l0, that number is called the bettor's "point." Once the point is established, the rule is, If the bettor rolls a 7 before the "point," the bettor loses; but if the "point" is rolled before a 7, the bettor wins.

(a) List the 36 outcomes in the sample space for the roll of a pair of dice. Assume that each of them has a probability of 1/36.

(b) Find the probability that the bettor wins on the first roll. That is, find the probability of rolling a 7 or 11, $P(7 \text{ or } 11)$.

(c) Given that 8 is the outcome on the first roll, find the probability that the bettor now rolls the point 8 before rolling a 7 and thus wins. Note that at this stage in the game the only outcomes of interest are 7 and 8. Thus find $P(8 \mid 7 \text{ or } 8)$.

(d) The probability that a bettor rolls an 8 on the first roll and then wins is given by $P(8)P(8 \mid 7 \text{ or } 8)$. Show that this probability is $(5/36)(5/11)$.

(e) Show that the total probability that a bettor wins in the game of craps is 0.49293. HINT: Note that the bettor can win in one of several mutually exclusive ways: by rolling a 7 or 11 on the first roll or by establishing one of the points 4, 5, 6, 8, 9, or 10 on the first roll and then obtaining that point before a 7 on successive rolls.

1.4-14 A single card is drawn at random from each of six well-shuffled decks of playing cards. Let A be the event that all six cards drawn are different.

(a) Find $P(A)$.

(b) Find the probability that at least two of the drawn cards match.

1.4-15 Consider the birthdays of the students in a class of size r. Assume that the year consists of 365 days.

(a) How many different ordered samples of birthdays are possible (r in sample) allowing repetitions (with replacement)?

(b) The same as part (a) except requiring that all the students have different birthdays (without replacement)?

(c) If we can assume that each ordered outcome in part (a) has the same probability, what is the probability that at least two students have the same birthday?

(d) For what value of r is the probability in part (c) about equal to 1/2? Is this number surprisingly small? HINT: Use a calculator or computer to find r.

1.4-16 You are a member of a class of 18 students. A bowl contains 18 chips, 1 blue and 17 red. Each student is to take 1 chip from the bowl without replacement. The student who draws the blue chip is guaranteed an A for the course.

    (a) If you have a choice of drawing first, fifth, or last, which position would you choose? Justify your choice using probability.

    (b) Suppose the bowl contains 2 blue and 16 red chips. What position would you now choose?

1.4-17 A drawer contains four black, six brown, and eight olive socks. Two socks are selected at random from the drawer.

    (a) Compute the probability that both socks are the same color.

    (b) Compute the probability that both socks are olive if it is known that they are the same color.

1.4-18 Bowl A contains three red and two white chips, and bowl B contains four red and three white chips. A chip is drawn at random from bowl A and transferred to bowl B. Compute the probability of then drawing a red chip from bowl B.

1.4-19 An urn contains four balls numbered 1 through 4. The balls are selected one at a time without replacement. A match occurs if ball numbered m is the mth ball selected. Let the event A_i denote a match on the ith draw, $i = 1, 2, 3, 4$. Show that

    (a) $P(A_i) = \dfrac{3!}{4!}$.

    (b) $P(A_i \cap A_j) = \dfrac{2!}{4!}$.

    (c) $P(A_i \cap A_j \cap A_k) = \dfrac{1!}{4!}$.

    (d) The probability of at least one match is

$$P(A_1 \cup A_2 \cup A_3 \cup A_4) = 1 - \frac{1}{2!} + \frac{1}{3!} - \frac{1}{4!}.$$

    (e) Extend this exercise so that there are n balls in the urn. Show that the probability of at least one match is

$$P(A_1 \cup A_2 \cup \cdots \cup A_n) = 1 - \frac{1}{2!} + \frac{1}{3!} - \frac{1}{4!} + \cdots + \frac{(-1)^{n+1}}{n!}$$

$$= 1 - \left(1 - \frac{1}{1!} + \frac{1}{2!} - \frac{1}{3!} + \cdots + \frac{(-1)^n}{n!}\right).$$

    (f) What is the limit of this probability as n increases without bound?

1.4-20 Paper, like butcher's paper, is often tested for "burst strength" and "tear strength." Say we classify these strengths as low, middle, high; and we find after examinations of 100 pieces of paper the following:

Tear Strength	Burst Strength		
	A_1 (low)	A_2 (middle)	A_3 (high)
B_1 (low)	7	11	13
B_2 (middle)	11	21	9
B_3 (high)	12	9	7

If we select one of the pieces at random, what are the probabilities of selecting one with the following characteristics:

(a) A_1,
(b) $A_3 \cap B_2$,
(c) $A_2 \cup B_3$,
(d) A_1, given it is B_2,
(e) B_1, given it is A_3?

1.5 INDEPENDENT EVENTS

For certain pairs of events, the occurrence of one of them may or may not change the probability of the occurrence of the other. In the latter case they are said to be **independent events**. However, before giving the formal definition of independence, let us consider an example.

EXAMPLE 1.5-1

Flip a coin twice and observe the sequence of heads and tails. The sample space is then

$$S = \{HH, HT, TH, TT\}.$$

It is reasonable to assign a probability of 1/4 to each of these four outcomes. Let

$$A = \{\text{heads on the first flip}\} = \{HH, HT\},$$
$$B = \{\text{tails on the second flip}\} = \{HT, TT\},$$
$$C = \{\text{tails on both flips}\} = \{TT\}.$$

Now $P(B) = 2/4 = 1/2$. However, if we are given that C has occurred, then $P(B \mid C) = 1$ because $C \subset B$. That is, the knowledge of the occurrence of C has changed the probability of B. On the other hand, if we are given that A has occurred,

$$P(B \mid A) = \frac{P(A \cap B)}{P(A)} = \frac{1/4}{2/4} = \frac{1}{2} = P(B).$$

So the occurrence of A has not changed the probability of B. Hence the probability of B does not depend upon knowledge about event A, so we say that A and B are independent events. That is, events A and B are independent if the occurrence of one of them does not affect the probability of the occurrence of the other. A more mathematical way of saying this is

$$P(B \mid A) = P(B) \qquad \text{or} \qquad P(A \mid B) = P(A),$$

provided that $P(A) > 0$ or, in the latter case, $P(B) > 0$. With the first of these equalities and the multiplication rule (Definition 1.4-2), we have

$$P(A \cap B) = P(A)P(B \mid A) = P(A)P(B).$$

The second of these equalities, namely $P(A \mid B) = P(A)$, gives us the same result

$$P(A \cap B) = P(B)P(A \mid B) = P(B)P(A).$$ ◄

This example motivates the following definition of independent events.

Definition 1.5-1   Events A and B are *independent* if and only if $P(A \cap B) = P(A)P(B)$. Otherwise A and B are called *dependent* events.

Events that are independent are sometimes called **statistically independent**, **stochastically independent**, or **independent in a probabilistic sense**, but in most instances we use independent without a modifier if there is no possibility of misunderstanding. It is interesting to note that the definition always holds if $P(A) = 0$ or $P(B) = 0$ because then $P(A \cap B) = 0$, since $(A \cap B) \subset A$ and $(A \cap B) \subset B$. Thus the left-hand and right-hand member of $P(A \cap B) = P(A)P(B)$ are both equal to zero and thus are equal to each other.

EXAMPLE 1.5-2   A red die and a white die are rolled. Let event $A = \{4$ on the red die$\}$ and event $B = \{$sum of dice is odd$\}$. Of the 36 equally likely outcomes, 6 are favorable to A, 18 are favorable to B, and 3 are favorable to $A \cap B$. Thus

$$P(A)P(B) = \frac{6}{36} \cdot \frac{18}{36} = \frac{3}{36} = P(A \cap B).$$

Hence A and B are independent by Definition 1.5-1.   ◄

EXAMPLE 1.5-3   A red die and a white die are rolled. Let event $C = \{5$ on red die$\}$ and event $D = \{$sum of dice is 11$\}$. Of the 36 equally likely outcomes, 6 are favorable to C, 2 are favorable to D, and 1 is favorable to $C \cap D$. Thus

$$P(C)P(D) = \frac{6}{36} \cdot \frac{2}{36} = \frac{1}{108} \neq \frac{1}{36} = P(C \cap D).$$

Hence C and D are dependent events by Definition 1.5-1.   ◄

Theorem 1.5-1   If A and B are independent events, then the following pairs of events are also independent:

(a) A and B'.
(b) A' and B.
(c) A' and B'.

Proof: We know that conditional probability satisfies the axioms for a probability function. Hence, if $P(A) > 0$, then $P(B'\,|\,A) = 1 - P(B\,|\,A)$. Thus

$$P(A \cap B') = P(A)P(B'\,|\,A) = P(A)[1 - P(B\,|\,A)]$$
$$= P(A)[1 - P(B)]$$
$$= P(A)P(B'),$$

since $P(B\,|\,A) = P(B)$ by hypothesis. Thus A and B' are independent events. The proofs for parts (b) and (c) are left as exercises. ○

Before extending the definition of independent events to more than two events, we present the following example.

EXAMPLE **1.5-4**   An urn contains four balls numbered 1, 2, 3, and 4. One ball is to be drawn at random from the urn. Let the events A, B, and C be defined by $A = \{1, 2\}$, $B = \{1, 3\}$, $C = \{1, 4\}$. Then $P(A) = P(B) = P(C) = 1/2$. Furthermore,

$$P(A \cap B) = \frac{1}{4} = P(A)P(B),$$

$$P(A \cap C) = \frac{1}{4} = P(A)P(C),$$

$$P(B \cap C) = \frac{1}{4} = P(B)P(C),$$

which implies that A, B, and C are independent in pairs (called **pairwise independence**). However, since $A \cap B \cap C = \{1\}$, we have

$$P(A \cap B \cap C) = \frac{1}{4} \neq \frac{1}{8} = P(A)P(B)P(C).$$

That is, something seems to be lacking for the complete independence of A, B, and C. ◄

This example illustrates the reason for the second condition in Definition 1.5-2.

Definition 1.5-2   Events A, B, and C are *mutually independent* if and only if the following two conditions hold:

(a) They are pairwise independent; that is,

$$P(A \cap B) = P(A)P(B), \qquad P(A \cap C) = P(A)P(C),$$

and

$$P(B \cap C) = P(B)P(C).$$

(b) $P(A \cap B \cap C) = P(A)P(B)P(C).$

Definition 1.5-2 can be extended to mutual independence of four or more events. In this extension, each pair, triple, quartet, and so on, must satisfy this type of multiplication rule. If there is no possibility of misunderstanding, *independent* is often used without the modifier *mutually* when considering several events.

EXAMPLE 1.5-5 A rocket has a built-in redundant system. In this system, if component K_1 fails, it is bypassed and component K_2 is used. If component K_2 fails, it is bypassed and component K_3 is used. (An example of such components is computer systems.) Suppose that the probability of failure of any one of these components is 0.15 and assume that the failures of these components are mutually independent events. Let A_i denote the event that component K_i fails for $i = 1, 2, 3$. Because the system fails if K_1 fails and K_2 fails and K_3 fails, the probability that the system does not fail is given by

$$P[(A_1 \cap A_2 \cap A_3)'] = 1 - P(A_1 \cap A_2 \cap A_3)$$

$$= 1 - P(A_1)P(A_2)P(A_3)$$

$$= 1 - (0.15)^3$$

$$= 0.9966.$$

One way to increase the reliability of such a system is to add more components (realizing that this also adds weight and takes up space). For example, if a fourth component K_4 were added to this system, the probability that the system does not fail is

$$P[(A_1 \cap A_2 \cap A_3 \cap A_4)'] = 1 - (0.15)^4 = 0.9995. \qquad \blacktriangleleft$$

The proof and illustration of the following results are left as exercises. If A, B, and C are mutually independent events, then the following events are also independent:

(a) A and $(B \cap C)$;
(b) A and $(B \cup C)$;
(c) A' and $(B \cap C')$.

In addition, A', B', and C' are mutually independent.

Many experiments consist of a sequence of n trials that are mutually independent. If the outcomes of the trials, in fact, do not have anything to do with one another, then events, such that each is associated with a different trial, should be independent in the probability sense. That is, if the event A_i is associated with the ith trial, $i = 1, 2, \ldots, n$, then

$$P(A_1 \cap A_2 \cap \cdots \cap A_n) = P(A_1)P(A_2) \cdots P(A_n).$$

EXAMPLE 1.5-6 A fair six-sided die is rolled six independent times. Let A_i be the event that side i is observed on the ith roll, called a match on the ith trial, $i = 1, 2, \ldots, 6$.

Thus $P(A_i) = 1/6$, and $P(A'_i) = 1 - 1/6 = 5/6$. If we let B denote the event that at least one match occurs, then B' is the event that no matches occur. Thus

$$P(B) = 1 - P(B') = 1 - P(A'_1 \cap A'_2 \cap \cdots \cap A'_6)$$

$$= 1 - \frac{5}{6} \cdot \frac{5}{6} \cdot \frac{5}{6} \cdot \frac{5}{6} \cdot \frac{5}{6} \cdot \frac{5}{6} = 1 - \left(\frac{5}{6}\right)^6$$

is the probability of B. ◄

The sample space for an experiment of n trials is a set of n-tuples, where the ith component denotes the outcome on the ith trial. For example, if a six-sided die is rolled five times,

$$S = \{(O_1, O_2, O_3, O_4, O_5) : O_i = 1, 2, 3, 4, 5, \text{ or } 6, \text{ for } i = 1, 2, 3, 4, 5\}.$$

That is, S is a set of five-tuples, where each component is one of the first six positive integers.

If a coin is tossed two times,

$$S = \{(O_1, O_2) : O_i = H \text{ or } T, i = 1, 2\}.$$

We often drop the commas and parentheses and let, for example, $(H, T) = HT$, as in Example 1.5-1.

EXAMPLE 1.5-7   The probability that a company's work force has at least one accident in a month is $(0.01)k$, where k is the number of days in that month (say February has 28 days). Assume the numbers of accidents is independent from month to month. If the company's year starts with January, the probability that the first accident is in April is

$P(\text{none in Jan., none in Feb., none in March, at least one in April}) =$
$$(1 - 0.31)(1 - 0.28)(1 - 0.31)(0.30) = (0.69)(0.72)(0.69)(0.30) = 0.103.$$

◄

EXAMPLE 1.5-8   Three inspectors look at a critical component of a product. Their probabilities of detecting a defect are different, namely 0.99, 0.98, and 0.96, respectively. If we assume independence the probability of at least one detecting the defect is

$$1 - (0.01)(0.02)(0.04) = 0.999992.$$

The probability of only one finding the defect is

$$(0.99)(0.02)(0.04) + (0.01)(0.98)(0.04) + (0.01)(0.02)(0.96) = 0.001576.$$

As an exercise, compute the probability exactly: **(a)** two find the defect, **(b)** three find the defect. ◄

EXAMPLE 1.5-9   Suppose that on five consecutive days an "instant winner" lottery ticket is purchased and the probability of winning is 1/5 on each day. Assuming independent trials,

$$P(WWLLL) = \left(\frac{1}{5}\right)^2 \left(\frac{4}{5}\right)^3,$$

$$P(LWLWL) = \frac{4}{5} \cdot \frac{1}{5} \cdot \frac{4}{5} \cdot \frac{1}{5} \cdot \frac{4}{5} = \left(\frac{1}{5}\right)^2 \left(\frac{4}{5}\right)^3.$$

In general, the probability of purchasing two winning tickets and three losing tickets is

$$\binom{5}{2} \left(\frac{1}{5}\right)^2 \left(\frac{4}{5}\right)^3 = \frac{5!}{2!3!} \left(\frac{1}{5}\right)^2 \left(\frac{4}{5}\right)^3 = 0.2048$$

because there are $\binom{5}{2}$ ways to select the positions (or the days) for the winning tickets, and each of these $\binom{5}{2}$ ways has the probability $(1/5)^2(4/5)^3$.   ◄

EXERCISES

1.5-1 Let A and B be independent events with $P(A) = 0.7$ and $P(B) = 0.2$. Compute **(a)** $P(A \cap B)$, **(b)** $P(A \cup B)$, and **(c)** $P(A' \cup B')$.

1.5-2 Let $P(A) = 0.3$ and $P(B) = 0.6$.

  (a) Find $P(A \cup B)$ when A and B are independent.
  (b) Find $P(A \mid B)$ when A and B are mutually exclusive.

1.5-3 Let A and B be independent events with $P(A) = 1/4$ and $P(B) = 2/3$. Compute: **(a)** $P(A \cap B)$, **(b)** $P(A \cap B')$, **(c)** $P(A' \cap B')$, **(d)** $P[(A \cup B)']$, and **(e)** $P(A' \cap B)$.

1.5-4 Prove parts (b) and (c) of Theorem 1.5-1.

1.5-5 If $P(A) = 0.8$, $P(B) = 0.5$, and $P(A \cup B) = 0.9$, are A and B independent events? Why?

1.5-6 If A, B, and C are mutually independent, show that the following pairs of events are independent: A and $(B \cap C)$, A and $(B \cup C)$, A' and $(B \cap C')$. Also show that A', B' and C' are mutually independent.

1.5-7 Each of three football players will attempt to kick a field goal from the 25-yard line. Let A_i denote the event that the field goal is made by player i, $i = 1, 2, 3$. Assume that A_1, A_2, A_3 are mutually independent and that $P(A_1) = 0.5$, $P(A_2) = 0.7$, $P(A_3) = 0.6$.

  (a) Compute the probability that exactly one player is successful.
  (b) Compute the probability that exactly two players make a field goal (i.e., one misses).

1.5-8 Die A has orange on one face and blue on five faces, Die B has orange on two faces and blue on four faces, Die C has orange on three faces and blue on three faces. These are unbiased dice. If the three dice are rolled, find the probability that exactly two of the three dice come up orange.

1.5-9 Suppose that A, B, and C are mutually independent events and that $P(A) = 0.5$, $P(B) = 0.8$, and $P(C) = 0.9$. Find the probabilities that **(a)** all three events occur, **(b)** exactly two of the three events occur, and **(c)** none of the events occur.

1.5-10 Let D_1, D_2, D_3 be three four-sided dice whose sides have been labeled as follows:

$$D_1: 0333 \qquad D_2: 2225 \qquad D_3: 1146$$

These dice are rolled at random. Let A, B, and C be the events that the outcome on die D_1 is larger than the outcome on D_2, the outcome on D_2 is larger than the outcome on D_3, the outcome on D_3 is larger than the outcome on D_1, respectively. Show that **(a)** $P(A) = 9/16$, **(b)** $P(B) = 9/16$, **(c)** $P(C) = 10/16$. Do you find it interesting that each of the probabilities that D_1 "beats" D_2, D_2 "beats" D_3, and D_3 "beats" D_1 is greater than 1/2? Thus it is difficult to determine the "best" die.

1.5-11 Let A and B be two events.

(a) If the events A and B are mutually exclusive, are A and B always independent? If the answer is no, can they ever be independent? Explain.

(b) If $A \subset B$, can A and B ever be independent events? Explain.

1.5-12 Flip an unbiased coin five independent times. Compute the probability of

(a) *HHTHT*.

(b) *THHHT*.

(c) *HTHTH*.

(d) Three heads occurring in the five trials.

1.5-13 An urn contains two red balls and four white balls. Sample successively five times at random and with replacement so that the trials are independent. Compute the probability of each of the two sequences $WWRWR$ and $RWWWR$.

1.5-14 In Example 1.5-5, suppose that the probability of failure of a component is $p = 0.4$. Find the probability that the system does not fail if the number of redundant components is

(a) 3.

(b) 8.

1.5-15 An urn contains 10 red and 10 white balls. The balls are drawn from the urn at random, one at a time. Find the probability that the fourth white ball is the sixth ball drawn if the sampling is done

(a) With replacement.

(b) Without replacement.

(c) In the World Series the American League (red) and National League (white) teams play until one team wins four games. Do you think that this urn model could be used to describe the probabilities of a 4-, 5-, 6-, or 7-game series? If your answer is yes, would you choose sampling with or without replacement in your model? (For your information, the numbers of 4-, 5-, 6-, and 7-game series, up to and including 2004, were 18, 21, 21, 36. The World Series was cancelled in 1994. In 1903 and 1919–1921, winners had to take 5 out of 9 games. Three of those series went eight, one went seven.)

1.5-16 An urn contains five balls, one marked WIN and four marked LOSE. You and another player take turns selecting a ball from the urn, one at a time. The first person to select the WIN ball is the winner. If you draw first, find the probability that you will win if the sampling is done

(a) With replacement.

(b) Without replacement.

1.5-17 Each of the 12 students in a class is given a fair 12-sided die. In addition, each student is numbered from 1 to 12.

(a) If the students roll their dice, what is the probability that there is at least one "match" like student 4 rolls a 4?

(b) If you are a member of this class, what is the probability that at least one of the other 11 students rolls the same number as you do?

1.5-18 An eight-team single elimination tournament is set up as follows:

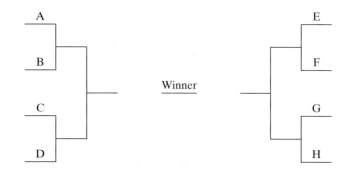

For example, eight students (called A–H) set up a tournament among themselves. The top listed student in each bracket calls Heads or Tails when their opponent flips a coin. If the call is correct, the student moves on to the next bracket.

(a) How many coin flips are required to determine the tournament winner?

(b) What is the probability that you can predict all of the winners?

(c) In NCAA Division I basketball, after a play-in game between the 64th and 65th seeds, 64 teams participate in a single elimination tournament to determine the national champion. Considering only the remaining 64 teams, how many games are required to determine the national champion?

(d) Assume that for any given game, either team has an equal chance of winning. (That is probably not true.) *Time*, on page 43 of the March 22, 1999 issue, claimed that the "mathematical odds of predicting all 63 NCAA games correctly is 1 in 75 million." Do you agree with this statement? If not, why not?

1.5-19 Extend Example 1.5-6 to an n-sided die. That is, suppose that a fair n-sided die is rolled n independent times. A match occurs if side i is observed on the ith trial, $i = 1, 2, \ldots, n$.

(a) Show that the probability of at least one match is

$$1 - \left(\frac{n-1}{n}\right)^n = 1 - \left(1 - \frac{1}{n}\right)^n.$$

(b) Find the limit of this probability as n increases without bound.

1.5-20 An urn contains n balls numbered from 1 through n. A random sample of n balls is selected from the urn, one at a time. A match occurs if ball numbered i is selected on the ith draw. For $n = 1$ to 15, find the probability of at least one match if the sampling is done

(a) With replacement (see Exercise 1.5-19).

(b) Without replacement (see Exercise 1.4-19).

(c) How much does n affect these probabilities?

(d) How does sampling with and without replacement affect these probabilities?

(e) Illustrate these probabilities empirically, either performing the experiments physically or simulating them on a computer.

1.6 BAYES'S THEOREM

We consider Bayes's theorem in this section, beginning with an example.

EXAMPLE 1.6-1 Bowl B_1 contains two red and four white chips; bowl B_2 contains one red and two white chips; and bowl B_3 contains five red and four white chips. Say that the probabilities for selecting the bowls are not the same but are given by $P(B_1) = 1/3$, $P(B_2) = 1/6$, and $P(B_3) = 1/2$, where B_1, B_2, and B_3 are the events that bowls B_1, B_2, and B_3 are chosen, respectively. The experiment consists of selecting a bowl with these probabilities and then drawing a chip at random from that bowl. Let us compute the probability of event R, drawing a red chip, say $P(R)$. Note that $P(R)$ is dependent first of all on which bowl is selected and then on the probability of drawing a red chip from the selected bowl. That is, the event R is the union of the mutually exclusive events $B_1 \cap R$, $B_2 \cap R$, and $B_3 \cap R$. Thus

$$P(R) = P(B_1 \cap R) + P(B_2 \cap R) + P(B_3 \cap R)$$
$$= P(B_1)P(R \mid B_1) + P(B_2)P(R \mid B_2) + P(B_3)P(R \mid B_3)$$
$$= \frac{1}{3} \cdot \frac{2}{6} + \frac{1}{6} \cdot \frac{1}{3} + \frac{1}{2} \cdot \frac{5}{9} = \frac{4}{9}.$$

Suppose now that the outcome of the experiment is a red chip, but we do not know from which bowl it was drawn. Accordingly, we compute the conditional probability that the chip was drawn from bowl B_1, namely, $P(B_1 \mid R)$. From the definition of conditional probability and the result above, we have that

$$P(B_1 \mid R) = \frac{P(B_1 \cap R)}{P(R)}$$
$$= \frac{P(B_1)P(R \mid B_1)}{P(B_1)P(R \mid B_1) + P(B_2)P(R \mid B_2) + P(B_3)P(R \mid B_3)}$$
$$= \frac{(1/3)(2/6)}{(1/3)(2/6) + (1/6)(1/3) + (1/2)(5/9)} = \frac{2}{8}.$$

Similarly, we have that

$$P(B_2 \mid R) = \frac{P(B_2 \cap R)}{P(R)} = \frac{(1/6)(1/3)}{4/9} = \frac{1}{8}$$

and

$$P(B_3 \mid R) = \frac{P(B_3 \cap R)}{P(R)} = \frac{(1/2)(5/9)}{4/9} = \frac{5}{8}.$$

Note that the conditional probabilities $P(B_1 \mid R)$, $P(B_2 \mid R)$, and $P(B_3 \mid R)$ have changed from the original probabilities $P(B_1)$, $P(B_2)$, and $P(B_3)$ in a way that agrees with your intuition. Namely, once the red chip has been observed, the probability concerning B_3 seems more favorable than originally because B_3 has a larger percentage of red chips than do B_1 and B_2. The conditional probabilities of B_1 and B_2 decrease from their original ones once the red chip is observed. Frequently, the original probabilities are called *prior probabilities*, and the conditional probabilities are the *posterior probabilities*.   ◀

We generalize the result of Example 1.6-1. Let $B_1, B_2, \ldots, B_m$ constitute a *partition* of the sample space S. That is,

$$S = B_1 \cup B_2 \cup \cdots \cup B_m \text{ and } B_i \cap B_j = \emptyset, i \neq j.$$

Of course, the events $B_1, B_2, \ldots, B_m$ are mutually exclusive and exhaustive (since the union of the disjoint sets equals the sample space S). Furthermore, suppose the **prior probability** of the event B_i is positive; that is $P(B_i) > 0$, $i = 1, \ldots, m$. If A is an event, then A is the union of m mutually exclusive events, namely,

$$A = (B_1 \cap A) \cup (B_2 \cap A) \cup \cdots \cup (B_m \cap A).$$

Thus

$$P(A) = \sum_{i=1}^{m} P(B_i \cap A)$$

$$= \sum_{i=1}^{m} P(B_i) P(A \mid B_i). \tag{1.6-1}$$

If $P(A) > 0$, we have that

$$P(B_k \mid A) = \frac{P(B_k \cap A)}{P(A)}, \qquad k = 1, 2, \ldots, m. \tag{1.6-2}$$

Using Equation 1.6-1 and replacing $P(A)$ in Equation 1.6-2, we have **Bayes's theorem**:

$$P(B_k \mid A) = \frac{P(B_k) P(A \mid B_k)}{\sum_{i=1}^{m} P(B_i) P(A \mid B_i)}, \qquad k = 1, 2, \ldots, m.$$

The conditional probability $P(B_k \mid A)$ is often called the **posterior probability** of B_k. The following example illustrates one application of Bayes's result.

EXAMPLE 1.6-2   In a certain factory, machines I, II, and III are all producing springs of the same length. Of their production, machines I, II, and III produce 2%, 1%, and 3% defective springs, respectively. Of the total production of springs in the factory, machine I produces 35%, machine II produces 25%, and machine III produces 40%. If one spring is selected at random from the total springs produced in a day, the probability that it is defective, in an obvious notation, equals

$$P(D) = P(I)P(D \mid I) + P(II)P(D \mid II) + P(III)P(D \mid III)$$

$$= \left(\frac{35}{100}\right)\left(\frac{2}{100}\right) + \left(\frac{25}{100}\right)\left(\frac{1}{100}\right) + \left(\frac{40}{100}\right)\left(\frac{3}{100}\right) = \frac{215}{10,000}.$$

If the selected spring is defective, the conditional probability that it was produced by machine III is, by Bayes's formula,

$$P(III \mid D) = \frac{P(III)P(D \mid III)}{P(D)} = \frac{(40/100)(3/100)}{215/10,000} = \frac{120}{215}.$$

Note how the posterior probability of III increased from the prior probability of III after the defective spring was observed because III produces a larger percentage of defectives than do I and II. ◄

EXAMPLE 1.6-3 A Pap smear is a screening procedure used to detect cervical cancer. For women with this cancer, there are about 16% *false negatives*; that is,

$$P(T^- = \text{test negative} \mid C = \text{cancer}) = 0.16.$$

Thus

$$P(T^+ = \text{test positive} \mid C = \text{cancer}) = 0.84.$$

For women without cancer, there are about 19% false positives; that is,

$$P(T^+ \mid C' = \text{not cancer}) = 0.19.$$

Hence,

$$P(T^- \mid C' = \text{not cancer}) = 0.81.$$

In the United States, there are about 8 women in 100,000 who have this cancer; that is,

$$P(C) = 0.00008; \quad \text{so} \quad P(C') = 0.99992.$$

By Bayes's theorem,

$$P(C \mid T^+) = \frac{P(C \text{ and } T^+)}{P(T^+)}$$

$$= \frac{(0.00008)(0.84)}{(0.00008)(0.84) + (0.99992)(0.19)}$$

$$= \frac{672}{672 + 1899848} = 0.0000354.$$

What this means is that for every million positive Pap smears, only 354 represent true cases of cervical cancer and makes one question the value of the procedure. The reason for this ineffective procedure is the percentage of women having that cancer is so small and the error rates of the procedure, namely 0.16 and 0.19, are so high. (These numbers are close to the true values.) ◄

EXERCISES

1.6-1 Bowl B_1 contains 2 white chips, bowl B_2 contains 2 red chips, bowl B_3 contains 2 white and 2 red chips, and bowl B_4 contains 3 white chips and 1 red chip. The probabilities of selecting bowl B_1, B_2, B_3, or B_4 are 1/2, 1/4, 1/8, and 1/8, respectively. A bowl is selected using these probabilities, and a chip is then drawn at random. Find

    (a) $P(W)$, the probability of drawing a white chip.

    (b) $P(B_1 \mid W)$, the conditional probability that bowl B_1 had been selected, given that a white chip was drawn.

1.6-2 Bean seeds from supplier A have an 85% germination rate and those from supplier B have a 75% germination rate. A seed packaging company purchases 40% of their bean seeds from supplier A and 60% from supplier B and mixes these seeds together.

    (a) Find the probability $P(G)$ that a seed selected at random from the mixed seeds will germinate.

    (b) Given that a seed germinates, find the probability that the seed was purchased from supplier A.

1.6-3 The Belgium 20-franc coin ($B20$), the Italian 500-lire coin ($I500$), and the Hong Kong 5-dollar coin ($HK5$) are approximately the same size. Coin purse 1 ($C1$) contains 6 of each of these coins. Coin purse 2 ($C2$) contains 9 $B20$'s, 6 $I500$'s, and 3 $HK5$'S. A fair four-sided die is rolled. If the outcome is {1}, a coin is selected randomly from $C1$. If the outcome belongs to {2, 3, 4}, a coin is selected randomly from $C2$. Find

    (a) $P(B20)$, the probability of selecting a Belgian coin.

    (b) $P(C1 \mid B20)$, the probability that the coin was selected from $C1$, given that it was a Belgian coin.

1.6-4 Assume that an insurance company knows the following probabilities relating to automobile accidents:

Age of Driver	Probability of Accident	Fraction of Company's Insured Drivers
16–25	0.05	0.10
26–50	0.02	0.55
51–65	0.03	0.20
66–90	0.04	0.15

A randomly selected driver from the company's insured drivers has a accident. What is the conditional probability that the driver is in the 16–25 age group?

1.6-5 At a hospital's emergency room, patients are classified and 20% of them are critical, 30% are serious, and 50% are stable. Of the critical ones, 30% die; of the serious, 10% die; and of the stable, 1% die. Given that a patient dies, what is the conditional probability that the patient was classified as critical.

1.6-6 A life insurance company issues standard, preferred, and ultra-preferred policies. Of the company's policyholders of a certain age, 60% are standard with a probability of 0.01 of dying in the next year, 30% preferred with a probability of 0.008 of dying in the next year, and 10% are ultra-preferred with a probability of 0.007 of dying in the next year. A policyholder of that age dies in the next year. What are the conditional probabilities of the deceased being standard, preferred, and ultra-preferred?

1.6-7 Among 60-year-old college professors, 10% are smokers and 90% are nonsmokers. The probability of a nonsmoker dying in the next year is 0.005 and the probability for smokers is 0.05. Given that one of this group of college professors dies in the next year, what is the conditional probability that the professor is a smoker?

1.6-8 A store sells four brands of VCRs. The least expensive brand B_1 accounts for 40% of the sales. The other brands (in order of their price) have the following percentages of sales: B_2, 30%; B_3, 20%; and B_4, 10%. The respective probabilities of needing repair during warranty are 0.10 for B_1, 0.05 for B_2, 0.03 for B_3, and 0.02 for B_4. A randomly selected purchaser has a VCR that needed repair under warranty. What are the four conditional probabilities of being brand B_i, $i = 1, 2, 3, 4$?

1.6-9 There is a new diagnostic test for a disease that occurs in about 0.05% of the population. The test is not perfect but will detect a person with the disease 99% of the time. It will, however, say that a person without the disease has the disease about 3% of the time. A person is selected at random from the population and the test indicates that this person has the disease. What are the conditional probabilities that

(a) the person has the disease?

(b) the person does not have the disease?

Discuss. HINT: Note the fraction 0.0005 of diseased persons in the population is much smaller than the error probabilities of 0.01 and 0.03.

1.6-10 Suppose we want to investigate the percentage of abused children in a certain population. To do this, doctors examine some of these children taken at random from this population. However, doctors are not perfect: They sometimes classify an abused child (A) as one not abused (ND) or they classify a nonabused child (N) as one that is abused (AD). Suppose these error rates are $P(ND \mid A) = 0.08$ and $P(AD \mid N) = 0.05$, respectively; thus $P(AD \mid A) = 0.92$ and $P(ND \mid N) = 0.95$ are probabilities of the correct decisions. Let us pretend that only two percent of all children are abused; that is, $P(A) = 0.02$ and $P(N) = 0.98$.

(a) Select a child at random. What is the probability that the doctor classifies this child as abused. That is, compute

$$P(AD) = P(A)P(AD \mid A) + P(N)P(AD \mid N).$$

(b) Given the child is classified by the doctor as abused, compute $P(N \mid AD)$ and $P(A \mid AD)$.

(c) Also compute $P(N \mid ND)$ and $P(A \mid ND)$.

(d) Are these probabilities in (b) and (c) alarming? This happens because the error rates of 0.08 and 0.05 are high relative to the fraction 0.02 of abused children in the population.

CHAPTER

2

DISCRETE DISTRIBUTIONS

2.1 RANDOM VARIABLES OF THE DISCRETE TYPE

An outcome space S may be difficult to describe if the elements of S are not numbers. We shall now discuss how we can use a rule by which each outcome of a random experiment, an element s of S may be associated with a real number x. We begin the discussion with an example.

EXAMPLE 2.1-1

A rat is selected at random from a cage and its sex is determined. The set of possible outcomes is female and male. Thus the outcome space is $S = \{\text{female, male}\} = \{F, M\}$. Let X be a function defined on S such that $X(F) = 0$ and $X(M) = 1$. Thus X is a real-valued function that has the outcome space S as its domain and the set of real numbers $\{x : x = 0, 1\}$ as its range. We call X a random variable and, in this example, the space associated with X is $\{x : x = 0, 1\}$. ◄

We now formulate the definition of a random variable.

Definition 2.1-1	Given a random experiment with an outcome space S, a function X that assigns to each element s in S one and only one real number $X(s) = x$ is called a *random variable*. The **space** of X is the set of real numbers $\{x : X(s) = x, s \in S\}$, where $s \in S$ means the element s belongs to the set S.

Remark As we give examples of random variables and their probability distributions, the reader soon recognizes that when observing a random experiment the experimenter must take some type of measurement (or measurements). This measurement can be thought of as the outcome of a random variable. We would simply like to know the probability of a measurement ending in A, a subset of the space of X. If this is known for all subsets A, then we know the probability distribution of the random variable. Obviously, in practice, we do not very often know this distribution exactly. Hence statisticians make conjectures about these distributions; that is, we construct probabilistic models for random variables. The ability of a statistician to model a real situation appropriately is a valuable trait. In this chapter we introduce some probability models in which the spaces of the random variables consist of sets of integers. ∎

It may be that the set S has elements that are themselves real numbers. In such an instance we could write $X(s) = s$ so that X is the identity function and the space of X is also S. This is illustrated in Example 2.1-2.

EXAMPLE 2.1-2	Let the random experiment be the cast of a die, observing the number of spots on the side facing up. The outcome space associated with this experiment is $S = \{1, 2, 3, 4, 5, 6\}$. For each $s \in S$, let $X(s) = s$. The space of the random variable X is then $\{1, 2, 3, 4, 5, 6\}$.

If we associate a probability of 1/6 with each outcome, then $P(X = 5) = 1/6$, $P(2 \leq X \leq 5) = 4/6$, and $P(X \leq 2) = 2/6$, for example, seem to be reasonable assignments, where $\{2 \leq X \leq 5\}$ means $\{X = 2, 3, 4, \text{ or } 5\}$ and $\{X \leq 2\}$ means $\{X = 1 \text{ or } 2\}$, in this example. ◄

The student will no doubt recognize two major difficulties here:

1. In many practical situations the probabilities assigned to the events are unknown.
2. Since there are many ways of defining a function X on S, which function do we want to use?

As a matter of fact, the solutions to these problems in particular cases are major concerns in applied statistics. In considering (2), statisticians try to determine what *measurement* (or measurements) should be taken on an outcome; that is, how best do we "mathematize" the outcome? These measurement problems are most difficult and can be answered only by getting involved in a practical project. For (1), we often need, through repeated observations (called sampling), to estimate these probabilities or percentages. For example, what percentage of newborn girls in the University of Iowa Hospital weigh less than 7 pounds? Here a newborn baby girl is the outcome, and we have measured her one way (by weight); but obviously there are many other ways of measuring her. If we let X be the weight in pounds, we are interested in the probability $P(X < 7)$, and we can only estimate this by repeated observations. One obvious way of estimating this is by use of the relative frequency of $\{X < 7\}$ after a number of observations. If it is reasonable to make additional assumptions, we will study, in this text, other ways of estimating this probability. It is this latter aspect with which mathematical statistics is concerned. That is, if we assume certain models, we find that the theory of statistics can explain how best to draw conclusions or make predictions.

In many instances, it is clear exactly what function X the experimenter wants to define on the outcome space. For example, the caster in the dice game "craps" is concerned about the sum of the spots, say X, that are up on the pair of dice. Hence we go directly to the space of X, which we shall denote by the same letter S. After all, in the dice game the caster is directly concerned only with the probabilities associated with X. Hence for convenience, the reader can, in many instances, think of the space of X as being the outcome space.

Let X denote a random variable with space S. Suppose that we know how the probability is distributed over the various subsets A of S; that is, we can compute $P(X \in A)$. In this sense, we speak of the distribution of the random variable X, meaning, of course, the distribution of probability associated with the space S of X.

Let X denote a random variable with one-dimensional space S, a subset of the real numbers. Suppose that the space S contains a countable number of points; that is, S contains either a finite number of points or the points of S can be put into a one-to-one correspondence with the positive integers. Such a set S is called a set of discrete points or simply a discrete outcome space. Furthermore, the random variable X is called a random variable of the **discrete type**, and X is said to have a distribution of the discrete type.

For a random variable X of the discrete type, the probability $P(X = x)$ is frequently denoted by $f(x)$, and this function $f(x)$ is called the **probability mass function**. Note that some authors refer to $f(x)$ as the probability function, the frequency function, or the probability density function. In the discrete case we shall use "probability mass function," and it is hereafter abbreviated p.m.f.

Let $f(x)$ be the p.m.f. of the random variable X of the discrete type, and let S be the space of X. Since $f(x) = P(X = x)$, for $x \in S$, $f(x)$ must be non-negative for $x \in S$ and we want all these probabilities to add to 1 because each $P(X = x)$ represents the fraction of times x can be expected to occur. Moreover, to determine the probability associated with the event $A \in S$, we would sum the probabilities of the x values in A. This leads us to the following definition.

Definition 2.1-2   The p.m.f. $f(x)$ of a discrete random variable X is a function that satisfies the following properties:

(a) $f(x) > 0, \qquad x \in S$;

(b) $\displaystyle\sum_{x \in S} f(x) = 1$;

(c) $\displaystyle P(X \in A) = \sum_{x \in A} f(x), \qquad$ where $A \subset S$.

Of course, we usually let $f(x) = 0$ when $x \notin S$ and thus the domain of $f(x)$ is the set of real numbers. When we define the p.m.f. $f(x)$ and do not say zero elsewhere, then we tacitly mean that $f(x)$ has been defined at all x's in the space S, and it is assumed that $f(x) = 0$ elsewhere, namely, $f(x) = 0$ when $x \notin S$. Since the probability $P(X = x) = f(x) > 0$ when $x \in S$ and since S contains all the probability associated with X, we sometimes refer to S as the **support** of X as well as the space of X.

When a p.m.f. is constant on the space or support, we say that the distribution is **uniform** over that space. For illustration, in Example 2.1-2, X has a discrete uniform distribution on $S = \{1, 2, 3, 4, 5, 6\}$ and its p.m.f. is

$$f(x) = \frac{1}{6}, \qquad x = 1, 2, 3, 4, 5, 6.$$

We can generalize this by letting X have a discrete uniform distribution over the first m positive integers so that its p.m.f. is

$$f(x) = \frac{1}{m}, \qquad x = 1, 2, 3, \ldots, m.$$

We now give an example in which X does not have a uniform distribution.

EXAMPLE 2.1-3   Roll a four-sided die twice and let X equal the larger of the two outcomes if they are different and the common value if they are the same. The outcome space for this experiment is $S_0 = \{(d_1, d_2) : d_1 = 1, 2, 3, 4; d_2 = 1, 2, 3, 4\}$, where we assume that each of these 16 points has probability 1/16. Then $P(X = 1) = P[(1, 1)] = 1/16, P(X = 2) = P[\{(1, 2), (2, 1), (2, 2)\}] = 3/16$, and similarly $P(X = 3) = 5/16$ and $P(X = 4) = 7/16$. That is, the p.m.f. of X can be written simply as

$$f(x) = P(X = x) = \frac{2x - 1}{16}, \qquad x = 1, 2, 3, 4. \qquad (2.1\text{-}1)$$

We could add that $f(x) = 0$ elsewhere; but if we do not, the reader should take $f(x)$ to equal zero when $x \notin S = \{1, 2, 3, 4\}$. ◄

A better understanding of a particular probability distribution can often be obtained with a graph that depicts the p.m.f. of X. Note that the graph of the p.m.f., when $f(x) > 0$, would be simply the set of points $\{[x, f(x)] : x \in S\}$, where S is the space of X. Two types of graphs can be used to give a better visual appreciation of the p.m.f., namely, a bar graph and a probability histogram. A **bar graph** of the p.m.f. $f(x)$ of the random variable X is a graph having a vertical line segment drawn from $(x, 0)$ to $[x, f(x)]$ at each x in S, the space of X. If X can only assume integer values, a **probability histogram** of the p.m.f. $f(x)$ is a graphical representation that has a rectangle of height $f(x)$ and a base of length 1, centered at x for each $x \in S$, the space of X. Thus the area of each rectangle is equal to the respective probability $f(x)$ and the total area of a probability histogram is one.

Figure 2.1-1 displays a bar graph and a probability histogram for the p.m.f. $f(x)$ defined in Equation 2.1-1.

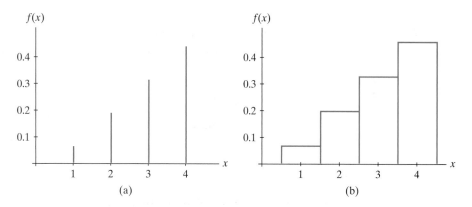

Figure 2.1-1: Bar graph and probability histogram

Our next probability model uses the material in Section 1.3 on methods of enumeration. Consider a collection of $N = N_1 + N_2$ similar objects, N_1 of them belonging to one of two dichotomous classes (red chips, say) and N_2 of them belonging to the second class (blue chips, say). A collection of n objects is selected from these N objects at random and without replacement. Find the probability that exactly x (where the nonnegative integer x satisfies $x \leq n$, $x \leq N_1$, and $n - x \leq N_2$) of these n objects are red (i.e., x belong to the first class and $n - x$ belong to the second). Of course, we can select x red chips in any one of $\binom{N_1}{x}$ ways and $n - x$ blue chips in any one of $\binom{N_2}{n-x}$ ways. By the multiplication principle, the product $\binom{N_1}{x}\binom{N_2}{n-x}$ equals the number of ways the joint operation can be performed. If we assume that each of the $\binom{N}{n}$ ways of selecting n objects from $N = N_1 + N_2$ objects has the same probability, we have that the probability of selecting exactly x red chips is

$$f(x) = P(X = x) = \frac{\binom{N_1}{x}\binom{N_2}{n-x}}{\binom{N}{n}},$$

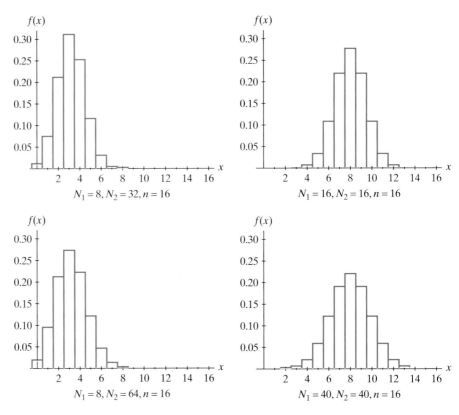

Figure 2.1-2: Hypergeometric probability histograms

where the space S is the collection of nonnegative integers x that satisfies the inequalities $x \leq n$, $x \leq N_1$, and $n - x \leq N_2$. We say that the random variable X has a **hypergeometric distribution**.

EXAMPLE 2.1-4 Some examples of hypergeometric probability histograms are given in Figure 2.1-2. The values of N_1, N_2, and n are given with each figure. ◄

EXAMPLE 2.1-5 In a small pond there are 50 fish, 10 of which have been tagged. If a fisherman's catch consists of 7 fish, selected at random and without replacement, and X denotes the number of tagged fish, the probability that exactly 2 tagged fish are caught is

$$P(X = 2) = \frac{\binom{10}{2}\binom{40}{5}}{\binom{50}{7}} = \frac{(45)(658,008)}{99,884,400} = \frac{246,753}{832,370} = 0.2964,$$

approximately. ◄

EXAMPLE 2.1-6   A lot, consisting of 100 fuses, is inspected by the following procedure. Five fuses are chosen at random and tested; if all 5 blow at the correct amperage, the lot is accepted. Suppose that the lot contains 20 defective fuses. If X is a random variable equal to the number of defective fuses in the sample of 5, the probability of accepting the lot is

$$P(X = 0) = \frac{\binom{20}{0}\binom{80}{5}}{\binom{100}{5}} = \frac{19{,}513}{61{,}110} = 0.3193,$$

approximately. More generally, the p.m.f. of X is

$$f(x) = P(X = x) = \frac{\binom{20}{x}\binom{80}{5-x}}{\binom{100}{5}}, \qquad x = 0, 1, 2, 3, 4, 5. \qquad \blacktriangleleft$$

EXAMPLE 2.1-7   A supplier ships parts to another company in lots of 25 parts, some of which could be defective. The receiving company has an *acceptance sampling plan* which states that $n = 5$ parts are to be taken at random and without replacement from each lot. If there are zero defectives among those 5 parts, the entire lot is accepted; otherwise, the lot is rejected. That is, if one or more defectives are found among the 5 in the sample, the lot is rejected. If X is the number of defectives among the 5 sampled parts, X must have a hypergeometric distribution; but here we do not know the values of N_1, the number of defectives in the lot, and N_2, the number of good parts in the lot, but only that $N_1 + N_2 = 25$. Clearly, we want the probability of accepting the lot, namely $P(X = 0)$, to be large if N_1 is very small. This probability is usually called the *operating characteristic curve* when treated as a function of N_1 or equivalently, the fraction defective, $p = N_1/25$, in the lot. That is, the operating characteristic curve is

$$OC(p) = P(X = 0) = \frac{\binom{N_1}{0}\binom{25 - N_1}{5}}{\binom{25}{5}},$$

where $p = N_1/25$. Letting $N_1 = 0, 1, 2, \ldots$ so that $p = 0.00, 0.04, 0.08, \ldots$, we find using Table I in the Appendix that

$$OC(0.00) = \frac{\binom{0}{0}\binom{25}{5}}{\binom{25}{5}} = 1.00,$$

$$OC(0.04) = \frac{\binom{1}{0}\binom{24}{5}}{\binom{25}{5}} = 0.800,$$

$$OC(0.08) = \frac{\binom{2}{0}\binom{23}{5}}{\binom{25}{5}} = 0.633,$$

$$OC(0.12) = \frac{\binom{3}{0}\binom{22}{5}}{\binom{25}{5}} = 0.496,$$

$$OC(0.16) = \frac{\binom{4}{0}\binom{21}{5}}{\binom{25}{5}} = 0.383,$$

and so on. Upon observing these probabilities, the company might find this acceptance sampling plan unsatisfactory because possibly, with $N_1 = (25)(0.04) = 1$, $OC(0.04) = P(X = 0) = 0.8$ is too low and, with $N_1 = (25)(0.16) = 4$, $OC(0.16) = P(X = 0) = 0.383$ is too high. Accordingly, if this is so, the plan must be changed. In industrial practice, usually lot sizes are much larger than 25, sometimes running into the thousands, and the sample sizes might be in the hundreds rather than $n = 5$. Here we kept the numbers small so we could use Table I, but we illustrate what is actually done in practice with those larger values in Section 2.6 on the Poisson distribution. ◀

In Section 1.2 we discussed the relationship between the probability $P(A)$ of an event A and the relative frequency $N(A)/n$ of occurrences of event A in n repetitions of an experiment. We shall now extend those ideas.

Suppose that a random experiment is repeated n independent times. Let $A = \{X = x\}$, the event that x is the outcome of the experiment. Then we would expect the relative frequency $N(A)/n$ to be close to $f(x)$. The following example illustrates this.

EXAMPLE 2.1-8 A tetrahedron (four-sided die with outcomes 1, 2, 3, 4) is rolled twice. Let X equal the sum of the two outcomes. Then the possible values that X can equal are 2, 3, 4, 5, 6, 7, and 8. The following argument suggests that the p.m.f. of X is given by $f(x) = (4 - |x - 5|)/16$, for $x = 2, 3, 4, 5, 6, 7, 8$. That is, $f(2) = 1/16$, $f(3) = 2/16$, $f(4) = 3/16$, $f(5) = 4/16$, $f(6) = 3/16$, $f(7) = 2/16$, and $f(8) = 1/16$. Intuitively, these probabilities seem correct if we think of the

x	Number of Observations of x	Relative Frequency of x	Probability of $\{X = x\}$, $f(x)$
		Table 2.1-1: Sum of Two Tetrahedral Dice	
2	71	0.071	0.0625
3	124	0.124	0.1250
4	194	0.194	0.1875
5	258	0.258	0.2500
6	177	0.177	0.1875
7	122	0.122	0.1250
8	54	0.054	0.0625

16 points (result on first roll, result on second roll), and assume that each has probability 1/16. Then note that $X = 2$ only for the point $(1, 1)$; $X = 3$ for the two points $(2, 1)$ and $(1, 2)$; and so on. This experiment was simulated 1000 times on a computer. Table 2.1-1 lists the results and compares the relative frequencies with the corresponding probabilities.

A graph can be used to display the results given in Table 2.1-1. The probability histogram of the p.m.f. $f(x)$ of X is given in Figure 2.1-3. It is superimposed over the shaded histogram that represents the observed relative frequencies of the corresponding x values. The shaded histogram is the relative frequency histogram. For random experiments of the discrete type, this relative frequency histogram of a set of data gives an estimate of the probability histogram of the associated random variable when the latter is unknown. Estimation is considered in detail later in this book.   ◄

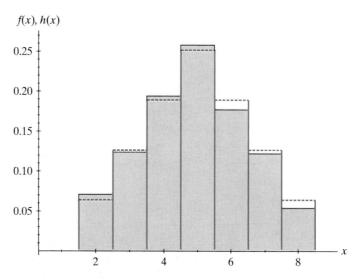

Figure 2.1-3: Sum of two tetrahedral dice

EXERCISES

2.1-1 Let the p.m.f. of X be defined by $f(x) = x/9$, $x = 2, 3, 4$.

(a) Draw a bar graph for this p.m.f.
(b) Draw a probability histogram for this p.m.f.

2.1-2 Let a chip be taken at random from a bowl that contains six white chips, three red chips, and one blue chip. Let the random variable $X = 1$ if the outcome is a white chip; let $X = 5$ if the outcome is a red chip; and let $X = 10$ if the outcome is a blue chip.

(a) Find the p.m.f. of X.
(b) Graph the p.m.f. as a bar graph.

2.1-3 For each of the following, determine the constant c so that $f(x)$ satisfies the conditions of being a p.m.f. for a random variable X. Depict each p.m.f. as a bar graph.

(a) $f(x) = x/c$, $\quad x = 1, 2, 3, 4$.
(b) $f(x) = cx$, $\quad x = 1, 2, 3, \ldots, 10$.
(c) $f(x) = c(1/4)^x$, $\quad x = 1, 2, 3, \ldots$.
(d) $f(x) = c(x + 1)^2$, $\quad x = 0, 1, 2, 3$.
(e) $f(x) = x/c$, $\quad x = 1, 2, 3, \ldots, n$.

2.1-4 The state of Michigan generates a three-digit number at random six days a week for their daily lottery. The numbers are generated one digit at a time. Consider the following set of 50 three-digit numbers as 150 one-digit integers that were generated at random:

169	938	506	757	594	656	444	809	321	545
732	146	713	448	861	612	881	782	209	752
571	701	852	924	766	633	696	023	601	789
137	098	534	826	642	750	827	689	979	000
933	451	945	464	876	866	236	617	418	988

Let X denote the outcome when a single digit is generated.

(a) With true random numbers, what is the p.m.f. of X? Draw the probability histogram.
(b) For the 150 observations, determine the relative frequencies of 0, 1, 2, 3, 4, 5, 6, 7, 8, and 9, respectively.
(c) Draw the relative frequency histogram of the observations on the same graph paper as that of the probability histogram. Use a colored or dashed line for the relative frequency histogram.

2.1-5 The p.m.f. of X is $f(x) = (5 - x)/10$, $x = 1, 2, 3, 4$.

(a) Graph the p.m.f. as a bar graph.
(b) Use the following independent observations of X, simulated on a computer, to construct a table like Table 2.1-1.

3	1	2	2	3	2	2	2	1	3	3	2	3	2	4	4	2	1	1	3
3	1	2	2	1	1	4	2	3	1	1	1	2	1	3	1	1	3	3	1
1	1	1	1	1	4	1	3	1	2	4	1	1	2	3	4	3	1	4	2
2	1	3	2	1	4	1	1	1	2	1	3	4	3	2	1	4	4	1	3
2	2	2	1	2	3	1	1	4	2	1	4	2	1	2	3	1	4	2	3

(c) Construct a probability histogram and a relative frequency histogram like Figure 2.1-3.

2.1-6 Let a random experiment be the cast of a pair of unbiased dice, each having six faces, and let the random variable X denote the sum of the dice.

(a) With reasonable assumptions, determine the p.m.f. $f(x)$ of X. HINT: Picture the sample space consisting of the 36 points, (result on first die, result on second die), and assume that each has probability 1/36. Find the probability for each possible outcome of X, namely, $x = 2, 3, 4, \ldots, 12$.

(b) Draw the bar graph for $f(x)$.

2.1-7 Let a random experiment be the cast of a pair of unbiased six-sided dice and let X equal the smaller of the outcomes if they are different and the common value if they are equal.

(a) With reasonable assumptions, find the p.m.f. of X.

(b) Draw a probability histogram.

(c) Let Y equal the range of the two outcomes (i.e., the absolute value of the difference of the largest and the smallest outcomes). Determine the p.m.f. $g(y)$ of Y for $y = 0, 1, 2, 3, 4, 5$.

(d) Draw the probability histogram for $g(y)$.

2.1-8 An unbiased six-sided die X has three faces numbered 0 and three faces numbered 2. An unbiased six-sided die Y has its faces numbered 0, 1, 4, 5, 8, and 9. The two dice are rolled. Let $W = X + Y$, the sum of the two outcomes.

(a) Determine the p.m.f. of W.

(b) Draw its probability histogram.

2.1-9 Let the p.m.f. of X be defined by $f(x) = (1 + |x - 3|)/11, x = 1, 2, 3, 4, 5$. Graph the p.m.f. of X as a bar graph.

2.1-10 Suppose there are 3 defective items in a lot of 50 items. A sample of size 10 is taken at random and without replacement. Let X denote the number of defective items in the sample. Find the probability that the sample contains

(a) Exactly 1 defective item.

(b) At most 1 defective item.

2.1-11 In a lot of 100 light bulbs, there are 5 bad bulbs. An inspector inspects 10 bulbs selected at random. Find the probability of finding at least one defective bulb. HINT: First compute the probability of finding no defectives in the sample.

2.1-12 A lot of $N_1 + N_2 = 25$ items is accepted if the number of defectives, X, among $n = 5$ items taken at random and without replacement from the lot is less than or equal to 1. The operating characteristic curve is defined as the probability of accepting the lot, namely,

$$OC(p) = P(X \leq 1) = P(X = 0) + P(X = 1),$$

where $p = N_1/25$. Determine the probabilities $OC(0.04)$, $OC(0.08)$, $OC(0.12)$, and $OC(0.16)$ (and compare them to those in Example 2.1-7).

2.1-13 Let X be the number of accidents in a factory per week having p.m.f.

$$f(x) = \frac{1}{(x + 1)(x + 2)}, \qquad x = 0, 1, 2, \ldots.$$

Find the conditional probability of $X \geq 4$, given that $X \geq 1$. HINT: Write $f(x) = 1/(x + 1) - 1/((x + 2))$.

2.1-14 Often in buying a product at a supermarket, there is a concern about that item being underweight. Suppose there are 20 "one-pound" packages of frozen ground turkey on display and 3 of them are underweight. A consumer group buys five of the 20 packages at random. What is the probability of at least one of the five being underweight?

2.1-15 A lot of 200 semiconductor chips is inspected by selecting five at random and without replacement. If at least one of the five is defective, the lot is rejected. Find the probability of rejecting the lot if in the 200,

(a) zero are defective,
(b) 10 are defective,
(c) 20 are defective.

2.2 MATHEMATICAL EXPECTATION

An extremely important concept in summarizing important characteristics of distributions of probability is that of mathematical expectation which we introduce with an example.

EXAMPLE 2.2-1

An enterprising young man who needs a little extra money devises a game of chance in which some of his friends might wish to participate. The game that he proposes is to let the participant cast an unbiased die and then receive a payment according to the following schedule: If the event $A = \{1, 2, 3\}$ occurs, he receives 1¢; if $B = \{4, 5\}$ occurs, he receives 5¢; and if $C = \{6\}$ occurs, he receives 35¢. The probabilities of the respective events are assumed to be 3/6, 2/6, and 1/6, since the die is unbiased. The problem that now faces the young man is the determination of the amount that should be charged for the opportunity of playing the game. He reasons, correctly, that if the game is played a large number of times about 3/6 of the trials will require a payment of 1¢, about 2/6 of them will require one of 5¢, and about 1/6 of them will require one of 35¢. Thus the approximate average payment is

$$(1)\left(\frac{3}{6}\right) + (5)\left(\frac{2}{6}\right) + (35)\left(\frac{1}{6}\right) = 8.$$

That is, he expects to pay 8¢ "on the average." Note that he never pays exactly 8¢; the payment is either 1¢, 5¢, or 35¢. However, the "weighted average" of 1, 5, and 35, in which the weights are the respective probabilities 3/6, 2/6, and 1/6, equals eight. Such a weighted average is called the mathematical expectation of payment. Thus, if the young man decides to charge 10¢ per play, he would make 2¢ per play "on the average." Since the most that a player would lose at the charge of 10¢ per play is 9¢, the young man might find that several players are attracted by the possible gain of 25¢. ◀

A more mathematical way of formulating the preceding example would be to let X be the random variable defined by the outcome of the cast of the die. Thus the p.m.f. of X is the uniform one given by

$$f(x) = \frac{1}{6}, \qquad x = 1, 2, 3, 4, 5, 6.$$

In terms of the observed value x, the payment is given by the function

$$u(x) = \begin{cases} 1, & x = 1, 2, 3, \\ 5, & x = 4, 5, \\ 35, & x = 6. \end{cases}$$

The mathematical expectation of payment is then equal to

$$\sum_{x=1}^{6} u(x)f(x) = (1)\left(\frac{1}{6}\right) + (1)\left(\frac{1}{6}\right) + (1)\left(\frac{1}{6}\right) + (5)\left(\frac{1}{6}\right) + (5)\left(\frac{1}{6}\right) + (35)\left(\frac{1}{6}\right)$$

$$= (1)\left(\frac{3}{6}\right) + (5)\left(\frac{2}{6}\right) + (35)\left(\frac{1}{6}\right)$$

$$= 8.$$

This discussion suggests the more general definition of mathematical expectation of a function of X.

Definition 2.2-1   If $f(x)$ is the p.m.f. of the random variable X of the discrete type with space S and if the summation

$$\sum_{x \in S} u(x)f(x), \qquad \text{which is sometimes written} \qquad \sum_{S} u(x)f(x),$$

exists, then the sum is called the ***mathematical expectation*** or the ***expected value*** of the function $u(X)$, and it is denoted by $E[u(X)]$. That is,

$$E[u(X)] = \sum_{x \in S} u(x)f(x).$$

We can think of the expected value $E[u(X)]$ as a weighted mean of $u(x)$, $x \in S$, where the weights are the probabilities $f(x) = P(X = x)$, $x \in S$.

Remark The usual definition of mathematical expectation of $u(X)$ requires that the sum converge absolutely; that is,

$$\sum_{x \in S} |u(x)|\, f(x)$$

converges and is finite. However, in this book, each $u(x)$ is such that the convergence is absolute, and we do not burden the student with this additional requirement. Moreover, sometimes $E[u(X)]$ is called, more simply, the expectation of $u(X)$. ∎

There is another important observation that must be made about the consistency of this definition. Certainly, this function $u(X)$ of the random variable X is itself a random variable, say Y. Suppose that we find the p.m.f. of Y to be $g(y)$ on the support S_1. Then $E(Y)$ is given by the summation

$$\sum_{y \in S_1} y\, g(y).$$

In general, it is true that

$$\sum_{x \in S} u(x)f(x) = \sum_{y \in S_1} y\, g(y).$$

That is, the same expectation is obtained by either formula. We do not prove this general result but only illustrate it in the following example.

EXAMPLE 2.2-2 Let the random variable X have the p.m.f.

$$f(x) = \frac{1}{3}, \qquad x \in S,$$

where $S = \{-1, 0, 1\}$. Let $u(X) = X^2$. Then

$$\sum_{x \in S} x^2 f(x) = (-1)^2 \left(\frac{1}{3}\right) + (0)^2 \left(\frac{1}{3}\right) + (1)^2 \left(\frac{1}{3}\right) = \frac{2}{3}.$$

However, the support of the random variable $Y = X^2$ is $S_1 = \{0, 1\}$ and

$$P(Y = 0) = P(X = 0) = \frac{1}{3},$$

$$P(Y = 1) = P(X = -1) + P(X = 1) = \frac{1}{3} + \frac{1}{3} = \frac{2}{3}.$$

That is,

$$g(y) = \begin{cases} \dfrac{1}{3}, & y = 0, \\ \dfrac{2}{3}, & y = 1; \end{cases}$$

and $S_1 = \{0, 1\}$. Hence

$$\sum_{y \in S_1} y\, g(y) = (0) \left(\frac{1}{3}\right) + (1) \left(\frac{2}{3}\right) = \frac{2}{3},$$

which illustrates the preceding observation. ◀

Before presenting additional examples, we list some useful facts about mathematical expectation in the following theorem.

Theorem 2.2-1 When it exists, mathematical expectation E satisfies the following properties:

(a) If c is a constant, $E(c) = c$.

(b) If c is a constant and u is a function,

$$E[c\, u(X)] = c E[u(X)].$$

(c) If c_1 and c_2 are constants and u_1 and u_2 are functions, then

$$E[c_1 u_1(X) + c_2 u_2(X)] = c_1 E[u_1(X)] + c_2 E[u_2(X)].$$

Proof: First, we have for the proof of (a) that

$$E(c) = \sum_{x \in S} cf(x) = c \sum_{x \in S} f(x) = c$$

because

$$\sum_{x \in S} f(x) = 1.$$

Next, to prove (b), we see that

$$E[c\,u(X)] = \sum_{x \in S} c\,u(x)f(x)$$

$$= c \sum_{x \in S} u(x)f(x)$$

$$= c\,E[u(X)].$$

Finally, the proof of (c) is given by

$$E[c_1 u_1(X) + c_2 u_2(X)] = \sum_{x \in S} [c_1 u_1(x) + c_2 u_2(x)]f(x)$$

$$= \sum_{x \in S} c_1 u_1(x)f(x) + \sum_{x \in S} c_2 u_2(x)f(x).$$

By applying (b), we obtain

$$E[c_1 u_1(X) + c_2 u_2(X)] = c_1 E[u_1(X)] + c_2 E[u_2(X)]. \qquad \bigcirc$$

Property (c) can be extended to more than two terms by mathematical induction; that is, we have

$$(c)' \quad E\left[\sum_{i=1}^{k} c_i\,u_i(X))\right] = \sum_{i=1}^{k} c_i\,E[u_i(X)].$$

Because of property $(c)'$, mathematical expectation E is often called a **linear** or **distributive** operator.

EXAMPLE 2.2-3   Let X have the p.m.f.

$$f(x) = \frac{x}{10}, \qquad x = 1, 2, 3, 4.$$

Then

$$E(X) = \sum_{x=1}^{4} x\left(\frac{x}{10}\right)$$

$$= (1)\left(\frac{1}{10}\right) + (2)\left(\frac{2}{10}\right) + (3)\left(\frac{3}{10}\right) + (4)\left(\frac{4}{10}\right) = 3,$$

$$E(X^2) = \sum_{x=1}^{4} x^2\left(\frac{x}{10}\right)$$

$$= (1)^2\left(\frac{1}{10}\right) + (2)^2\left(\frac{2}{10}\right) + (3)^2\left(\frac{3}{10}\right) + (4)^2\left(\frac{4}{10}\right) = 10,$$

and

$$E[X(5 - X)] = 5E(X) - E(X^2) = (5)(3) - 10 = 5. \qquad \blacktriangleleft$$

EXAMPLE 2.2-4   Let $u(x) = (x - b)^2$, where b is not a function of X, and suppose $E[(X - b)^2]$ exists. To find that value of b for which $E[(X - b)^2]$ is a minimum, we write

$$g(b) = E[(X - b)^2] = E[X^2 - 2bX + b^2]$$

$$= E(X^2) - 2bE(X) + b^2$$

because $E(b^2) = b^2$. To find the minimum, we differentiate $g(b)$ with respect to b, set $g'(b) = 0$ and solve for b as follows:

$$g'(b) = -2E(X) + 2b = 0,$$

$$b = E(X).$$

Since $g''(b) = 2 > 0$, $E(X)$ is the value of b that minimizes $E[(X - b)^2]$. $\qquad \blacktriangleleft$

EXAMPLE 2.2-5   Let X have a hypergeometric distribution in which n objects are selected from $N = N_1 + N_2$ objects as described in Section 2.1. Then

$$E(X) = \sum_{x \in S} x \, \frac{\binom{N_1}{x}\binom{N_2}{n - x}}{\binom{N}{n}}.$$

Since the first term of this summation equals zero when $x = 0$ and

$$\binom{N}{n} = \left(\frac{N}{n}\right)\binom{N-1}{n-1},$$

we can write

$$E(X) = \sum_{0 < x \in S} x \, \frac{N_1!}{x!(N_1 - x)!} \frac{\binom{N_2}{n-x}}{\binom{N}{n}\binom{N-1}{n-1}}.$$

Of course, $x/x! = 1/(x-1)!$ when $x \neq 0$; thus

$$E(X) = \left(\frac{n}{N}\right) \sum_{0 < x \in S} \frac{(N_1)(N_1 - 1)!}{(x-1)!(N_1 - x)!} \frac{\binom{N_2}{n-x}}{\binom{N-1}{n-1}}$$

$$= n \left(\frac{N_1}{N}\right) \sum_{0 < x \in S} \frac{\binom{N_1 - 1}{x-1}\binom{N_2}{n-1-(x-1)}}{\binom{N-1}{n-1}}.$$

However, the summand of this last expression represents, when $x > 0$, the probability of obtaining $x - 1$ red chips if $n - 1$ chips are selected from $N_1 - 1$ red chips and N_2 blue chips. Since the summation is over all possible values of $x - 1$, it must sum to one as it is the sum of all possible probabilities of $x - 1$. Thus

$$E(X) = n \left(\frac{N_1}{N}\right),$$

which is a result that agrees with our intuition: We expect the number X of red chips to equal the product of the number of selections, n, and the fraction, N_1/N, of red chips in the original collection. ◀

EXERCISES

2.2-1 Let the random variable X be the number of days that a certain patient needs to be in the hospital. Say X has the p.m.f.

$$f(x) = \frac{5 - x}{10}, \qquad x = 1, 2, 3, 4.$$

If the patient is to receive from an insurance company \$200 for each of the first two days in the hospital and \$100 for each day after the first two days, what is the expected payment for the hospitalization?

2.2-2 An insurance company sells an automobile policy with a deductible of one unit. Let X be the amount of the loss having p.m.f.

$$f(x) = \begin{cases} 0.9, & x = 0, \\ \dfrac{c}{x}, & x = 1, 2, 3, 4, 5, 6, \end{cases}$$

where c is a constant. Determine c and the expected value of the amount the insurance company must pay.

2.2-3 Find $E(X)$ for each of the distributions given in Exercise 2.1-3. HINT: Part (c). Note that the difference $E(X) - (1/4)E(X)$ is equal to the sum of a geometric series.

2.2-4 Let the random variable X have the p.m.f.

$$f(x) = \frac{(|x| + 1)^2}{9}, \qquad x = -1, 0, 1.$$

Compute $E(X)$, $E(X^2)$, and $E(3X^2 - 2X + 4)$.

2.2-5 In a particular lottery 3,000,000 tickets are sold each week for 50¢ apiece. Out of the 3,000,000 tickets, 12,006 are drawn at random and without replacement and awarded prizes: twelve thousand $25 prizes, four $10,000 prizes, one $50,000 prize, and one $200,000 prize. If you purchased a single ticket each week, what is the expected value of this game to you?

2.2-6 In a state lottery a three-digit integer is selected at random. A player bets $1 on a particular number and if that number is selected, the payoff is $500 minus the $1 paid for the ticket. Let X equal the payoff to the bettor, namely $-$1 or $499, and find $E(X)$.

2.2-7 In the gambling game chuck-a-luck, for a $1 bet it is possible to win $1, $2, or $3 with respective probabilities 75/216, 15/216, and 1/216. One dollar is lost with probability 125/216. Let X equal the payoff for this game and find $E(X)$. Note that when a bet is won, the $1 that was bet is returned to the bettor in addition to the $1, $2, or $3 that is won.

2.2-8 Let the p.m.f. of X be defined by $f(x) = 6/(\pi^2 x^2)$, $x = 1, 2, 3, \ldots$. Show that $E(X)$ does not exist in this case.

2.2-9 Let us select at random a number from the first n positive integers. If the payment is equal to the reciprocal of the number, find an expression for the expected payment. Evaluate this number when $n = 5$. Approximate this number when $n = 100$ or find the exact value using the computer. HINT: For the approximation, use a modification of the integral test for testing convergence of series.

2.2-10 Let X be a random variable with support $\{1, 2, 3, 5, 15, 25, 50\}$, each point of which has the same probability 1/7. Argue that $c = 5$ is the value that minimizes $h(c) = E(|X - c|)$. Compare this to the value of b that minimizes $g(b) = E[(X - b)^2]$.

2.2-11 A roulette wheel used in a United States casino has 38 slots of which 18 are red, 18 are black, and 2 are green. A roulette wheel used in a French casino has 37 slots of which 18 are red, 18 are black, and 1 is green. A ball is rolled around the wheel and ends up in one of the slots with equal probability. Suppose that a player bets on red. If a $1 bet is placed, the player wins $1 if the ball ends up in a red slot (the player's $1 bet is returned). If the ball ends up in a black or green slot, the player loses $1. Find the expected value of this game to the player in

(a) The United States.
(b) France.

2.2-12 In the casino game called **High-Low**, there are three possible bets. Assume that $1 is the size of the bet. A pair of fair six-sided dice is rolled and their sum is calculated. If you bet **Low**, you win $1 if the sum of the dice is {2, 3, 4, 5, 6}. If you bet **High**, you win $1 if the sum of the dice is {8, 9, 10, 11, 12}. If you bet on {7}, you win $4 if a sum of 7 is rolled. Otherwise you lose on each of the three bets. In all three cases, your original dollar is returned if you win. Find the expected value of the game to the bettor for each of these three bets.

2.2-13 In the gambling game craps (see Exercise 1.4-13), the player wins $1 with probability 0.49293 and loses $1 with probability 0.50707 for each $1 bet. What is the expected value of the game to the player?

2.2-14 Suppose that a school has 20 classes: 16 with 25 students in each, three with 100 students in each, and one with 300 students for a total of 1000 students.

(a) What is the average class size?
(b) Select a student randomly out of the 1000 students. Let the random variable X equal the size of the class to which this student belongs and define the p.m.f. of X.
(c) Find $E(X)$, the expected value of X. Does this answer surprise you?

2.3 THE MEAN, VARIANCE, AND STANDARD DEVIATION

Let us consider an example in which $x \in \{1, 2, 3\}$ and the p.m.f. is given by $f(1) = 3/6$, $f(2) = 2/6$, $f(3) = 1/6$. That is, the probability that the random variable X equals one, denoted by $P(X = 1)$, is $f(1) = 3/6$. Likewise $P(X = 2) = f(2) = 2/6$ and $P(X = 3) = f(3) = 1/6$. Of course, $f(x) > 0$ when $x \in S$ and

$$\sum_{x \in S} f(x) = f(1) + f(2) + f(3) = 1.$$

We can think of the points 1, 2, 3 having respective weights (probabilities) 3/6, 2/6, 1/6 and their weighted mean (weighted average) is

$$1 \cdot \frac{3}{6} + 2 \cdot \frac{2}{6} + 3 \cdot \frac{1}{6} = \frac{10}{6},$$

which, in this illustration, does not equal one of the x values in S. As a matter of fact, it is two-thirds of the way between $x = 1$ and $x = 2$. We denote this weighted average or mean by the Greek m, namely mu, having symbol μ. In our example,

$$\mu = \sum_{x \in S} x f(x) = \frac{10}{6}.$$

More generally, suppose the random variable X has the space $S = \{u_1, u_2, \ldots, u_k\}$ and these points have respective probabilities $P(X = u_i) = f(u_i) > 0$, where $f(x)$ is the p.m.f. Of course,

$$\sum_{x \in S} f(x) = 1$$

and the **mean** of the random variable X (or of its distribution) is

$$\mu = \sum_{x \in S} x f(x) = u_1 f(u_1) + u_2 f(u_2) + \cdots + u_k f(u_k).$$

Now u_i is the distance of that ith point from the origin. In mechanics, the product of a distance and its weight is called a moment; so $u_i f(u_i)$ is a moment having a moment arm of length u_i. The sum of such products would be the moment of the system of distances and weights. Actually, it is called the first moment about the origin since the distances are simply to the first power and the lengths of the arms (distances) are measured from the origin. However, if we computed the first moment about the mean μ, we have, since here a moment arm equals $(x - \mu)$,

$$\sum_{x \in S} (x - \mu) f(x) = \sum_{x \in S} x f(x) - \sum_{x \in S} \mu f(x)$$

$$= \sum_{x \in S} x f(x) - \mu \sum_{x \in S} f(x) = \mu - \mu \cdot 1 = 0.$$

That is, that first moment about μ is equal to zero. In mechanics μ is called the centroid. The last equation implies that if a fulcrum is placed at the centroid μ, the system of weights would balance as the sum of the positive moments (when $x > \mu$) about μ equals the sum of the negative moments (when $x < \mu$). In our first illustration, $\mu = 10/6$ is the centroid so that the negative moment

$$\left(1 - \frac{10}{6}\right) \cdot \frac{3}{6} = -\frac{12}{36}$$

equals the sum of the two positive moments

$$\left(2 - \frac{10}{6}\right) \cdot \frac{2}{6} + \left(3 - \frac{10}{6}\right) \cdot \frac{1}{6} = \frac{12}{36}.$$

In the notation of Section 2.2, we see that $\mu = E(X)$. Also from Example 2.2-4 we note that $b = \mu$ minimizes $E[(X - b)^2]$ and Example 2.2-5 shows that

$$\mu = n\left(\frac{N_1}{N}\right)$$

is the mean of the hypergeometric distribution.

Statisticians often find it valuable to compute the second moment about the mean μ. It is called the second moment because the distances are raised to the second power and is equal to

$$\sum_{x \in S} (x - \mu)^2 f(x) = (u_1 - \mu)^2 f(u_1) + (u_2 - \mu)^2 f(u_2) + \cdots + (u_k - \mu)^2 f(u_k).$$

This weighted mean of the squares of those distances is called the **variance** of the random variable X (or of its distribution). The positive square root of the variance is called the **standard deviation** of X and is denoted by the Greek s, namely sigma, having symbol σ. Thus the variance is σ^2, sometimes denoted by Var(X). In our first illustration, since $\mu = 10/6$, we have that the variance equals

$$\sigma^2 = \text{Var}(X) = \left(1 - \frac{10}{6}\right)^2 \cdot \frac{3}{6} + \left(2 - \frac{10}{6}\right)^2 \cdot \frac{2}{6} + \left(3 - \frac{10}{6}\right)^2 \cdot \frac{1}{6} = \frac{120}{216}.$$

Thus the standard deviation is

$$\sigma = \sqrt{\sigma^2} = \sqrt{\frac{120}{216}} = 0.745.$$

It is worth noting that the variance can be computed in another way, because

$$\sigma^2 = \sum_{x \in S} (x - \mu)^2 f(x) = \sum_{x \in S} (x^2 - 2\mu x + \mu^2) f(x)$$

$$= \sum_{x \in S} x^2 f(x) - 2\mu \sum_{x \in S} x f(x) + \mu^2 \sum_{x \in S} f(x)$$

$$= \sum_{x \in S} x^2 f(x) - 2\mu \cdot \mu + \mu^2 \cdot 1 = \sum_{x \in S} x^2 f(x) - \mu^2.$$

That is, the variance σ^2 equals the difference of the second moment about the origin and the square of the mean. For our first illustration,

$$\sigma^2 = \sum_{x=1}^{3} x^2 f(x) - \mu^2$$

$$= 1^2 \left(\frac{3}{6}\right) + 2^2 \left(\frac{2}{6}\right) + 3^2 \left(\frac{1}{6}\right) - \left(\frac{10}{6}\right)^2 = \frac{20}{6} - \frac{100}{36} = \frac{120}{216},$$

which agrees with our previous computation.

In the notation of Section 2.2, we observe that

$$\sigma^2 = E[(X - \mu)^2] = E(X^2) - \mu^2.$$

EXAMPLE 2.3-1   Let X equal the number of spots on the side that is up after a six-sided die (one of a pair of dice) is cast at random. If everything is fair about this experiment, a reasonable probability model is given by the p.m.f.

$$f(x) = P(X = x) = \frac{1}{6}, \qquad x = 1, 2, 3, 4, 5, 6.$$

The mean of X is

$$\mu = E(X) = \sum_{x=1}^{6} x \left(\frac{1}{6}\right) = \frac{1 + 2 + 3 + 4 + 5 + 6}{6} = \frac{7}{2}.$$

The second moment about the origin is

$$E(X^2) = \sum_{x=1}^{6} x^2 \left(\frac{1}{6}\right) = \frac{1^2 + 2^2 + 3^2 + 4^2 + 5^2 + 6^2}{6} = \frac{91}{6}.$$

Thus the variance equals

$$\sigma^2 = \frac{91}{6} - \left(\frac{7}{2}\right)^2 = \frac{182 - 147}{12} = \frac{35}{12}.$$

The standard deviation is $\sigma = \sqrt{35/12} = 1.708$. ◀

EXAMPLE 2.3-2 Let the p.m.f. of X be defined by $f(x) = x/6$, $x = 1, 2, 3$. The mean of X is

$$\mu = E(X) = 1\left(\frac{1}{6}\right) + 2\left(\frac{2}{6}\right) + 3\left(\frac{3}{6}\right) = \frac{7}{3}.$$

To find the variance and standard deviation of X we first find

$$E(X^2) = 1^2\left(\frac{1}{6}\right) + 2^2\left(\frac{2}{6}\right) + 3^2\left(\frac{3}{6}\right) = \frac{36}{6} = 6.$$

Thus the variance of X is

$$\sigma^2 = E(X^2) - \mu^2 = 6 - \left(\frac{7}{3}\right)^2 = \frac{5}{9}$$

and the standard deviation of X is

$$\sigma = \sqrt{5/9} = 0.745.$$ ◀

Although most students understand that $\mu = E(X)$ is, in some sense, a measure of the middle of the distribution of X, it is difficult to get much of a feeling for the variance and the standard deviation. The following example illustrates that the standard deviation is a measure of dispersion or spread of the points belonging to the space S.

EXAMPLE 2.3-3 Let X have the p.m.f. $f(x) = 1/3$, $x = -1, 0, 1$. Here the mean is

$$\mu = \sum_{x=-1}^{1} xf(x) = (-1)\left(\frac{1}{3}\right) + (0)\left(\frac{1}{3}\right) + (1)\left(\frac{1}{3}\right) = 0.$$

Accordingly, the variance, denoted by σ_X^2, is

$$\sigma_X^2 = E[(X - 0)^2]$$

$$= \sum_{x=-1}^{1} x^2 f(x)$$

$$= (-1)^2\left(\frac{1}{3}\right) + (0)^2\left(\frac{1}{3}\right) + (1)^2\left(\frac{1}{3}\right)$$

$$= \frac{2}{3},$$

so the standard deviation is $\sigma_X = \sqrt{2/3}$. Next let another random variable Y have the p.m.f. $g(y) = 1/3$, $y = -2, 0, 2$. Its mean is also zero, and it is easy to show that $\text{Var}(Y) = 8/3$, so the standard deviation of Y is $\sigma_Y = 2\sqrt{2/3}$. Here the standard deviation of Y is twice that of the standard deviation of X, reflecting the fact that the probability of Y is spread out twice as much as that of X. ◄

EXAMPLE 2.3-4 The mean of X, which has a uniform distribution on the first m positive integers, is given by

$$\mu = E(X) = \sum_{x=1}^{m} x\left(\frac{1}{m}\right) = \frac{1}{m}\sum_{x=1}^{m} x$$

$$= \left(\frac{1}{m}\right)\frac{m(m+1)}{2} = \frac{m+1}{2}.$$

To find the variance of X, we first find

$$E(X^2) = \sum_{x=1}^{m} x^2\left(\frac{1}{m}\right) = \frac{1}{m}\sum_{x=1}^{m} x^2$$

$$= \left(\frac{1}{m}\right)\frac{m(m+1)(2m+1)}{6} = \frac{(m+1)(2m+1)}{6}.$$

Thus the variance of X is

$$\sigma^2 = \text{Var}(X) = E[(X-\mu)^2]$$

$$= E(X^2) - \mu^2 = \frac{(m+1)(2m+1)}{6} - \left(\frac{m+1}{2}\right)^2$$

$$= \frac{m^2 - 1}{12}.$$

For example, we find that if X equals the outcome when rolling a fair six-sided die, the p.m.f. of X is

$$f(x) = \frac{1}{6}, \qquad x = 1, 2, 3, 4, 5, 6;$$

the respective mean and variance of X are

$$\mu = \frac{6+1}{2} = 3.5 \qquad \text{and} \qquad \sigma^2 = \frac{6^2 - 1}{12} = \frac{35}{12},$$

which agrees with calculations of Example 2.3-1. ◄

Now let X be a random variable with mean μ_X and variance σ_X^2. Of course, $Y = aX + b$, where a and b are constants, is a random variable, too. The mean of Y is

$$\mu_Y = E(Y) = E(aX + b) = aE(X) + b = a\mu_X + b.$$

Moreover, the variance of Y is

$$\sigma_Y^2 = E[(Y - \mu_Y)^2] = E[(aX + b - a\mu_X - b)^2] = E[a^2(X - \mu_X)^2] = a^2\sigma_X^2.$$

Thus $\sigma_Y = |a|\sigma_X$. To illustrate, in Example 2.3-3 we note that the relationship between the two distributions could be explained by $Y = 2X$ so that $\sigma_Y^2 = 4\sigma_X^2$ and thus $\sigma_Y = 2\sigma_X$, which we had observed there. In addition, we see that adding or subtracting a constant from X does not change the variance. In our example, $\text{Var}(X - 1) = \text{Var}(X)$, because $a = 1$ and $b = -1$.

Let r be a positive integer. If

$$E(X^r) = \sum_{x \in S} x^r f(x)$$

is finite, it is called the rth **moment** of the distribution about the origin. In addition, the expectation

$$E[(X - b)^r] = \sum_{x \in S} (x - b)^r f(x)$$

is called the rth moment of the distribution about b.

For a given positive integer r,

$$E[(X)_r] = E[X(X - 1)(X - 2) \cdots (X - r + 1)]$$

is called the rth factorial moment. We note that the second factorial moment is equal to the difference of the second and first moments:

$$E[X(X - 1)] = E(X^2) - E(X).$$

There is another formula that can be used for computing the variance that uses the second factorial moment and sometimes simplifies the calculations. First find the values of $E(X)$ and $E[X(X - 1)]$. Then

$$\sigma^2 = E[X(X - 1)] + E(X) - [E(X)]^2,$$

since, using the distributive property of E, this becomes

$$\sigma^2 = E(X^2) - E(X) + E(X) - [E(X)]^2 = E(X^2) - \mu^2.$$

EXAMPLE 2.3-5   In Example 2.2-5 concerning the hypergeometric distribution, we found that the mean of that distribution is

$$\mu = E(X) = n\left(\frac{N_1}{N}\right) = np,$$

where $p = N_1/N$, the fraction of red chips in the N chips. In Exercise 2.3-9 it is determined that

$$E[X(X-1)] = \frac{(n)(n-1)(N_1)(N_1-1)}{N(N-1)}.$$

Thus the variance of X is $E[X(X-1)] + E(X) - [E(X)]^2$, namely,

$$\sigma^2 = \frac{n(n-1)(N_1)(N_1-1)}{N(N-1)} + \frac{nN_1}{N} - \left(\frac{nN_1}{N}\right)^2.$$

After some straightforward algebra, we find that

$$\sigma^2 = n\left(\frac{N_1}{N}\right)\left(\frac{N_2}{N}\right)\left(\frac{N-n}{N-1}\right) = np(1-p)\left(\frac{N-n}{N-1}\right). \qquad \blacktriangleleft$$

Suppose that we now consider the situation in which we actually perform a certain random experiment n times obtaining n observed values of the random variable, say $x_1, x_2, \ldots, x_n$. Often the collection is referred to as a **sample**. It is possible that some of these values might be the same, but we do not worry about this at this time. We artificially create a probability distribution by placing the weight $1/n$ on each of these x values. Note that these weights are positive and sum to one; so we have a distribution which we call the **empirical distribution** since it is determined by the data $x_1, x_2, \ldots, x_n$. The mean of the empirical distribution is

$$\sum_{i=1}^{n} x_i\left(\frac{1}{n}\right) = \frac{1}{n}\sum_{i=1}^{n} x_i,$$

which is the arithmetic mean of the observations $x_1, x_2, \ldots, x_n$. We denote this mean by $\overline{x}$ and call it the **sample mean** (or mean of the sample $x_1, x_2, \ldots, x_n$). That is, the sample mean is

$$\overline{x} = \frac{1}{n}\sum_{i=1}^{n} x_i,$$

which is, in some sense, an estimate of μ, if the latter is unknown.

Likewise, the **variance of the empirical distribution** is

$$v = \sum_{i=1}^{n} (x_i - \overline{x})^2\left(\frac{1}{n}\right) = \frac{1}{n}\sum_{i=1}^{n} (x_i - \overline{x})^2,$$

which can be written as

$$v = \sum_{i=1}^{n} x_i^2 \left(\frac{1}{n}\right) - \overline{x}^2 = \frac{1}{n} \sum_{i=1}^{n} x_i^2 - \overline{x}^2;$$

that is, the second moment about the origin minus the square of the mean. However, v is not called the sample variance, but

$$s^2 = \left[\frac{n}{n-1}\right] v = \frac{1}{n-1} \sum_{i=1}^{n} (x_i - \overline{x})^2$$

is, because we will see later that, in some sense, s^2 is a better estimate of an unknown σ^2 than is v. Thus the **sample variance** is

$$s^2 = \frac{1}{n-1} \sum_{i=1}^{n} (x_i - \overline{x})^2.$$

The **sample standard deviation**, $s = \sqrt{s^2} \geq 0$, gives a measure of how dispersed the data are from the sample mean. At this stage of the study of statistics it is difficult to get a good understanding or meaning of the standard deviation s, but you can roughly think of it as the average distance of the values $x_1, x_2, \ldots, x_n$ from the mean $\overline{x}$. This is not true exactly for, in general,

$$s > \frac{1}{n} \sum_{i=1}^{n} |x_i - \overline{x}|,$$

but it is fair to say that s is somewhat larger yet of the same magnitude as the average of the distances of $x_1, x_2, \ldots, x_n$ from $\overline{x}$.

EXAMPLE 2.3-6 Rolling a fair six-sided die five times could result in the following sample of $n = 5$ observations:

$$x_1 = 3, \quad x_2 = 1, \quad x_3 = 2, \quad x_4 = 6, \quad x_5 = 3.$$

In this case

$$\overline{x} = \frac{3 + 1 + 2 + 6 + 3}{5} = 3$$

and

$$s^2 = \frac{(3-3)^2 + (1-3)^2 + (2-3)^2 + (6-3)^2 + (3-3)^2}{4} = \frac{14}{4} = 3.5.$$

It follows that $s = \sqrt{14/4} = 1.87$. Roughly, s can be thought of as the average distance that the x-values are away from the sample mean $\overline{x}$. Clearly this is not exact because in this example the distances from $\bar{x} = 3$ are 0, 2, 1, 3, 0, with an average of 1.2. In general s will be somewhat larger than this average distance. ◄

There is an alternative way of computing s^2, because $s^2 = [n/(n-1)]v$ and

$$v = \frac{1}{n}\sum_{i=1}^{n}(x_i - \bar{x})^2 = \frac{1}{n}\sum_{i=1}^{n}x_i^2 - \bar{x}^2.$$

It follows that

$$s^2 = \frac{\displaystyle\sum_{i=1}^{n}x_i^2 - n\bar{x}^2}{n-1} = \frac{\displaystyle\sum_{i=1}^{n}x_i^2 - \frac{1}{n}\left(\sum_{i=1}^{n}x_i\right)^2}{n-1}.$$

COMPUTATIONAL COMMENTS (CAS) When it comes to computing probabilities, means, and variances in the discrete case, we need to evaluate summations and could use a computer algebra system (CAS) such as *Maple* or *Mathematica*. The same statement could be made about the continuous case in Chapter 3 with integrals replacing summations.

For example, let us consider the hypergeometric distribution.

```
>   f:=binomial(N[1],x)*binomial(N[2],n-x)/binomial(N[1]+N[2],n);
```

$$f := \frac{\text{binomial}(N_1, x)\,\text{binomial}(N_2, n - x)}{\text{binomial}(N_1 + N_2, n)}$$

```
>   simplify(sum(f, x = 0 .. N[1]));
```

$$1$$

```
>   mu := simplify(sum(x*f, x = 0 .. N[1]));
```

$$\mu := \frac{N_1\, n}{N_1 + N_2}$$

```
>   var := simplify(sum(x^2*f, x = 0 .. N[1]) - mu^2);
```

$$var := -\frac{N_2\, N_1\,(n - N_1 - N_2)\, n}{(N_1 + N_2)^2\,(N_1 - 1 + N_2)}$$

```
>   N[1] := 20: N[2] := 30: n := 10:
>   mu := sum(x*f, x = 0 .. N[1]);
```

$$\mu := 4$$

```
>   var := sum((x - mu)^2*f, x = 0 .. N[1]);
```

$$var := \frac{96}{49}$$

```
>   sum(f, x = 3 .. 6);
```

$$\frac{8469282699}{10272278170}$$

```
>   evalf(%);
```

$$0.8244794931$$

Maple does not always write the output in the standard way. So we note that usually we write

$$\mu = n\left(\frac{N_1}{N}\right) = np$$

and

$$\sigma^2 = n\left(\frac{N_1}{N}\right)\left(\frac{N_2}{N}\right)\left(\frac{N-n}{N-1}\right) = np(1-p)\left(\frac{N-n}{N-1}\right).$$

The last two commands in the *Maple* output give the values of $P(3 \leq X \leq 6)$, as a fraction and as a decimal. It is also possible to find these probabilities using a program like Minitab or even some calculators can do this. You are encouraged to use whatever computing facilities are available to you. ■

Remark Note that we have introduced two types of means and standard deviations. The first pair is associated with the probability distribution and is denoted by the Greek symbols (μ, σ). The second pair is associated with data and is denoted by the Roman letters $(\bar{x}, s)$. ■

EXERCISES

2.3-1 Find the mean and variance for the following discrete distributions:

(a) $f(x) = \frac{1}{5}$, $x = 5, 10, 15, 20, 25$.

(b) $f(x) = 1$, $x = 5$.

(c) $f(x) = \frac{4-x}{6}$, $x = 1, 2, 3$.

2.3-2 For each of the following distributions, find $\mu = E(X)$, $E[X(X-1)]$, and $\sigma^2 = E[X(X-1)] + E(X) - \mu^2$:

(a) $f(x) = \frac{3!}{x!(3-x)!}\left(\frac{1}{4}\right)^x\left(\frac{3}{4}\right)^{3-x}$, $x = 0, 1, 2, 3$.

(b) $f(x) = \frac{4!}{x!(4-x)!}\left(\frac{1}{2}\right)^4$, $x = 0, 1, 2, 3, 4$.

2.3-3 Given $E(X+4) = 10$ and $E[(X+4)^2] = 116$, determine (a) $\text{Var}(X+4)$, (b) μ, and (c) σ^2.

2.3-4 Let μ and σ^2 denote the mean and variance of the random variable X. Determine $E[(X-\mu)/\sigma]$ and $E\{[(X-\mu)/\sigma]^2\}$.

2.3-5 A measure of **skewness** is defined by

$$\frac{E[(X - \mu)^3]}{\{E[(X - \mu)^2]\}^{3/2}} = \frac{E[(X - \mu)^3]}{(\sigma^2)^{3/2}} = \frac{E[(X - \mu)^3]}{\sigma^3}.$$

When a distribution is symmetrical about the mean, the skewness is equal to zero. If the probability histogram has a longer "tail" to the right than to the left, the measure of skewness is positive, and we say that the distribution is skewed positively or to the right. If the probability histogram has a longer tail to the left than to the right, the measure of skewness is negative, and we say that the distribution is skewed negatively or to the left. If the p.m.f. of X is given by $f(x)$, **(i)** depict the p.m.f. as a probability histogram and find the values of **(ii)** the mean, **(iii)** the standard deviation and **(iv)** skewness.

(a)
$$f(x) = \begin{cases} \dfrac{2^{6-x}}{64}, & x = 1, 2, 3, 4, 5, 6, \\[2mm] \dfrac{1}{64}, & x = 7. \end{cases}$$

(b)
$$f(x) = \begin{cases} \dfrac{1}{64}, & x = 1, \\[2mm] \dfrac{2^{x-2}}{64}, & x = 2, 3, 4, 5, 6, 7. \end{cases}$$

2.3-6 In LOTTO 49, Michigan's lottery game, a player selects 6 integers out of the first 49 positive integers. The state then randomly selects 6 out of the first 49 integers. Cash prizes are given to a player who matches 4, 5, or 6 integers. Let X equal the number of integers selected by a player that match integers selected by the state.

(a) State the p.m.f. of X.
(b) Calculate the mean, variance, and standard deviation of X.
(c) What value of X is most likely to occur?
(d) On February 25, 1995, the jackpot was worth $45,000,000. When the prize is this large, many bets are placed. Out of the 25,000,000 bets that were placed, 3 people matched all 6 numbers with each winning $15,000,000 (most of this paid by losers during preceding games), 390 matched 5 numbers to win $2500, and 22,187 matched 4 numbers to win $100. Are these numbers of winners consistent with the probability model?
(e) A mathematics professor convinced some colleagues to pool their lotto bets so that they were able to purchase 138 tickets together. They let the state computer randomly select the numbers on which they placed their bets. Among their 138 bets, 65 matched 0 of the winning LOTTO numbers, {3, 20, 33, 34, 43, 46}, 55 matched 1, 16 matched 2, and 2 matched 3 numbers. How do these results compare with what they could have expected?

2.3-7 Let X equal the larger outcome when a pair of four-sided dice is rolled. The p.m.f. of X is

$$f(x) = \frac{2x - 1}{16}, \qquad x = 1, 2, 3, 4.$$

(a) Find the mean, variance, and standard deviation of X.

(b) Calculate the sample mean, sample variance, and sample standard deviation of the following 100 simulated observations of X and compare this answer to $E(X)$ found in part (a):

4	4	4	4	2	2	2	3	1	4	3	3	2	3	2	4	4	2	3	4
4	3	4	3	4	3	3	2	4	4	3	2	3	3	3	2	4	4	3	4
1	4	3	4	4	4	3	2	4	4	4	3	1	3	2	4	4	4	4	1
3	4	3	2	4	4	3	3	1	3	3	3	3	2	2	2	3	4	3	3
2	4	2	3	3	2	4	4	3	4	4	4	4	3	4	4	4	4	4	4

(c) Draw the graphs of the probability histogram and the relative frequency histogram on the same figure.

(d) Generate your own data, either rolling dice (four-sided or six-sided) or simulating this experiment using a computer.

2.3-8 A Bingo card has 25 squares with numbers on 24 of them, the center being a free square. The integers that are placed on the Bingo card are selected randomly and without replacement from 1 to 75, inclusive. When a game called "cover-up" is played, balls numbered from 1 to 75, inclusive, are selected randomly and without replacement until a player covers each of the numbers on a card. Let X equal the number of balls that must be drawn to cover all the numbers on a single card.

(a) Argue that the p.m.f. of X, for $x = 24, 25, \ldots, 75$, is

$$f(x) = \frac{\binom{24}{23}\binom{51}{x-24}}{\binom{75}{x-1}} \cdot \frac{1}{75 - (x-1)} = \frac{\binom{51}{x-24}}{\binom{75}{x}} \cdot \frac{24}{x} = \frac{\binom{x}{24}}{\binom{75}{24}} \cdot \frac{24}{x}.$$

(b) What value of X is most likely to occur? In other words, what is the mode of this distribution?

(c) To show that the mean of X is $(24)(76)/25 = 72.96$ use the combinatorial identity

$$\binom{k+n+1}{k+1} = \sum_{x=k}^{n+k}\binom{x}{k}.$$

(d) Show that $E[X(X+1)] = \dfrac{24 \cdot 77 \cdot 76}{26} = 5401.8462.$

(e) Calculate $\text{Var}(X) = E[X(X+1)] - E[X] - [E(X)]^2 = \dfrac{46{,}512}{8125} = 5.7246$ and $\sigma = 2.39$.

(f) The following 100 observations of X were simulated on the computer.

75	73	74	74	73	75	75	71	71	74	75	74	74	73	75
75	75	71	68	75	74	72	73	75	74	67	73	71	74	74
73	75	71	73	71	68	74	75	66	75	75	74	75	71	75
72	75	74	75	75	72	75	75	74	73	75	62	64	75	72
74	74	75	73	75	75	73	74	75	68	75	69	74	75	61
73	73	73	72	74	68	74	71	73	75	73	74	72	74	73
75	73	71	70	62	74	74	72	75	74					

Compare these data with the probability model. In particular, make the following comparisons: **(i)** $\bar{x}$ with μ, **(ii)** s^2 with σ^2, **(iii)** s with σ, **(iv)** $(1/100)\sum_{i=1}^{100} x_i(x_i+1)$ with $E[X(X+1)]$.

(g) Compare the probability histogram with the relative frequency histogram.

2.3-9 To find the variance of a hypergeometric random variable in Example 2.3-5 we used the fact that

$$E[X(X-1)] = \frac{N_1(N_1-1)(n)(n-1)}{N(N-1)}.$$

Prove this result by making a change of variables $k = x - 2$ and noting that

$$\binom{N}{n} = \frac{N(N-1)}{n(n-1)}\binom{N-2}{n-2}.$$

2.3-10 A deck of $n = 10$ cards is numbered from 1 to 10. The cards are shuffled and laid down from left to right, face up. Order each of the five successive pairs of cards. Each of these five pairs determines a random interval. Let X equal the number of these five random intervals that intersects each of the other four intervals. It can be shown that

$$P(X \geq k) = \frac{2^k}{\binom{2k+1}{k}}, \qquad k = 0, 1, 2, 3, 4, 5.$$

(a) Find the probability that at least one interval intersects all the other intervals. (You could also simulate this.)
(b) Prove that $\mu = E(X) = \sum_{k=1}^{5} P(X \geq k)$.
(c) Find the value of μ for $n = 10$ cards.
(d) Let the number of cards increase without bound. Find the value of

$$\mu = \sum_{k=0}^{\infty} \frac{2^k}{\binom{2k+1}{k}}.$$

2.3-11 Calculate the mean $\bar{x}$ and the variance s^2 of the samples given in
(a) Exercise 1.1-3,
(b) Exercise 1.1-4,
(c) Exercise 1.1-5,
(d) Exercise 1.1-7.

2.3-12 Referring to Exercise 1.1-6, compute the sample mean $\bar{x}$. Could a statistician tell that family about how many boxes of cereal they could expect to purchase to obtain all four prizes?

2.3-13 Consider an experiment that consists of selecting at random a card from an ordinary deck of cards. Let the random variable X equal the value of the selected card, where Ace = 1, Jack = 11, Queen = 12, and King = 13. Thus the space of X is $S = \{1, 2, 3, \ldots, 13\}$. If the experiment is performed in an unbiased manner, assign probabilities to these 13 outcomes and compute the mean μ of this probability distribution.

2.3-14 Place eight chips in a bowl: three of which have the number one, two of which have the number two, and three of which have the number three. Say each chip has the probability of 1/8 of being drawn at random. Let the random variable X equal the number on the chip that is selected so that the space of X is $S = \{1, 2, 3\}$. Make reasonable probability assignments to each of these three outcomes and compute the mean μ and the variance σ^2 of this probability distribution.

2.3-15 A fair coin is flipped successively at random until the first head is observed. Let the random variable X denote the number of flips of the coin that are required. Then the space of X is $S = \{x : x = 1, 2, 3, 4, \ldots\}$. Later we learn that under certain conditions, we can assign probabilities to these outcomes in S with the function $f(x) = (1/2)^x$, $x = 1, 2, 3, 4, \ldots$. Compute the mean μ. HINT: Write out the series for μ and then construct the series for $(1/2)\mu$ and take the difference. An alternative method would be to compare the series for μ to that of the negative binomial $(1 - z)^{-2}$, with $z = 1/2$.

2.3-16 Let X equal the number of calls per hour received by 911 between midnight and noon and reported in the *Holland Sentinel*. On October 29 and October 30, the following numbers of call were reported:

$$0\ 1\ 1\ 1\ 0\ 1\ 2\ 1\ 4\ 1\ 2\ 3\ 0\ 3\ 0\ 1\ 0\ 1\ 1\ 2\ 3\ 0\ 2\ 2$$

(a) Find the sample mean.
(b) Find the sample variance.

2.3-17 For determining half-lives of radioactive isotopes, it is important to know what the background radiation is in a given detector over a period of time. Data were collected in a γ-ray detection experiment over 98 ten-second intervals (observations of a random variable X) yielding the following data:

58	50	57	58	64	63	54	64	59	41	43	56	60	50
46	59	54	60	59	60	67	52	65	63	55	61	68	58
63	36	42	54	58	54	40	60	64	56	61	51	48	50
60	42	62	67	58	49	66	58	57	59	52	54	53	53
57	43	73	65	45	43	57	55	73	62	68	55	51	55
53	68	58	53	51	73	44	50	53	62	58	47	63	59
59	56	60	59	50	52	62	51	66	51	56	53	59	57

(a) Find the sample mean.
(b) Find the sample variance.

2.3-18 Students in a statistics class were asked to report the number of pets in their family. The following data were collected:

3	3	2	4	4	4	3	2	3	2	2	3	4	2	2	3
2	3	4	4	2	3	2	3	3	2	4	3	3	3	2	2
2	2	1	4	4	2	3	3	1	2	2	2	2	3	1

(a) Find the frequencies of 1, 2, 3, 4.
(b) Calculate the sample mean and the sample standard deviation.
(c) Construct a histogram and locate the mean on the histogram.

2.3-19 A warranty is written on a product worth $10,000 so that the buyer is given $8000 if it fails in the first year, $6000 if it fails in the second, $4000 if it fails in the third, $2000 if it fails in the fourth, and zero after that. Its probability of failing in a year is 0.1; failures are independent of those of other years. What is the expected value of the warranty?

2.4 BERNOULLI TRIALS AND THE BINOMIAL DISTRIBUTION

The probability models for random experiments that will be described in this section occur frequently in applications.

A **Bernoulli experiment** is a random experiment, the outcome of which can be classified in but one of two mutually exclusive and exhaustive ways, say, success or failure (e.g., female or male, life or death, nondefective or defective). A sequence of **Bernoulli trials** occurs when a Bernoulli experiment is performed several *independent* times so that the probability of success, say

p, remains the *same* from trial to trial. That is, in such a sequence we let p denote the probability of success on each trial. In addition, we shall frequently let $q = 1 - p$ denote the probability of failure; that is, we shall use q and $1 - p$ interchangeably.

EXAMPLE 2.4-1 Suppose that the probability of germination of a beet seed is 0.8 and the germination of a seed is called a success. If we plant 10 seeds and can assume that the germination of one seed is independent of the germination of another seed, this would correspond to 10 Bernoulli trials with $p = 0.8$. ◄

EXAMPLE 2.4-2 In the Michigan daily lottery the probability of winning when placing a six-way boxed bet is 0.006. A bet placed on each of 12 successive days would correspond to 12 Bernoulli trials with $p = 0.006$. ◄

Let X be a random variable associated with a Bernoulli trial by defining it as follows:

$$X(\text{success}) = 1 \quad \text{and} \quad X(\text{failure}) = 0.$$

That is, the two outcomes, success and failure, are denoted by one and zero, respectively. The p.m.f. of X can be written as

$$f(x) = p^x (1 - p)^{1-x}, \qquad x = 0, 1,$$

and we say that X has a **Bernoulli distribution**. The expected value of X is

$$\mu = E(X) = \sum_{x=0}^{1} x\, p^x (1 - p)^{1-x} = (0)(1 - p) + (1)(p) = p,$$

and the variance of X is

$$\sigma^2 = \text{Var}(X) = \sum_{x=0}^{1} (x - p)^2 p^x (1 - p)^{1-x}$$

$$= (0 - p)^2 (1 - p) + (1 - p)^2 p = p(1 - p) = pq.$$

It follows that the standard deviation of X is

$$\sigma = \sqrt{p(1 - p)} = \sqrt{pq}.$$

In a sequence of n Bernoulli trials, we shall let X_i denote the Bernoulli random variable associated with the ith trial. An observed sequence of n Bernoulli trials will then be an n-tuple of zeros and ones, and we often call this collection a **random sample** of size n from a Bernoulli distribution.

EXAMPLE 2.4-3   Out of millions of instant lottery tickets, suppose that 20% are winners. If five such tickets are purchased, $(0, 0, 0, 1, 0)$ is a possible observed sequence in which the fourth ticket is a winner and the other four are losers. Assuming independence among winning and losing tickets, the probability of this outcome is

$$(0.8)(0.8)(0.8)(0.2)(0.8) = (0.2)(0.8)^4.$$ ◄

EXAMPLE 2.4-4   If five beet seeds are planted in a row, a possible observed sequence would be $(1, 0, 1, 0, 1)$ in which the first, third, and fifth seeds germinated and the other two did not. If the probability of germination is $p = 0.8$, the probability of this outcome is, assuming independence,

$$(0.8)(0.2)(0.8)(0.2)(0.8) = (0.8)^3(0.2)^2.$$ ◄

In a sequence of Bernoulli trials, we are often interested in the total number of successes and not in the order of their occurrence. If we let the random variable X equal the number of observed successes in n Bernoulli trials, the possible values of X are $0, 1, 2, \ldots, n$. If x successes occur, where $x = 0, 1, 2, \ldots, n$, then $n - x$ failures occur. The number of ways of selecting x positions for the x successes in the n trials is

$$\binom{n}{x} = \frac{n!}{x!(n-x)!}.$$

Since the trials are independent and since the probabilities of success and failure on each trial are, respectively, p and $q = 1 - p$, the probability of each of these ways is $p^x(1-p)^{n-x}$. Thus $f(x)$, the p.m.f. of X, is the sum of the probabilities of these $\binom{n}{x}$ mutually exclusive events; that is,

$$f(x) = \binom{n}{x}p^x(1-p)^{n-x}, \qquad x = 0, 1, 2, \ldots, n.$$

These probabilities are called binomial probabilities, and the random variable X is said to have a **binomial distribution**.

Summarizing, a binomial experiment satisfies the following properties:

1. A Bernoulli (success-failure) experiment is performed n times.
2. The trials are independent.
3. The probability of success on each trial is a constant p; the probability of failure is $q = 1 - p$.
4. The random variable X equals the number of successes in the n trials.

A binomial distribution will be denoted by the symbol $b(n, p)$ and we say that the distribution of X is $b(n, p)$. The constants n and p are called the **parameters** of the binomial distribution; they correspond to the number n of independent trials and the probability p of success on each trial. Thus, if we say that the distribution of X is $b(12, 1/4)$, we mean that X is the number of successes in a random sample of size $n = 12$ from a Bernoulli distribution with $p = 1/4$.

EXAMPLE 2.4-5 In the instant lottery with 20% winning tickets, if X is equal to the number of winning tickets among $n = 8$ that are purchased, the probability of purchasing 2 winning tickets is

$$f(2) = P(X = 2) = \binom{8}{2}(0.2)^2(0.8)^6 = 0.2936.$$

The distribution of the random variable X is $b(8, 0.2)$. ◀

EXAMPLE 2.4-6 In order to obtain a better feeling for the effect of the parameters n and p on the distribution of probabilities, four probability histograms are displayed in Figure 2.4-1. ◀

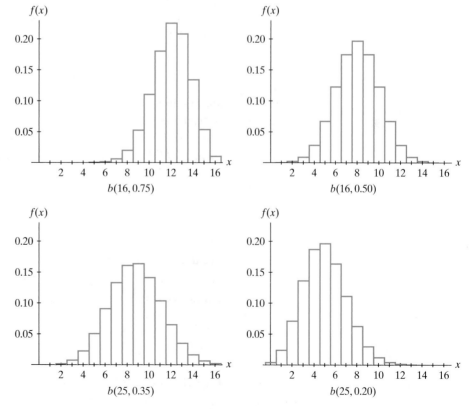

Figure 2.4-1: Binomial probability histograms

EXAMPLE 2.4-7 In Example 2.4-1, the number X of seeds that germinate in $n = 10$ independent trials is $b(10, 0.8)$: that is,

$$f(x) = \binom{10}{x}(0.8)^x(0.2)^{10-x}, \qquad x = 0, 1, 2, \ldots, 10.$$

In particular,

$$P(X \leq 8) = 1 - P(X = 9) - P(X = 10)$$
$$= 1 - 10(0.8)^9(0.2) - (0.8)^{10} = 0.6242.$$

Also, we could compute, with a little more work,

$$P(X \leq 6) = \sum_{x=0}^{6}\binom{10}{x}(0.8)^x(0.2)^{10-x}. \qquad \blacktriangleleft$$

Cumulative probabilities like those in the previous example are often of interest. We call the function defined by

$$F(x) = P(X \leq x)$$

the **cumulative distribution function**, or more simply, the **distribution function** of the random variable X. Values of the distribution function of a random variable X that is $b(n, p)$ are given in Table II in the Appendix for selected values of n and p. The use of this table is illustrated in the next example.

EXAMPLE 2.4-8 Leghorn chickens are raised for laying eggs. Let $p = 0.5$ be the probability of a female chick hatching. Assuming independence, let X equal the number of female chicks out of 10 newly hatched chicks selected at random. Then the distribution of X is $b(10, 0.5)$. The probability of 5 or fewer female chicks is, using Table II in the Appendix,

$$P(X \leq 5) = 0.6230.$$

The probability of exactly 6 female chicks is

$$P(X = 6) = \binom{10}{6}\left(\frac{1}{2}\right)^6\left(\frac{1}{2}\right)^4$$
$$= P(X \leq 6) - P(X \leq 5)$$
$$= 0.8281 - 0.6230 = 0.2051,$$

since $P(X \leq 6) = 0.8281$. The probability of at least 6 female chicks is

$$P(X \geq 6) = 1 - P(X \leq 5) = 1 - 0.6230 = 0.3770. \qquad \blacktriangleleft$$

Although probabilities for the binomial distribution $b(n, p)$ are given in Table II in the Appendix for selected values of p that are less than or equal to 0.5, the next example demonstrates that this table can also be used for values of p that are greater than 0.5. (In later sections we learn how to approximate certain binomial probabilities with those of other distributions.)

EXAMPLE 2.4-9   Suppose that we are in one of those rare times when 65% of the American public approve of the way the President of the United States is handling his job. Take a random sample of $n = 8$ Americans and let Y equal the number who give approval. Then the distribution of Y is $b(8, 0.65)$. To find $P(Y \geq 6)$, note that

$$P(Y \geq 6) = P(8 - Y \leq 8 - 6) = P(X \leq 2),$$

where $X = 8 - Y$ counts the number who disapprove. Since $q = 1 - p = 0.35$ equals the probability of disapproval by each person selected, the distribution of X is $b(8, 0.35)$ (see Figure 2.4-2). From Table II in the Appendix, since $P(X \leq 2) = 0.4278$, it follows that $P(Y \geq 6) = 0.4278$.

Similarly,

$$P(Y \leq 5) = P(8 - Y \geq 8 - 5)$$
$$= P(X \geq 3) = 1 - P(X \leq 2)$$
$$= 1 - 0.4278 = 0.5722$$

and

$$P(Y = 5) = P(8 - Y = 8 - 5)$$
$$= P(X = 3) = P(X \leq 3) - P(X \leq 2)$$
$$= 0.7064 - 0.4278 = 0.2786.$$ ◄

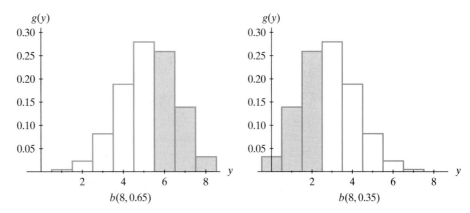

Figure 2.4-2: Presidential approval histogram

Recall that if n is a positive integer, then

$$(a+b)^n = \sum_{x=0}^{n} \binom{n}{x} b^x a^{n-x}.$$

Thus the sum of the binomial probabilities, if we use the above binomial expansion with $b = p$ and $a = 1 - p$, is

$$\sum_{x=0}^{n} \binom{n}{x} p^x (1-p)^{n-x} = [(1-p)+p]^n = 1,$$

a result that had to follow from the fact that $f(x)$ is a p.m.f.

Since the summations to determine the mean and the variance of the binomial distribution are rather difficult to evaluate, we find the mean and variance by a different way in the next section (see Example 2.5-3) to be

$$\mu = np \qquad \text{and} \qquad \sigma^2 = np(1-p) = npq.$$

Note that when p is the probability of success on each trial, the expected number of successes in n trials is np, a result that agrees with our intuition. Of course, the standard deviation is

$$\sigma = \sqrt{np(1-p)}.$$

EXAMPLE 2.4-10 Suppose that observation over a long period of time has disclosed that, on the average, one out of 10 items produced by a process is defective. Select five items independently from the production line and test them. Let X denote the number of defective items among the $n = 5$ items. Then X is $b(5, 0.1)$. Furthermore,

$$E(X) = 5(0.1) = 0.5, \quad \text{Var}(X) = 5(0.1)(0.9) = 0.45.$$

For example, the probability of observing at most one defective item is

$$P(X \le 1) = \binom{5}{0}(0.1)^0(0.9)^5 + \binom{5}{1}(0.1)^1(0.9)^4 = 0.9185. \qquad \blacktriangleleft$$

The next example shows the relationship between the binomial probability model and a set of observed data. Replications of this example would of course yield different results, in which the fits could be better or worse than the given one.

EXAMPLE 2.4-11 Consider the simple experiment of flipping a fair coin five independent times. If X equals the number of heads that is observed, then X is $b(5, 0.5)$, $\mu = 2.5$, $\sigma^2 = 1.25$, and $\sigma = 1.118$. This experiment was simulated 100 times, yielding the following data:

2	3	2	4	1	2	1	1	4	2	4	2	0	4	4	2	4	4	3	4
2	2	4	4	1	1	3	3	1	4	2	3	1	2	4	1	2	5	3	2
4	3	2	2	2	3	5	2	0	3	2	1	3	4	2	2	4	0	2	1
3	3	2	3	2	1	3	2	2	2	1	1	3	3	1	1	4	2	1	5
3	2	3	0	3	5	3	2	4	3	3	5	2	3	3	1	3	2	1	1

For these data $\bar{x} = 2.47$, $s^2 = 1.5243$, and $s = 1.235$. In Figure 2.4-3 the probability histogram and the relative frequency histogram (shaded) are given. ◀

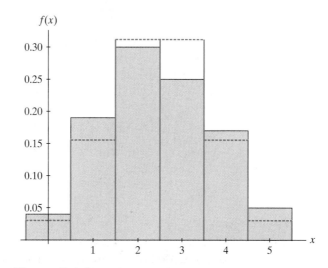

Figure 2.4-3: Number of heads when five coins are flipped

Suppose that an urn contains N_1 success balls and N_2 failure balls and we let $p = N_1/(N_1 + N_2)$. Let X equal the number of success balls in a random sample of size n that is taken from this urn. If the sampling is done one at a time with replacement, the distribution of X is $b(n, p)$; and if the sampling is done without replacement, then X has a hypergeometric distribution with p.m.f.

$$f(x) = \frac{\binom{N_1}{x}\binom{N_2}{n - x}}{\binom{N_1 + N_2}{n}},$$

where x is a nonnegative integer such that $x \leq n$, $x \leq N_1$, and $n - x \leq N_2$. When $N_1 + N_2$ is large and n is relatively small, it makes little difference if the sampling is done with or without replacement. In Figure 2.4-4 the probability histograms are compared for different combinations of n, N_1, and N_2. You are asked to compute some of these probabilities in Exercise 2.4-19.

STATISTICAL COMMENTS (Ordered Restricted Estimates) As indicated earlier, the relative frequency, X/n, in the notation of this section, is an

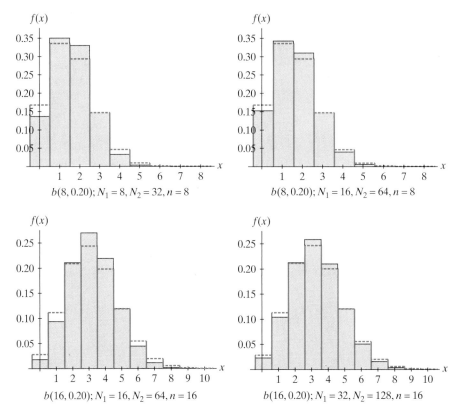

Figure 2.4-4: Binomial and hypergeometric (shaded) probability histograms

estimator of the unknown p. Suppose we are interested in knowing how well professional golfers putt. In particular, what is the probability, p_r, of sinking a putt of r feet on a relatively flat green? Say we observe many independent putts of r feet by ranked professionals and get the following data (these are not exact as they have been rounded off for easier computations, but these statistics reflect the abilities of professional golfers).

Length (r)	n_i	x_i	x_i/n_i	Length (r)	n_i	x_i	x_i/n_i
2	1000	946	0.946	12	200	67	0.335
3	800	676	0.845	13	200	61	0.305
4	500	376	0.752	14	200	66	0.330
5	400	236	0.590	15	200	46	0.230
6	300	162	0.540	16	200	41	0.205
7	250	134	0.536	17	200	35	0.175
8	250	123	0.492	18	100	17	0.170
9	200	66	0.330	19	100	14	0.140
10	200	73	0.375	20	100	15	0.150
11	200	69	0.345				

So we might use $376/500 = 0.752$ as an estimate of p_4, the probability of a professional golfer sinking a four foot putt on a relatively flat green. However, we get into trouble further down the list. While we would assume that $p_r \geq p_{r+1}$, we get reversals in our estimates. For example, the estimate of p_{10} is 0.375, which is greater than that of p_9, which is 0.330, and this does not agree with our intuition.

As a matter of fact, if we wanted to be dishonest statisticians, we might give these estimates of p_9, p_{12}, p_{14}, namely, 0.330, 0.335, 0.330 as evidence that the probability of sinking a putt does not change much for putts between 9 and 14 feet. This simply is not true because this total data set clearly indicates that the probabilities tend to decrease as the length increases. Hence we really need **ordered restricted estimates**, which is a topic included in a more advanced course in statistics. However, one solution is called the **pool-adjacent-violators-algorithm**, which would correct the estimates of p_9 and p_{10} to be 0.3475 each; p_{13} and p_{14} to be 0.3175 each; and p_{19} and p_{20} to be 0.145 each. The others are in the correct order and they remain as given.   ■

EXERCISES

2.4-1 An urn contains 7 red and 11 white balls. Draw one ball at random from the urn. Let $X = 1$ if a red ball is drawn, and let $X = 0$ if a white ball is drawn. Give the p.m.f., mean, and variance of X.

2.4-2 Suppose that in Exercise 2.4-1, $X = 1$ if a red ball is drawn, and $X = -1$ if a white ball is drawn. Give the p.m.f., mean, and variance of X.

2.4-3 On a six-question multiple-choice test there are five possible answers for each question, of which one is correct (C) and four are incorrect (I). If a student guesses randomly and independently, find the probability of

   (a) Being correct only on questions 1 and 4 (i.e., scoring C, I, I, C, I, I).
   (b) Being correct on two questions.

2.4-4 A recent national study showed that approximately 45% of college students binge drink. Let X equal the number of students in a random sample of size $n = 12$ who binge drink. Find the probability that

   (a) X is at most 5.
   (b) X is at least 6.
   (c) X is equal to 7.
   (d) Give the mean, variance, and standard deviation of X.

2.4-5 According to a *CNN/USA Today* poll, approximately 70% of Americans believe the IRS abuses its power. Let X equal the number of people who believe the IRS abuses its power in a random sample of $n = 25$ Americans. Assuming that the poll results are still valid, find the probability that

   (a) X is at least 13.
   (b) X is at most 11.
   (c) X is equal to 12.
   (d) Give the mean, variance, and standard deviation of X.

2.4-6 It is claimed that 15% of the ducks in a particular region have patent schistosome infection. Suppose that seven ducks are selected at random. Let X equal the number of ducks that are infected.

   (a) Assuming independence, how is X distributed?
   (b) Find **(i)** $P(X \geq 2)$, **(ii)** $P(X = 1)$, and **(iii)** $P(X \leq 3)$.

2.4-7 Suppose that 2000 points are selected independently and at random from the unit square $\{(x, y): 0 \leq x < 1, 0 \leq y < 1\}$. Let W equal the number of points that fall in $A = \{(x, y): x^2 + y^2 < 1\}$.

(a) How is W distributed?
(b) Give the mean, variance, and standard deviation of W.
(c) What is the expected value of $W/500$?
(d) Use the computer to select 2000 pairs of random numbers. Determine the value of W and use that value to find an estimate for π. (Of course, we know the real value of π, and more will be said about estimation later on in this text.)
(e) How could you extend part (d) to estimate the volume of a ball of radius 1 in 3-space, namely $(4/3)\pi$?
(f) How could you extend these techniques to estimate the "volume" of a ball of radius 1 in n-space?

2.4-8 It is believed that 20% of Americans do not have any health insurance. Suppose this is true and let X equal the number with no health insurance in a random sample of $n = 15$ Americans.

(a) How is X distributed?
(b) Give the mean, variance, and standard deviation of X.
(c) Find $P(X \geq 5)$.

2.4-9 Let p equal the proportion of all college and university students who would say yes to the question, "Would you drink from the same glass as your friend if you suspected that this friend were an AIDS virus carrier?" Assume that $p = 0.10$. Let X equal the number of students out of a random sample of size $n = 9$ who would say yes to this question.

(a) How is X distributed?
(b) Give the values of the mean, variance, and standard deviation of X.
(c) Find (i) $P(X = 2)$ and (ii) $P(X \geq 2)$.

2.4-10 Suppose that, in manufacturing, the probability of a certain item being defective is $p = 0.05$. Suppose further that the quality of an item is independent of the quality of the other manufactured items. An inspector selects six items at random. Let X equal the number of defective items in the sample.

(a) How is X distributed?
(b) Give the values of $E(X)$ and $Var(X)$.
(c) Find (i) $P(X = 0)$, (ii) $P(X \leq 1)$, and (iii) $P(X \geq 2)$.

2.4-11 According to a Des Moines Register poll, 40% of Iowa farmers support an independent Palestinian state. Let X equal the number of Iowa farmers out of a random sample of $n = 20$ who support an independent Palestinian state. Find

(a) $P(X \leq 10)$.
(b) $P(X \geq 8)$.
(c) $P(X = 9)$.

2.4-12 One of the 12 courses in a Chinese vegetarian banquet at a Buddhist temple was a platter of mushrooms. When attempting to pick up a round, slippery mushroom with chopsticks, an American tourist was successful 55% of the time on the first try. Let X equal the number of times that the tourist was successful on the first try at picking up 14 mushrooms. Assume that the "first tries" are independent trials.

(a) Give the mean and variance of X.
(b) Find $P(X < 8)$ and $P(X > 6)$.

2.4-13 A random variable X has a binomial distribution with mean 6 and variance 3.6. Find $P(X = 4)$.

2.4-14 A certain type of mint has a label weight of 20.4 grams. Suppose that the probability is 0.90 that a mint weighs more than 20.7 grams. Let X equal the number of mints that weigh more than 20.7 grams in a sample of 8 mints selected at random.

    (a) How is X distributed if we assume independence?
    (b) Find (i) $P(X = 8)$, (ii) $P(X \leq 6)$, and (iii) $P(X \geq 6)$.

2.4-15 It is given that the probability of germination of each beet seed is 0.8. Thus if X denotes the number of seeds that germinate in a set of nine seeds, then X is $b(9, 0.8)$, provided that the assumption of independence is valid.

    (a) Give the values of μ, σ^2, and σ.
    (b) Draw the probability histogram for the p.m.f. of X.
    (c) If 100 sets of nine beet seeds were planted, calculate $\bar{x}, s^2$, and s for the following 100 observations.

7	9	7	7	9	7	9	8	5	8	8	6	8	7	8	8	9	7	8	8
8	7	8	7	7	7	6	7	9	7	6	7	6	7	6	5	7	8	7	4
8	8	7	6	9	7	8	7	7	6	7	9	6	6	8	8	8	7	9	5
7	4	7	8	8	9	6	7	6	8	7	5	7	7	9	7	7	8	7	9
7	5	9	8	7	7	8	9	8	5	9	9	8	6	9	8	8	8	9	7

    (d) Superimpose the relative frequency histogram of these observations on the graph of the probability histogram.

2.4-16 In the casino game chuck-a-luck, three unbiased six-sided dice are rolled. One possible bet is $1 on fives, and the payoff is equal to $1 for each five on that roll. In addition, the dollar bet is returned if at least one five is rolled. The dollar that was bet is lost only if no fives are rolled. Let X denote the payoff for this game. Then X can equal $-1, 1, 2$, or 3.

    (a) Determine the p.m.f. $f(x)$.
    (b) Calculate μ, σ^2, and σ.
    (c) Depict the p.m.f. as a probability histogram.
    (d) Calculate $\bar{x}, s^2$, and s using the 100 following simulated observations of X.

-1	1	-1	-1	1	-1	-1	-1	-1	1
-1	1	-1	2	1	-1	-1	-1	1	-1
1	-1	-1	1	1	-1	-1	-1	-1	1
-1	-1	2	1	1	-1	-1	1	-1	1
-1	-1	1	2	-1	-1	1	-1	-1	-1
-1	1	-1	-1	-1	1	-1	1	-1	2
-1	1	-1	1	-1	2	1	-1	-1	1
-1	1	1	-1	1	1	-1	3	1	-1
-1	-1	1	1	1	1	-1	-1	2	-1
1	-1	1	-1	1	2	-1	1	1	-1

    (e) Superimpose the relative frequency histogram of these observations on the probability histogram.

2.4-17 It is claimed that for a particular lottery, 1/10 of the 50,000,000 tickets will win a prize. What is the probability of winning at least one prize if you purchase (a) 10 tickets or (b) 15 tickets?

2.4-18 For the lottery described in Exercise 2.4-17, find the smallest number of tickets that must be purchased so that the probability of winning at least one prize is greater than (a) 0.50, (b) 0.95.

2.4-19 Construct a table that gives the probabilities for the $b(8, 0.2)$ distribution, the hypergeometric distribution with $N_1 = 8, N_2 = 32$, and $n = 8$, and the hypergeometric distribution with $N_1 = 16, N_2 = 64$, and $n = 8$ (see Figure 2.1-2).

2.4-20 A hospital obtains 40% of its flu vaccine from Company A, 50% from Company B, and 10% from Company C. From past experience it is known that 3% of the vials from A are ineffective, 2% from B are ineffective, and 5% from C are ineffective. The hospital tests five vials from each shipment. If at least one of the five is ineffective, find the conditional probability of that shipment coming from C.

2.4-21 Many products can operate if out of n parts, k of them are working. Say $n = 20$ and $k = 17$. Say $p = 0.05$ is the probability that a part fails. Assuming independence, what is the probability that the product operates?

2.4-22 A boiler has four relief valves. The probability that each opens properly is 0.99.

    (a) Find the probability that at least one opens properly.

    (b) Find the probability that all four open properly.

2.4-23 A business man has an important meeting to attend, but he is running a little late. He can take one route to work that has six stop lights and another longer route that has two stop lights. He figures if he stops at more than half of the lights on either route, he will be late for the meeting. Assuming independence and the chance of stopping at each light is $p = 0.5$, which route should he take?

2.5 THE MOMENT-GENERATING FUNCTION

The mean, variance, and standard deviation are important characteristics of a distribution. For some distributions, the binomial for instance, it is fairly difficult to directly compute $E(X)$ and $E(X^2)$ to find the mean and the variance. In this section, we define a function of t that will help us generate the moments of a distribution, and thus it is called the moment-generating function. While this generating characteristic is extremely important, there is a uniqueness property that is even more important. We first define this new function of t and then explain this uniqueness property before showing how it can be used to compute the moments of X.

Definition 2.5-1

Let X be a random variable of the discrete type with p.m.f. $f(x)$ and space S. If there is a positive number h such that

$$E(e^{tX}) = \sum_{x \in S} e^{tx} f(x)$$

exists and is finite for $-h < t < h$, then the function of t defined by

$$M(t) = E(e^{tX})$$

is called the *moment-generating function of X* (or of the distribution of X). This is often abbreviated as m.g.f.

First, it is evident that if we set $t = 0$, we have $M(0) = 1$. Moreover, if the space of S is $\{b_1, b_2, b_3, \ldots\}$, the moment-generating function is given by the expansion

$$M(t) = e^{tb_1} f(b_1) + e^{tb_2} f(b_2) + e^{tb_3} f(b_3) + \cdots.$$

Thus we see that the coefficient of e^{tb_i} is the probability

$$f(b_i) = P(X = b_i).$$

Accordingly if two random variables (or two distributions of probability) have the same moment-generating function, they must have the same distribution of probability. That is, if the two had the two probability mass functions $f(x)$ and $g(y)$, and the same space $S = \{b_1, b_2, b_3, \ldots\}$ and

$$e^{tb_1} f(b_1) + e^{tb_2} f(b_2) + \cdots = e^{tb_1} g(b_1) + e^{tb_2} g(b_2) + \cdots \qquad (2.5\text{-}1)$$

for all t, $-h < t < h$, mathematical transform theory requires that

$$f(b_i) = g(b_i), \qquad i = 1, 2, 3, \ldots.$$

So we see that the moment-generating function uniquely determines the distribution of a random variable. In other words, if the m.g.f. exists, there is one and only one distribution of probability associated with that m.g.f.

Remark From elementary algebra, we can get some understanding of why Equation 2.5-1 requires $f(b_i) = g(b_i)$. In that equation, let $e^t = w$ and say the points in the support, namely $b_1, b_2, \ldots, b_k$, are positive integers, the largest of which is m. Thus Equation 2.5-1 provides the equality of two mth degree polynomials in w for an uncountable number of values of w. A fundamental theorem of algebra requires that the corresponding coefficients of the two polynomials be equal; that is, $f(b_i) = g(b_i), i = 1, 2, \ldots, k$. ∎

EXAMPLE 2.5-1   If X has the m.g.f.

$$M(t) = e^t \left(\frac{3}{6}\right) + e^{2t} \left(\frac{2}{6}\right) + e^{3t} \left(\frac{1}{6}\right),$$

then the probabilities are

$$P(X = 1) = \frac{3}{6}, \qquad P(X = 2) = \frac{2}{6}, \qquad P(X = 3) = \frac{1}{6}.$$

We could write this, if we choose to do so, by saying X has the p.m.f.

$$f(x) = \frac{4 - x}{6}, \qquad x = 1, 2, 3. \qquad \blacktriangleleft$$

EXAMPLE 2.5-2   Suppose we are given that the m.g.f. of X is

$$M(t) = \frac{e^t/2}{1 - e^t/2}, \qquad t < \ln 2.$$

Until we expand $M(t)$, we cannot detect the coefficients of $e^{b_i t}$. Recalling

$$(1 - z)^{-1} = 1 + z + z^2 + z^3 + \cdots, \qquad -1 < z < 1,$$

we have that

$$\frac{e^t}{2}\left(1 - \frac{e^t}{2}\right)^{-1} = \frac{e^t}{2}\left(1 + \frac{e^t}{2} + \frac{e^{2t}}{2^2} + \frac{e^{3t}}{2^3} + \cdots\right)$$

$$= \left(e^t\right)\left(\frac{1}{2}\right) + \left(e^{2t}\right)\left(\frac{1}{2}\right)^2 + \left(e^{3t}\right)\left(\frac{1}{2}\right)^3 + \cdots,$$

when $e^t/2 < 1$ and thus $t < \ln 2$. That is,

$$P(X = x) = \left(\frac{1}{2}\right)^x,$$

when x is a positive integer, or, equivalently, the p.m.f. of X is

$$f(x) = \left(\frac{1}{2}\right)^x, \qquad x = 1, 2, 3, \dots. \qquad \blacktriangleleft$$

From the theory of Laplace transforms, it can be shown that the existence of $M(t)$, for $-h < t < h$, implies that derivatives of $M(t)$ of all orders exist at $t = 0$; moreover, it is permissible to interchange differentiation and summation as the series converges uniformly. Thus

$$M'(t) = \sum_{x \in S} x e^{tx} f(x),$$

$$M''(t) = \sum_{x \in S} x^2 e^{tx} f(x),$$

and, for each positive integer r,

$$M^{(r)}(t) = \sum_{x \in S} x^r e^{tx} f(x).$$

Setting $t = 0$, we see that

$$M'(0) = \sum_{x \in S} x f(x) = E(X),$$

$$M''(0) = \sum_{x \in S} x^2 f(x) = E(X^2),$$

and, in general,

$$M^{(r)}(0) = \sum_{x \in S} x^r f(x) = E(X^r).$$

In particular, if the moment-generating function exists,

$$\mu = M'(0) \qquad \text{and} \qquad \sigma^2 = M''(0) - [M'(0)]^2.$$

The above argument shows that we can find the moments of X by differentiating $M(t)$. In using this technique, it must be emphasized that in use, we first evaluate the summation representing $M(t)$ to obtain a closed form solution, and then we differentiate that solution to obtain the moments of X. The following illustrates the use of the moment-generating function for finding the first and second moments and then the mean and variance of the important binomial distribution.

EXAMPLE 2.5-3   The p.m.f. of the binomial distribution is

$$f(x) = \binom{n}{x} p^x (1 - p)^{n-x}, \qquad x = 0, 1, 2, \ldots, n.$$

Thus the m.g.f. is

$$M(t) = E(e^{tX}) = \sum_{x=0}^{n} e^{tx} \binom{n}{x} p^x (1 - p)^{n-x}$$

$$= \sum_{x=0}^{n} \binom{n}{x} (pe^t)^x (1 - p)^{n-x}$$

$$= [(1 - p) + pe^t]^n, \qquad -\infty < t < \infty,$$

from the expansion of $(a + b)^n$ with $a = 1 - p$ and $b = pe^t$. It is interesting to note that here and elsewhere that the m.g.f. is usually rather easy to compute if the p.m.f. has a factor involving an exponential, like p^x in the binomial p.m.f.

The first two derivatives of $M(t)$ are

$$M'(t) = n[(1 - p) + pe^t]^{n-1}(pe^t)$$

and

$$M''(t) = n(n - 1)[(1 - p) + pe^t]^{n-2}(pe^t)^2 + n[(1 - p) + pe^t]^{n-1}(pe^t).$$

Thus

$$\mu = E(X) = M'(0) = np$$

and

$$\sigma^2 = E(X^2) - [E(X)]^2 = M''(0) - [M'(0)]^2$$
$$= n(n - 1)p^2 + np - (np)^2 = np(1 - p),$$

as was claimed in Section 2.4.

In the special case when $n = 1$, X has a Bernoulli distribution and

$$M(t) = (1 - p) + pe^t$$

for all real values of t, $\mu = p$, and $\sigma^2 = p(1 - p)$. ◀

We turn now to the problem of observing a sequence of Bernoulli trials until exactly r successes occur, where r is a fixed positive integer. Let the random variable X denote the number of trials needed to observe the rth success. That is, X is the trial number on which the rth success is observed. By the multiplication rule of probabilities, the p.m.f. of X, say $g(x)$, equals the product of the probability

$$\binom{x-1}{r-1} p^{r-1}(1 - p)^{x-r} = \binom{x-1}{r-1} p^{r-1} q^{x-r}$$

of obtaining exactly $r - 1$ successes in the first $x - 1$ trials and the probability p of a success on the rth trial. Thus the p.m.f. of X is

$$g(x) = \binom{x-1}{r-1} p^r (1 - p)^{x-r} = \binom{x-1}{r-1} p^r q^{x-r}, \quad x = r, r+1, \dots.$$

We say that X has a **negative binomial distribution**.

Remark The reason for calling this distribution the negative binomial distribution is the following. Consider $h(w) = (1 - w)^{-r}$, the binomial $(1 - w)$ with the negative exponent $-r$. Using Maclaurin's series expansion (see Appendix A.4), we have

$$(1 - w)^{-r} = \sum_{k=0}^{\infty} \frac{h^{(k)}(0)}{k!} w^k = \sum_{k=0}^{\infty} \binom{r+k-1}{r-1} w^k, \quad -1 < w < 1.$$

If we let $x = k + r$ in the summation, then $k = x - r$ and

$$(1 - w)^{-r} = \sum_{x=r}^{\infty} \binom{r+x-r-1}{r-1} w^{x-r} = \sum_{x=r}^{\infty} \binom{x-1}{r-1} w^{x-r},$$

the summand of which is, except for the factor p^r, the negative binomial probability when $w = q$. In particular, we see that the sum of the probabilities for the negative binomial distribution is 1 because

$$\sum_{x=r}^{\infty} g(x) = \sum_{x=r}^{\infty} \binom{x-1}{r-1} p^r q^{x-r} = p^r (1 - q)^{-r} = 1. \qquad \blacksquare$$

If $r = 1$ in the negative binomial distribution, we say that X has a **geometric distribution** since the p.m.f. consists of terms of a geometric series, namely

$$g(x) = p(1 - p)^{x-1}, \qquad x = 1, 2, 3, \dots.$$

Recall that for a geometric series (see Appendix A.4 for a review), the sum is given by

$$\sum_{k=0}^{\infty} ar^k = \sum_{k=1}^{\infty} ar^{k-1} = \frac{a}{1 - r}$$

when $|r| < 1$. Thus, for the geometric distribution,

$$\sum_{x=1}^{\infty} g(x) = \sum_{x=1}^{\infty} (1 - p)^{x-1} p = \frac{p}{1 - (1 - p)} = 1,$$

so that $g(x)$ does satisfy the properties of a p.m.f.

From the sum of a geometric series we also note that, when k is an integer,

$$P(X > k) = \sum_{x=k+1}^{\infty} (1 - p)^{x-1} p = \frac{(1 - p)^k p}{1 - (1 - p)} = (1 - p)^k = q^k,$$

and thus the value of the distribution function at a positive integer k is

$$P(X \le k) = \sum_{x=1}^{k} (1 - p)^{x-1} p = 1 - P(X > k) = 1 - (1 - p)^k = 1 - q^k.$$

EXAMPLE 2.5-4 Some biology students were checking the eye color for a large number of fruit flies. For the individual fly, suppose that the probability of white eyes is 1/4 and the probability of red eyes is 3/4, and that we may treat these observations as having independent Bernoulli trials. The probability that at least four flies have to be checked for eye color to observe a white-eyed fly is given by

$$P(X \ge 4) = P(X > 3) = q^3 = \left(\frac{3}{4}\right)^3 = \frac{27}{64} = 0.4219.$$

The probability that at most four flies have to be checked for eye color to observe a white-eyed fly is given by

$$P(X \le 4) = 1 - q^4 = 1 - \left(\frac{3}{4}\right)^4 = \frac{175}{256} = 0.6836.$$

The probability that the first fly with white eyes is the fourth fly considered is

$$P(X = 4) = q^{4-1}p = \left(\frac{3}{4}\right)^3\left(\frac{1}{4}\right) = \frac{27}{256} = 0.1055.$$

It is also true that

$$P(X = 4) = P(X \le 4) - P(X \le 3)$$
$$= [1 - (3/4)^4] - [1 - (3/4)^3]$$
$$= \left(\frac{3}{4}\right)^3\left(\frac{1}{4}\right).$$

◀

We now show that the mean and the variance of a negative binomial random variable X are

$$\mu = E(X) = \frac{r}{p} \quad \text{and} \quad \sigma^2 = \frac{rq}{p^2} = \frac{r(1-p)}{p^2}.$$

In particular, if $r = 1$ so that X has a geometric distribution, then

$$\mu = \frac{1}{p} \quad \text{and} \quad \sigma^2 = \frac{q}{p^2} = \frac{1-p}{p^2}.$$

The mean $\mu = 1/p$ agrees with our intuition. Let's check: if $p = 1/6$, then we would expect, on the average, $1/(1/6) = 6$ trials before the first success.

To find these moments, we determine the m.g.f. of the negative binomial distribution. It is

$$M(t) = \sum_{x=r}^{\infty} e^{tx}\binom{x-1}{r-1}p^r(1-p)^{x-r}$$
$$= (pe^t)^r \sum_{x=r}^{\infty}\binom{x-1}{r-1}[(1-p)e^t]^{x-r}$$
$$= \frac{(pe^t)^r}{[1 - (1-p)e^t]^r}, \quad \text{where } (1-p)e^t < 1,$$

(or, equivalently, when $t < -\ln(1-p)$). Thus

$$M'(t) = (pe^t)^r(-r)[1 - (1-p)e^t]^{-r-1}[-(1-p)e^t] +$$
$$r(pe^t)^{r-1}(pe^t)[1 - (1-p)e^t]^{-r}$$
$$= r(pe^t)^r[1 - (1-p)e^t]^{-r-1}$$

and

$$M''(t) = r(pe^t)^r(-r-1)[1-(1-p)e^t]^{-r-2}[-(1-p)e^t] + r^2(pe^t)^{r-1}(pe^t)[1-(1-p)e^t]^{-r-1}.$$

Accordingly,

$$M'(0) = rp^r p^{-r-1} = rp^{-1}$$

and

$$M''(0) = r(r+1)p^r p^{-r-2}(1-p) + r^2 p^r p^{-r-1}$$
$$= rp^{-2}[(1-p)(r+1) + rp] = rp^{-2}(r+1-p).$$

Thus we have

$$\mu = \frac{r}{p} \quad \text{and} \quad \sigma^2 = \frac{r(r+1-p)}{p^2} - \frac{r^2}{p^2} = \frac{r(1-p)}{p^2}.$$

Even these calculations are a little messy, and a somewhat easier way is given in Exercise 2.5-20.

EXAMPLE 2.5-5 Suppose that during practice, a basketball player can make a free throw 80% of the time. Furthermore, assume that a sequence of free throw shooting can be thought of as independent Bernoulli trials. Let X equal the minimum number of free throws that this player must attempt to make a total of 10 shots. The p.m.f. of X is

$$g(x) = \binom{x-1}{10-1}(0.80)^{10}(0.20)^{x-10}, \qquad x = 10, 11, 12, \ldots.$$

The mean, variance, and standard deviation of X are

$$\mu = 10\left(\frac{1}{0.80}\right) = 12.5, \qquad \sigma^2 = \frac{10(0.20)}{0.80^2} = 3.125, \qquad \text{and} \qquad \sigma = 1.768.$$

And we have, for example,

$$P(X = 12) = g(12) = \binom{11}{9}(0.80)^{10}(0.20)^2 = 0.2362. \qquad \blacktriangleleft$$

EXAMPLE 2.5-6 To consider the effect of p and r on the negative binomial distribution, Figure 2.5-1 gives the probability histograms for four combinations of p and r. Note that since $r = 1$ in the first of these, it represents a geometric p.m.f. $\qquad \blacktriangleleft$

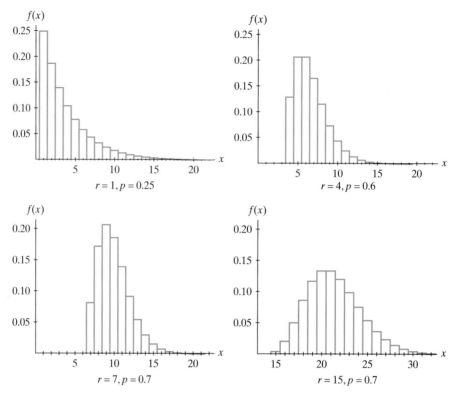

Figure 2.5-1: Negative binomial probability histograms

When the moment-generating function exists, derivatives of all orders exist at $t = 0$. Thus it is possible to represent $M(t)$ as a Maclaurin's series, namely,

$$M(t) = M(0) + M'(0)\left(\frac{t}{1!}\right) + M''(0)\left(\frac{t^2}{2!}\right) + M'''(0)\left(\frac{t^3}{3!}\right) + \cdots.$$

If the Maclaurin's series expansion of $M(t)$ exists and the moments are given, we can frequently sum the Maclaurin's series to obtain the closed form of $M(t)$. This is illustrated in the next example.

EXAMPLE 2.5-7 Let the moments of X be defined by

$$E(X^r) = 0.8, \qquad r = 1, 2, 3, \dots.$$

The moment-generating function of X is then

$$M(t) = M(0) + \sum_{r=1}^{\infty} 0.8\left(\frac{t^r}{r!}\right) = 1 + 0.8 \sum_{r=1}^{\infty} \frac{t^r}{r!}$$

$$= 0.2 + 0.8 \sum_{r=0}^{\infty} \frac{t^r}{r!} = 0.2e^{0t} + 0.8e^{1t}.$$

Thus

$$P(X = 0) = 0.2 \quad \text{and} \quad P(X = 1) = 0.8. \quad \blacktriangleleft$$

COMPUTATIONAL COMMENTS (CAS) We now illustrate how *Maple* can be used to find the moment-generating function for the negative binomial distribution. It is then used to find the mean and the variance.

```
> f := binomial(x - 1, r - 1)*p^r*(1 - p)^(x - r);
```

$$f := binomial(x - 1, r - 1) \, p^r \, (1 - p)^{(x-r)}$$

```
> mu := sum(x*f, x = r .. infinity);
```

$$\mu := \frac{r}{p}$$

```
> var := sum((x - mu)^2*f, x = r .. infinity);
```

$$var := -\frac{r(-1 + p)}{p^2}$$

```
> M := sum(exp(t*x)*f, x = r .. infinity);
```

$$M := \frac{e^{(t\,r)} \, p^r}{(1 - e^t + e^t \, p)^r}$$

```
> Mprime := diff(M, t):
> Mdoubleprime := diff(Mprime, t):
> mu := subs(t = 0, Mprime):
> mu := factor(%);
```

$$\mu := \frac{r}{p}$$

```
> var := subs(t = 0, Mdoubleprime) - mu^2:
> var := factor(%);
```

$$var := -\frac{r(-1 + p)}{p^2}$$

Note that although the result for the variance is correct, we would probably write it as

$$\sigma^2 = \frac{r(1 - p)}{p^2}. \quad \blacksquare$$

EXERCISES

2.5-1 Define the p.m.f. and give the values of μ, σ^2, and σ when the moment-generating function of X is defined by

    (a) $M(t) = 1/3 + (2/3)e^t$.

    (b) $M(t) = (0.25 + 0.75e^t)^{12}$.

2.5-2 (i) Give the name of the distribution of X (if it has a name); (ii) find the values of μ and σ^2; and (iii) calculate $P(1 \le X \le 2)$ when the moment-generating function of X is given by

    (a) $M(t) = (0.3 + 0.7e^t)^5$.

    (b) $M(t) = \dfrac{0.3e^t}{1 - 0.7e^t}$, $t < -\ln(0.7)$.

    (c) $M(t) = 0.45 + 0.55e^t$.

    (d) $M(t) = 0.3e^t + 0.4e^{2t} + 0.2e^{3t} + 0.1e^{4t}$.

    (e) $M(t) = (0.6e^t)^2(1 - 0.4e^t)^{-2}$, $t < -\ln(0.4)$.

    (f) $M(t) = \sum_{x=1}^{10} (0.1)e^{tx}$.

2.5-3 If the moment-generating function of X is

$$M(t) = \frac{2}{5} e^t + \frac{1}{5} e^{2t} + \frac{2}{5} e^{3t},$$

find the mean, variance, and p.m.f. of X.

2.5-4 Let X equal the number of people selected at random that you must ask in order to find someone with the same birthday as yours. Assuming each day of the year is equally likely (and ignoring February 29),

    (a) What is the p.m.f. of X?

    (b) Give the values of the mean, variance, and standard deviation of X.

    (c) Find $P(X > 400)$ and $P(X < 300)$.

2.5-5 For each question on a multiple-choice test, there are five possible answers of which exactly one is correct. If a student selects answers at random, give the probability that the first question answered correctly is question 4.

2.5-6 The probability that a machine produces a defective item is 0.01. Each item is checked as it is produced. Assume that these are independent trials and compute the probability that at least 100 items must be checked to find one that is defective.

2.5-7 Apples are packaged automatically in 3-pound bags. Suppose that 4% of the time the bag of apples weighs less than 3 pounds. If you select bags randomly and weigh them in order to find one underweight bag of apples, find the probability that the number of bags that must be selected is

    (a) At least 20.

    (b) At most 20.

    (c) Exactly 20.

2.5-8 Show that 63/512 is the probability that the fifth head is observed on the tenth independent flip of an unbiased coin.

2.5-9 A CD player has a magazine that holds six CDs. The machine is capable of randomly selecting a CD at random and then selecting a song randomly from that CD. Suppose that five CDs are albums by Paul McCartney and one is by Billy Joel. The player selects songs until a song by Billy Joel is played after which the machine is turned off. Give the probability that the machine is turned off after

    (a) The sixth song.

    (b) At least five songs have been played.

    (c) Suppose now that the songs continue to be played until the second song by Billy Joel has been played. Find the probability that at most four songs are played.

2.5-10 Suppose that a basketball player, different from the one in Example 2.5-5, can make a free throw 60% of the time. Let X equal the minimum number of free throws that this player must attempt to make a total of 10 shots.

  (a) Give the mean, variance, and standard deviation of X.
  (b) Find $P(X = 16)$.

2.5-11 Let X equal the number of flips of a fair coin that are required to observe the same face on consecutive flips.

  (a) Find the p.m.f. of X.
  (b) Give the values of the mean, variance, and standard deviation of X.
  (c) Find the values of **(i)** $P(X \leq 3)$, **(ii)** $P(X \geq 5)$, and **(iii)** $P(X = 3)$.

2.5-12 Let X have a geometric distribution. Show that

$$P(X > k + j \,|\, x > k) = P(X > j),$$

where k and j are nonnegative integers. NOTE: We sometimes say that in this situation there has been loss of memory.

2.5-13 Let X equal the number of flips of a fair coin that are required to observe heads-tails on consecutive flips.

  (a) Find the p.m.f. of X.
  (b) Determine the values of the mean, variance, and standard deviation of X.
  (c) Find the values of **(i)** $P(X \leq 3)$, **(ii)** $P(X \geq 5)$, and **(iii)** $P(X = 3)$.

2.5-14 Let X equal the number of flips of a fair coin that are required to observe heads on consecutive flips. Let f_n equal the nth Fibonacci number where $f_1 = 1$, $f_2 = 1$, and $f_n = f_{n-1} + f_{n-2}, n = 3, 4, 5, \ldots$.

  (a) Show that the p.m.f. of X is

$$f(x) = \frac{f_{x-1}}{2^x}, \qquad x = 2, 3, 4, \ldots.$$

  (b) Use the fact that

$$f_x = \frac{1}{\sqrt{5}} \left[\left(\frac{1 + \sqrt{5}}{2} \right)^x - \left(\frac{1 - \sqrt{5}}{2} \right)^x \right]$$

  to show that $\sum_{x=2}^{\infty} f(x) = 1$.
  (c) Show that $\mu = E(X) = 6$.
  (d) Show that $E[X(X - 1)] = 52$ so that the variance of X is $\sigma^2 = 22$ and the standard deviation is $\sigma = 4.690$.
  (e) Either simulate this experiment on the computer or physically toss a coin to support the theoretical characteristics of this distribution.

2.5-15 One of four different prizes was randomly put into each box of a cereal. If a family decided to buy this cereal until it obtained at least one of each of the four different prizes, what is the expected number of boxes of cereal that must be purchased?

2.5-16 A fair six-sided die is rolled until a 3 is observed.

  (a) What is the expected number of rolls needed?
  (b) A pair of fair six-sided dice is rolled. What is the expected number of rolls needed to observe at least one 3 on at least one of the two dice?
  (c) What is the smallest number of dice required so that the expected number of rolls needed to observe at least one 3 does not exceed 2?

2.5-17 From 1999-2002 Red Rose Tea randomly placed one of 10 English porcelain miniature animals in a l00-bag box of Red Rose Tea, selecting from 10 endangered North American animals.

  (a) On the average, how many boxes of tea must be purchased by a customer to obtain a complete collection consisting of 10 different animals?

(b) If the customer uses one tea bag per day, how long will it take, approximately, to obtain a complete collection?

(c) Beginning in 2002 new figurines are part of the Noah's Ark Animal Series which contains 15 pieces. Again assuming that these figurines are selected randomly, how many boxes of tea must be purchased on the average to obtain the complete collection?

2.5-18 If $E(X^r) = 5^r, r = 1, 2, 3, \ldots$, find $M(t)$, the moment-generating function of X, and the p.m.f. of X.

2.5-19 Let the moment-generating function $M(t)$ of X exist for $-h < t < h$. Consider the function $R(t) = \ln M(t)$. The first two derivatives of $R(t)$ are, respectively,

$$R'(t) = \frac{M'(t)}{M(t)} \quad \text{and} \quad R''(t) = \frac{M(t)M''(t) - [M'(t)]^2}{[M(t)]^2}.$$

Setting $t = 0$, show that

(a) $\mu = R'(0)$.

(b) $\sigma^2 = R''(0)$.

2.5-20 Use the result of Exercise 2.5-19 to find the mean and variance of the

(a) Bernoulli distribution.

(b) Binomial distribution.

(c) Geometric distribution.

(d) Negative binomial distribution.

2.5-21 Given a random permutation of the integers in the set $\{1, 2, 3, 4, 5\}$, let X equal the number of integers that are in their natural position. The moment-generating function of X is

$$M(t) = \frac{44}{120} + \frac{45}{120}e^t + \frac{20}{120}e^{2t} + \frac{10}{120}e^{3t} + \frac{1}{120}e^{5t}.$$

(a) Find the mean and variance of X.

(b) Find the probability that at least one integer is in its natural position.

(c) Draw a graph of the probability histogram of the p.m.f. of X.

2.5-22 The probability that a company's work force has no accidents in a given month is 0.7. The numbers of accidents from month to month are independent. What is the probability that the third month in a year is the first month at least one accident occurs?

2.5-23 Suppose an airport metal detector catches a person with a metal 99% of the time. That is, it misses detecting a person with metal 1% of the time. Assuming independence of people carrying metal, what is the probability that the first person missed (not detected) is among the first 50 persons?

2.5-24 An excellent free throw shooter attempts several free throws until she misses.

(a) If $p = 0.9$ is her probability of making a free throw, what is the probability of having the first miss after 12 attempts?

(b) If she continues shooting until she misses three, what is the probability that the third miss occurs on the 30th attempt?

2.6 **THE POISSON DISTRIBUTION**

Some experiments result in counting the number of times particular events occur in given times or on given physical objects. For example, we could count the number of phone calls arriving at a switchboard between 9 and 10 A.M., the number of flaws in 100 feet of wire, the number of customers that arrive at a ticket window between 12 noon and 2 P.M., or the number of defects in a 100-foot roll of aluminum screen that is 2 feet wide. Each count can be looked upon as a random variable associated with an approximate Poisson process provided the conditions in Definition 2.6-1 are satisfied.

Definition 2.6-1

Let the number of changes that occur in a given continuous interval be counted. We have an *approximate Poisson process* with parameter $\lambda > 0$ if the following are satisfied:

(a) The numbers of changes occurring in non-overlapping intervals are independent.
(b) The probability of exactly one change in a sufficiently short interval of length h is approximately λh.
(c) The probability of two or more changes in a sufficiently short interval is essentially zero.

Remark In this definition, we have modified the usual requirements of a Poisson process by using the words *approximately* and *essentially* in (b) and (c). We do this to avoid some advanced mathematics. Thus we refer to this as the *approximate* Poisson process. ■

Suppose that an experiment satisfies the three points of an approximate Poisson process. Let X denote the number of changes in an interval of "length 1" (where "length 1" represents one unit of the quantity under consideration). We would like to find an approximation for $P(X = x)$, where x is a nonnegative integer. To achieve this, we partition the unit interval into n subintervals of equal length $1/n$. If n is sufficiently large (i.e., much larger than x), we shall approximate the probability that x changes occur in this unit interval by finding the probability that one change occurs in each of exactly x of these n subintervals. The probability of one change occurring in any one subinterval of length $1/n$ is approximately $\lambda(1/n)$ by condition (b). The probability of two or more changes in any one subinterval is essentially zero by condition (c). So for each subinterval, exactly one change occurs with a probability of approximately $\lambda(1/n)$. Consider the occurrence or nonoccurrence of a change in each subinterval as a Bernoulli trial. By condition (a) we have a sequence of n Bernoulli trials with probability p approximately equal to $\lambda(1/n)$. Thus an approximation for $P(X = x)$ is given by the binomial probability

$$\frac{n!}{x!\,(n-x)!} \left(\frac{\lambda}{n}\right)^x \left(1 - \frac{\lambda}{n}\right)^{n-x}.$$

If n increases without bound, we have that

$$\lim_{n\to\infty} \frac{n!}{x!\,(n-x)!} \left(\frac{\lambda}{n}\right)^x \left(1-\frac{\lambda}{n}\right)^{n-x}$$

$$= \lim_{n\to\infty} \frac{n(n-1)\cdots(n-x+1)}{n^x}\frac{\lambda^x}{x!}\left(1-\frac{\lambda}{n}\right)^n \left(1-\frac{\lambda}{n}\right)^{-x}.$$

Now, for fixed x, we have (see Appendix A.3)

$$\lim_{n\to\infty} \frac{n(n-1)\cdots(n-x+1)}{n^x} = \lim_{n\to\infty}\left[(1)\left(1-\frac{1}{n}\right)\cdots\left(1-\frac{x-1}{n}\right)\right] = 1,$$

$$\lim_{n\to\infty}\left(1-\frac{\lambda}{n}\right)^n = e^{-\lambda},$$

$$\lim_{n\to\infty}\left(1-\frac{\lambda}{n}\right)^{-x} = 1.$$

Thus

$$\lim_{n\to\infty} \frac{n!}{x!\,(n-x)!}\left(\frac{\lambda}{n}\right)^x\left(1-\frac{\lambda}{n}\right)^{n-x} = \frac{\lambda^x e^{-\lambda}}{x!} = P(X=x).$$

The distribution of probability associated with this process has a special name. We say that the random variable X has a **Poisson distribution** if its p.m.f. is of the form

$$f(x) = \frac{\lambda^x e^{-\lambda}}{x!}, \qquad x = 0, 1, 2, \ldots,$$

where $\lambda > 0$.

It is easy to see that $f(x)$ enjoys the properties of a p.m.f. because clearly $f(x) \geq 0$ and, from the Maclaurin's series expansion of e^{λ} (see Appendix A.4), we have

$$\sum_{x=0}^{\infty} \frac{\lambda^x e^{-\lambda}}{x!} = e^{-\lambda}\sum_{x=0}^{\infty}\frac{\lambda^x}{x!} = e^{-\lambda}e^{\lambda} = 1.$$

To discover the exact role of the parameter $\lambda > 0$, let us find some of the characteristics of the Poisson distribution.

The moment-generating function of X is

$$M(t) = E(e^{tX}) = \sum_{x=0}^{\infty} e^{tx}\frac{\lambda^x e^{-\lambda}}{x!} = e^{-\lambda}\sum_{x=0}^{\infty}\frac{(\lambda e^t)^x}{x!}.$$

From the series representation of the exponential function, we have that

$$M(t) = e^{-\lambda} e^{\lambda e^t} = e^{\lambda(e^t - 1)}$$

for all real values of t. Now the first and second derivatives are

$$M'(t) = \lambda e^t e^{\lambda(e^t - 1)},$$

$$M''(t) = (\lambda e^t)^2 e^{\lambda(e^t - 1)} + \lambda e^t e^{\lambda(e^t - 1)}.$$

The values of the mean and variance of X are

$$\mu = M'(0) = \lambda,$$

$$\sigma^2 = M''(0) - [M'(0)]^2 = (\lambda^2 + \lambda) - \lambda^2 = \lambda.$$

That is, for the Poisson distribution, $\mu = \sigma^2 = \lambda$.

Remark   It is also possible to find the mean and the variance for the Poisson distribution directly without using the moment-generating function. The mean for the Poisson distribution is given by

$$E(X) = \sum_{x=0}^{\infty} x \frac{\lambda^x e^{-\lambda}}{x!} = e^{-\lambda} \sum_{x=1}^{\infty} \frac{\lambda^x}{(x-1)!}$$

because $(0) f(0) = 0$ and $x/x! = 1/(x-1)!$ when $x > 0$. If we let $k = x - 1$, then

$$E(X) = e^{-\lambda} \sum_{k=0}^{\infty} \frac{\lambda^{k+1}}{k!} = \lambda e^{-\lambda} \sum_{k=0}^{\infty} \frac{\lambda^k}{k!} = \lambda e^{-\lambda} e^{\lambda} = \lambda.$$

To find the variance, we first determine the second factorial moment, namely, $E[X(X-1)]$. We have

$$E[X(X-1)] = \sum_{x=0}^{\infty} x(x-1) \frac{\lambda^x e^{-\lambda}}{x!} = e^{-\lambda} \sum_{x=2}^{\infty} \frac{\lambda^x}{(x-2)!}$$

because $(0)(0-1) f(0) = 0$, $(1)(1-1) f(1) = 0$, and $x(x-1)/x! = 1/(x-2)!$ when $x > 1$ If we let $k = x - 2$, then

$$E[X(X-1)] = e^{-\lambda} \sum_{k=0}^{\infty} \frac{\lambda^{k+2}}{k!} = \lambda^2 e^{-\lambda} \sum_{k=0}^{\infty} \frac{\lambda^k}{k!} = \lambda^2 e^{-\lambda} e^{\lambda} = \lambda^2.$$

Thus

$$\text{Var}(X) = E(X^2) - [E(X)]^2 = E[X(X-1)] + E(X) - [E(X)]^2$$

$$= \lambda^2 + \lambda - \lambda^2 = \lambda.$$

We again see that, for the Poisson distribution, $\mu = \sigma^2 = \lambda$.   ■

Table III in the Appendix gives values of the distribution function of a Poisson random variable for selected values of λ. This table is illustrated in the next example.

EXAMPLE 2.6-1 Let X have a Poisson distribution with a mean of $\lambda = 5$. Then using Table III in the Appendix,

$$P(X \leq 6) = \sum_{x=0}^{6} \frac{5^x e^{-5}}{x!} = 0.762,$$

$$P(X > 5) = 1 - P(X \leq 5) = 1 - 0.616 = 0.384,$$

and

$$P(X = 6) = P(X \leq 6) - P(X \leq 5) = 0.762 - 0.616 = 0.146. \quad \blacktriangleleft$$

EXAMPLE 2.6-2 To see the effect of λ on the p.m.f. $f(x)$ of X, Figure 2.6-1 shows the probability histograms of $f(x)$ for four different values of λ. $\quad \blacktriangleleft$

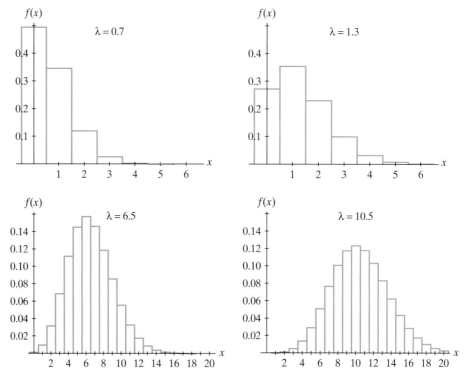

Figure 2.6-1: Poisson probability histograms

EXAMPLE 2.6-3   Let X denote the number of alpha particles emitted by barium-133 in 1/10 of a second and counted by a Geiger counter. One hundred observations of X produced the data in Table 2.6-1. For these data, $\bar{x} = 559/100 = 5.59$ and $s^2 = 3,619/99 - (559)^2/9900 = 4.992$. Thus we see that the sample mean and sample variance are fairly close to each other. In Figure 2.6-2 the probability histogram for a Poisson distribution with $\lambda = 5.59$ is drawn. The relative frequency histogram of the data is shaded. Accordingly, it looks as if the Poisson distribution provides a fairly reasonable probability model for this random variable X. Later in the text, formal tests of the goodness of fit of various models are given. ◄

Table 2.6-1: Barium 133			
Outcome (x)	**Frequency (f)**	**fx**	**fx²**
1	1	1	1
2	4	8	16
3	13	39	117
4	19	76	304
5	16	80	400
6	15	90	540
7	9	63	441
8	12	96	768
9	7	63	567
10	2	20	200
11	1	11	121
12	1	12	144
	100	559	3619

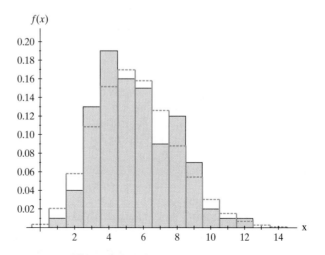

Figure 2.6-2: Barium histograms

If events in a Poisson process occur at a mean rate of λ per unit, then the expected number of occurrences in an interval of length t is λt. For example, if phone calls arrive at a switchboard following a Poisson process at a mean rate of three per minute, then the expected number of phone calls in a 5-minute period is $(3)(5) = 15$. Or if calls arrive at a mean rate of 22 in a 5-minute period, the expected number of calls per minute is $\lambda = 22(1/5) = 4.4$. Moreover, the number of occurrences, say X, in the interval of length t has the Poisson p.m.f.

$$f(x) = \frac{(\lambda t)^x e^{-\lambda t}}{x!}, \qquad x = 0, 1, 2, \dots.$$

This follows by treating the interval of length t as if it were the "unit interval" with mean λt instead of λ.

EXAMPLE 2.6-4   A disk is often used when backing up a computer. However, in the past a less reliable backup system that was used was a computer tape and flaws occurred on these tapes. In a particular situation, flaws (bad records) on a used computer tape occurred on the average of one flaw per 1200 feet. If one assumes a Poisson distribution, what is the distribution of X, the number of flaws in a 4800-foot roll? The expected number of flaws in $4800 = 4(1200)$ feet is 4; that is, $E(X) = 4$. Thus the p.m.f. of X is

$$f(x) = \frac{4^x e^{-4}}{x!}, \qquad x = 0, 1, 2, \dots,$$

and, in particular,

$$P(X = 0) = \frac{4^0 e^{-4}}{0!} = e^{-4} = 0.018,$$

$$P(X \le 4) = 0.629,$$

by Table III in the Appendix.   ◄

EXAMPLE 2.6-5   Telephone calls enter a college switchboard on the average of two every 3 minutes. If one assumes an approximate Poisson process, what is the probability of five or more calls arriving in a 9-minute period? Let X denote the number of calls in a 9-minute period. We see that $E(X) = 6$; that is, on the average, six calls will arrive during a 9-minute period. Thus,

$$P(X \ge 5) = 1 - P(X \le 4) = 1 - \sum_{x=0}^{4} \frac{6^x e^{-6}}{x!}$$

$$= 1 - 0.285 = 0.715,$$

using Table III in the Appendix.   ◄

Not only is the Poisson distribution important in its own right, but it can also be used to approximate probabilities for a binomial distribution. If X has a Poisson distribution with parameter λ, we saw that with n large,

$$P(X = x) \approx \binom{n}{x}\left(\frac{\lambda}{n}\right)^x\left(1 - \frac{\lambda}{n}\right)^{n-x},$$

where $p = \lambda/n$ so that $\lambda = np$ in the above binomial probability. That is, if X has the binomial distribution $b(n, p)$ with large n and small p, then

$$\frac{(np)^x e^{-np}}{x!} \approx \binom{n}{x}p^x(1 - p)^{n-x}.$$

This approximation is reasonably good if n is large. But since λ was a fixed constant in that earlier argument, p should be small since $np = \lambda$. In particular, the approximation is quite accurate if $n \geq 20$ and $p \leq 0.05$ or if $n \geq 100$ and $p \leq 0.10$, but it is not bad in other situations violating these bounds somewhat, like $n = 50$ and $p = 0.12$.

EXAMPLE 2.6-6   A manufacturer of Christmas tree light bulbs knows that 2% of its bulbs are defective. Assuming independence, we have a binomial distribution with parameters $p = 0.02$ and $n = 100$. To approximate the probability that a box of 100 of these bulbs contains at most three defective bulbs we use the Poisson distribution with $\lambda = 100(0.02) = 2$, which gives

$$\sum_{x=0}^{3} \frac{2^x e^{-2}}{x!} = 0.857$$

from Table III in the Appendix. Using the binomial distribution, we obtain, after some tedious calculations,

$$\sum_{x=0}^{3} \binom{100}{x}(0.02)^x(0.98)^{100-x} = 0.859.$$

Hence, in this case, the Poisson approximation is extremely close to the true value, but much easier to find.   ◄

Remark   With the availability of better and faster computing power, it is often very easy to find binomial probabilities. So do not use the Poisson approximation if you are able to find the probability exactly.   ∎

EXAMPLE 2.6-7   In Figure 2.6-3 Poisson probability histograms have been superimposed on shaded binomial probability histograms so that we can see whether or not these are close to each other. If the distribution of X is $b(n, p)$, the approximating

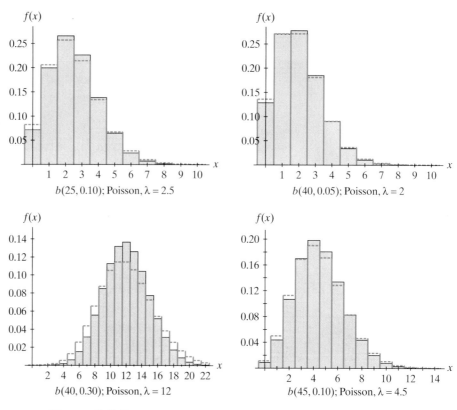

Figure 2.6-3: Binomial (shaded) and Poisson probability histograms

Poisson distribution has a mean of $\lambda = np$. Note that the approximation is not good when p is large (e.g., $p = 0.30$).   ◀

EXAMPLE 2.6-8   A lot of 1000 parts is shipped to a company. A sampling plan dictates that $n = 100$ parts are to be taken at random and without replacement and the lot accepted if no more than two of these 100 parts are defective. Here $Ac = 2$ is usually called the **acceptance number**. The operating characteristic curve

$$OC(p) = P(X \leq 2),$$

where p is the fraction defective in the lot, is really the sum of the three hypergeometric probabilities

$$P(X = x) = \frac{\binom{N_1}{x}\binom{1000 - N_1}{100 - x}}{\binom{1000}{100}}, \qquad x = 0, 1, 2,$$

where $N_1 = (1000)(p)$. However, we have seen that the hypergeometric distribution can be approximated by the binomial distribution, which in turn can be approximated by the Poisson distribution when n is large and p is small. This is exactly our situation since $n = 100$ and since we are interested in values of p in the range 0.00 to 0.10. Thus

$$OC(p) = P(X \le 2) \approx \sum_{x=0}^{2} \frac{(100p)^x e^{-100p}}{x!}.$$

Using Table III in the Appendix, we find that some approximate values of the operating characteristic curve are $OC(0.01) = 0.920$, $OC(0.02) = 0.677$, $OC(0.03) = 0.423$, $OC(0.05) = 0.125$, and $OC(0.10) = 0.003$ and it is plotted in Figure 2.6-4. If this is not a satisfactory operating characteristic curve for the type of parts involved, the sample size n and the acceptance number Ac must be changed. ◀

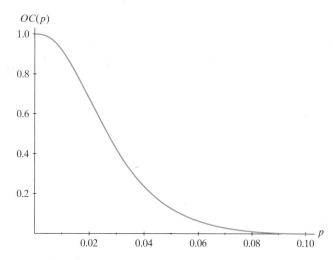

Figure 2.6-4: Operating characteristic curve

EXERCISES

2.6-1 Let X have a Poisson distribution with a mean of 4. Find
    (a) $P(2 \le X \le 5)$.
    (b) $P(X \ge 3)$.
    (c) $P(X \le 3)$.

2.6-2 Let X have a Poisson distribution with a variance of 3. Find $P(X = 2)$.

2.6-3 Customers arrive at a travel agency at a mean rate of 11 per hour. Assuming that the number of arrivals per hour has a Poisson distribution, give the probability that more than 10 customers arrive in a given hour.

2.6-4 If X has a Poisson distribution so that $3P(X = 1) = P(X = 2)$, find $P(X = 4)$.

2.6-5 Flaws in a certain type of drapery material appear on the average of one in 150 square feet. If we assume the Poisson distribution, find the probability of at most one flaw in 225 square feet.

2.6-6 A certain type of aluminum screen that is 2 feet wide has on the average one flaw in a 100-foot roll. Find the probability that a 50-foot roll has no flaws.

2.6-7 With probability 0.001, a prize of $499 is won in the Michigan Daily Lottery when a $1 straight bet is placed. Let Y equal the number of $499 prizes won by a gambler after placing n straight bets. Note that Y is $b(n, 0.001)$. After placing $n = 2000$ $1 bets, a gambler is behind if $\{Y \leq 4\}$. Use the Poisson distribution to approximate $P(Y \leq 4)$ when $n = 2000$.

2.6-8 Suppose that the probability of suffering a side effect from a certain flu vaccine is 0.005. If 1000 persons are inoculated, find approximately the probability that

(a) At most 1 person suffers.
(b) 4, 5, or 6 persons suffer.

2.6-9 A roll of a biased die results in a two only 1/10 of the time. Let X denote the number of twos in 100 rolls of this die. Approximate.

(a) $P(3 \leq X \leq 7)$.
(b) $P(X \geq 5)$.

2.6-10 The mean of a Poisson random variable X is $\mu = 9$. Compute

$$P(\mu - 2\sigma < X < \mu + 2\sigma).$$

2.6-11 A Geiger counter was set up in the physics laboratory to record the number of alpha particle emissions of carbon-14 in 0.5 second. The following are 50 observations.

4	6	6	12	11	11	10	5	10	7
9	6	9	11	9	6	4	9	11	8
10	11	7	4	5	8	6	5	8	7
6	3	4	6	4	12	7	14	5	9
7	10	9	6	6	12	10	7	12	13

(a) Calculate the sample mean, $\bar{x}$, and sample variance, s^2, for these data. Are these two measurements close to each other?
(b) On the same graph depict the probability histogram for a Poisson distribution with $\lambda = 8$ (we use 8 only because it is the closest value of λ to $\bar{x}$ in Table III in the Appendix), with the relative frequency histogram of the data superimposed.

2.6-12 For determining the half-lives of radioactive isotopes, it is important to know what the background radiation is in a given detector over a period of time. Data taken in a γ-ray detection experiment over 300 one-second intervals yielded the following data:

0	2	4	6	6	1	7	4	6	1	1	2	3	6	4	2	7	4	4	2
2	5	4	4	4	1	2	4	3	2	2	5	0	3	1	1	0	0	5	2
7	1	3	3	3	2	3	1	4	1	3	5	3	5	1	3	3	0	3	2
6	1	1	4	6	3	6	4	4	2	2	4	3	3	6	1	6	2	5	0
6	3	4	3	1	1	4	6	1	5	1	1	4	1	4	1	1	1	3	3
4	3	3	2	5	2	1	3	5	3	2	7	0	4	2	3	3	5	6	1
4	2	6	4	2	0	4	4	7	3	5	2	2	3	1	3	1	3	6	5
4	8	2	2	4	2	2	1	4	7	5	2	1	1	4	1	4	3	6	2
1	1	2	2	2	2	3	5	4	3	2	2	3	3	2	4	4	3	2	2
3	6	1	1	3	3	2	1	4	5	5	1	2	3	3	1	3	7	2	5
4	2	0	6	2	3	2	3	0	4	4	5	2	5	3	0	4	6	2	2
2	2	2	5	2	2	3	4	2	3	7	1	1	7	1	3	6	0	5	3
0	0	3	3	0	2	4	3	1	2	3	3	3	4	3	2	2	7	5	3
5	1	1	2	2	6	1	3	1	4	4	2	3	4	5	1	3	4	3	1
0	3	7	4	0	5	2	5	4	4	2	2	3	2	4	6	5	5	3	4

Do these look like observations of a Poisson random variable with mean $\lambda = 3$? To help answer this question, do the following.

(a) Find the frequencies of $0, 1, 2, \ldots, 8$.

(b) Calculate the sample mean and sample variance. Are they approximately equal to each other?

(c) Construct a probability histogram with $\lambda = 3$ and a relative frequency histogram on the same graph.

(d) What is your conclusion? (Later in this book we consider tests of hypotheses which help give a statistical answer.)

2.6-13 Let X equal the number of telephone calls per hour that are received by 911 between midnight and noon and reported in the *Holland Sentinel*. On October 29 and October 30, the following numbers of calls were reported:

$$0 \quad 1 \quad 1 \quad 1 \quad 0 \quad 1 \quad 2 \quad 1 \quad 4 \quad 1 \quad 2 \quad 3$$
$$0 \quad 3 \quad 0 \quad 1 \quad 0 \quad 1 \quad 1 \quad 2 \quad 3 \quad 0 \quad 2 \quad 2$$

(a) Calculate the sample mean and sample variance for these data. Are they approximately equal to each other?

(b) Assume that $\lambda = 1.3$ so that Table III can be used. Draw a probability histogram for the Poisson distribution and a relative frequency histogram of the data on the same graph.

(c) Does it look like the Poisson distribution with $\lambda = 1.3$ could be a reasonable probability model based on these limited data?

2.6-14 Let X equal the number of green peanut m&m's in packages of size 22. Forty-five observations of X yielded the following frequencies for the possible outcomes of X:

Outcome (x):	0	1	2	3	4	5	6	7	8	9
Frequency:	0	2	4	5	7	9	8	5	3	2

(a) Use these data to construct a relative frequency histogram and superimpose over it a Poisson probability histogram with a mean of $\lambda = 5$.

(b) Do these data appear to be observations of a Poisson random variable?

2.6-15 Let X equal the number of chocolate chips in a chocolate-chip cookie. Sixty-two observations of X yielded the following frequencies for the possible outcomes of X:

Outcome (x):	0	1	2	3	4	5	6	7	8	9	10
Frequency:	0	0	2	8	7	13	13	10	4	4	1

(a) Use these data to graph the relative frequency histogram and the Poisson probability histogram with $\lambda = 5.6$ on the same figure.

(b) Do these data seem to be observations of a Poisson random variable with a mean of $\lambda = 5.6$?

2.6-16 While testing a used computer tape for bad records, a computer operator counted the number of flaws per 100 feet of tape. Let X equal this random variable. Given that 40 observations of X yielded 5 zeros, 7 ones, 12 twos, 9 threes, 5 fours, 1 five, and 1 six,

(a) Construct a relative frequency histogram and superimpose a Poisson p.m.f. with $\lambda = 2.2$.

(b) Does the Poisson distribution seem to be a possible probability model for this experiment?

2.6-17 A student trainer worked in the training room in the physical education building. The number of students coming in to be taped or for treatment of other injuries was recorded each half-hour, yielding the following data:

$$3 \quad 3 \quad 2 \quad 0 \quad 4 \quad 5 \quad 6 \quad 4 \quad 4 \quad 3 \quad 2 \quad 1 \quad 2 \quad 3 \quad 0$$
$$5 \quad 5 \quad 3 \quad 2 \quad 3 \quad 5 \quad 4 \quad 1 \quad 2 \quad 0 \quad 3 \quad 2 \quad 4 \quad 2 \quad 6$$

   (a) Find the sample mean, $\bar{x}$.
   (b) Find the sample variance, s^2.
   (c) Find the sample standard deviation, s.
   (d) Draw a relative frequency histogram of these data with an appropriate Poisson probability histogram superimposed.

2.6-18 A lot of 10,000 items is accepted if a sample of $n = 400$ has no more than $Ac = 3$ defective items. Determine the approximate values of $OC(0.002)$, $OC(0.004)$, $OC(0.006)$, $OC(0.01)$, and $OC(0.02)$ and plot the operating characteristic curve.

2.6-19 Consider a lot of 5000 items. Design a sampling plan (i.e., find n and Ac) so that $OC(0.02) = 0.95$ and $OC(0.08) = 0.10$ approximately.
   HINT: Start with $n = 50$ and $Ac = 2$ and modify as necessary, noting that increasing n creates a steeper OC curve.

2.6-20 A baseball team loses $10,000 for each consecutive day it rains. Say X, the number of consecutive days it rains at the beginning of the season, has a Poisson distribution with mean 0.2. What is the expected loss before the opening game?

2.6-21 An airline always overbooks, if possible. A particular plane has 95 seats on a flight in which a ticket sells for $300. The airline sells 100 such tickets for this flight.

   (a) If the probability of an individual not showing is 0.05, assuming independence, what is the probability that the airline can accommodate all the passengers who do show up?
   (b) If the airline must return the $300 price plus a penalty of $400 to each passenger that cannot get on the flight, what is the expected penalty that the airline will pay?

2.6-22 A quality control expert suggests the use of an acceptance sampling plan on large incoming lots of products. The plan is this: Observe n of these products at random and if k or less are defective, the lot is accepted. Otherwise it is rejected. If $p = 0.01$ is the fraction defective in the lot, he wants to accept the lot with probability of at least 0.98. If, however, $p = 0.05$, he wants to accept the lot with a small probability less that 0.07. Find n and k to achieve these probabilities.

2.6-23 A store selling newspapers only orders $n = 4$ of a certain newspaper because the manager does not get many calls for that publication. If the number of requests per day follows a Poisson distribution with mean three,

   (a) What is the expected value of the number sold?
   (b) How many should the manager order so that the chance of running out is less that 0.05?

2.6-24 In the decade of the 1980s, the number X of cases of diphtheria reported each year in the United States followed a Poisson distribution with mean 2.5. What is the probability that

   (a) No cases were reported in a given year?
   (b) Exactly three were reported?
   (c) At least three were reported?
   (d) At most three were reported?

CHAPTER

3

CONTINUOUS DISTRIBUTIONS

3.1 CONTINUOUS-TYPE DATA

In Chapter 2 we considered probability distributions of random variables whose space S contains a countable number of outcomes: either a finite number of outcomes or outcomes that can be put into a one-to-one correspondence with the positive integers. Such a random variable is said to be of the **discrete type** and its distribution of probabilities is of the discrete type.

Many experiments or observations of random phenomena do not have integers as outcomes but instead are measurements selected from an interval of numbers. For example, you could find the length of time that it takes when waiting in line to buy frozen yogurt. Or, the weight of a "one-pound" package of hot dogs could be any number between 0.94 and 1.25 pounds. The weight of a miniature Baby Ruth candy bar could be any number between 20 and 27 grams. Even though such times and weights could be selected from an interval of values, times and weights are generally rounded off so that the data often look like discrete data. If conceptually the measurements could come from an interval of possible outcomes, we call them data from a continuous-type population or, more simply, **continuous-type data**.

Given a set of continuous-type data, we shall group the data into classes and then construct a histogram of the grouped data. This will help us better visualize the data. The following guidelines and terminology will be used to group continuous-type data into classes of equal length. These guidelines can also be used for sets of discrete data that have a large range.

1. Determine the largest (maximum) and smallest (minimum) observations. The **range** is the difference, $R = \text{maximum} - \text{minimum}$.
2. Generally select from $k = 5$ to $k = 20$ classes which are usually nonoverlapping intervals of equal length, so that these classes cover the interval from the minimum to the maximum.
3. Each interval begins and ends halfway between two possible values of the measurements which have been rounded off to a given number of decimal places.
4. The first interval should begin about as much below the smallest value as the last interval ends above the largest.
5. The intervals are called **class intervals** and the boundaries are called **class boundaries** or **cutpoints**. We shall denote these k class intervals by

$$(c_0, c_1), (c_1, c_2), \ldots, (c_{k-1}, c_k).$$

6. The **class limits** are the smallest and the largest possible observed (recorded) values in a class.
7. The **class mark** is the midpoint of a class.

A frequency table is constructed that lists the class intervals, the class limits, a tabulation of the measurements in the various classes, the frequency f_i of each class, the class marks, and (optionally) a column that is used to construct a relative frequency (density) histogram. With class intervals of equal length, a frequency histogram is constructed by drawing a rectangle for each class having as its base the class interval and height equal to the frequency of the class. For the relative frequency histogram, each rectangle has an **area** equal to the relative frequency f_i/n of the observations for the class. That is, the function defined by

$$h(x) = \frac{f_i}{(n)(c_i - c_{i-1})}, \qquad \text{for } c_{i-1} < x \le c_i, \quad i = 1, 2, \ldots, k,$$

is called a **relative frequency histogram** or **density histogram**, where f_i is the frequency of the ith class and n is the total number of observations. Clearly, if the class intervals are of equal length, the relative frequency histogram, $h(x)$, is proportional to the **frequency histogram** f_i, for $c_{i-1} < x \le c_i, i = 1, 2, \ldots, k$. The frequency histogram should be used only in those situations in which the class intervals are of equal length.

EXAMPLE 3.1-1   The weights in grams of 40 miniature Baby Ruth candy bars, with the weights ordered, are given in Table 3.1-1.
    We shall group these data and then construct a histogram to visualize the distribution of weights.
    The range of the data is

$$R = 26.7 - 20.5 = 6.2.$$

Table 3.1-1: Candy Bar Weights									
20.5	20.7	20.8	21.0	21.0	21.4	21.5	22.0	22.1	22.5
22.6	22.6	22.7	22.7	22.9	22.9	23.1	23.3	23.4	23.5
23.6	23.6	23.6	23.9	24.1	24.3	24.5	24.5	24.8	24.8
24.9	24.9	25.1	25.1	25.2	25.6	25.8	25.9	26.1	26.7

The interval $(20.5, 26.7)$ could be covered with $k = 8$ classes of width 0.8, or with $k = 9$ classes of width 0.7, (there are other possibilities). We shall use $k = 7$ classes of width 0.9. The first class interval will be $(20.45, 21.35)$ and the last class interval will be $(25.85, 26.75)$. The data are grouped in Table 3.1-2.

Table 3.1-2: Frequency Table of Candy Bar Weights					
Class Interval	Class Limits	Tabulation	Frequency (f_i)	$h(x)$	Class Marks
$(20.45, 21.35)$	20.5-21.3	卌	5	5/36	20.9
$(21.35, 22.25)$	21.4-22.2	||||	4	4/36	21.8
$(22.25, 23.15)$	22.3-23.1	卌 |||	8	8/36	22.7
$(23.15, 24.05)$	23.2-24.0	卌 ||	7	7/36	23.6
$(24.05, 24.95)$	24.1-24.9	卌 |||	8	8/36	24.5
$(24.95, 25.85)$	25.0-25.8	卌	5	5/36	25.4
$(25.85, 26.75)$	25.9-26.7	|||	3	3/36	26.3

A relative frequency histogram of these data is given in Figure 3.1-1. Note that the total area of this histogram is equal to one. We could also construct a frequency histogram in which the heights of the rectangles would be equal

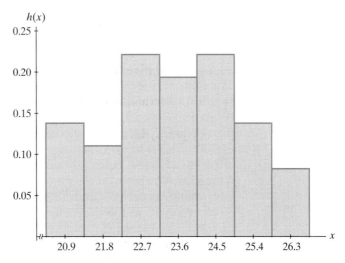

Figure 3.1-1: Relative frequency histogram of weights of candy bars

to the frequencies of the classes. The shape of the two histograms is the same. Later we will see the reason for preferring the relative frequency histogram. In particular, we will be superimposing on the relative frequency histogram the graph of a continuous function, called a probability density function, that corresponds to a probability model. ◄

There is a refinement of the relative frequency histogram that can be made when the class intervals are of equal length. The **relative frequency polygon** smooths out the corresponding histogram somewhat. To form such a polygon mark the midpoints at the top of each "bar" of the histogram. Connect adjacent midpoints with straight line segments. On each of the two end bars draw a line segment from the top middle mark through the middle point of the outer vertical line of the bar. Of course, if the area underneath the tops of the relative frequency histogram is equal to one, which it should be, the area underneath the relative frequency polygon is equal to one because the areas lost and gained cancel out by consideration of congruent triangles. This is clear in the next example taken from the "great home run race of 1998."

EXAMPLE 3.1-2   Mark McGwire and Sammy Sosa battled through the 1998 season, particularly the last two months, for the home run championship and a chance to beat Roger Maris's total of 61 home runs for the season. Both actually beat that record with 70 and 66, respectively. In this example, we want to consider the distances of these home runs, and we concentrate on McGwire's distances. Sosa's distances are given in Exercise 3.1-11. There probably is some measurement errors in these distances, but at least these are approximately correct. In McGwire's case, $n = 70$, and the measurements in feet are the following:

364	368	364	419	424	347	462	419	437	419
371	362	358	527	381	545	478	440	471	451
425	366	477	397	433	388	423	409	356	409
425	366	477	397	433	388	423	409	356	409
438	437	449	433	461	431	472	485	405	415
511	425	458	452	408	374	464	398	409	369
385	477	393	509	501	450	472	497	458	381
430	341	385	417	423	375	403	435	377	370

The minimum value equals 341 (which was his record-breaking 62nd home run down the left field line) and the maximum value equals 545. The range is $545 - 341 = 204$. With only $n = 70$ values, we do not want to take too many classes, and we note that $k = 7$ classes of length 30 covers the data; that is, $(7)(30) = 210 > 204$. We decided to start at 337.5; thus we end at $337.5 + 210 = 547.5$. These boundaries are about the same distance below the minimum and above the maximum, and by starting at 337.5 there is no question into what class each measurement goes. See Table 3.1-3.

The plots of the relative frequency histogram and polygon are given in Figure 3.1-2. We must note that it does not make a great deal of difference in these graphs if we select different numbers and lengths of class intervals and the start of the first one. Later in this section we use a slightly different approach and will see that to be true. ◄

		Table 3.1-3: Frequency Table of Home Run Distances			
Class Interval	Class Mark (u_i)	Tabulation	Frequency (f_i)	Relative Frequency	$h(x) = f_i/2100$
(337.5, 367.5)	352.5	⸽⸽⸽⸽⸽ ⸽⸽⸽	8	8/70	0.0038
(367.5, 397.5)	382.5	⸽⸽⸽⸽⸽ ⸽⸽⸽⸽⸽ ⸽⸽⸽⸽	14	14/70	0.0067
(397.5, 427.5)	412.5	⸽⸽⸽⸽⸽ ⸽⸽⸽⸽⸽ ⸽⸽	17	17/70	0.0081
(427.5, 457.5)	442.5	⸽⸽⸽⸽⸽ ⸽⸽⸽⸽⸽ ⸽⸽⸽	13	13/70	0.0062
(457.5, 487.5)	472.5	⸽⸽⸽⸽⸽ ⸽⸽⸽⸽⸽ ⸽⸽	12	12/70	0.0057
(487.5, 517.5)	502.5	⸽⸽⸽⸽	4	4/70	0.0019
(517.5, 547.5)	532.5	⸽⸽	2	2/70	0.0010

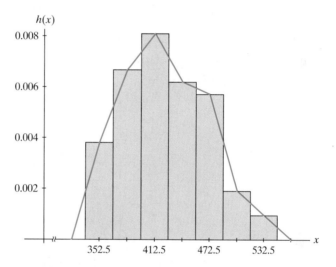

Figure 3.1-2: Distances of McGwire's home runs

In some situations it is not necessarily desirable to use class intervals of equal widths in the construction of the frequency distribution and histogram. This is particularly true if the data are skewed with a very long tail. We now present such an illustration in which it seems desirable to use class intervals of unequal lengths, and thus we cannot use the relative frequency polygon.

EXAMPLE 3.1-3   The following 40 losses, due to wind-related catastrophes, were recorded to the nearest one million dollars. These data include only those losses of $2 million or more; and, for convenience, they have been ordered and recorded in millions.

$$
\begin{array}{cccccccccc}
2 & 2 & 2 & 2 & 2 & 2 & 2 & 2 & 2 & 2 \\
2 & 2 & 3 & 3 & 3 & 3 & 4 & 4 & 4 & 5 \\
5 & 5 & 5 & 6 & 6 & 6 & 6 & 8 & 8 & 9 \\
15 & 17 & 22 & 23 & 24 & 24 & 25 & 27 & 32 & 43
\end{array}
$$

The selection of class boundaries is more subjective in this case. It makes sense to let $c_0 = 1.5$ and $c_1 = 2.5$ because only values of \$2 million or more are recorded and there are 12 observations equal to 2. We could let $c_2 = 6.5$, $c_3 = 29.5$, and $c_4 = 49.5$. This would yield the following relative frequency histogram:

$$h(x) = \begin{cases} \dfrac{12}{40}, & 1.5 < x \leq 2.5, \\[2mm] \dfrac{15}{(40)(4)}, & 2.5 < x \leq 6.5, \\[2mm] \dfrac{11}{(40)(23)}, & 6.5 < x \leq 29.5, \\[2mm] \dfrac{2}{(40)(20)}, & 29.5 < x \leq 49.5. \end{cases}$$

It is displayed in Figure 3.1-3. It takes some experience before a person can display a relative frequency histogram that is most meaningful.

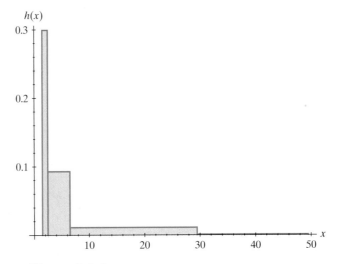

Figure 3.1-3: Relative frequency histogram of losses

The areas of the four rectangles, $0.300, 0.375, 0.275, 0.050$, are the respective relative frequencies. It is important to note in the case of unequal lengths among class intervals that the areas, not the heights, of the rectangles are proportional to the frequencies. In particular, the first and second classes have frequencies $f_1 = 12$ and $f_2 = 15$, and yet the height of the first is greater than the height of the second while here $f_1 < f_2$. If we have equal lengths among the class intervals, then the heights are proportional to the frequencies. ◄

For continuous-type data, the interval with the largest class height is called the **modal class** and the respective class mark is called the **mode**. Hence in the last example, $x = 2$ is the mode and $(1.5, 2.5)$ the modal class.

To explore the characteristics of an unknown distribution, we need to take a sample of n observations, $x_1, x_2, \ldots, x_n$, from that distribution and often need to order them from the smallest to the largest. One convenient way of doing this is using a stem-and-leaf display, a method that was started by John W. Tukey. For more details, see the books by Tukey (1977) and Velleman and Hoaglin (1981).

Possibly the easiest way to begin is with an example to which all of us can relate. Say we have the following 50 test scores on a statistics examination:

93	77	67	72	52	83	66	84	59	63
75	97	84	73	81	42	61	51	91	87
34	54	71	47	79	70	65	57	90	83
58	69	82	76	71	60	38	81	74	69
68	76	85	58	45	73	75	42	93	65

We can do much the same thing as a frequency table and histogram, but keep the original values, through a **stem-and-leaf display**. For this particular data set, we could use the following procedure. The first number in the set, 93, is recorded as follows: The 9 (in the "tens" place) is treated as the stem, and the 3 (in the "units" place) is the corresponding leaf. Note that this leaf of 3 is the first digit after the stem of 9 in Table 3.1-4. The second number, 77, is given by the leaf of 7 after the stem of 7; the third number, 67, by the leaf of 7 after the stem of 6; the fourth number, 72, as the leaf of 2 after the stem of 7 (note that this is the second leaf on the 7 stem); and so on. Table 3.1-4 is called a stem-and-leaf display. If the leaves are carefully aligned vertically, this table has the same effect as a histogram, but the original numbers are not lost.

Table 3.1-4: Stem-and-leaf Display of Scores from 50 Statistics Examinations			
Stems	**Leaves**	**Frequency**	**Depths**
3	4 8	2	2
4	2 7 5 2	4	6
5	2 9 1 4 7 8 8	7	13
6	7 6 3 1 5 9 0 9 8 5	10	23
7	7 2 5 3 1 9 0 6 1 4 6 3 5	13	(13)
8	3 4 4 1 7 3 2 1 5	9	14
9	3 7 1 0 3	5	5

An additional new term in these stem-and-leaf displays is **depths**. In the "depths" column the frequencies are added together (accumulated) from the low end and the high end until the row is reached that contains the middle value in the ordered display. Later we call that middle value the median of the sample, and the frequency of that row is simply placed in parentheses. That is, the frequency of the row containing this median is not included in either the sum from the low end or the high end.

It is useful to modify the stem-and-leaf display by ordering the leaves in each row from smallest to largest. The resulting stem-and-leaf diagram is called an **ordered stem-and-leaf display**. Table 3.1-5 gives an ordered stem-and-leaf using the data in Table 3.1-4.

Table 3.1-5: Ordered Stem-and-leaf Display of Statistics Examinations			
Stems	**Leaves**	**Frequency**	**Depths**
3	4 8	2	2
4	2 2 5 7	4	6
5	1 2 4 7 8 8 9	7	13
6	0 1 3 5 5 6 7 8 9 9	10	23
7	0 1 1 2 3 3 4 5 5 6 6 7 9	13	(13)
8	1 1 2 3 3 4 4 5 7	9	14
9	0 1 3 3 7	5	5

There is another modification that can also be helpful. Suppose that we want two rows of leaves with each original stem. We can do this by recording leaves 0, 1, 2, 3, and 4 with a stem adjoined with an asterisk ($*$) and leaves 5, 6, 7, 8, and 9 with a stem adjoined with a dot ($\bullet$). Of course, in our example, by going from seven original classes to 14 classes, we lose a certain amount of smoothness with this particular data set, as illustrated in Table 3.1-6, which is also ordered.

Table 3.1-6: Ordered Stem-and-leaf Display of Statistics Examinations			
Stems	**Leaves**	**Frequency**	**Depths**
3$*$	4	1	1
3$\bullet$	8	1	2
4$*$	2 2	2	4
4$\bullet$	5 7	2	6
5$*$	1 2 4	3	9
5$\bullet$	7 8 8 9	4	13
6$*$	0 1 3	3	16
6$\bullet$	5 5 6 7 8 9 9	7	23
7$*$	0 1 1 2 3 3 4	7	(7)
7$\bullet$	5 5 6 6 7 9	6	20
8$*$	1 1 2 3 3 4 4	7	14
8$\bullet$	5 7	2	7
9$*$	0 1 3 3	4	5
9$\bullet$	7	1	1

Tukey suggested another modification used in the next example.

EXAMPLE 3.1-4   The following numbers represent ACT composite scores for 60 entering freshmen at a certain college:

26	19	22	28	31	29	25	23	20	33	23	26
30	27	26	29	20	23	18	24	29	27	32	24
25	26	22	29	21	24	20	28	23	26	30	19
27	21	32	28	29	23	25	21	28	22	25	24
19	24	35	26	25	20	31	27	23	26	30	29

Table 3.1-7: Ordered Stem-and-leaf Display of 60 ACT Scores

Stems	Leaves	Frequency	Depths
1•	8 9 9 9	4	4
2*	0 0 0 0 1 1 1	7	11
2t	2 2 2 3 3 3 3 3 3	9	20
2f	4 4 4 4 4 5 5 5 5 5	10	30
2s	6 6 6 6 6 6 7 7 7 7	11	30
2•	8 8 8 8 9 9 9 9 9 9	10	19
3*	0 0 0 1 1	5	9
3t	2 2 3	3	4
3f	5	1	1

An ordered stem-and-leaf display is given in Table 3.1-7. Here leaves are recorded as zeros and ones with a stem adjoined with an asterisk (*), twos and threes with a stem adjoined with t, fours and fives with a stem adjoined with f, sixes and sevens with a stem adjoined with s, and eights and nines with a stem adjoined with a dot (•).   ◄

There is a reason for constructing ordered stem-and-leaf diagrams. For a sample of n observations, $x_1, x_2, \ldots, x_n$, when the observations are ordered from small to large, the resulting ordered data are called the **order statistics** of the sample. For years statisticians have found that order statistics and certain functions of the order statistics are extremely valuable. It is very easy to determine the values of the sample in order from an ordered stem-and-leaf display. For illustration, consider the values in Table 3.1-5 or Table 3.1-6. The order statistics of the 50 test scores are given in Table 3.1-8.

Table 3.1-8: Order Statistics of 50 Exam Scores

34	38	42	42	45	47	51	52	54	57
58	58	59	60	61	63	65	65	66	67
68	69	69	70	71	71	72	73	73	74
75	75	76	76	77	79	81	81	82	83
83	84	84	85	87	90	91	93	93	97

Sometimes we give ranks to these order statistics and use the rank as the subscript on y. The first order statistic $y_1 = 34$ has rank 1; the second order statistic $y_2 = 38$ has rank 2; the third order statistic $y_3 = 42$ has rank 3; the fourth order statistic $y_4 = 42$ has rank 4, ... ; and the 50th order statistic $y_{50} = 97$ has rank 50. It is also about as easy to determine these values from the ordered stem-and-leaf display. We see that $y_1 \le y_2 \le \cdots \le y_{50}$.

From either these order statistics or the corresponding ordered stem-and-leaf display, it is rather easy to find the **sample percentiles**. If $0 < p < 1$, the $(100p)$th sample percentile has *approximately* np sample observations less than it and also $n(1 - p)$ sample observations greater than it. One way of achieving this is to take the $(100p)$th sample percentile as the $(n + 1)p$th order statistic, provided that $(n + 1)p$ is an integer. If $(n + 1)p$ is not an integer but is equal to r plus some proper fraction, say a/b, use a weighted average of the rth and the $(r + 1)$st order statistics. That is, define the $(100p)$th sample percentile as

$$\tilde{\pi}_p = y_r + (a/b)(y_{r+1} - y_r) = (1 - a/b)y_r + (a/b)y_{r+1}.$$

This is simply a linear interpolation between y_r and y_{r+1}. (If $p < 1/(n + 1)$ or $p > n/(n + 1)$, that sample percentile is not defined.)

For illustration, consider the 50 ordered test scores. With $p = 1/2$, we find the 50th percentile by averaging the 25th and 26th order statistics, since $(n + 1)p = (51)(1/2) = 25.5$. Thus the 50th percentile is

$$\tilde{\pi}_{0.50} = (1/2)y_{25} + (1/2)y_{26} = (71 + 71)/2 = 71.$$

With $p = 1/4$, we have $(n + 1)p = (51)(1/4) = 12.75$; and thus the 25th sample percentile is

$$\tilde{\pi}_{0.25} = (1 - 0.75)y_{12} + (0.75)y_{13} = (0.25)(58) + (0.75)(59) = 58.75.$$

With $p = 3/4$, so that $(n+1)p = (51)(3/4) = 38.25$, the 75th sample percentile is

$$\tilde{\pi}_{0.75} = (1 - 0.25)y_{38} + (0.25)y_{39} = (0.75)(81) + (0.25)(82) = 81.25.$$

Note that *approximately* 50%, 25%, and 75% of the sample observations are less than 71, 58.75, and 81.25, respectively.

Special names are given to certain percentiles. The 50th percentile is the **median** of the sample. The 25th, 50th, and 75th percentiles are the **first, second,** and **third quartiles** of the sample. For notation we let $\tilde{q}_1 = \tilde{\pi}_{0.25}, \tilde{q}_2 = \tilde{m} = \tilde{\pi}_{0.50}$, and $\tilde{q}_3 = \tilde{\pi}_{0.75}$. The 10th, 20th, ..., and 90th percentiles are the **deciles** of the sample. So note that the 50th percentile is also the median, the second quartile, and the fifth decile. By using the set of 50 test scores, since $(51)(2/10) = 10.2$ and $(51)(9/10) = 45.9$, the second and ninth deciles are

$$\tilde{\pi}_{0.20} = (0.8)y_{10} + (0.2)y_{11} = (0.8)(57) + (0.2)(58) = 57.2$$

and

$$\tilde{\pi}_{0.90} = (0.1)y_{45} + (0.9)y_{46} = (0.1)(87) + (0.9)(90) = 89.7,$$

commonly called the 20th and 90th percentiles, respectively.

EXAMPLE **3.1-5**   Let us return to the data given on the distances of Mark McGwire's $n = 70$ home runs in Example 3.1-2. For convenience, we use 20 as the length of a class interval using Tukey's suggestion and leaves with two integers. That is, the first home run of length of 371 follows the stem $3s$ with leaf 71. The ordered stem-and-leaf display is given in Table 3.1-9. The corresponding relative frequency histogram and polygon are in Figure 3.1-4.   ◄

Stems	Leaves	Frequency	Depths
$3f$	41 47 56 58	4	4
$3s$	62 64 64 66 68 69 70 71 74 75 77	11	15
$3\bullet$	81 81 85 85 88 93 97 98	8	23
$4*$	03 05 08 09 09 09 15 17 19 19 19	11	34
$4t$	23 23 24 25 25 30 31 33 33 35 37 37 38	13	(13)
$4f$	40 49 50 51 52 58 58	7	23
$4s$	61 62 64 71 72 72 77 77 78	9	16
$4\bullet$	85 97	2	7
$5*$	01 09 11	3	5
$5t$	27	1	2
$5f$	45	1	1

Table 3.1-9: Ordered Stem-and-leaf of Distances of McGwire's Home Runs

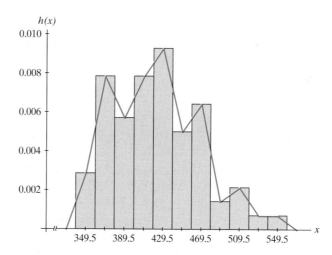

Figure 3.1-4: Histogram and polygon of distances of McGwire's home runs

Due to the fact that we have more and shorter class intervals, the relative frequency histogram and polygon are not as smooth as the previous ones. Also it picks up the fact that there might be two modes. The first, having the smaller frequency, is around 370 feet while the second is about 430. Those homers of distances around 370 or less have been hit to left or right field because they would not be home runs if hit toward center field. This accounts for the first mode. Clearly hits of 430 feet will be homers in almost all ball parks no matter what direction they go. However, if we were taking data on all long batted balls (some of which would be outs), these would probably show a unimodal pattern. You should compare McGwire's results with those of Sosa's in Exercise 3.1-19.

EXERCISES

3.1-1 One characteristic of a car's storage console that is checked by the manufacturer is the time in seconds that it takes for the lower storage compartment door to open completely. A random sample of size $n = 5$ yielded the following times:

$$1.1 \quad 0.9 \quad 1.4 \quad 1.1 \quad 1.0$$

(a) Find the sample mean, $\bar{x}$.
(b) Find the sample variance, s^2.
(c) Find the sample standard deviation, s.

3.1-2 A leakage test was conducted to determine the effectiveness of a seal designed to keep the inside of a plug air tight. An air needle was inserted into the plug and this was placed underwater. The pressure was then increased until leakage was observed. The magnitude of this pressure was recorded for 10 trials in psi:

$$3.1 \quad 3.5 \quad 3.3 \quad 3.7 \quad 4.5 \quad 4.2 \quad 2.8 \quad 3.9 \quad 3.5 \quad 3.3$$

Find the sample mean and sample standard deviation for these 10 measurements.

3.1-3 The costs of 28 "300 level" textbooks (rounded to the nearest dollar) for spring 2004, selected from all major disciplines, were as follows:

128	23	108	104	44	77	25
112	45	65	116	53	19	84
20	115	93	55	80	73	108
98	18	58	75	45	19	25

(a) Group these data into six classes using as class intervals (9.50, 29.50), (29.50, 49.50), and so on.
(b) Construct a histogram and interpret the histogram.
(c) Find the mean, $\bar{x}$ and locate it on your histogram.
(d) Find the standard deviation, s, and locate the points $\bar{x} \pm s$ and $\bar{x} \pm 2s$ on your histogram.

3.1-4 Decks have become very popular items for American homeowners. Lumber yards sell many styles of decks during the summer months, but they all use the same type of nail. The question that always arises is how many nails to buy. If 800 nails are needed, how many grams of nails should be purchased? The weights (in grams) of the following 50 nails should help to answer the question.

9.42	8.69	8.93	8.27	8.82	8.66	8.90	8.31	9.15	9.63
9.41	8.56	8.82	8.58	8.43	8.05	8.56	8.55	8.88	8.73
8.29	8.79	8.51	8.85	9.34	9.21	8.38	8.51	8.41	8.98
8.58	9.21	8.27	8.76	9.26	8.59	8.36	8.71	8.51	8.88
9.20	8.24	8.57	8.85	8.69	8.85	9.08	9.40	9.25	8.79

(a) Group these data into seven classes using as class intervals (7.995, 8.245), (8.245, 8.495), and so on.
(b) Construct a histogram.
(c) Calculate the sample mean and sample standard deviation.
(d) How many grams of nails would you recommend?

3.1-5 Before Halloween two bags of miniature Clark bars were purchased and each bar was weighed (in grams), 19 in the first bag and 23 in the second bag, yielding the following weights.

19.8	20.3	20.2	19.6	20.0	19.4	15.4	19.9	20.5	20.5	19.8
15.5	21.8	20.0	15.4	20.6	15.7	20.4	21.3	16.7	15.6	14.2
15.9	16.8	15.2	15.0	16.0	16.4	15.4	16.0	16.1	17.0	16.0
15.1	14.1	15.8	15.7	16.6	15.2	15.5	14.9	15.3		

(a) Group these weights using as class boundaries 13.95–14.95, 14.95–15.95, and so on and construct a histogram.

(b) Calculate the values of the sample mean and sample standard deviation.

(c) Locate on your histogram $\bar{x}, \bar{x} \pm s_x, \bar{x} \pm 2s_x$.

(d) Give an interpretation.

3.1-6 In the Old Kent River Bank Run (a 25K race) the times are listed separately for males and females and also by age groupings. A 53-year-old male runner was interested in the distribution of race times for males, 50–54 years old. There were 125 runners in this category and their race times are listed in order in minutes and seconds.

96:21	98:24	100:26	101:23	101:29	101:38	103:11	103:58	104:28
105:02	105:14	106:00	106:14	108:49	109:28	109:32	109:42	110:49
110:54	111:48	111:54	112:17	112:33	112:48	112:54	112:55	113:09
113:10	113:15	113:43	114:30	114:36	115:28	115:39	115:44	116:11
116:34	116:56	117:45	118:10	118:11	118:12	118:28	118:30	118:44
119:10	119:27	119:53	120:17	120:19	121:12	122:15	122:17	122:46
122:57	122:59	123:46	124:04	124:06	124:49	124:55	125:14	125:15
125:16	126:09	126:12	126:29	126:43	127:21	127:25	127:57	128:02
128:23	129:13	129:21	129:39	130:01	130:07	130:29	130:34	131:07
131:22	131:58	132:27	132:51	132:58	133:51	134:04	134:40	134:53
135:05	135:41	136:27	136:51	136:52	136:53	137:09	137:13	137:31
137:48	138:01	138:13	138:44	139:34	139:51	140:03	140:05	141:20
142:28	143:08	143:46	144:10	144:35	145:40	147:05	148:12	151:18
151:39	152:00	156:13	160:43	161:12	162:29	166:42	169:54	

(a) Group these data into 10 classes with a class width of eight minutes.

(b) Draw a histogram of the grouped data.

(c) Describe the shape of the histogram. Is it what you would have expected? (Similar data can be found following most road races.)

3.1-7 In Example 1.1-7 the sample space for underwater weights was given. These weights for 82 male students were

3.7	3.6	4.0	4.3	3.8	3.4	4.1	4.0	3.7	3.4	3.5	3.8	3.7	4.9
3.5	3.8	3.3	4.8	3.4	4.6	3.5	5.3	4.4	4.2	2.5	3.1	5.2	3.8
3.3	3.4	4.1	4.6	4.0	1.4	4.3	3.8	4.7	4.4	5.0	3.2	3.1	4.2
4.9	4.5	3.8	4.2	2.7	3.8	3.8	2.0	3.4	4.9	3.3	4.3	5.6	3.2
4.7	4.5	5.2	5.0	5.0	4.0	3.8	5.3	4.5	3.8	3.8	3.4	3.6	3.3
4.2	5.1	4.0	4.7	6.5	4.4	3.6	4.7	4.5	2.3	4.0	3.7		

(a) Group these data into 11 classes with a class width of 0.5 kilograms, in which the class marks are 1.5, 2.0, 2.5, . . . , 6.5, respectively.

(b) Draw a histogram of the grouped data.

(c) Describe the shape of the histogram. Is it what you would have expected?

(d) Find the sample mean, $\bar{x}$, and locate it on your histogram.

(e) Find the sample standard deviation, s_x, and locate the points one standard deviation above and below the mean on your histogram.

3.1-8 A small part for an automobile rear view mirror was produced on two different punch presses. In order to describe the distribution of the weights of those parts, a random sample was selected, and each piece was weighed in grams, resulting in the following data set.

3.968	3.534	4.032	3.912	3.572	4.014	3.682	3.608
3.669	3.705	4.023	3.588	3.945	3.871	3.744	3.711
3.645	3.977	3.888	3.948	3.551	3.796	3.657	3.667
3.799	4.010	3.704	3.642	3.681	3.554	4.025	4.079
3.621	3.575	3.714	4.017	4.082	3.660	3.692	3.905
3.977	3.961	3.948	3.994	3.958	3.860	3.965	3.592
3.681	3.861	3.662	3.995	4.010	3.999	3.993	4.004
3.700	4.008	3.627	3.970	3.647	3.847	3.628	3.646
3.674	3.601	4.029	3.603	3.619	4.009	4.015	3.615
3.672	3.898	3.959	3.607	3.707	3.978	3.656	4.027
3.645	3.643	3.898	3.635	3.865	3.631	3.929	3.635
3.511	3.539	3.830	3.925	3.971	3.646	3.669	3.931
4.028	3.665	3.681	3.984	3.664	3.893	3.606	3.699
3.997	3.936	3.976	3.627	3.536	3.695	3.981	3.587
3.680	3.888	3.921	3.953	3.847	3.645	4.042	3.692
3.910	3.672	3.957	3.961	3.950	3.904	3.928	3.984
3.721	3.927	3.621	4.038	4.047	3.627	3.774	3.983
3.658	4.034	3.778					

(a) Construct a frequency distribution using about 10 (say, 8 to 12) classes.
(b) Draw a histogram of the data.
(c) Describe the shape of this distribution.

3.1-9 In the casino game roulette, if a player bets $1 on red, the probability of winning $1 is 18/38 and the probability of losing $1 is 20/38. Let X equal the number of successive $1 bets that a player makes before losing $5. One hundred observations of X were simulated on a computer, yielding the following data:

23	127	877	65	101	45	61	95	21	43
53	49	89	9	75	93	71	39	25	91
15	131	63	63	41	7	37	13	19	413
65	43	35	23	135	703	83	7	17	65
49	177	61	21	9	27	507	7	5	87
13	213	85	83	75	95	247	1815	7	13
71	67	19	615	11	15	7	131	47	25
25	5	471	11	5	13	75	19	307	33
57	65	9	57	35	19	9	33	11	51
27	9	19	63	109	515	443	11	63	9

(a) Find the sample mean and sample standard deviation of these data.
(b) Construct a relative frequency histogram using about 10 classes. These do not need to be of the same length.
(c) Locate on your histogram $\bar{x}, \bar{x} \pm s, \bar{x} \pm 2s, \bar{x} \pm 3s$.
(d) If a friend asked you how many bets could typically be placed before losing $5, would $\bar{x}$ be a good answer? Why?

3.1-10 An insurance company experienced the following mobile home losses in 10,000's of dollars for 50 catastrophic events:

1	2	2	3	3	4	4	5	5	5
5	6	7	7	9	9	9	10	11	12
22	24	28	29	31	33	36	38	38	38
39	41	48	49	53	55	74	82	117	134
192	207	224	225	236	280	301	308	351	527

(a) Group these data into four to eight classes. As in Example 3.1-3, the classes do not have the same width.
(b) Construct a relative frequency histogram.
(c) Describe the distribution of losses.

3.1-11 The distances in feet of Sammy Sosa's $n = 66$ home runs in 1998 are the following.

371	350	430	420	430	434	370	420	440	410
420	460	400	430	410	370	370	410	380	340
410	420	410	415	430	380	380	366	500	380
390	400	364	432	428	440	365	420	347	438
390	375	374	400	361	480	360	368	430	440
380	438	414	482	364	363	374	417	464	430
480	480	434	344	410	462				

The minimum equals 340 and the maximum is 500; thus the range is $500 - 340 = 160$. There are a number of possibilities for k, but so each member of the class gets the same result, take $k = 7$ class intervals each with length 23. That is, $(7)(23) = 161 > 160$. We start with 339.5 and accordingly end at 500.5.

(a) Construct a frequency table.
(b) Plot the relative frequency histogram and polygon.
(c) Compute $\bar{x}$ and s_x^2 using the original data.
(d) Compare McGwire's and Sosa's results.
(e) Write and answer some questions about the combined sets of distances for the 136 home runs.

3.1-12 Mark McGwire and Sammy Sosa continued to hit home runs in 1999, although not quite as many as in 1998. The following data give not only the distances for the home runs but they are also coded for the direction as follows: 1–left field, 2–left center, 3–center, 4–right center, and 5–right field.

Mark McGwire's Distances

400, 4	390, 1	410, 2	430, 3	420, 3	420, 4	398, 2	422, 2	417, 2
373, 1	420, 3	500, 3	412, 1	427, 1	368, 2	400, 2	435, 2	340, 1
434, 1	410, 3	421, 5	473, 1	440, 2	392, 1	432, 1	424, 3	431, 1
360, 1	425, 3	390, 2	400, 3	390, 4	365, 5	423, 3	425, 3	400, 1
420, 1	395, 5	372, 1	347, 5	368, 1	420, 2	451, 3	479, 2	389, 5
419, 2	393, 2	410, 2	502, 3	401, 2	440, 3	450, 2	410, 1	420, 2
419, 1	422, 2	400, 4	423, 2	410, 3	390, 1	450, 1	462, 3	467, 3
424, 3	461, 3							

Sammy Sosa's Distances

440, 3	370, 1	450, 2	390, 4	430, 1	380, 2	450, 1	410, 4	380, 4
450, 3	460, 3	460, 2	345, 5	420, 2	390, 2	430, 2	370, 2	405, 2
400, 2	400, 1	440, 2	410, 4	370, 5	450, 2	460, 4	420, 2	465, 3
416, 1	429, 2	420, 1	440, 3	350, 1	430, 4	430, 4	385, 4	410, 4
410, 3	465, 3	423, 5	380, 2	355, 1	380, 2	380, 2	401, 2	390, 2
386, 5	440, 1	430, 2	380, 2	373, 2	389, 4	480, 1	390, 1	410, 5
372, 1	431, 3	400, 3	395, 1	430, 3	403, 3	500, 1	420, 2	389, 1

(a) Compare the sample means and sample standard deviations of these distances with those for 1998.
(b) Using histograms, compare the distributions of distances for 1998 and 1999.

3.1-13 For determining half-lives of radioactive isotopes, it is important to know what the background radiation is in a given detector over a period of time. Data were collected in a γ-ray detection experiment over 98 ten-second intervals (observations of a random variable X), yielding the following data:

58	50	57	58	64	63	54	64	59	41	43	56	60	50
46	59	54	60	59	60	67	52	65	63	55	61	68	58
63	36	42	54	58	54	40	60	64	56	61	51	48	50
60	42	62	67	58	49	66	58	57	59	52	54	53	53
57	43	73	65	45	43	57	55	73	62	68	55	51	55
53	68	58	53	51	73	44	50	53	62	58	47	63	59
59	56	60	59	50	52	62	51	66	51	56	53	59	57

(a) Find the sample mean and the sample variance.

(b) Construct a histogram by grouping these data into 10 classes using as "cut points" 34.5, 38.5, 42.5, . . . , 74.5.

(c) Locate on your histogram the points $\bar{x}$, $\bar{x} \pm s$, and $\bar{x} \pm 2s$.

3.1-14 When researching ground water it is often important to know the characteristics of the soil at a certain site. Many of these characteristics, such as porosity, are at least partially dependent upon the grain size. The diameter of individual grains of soil can be measured. Here are the diameters (in mm) of 30 grains.

1.24	1.36	1.28	1.31	1.35	1.20	1.39	1.35	1.41	1.31
1.28	1.26	1.37	1.49	1.32	1.40	1.33	1.28	1.25	1.39
1.38	1.34	1.40	1.27	1.33	1.36	1.43	1.33	1.29	1.34

(a) Calculate the sample mean and the sample variance for these data.

(b) Construct a histogram of these data.

3.1-15 Old Faithful Geyser is a geyser in Yellowstone National Park. Tourists always want to know when the next eruption will occur, so data have been collected to help make those predictions. In the following data set, observations were made on several consecutive days and the data recorded give the starting time of the eruption (STE), the duration of the eruption in seconds (DIS), the predicted time until the next eruption in minutes (PTM), the actual time until the next eruption in minutes (ATM), and the duration of the eruption in minutes (DIM).

STE	DIS	PTM	ATM	DIM	STE	DIS	PTM	ATM	DIM
706	150	65	72	2.500	1411	110	55	65	1.833
818	268	89	88	4.467	616	289	89	97	4.817
946	140	65	62	2.333	753	114	58	52	1.900
1048	300	95	87	5.000	845	271	89	94	4.517
1215	101	55	57	1.683	1019	120	58	60	2.000
1312	270	89	94	4.500	1119	279	89	84	4.650
651	270	89	91	4.500	1253	109	55	63	1.817
822	125	59	51	2.083	1356	295	95	91	4.917
913	262	89	98	4.367	608	240	85	83	4.000
1051	95	55	59	1.583	731	259	86	84	4.317
1150	270	89	93	4.500	855	128	60	71	2.133
637	273	89	86	4.550	1006	287	92	83	4.783
803	104	55	70	1.733	1129	253	65	70	4.217
913	129	62	63	2.150	1239	284	89	81	4.733
1016	264	89	91	4.400	608	120	58	60	2.000
1147	239	82	82	3.983	708	283	92	91	4.717
1309	106	55	58	1.767	839	115	58	51	1.917
716	259	85	97	4.317	930	254	85	85	4.233
853	115	55	59	1.917	1055	94	55	55	1.567
952	275	89	90	4.583	1150	274	89	98	4.567
1122	110	55	58	1.833	1328	128	64	49	2.133
1220	286	92	98	4.767	557	270	93	85	4.500
735	115	55	55	1.917	722	103	58	65	1.717
830	266	89	107	4.433	827	287	89	102	4.783
1017	105	55	61	1.750	1009	111	55	56	1.850
1118	275	89	82	4.583	1105	275	89	86	4.583
1240	226	79	91	3.767	1231	104	55	62	1.733

(a) Construct a histogram of the lengths of the eruptions given in seconds. Use 10 to 12 classes.

(b) Calculate the sample mean and locate it on your histogram. Does it give a good measure of the average length of an eruption? Why?

(c) Construct a histogram of the lengths of the times between eruptions. Use 10 to 12 classes.

(d) Calculate the sample mean and locate it on your histogram. Does it give a good measure of the average length of the times between eruptions?

3.1-16 A company manufactures mints that have a label weight of 20.4 grams. It regularly samples from the production line and weighs the selected mints. During two mornings of production it sampled 81 mints with the following weights:

21.8	21.7	21.7	21.6	21.3	21.6	21.5	21.3	21.2
21.0	21.6	21.6	21.6	21.5	21.4	21.8	21.7	21.6
21.6	21.3	21.9	21.9	21.6	21.0	20.7	21.8	21.7
21.7	21.4	20.9	22.0	21.3	21.2	21.0	21.0	21.9
21.7	21.5	21.5	21.1	21.3	21.3	21.2	21.0	20.8
21.6	21.6	21.5	21.5	21.2	21.5	21.4	21.4	21.3
21.2	21.8	21.7	21.7	21.6	20.5	21.8	21.7	21.5
21.4	21.4	21.9	21.8	21.7	21.4	21.3	20.9	21.9
20.7	21.1	20.8	20.6	20.6	22.0	22.0	21.7	21.6

(a) Construct an ordered stem-and-leaf display using stems of $20f$, $20s$, $20\bullet$, $21*, \ldots, 22*$.

(b) Find (i) the three quartiles, (ii) the 60th percentile, and (iii) the 15th percentile.

3.1-17 Let X denote the concentration of $CaCO_3$ in milligrams per liter. Twenty observations of X are

130.8	129.9	131.5	131.2	129.5
132.7	131.5	127.8	133.7	132.2
134.8	131.7	133.9	129.8	131.4
128.8	132.7	132.8	131.4	131.3

(a) Construct an ordered stem-and-leaf display, using stems of 127, 128, ... , 134.

(b) Find the midrange, range, interquartile range, median, sample mean, and sample variance.

3.1-18 *PC Magazine* in February 1994 includes "Going Mainstream" by John R. Quain, which reports tests of 26 CD-ROM drives. One of the tests was a random read of a 4K (4,000 byte) block of data. The read rates, recorded in kilobytes (K) per second, were

29.21	14.50	25.10	22.90	30.80	30.62	27.88	20.65	30.90
30.60	6.12	19.94	19.61	21.51	28.80	21.72	21.95	18.89
15.60	29.38	25.45	30.41	30.11	30.51	28.17	21.42	

Construct an ordered stem-and-leaf display using three-digit leaves. That is, 29.21 would have the stem of 2 and leaf of (921). Would much be lost if, before constructing the stem-and-leaf display, each number was rounded to the nearest integer?

3.1-19 Use the data for the distances of Sosa's homers that is given in Exercise 3.1-11. Determine the stem-and-leaf display, the relative frequency histogram, and the relative frequency polygon, beginning with the stem $3f$ as was done with McGwire's data. Compare the results of these two home run hitters.

3.2   RANDOM VARIABLES OF THE CONTINUOUS TYPE

The relative frequency (or density) histogram $h(x)$ associated with n observations of a random variable of the continuous type is a nonnegative function defined so that the total area between its graph and the x axis equals one. This statement is also true of the function created by the relative frequency polygon. In addition, $h(x)$ is constructed so that the integral

$$\int_a^b h(x)\, dx$$

is approximately equal to the relative frequency of the interval $\{x : a < x < b\}$ and hence can be thought of as an estimate of the probability that the random variable X falls in the interval (a, b). This probability is often written $P(a < X < b)$.

Let us now consider what happens to the function $h(x)$ in the limit, or what happens to the relative frequency polygon as n increases without bound and as the lengths of the class intervals decrease to zero. It is to be hoped that $h(x)$ and the relative frequency polygon will become closer and closer to some function, say $f(x)$, that gives the true probabilities, such as $P(a < X < b)$, through the integral

$$P(a < X < b) = \int_a^b f(x)\, dx.$$

That is, $f(x)$ should be a nonnegative function such that the total area between its graph and the x axis equals one. Moreover, the probability $P(a < X < b)$ is the area bounded by the graph of $f(x)$, the x axis, and the lines $x = a$ and $x = b$. Thus we say that the **probability density function (p.d.f.)** of a random variable X of the **continuous type**, with space S that is an interval or union of intervals, is an integrable function $f(x)$ satisfying the following conditions:

(a) $f(x) > 0, \qquad x \in S.$

(b) $\int_S f(x)\, dx = 1.$

(c) If $(a, b) \subseteq S$, the probability of the event $\{a < X < b\}$ is

$$P(a < X < b) = \int_a^b f(x)\, dx.$$

The corresponding distribution of probability is said to be one of the continuous type.

EXAMPLE 3.2-1   Let the random variable X be the lengths of time **in minutes** between calls to 911 in a small city that were reported in the newspaper on February 26 and 27. Suppose that a reasonable probability model for X is given by the p.d.f.

$$f(x) = \frac{1}{20} e^{-x/20}, \qquad 0 \le x < \infty.$$

Note that $S = \{x : 0 \leq x < \infty\}$ and $f(x) > 0$ for $x \in S$. Also,

$$\int_S f(x)\,dx = \int_0^\infty \frac{1}{20} e^{-x/20}\,dx$$

$$= \lim_{b \to \infty} \left[-e^{-x/20} \right]_0^b$$

$$= 1 - \lim_{b \to \infty} e^{-b/20} = 1.$$

The probability that the time between calls is greater than 20 minutes is

$$P(X > 20) = \int_{20}^\infty \frac{1}{20} e^{-x/20}\,dx = e^{-1} = 0.368.$$

The p.d.f. and the probability of interest are depicted in Figure 3.2-1.   ◀

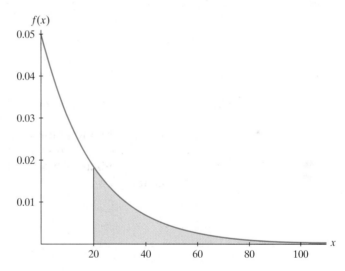

Figure 3.2-1: Times between calls to 911

So that we can avoid repeated references to the space (or support) S of the random variable X, we shall adopt the same convention when describing probability density functions of the continuous type as we did in the discrete case. We extend the definition of the p.d.f. $f(x)$ to the entire set of real numbers by letting it equal zero when $x \notin S$. For example,

$$f(x) = \begin{cases} \dfrac{1}{20} e^{-x/20}, & 0 \leq x < \infty, \\[2mm] 0, & \text{elsewhere,} \end{cases}$$

has the properties of a p.d.f. of a continuous-type random variable X having support $\{x : 0 \leq x < \infty\}$. It will always be understood that $f(x) = 0$ when $x \notin S$, even when this is not explicitly written out.

The **distribution function** of a random variable X of the continuous type, defined in terms of the p.d.f. of X, is given by

$$F(x) = P(X \leq x) = \int_{-\infty}^{x} f(t)\, dt.$$

Here, again, $F(x)$ accumulates (or more simply, cumulates) all of the probability less than or equal to x. (This function is sometimes called a **cumulative distribution function** or c.d.f.). From the fundamental theorem of calculus we have, for x values for which the derivative $F'(x)$ exists, that $F'(x) = f(x)$.

EXAMPLE 3.2-2   Continuing with Example 3.2-1, if the p.d.f. of X is

$$f(x) = \begin{cases} 0, & -\infty < x < 0, \\ \dfrac{1}{20}e^{-x/20}, & 0 \leq x < \infty, \end{cases}$$

the distribution function of X is $F(x) = 0$ for $x < 0$ and, for $x \geq 0$,

$$F(x) = \int_{-\infty}^{x} f(t)\, dt = \int_{0}^{x} \frac{1}{20}e^{-t/20}\, dt$$

$$= \left[-e^{-t/20}\right]_{0}^{x} = 1 - e^{-x/20}.$$

Note that

$$F'(x) = \begin{cases} 0, & -\infty < x < 0 \\ \dfrac{1}{20}e^{-x/20}, & 0 < x < \infty. \end{cases}$$

Also, $F'(0)$ does not exist. (Sketch a graph of $y = F(x)$ to see why this is true.)   ◀

Since there are no steps or jumps in $F(x)$, a distribution function of the continuous type, it must be true that

$$P(X = b) = 0$$

for all real values of b. This agrees with the fact that the integral $\int_{b}^{b} f(x)\, dx$ is taken to be zero in calculus. In general we see that

$$P(a \leq X \leq b) = P(a < X < b) = P(a \leq X < b)$$

$$= P(a < X \leq b) = F(b) - F(a),$$

provided that X is a random variable of the continuous type. Moreover, we can change the definition of a p.d.f. of a random variable of the continuous type at a finite (actually countable) number of points without altering the distribution of probability. For illustration,

$$f(x) = \begin{cases} 0, & -\infty < x < 0 \\ \dfrac{1}{20}e^{-x/20}, & 0 \le x < \infty, \end{cases}$$

and

$$f(x) = \begin{cases} 0, & -\infty < x \le 0 \\ \dfrac{1}{20}e^{-x/20}, & 0 < x < \infty, \end{cases}$$

are equivalent in the computation of probabilities involving this random variable.

EXAMPLE 3.2-3   The following table lists 105 observations of X, the times between calls to 911. To help determine visually whether the exponential model in Example 3.2-1 is perhaps appropriate for this situation, we shall look at two graphs.

30	17	65	8	38	35	4	19	7	14	12	4	5	4	2
7	5	12	50	33	10	15	2	10	1	5	30	41	21	31
1	18	12	5	24	7	6	31	1	3	2	22	1	30	2
1	3	12	12	9	28	6	50	63	5	17	11	23	2	46
90	13	21	55	43	5	19	47	24	4	6	27	4	6	37
16	41	68	9	5	28	42	3	42	8	52	2	11	41	4
35	21	3	17	10	16	1	68	105	45	23	5	10	12	17

We have constructed a relative frequency histogram, $h(x)$, of these data with $f(x) = (1/20)e^{-x/20}$ superimposed in Figure 3.2-2(a). We have also constructed the empirical distribution function of these data in Figure 3.2-2(b) with the theoretical distribution function superimposed. Note that $F_n(x)$, the **empirical distribution function**, is a step function with a step of size $1/n$ at each observation of X. If k observations are equal, the step at that value is k/n. Later in this text we give statistical tests to help us decide whether this probability model provides a good fit for these data. ◀

EXAMPLE 3.2-4   Let Y be a continuous random variable with p.d.f. $g(y) = 2y, 0 < y < 1$. The distribution function of Y is defined by

$$G(y) = \begin{cases} 0, & y < 0, \\ \displaystyle\int_0^y 2t\,dt = y^2, & 0 \le y < 1, \\ 1, & 1 \le y. \end{cases}$$

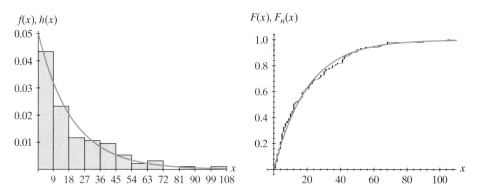

Figure 3.2-2: Times between calls to 911

Figure 3.2-3 gives the graph of the p.d.f. $g(y)$ and the graph of the distribution function $G(y)$. For examples of computations of probabilities, consider

$$P\left(\frac{1}{2} < Y \le \frac{3}{4}\right) = G\left(\frac{3}{4}\right) - G\left(\frac{1}{2}\right) = \left(\frac{3}{4}\right)^2 - \left(\frac{1}{2}\right)^2 = \frac{5}{16}$$

and

$$P\left(\frac{1}{4} \le Y < 2\right) = G(2) - G\left(\frac{1}{4}\right) = 1 - \left(\frac{1}{4}\right)^2 = \frac{15}{16}.$$ ◀

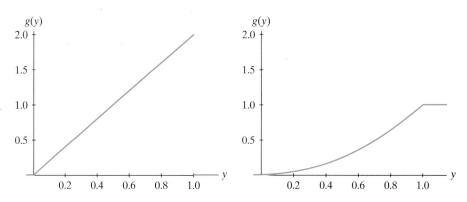

Figure 3.2-3: Continuous distribution p.d.f. and c.d.f.

Recall that the p.m.f. $f(x)$ of a random variable of the discrete type is bounded by one because $f(x)$ gives a probability, namely

$$f(x) = P(X = x).$$

For random variables of the continuous type, the p.d.f. does not have to be bounded (see Exercises 3.2-2(c) and 3.2-3(c)). The restriction is that *the area between the p.d.f. and the x axis must equal one.* Furthermore, it should be noted that the p.d.f. of a random variable X of the continuous type does not need to be a continuous function. For example,

$$f(x) = \begin{cases} \dfrac{1}{2}, & 0 < x < 1 \quad \text{or} \quad 2 < x < 3, \\ 0, & \text{elsewhere,} \end{cases}$$

enjoys the properties of a p.d.f. of a distribution of the continuous type, and yet $f(x)$ has discontinuities at $x = 0, 1, 2$, and 3. However, the distribution function associated with a distribution of the continuous type is always a continuous function.

For continuous-type random variables, the definitions associated with mathematical expectation are the same as those in the discrete case except that integrals replace summations. For illustration, let X be a continuous random variable with a p.d.f. $f(x)$.

The **expected value of** X or **mean of** X is

$$\mu = E(X) = \int_{-\infty}^{\infty} x f(x)\, dx.$$

The **variance of** X is

$$\sigma^2 = \text{Var}(X) = E[(X - \mu)^2] = \int_{-\infty}^{\infty} (x - \mu)^2 f(x)\, dx.$$

The **standard deviation of** X is

$$\sigma = \sqrt{\text{Var}(X)}.$$

The **moment-generating function**, if it exists, is

$$M(t) = \int_{-\infty}^{\infty} e^{tx} f(x)\, dx, \qquad -h < t < h.$$

Moreover, results such as $\sigma^2 = E(X^2) - \mu^2$ and $\mu = M'(0)$, $\sigma^2 = M''(0) - [M'(0)]^2$ are still valid. Again it is important to note that the moment-generating function, if it is finite for $-h < t < h$ for some $h > 0$, completely determines the distribution.

Remark In both the discrete and continuous cases, it should be noted that if the rth moment, $E(X^r)$, exists and is finite, then the same is true of all lower order moments, $E(X^k)$, $k = 1, 2, \ldots, r - 1$. However, the converse is not true; for illustration the first moment can exist and be finite, but the second moment is not necessarily finite (see Exercise 3.2-9). Moreover, if $E(e^{tX})$ exists and is finite for $-h < t < h$, then all moments exist and are finite, but the converse is not necessarily true. ∎

EXAMPLE 3.2-5 For the random variable Y in Example 3.2-4,

$$\mu = E(Y) = \int_0^1 y\,(2y)\,dy = \left[\left(\frac{2}{3}\right) y^3\right]_0^1 = \frac{2}{3}$$

and

$$\sigma^2 = \text{Var}(Y) = E(Y^2) - \mu^2$$

$$= \int_0^1 y^2 (2y)\,dy - \left(\frac{2}{3}\right)^2 = \left[\left(\frac{1}{2}\right) y^4\right]_0^1 - \frac{4}{9} = \frac{1}{18},$$

are the mean and variance, respectively, of Y. ◀

EXAMPLE 3.2-6 Let X have the p.d.f.

$$f(x) = \begin{cases} xe^{-x}, & 0 \le x < \infty, \\ 0, & \text{elsewhere.} \end{cases}$$

Then

$$M(t) = \int_0^\infty e^{tx} xe^{-x}\,dx = \lim_{b\to\infty} \int_0^b xe^{-(1-t)x}\,dx$$

$$= \lim_{b\to\infty} \left[-\frac{xe^{-(1-t)x}}{1-t} - \frac{e^{-(1-t)x}}{(1-t)^2} \right]_0^b$$

$$= \lim_{b\to\infty} \left[-\frac{be^{-(1-t)b}}{1-t} - \frac{e^{-(1-t)b}}{(1-t)^2} \right] + \frac{1}{(1-t)^2}$$

$$= \frac{1}{(1-t)^2},$$

provided that $t < 1$. Note that $M(0) = 1$, which is true for every moment-generating function. Now

$$M'(t) = \frac{2}{(1-t)^3} \qquad \text{and} \qquad M''(t) = \frac{6}{(1-t)^4}.$$

Thus

$$\mu = M'(0) = 2$$

and

$$\sigma^2 = M''(0) - [M'(0)]^2 = 6 - 2^2 = 2.$$ ◀

The $(100p)$th **percentile** is a number π_p such that the area under $f(x)$ to the left of π_p is p. That is,

$$p = \int_{-\infty}^{\pi_p} f(x)\,dx = F(\pi_p).$$

The 50th percentile is called the **median**. We let $m = \pi_{0.50}$. The 25th and 75th percentiles are called the **first** and **third quartiles**, respectively, denoted by $q_1 = \pi_{0.25}$ and $q_3 = \pi_{0.75}$. Of course, the median $m = \pi_{0.50} = q_2$ is also called the second quartile.

The $(100p)$th percentile of a distribution is often called the quantile of order p. So if $y_1 \le y_2 \le \cdots \le y_n$ are the order statistics associated with the sample $x_1, x_2, \ldots, x_n$, then y_r is called the **quantile of order** $r/(n+1)$ as well as the $100r/(n+1)$ **percentile**. Also, the percentile π_p of a theoretical distribution is the quantile of order p. Say the theoretical distribution is a good model for the observations. If we plot (y_r, π_p), where $p = r/(n+1)$, for several values of r (possibly even for all r values, $r = 1, 2, \ldots, n$), we would expect these points (y_r, π_p) to lie close to a line through the origin with slope equal to one because $y_r \approx \pi_p$. If they are not close to that line, then we would doubt that the theoretical distribution is a good model for the observations. The plot of (y_r, π_p) for several values of r is called the **quantile-quantile plot** or, more simply, the **q-q plot**. (See Figure 3.5-3 for an example.)

EXAMPLE 3.2-7   The time X in months until failure of a certain product has the p.d.f. (of the Weibull type)

$$f(x) = \frac{3x^2}{4^3}\, e^{-(x/4)^3}, \qquad 0 < x < \infty.$$

Its distribution function is

$$F(x) = 1 - e^{-(x/4)^3}, \qquad 0 \le x < \infty.$$

Thus, for example, the 30th percentile, $\pi_{0.3}$, is given by

$$F(\pi_{0.3}) = 0.3$$

or, equivalently,

$$1 - e^{-(\pi_{0.3}/4)^3} = 0.3$$

$$\ln(0.7) = -(\pi_{0.3}/4)^3$$

$$\pi_{0.3} = -4\sqrt[3]{\ln(0.7)} = 2.84.$$

Likewise, $\pi_{0.9}$ is found by

$$F(\pi_{0.9}) = 0.9;$$

so

$$\pi_{0.9} = -4\sqrt[3]{\ln(0.1)} = 5.28.$$

Thus

$$P(2.84 < X < 5.28) = 0.6.$$

The 30th and 90th percentiles are shown in Figure 3.2-4. ◄

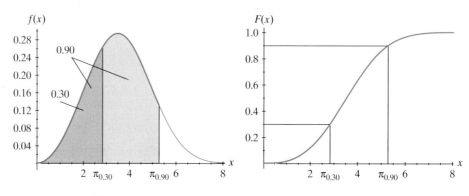

Figure 3.2-4: Illustration of percentiles $\pi_{0.30}$ and $\pi_{0.90}$

We conclude this section by reviewing, through the next example, the important ideas that have been considered in this section.

EXAMPLE 3.2-8   Let X have the p.d.f.

$$f(x) = e^{-x-1}, \qquad -1 < x < \infty.$$

Then

$$P(1 \le X) = \int_{1}^{\infty} e^{-x-1}\,dx = e^{-1}\left[-e^{-x}\right]_{1}^{\infty} = e^{-2} = 0.135.$$

Also we have

$$M(t) = E(e^{tX}) = \int_{-1}^{\infty} e^{tx}e^{-x-1}\,dx$$

$$= e^{-1}\left[\frac{-e^{-(1-t)x}}{1-t}\right]_{-1}^{\infty}$$

$$= e^{-t}(1-t)^{-1}, \qquad t < 1.$$

Thus, since

$$M'(t) = (e^{-t})(1-t)^{-2} - e^{-t}(1-t)^{-1}$$

and

$$M''(t) = (e^{-t})(2)(1-t)^{-3} - 2e^{-t}(1-t)^{-2} + e^{-t}(1-t)^{-1},$$

we have

$$\mu = M'(0) = 0 \quad \text{and} \quad \sigma^2 = M''(0) - [M'(0)]^2 = 1.$$

The distribution function is

$$F(x) = \int_{-1}^{x} e^{-w-1} \, dw = e^{-1} \left[-e^{-w} \right]_{-1}^{x}$$

$$= e^{-1} \left[e^1 - e^{-x} \right] = 1 - e^{-x-1}, \quad -1 < x < \infty,$$

and zero for $x \leq -1$.

For illustration, the median $\pi_{0.5}$ is found by solving

$$F(\pi_{0.5}) = 0.5,$$

which is equivalent to

$$-\pi_{0.5} - 1 = \ln(0.5);$$

so

$$\pi_{0.5} = \ln 2 - 1 = -0.307. \quad \blacktriangleleft$$

EXERCISES

3.2-1 Let the random variable X have the p.d.f. $f(x) = 2(1-x), 0 \leq x \leq 1$, zero elsewhere.
   (a) Sketch the graph of this p.d.f.
   (b) Determine and sketch the graph of the distribution function of X.
   (c) Find **(i)** $P(0 \leq X \leq 1/2)$, **(ii)** $P(1/4 \leq X \leq 3/4)$, **(iii)** $P(X = 3/4)$, and **(iv)** $P(X \geq 3/4)$.

3.2-2 For each of the following functions, **(i)** find the constant c so that $f(x)$ is a p.d.f. of a random variable X, **(ii)** find the distribution function, $F(x) = P(X \leq x)$, and **(iii)** sketch graphs of the p.d.f. $f(x)$ and the distribution function $F(x)$.
   (a) $f(x) = x^3/4$, $\quad 0 < x < c$;
   (b) $f(x) = (3/16)x^2$, $\quad -c < x < c$;
   (c) $f(x) = c/\sqrt{x}$, $\quad 0 < x < 1$. Is this p.d.f. bounded?

3.2-3 For each of the following functions, (i) find the constant c so that $f(x)$ is a p.d.f. of a random variable X, (ii) find the distribution function, $F(x) = P(X \leq x)$, and (iii) sketch graphs of the p.d.f. $f(x)$ and the distribution function $F(x)$.
   (a) $f(x) = 4x^c$, $\quad 0 \leq x \leq 1$,
   (b) $f(x) = c\sqrt{x}$, $\quad 0 \leq x \leq 4$,
   (c) $f(x) = c/x^{3/4}$, $\quad 0 < x < 1$.

3.2-4 For each of the distributions in Exercise 3.2-2, find μ, σ^2, and σ.

3.2-5 For each of the distributions in Exercise 3.2-3, find μ, σ^2, and σ.

3.2-6 Let $f(x) = (1/2)x^2 e^{-x}$, $0 < x < \infty$ be the p.d.f. of X.
   (a) Find the m.g.f. $M(t)$.
   (b) Find the values of μ and σ^2.

3.2-7 Let $f(x) = (1/2) \sin x, 0 \le x \le \pi$, be the p.d.f. of X.

(a) Find μ and σ^2.

(b) Sketch the graph of the p.d.f. of X.

(c) Determine and sketch the graph of the distribution function of X.

3.2-8 The p.d.f. of X is $f(x) = c/x^2, 1 < x < \infty$.

(a) Calculate the value of c so that $f(x)$ is a p.d.f.

(b) Show that $E(X)$ is not finite.

3.2-9 The p.d.f. of Y is $g(y) = d/y^3, 1 < y < \infty$.

(a) Calculate the value of d so that $g(y)$ is a p.d.f.

(b) Find $E(Y)$.

(c) Show that Var(Y) is not finite.

3.2-10 Sketch the graphs of the following probability density functions. Also find and sketch the graphs of the distribution functions associated with these distributions. Note carefully the relationship between the shape of the graph of the p.d.f. and the concavity of the graph of the distribution function.

(a) $f(x) = \left(\dfrac{3}{2}\right)x^2, \quad -1 < x < 1.$

(b) $f(x) = \dfrac{1}{2}, \quad -1 < x < 1.$

(c) $f(x) = \begin{cases} x + 1, & -1 < x < 0, \\ 1 - x, & 0 \le x < 1. \end{cases}$

3.2-11 Find the mean and variance for each of the distributions in Exercise 3.2-10.

3.2-12 Let $R(t) = \ln M(t)$, where $M(t)$ is the moment-generating function of a random variable of the continuous type. Show that

(a) $\mu = R'(0)$.

(b) $\sigma^2 = R''(0)$.

3.2-13 If $M(t) = e^{-t}(1 - t)^{-1}, t < 1$, use $R(t) = \ln M(t)$ and the result in Exercise 3.2-12 to find (a) μ and (b) σ^2.

3.2-14 Find the moment-generating function $M(t)$ of the distribution with p.d.f. $f(x) = (1/10)e^{-x/10}, 0 < x < \infty$. Use $M(t)$ or $R(t) = \ln M(t)$ to determine the mean μ and the variance σ^2.

3.2-15 The logistic distribution is associated with the distribution function

$$F(x) = (1 + e^{-x})^{-1}, \quad -\infty < x < \infty.$$

Find the p.d.f. of the logistic distribution and show that its graph is symmetric about $x = 0$.

3.2-16 Let $f(x) = 1/2, 0 < x < 1$ or $2 < x < 3$, zero elsewhere, be the p.d.f. of X.

(a) Sketch the graph of this p.d.f.

(b) Define the distribution function of X and sketch its graph.

(c) Find $q_1 = \pi_{0.25}$.

(d) Find $m = \pi_{0.50}$. Is it unique?

(e) Find $q_3 = \pi_{0.75}$.

3.2-17 Let $f(x) = 1/2, -1 < x < 1$. Find (a) $m = \pi_{0.50}$, (b) $q_1 = \pi_{0.25}$, and (c) $\pi_{0.90}$.

3.2-18 Let $f(x) = (x + 1)/2, -1 < x < 1$. Find (a) $\pi_{0.64}$, (b) $q_1 = \pi_{0.25}$, and (c) $\pi_{0.81}$.

3.2-19 Let the random variable X_n have the p.d.f. $f(x_n) = n, 0 < x_n < 1/n$.

(a) Define the distribution function of X_n, say $F_n(x_n)$.

(b) Graph the p.d.f. and the distribution function for $n = 1, 5$, and 10.

3.2-20 One hundred observations of X having a certain distribution were taken and then ordered, yielding

0.19	0.25	0.26	0.26	0.28	0.31	0.39	0.42	0.44	0.45
0.48	0.53	0.54	0.57	0.66	0.69	0.70	0.72	0.73	0.79
0.83	0.85	0.87	0.90	0.91	0.94	0.96	0.99	1.01	1.02
1.02	1.03	1.05	1.07	1.13	1.13	1.14	1.20	1.22	1.27
1.28	1.30	1.34	1.34	1.37	1.39	1.40	1.43	1.43	1.46
1.49	1.50	1.51	1.53	1.54	1.56	1.56	1.57	1.60	1.60
1.61	1.62	1.62	1.62	1.63	1.64	1.66	1.67	1.70	1.71
1.71	1.71	1.72	1.73	1.73	1.74	1.75	1.76	1.77	1.80
1.81	1.84	1.85	1.85	1.86	1.87	1.87	1.88	1.90	1.91
1.91	1.92	1.92	1.93	1.93	1.93	1.94	1.95	1.98	1.99

(a) Calculate the sample mean $\bar{x}$.
(b) Calculate the sample standard deviation s.
(c) Group these data into 10 classes: $(0.00, 0.20], (0.20, 0.40], \ldots, (1.80, 2.00]$.
(d) Construct a relative frequency histogram with the p.d.f. $f(x) = x/2, 0 < x < 2$, superimposed. Is there a good fit?
(e) Find μ and σ, assuming that the p.d.f. of X is $f(x) = x/2, 0 < x \leq 2$. Is $\bar{x}$ close to μ? Is s close to σ?

3.2-21 The lifetime X (in years) of a machine has a p.d.f.

$$f(x) = \frac{2x}{\theta^2} e^{-(x/\theta)^2}, \qquad 0 < x < \infty.$$

If $P(X > 5) = 0.01$, determine θ.

3.2-22 The weekly demand X for propane gas (in thousands of gallons) has the p.d.f.

$$f(x) = 4x^3 e^{-x^4}, \qquad 0 < x < \infty.$$

If two thousand gallons is in stock at the beginning of each week (and there is none extra during the week), what is the probability of not being able to meet the demand during a given week?

3.2-23 An insurance agent receives a bonus if the loss ratio L on his business is less than 0.5, where L is the total losses, say X, divided by the total premiums, say T. The bonus equals $(0.5 - L)(T/30)$ if $L < 0.5$ and equals zero otherwise. If X (in \$100,000) has the p.d.f.

$$f(x) = \frac{3}{x^4}, \qquad x > 1,$$

and if T (in \$100,000) equals 3, determine the expected value of the bonus.

3.2-24 The p.d.f. of time X to failure of an electronic component is

$$f(x) = \frac{2x}{1000^2} e^{-(x/1000)^2}, \qquad 0 < x < \infty.$$

(a) Compute $P(X > 2000)$.
(b) Determine the 75th percentile, $\pi_{0.75}$, of the distribution.
(c) Find the 10th and 60th percentiles, $\pi_{0.10}$ and $\pi_{0.60}$.

3.2-25 The life X in years of a voltage regulator of a car has the p.d.f.

$$f(x) = \frac{3x^2}{7^3} e^{-(x/7)^3}, \qquad 0 < x < \infty.$$

(a) What is the probability that this regulator will last at least 7 years?
(b) Given that it has lasted at least 7 years, what is the conditional probability it will last at least another 3.5 years?

3.3 **THE UNIFORM AND EXPONENTIAL DISTRIBUTIONS**

Let the random variable X denote the outcome when a point is selected at random from an interval $[a, b]$, $-\infty < a < b < \infty$. If the experiment is performed in a fair manner, it is reasonable to assume that the probability that the point is selected from the interval $[a, x]$, $a \leq x < b$ is $(x - a)/(b - a)$. That is, the probability is proportional to the length of the interval so that the distribution function of X is

$$
F(x) = \begin{cases} 0, & x < a, \\ \dfrac{x - a}{b - a}, & a \leq x < b, \\ 1, & b \leq x. \end{cases}
$$

Because X is a continuous-type random variable, $F'(x)$ is equal to the p.d.f. of X whenever $F'(x)$ exists; thus when $a < x < b$, we have $f(x) = F'(x) = 1/(b - a)$.

The random variable X has a **uniform distribution** if its p.d.f. is equal to a constant on its support. In particular, if the support is the interval $[a, b]$, then

$$
f(x) = \frac{1}{b - a}, \qquad a \leq x \leq b.
$$

Moreover, we shall say that X **is** $U(a, b)$. This distribution is also referred to as **rectangular** because the graph of $f(x)$ suggests that name. See Figure 3.3-1 for the graph of $f(x)$ and the distribution function $F(x)$ when $a = 0.30$ and $b = 1.55$. Note that we could have taken $f(a) = 0$ or $f(b) = 0$ without altering the probabilities, since this is a continuous-type distribution, and we shall do this in some cases.

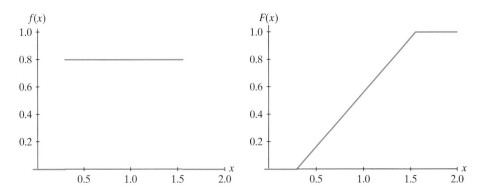

Figure 3.3-1: Uniform p.d.f. and distribution function

The mean, variance, and moment-generating function of X are not difficult to calculate (see Exercise 3.3-1). They are

$$\mu = \frac{a + b}{2}, \qquad \sigma^2 = \frac{(b - a)^2}{12},$$

$$M(t) = \begin{cases} \dfrac{e^{tb} - e^{ta}}{t(b - a)}, & t \neq 0, \\ 1, & t = 0. \end{cases}$$

An important uniform distribution is that for which $a = 0$ and $b = 1$, namely $U(0, 1)$. If X is $U(0, 1)$, approximate values of X can be simulated on most computers using a random number generator. In fact, it should be called a **pseudo-random number generator** because the programs that produce the random numbers are usually such that if the starting number (the seed number) is known, all subsequent numbers in the sequence may be determined by simple arithmetical operations. Yet, despite their deterministic origin, these computer-produced numbers do behave as if they were truly randomly generated, and we shall not encumber our terminology by adding *pseudo*. Examples of computer-produced random numbers are given in the Appendix in Table IX. Place a decimal point in front of each of these four digits so that each is a number between 0 and 1.

EXAMPLE 3.3-1   Let X have the p.d.f.

$$f(x) = \frac{1}{100}, \qquad 0 < x < 100,$$

so that X is $U(0, 100)$. The mean and the variance are

$$\mu = \frac{0 + 100}{2} = 50 \quad \text{and} \quad \sigma^2 = \frac{(100 - 0)^2}{12} = \frac{10,000}{12}.$$

The standard deviation is $\sigma = 100/\sqrt{12}$, which is 100 times that of the $U(0, 1)$ distribution. This agrees with our intuition since the standard deviation is a measure of spread and $U(0, 100)$ is clearly spread out 100 times more than $U(0, 1)$. ◀

We turn now to a continuous distribution that is related to the Poisson distribution. When previously observing a process of the (approximate) Poisson type, we counted the number of changes occurring in a given interval. This number was a discrete-type random variable with a Poisson distribution. But not only is the number of changes a random variable; the waiting times between successive changes are also random variables. However, the latter are of the continuous type, since each of them can assume any positive value. In particular, let W denote the waiting time until the first change occurs when observing a Poisson process in which the mean number of changes in the unit interval is λ. Then W is a continuous-type random variable, and we proceed to find its distribution function.

Because this waiting time is nonnegative, the distribution function $F(w) = 0$, $w < 0$. For $w \geq 0$,

$$F(w) = P(W \leq w) = 1 - P(W > w)$$

$$= 1 - P(\text{no changes in } [0, w])$$

$$= 1 - e^{-\lambda w},$$

since we previously discovered that $e^{-\lambda w}$ equals the probability of no changes in an interval of length w. That is, if the mean number of changes per unit interval is λ, then the mean number of changes in an interval of length w is proportional to w, namely, λw. Thus, when $w > 0$, the p.d.f. of W is given by

$$F'(w) = f(w) = \lambda e^{-\lambda w}.$$

We often let $\lambda = 1/\theta$ and say that the random variable X has an **exponential distribution** if its p.d.f. is defined by

$$f(x) = \frac{1}{\theta} e^{-x/\theta}, \qquad 0 \leq x < \infty,$$

where the parameter $\theta > 0$. Accordingly, the waiting time W until the first change in a Poisson process has an exponential distribution with $\theta = 1/\lambda$. To determine the exact meaning of the parameter θ, we first find the moment-generating function of X. It is

$$M(t) = \int_0^\infty e^{tx} \left(\frac{1}{\theta}\right) e^{-x/\theta} \, dx = \lim_{b \to \infty} \int_0^b \left(\frac{1}{\theta}\right) e^{-(1-\theta t)x/\theta} \, dx$$

$$= \lim_{b \to \infty} \left[-\frac{e^{-(1-\theta t)x/\theta}}{1 - \theta t} \right]_0^b = \frac{1}{1 - \theta t}, \qquad t < \frac{1}{\theta}.$$

Thus

$$M'(t) = \frac{\theta}{(1 - \theta t)^2}$$

and

$$M''(t) = \frac{2\theta^2}{(1 - \theta t)^3}.$$

Hence, for an exponential distribution, we have

$$\mu = M'(0) = \theta \qquad \text{and} \qquad \sigma^2 = M''(0) - [M'(0)]^2 = \theta^2.$$

So if λ is the mean number of changes in the unit interval, then $\theta = 1/\lambda$ is the mean waiting time for the first change. In particular, suppose that $\lambda = 7$ is the mean number of changes per minute; then the mean waiting time for the first change is 1/7 of a minute, a result that agrees with our intuition.

EXAMPLE 3.3-2   Let X have an exponential distribution with a mean of $\theta = 20$. So the p.d.f. of X is that of Example 3.2-1, namely

$$f(x) = \frac{1}{20}e^{-x/20}, \qquad 0 \le x < \infty.$$

See Figure 3.2-1 for the graph of this p.d.f. The probability that X is less than 18 is

$$P(X < 18) = \int_0^{18} \frac{1}{20}e^{-x/20}\,dx = 1 - e^{-18/20} = 0.593. \qquad \blacktriangleleft$$

Let X have an exponential distribution with mean $\mu = \theta$. Then the distribution function of X is

$$F(x) = \begin{cases} 0, & -\infty < x < 0, \\ 1 - e^{-x/\theta}, & 0 \le x < \infty. \end{cases}$$

For example, the p.d.f. and distribution function are graphed in Figure 3.3-2 for $\theta = 5$. The median, m, is found by solving $F(m) = 0.5$. That is,

$$1 - e^{-m/\theta} = 0.5.$$

Thus

$$m = -\theta \ln(0.5) = \theta \ln(2).$$

So with $\theta = 5$, the median is $m = -5\ln(0.5) = 3.466$. Both the median and the mean $\theta = 5$ are indicated on the graphs.

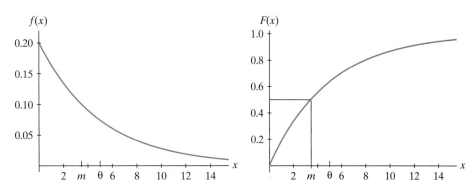

Figure 3.3-2: Exponential p.d.f. and distribution function

It is useful to note that for an exponential random variable, X, we have that

$$P(X > x) = 1 - F(x) = 1 - (1 - e^{-x/\theta})$$

$$= e^{-x/\theta}, \qquad \text{when } x > 0.$$

EXAMPLE 3.3-3 Customers arrive in a certain shop according to an approximate Poisson process at a mean rate of 20 per hour. What is the probability that the shopkeeper will have to wait more than 5 minutes for the arrival of the first customer? Let X denote the waiting time in minutes until the first customer arrives and note that $\lambda = 1/3$ is the expected number of arrivals per minute. Thus

$$\theta = \frac{1}{\lambda} = 3$$

and

$$f(x) = \frac{1}{3} e^{-(1/3)x}, \qquad 0 \le x < \infty.$$

Hence

$$P(X > 5) = \int_5^\infty \frac{1}{3} e^{-(1/3)x} \, dx = e^{-5/3} = 0.1889.$$

The median time until the first arrival is

$$m = -3 \ln(0.5) = 2.0794. \qquad \blacktriangleleft$$

EXAMPLE 3.3-4 Suppose that the life of a certain type of electronic component has an exponential distribution with a mean life of 500 hours. If X denotes the life of this component (or the time to failure of this component), then

$$P(X > x) = \int_x^\infty \frac{1}{500} e^{-t/500} \, dt = e^{-x/500}.$$

Suppose that the component has been in operation for 300 hours. The conditional probability that it will last for another 600 hours is

$$P(X > 900 \mid X > 300) = \frac{P(X > 900)}{P(X > 300)} = \frac{e^{-900/500}}{e^{-300/500}} = e^{-6/5}.$$

It is important to note that this conditional probability is exactly equal to $P(X > 600) = e^{-6/5}$. That is, the probability that it will last an additional 600 hours, given that it has operated 300 hours, is the same as the probability that it would last 600 hours when first put into operation. Thus, for such components, an old component is as good as a new one, and we say that the failure rate is constant. Certainly, with constant failure rate, there is no advantage in replacing components that are operating satisfactorily. Obviously, this is not true in practice because most would have an increasing failure rate with time; hence the exponential distribution is probably not the best model for the probability distribution of such a life.   ◀

Remark   In Exercise 3.3-11, the result of Example 3.3-4 is generalized; namely, if the length of life has an exponential distribution, the probability that it will last a time of at least $x + y$ units, given that it has lasted at least x units, is exactly the same as the probability that it will last at least y units when first put into operation. In effect, this states that the exponential distribution has a forgetfulness (or no memory) property. It is also interesting to observe that, for continuous random variables whose support is $(0, \infty)$, the exponential distribution is the only distribution with this forgetfulness property. Recall, however, that when we considered distributions of the discrete type, we noted that the geometric distribution has this property (see Exercise 2.5-12).   ■

We often want to check to see if the order statistics $y_1 \leq y_2 \leq \cdots \leq y_n$ are those of a random sample from some exponential distribution with mean θ. If we knew θ, say $\theta = \theta_0$, we could construct the q-q plot (y_r, π_p), where $p = r/(n+1)$ and

$$F(\pi_p) = 1 - e^{-\pi_p/\theta_0} = p$$

or, equivalently,

$$\pi_p = -\theta_0 \ln(1 - p).$$

If θ_0 is the correct value of θ and if the exponential distribution is a suitable model, then $(y_r, \pi_p), r = 1, 2, \ldots, n$, would lie close to a straight line through the origin with slope one. However, we do not know θ. Fortunately, there is a way out of our dilemma. Simply plot (y_r, π_p), with $\theta = 1$; that is,

$$\pi_p = -\ln(1 - p).$$

If these points (y_r, π_p) fall close to a straight line through the origin, then the exponential model is appropriate. Moreover, its slope, say s_0, provides an estimate of θ through $s_0 = 1/\theta$. That is, with $\theta = 1/s_0$, the points

$$[y_r, \pi_p = -\theta \ln(1 - p)], \qquad r = 1, 2, \ldots, n,$$

then lie close to a straight line through the origin with slope one.

EXERCISES

3.3-1   Show that the mean, variance, and moment-generating function of the uniform distribution are as given in this section.

3.3-2   Let $f(x) = 1/2, -1 \leq x \leq 1$, be the p.d.f. of X. Graph the p.d.f. and distribution function, and record the mean and variance of X.

3.3-3   Customers arrive randomly at a bank teller's window. Given that one customer arrived during a particular 10-minute period, let X equal the time within the 10 minutes that the customer arrived. If X is $U(0, 10)$, find:

(a)   the p.d.f. of X,

(b) $P(X \geq 8)$,

(c) $P(2 \le X < 8)$,

(d) $E(X)$, and

(e) $\text{Var}(X)$.

3.3-4 If the moment-generating function of X is

$$M(t) = \frac{e^{5t} - e^{4t}}{t}, \quad t \ne 0, \qquad \text{and} \qquad M(0) = 1,$$

find **(a)** $E(X)$, **(b)** $\text{Var}(X)$, and **(c)** $P(4.2 < X \le 4.7)$.

3.3-5 Let Y have a uniform distribution $U(0, 1)$, and let

$$W = a + (b - a)Y, \qquad a < b.$$

(a) Find the distribution function of W.
   HINT: Find $P[a + (b - a)Y \le w]$.

(b) How is W distributed?

3.3-6 Let X have an exponential distribution with a mean of $\theta = 20$. Compute

(a) $P(10 < X < 30)$.

(b) $P(X > 30)$.

(c) $P(X > 40 \mid X > 10)$.

(d) What are the variance and the moment-generating function of X?

(e) The following ordered data were simulated from an exponential distribution with $\theta = 20$. Compare the relative frequencies of the appropriate events in the data with the respective probabilities in parts (a), (b), and (c).

0.45	0.65	0.66	0.70	0.94	1.05	1.17	1.28	1.32	1.35
1.45	1.52	1.59	1.76	1.92	2.05	2.75	2.88	3.48	3.63
4.01	4.37	4.54	4.84	5.39	5.53	5.95	6.26	6.44	6.79
6.85	7.45	7.51	7.75	8.23	8.41	9.42	9.45	9.57	9.72
9.85	9.97	10.26	10.27	10.34	11.05	11.37	11.42	11.45	11.78
12.26	12.90	13.09	13.19	13.45	16.29	16.55	17.07	17.18	17.32
19.40	19.49	20.04	20.95	21.11	21.13	21.37	22.49	23.34	23.88
25.06	25.41	27.86	27.91	28.32	28.60	29.41	30.84	31.74	33.64
35.42	35.60	36.26	37.06	37.98	39.49	40.13	41.34	44.13	45.11
46.10	47.67	47.68	54.36	55.98	62.81	72.92	76.08	92.16	120.54

3.3-7 The *Holland Sentinel* reported the following numbers of calls per hour received by 911 between noon, February 26, and all day February 27.

3	0	3	4	9	1	6	2	2	5	7	6	4	2	2	4	1	0
3	1	3	3	4	2	1	3	3	2	3	2	5	2	1	0	2	4

They also reported the following lengths of time per minute between calls.

30	17	65	8	38	35	4	19	7	14	12	4	5	4	2
7	5	12	50	33	10	15	2	10	1	5	30	41	21	31
1	18	12	5	24	7	6	31	1	3	2	22	1	30	2
1	3	12	12	9	28	6	50	63	5	17	11	23	2	46
90	13	21	55	43	5	19	47	24	4	6	27	4	6	37
16	41	68	9	5	28	42	3	42	8	52	2	11	41	4
35	21	3	17	10	16	1	68	105	45	23	5	10	12	17

(a) Show graphically that the numbers of calls per hour have an approximate Poisson distribution with a mean of $\lambda = 3$.

(b) Show that the sample mean and the sample standard deviation of the times between calls are both approximately equal to 20.

(c) If X is an exponential random variable with mean $\theta = 20$, compare the probability $P(X > 15)$ with the proportion of times that are greater than 15.

(d) Compare $P(X > 45.5 \mid X > 30.5)$ with the proportion of observations that satisfy this condition.

3.3-8 Telephone calls enter a college switchboard according to a Poisson process on the average of two every 3 minutes. Let X denote the waiting time until the first call that arrives after 10 A.M.

(a) What is the p.d.f. of X?

(b) Find $P(X > 2)$.

3.3-9 What are the p.d.f., the mean, and the variance of X if the moment-generating function of X is given by the following?

(a) $M(t) = \dfrac{1}{1 - 3t}$, $t < 1/3$.

(b) $M(t) = \dfrac{3}{3 - t}$, $t < 3$.

3.3-10 Let X equal the number of students who use a college card catalog every 15 minutes. Assume that X has a Poisson distribution with mean 5. Let W equal the time in minutes between two student arrivals. Then W has an exponential distribution with mean $\theta = 3$. Find

(a) $P(W > 6)$.

(b) $P(W > 12 \mid W > 6)$.

3.3-11 Let X have an exponential distribution with mean $\theta > 0$. Show that

$$P(X > x + y \mid X > x) = P(X > y).$$

3.3-12 Let $F(x)$ be the distribution function of the continuous-type random variable X, and assume that $F(x) = 0$ for $x \leq 0$ and $0 < F(x) < 1$ for $0 < x$. Prove that if

$$P(X > x + y \mid X > x) = P(X > y),$$

then

$$F(x) = 1 - e^{-\lambda x}, \qquad 0 < x.$$

HINT: Show that $g(x) = 1 - F(x)$ satisfies the functional equation

$$g(x + y) = g(x)g(y),$$

which implies that $g(x) = a^{cx}$.

3.3-13 Let X equal the number of bad records in each 100 feet of a used computer tape. Assume that X has a Poisson distribution with mean 2.5. Let W equal the number of feet before the first bad record is found.

(a) Give the mean number of flaws per foot.

(b) How is W distributed?

(c) Give the mean and variance of W.

(d) Find **(i)** $P(W \leq 20)$, **(ii)** $P(W > 40)$, and **(iii)** $P(W > 60 \mid W > 20)$.

3.3-14 The initial value of an appliance is 700 dollars and its value in the future is given by

$$v(t) = 100 \, (2^{3-t} - 1), \qquad 0 \leq t \leq 3,$$

where t is time in years. Thus after the first 3 years the appliance is worth nothing as far as the warranty is concerned. If it fails in the first three years, the warranty pays $v(t)$. Compute the expected value of the payment on the warranty if T has a exponential distribution with mean five.

3.3-15 Let X equal the time (in minutes) between calls that are made over the public safety radio. On four different days (February 14, 21, and 28, and March 6) and during a period of one hour on each day, the following observations of X were made:

5	7	8	20	17	2	24	8	8	6	4
3	42	10	18	5	7	8	4	5	10	

If calls arrive randomly in accordance with an approximate Poisson process, then the distribution of X should be approximately exponential.

(a) Calculate the values of the sample mean and sample standard deviation. Are they close to each other in value?

(b) Construct a q-q plot of the ordered observations versus the respective quartiles of the exponential distribution with a mean of $\theta = 1$. If this is approximately linear, the exponential model is supported. Since the mean of these data is not close to 1, this line will not have slope 1, but a linear fit will still indicate an exponential model. What is your conclusion?

3.3-16 A grocery store has available n watermelons to sell and makes \$1.00 on each sale. Say the number of consumers of these watermelons is a random variable that has a distribution that can be approximated by

$$f(x) = \frac{1}{200}, \qquad 0 < x < 200,$$

a p.d.f. of the continuous type. If the grocer does not have enough watermelons to sell to all consumers, she figures that she loses \$5.00 in "good-will" from each unhappy customer. But if she has surplus watermelons, she loses 50 cents on each extra watermelon. What should n be to maximize "profit"? HINT: If $X \leq n$, then her profit is $(1.00)X + (-0.50)(n - X)$; but if $X > n$, her profit is $(1.00)n + (-5.00)(X - n)$. Find the expected value of profit as a function of n and then select n to maximize this.

3.3-17 There are times when a shifted exponential model is appropriate. That is, let the p.d.f. of X be

$$f(x) = \frac{1}{\theta} e^{-(x-\delta)/\theta}, \qquad \delta < x < \infty.$$

(a) Define the distribution function of X.

(b) Calculate the mean and variance of X.

3.3-18 A certain type of aluminum screen 2 feet in width has on the average three flaws in a 100-foot roll.

(a) What is the probability that the first 40 feet in a roll contain no flaws?

(b) What assumption did you make to solve part (a)?

3.3-19 Let X have an exponential distribution with mean θ.

(a) Find the first quartile, q_1.

(b) How far is the first quartile below the mean?

(c) Find the third quartile, q_3.

(d) How far is the third quartile above the mean?

3.3-20 Let X have a logistic distribution with p.d.f.

$$f(x) = \frac{e^{-x}}{(1 + e^{-x})^2}, \qquad -\infty < x < \infty.$$

Show that

$$Y = \frac{1}{1 + e^{-X}}$$

has a $U(0, 1)$ distribution.

HINT: Find $G(y) = P(Y \leq y) = P\left(\frac{1}{1 + e^{-X}} \leq y\right)$, when $0 < y < 1$.

THE GAMMA AND CHI-SQUARE DISTRIBUTIONS

In the (approximate) Poisson process with mean λ, we have seen that the waiting time until the first change has an exponential distribution. We now let W denote the waiting time until the αth change occurs and find the distribution of W.

The distribution function of W, when $w \geq 0$, is given by

$$F(w) = P(W \leq w) = 1 - P(W > w)$$

$$= 1 - P(\text{fewer than } \alpha \text{ changes occur in } [0, w])$$

$$= 1 - \sum_{k=0}^{\alpha-1} \frac{(\lambda w)^k e^{-\lambda w}}{k!} \tag{3.4-1}$$

since the number of changes in the interval $[0, w]$ has a Poisson distribution with mean λw. Because W is a continuous-type random variable, $F'(w)$ is equal to the p.d.f. of W whenever this derivative exists. We have, provided $w > 0$, that

$$F'(w) = \lambda e^{-\lambda w} - e^{-\lambda w} \sum_{k=1}^{\alpha-1} \left[\frac{k(\lambda w)^{k-1} \lambda}{k!} - \frac{(\lambda w)^k \lambda}{k!} \right]$$

$$= \lambda e^{-\lambda w} - e^{-\lambda w} \left[\lambda - \frac{\lambda(\lambda w)^{\alpha-1}}{(\alpha - 1)!} \right]$$

$$= \frac{\lambda(\lambda w)^{\alpha-1}}{(\alpha - 1)!} e^{-\lambda w}.$$

If $w < 0$, then $F(w) = 0$ and $F'(w) = 0$. A p.d.f. of this form is said to be one of the gamma type, and the random variable W is said to have a **gamma distribution**.

Before determining the characteristics of the gamma distribution, let us consider the gamma function for which the distribution is named. The **gamma function** is defined by

$$\Gamma(t) = \int_0^\infty y^{t-1} e^{-y} \, dy, \qquad 0 < t.$$

This integral is positive for $0 < t$ because the integrand is positive. Values of it are often given in a table of integrals. If $t > 1$, integration of the gamma function of t by parts yields

$$\Gamma(t) = \left[-y^{t-1} e^{-y} \right]_0^\infty + \int_0^\infty (t - 1) y^{t-2} e^{-y} \, dy$$

$$= (t - 1) \int_0^\infty y^{t-2} e^{-y} \, dy = (t - 1)\Gamma(t - 1).$$

For example, $\Gamma(6) = 5\Gamma(5)$ and $\Gamma(3) = 2\Gamma(2) = (2)(1)\Gamma(1)$. Whenever $t = n$, a positive integer, we have, by repeated application of $\Gamma(t) = (t-1)\Gamma(t-1)$, that

$$\Gamma(n) = (n-1)\Gamma(n-1) = (n-1)(n-2)\cdots(2)(1)\Gamma(1).$$

However,

$$\Gamma(1) = \int_0^\infty e^{-y}\,dy = 1.$$

Thus, when n is a positive integer, we have that

$$\Gamma(n) = (n-1)!$$

and, for this reason, the gamma function is called the generalized factorial. Incidentally, $\Gamma(1)$ corresponds to $0!$, and we have noted that $\Gamma(1) = 1$, which is consistent with earlier discussions.

Let us now formally define the p.d.f. of the gamma distribution and find its characteristics. The random variable X has a **gamma distribution** if its p.d.f. is defined by

$$f(x) = \frac{1}{\Gamma(\alpha)\theta^\alpha} x^{\alpha-1} e^{-x/\theta}, \qquad 0 \le x < \infty.$$

Hence, W, the waiting time until the αth change in a Poisson process, has a gamma distribution with parameters α and $\theta = 1/\lambda$. To see that $f(x)$ actually has the properties of a p.d.f., note that $f(x) \ge 0$ and

$$\int_{-\infty}^\infty f(x)\,dx = \int_0^\infty \frac{x^{\alpha-1}e^{-x/\theta}}{\Gamma(\alpha)\theta^\alpha}\,dx$$

which, by the change of variables $y = x/\theta$, equals

$$\int_0^\infty \frac{(\theta y)^{\alpha-1}e^{-y}}{\Gamma(\alpha)\theta^\alpha}\theta\,dy = \frac{1}{\Gamma(\alpha)}\int_0^\infty y^{\alpha-1}e^{-y}\,dy = \frac{\Gamma(\alpha)}{\Gamma(\alpha)} = 1.$$

The moment-generating function of X is (Exercise 3.4-3)

$$M(t) = \frac{1}{(1-\theta t)^\alpha}, \qquad t < 1/\theta.$$

The mean and variance are (Exercise 3.4-4)

$$\mu = \alpha\theta \qquad \text{and} \qquad \sigma^2 = \alpha\theta^2.$$

EXAMPLE 3.4-1 Suppose that an average of 30 customers per hour arrive at a shop in accordance with a Poisson process. That is, if a minute is our unit, then $\lambda = 1/2$. What is the probability that the shopkeeper will wait more than 5 minutes before both of the first two customers arrive? If X denotes the waiting time in minutes until the second customer arrives, then X has a gamma distribution with $\alpha = 2$, $\theta = 1/\lambda = 2$. Hence

$$P(X > 5) = \int_5^\infty \frac{x^{2-1}e^{-x/2}}{\Gamma(2)2^2}\, dx = \int_5^\infty \frac{xe^{-x/2}}{4}\, dx$$

$$= \frac{1}{4}\left[(-2)xe^{-x/2} - 4e^{-x/2}\right]_5^\infty$$

$$= \frac{7}{2}e^{-5/2} = 0.287.$$

We could also have used Equation 3.4-1 with $\lambda = 1/\theta$ because α is an integer. From Equation 3.4-1 we have

$$P(X > x) = \sum_{k=0}^{\alpha-1} \frac{(x/\theta)^k e^{-x/\theta}}{k!}.$$

Thus, with $x = 5$, $\alpha = 2$, and $\theta = 2$, this is equal to

$$P(X > 5) = \sum_{k=0}^{2-1} \frac{(5/2)^k e^{-5/2}}{k!}$$

$$= e^{-5/2}\left(1 + \frac{5}{2}\right) = \left(\frac{7}{2}\right)e^{-5/2}. \quad \blacktriangleleft$$

EXAMPLE 3.4-2   Telephone calls arrive at a switchboard at a mean rate of $\lambda = 2$ per minute according to a Poisson process. Let X denote the waiting time in minutes until the fifth call arrives. The p.d.f. of X, with $\alpha = 5$ and $\theta = 1/\lambda = 1/2$, is

$$f(x) = \frac{2^5 x^4}{4!}e^{-2x}, \qquad 0 \le x < \infty.$$

The mean and the variance of X are, respectively, $\mu = 5/2$ and $\sigma^2 = 5/4$. $\blacktriangleleft$

In order to see the effect of the parameters on the shape of the p.d.f., several combinations of α and θ have been used for graphs that are displayed in Figure 3.4-1. Note that for a fixed θ, as α increases, the probability moves to the right. The same is true for increasing θ, with fixed α. Since $\theta = 1/\lambda$, as θ increases, λ decreases. That is, if $\theta_2 > \theta_1$, then $\lambda_2 = 1/\theta_2 < \lambda_1 = 1/\theta_1$. So if the mean number of changes per unit decreases, the waiting time to observe α changes can be expected to increase.

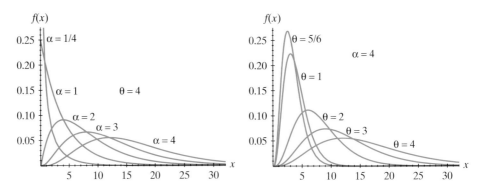

Figure 3.4-1: Gamma p.d.f.s: $\theta = 4$ with $\alpha = 1, 2, 3, 4$; $\alpha = 4$ with $\theta = 1, 2, 3, 4$

We now consider a special case of the gamma distribution that plays an important role in statistics. Let X have a gamma distribution with $\theta = 2$ and $\alpha = r/2$, where r is a positive integer. The p.d.f. of X is

$$f(x) = \frac{1}{\Gamma(r/2)2^{r/2}} x^{r/2-1} e^{-x/2}, \qquad 0 \le x < \infty.$$

We say that X has a **chi-square distribution with r degrees of freedom**, which we abbreviate by saying X is $\chi^2(r)$. The mean and the variance of this chi-square distribution are

$$\mu = \alpha\theta = \left(\frac{r}{2}\right)2 = r \qquad \text{and} \qquad \sigma^2 = \alpha\theta^2 = \left(\frac{r}{2}\right)2^2 = 2r.$$

That is, the mean equals the number of degrees of freedom, and the variance equals twice the number of degrees of freedom. An explanation of "number of degrees of freedom" is given later. From the results concerning the more general gamma distribution, we see that its moment-generating function is

$$M(t) = (1 - 2t)^{-r/2}, \qquad t < \frac{1}{2}.$$

In Figure 3.4-2 the graphs of chi-square p.d.f.s for $r = 2, 3, 5,$ and 8 are given. Note the relationship between the mean, $\mu = r$, and the point at which the p.d.f. obtains its maximum (see Exercise 3.4-13).

Because the chi-square distribution is so important in applications, tables have been prepared giving the values of the distribution function

$$F(x) = \int_0^x \frac{1}{\Gamma(r/2)2^{r/2}} w^{r/2-1} e^{-w/2} \, dw$$

for selected values of r and x. For an example, see Table IV in the Appendix.

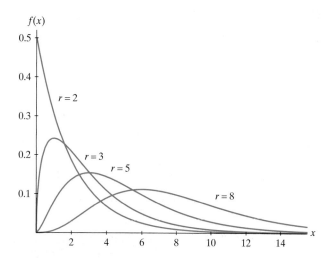

Figure 3.4-2: Chi-square p.d.f.s with $r = 2, 3, 5, 8$

EXAMPLE 3.4-3 Let X have a chi-square distribution with $r = 5$ degrees of freedom. Then, using Table IV in the Appendix,

$$P(1.145 \leq X \leq 12.83) = F(12.83) - F(1.145) = 0.975 - 0.050 = 0.925$$

and

$$P(X > 15.09) = 1 - F(15.09) = 1 - 0.99 = 0.01.$$ ◄

EXAMPLE 3.4-4 If X is $\chi^2(7)$, two constants, a and b, such that

$$P(a < X < b) = 0.95$$

are $a = 1.690$ and $b = 16.01$. Other constants a and b can be found, and we are only restricted in our choices by the limited table. ◄

Probabilities like that of Example 3.4-4 are so important in statistical applications that we use special symbols for a and b. Let α be a positive probability (that is usually less than 0.5) and let X have a chi-square distribution with r degrees of freedom. Then $\chi_\alpha^2(r)$ is a number such that

$$P[X \geq \chi_\alpha^2(r)] = \alpha.$$

That is, $\chi_\alpha^2(r)$ is the $100(1 - \alpha)$ percentile (or upper 100α percent point) of the chi-square distribution with r degrees of freedom. Then the 100α percentile is the number $\chi_{1-\alpha}^2(r)$ such that

$$P[X \leq \chi_{1-\alpha}^2(r)] = \alpha.$$

That is, the probability to the right of $\chi_{1-\alpha}^2(r)$ is $1 - \alpha$ (see Figure 3.4-3).

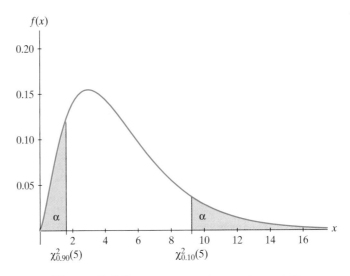

Figure 3.4-3: Chi-square tails, $r = 5, \alpha = 0.10$

EXAMPLE 3.4-5 Let X have a chi-square distribution with five degrees of freedom. Then, using Table IV in the Appendix, $\chi^2_{0.10}(5) = 9.236$ and $\chi^2_{0.90}(5) = 1.610$. These are the points that are indicated on Figure 3.4-3. ◀

EXAMPLE 3.4-6 If customers arrive at a shop on the average of 30 per hour in accordance with a Poisson process, what is the probability that the shopkeeper will have to wait longer than 9.390 minutes for the first nine customers to arrive? Note that the mean rate of arrivals per minute is $\lambda = 1/2$. Thus, $\theta = 2$ and $\alpha = r/2 = 9$. If X denotes the waiting time until the ninth arrival, then X is $\chi^2(18)$. Hence

$$P(X > 9.390) = 1 - 0.05 = 0.95.$$ ◀

EXAMPLE 3.4-7 If X has an exponential distribution with a mean of 2, the p.d.f. of X is

$$f(x) = \frac{1}{2} e^{-x/2} = \frac{x^{2/2-1} e^{-x/2}}{\Gamma(2/2) 2^{2/2}}, \qquad 0 \le x < \infty.$$

That is, X is $\chi^2(2)$. Thus, for illustration,

$$P(0.051 < X < 7.378) = 0.975 - 0.025 = 0.95.$$ ◀

COMPUTATIONAL COMMENTS (CAS) When finding the mean, variance, and moment-generating function, a CAS can often be helpful. We use *Maple* to illustrate this with the gamma distribution. In the following *Maple* code, we first must let *Maple* know the possible values of the parameters. Usually *Maple* denotes assumed variables with a tilde ($\sim$). The first line suppresses this symbol. The mean and variance are calculated using integration. And then the moment-generating function is calculated and it is used to find the mean and the variance.

```
>   interface(showassumed=0);
>   assume(alpha > 0): additionally(theta > 0):
>   f := x^(alpha - 1)*exp(-x/theta)/GAMMA(alpha)/theta^alpha;
```

$$f := \frac{x^{(\alpha-1)}\, e^{(-\frac{x}{\theta})}}{\Gamma(\alpha)\,\theta^{\alpha}}$$

```
>   mu := int(x*f, x = 0 .. infinity);
```

$$\mu := \theta\,\alpha$$

```
>   var := factor(int((x - mu)^2*f, x = 0 .. infinity));
```

$$var := \theta^{2}\,\alpha$$

$$\sigma := \theta\,\sqrt{\alpha}$$

```
>   M := int(exp(t*x)*f, x = 0 .. infinity);
```

$$M := (1 - t\theta)^{(-\alpha)}$$

```
>   Mprime := diff(M, t);
```

$$Mprime := \frac{(1 - t\theta)^{(-\alpha)}\,\alpha\,\theta}{1 - t\theta}$$

```
>   Mdoubleprime := diff(Mprime, t);
```

$$Mdoubleprime := \frac{(1 - t\theta)^{(-\alpha)}\,\alpha^{2}\,\theta^{2}}{(1 - t\theta)^{2}} + \frac{(1 - t\theta)^{(-\alpha)}\,\alpha\,\theta^{2}}{(1 - t\theta)^{2}}$$

```
>   mu := subs(t = 0, Mprime);
```

$$\mu := \theta\,\alpha$$

```
>   var := subs(t = 0, Mdoubleprime) - mu^2;
```

$$var := \theta^{2}\,\alpha$$

It is very easy to find probabilities for the gamma distribution for given values of the parameters.   ■

EXERCISES

3.4-1 Telephone calls enter a college switchboard at a mean rate of 2/3 call per minute according to a Poisson process. Let X denote the waiting time until the tenth call arrives.

(a) What is the p.d.f. of X?

(b) What are the moment-generating function, mean, and variance of X?

3.4-2 If X has a gamma distribution with $\theta = 4$ and $\alpha = 2$, find $P(X < 5)$.

3.4-3 Find the moment-generating function for the gamma distribution with parameters α and θ.

HINT: In the integral representing $E(e^{tX})$, change variables by letting $y = (1 - \theta t)x/\theta$, where $1 - \theta t > 0$.

3.4-4 Use the moment-generating function of a gamma distribution to show that $E(X) = \alpha\theta$ and $\text{Var}(X) = \alpha\theta^2$.

3.4-5 If the moment-generating function of a random variable W is

$$M(t) = (1 - 7t)^{-20},$$

find the p.d.f., mean, and variance of W.

3.4-6 Let X denote the number of alpha particles emitted by barium-133 and observed by a Geiger counter in a fixed position. Assume X has a Poisson distribution and $\lambda = 14.7$ is the mean number of counts per second. Let W denote the waiting time to observe 100 counts. Twenty-five independent observations of W are

6.9	7.3	6.7	6.4	6.3
5.9	7.0	7.1	6.5	7.6
7.2	7.1	6.1	7.3	7.6
7.6	6.7	6.3	5.7	6.7
7.5	5.3	5.4	7.4	6.9

(a) Give the p.d.f., mean, and variance of W.
(b) Calculate the sample mean and sample variance of the 25 observations of W.
(c) Use the relative frequency of event $\{W \le 6.6\}$ to approximate $P(W \le 6.6)$.

3.4-7 The waiting times in minutes until two calls to 911, as reported by the *Holland Sentinel* on November 13 between noon and midnight, were

20	28	81	4	9	41	9	11	10	24	20
44	18	30	16	53	15	38	50	84	44	69

Could these times represent a random sample from a gamma distribution with $\alpha = 2$ and $\theta = 120/7$?

(a) Compare the distribution and sample means.
(b) Compare the distribution and sample variances.
(c) Compare $P(X < 35)$ with the proportion of times that are less than 35.
(d) If possible, make some graphical comparisons.
(e) What is your conclusion?

3.4-8 Let X equal the number of alpha particle emissions of carbon-14 that are counted by a Geiger counter per second. Assume that the distribution of X is Poisson with mean 16. Let W equal the time in seconds before the seventh count is made.

(a) Give the distribution of W.
(b) Find $P(W \le 0.5)$. HINT: Use formula 3.4-1 with $\lambda w = 16$.

3.4-9 If X is $\chi^2(17)$, find

(a) $P(X < 7.564)$.
(b) $P(X > 27.59)$.
(c) $P(6.408 < X < 27.59)$.
(d) $\chi^2_{0.95}(17)$.
(e) $\chi^2_{0.025}(17)$.

3.4-10 If X is $\chi^2(12)$, find constants a and b such that

$$P(a < X < b) = 0.90 \text{ and } P(X < a) = 0.05.$$

3.4-11 If X is $\chi^2(23)$, find the following:

(a) $P(14.85 < X < 32.01)$.

(b) Constants a and b such that $P(a < X < b) = 0.95$ and $P(X < a) = 0.025$.

(c) The mean and variance of X.

(d) $\chi_{0.05}^2(23)$ and $\chi_{0.95}^2(23)$.

3.4-12 If the moment-generating function of X is $M(t) = (1 - 2t)^{-12}, t < 1/2$, find

(a) $E(X)$.

(b) $\text{Var}(X)$.

(c) $P(15.66 < X < 42.98)$.

3.4-13 Let the distribution of X be $\chi^2(r)$.

(a) Find the point at which the p.d.f. of X obtains its maximum when $r \geq 2$. This is the mode of a $\chi^2(r)$ distribution.

(b) Find the points of inflection for the p.d.f. of X.

(c) Use the results of parts (a) and (b) to sketch the p.d.f. of X when $r = 4$ and when $r = 10$.

3.4-14 Cars arrive at a toll booth at a mean rate of five cars every 10 minutes according to a Poisson process. Find the probability that the toll collector will have to wait longer than 26.30 minutes before collecting the eighth toll.

3.4-15 If 15 observations are taken independently from a chi-square distribution with four degrees of freedom, find the probability that at most three of the 15 observations exceed 7.779.

3.4-16 If 10 observations are taken independently from a chi-square distribution with 19 degrees of freedom, find the probability that exactly 2 of the 10 sample items exceed 30.14.

3.5 DISTRIBUTIONS OF FUNCTIONS OF A RANDOM VARIABLE

Let X be a random variable of the continuous type. If we consider a function of X, say $Y = u(X)$, then Y must also be a random variable that has its own distribution. If we can find its distribution function, say

$$G(y) = P(Y \leq y) = P[u(X) \leq y],$$

then its p.d.f. is given by $g(y) = G'(y)$. This **distribution function technique** is now illustrated by two examples.

EXAMPLE 3.5-1   Let X have a gamma distribution with p.d.f.

$$f(x) = \frac{1}{\Gamma(\alpha)\theta^\alpha} x^{\alpha-1} e^{-x/\theta}, \qquad 0 < x < \infty,$$

where $\alpha > 0, \theta > 0$. Let $Y = e^X$ so that the support of Y is $1 < y < \infty$. For each y in the support, the distribution function of Y is

$$G(y) = P(Y \leq y) = P(e^X \leq y) = P(X \leq \ln y).$$

That is,

$$G(y) = \int_0^{\ln y} \frac{1}{\Gamma(\alpha)\theta^\alpha} x^{\alpha-1} e^{-x/\theta} dx$$

and thus the p.d.f. $g(y) = G'(y)$ of Y is

$$g(y) = \frac{1}{\Gamma(\alpha)\theta^\alpha} (\ln y)^{\alpha-1} e^{-(\ln y)/\theta} \left(\frac{1}{y}\right), \qquad 1 < y < \infty.$$

Equivalently, we have

$$g(y) = \frac{1}{\Gamma(\alpha)\theta^\alpha} \frac{(\ln y)^{\alpha-1}}{y^{1+1/\theta}}, \qquad 1 < y < \infty,$$

which is called a **loggamma** p.d.f. See Figure 3.5-1 for some graphs. Note that $\alpha\theta$ and $\alpha\theta^2$ are not the mean and the variance of Y, but of the original random variable $X = \ln Y$. For the loggamma distribution,

$$\mu = \frac{1}{(1-\theta)^\alpha}, \qquad \theta < 1,$$

$$\sigma^2 = \frac{1}{(1-2\theta)^\alpha} - \frac{1}{(1-\theta)^{2\alpha}}, \qquad \theta < \frac{1}{2}. \qquad \blacktriangleleft$$

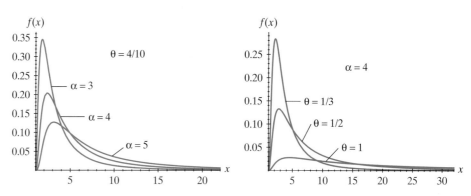

Figure 3.5-1: Loggamma probability density functions

There is another interesting distribution, this one involving a transformation of a uniform random variable.

EXAMPLE 3.5-2   A spinner is mounted at the point $(0, 1)$. Let w be the smallest angle between the y axis and the spinner (see Figure 3.5-2). Assume that w is the value of a random variable W that has a uniform distribution on the interval $(-\pi/2, \pi/2)$. That is, W is $U(-\pi/2, \pi/2)$, and the distribution function of W is

$$P(W \le w) = F(w) = \begin{cases} 0, & -\infty < w < -\dfrac{\pi}{2}, \\[2mm] \left(w + \dfrac{\pi}{2}\right)\left(\dfrac{1}{\pi}\right), & -\dfrac{\pi}{2} \le w < \dfrac{\pi}{2}, \\[2mm] 1, & \dfrac{\pi}{2} \le w < \infty. \end{cases}$$

The relationship between x and w is given by $x = \tan w$; that is, x is the point on the x-axis which is the intersection of that axis and the linear extension of the spinner. To find the distribution of the random variable $X = \tan W$, we see that the distribution function of X is given by

$$G(x) = P(X \le x) = P(\tan W \le x) = P(W \le \arctan x)$$

$$= F(\arctan x) = \left(\arctan x + \frac{\pi}{2}\right)\left(\frac{1}{\pi}\right), \qquad -\infty < x < \infty.$$

The last equality follows because $-\pi/2 < w = \arctan x < \pi/2$. The p.d.f. of X is given by

$$g(x) = G'(x) = \frac{1}{\pi(1 + x^2)}, \qquad -\infty < x < \infty.$$

In Figure 3.5-2 the graph of this **Cauchy p.d.f.** is given. In Exercise 3.5-12 you will be asked to show that $E(X)$ does not exist because the tails of the Cauchy p.d.f. contain too much probability for this p.d.f. to "balance" at $x = 0$.   ◀

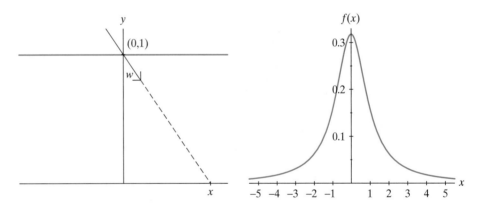

Figure 3.5-2: Spinner and Cauchy p.d.f.

Thus far the examples have illustrated the use of the distribution function technique. By making a simple observation, we can sometimes "short cut" the distribution function technique by using what is frequently called the **change-of-variable technique.** Let X be a continuous-type random variable with p.d.f. $f(x)$ with support $c_1 < x < c_2$. We begin this discussion by taking $Y = u(X)$ as a continuous increasing function of X with the inverse function $X = v(Y)$. Say the support of X, namely $c_1 < x < c_2$, maps onto $d_1 = u(c_1) < y < d_2 = u(c_2)$, the support of Y. Thus the distribution function of Y is

$$G(y) = P(Y \le y) = P[u(X) \le y] = P[X \le v(y)], \qquad d_1 < y < d_2,$$

since u and v are continuous increasing functions. Of course, $G(y) = 0$, $y \leq d_1$, and $G(y) = 1$, $y \geq d_2$. Thus,

$$G(y) = \int_{c_1}^{v(y)} f(x)\,dx, \qquad d_1 < y < d_2.$$

Recall from calculus that the derivative, $G'(y) = g(y)$, of such an expression is given by

$$G'(y) = g(y) = f[v(y)][v'(y)], \qquad d_1 < y < d_2.$$

Of course, $G'(y) = g(y) = 0$ if $y < d_1$ or $y > d_2$. We may let $g(d_1) = g(d_2) = 0$.

For illustration of this change-of-variable technique, let us reconsider Example 3.5-1 with $= e^X$, where X has p.d.f.

$$f(x) = \frac{1}{\Gamma(\alpha)\theta^\alpha}\, x^{\alpha-1} e^{-x/\theta}, \qquad 0 < x < \infty.$$

Here $c_1 = 0$ and $c_2 = \infty$ and thus $d_1 = 1$ and $d_2 = \infty$; also $X = \ln Y = v(Y)$. Since $v'(y) = 1/y$, the p.d.f. of Y is

$$g(y) = \frac{1}{\Gamma(\alpha)\theta^\alpha}\, (\ln y)^{\alpha-1} e^{-(\ln y)/\theta} \left(\frac{1}{y}\right), \qquad 1 < y < \infty,$$

which is the same result as obtained in that example.

Suppose now the function $Y = u(X)$ and its inverse $X = v(Y)$ were continuous decreasing functions. Thus the mapping of $c_1 < x < c_2$ would be $d_1 = u(c_1) > y > d_2 = u(c_2)$.

Since u and v are decreasing functions, we have

$$G(y) = P(Y \leq y) = P[u(X) \leq y] = P[X \geq v(y)] = \int_{v(y)}^{c_2} f(x)\,dx, \ d_2 < y < d_1.$$

Accordingly, from calculus, we have

$$G'(y) = g(y) = f[v(y)][-v'(y)], \qquad d_2 < y < d_1,$$

and $G'(y) = g(y) = 0$, elsewhere. Note in both the increasing and decreasing cases, we could write

$$g(y) = f[v(y)]\,|v'(y)|, \qquad y \in S_y,$$

where S_y is the support of Y found by mapping the support of X, say S_x, onto it. The absolute value $|v'(y)|$ assures that $g(y)$ is nonnegative.

EXAMPLE 3.5-3 Let X have the p.d.f.

$$f(x) = 3(1 - x)^2, \qquad 0 < x < 1.$$

Say $Y = (1-X)^3 = u(X)$, a decreasing function of X. Thus $X = 1 - Y^{1/3} = v(Y)$ and $0 < x < 1$ is mapped onto $0 < y < 1$. Since

$$v'(y) = -\frac{1}{3y^{2/3}},$$

we have

$$g(y) = 3[1 - (1 - y^{1/3})]^2 \left| \frac{-1}{3y^{2/3}} \right|, \qquad 0 < y < 1.$$

That is,

$$g(y) = 3y^{2/3}\left(\frac{1}{3y^{2/3}}\right) = 1, \qquad 0 < y < 1;$$

so $Y = (1-X)^3$ has the uniform distribution $U(0, 1)$.   ◀

As we have seen, it is sometimes easier to use the change-of-variable technique than the distribution function technique. However, there are many occasions that the latter is more convenient to use. As a matter of fact, we had to use the distribution function in finding the gamma distribution from the Poisson process; see Section 3.4. We again use the distribution function technique to prove two theorems involving the uniform distribution.

Theorem 3.5-1   Let Y have a distribution that is $U(0, 1)$. Let $F(x)$ have the properties of a distribution function of the continuous type with $F(a) = 0$, $F(b) = 1$, and suppose that $F(x)$ is strictly increasing on the support $a < x < b$, where a and b could be $-\infty$ and ∞, respectively. Then the random variable X defined by $X = F^{-1}(Y)$ is a continuous-type random variable with distribution function $F(x)$.

Proof: The distribution function of X is

$$P(X \le x) = P[F^{-1}(Y) \le x], \qquad a < x < b.$$

However, since $F(x)$ is strictly increasing, $\{F^{-1}(Y) \le x\}$ is equivalent to $\{Y \le F(x)\}$ and hence

$$P(X \le x) = P[Y \le F(x)], \qquad a < x < b.$$

But Y is $U(0, 1)$; so $P(Y \le y) = y$ for $0 < y < 1$, and accordingly

$$P(X \le x) = P[Y \le F(x)] = F(x), \qquad 0 < F(x) < 1.$$

That is, the distribution function of X is $F(x)$.   ○

We now illustrate how Theorem 3.5-1 can be used to simulate observations from a given distribution in the next two examples.

EXAMPLE 3.5-4
To see how we can simulate observations from an exponential distribution with a mean of $\theta = 10$, note that the distribution function of X is $F(x) = 1 - e^{-x/10}$ when $0 \le x < \infty$. Solving $y = F(x)$ for x yields $x = F^{-1}(y) = -10\ln(1 - y)$. So given independent random numbers $y_1, y_2, \ldots, y_n$ from $U(0, 1)$, we would expect $x_i = -10\ln(1 - y_i)$, $i = 1, 2, \ldots, n$, to represent n observations of an exponential random variable X with mean $\theta = 10$. Table 3.5-1 gives the values of 15 random numbers, y_i, along with the values of $x_i = -10\ln(1 - y_i)$.

Table 3.5-1: Random Exponential Observations	
y	$x = -10\ln(1 - y)$
0.1514	1.6417
0.6697	11.0775
0.0527	0.5414
0.4749	6.4417
0.2900	3.4249
0.2354	2.6840
0.9662	33.8729
0.0043	0.0431
0.1003	1.0569
0.9192	25.1578
0.4971	6.8736
0.7293	13.0674
0.9118	24.2815
0.8225	17.2878
0.5915	8.9526

As a rough check on whether these 15 observations seem to come from an exponential distribution with mean 10, let us construct a q-q plot. We must order these 15 observations and plot them against the corresponding quantiles q_i of the exponential distribution with mean $\theta = 10$. These are the solution of

$$F(q_i) = 1 - e^{-q_i/10} = \frac{i}{16}$$

or, equivalently,

$$q_i = -10\ln\left(1 - \frac{i}{16}\right).$$

The ordered x_i's and the respective quantiles, q_i, are given in Table 3.5-2.

Figure 3.5-3 shows the q-q plot for these data. The linearity of this q-q plot confirms the possibility of the exponential distribution being the correct model. The sample mean is $\bar{x} = 10.4270$ and the sample standard deviation is

Table 3.5-2: Exponential Quantiles	
Ordered x's	**Quantiles**
0.0431	0.6454
0.5414	1.3353
1.0569	2.0764
1.6417	2.8768
2.6840	3.7469
3.4249	4.7000
6.4417	5.7536
6.8736	6.9315
8.9526	8.2668
11.0775	9.8083
13.0674	11.6315
17.2878	13.8629
24.2815	16.7398
25.1578	20.7944
33.8729	27.7259

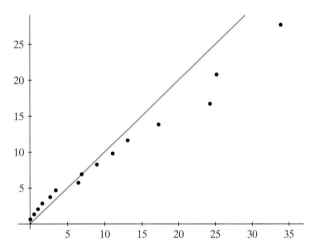

Figure 3.5-3: Exponential q-q plot

$s = 10.4311$, both of which are close to $\theta = 10$. We caution the reader that in other simulations, the fit is sometimes better and sometimes not as good. ◄

EXAMPLE 3.5-5 To help appreciate the large probability in the tails of the Cauchy distribution, it is useful to simulate some observations of a Cauchy random variable. We can first begin with a random number, Y, that is an observation from the $U(0, 1)$ distribution. From the distribution function of X, namely $G(x)$, which is that of the Cauchy distribution given in Example 3.5-2, we have

$$y = G(x) = \left(\arctan x + \frac{\pi}{2}\right)\left(\frac{1}{\pi}\right), \qquad -\infty < x < \infty,$$

or, equivalently,

$$x = \tan\left(\pi y - \frac{\pi}{2}\right). \tag{3.5-1}$$

This latter expression provides an observations of X.

In Table 3.5-3 the values of y are the first 10 random numbers in the last column of Table IX in the Appendix. The corresponding values of x are given by Equation 3.5-1. Although most of these observations from the Cauchy distribution are relatively small in magnitude, we see that a very large value (in magnitude) occurs occasionally. Another way of looking at this situation is by considering sightings (or firing of a gun) from an observation tower, here with coordinates $(0, 1)$, at independent random angles, each with the uniform distribution $U(-\pi/2, \pi/2)$; the target points would then be at Cauchy observations. ◄

Table 3.5-3: Cauchy Observations	
y	x
0.1514	−1.9415
0.6697	0.5901
0.0527	−5.9847
0.4749	−0.0790
0.2900	−0.7757
0.2354	−1.0962
0.9662	9.3820
0.0043	−74.0211
0.1003	−3.0678
0.9192	3.8545

The following probability integral transformation theorem is the converse of Theorem 3.5-1.

Theorem 3.5-2   Let X have the distribution function $F(x)$ of the continuous type that is strictly increasing on the support $a < x < b$. Then the random variable Y, defined by $Y = F(X)$, has a distribution that is $U(0, 1)$.

Proof: Since $F(a) = 0$ and $F(b) = 1$, the distribution function of Y is

$$P(Y \le y) = P[F(X) \le y], \qquad 0 < y < 1.$$

However, $\{F(X) \le y\}$ is equivalent to $\{X \le F^{-1}(y)\}$; thus

$$P(Y \le y) = P[X \le F^{-1}(y)], \qquad 0 < y < 1.$$

Since $P(X \le x) = F(x)$, we have that

$$P(Y \le y) = P[X \le F^{-1}(y)] = F[F^{-1}(y)] = y, \qquad 0 < y < 1,$$

which is the distribution function of a $U(0, 1)$ random variable. $\bigcirc$

Although in our statements and proofs of Theorems 3.5-1 and 3.5-2 we required $F(x)$ to be strictly increasing, this restriction can be dropped, and both theorems are still true. In our exposition we did not want to bother students with certain difficulties that are experienced if $F(x)$ is not strictly increasing.

The change-of-variable technique can be used for a variable X of the discrete type, but there is one major difference since the p.m.f. $f(x) = P(X = x), x \in S_x$, represents probability. Note that the support S_x consists of a countable number of points, say $c_1, c_2, c_3, \ldots$. Let $Y = u(X)$ be a one-to-one transformation with inverse $X = v(Y)$. The function $y = u(x)$ maps S_x onto $d_1 = u(c_1), d_2 = u(c_2), d_3 = u(c_3), \ldots$, which we denote by S_y. Hence the p.m.f. of Y is

$$g(y) = P(Y = y) = P[u(X) = y] = P[X = v(y)], \qquad y \in S_y.$$

Since $P(X = x) = f(x)$, we have $g(y) = f[v(y)], y \in S_y$. Note that, in this discrete case, the value of the derivative, namely $|v'(y)|$, is not needed.

EXAMPLE 3.5-6   Let X have a Poisson distribution with $\lambda = 4$ and thus the p.m.f. is

$$f(x) = \frac{4^x e^{-4}}{x!}, \qquad x = 0, 1, 2, \ldots.$$

If $Y = \sqrt{X}$, then, since $X = Y^2$, we have

$$g(y) = \frac{4^{y^2} e^{-4}}{(y^2)!}, \qquad y = 0, 1, \sqrt{2}, \sqrt{3}, \ldots. \qquad \blacktriangleleft$$

One final observation concerns the situation in which the transformation $Y = u(X)$ is not one-to-one, as it has been up to this point in this section. For illustration let $Y = X^2$, where X is Cauchy. Here $-\infty < x < \infty$ maps onto $0 \le y < \infty$; so

$$G(y) = P(X^2 \le y) = P(-\sqrt{y} \le X \le \sqrt{y})$$

$$= \int_{-\sqrt{y}}^{\sqrt{y}} f(x)\,dx, \, 0 \le y < \infty,$$

where

$$f(x) = \frac{1}{\pi(1 + x^2)}, \qquad -\infty < x < \infty.$$

Thus

$$G'(y) = g(y) = f(\sqrt{y}) \left| \frac{1}{2\sqrt{y}} \right| + f(-\sqrt{y}) \left| \frac{-1}{2\sqrt{y}} \right|$$

$$= \frac{1}{\pi(1 + y)\sqrt{y}}, \qquad 0 \le y < \infty.$$

That is, in this case of a two-to-one transformation, there is a need to sum two terms, each of which are similar to those of the one-to-one case; but here $x_1 = \sqrt{y}$ and $x_2 = -\sqrt{y}, 0 < y < \infty$, give the two inverse functions, respectively.

With careful thought, we can handle many situations which generalize this particular illustration.

EXAMPLE 3.5-7 Let X have the p.d.f.

$$f(x) = \frac{x^2}{3}, \qquad -1 < x < 2.$$

The random variable $Y = X^2$ will have the support $0 \le y < 4$. However, for $0 < y < 1$, we obtain the two-to-one transformation represented by $x_1 = -\sqrt{y}$ for $-1 < x_1 < 0$, and $x_2 = \sqrt{y}$ for $0 < x_2 < 1$. On the other hand, if $1 < y < 4$, the one-to-one transformation is represented by $x_2 = \sqrt{y}, 1 < x_2 < 2$. Since

$$\frac{dx_1}{dy} = \frac{-1}{2\sqrt{y}} \qquad \text{and} \qquad \frac{dx_2}{dy} = \frac{1}{2\sqrt{y}},$$

we have that the p.d.f. of $Y = X^2$ is

$$g(y) = \begin{cases} \dfrac{(-\sqrt{y})^2}{3} \left| \dfrac{-1}{2\sqrt{y}} \right| + \dfrac{(\sqrt{y})^2}{3} \left| \dfrac{1}{2\sqrt{y}} \right| = \dfrac{\sqrt{y}}{3}, & 0 < y < 1, \\[4mm] \dfrac{(\sqrt{y})^2}{3} \left| \dfrac{1}{2\sqrt{y}} \right| = \dfrac{\sqrt{y}}{6}, & 1 < y < 4. \end{cases}$$

Note in case $0 < y < 1$, the p.d.f. is the sum of two terms but only one in case $1 < y < 4$; these different expressions in $g(y)$ correspond to the two types of transformations, namely the two-to-one and the one-to-one transformations, respectively. ◀

EXERCISES

3.5-1 Let X have the p.d.f. $f(x) = 4x^3, 0 < x < 1$. Find the p.d.f. of $Y = X^2$.

3.5-2 Let X have the p.d.f. $f(x) = xe^{-x^2/2}, 0 < x < \infty$. Find the p.d.f. of $Y = X^2$.

3.5-3 Let X have a gamma distribution with $\alpha = 3$ and $\theta = 2$. Determine the p.d.f. of $Y = \sqrt{X}$.

3.5-4 The p.d.f. of X is $f(x) = 2x, 0 < x < 1$.

    (a) Find the distribution function of X.
    (b) Describe how an observation of X can be simulated.
    (c) Simulate 10 observations of X.
    (d) Construct a q-q plot using the ordered observations and the corresponding quantiles for the given distribution.

3.5-5 Simulate 10 independent observations from an exponential distribution with mean $\theta = 3$. Construct a q-q plot of your data using the ordered observations and the corresponding quantiles for the exponential distribution with $\theta = 3$.

3.5-6 Let X have a **logistic distribution** with p.d.f.

$$f(x) = \frac{e^{-x}}{(1 + e^{-x})^2}, \qquad -\infty < x < \infty.$$

Show that

$$Y = \frac{1}{1 + e^{-X}}$$

has a $U(0, 1)$ distribution.

3.5-7 A sum of $50,000 is invested at a rate R, which is selected from a uniform distribution on the interval $(0.03, 0.07)$. Once R is selected, the sum is compounded instantaneously for a year so that $X = 50000\, e^R$ dollars is the amount at the end of that year.

    (a) Find the distribution function and p.d.f. of X.
    (b) Verify that $X = 50000\, e^R$ is defined correctly. HINT: Divide the year into n equal parts, calculate the value of the amount at the end of each part, and then take the limit as $n \to \infty$.

3.5-8 The lifetime (in years) of a product is $Y = 5X^{0.7}$, where X has an exponential distribution with mean one. Find the distribution function and p.d.f. of Y.

3.5-9 Frequently statisticians use the **extreme value distribution** given by the distribution function

$$F(x) = 1 - \exp\left[-e^{(x - \theta_1)/\theta_2}\right], \qquad -\infty < x < \infty.$$

A simple case is when $\theta_1 = 0$ and $\theta_2 = 1$, giving

$$F(x) = 1 - \exp\left[-e^x\right], \qquad -\infty < x < \infty.$$

Let $Y = e^X$ or $X = \ln Y$; so the support of Y is $0 < y < \infty$.

(a) Show that the distribution of Y is exponential.

(b) Find the distribution function and the p.d.f. of Y when $\theta_1 \neq 0$ and $\theta_2 \neq 1$. What is this distribution?

(c) As suggested by its name, the extreme value distribution can be used to model the longest home run, the deepest mine, the greatest flood, and so on. Suppose the length X (in feet) of the maximum of someone's home runs was modeled by an extreme value distribution with $\theta_1 = 550$ and $\theta_2 = 25$. What is the probability that X exceeds 500 feet?

3.5-10 Let X have the uniform distribution $U(-1, 3)$. Find the p.d.f. of $Y = X^2$.

3.5-11 The p.d.f. of X is $f(x) = \theta\, x^{\theta-1}, 0 < x < 1, 0 < \theta < \infty$. Let $Y = -2\theta \ln X$. How is Y distributed?

3.5-12 Let $f(x) = 1/[\pi(1+x^2)], -\infty < x < \infty$, be the p.d.f. of the Cauchy random variable X. Show that $E(X)$ does not exist.

3.5-13 Let X have a Cauchy distribution. Find

(a) $P(X > 1)$.

(b) $P(X > 5)$.

(c) $P(X > 10)$.

3.5-14 Using Table IX in the Appendix or a computer,

(a) Simulate nine observations of a Cauchy random variable.

(b) Find the sample mean and the sample median for your nine observations. Which of these statistics seems to give the better estimate of the center (here zero) of the distribution? Compare your results with those of other students.

3.6* ADDITIONAL MODELS

The binomial, Poisson, gamma, and chi-square models are frequently used in statistics. However, many other interesting and very useful models can be found. We begin with a modification of one of the postulates of an approximate Poisson process as given in Section 2.6. In that definition, the numbers of changes occurring in nonoverlapping intervals are independent, and the probability of at least two changes in a sufficiently small interval is essentially zero. We continue to use these postulates, but now we say that the probability of exactly one change in a sufficiently short interval of length h is approximately λh, *where λ is a nonnegative function of the position of this interval*. To be explicit, say $p(x, w)$ is the probability of x changes in the interval $(0, w), 0 \leq w$. Then the last postulate, in more formal terms, becomes

$$p(x + 1, w + h) - p(x, w) \approx \lambda(w)h,$$

where $\lambda(w)$ is a nonnegative function of w. This means that if we want the approximate probability of zero changes in the interval $(0, w + h)$, we could take from the independence, the probability of zero changes in the interval $(0, w)$ times that of zero changes in the interval $(w, w + h)$. That is,

$$p(0, w + h) \approx p(0, w)[1 - \lambda(w)h]$$

because the probability of one or more changes in $(w, w + h)$ is about equal to $\lambda(w)h$. Equivalently,

$$\frac{p(0, w + h) - p(0, w)}{h} \approx -\lambda(w)p(0, w).$$

Taking limits as $h \to 0$, we have

$$D_w[p(0, w)] = -\lambda(w)p(0, w).$$

That is, the resulting differential equation is

$$\frac{D_w[p(0, w)]}{p(0, w)} = -\lambda(w);$$

and thus

$$\ln p(0, w) = -\int \lambda(w)\, dw + c_1.$$

Therefore,

$$p(0, w) = \exp\left[-\int \lambda(w)\, dw + c_1\right] = c_2 \exp\left[-\int \lambda(w)\, dw\right],$$

where $c_2 = e^{c_1}$. However, the boundary condition of the probability of zero changes in an interval of length zero must be one; that is, $p(0, 0) = 1$. So if we select

$$H(w) = \int \lambda(w)\, dw$$

to be such that $H(0) = 0$, then $c_2 = 1$. That is,

$$p(0, w) = e^{-H(w)},$$

where $H'(w) = \lambda(w)$ and $H(0) = 0$. Hence

$$H(w) = \int_0^w \lambda(t)\, dt.$$

Suppose that we now let the continuous-type random variable W be the interval necessary to produce the first change, then the distribution function of W is

$$G(w) = P(W \le w) = 1 - P(W > w), \qquad 0 \le w.$$

Because zero changes in the interval $(0, w)$ are the same as $W > w$, then

$$G(w) = 1 - p(0, w) = 1 - e^{-H(w)}, \qquad 0 \le w.$$

The p.d.f. of W is

$$g(w) = G'(w) = H'(w)e^{-H(w)} = \lambda(w) \exp\left[-\int_0^w \lambda(t)\, dt\right], \qquad 0 \le w.$$

From this, we see immediately that $\lambda(w)$ in terms of $g(w)$ and $G(w)$ is

$$\lambda(w) = \frac{g(w)}{1 - G(w)}.$$

In many applications of this result, W can be thought of as a random time interval. For example, if one change means "death" or "failure" of the item under consideration, then W is actually the length of life of the item. Usually $\lambda(w)$, which is commonly called the **failure rate** or **force of mortality**, is an increasing function of w. That is, the larger w (the older the item) is, the better chance of failure within a short interval of length h, namely $\lambda(w)h$. As we review the exponential distribution of Section 3.3, we note that there $\lambda(w)$ is a constant; that is, the failure rate or force of mortality does not increase as the item gets older. If this were true in human populations, it would mean that a person 80 years old would have as much chance of living another year as would a person 20 years old (sort of a mathematical "fountain of youth"). However, a constant failure rate (force of mortality) is not the case in most human populations nor in most populations of manufactured items. That is, the failure rate $\lambda(w)$ is usually an increasing function of w. We give two important examples of useful probabilistic models.

EXAMPLE 3.6-1   Let

$$H(w) = \left(\frac{w}{\beta}\right)^{\alpha}, \qquad 0 \le w,$$

so that the failure rate is

$$\lambda(w) = H'(w) = \frac{\alpha w^{\alpha - 1}}{\beta^{\alpha}},$$

where $\alpha > 0$, $\beta > 0$. Then the p.d.f. of W is

$$g(w) = \frac{\alpha w^{\alpha - 1}}{\beta^{\alpha}} \exp\left[-\left(\frac{w}{\beta}\right)^{\alpha}\right], \qquad 0 \le w.$$

Frequently, in engineering, this distribution, with appropriate values of α and β, is excellent for describing the life of a manufactured item. Often α is greater than one but less than five. This p.d.f. is frequently called that of the **Weibull distribution** and, in model fitting, is a strong competitor to the gamma p.d.f. The mean and variance of the Weibull distribution are

$$\mu = \beta \, \Gamma\left(1 + \frac{1}{\alpha}\right),$$

$$\sigma^2 = \beta^2 \left[\Gamma\left(1 + \frac{2}{\alpha}\right) - \left\{\Gamma\left(1 + \frac{1}{\alpha}\right)\right\}^2\right].$$

Some graphs of Weibull p.d.f.s are shown in Figure 3.6-1.   ◀

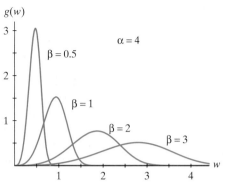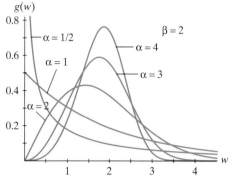

Figure 3.6-1: Weibull probability density functions

EXAMPLE 3.6-2   Persons are often shocked to learn that human mortality increases almost exponentially once a person reaches 25 years of age. Depending on which mortality table is used, one finds that this increase is about 10% each year, which means that the rate of mortality will double about every 7 years. Although this fact can be shocking, we can be thankful that the force of mortality starts very low. The probability that a person in reasonably good health at age 63 dies within the next year is only about 1%. Now, assuming an exponential force of mortality, we have

$$\lambda(w) = H'(w) = ae^{bw}, \qquad a > 0, \quad b > 0.$$

Thus

$$H(w) = \int_0^w ae^{bt}\, dt = \frac{a}{b}e^{bw} - \frac{a}{b}.$$

Thus

$$G(w) = 1 - \exp\left[-\frac{a}{b}e^{bw} + \frac{a}{b}\right], \qquad 0 \le w$$

and

$$g(w) = ae^{bw}\exp\left[-\frac{a}{b}e^{bw} + \frac{a}{b}\right], \qquad 0 \le w$$

are the distribution function and the p.d.f. associated with the famous **Gompertz law** found in actuarial science. Some graphs of p.d.f.s associated with the Gompertz law are shown in Figure 3.6-2. Note that the mode of the Gompertz distribution is $\ln(b/a)/b$. ◀

Both the gamma and Weibull distributions are skewed. In many studies (life testing, response times, incomes, etc.) these are valuable distributions for model selection. Another attractive one is called the loggamma distribution given in Example 3.5-1.

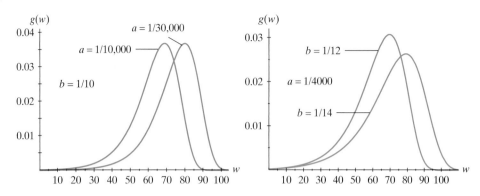

Figure 3.6-2: Gompertz law probability density functions

Thus far we have considered random variables that are either discrete or continuous. In most applications these are the types that are encountered. However, on some occasions, combinations of the two types of random variables are found. That is, in some experiments, positive probability is assigned to each of certain points and also is spread over an interval of outcomes, each point of which has zero probability. An illustration will help clarify these remarks.

EXAMPLE 3.6-3   A bulb for a slide projector is tested by turning it on, letting it burn for 1 hour, and then turning it off. Let X equal the length of time that the bulb performs satisfactorily during this test. There is positive probability that the bulb will burn out when it is turned on; hence

$$0 < P(X = 0) < 1,$$

It could also burn out during the 1-hour time period; thus

$$P(0 < X < 1) > 0$$

with $P(X = x) = 0$ when $x \in (0, 1)$. In addition, $P(X = 1) > 0$. The act of turning it off after 1 hour so that the actual failure time beyond 1 hour is not observed is called censoring, which is considered later in this section.   ◄

The distribution function for a distribution of the **mixed type** will be a combination of those for the discrete and continuous types. That is, at each point of positive probability the distribution function will be discontinuous so that the height of the step there equals the corresponding probability; at all other points the distribution function will be continuous.

EXAMPLE **3.6-4**   Let X have a distribution function $F(x)$ defined by

$$F(x) = \begin{cases} 0, & x < 0, \\[2mm] \dfrac{x^2}{4}, & 0 \le x < 1, \\[2mm] \dfrac{1}{2}, & 1 \le x < 2, \\[2mm] \dfrac{x}{3}, & 2 \le x < 3, \\[2mm] 1, & 3 \le x. \end{cases}$$

This distribution function is depicted in Figure 3.6-3. Probabilities can be computed using $F(x)$; for illustration consider

$$P(0 < X < 1) = \frac{1}{4},$$

$$P(0 < X \le 1) = \frac{1}{2},$$

$$P(X = 1) = \frac{1}{4}. \qquad \blacktriangleleft$$

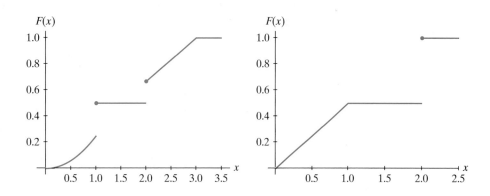

Figure 3.6-3: Mixed distribution functions

EXAMPLE **3.6-5**   Consider the following game. An unbiased coin is tossed. If the outcome is heads, the player receives \$2. If the outcome is tails, the player spins a balanced spinner that has a scale from 0 to 1 and receives that fraction of a dollar associated with the point selected by the spinner. If X denotes the amount received, the space of X is $S = [0, 1) \cup \{2\}$. The distribution function of X is defined by

$$F(x) = \begin{cases} 0, & x < 0, \\ \dfrac{x}{2}, & 0 \le x < 1, \\ \dfrac{1}{2}, & 1 \le x < 2, \\ 1, & 2 \le x. \end{cases}$$

The graph of the distribution function $F(x)$ is given in Figure 3.6-3. ◀

Suppose that the random variable X has a distribution of the mixed type. To find the expectation of the function $u(X)$ of X, a combination of a sum and a Riemann integral is used, as shown in Example 3.6-6.

EXAMPLE 3.6-6 We shall find the mean and variance for the random variable given in Example 3.6-4. Note that there $F'(x) = x/2$ when $0 < x < 1$, and $F'(x) = 1/3$ when $2 < x < 3$; also $P(X = 1) = 1/4$ and $P(X = 2) = 1/6$. Accordingly, we have

$$\mu = E(X) = \int_0^1 x\left(\frac{x}{2}\right) dx + 1\left(\frac{1}{4}\right) + 2\left(\frac{1}{6}\right) + \int_2^3 x\left(\frac{1}{3}\right) dx$$

$$= \left[\frac{x^3}{6}\right]_0^1 + \frac{1}{4} + \frac{1}{3} + \left[\frac{x^2}{6}\right]_2^3 = \frac{19}{12}$$

and

$$\sigma^2 = E(X^2) - [E(X)]^2$$

$$= \int_0^1 x^2\left(\frac{x}{2}\right) dx + 1^2\left(\frac{1}{4}\right) + 2^2\left(\frac{1}{6}\right) + \int_2^3 x^2\left(\frac{1}{3}\right) dx - \left(\frac{19}{12}\right)^2 = \frac{31}{48}. \quad ◀$$

Frequently, in life testing, we know that the length of life, say X, exceeds the number b, but the exact value is unknown. This is called **censoring**. For instance, this can happen when a subject in a cancer study simply disappears; the investigator knows that the subject has lived a certain number of months, but the exact length of life is unknown. Or it might happen when an investigator does not have enough time in an investigation to observe the moments of deaths of all the animals, say rats, in some study. Censoring can also occur in the insurance industry; in particular, consider a loss with a limited-pay policy in which the top amount is exceeded but it is not known by how much.

EXAMPLE 3.6-7 Reinsurance companies are concerned with large losses because they might agree, for illustration, to cover losses due to wind damages that are between

$2 million and $10 million. Say that X equals the size of a wind loss in millions of dollars, and suppose that it has the distribution function

$$F(x) = 1 - \left(\frac{10}{10+x}\right)^3, \qquad 0 \le x < \infty.$$

If losses beyond $10 million are reported only as 10, then $Y = X$, $X \le 10$, and $Y = 10$, $X > 10$, and the distribution function of this censored distribution is

$$G(y) = \begin{cases} 1 - \left(\dfrac{10}{10+y}\right)^3, & 0 \le y < 10, \\[2mm] 1, & 10 \le y < \infty, \end{cases}$$

which has a jump of $[10/(10+10)]^3 = 1/8$ at $y = 10$. ◀

EXAMPLE 3.6-8 A car worth 24 units (unit = $1,000) is insured for a year with a one unit deductible policy. The probability of no damage in a year is 0.95 and the probability of being totaled is 0.01. If the damage is partial, with probability 0.04, this damage follows the p.d.f.

$$f(x) = \frac{25}{24} \frac{1}{(x+1)^2}, \qquad 0 < x < 24.$$

In computing the expected payment, the insurance company recognizes that it will make zero payment if $X \le 1$, $24 - 1 = 23$ if it is totaled, and $X - 1$ if $1 < X < 24$. Thus the expected payment is

$$(0)(0.95) + (0)(0.04) \int_0^1 \frac{25}{24} \frac{1}{(x+1)^2}\, dx + (23)(0.01)$$

$$+ (0.04) \int_1^{24} (x - 1) \frac{25}{24} \frac{1}{(x+1)^2}\, dx.$$

That is, the answer is

$$0.23 + (0.04)(1.67) = 0.297,$$

because the last integral is equal to

$$\int_1^{24} (x + 1 - 2) \frac{25}{24} \frac{1}{(x+1)^2}\, dx = (-2) \int_1^{24} \frac{25}{24} \frac{1}{(x+1)^2}\, dx + \int_1^{24} \frac{25}{24} \frac{1}{(x+1)}\, dx$$

$$= (-2) \left[\frac{25}{24} \frac{-1}{(x+1)} \right]_1^{24} + \frac{25}{24} [\ln(x+1)]_1^{24}$$

$$= 1.67.$$ ◀

EXERCISES

3.6-1 Let the life W (in years) of the usual family car have a Weibull distribution with $\alpha = 2$.

    (a) Show that β must equal 10 for $P(W > 5) = e^{1/4} \approx 0.7788$.
    HINT: $P(W > 5) = e^{-H(5)}$.

    (b) Take a sample of size 8 from the uniform distribution $(0, 1)$ and simulate eight values from the Weibull $(\alpha = 2, \beta = 10)$ distribution. HINT: You may use the random number table. Also recall that if U has that uniform distribution, then $F^{-1}(U)$ has a distribution with distribution function F.

3.6-2 Suppose that the length W of a human life does follow the Gompertz distribution with $\lambda(w) = a(1.1)^w = ae^{(\ln 1.1)w}$, $P(63 < W < 64) = 0.01$. Determine the constant a and $P(W \leq 71 \mid 70 < W)$.

3.6-3 Let Y_1 be the smallest observation of three independent random variables W_1, W_2, W_3, each with a Weibull distribution with parameters α and β. Show that Y_1 has a Weibull distribution. What are the parameters of this latter distribution? HINT:

$$G(y_1) = P(Y_1 \leq y_1) = 1 - P(y_1 < W_i, i = 1, 2, 3) = 1 - [P(y_1 < W_1)]^3.$$

3.6-4 A frequent force of mortality used in actuarial science is $\lambda(w) = ae^{bw} + c$. Find the distribution function and p.d.f. associated with this **Makeham's law**.

3.6-5 From the graph of the first distribution function of X in Figure 3.6-4, determine the indicated probabilities.

    (a) $P(X < 0)$.   **(b)** $P(X < -1)$.   **(c)** $P(X \leq -1)$.

    (d) $P(X < 1)$.   **(e)** $P\left(-1 \leq X < \frac{1}{2}\right)$.   **(f)** $P(-1 < X \leq 1)$.

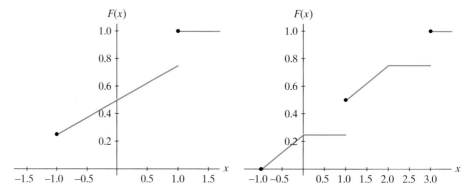

Figure 3.6-4: Mixed distribution functions

3.6-6 Determine the indicated probabilities from the graph of the second distribution function of X in Figure 3.6-4.

    (a) $P\left(-\frac{1}{2} \leq X \leq \frac{1}{2}\right)$.   **(b)** $P\left(\frac{1}{2} < X < 1\right)$.   **(c)** $P\left(\frac{3}{4} < X < 2\right)$.

    (d) $P(X > 1)$.   **(e)** $P(2 < X < 3)$.   **(f)** $P(2 < X \leq 3)$.

3.6-7 Let X be a random variable of the mixed type having the distribution function

$$F(x) = \begin{cases} 0, & x < 0, \\ \dfrac{x^2}{4}, & 0 \le x < 1, \\ \dfrac{x+1}{4}, & 1 \le x < 2, \\ 1, & 2 \le x. \end{cases}$$

(a) Carefully sketch the graph of $F(x)$.
(b) Find the mean and the variance of X.
(c) Find $P(1/4 < X < 1)$, $P(X = 1)$, $P(X = 1/2)$, and $P(1/2 \le X < 2)$.

3.6-8 Find the mean and variance of X if the distribution function of X is

$$F(x) = \begin{cases} 0, & x < 0, \\ 1 - \left(\dfrac{2}{3}\right)e^{-x}, & 0 \le x, \end{cases}$$

3.6-9 Consider the following game. A fair die is rolled. If the outcome is even, the player receives a number of dollars equal to the outcome on the die. If the outcome is odd, a number is selected at random from the interval $[0, 1)$ with a balanced spinner, and the player receives that fraction of a dollar associated with the point selected.

(a) Define and sketch the distribution function of X, the amount received.
(b) Find the expected value of X.
(c) Simulate 10 plays of this game.
(d) Is the average of your 10 plays close to the expected value calculated in part (b)?

3.6-10 Compute the means of both the uncensored and the censored distributions in Example 3.6-7.
HINT: Use integration by parts.

3.6-11 The lifetime X of a certain device has an exponential distribution with mean five years. However, it is not observed on a continuous basis until after three years. Hence we actually observe $Y = \max(X, 3)$. Compute $E(Y)$.

3.6-12 The weekly gravel demand X (in tons) follows the p.d.f.

$$f(x) = \left(\frac{1}{5}\right)e^{-x/5}, \qquad 0 < x < \infty.$$

However, the owner of the gravel pit can only produce at most 4 tons of gravel per week. Compute the expected value of the tons sold per week by the owner.

3.6-13 A loss X on a car has a mixed distribution with $p = 0.95$ on zero and $p = 0.05$ on an exponential distribution with mean of 5000 dollars. If the loss X on a car is greater than the deductible of 500 dollars, the difference $X - 500$ is paid to the owner of the car. Considering zero (if $X \le 500$) as a possible payment, determine the mean and the standard deviation of the payment.

3.6-14 Let X have an exponential distribution with $\theta = 1$; that is, the p.d.f. of X is $f(x) = e^{-x}, 0 < x < \infty$. Let T be defined by $T = \ln X$.

(a) Show that the p.d.f. of T is

$$g(t) = e^t e^{-e^t}, \qquad -\infty < x < \infty,$$

which is the p.d.f. of an extreme value distribution.

(b) Let W be defined by $T = \alpha + \beta \ln W$, where $-\infty < \alpha < \infty$ and $\beta > 0$. Show that W has a Weibull distribution.

CHAPTER

4

MULTIVARIATE DISTRIBUTIONS

4.1 DISTRIBUTIONS OF TWO RANDOM VARIABLES

So far we have taken only one measurement on a single item under observation. However, it is clear in many practical cases that it is possible, and often very desirable, to take more than one measurement of a random observation. Suppose, for illustration, we are observing female college students for some physical characteristics such as height, x, and weight, y, because we are trying to determine a relationship between those two characteristics. For instance, there may be some pattern between height and weight that can be described by an appropriate curve $y = u(x)$. Certainly not all the points observed will be on this curve, but we want to attempt to find the "best" curve to describe the relationship and then say something about the variation of the points around the curve.

Another example might concern high school rank, say x, and the ACT (or SAT) score, say y, of incoming college students. What is the relationship between these two? More important, how can we use these two measurements to predict a third one like first-year college GPA, say z, with a function $z = v(x, y)$? This is a very important problem for college admission offices, particularly when it comes to awarding an athletic scholarship, because the incoming student-athlete must satisfy certain conditions before receiving such an award.

EXAMPLE 4.1-1 Often when we collect data on a pair (x, y), such as $x =$ height (in inches) and $y =$ weight (in pounds) of female college students, we find that the n bivariate observations, $(x_1, y_1), (x_2, y_2), \ldots, (x_n, y_n)$, which frequently plot as in Figure 4.1-1. ◀

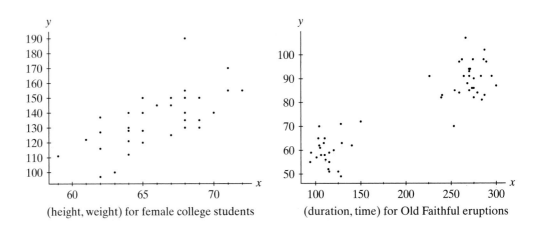

(height, weight) for female college students   (duration, time) for Old Faithful eruptions

Figure 4.1-1: Plots of bivariate data

EXAMPLE 4.1-2 It is always important, as in the univariate case, to plot the data, for often we find interesting patterns. For illustration, refer back to the data in Exercise 3.1-15 on Old Faithful Geyser. Consider $x =$ duration of eruption in seconds and $y =$ time in minutes until the next eruption. We plotted $n = 54$ pairs of such data in Figure 4.1-1. Here we see the interesting pattern in which we obtain two clusters of points: If the eruption is short, then the time until the next one is short; but a long eruption is followed by a long period until the next one. ◀

We can only touch on the consideration of relationships of several random variables in this chapter. There are many advanced courses in statistics covering this area. We hope that you become interested enough that you study some of these multivariate techniques. However, in this and the next few sections, we restrict ourselves to consideration of two random variables and later in the chapter we mention simple extensions to several random variables. We begin with two random variables of the discrete type.

Definition 4.1-1

Let X and Y be two random variables defined on a discrete probability space. Let S denote the corresponding two-dimensional space of X and Y, the two random variables of the discrete type. The probability that $X = x$ and $Y = y$ is denoted by $f(x, y) = P(X = x, Y = y)$. The function $f(x, y)$ is called the **joint probability mass function** (joint p.m.f.) of X and Y and has the following properties:

(a) $0 \leq f(x, y) \leq 1$.

(b) $\displaystyle\sum_{(x,y)\in S}\sum f(x, y) = 1$.

(c) $P[(X, Y) \in A] = \displaystyle\sum_{(x,y)\in A}\sum f(x, y)$, where A is a subset of the space S.

The following example will make this definition more meaningful.

EXAMPLE 4.1-3

Roll a pair of unbiased dice. For each of the 36 sample points with probability 1/36, let X denote the smaller and Y the larger outcome on the dice. For example, if the outcome is $(3, 2)$ then the observed values are $X = 2$, $Y = 3$. The event $\{X = 2, Y = 3\}$ could occur in two ways: $(3, 2)$ or $(2, 3)$; so its probability is

$$\frac{1}{36} + \frac{1}{36} = \frac{2}{36}.$$

If the outcome is $(2, 2)$, then the observed values are $X = 2$, $Y = 2$. Since the event $\{X = 2, Y = 2\}$ can occur in only one way, $P(X = 2, Y = 2) = 1/36$. The joint p.m.f. of X and Y is given by the probabilities

$$f(x, y) = \begin{cases} \dfrac{1}{36}, & 1 \leq x = y \leq 6, \\[2mm] \dfrac{2}{36}, & 1 \leq x < y \leq 6, \end{cases}$$

when x and y are integers. Figure 4.1-2 depicts the probabilities of the various points of the space S. ◀

Notice that certain numbers have been recorded in the bottom and left-hand margins of Figure 4.1-2. These numbers are the respective column and row totals of the probabilities. The column totals are the respective probabilities that X will assume the values in the x space $S_1 = \{1, 2, 3, 4, 5, 6\}$, and the row totals are the respective probabilities that Y will assume the values in the y space $S_2 = \{1, 2, 3, 4, 5, 6\}$. That is, the totals describe probability mass functions of X and Y, respectively. Since each collection of these probabilities is frequently recorded in the margins and satisfies the properties of a p.m.f. of one random variable, each is called a **marginal p.m.f.**

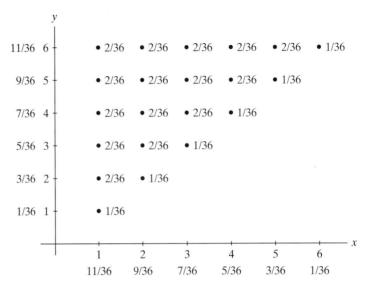

Figure 4.1-2: Discrete joint p.m.f.

Definition 4.1-2   Let X and Y have the joint probability mass function $f(x, y)$ with space S. The probability mass function of X alone, which is called the **marginal probability mass function of X**, is defined by

$$f_1(x) = \sum_y f(x, y) = P(X = x), \qquad x \in S_1,$$

where the summation is taken over all possible y values for each given x in the x space S_1. That is, the summation is over all (x, y) in S with a given x value. Similarly, the **marginal probability mass function of Y** is defined by

$$f_2(y) = \sum_x f(x, y) = P(Y = y), \qquad y \in S_2,$$

where the summation is taken over all possible x values for each given y in the y space S_2. The random variables X and Y are **independent** if and only if

$$P(X = x, Y = y) \equiv P(X = x)P(Y = y)$$

or, equivalently,

$$f(x, y) \equiv f_1(x)f_2(y), \qquad x \in S_1, \qquad y \in S_2;$$

otherwise X and Y are said to be **dependent**.

We note in Example 4.1-2 that X and Y are dependent because there are many x and y values for which $f(x, y) \neq f_1(x) f_2(y)$. For instance,

$$f_1(1) f_2(1) = \left(\frac{11}{36}\right)\left(\frac{1}{36}\right) \neq \frac{1}{36} = f(1, 1).$$

EXAMPLE 4.1-4 Let the joint p.m.f. of X and Y be defined by

$$f(x, y) = \frac{x+y}{21}, \qquad x = 1, 2, 3, \qquad y = 1, 2.$$

Then

$$f_1(x) = \sum_y f(x, y) = \sum_{y=1}^{2} \frac{x+y}{21}$$

$$= \frac{x+1}{21} + \frac{x+2}{21} = \frac{2x+3}{21}, \qquad x = 1, 2, 3;$$

and

$$f_2(y) = \sum_x f(x, y) = \sum_{x=1}^{3} \frac{x+y}{21} = \frac{6+3y}{21}, \qquad y = 1, 2.$$

Note that both $f_1(x)$ and $f_2(y)$ satisfy the properties of a probability mass function. Since $f(x, y) \neq f_1(x) f_2(y)$, X and Y are dependent. ◀

EXAMPLE 4.1-5 Let the joint p.m.f. of X and Y be

$$f(x, y) = \frac{xy^2}{30}, \qquad x = 1, 2, 3, \qquad y = 1, 2.$$

The marginal probability mass functions are

$$f_1(x) = \sum_{y=1}^{2} \frac{xy^2}{30} = \frac{x}{6}, \qquad x = 1, 2, 3,$$

and

$$f_2(y) = \sum_{x=1}^{3} \frac{xy^2}{30} = \frac{y^2}{5}, \qquad y = 1, 2.$$

Then $f(x, y) \equiv f_1(x) f_2(y)$ for $x = 1, 2, 3$ and $y = 1, 2$; thus X and Y are independent. See Figure 4.1-3. ◀

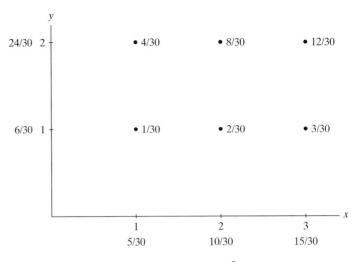

Figure 4.1-3: Joint p.m.f. $f(x, y) = \dfrac{xy^2}{30}, x = 1, 2, 3; y = 1, 2$

EXAMPLE 4.1-6   Let the joint p.m.f. of X and Y be

$$f(x, y) = \frac{xy^2}{13}, \qquad (x, y) = (1, 1), (1, 2), (2, 2).$$

Then the p.m.f. of X is

$$f_1(x) = \begin{cases} \dfrac{5}{13}, & x = 1, \\[2mm] \dfrac{8}{13}, & x = 2, \end{cases}$$

and that of Y is

$$f_2(y) = \begin{cases} \dfrac{1}{13}, & y = 1, \\[2mm] \dfrac{12}{13}, & y = 2. \end{cases}$$

X and Y are dependent because $f(x, y) \neq f_1(x) f_2(y)$ for $x = 1, 2$ and $y = 1, 2$. ◀

Note that in Example 4.1-6 the support S of X and Y is "triangular." Whenever this support S is not "rectangular," the random variables must be dependent because S cannot then equal the product set $\{(x, y): x \in S_1, y \in S_2\}$. That is, if we observe that the support S of X and Y is not a product set, then X and Y must be dependent. For example, in Example 4.1-6, X and Y are dependent because $S = \{(1, 1), (1, 2), (2, 2)\}$ is not a product set. On the other hand, if S equals the product set $\{(x, y): x \in S_1, y \in S_2\}$ and if the formula for

$f(x, y)$ is the product of an expression in x alone and an expression in y alone, then X and Y are independent, as shown in Example 4.1-5. Example 4.1-4 illustrates the fact that the support can be rectangular but the formula for $f(x, y)$ is not such a product and thus X and Y are dependent.

It is possible to define a probability histogram for a joint p.m.f. just like we did for a p.m.f. for a single random variable. Suppose that X and Y have a joint p.m.f. $f(x, y)$ with space S, where S is a set of pairs of integers. At a point (x, y) in S construct a "rectangular column" that is centered at (x, y), has a one unit by one unit base and a height equal to $f(x, y)$. Note that $f(x, y)$ is equal to the "volume" of this rectangular column. Furthermore, the sum of the volumes of the rectangular columns in this probability histogram is equal to one.

EXAMPLE 4.1-7   Let the joint p.m.f. of X and Y be

$$f(x, y) = \frac{xy^2}{30}, \qquad x = 1, 2, 3, \qquad y = 1, 2.$$

The probability histogram is shown in Figure 4.1-4.   ◀

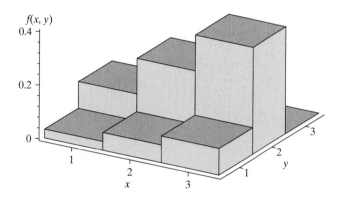

Figure 4.1-4: Joint p.m.f. $f(x, y) = \dfrac{xy^2}{30}, x = 1, 2, 3; y = 1, 2$

Sometimes it is convenient to replace the symbols X and Y representing random variables by X_1 and X_2. This is particularly true in the situations in which we have more than two random variables; so we use X and Y sometimes and then X_1 and X_2 at other times. The reader will see the advantage of the use of subscripts as we go further in the text.

Let X_1 and X_2 be random variables of the discrete type with the joint p.m.f. $f(x_1, x_2)$ on the space S. If $u(X_1, X_2)$ is a function of these two random variables, then

$$E[u(X_1, X_2)] = \sum \sum_{(x_1, x_2) \in S} u(x_1, x_2) f(x_1, x_2),$$

if it exists, is called the **mathematical expectation** (or **expected value**) of $u(X_1, X_2)$.

Remark The same remarks can be made here that were made in the univariate case. Namely,

$$\sum\sum_{(x_1, x_2) \in S} |u(x_1, x_2)| f(x_1, x_2)$$

must converge and be finite. Also $Y = u(X_1, X_2)$ is a random variable, say with p.m.f. $g(y)$ on space S_1, and it is true that

$$\sum\sum_{(x_1, x_2) \in S} u(x_1, x_2) f(x_1, x_2) = \sum_{y \in S_1} y g(y).$$ ■

EXAMPLE 4.1-8   There are eight similar chips in a bowl: three marked $(0, 0)$, two marked $(1, 0)$, two marked $(0, 1)$, and one marked $(1, 1)$. A player selects a chip at random and is given the sum of the two coordinates in dollars. If X_1 and X_2 represent those two coordinates, respectively, their joint p.m.f. is

$$f(x_1, x_2) = \frac{3 - x_1 - x_2}{8}, \qquad x_1 = 0, 1 \qquad \text{and} \qquad x_2 = 0, 1.$$

Thus

$$E(X_1 + X_2) = \sum_{x_2=0}^{1} \sum_{x_1=0}^{1} (x_1 + x_2) \frac{3 - x_1 - x_2}{8}$$

$$= (0)\left(\frac{3}{8}\right) + (1)\left(\frac{2}{8}\right) + (1)\left(\frac{2}{8}\right) + (2)\left(\frac{1}{8}\right) = \frac{3}{4}.$$

That is, the expected payoff is 75¢.   ◄

The following mathematical expectations, subject to their existence, have special names:

   (a) If $u_1(X_1, X_2) = X_i$, then

$$E[u_1(X_1, X_2)] = E(X_i) = \mu_i$$

is called the **mean** of X_i, $i = 1, 2$.

(b) If $u_2(X_1, X_2) = (X_i - \mu_i)^2$, then

$$E[u_2(X_1, X_2)] = E[(X_i - \mu_i)^2] = \sigma_i^2 = \text{Var}(X_i)$$

is called the **variance** of X_i, $i = 1, 2$.

The mean μ_i and the variance σ_i^2 can be computed using the joint p.m.f. $f(x_1, x_2)$ or the marginal p.m.f. $f_i(x_i)$, $i = 1, 2$.

The idea of joint distributions of two random variables of the discrete type can be extended to that of two random variables of the continuous type. The definitions are really the same except that integrals replace summations. The **joint probability density function** (joint p.d.f.) of two continuous-type random variables is an integrable function $f(x, y)$ with the following properties:

(a) $f(x, y) \geq 0$.

(b) $\int_{-\infty}^{\infty} \int_{-\infty}^{\infty} f(x, y) \, dx \, dy = 1$.

(c) $P[(X, Y) \in A] = \iint_A f(x, y) \, dx \, dy$, where $\{(X, Y) \in A\}$ is an event defined in the plane.

Property (c) implies that $P[(X, Y) \in A]$ is the volume of the solid over the region A in the xy plane and bounded by the surface $z = f(x, y)$.

EXAMPLE 4.1-9   Let X and Y have the joint p.d.f.

$$f(x, y) = \frac{3}{2} x^2 (1 - |y|), \qquad -1 < x < 1, \qquad -1 < y < 1.$$

The graph of $z = f(x, y)$ is given in Figure 4.1-5. Let $A = \{(x, y) : 0 < x < 1, 0 < y < x\}$. The probability that (X, Y) falls in A is given by

$$P[(X, Y) \in A] = \int_0^1 \int_0^x \frac{3}{2} x^2 (1 - y) \, dy \, dx = \int_0^1 \frac{3}{2} x^2 \left[y - \frac{y^2}{2} \right]_0^x dx$$

$$= \int_0^1 \frac{3}{2} \left(x^3 - \frac{x^4}{2} \right) dx = \frac{3}{2} \left[\frac{x^4}{4} - \frac{x^5}{10} \right]_0^1$$

$$= \frac{9}{40}. \qquad\qquad \blacktriangleleft$$

The respective **marginal p.d.f.s** of continuous-type random variables X and Y are given by

$$f_1(x) = \int_{-\infty}^{\infty} f(x, y) \, dy, \qquad x \in S_1,$$

and

$$f_2(y) = \int_{-\infty}^{\infty} f(x, y) \, dx, \qquad y \in S_2,$$

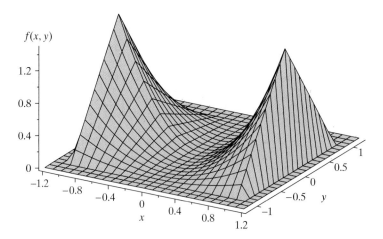

Figure 4.1-5: Joint p.d.f. $f(x, y) = \dfrac{3}{2} x^2 (1 - |y|)$

where S_1 and S_2 are the spaces of X and Y. The definitions associated with mathematical expectations are the same as those associated with the discrete case after replacing the summations with integrations.

For the random variables X and Y in Example 4.1-9, the marginal p.d.f.s are

$$f_1(x) = \int_{-1}^{1} \frac{3}{2} x^2 (1 - |y|)\, dy = \frac{3}{2} x^2, \qquad -1 < x < 1,$$

and

$$f_2(y) = \int_{-1}^{1} \frac{3}{2} x^2 (1 - |y|)\, dx = 1 - |y|, \qquad -1 < y < 1.$$

EXAMPLE 4.1-10   Let X and Y have the joint p.d.f.

$$f(x, y) = 2, \qquad 0 \le x \le y \le 1.$$

Then $S = \{(x, y) : 0 \le x \le y \le 1\}$ is the support and, for illustration,

$$P\left(0 \le X \le \frac{1}{2},\, 0 \le Y \le \frac{1}{2}\right) = P\left(0 \le X \le Y,\, 0 \le Y \le \frac{1}{2}\right)$$

$$= \int_{0}^{1/2} \int_{0}^{y} 2\, dx\, dy = \int_{0}^{1/2} 2y\, dy = \frac{1}{4}.$$

You should draw a figure to illustrate the set of points for which $f(x, y) > 0$ and then shade the region over which the integral is taken. The given probability is the volume above this shaded region under the surface $z = 2$. The marginal p.d.f.s are given by

$$f_1(x) = \int_x^1 2 \, dy = 2(1 - x), \qquad 0 \le x \le 1,$$

and

$$f_2(y) = \int_0^y 2 \, dx = 2y, \qquad 0 \le y \le 1.$$

Three illustrations of expected values are

$$E(X) = \int_0^1 \int_x^1 2x \, dy \, dx = \int_0^1 2x(1 - x) \, dx = \frac{1}{3},$$

$$E(Y) = \int_0^1 \int_0^y 2y \, dx \, dy = \int_0^1 2y^2 \, dy = \frac{2}{3},$$

$$E(Y^2) = \int_0^1 \int_0^y 2y^2 \, dx \, dy = \int_0^1 2y^3 \, dy = \frac{1}{2}.$$

From these calculations we see that $E(X)$, $E(Y)$, and $E(Y^2)$ could be calculated using the marginal p.d.f.s as well as the joint one.   ◄

The definition of independent random variables of the continuous type carries over naturally from the discrete case. That is, X and Y are **independent** if and only if the joint p.d.f. factors into the product of their marginal p.d.f.s; namely,

$$f(x, y) \equiv f_1(x) f_2(y), \qquad x \in S_1, \ y \in S_2.$$

Thus the random variables X and Y in Example 4.1-9 are independent. In addition, the rules that allow us to easily determine dependent and independent random variables are also valid here. So X and Y in Example 4.1-10 are dependent because the support S is not a product space, since it is bounded by the diagonal line $y = x$.

We give extensions of two important univariate distributions, the hypergeometric and the binomial, through examples.

EXAMPLE 4.1-11 Consider a population of 200 students who have just finished a first course in calculus. Of these 200, 40 have earned As, 60 Bs, and 100 Cs, Ds, or Fs. A sample of size 25 is taken at random and without replacement from this population in a way that each possible sample has probability

$$\frac{1}{\binom{200}{25}}$$

of being selected. Within the sample of 25, let X_1 be the number of A students, X_2 the number of B students, $25 - X_1 - X_2$ the number of other students. The space S of (X_1, X_2) is defined by the collection of nonnegative integers (x_1, x_2) such that $x_1 + x_2 \le 25$. The joint p.m.f. of X_1, X_2 is

$$f(x_1, x_2) = \frac{\binom{40}{x_1}\binom{60}{x_2}\binom{100}{25 - x_1 - x_2}}{\binom{200}{25}},$$

for $(x_1, x_2) \in S$, where it is understood that $\binom{k}{j} = 0$ if $j > k$. Without actually summing, we know that the marginal p.m.f. of X_1 is

$$f_1(x_1) = \frac{\binom{40}{x_1}\binom{160}{25 - x_1}}{\binom{200}{25}}, \qquad x_1 = 0, 1, 2, \ldots, 25,$$

since X_1 alone has a hypergeometric distribution. Of course, the function $f_2(x_2)$ is a hypergeometric p.m.f. and

$$f(x_1, x_2) \ne f_1(x_1) f_2(x_2);$$

so X_1 and X_2 are dependent. Note also that the space S is not "rectangular," which would also imply that the random variables are dependent.   ◄

We now extend the binomial distribution to the trinomial distribution. Here we have three mutually exclusive and exhaustive ways for an experiment to terminate: say, perfect, "seconds," and defective. We repeat the experiment n independent times and the probabilities $p_1, p_2, p_3 = 1 - p_1 - p_2$ of perfect, seconds, and defective remain the same from trial to trial. In the n trials, let $X_1 =$ number of perfect items, $X_2 =$ number of seconds, and $X_3 = n - X_1 - X_2 =$ number of defectives. If x_1 and x_2 are nonnegative integers such that $x_1 + x_2 \le n$, then the probability of having x_1 perfects, x_2 seconds, and $n - x_1 - x_2$ defectives in that order is

$$p_1^{x_1} p_2^{x_2} (1 - p_1 - p_2)^{n - x_1 - x_2}.$$

However, if we want $P(X_1 = x_1, X_2 = x_2)$, then we recognize that $X_1 = x_1, X_2 = x_2$ can be achieved in

$$\binom{n}{x_1, x_2, n - x_1 - x_2} = \frac{n!}{x_1! x_2! (n - x_1 - x_2)!}$$

different ways. Hence, the **trinomial** p.m.f. is

$$f(x_1, x_2) = P(X_1 = x_1, X_2 = x_2)$$

$$= \frac{n!}{x_1! x_2! (n - x_1 - x_2)!} \, p_1^{x_1} \, p_2^{x_2} (1 - p_1 - p_2)^{n - x_1 - x_2}.$$

Without summing, we know that X_1 is $b(n, p_1)$ and X_2 is $b(n, p_2)$, and thus X_1 and X_2 are dependent as the product of these marginal probability mass functions is not equal to $f(x_1, x_2)$.

EXAMPLE 4.1-12   Let X and Y have a trinomial distribution with parameters $p_1 = 1/5$, $p_2 = 2/5$, and $n = 5$. The probability histogram for the joint p.m.f. of X and Y is shown in Figure 4.1-6. ◀

EXAMPLE 4.1-13   In manufacturing a certain item, it is found that in normal production about 95% of the items are good ones, 4% are "seconds," and 1% are defective. This particular company has a program of quality control by statistical method; and each hour an on-line inspector observes 20 items selected at random, counting the number X of seconds, and the number Y of defectives. If, in fact, the production is normal, we shall find the probability that in this sample of size $n = 20$, at least two seconds or at least two defective items are found. If we let $A = \{(x, y): x \geq 2 \text{ or } y \geq 2\}$, then

$$P(A) = 1 - P(A')$$

$$= 1 - P(X = 0 \text{ or } 1 \text{ and } Y = 0 \text{ or } 1)$$

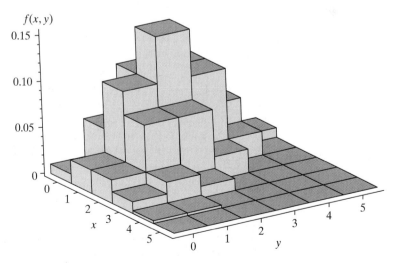

Figure 4.1-6: Trinomial distribution, $p_1 = 1/5$, $p_2 = 2/5$, and $n = 5$

$$= 1 - \frac{20!}{0!0!20!}(0.04)^0(0.01)^0(0.95)^{20} - \frac{20!}{1!0!19!}(0.04)^1(0.01)^0(0.95)^{19}$$

$$- \frac{20!}{0!1!19!}(0.04)^0(0.01)^1(0.95)^{19} - \frac{20!}{1!1!18!}(0.04)^1(0.01)^1(0.95)^{18}$$

$$= 0.204. \qquad \blacktriangleleft$$

EXERCISES

4.1-1 Let the joint p.m.f. of X and Y be defined by

$$f(x, y) = \frac{x + y}{32}, \qquad x = 1, 2, \ y = 1, 2, 3, 4.$$

Find

(a) $f_1(x)$, the marginal p.m.f. of X.
(b) $f_2(y)$, the marginal p.m.f. of Y.
(c) $P(X > Y)$.
(d) $P(Y = 2X)$.
(e) $P(X + Y = 3)$.
(f) $P(X \leq 3 - Y)$.
(g) Are X and Y independent or dependent?

4.1-2 Roll a pair of four-sided dice, one red and one black. Let X equal the outcome on the red die, and let Y equal the outcome on the black die.

(a) On graph paper, show the space of X and Y.
(b) Define the joint p.m.f. on the space (similar to Figure 4.1-2).
(c) Give the marginal p.m.f. of X in the margin.
(d) Give the marginal p.m.f. of Y in the margin.
(e) Are X and Y dependent or independent? Why?

4.1-3 Roll a pair of four-sided dice, one red and one black. Let X equal the outcome on the red die and let Y equal the sum of the two dice.

(a) On graph paper, describe the space of X and Y.
(b) Define the joint p.m.f. on the space (similar to Figure 4.1-2).
(c) Give the marginal p.m.f. of X in the margin.
(d) Give the marginal p.m.f. of Y in the margin.
(e) Are X and Y dependent or independent? Why?

4.1-4 The torque to remove bolts in a steel plate is rated as very high, high, average, and low and these occur about 30%, 40%, 20%, and 10% of the time, respectively. Suppose $n = 25$ bolts are rated; what is the probability of rating 7 very high, 8 high, 6 average, and 4 low? Assume independence.

4.1-5 Two construction companies make bids of X and Y (in $100,000's) on a remodeling project. The joint p.d.f. of X and Y is uniform on the space $2 < x < 2.5, 2 < y < 2.3$. If X and Y are within 0.1 of each other, the companies will be asked to rebid; otherwise the low bidder will be awarded the contract. What is the probability that they will be asked to rebid?

4.1-6 In a smoking survey among boys between the ages of 12 and 17, 78% prefer to date nonsmokers, 1% prefer to date smokers, and 21% don't care. Suppose seven such boys are selected randomly and X equals the number who prefer to date nonsmokers and Y equals the number who prefer to date smokers.

(a) Determine the joint p.m.f. of X and Y. Be sure to include the support of the p.m.f.

(b) Find the marginal p.m.f. of X. Again include the support.

4.1-7 A manufactured item is classified as good, a "second," or defective, with probabilities $6/10$, $3/10$, and $1/10$, respectively. Fifteen such items are selected at random from the production line. Let X denote the number of good items, Y the number of seconds, and $15 - X - Y$ the number of defective items.

(a) Give the joint p.m.f. of X and Y, $f(x, y)$.

(b) Sketch the set of points for which $f(x, y) > 0$. From the shape of this region, can X and Y be independent? Why?

(c) Find $P(X = 10, Y = 4)$.

(d) Give the marginal p.m.f. of X.

(e) Find $P(X \le 11)$.

4.1-8 Let $f(x, y) = 3/2, x^2 \le y \le 1, 0 \le x \le 1$, be the joint p.d.f. of X and Y. Find

(a) $P(0 \le X \le 1/2)$.

(b) $P(1/2 \le Y \le 1)$.

(c) $P(1/2 \le X \le 1, 1/2 \le Y \le 1)$.

(d) $P(X \ge 1/2, Y \ge 1/2)$.

(e) Are X and Y independent?

4.1-9 Let $f(x, y) = 2e^{-x-y}, 0 \le x \le y < \infty$, be the joint p.d.f. of X and Y. Find $f_1(x)$ and $f_2(y)$, the marginal p.d.f.s of X and Y, respectively. Are X and Y independent?

4.1-10 Let X and Y have the joint p.d.f. $f(x, y) = x + y, 0 \le x \le 1, 0 \le y \le 1$.

(a) Find the marginal p.d.f.s $f_1(x)$ and $f_2(y)$ and show that $f(x, y) \neq f_1(x)f_2(y)$. Thus X and Y are dependent. Compute

(b) μ_X.

(c) μ_Y.

(d) σ_X^2.

(e) σ_Y^2.

4.1-11 Let $f(x, y) = (3/16)xy^2, 0 \le x \le 2, 0 \le y \le 2$, be the joint p.d.f. of X and Y. Find $f_1(x)$ and $f_2(y)$, the marginal probability density functions. Are the two random variables independent?

4.1-12 Let T_1 and T_2 be random times for a company to complete two steps in a certain process. Say T_1 and T_2 are measured in days and they have the joint p.d.f. that is uniform over the space $1 < t_1 < 10, 2 < t_2 < 6, t_1 + 2t_2 < 14$. What is the $P(T_1 + T_2 > 10)$?

4.2 THE CORRELATION COEFFICIENT

In Section 4.1 we introduced the mathematical expectation of a function of two random variables, say X_1, X_2. We gave the respective special names of mean and variance of X_i to

$$\mu_i = E(X_i) \qquad \text{and} \qquad \sigma_i^2 = E[(X_i - \mu_i)^2], \qquad i = 1, 2.$$

We introduce two more special names:

(a) If $u_3(X_1, X_2) = (X_1 - \mu_1)(X_2 - \mu_2)$, then

$$E[u_3(X_1, X_2)] = E[(X_1 - \mu_1)(X_2 - \mu_2)] = \sigma_{12} = \text{Cov}(X_1, X_2)$$

is called the **covariance** of X_1 and X_2.

(b) If the standard deviations σ_1 and σ_2 are positive, then

$$\rho = \frac{\text{Cov}(X_1, X_2)}{\sigma_1 \sigma_2} = \frac{\sigma_{12}}{\sigma_1 \sigma_2}$$

is called the **correlation coefficient** of X_1 and X_2.

It is convenient that the mean and the variance of X_1 can be computed from either the joint p.m.f. or the marginal p.m.f. of X_1. For example, in the discrete case,

$$\mu_1 = E(X_1) = \sum_{x_1} \sum_{x_2} x_1 f(x_1, x_2)$$

$$= \sum_{x_1} x_1 \left[\sum_{x_2} f(x_1, x_2) \right] = \sum_{x_1} x_1 f_1(x_1).$$

However, to compute the covariance, we need the joint p.m.f.

Before considering the meaning of the covariance and the correlation coefficient, let us note a few simple facts. First,

$$E[(X_1 - \mu_1)(X_2 - \mu_2)] = E(X_1 X_2 - \mu_1 X_2 - \mu_2 X_1 + \mu_1 \mu_2)$$

$$= E(X_1 X_2) - \mu_1 E(X_2) - \mu_2 E(X_1) + \mu_1 \mu_2$$

because it is true that even in the multivariate situation, E is still a linear or distributive operator (see Exercise 4.2-4). Thus

$$\text{Cov}(X_1, X_2) = E(X_1 X_2) - \mu_1 \mu_2 - \mu_2 \mu_1 + \mu_1 \mu_2 = E(X_1 X_2) - \mu_1 \mu_2.$$

Since $\rho = \text{Cov}(X_1, X_2)/\sigma_1 \sigma_2$, we also have

$$E(X_1 X_2) = \mu_1 \mu_2 + \rho \sigma_1 \sigma_2.$$

That is, the expected value of the product of two random variables is equal to the product $\mu_1 \mu_2$ of their expectations plus their covariance $\rho \sigma_1 \sigma_2$.

A simple example at this point would be helpful.

EXAMPLE 4.2-1 Let X_1 and X_2 have the joint p.m.f.

$$f(x_1, x_2) = \frac{x_1 + 2x_2}{18}, \qquad x_1 = 1, 2, \qquad x_2 = 1, 2.$$

The marginal probability mass functions are, respectively,

$$f_1(x_1) = \sum_{x_2=1}^{2} \frac{x_1 + 2x_2}{18} = \frac{2x_1 + 6}{18}, \qquad x_1 = 1, 2,$$

and

$$f_2(x_2) = \sum_{x_1=1}^{2} \frac{x_1 + 2x_2}{18} = \frac{3 + 4x_2}{18}, \qquad x_2 = 1, 2.$$

Since $f(x_1, x_2) \neq f_1(x_1)f_2(x_2)$, X_1 and X_2 are dependent. The mean and the variance of X_1 are

$$\mu_1 = \sum_{x_1=1}^{2} x_1 \frac{2x_1 + 6}{18} = (1)\left(\frac{8}{18}\right) + (2)\left(\frac{10}{18}\right) = \frac{14}{9},$$

and

$$\sigma_1^2 = \sum_{x_1=1}^{2} x_1^2 \frac{2x_1 + 6}{18} - \left(\frac{14}{9}\right)^2 = \frac{24}{9} - \frac{196}{81} = \frac{20}{81}.$$

The mean and the variance of X_2 are

$$\mu_2 = \sum_{x_2=1}^{2} x_2 \frac{3 + 4x_2}{18} = (1)\left(\frac{7}{18}\right) + (2)\left(\frac{11}{18}\right) = \frac{29}{18},$$

and

$$\sigma_2^2 = \sum_{x_2=1}^{2} x_2^2 \frac{3 + 4x_2}{18} - \left(\frac{29}{18}\right)^2 = \frac{51}{18} - \frac{841}{324} = \frac{77}{324}.$$

The covariance of X_1 and X_2 is

$$\mathrm{Cov}(X_1, X_2) = \sum_{x_2=1}^{2} \sum_{x_1=1}^{2} x_1 x_2 \frac{x_1 + 2x_2}{18} - \left(\frac{14}{9}\right)\left(\frac{29}{18}\right)$$

$$= (1)(1)\left(\frac{3}{18}\right) + (2)(1)\left(\frac{4}{18}\right) + (1)(2)\left(\frac{5}{18}\right)$$

$$+ (2)(2)\left(\frac{6}{18}\right) - \left(\frac{14}{19}\right)\left(\frac{29}{18}\right)$$

$$= \frac{45}{18} - \frac{406}{162} = -\frac{1}{162}.$$

Hence the correlation coefficient is

$$\rho = \frac{-1/162}{\sqrt{(20/81)(77/324)}} = \frac{-1}{\sqrt{1540}} = -0.025.$$

◀

Insight into the correlation coefficient ρ of two discrete random variables X and Y may be gained by thoughtfully examining its definition

$$\rho = \frac{\sum_x \sum_y (x - \mu_X)(y - \mu_Y)f(x, y)}{\sigma_X \sigma_Y},$$

where $\mu_X, \mu_Y, \sigma_X,$ and σ_Y denote the respective means and standard deviations. If positive probability is assigned to pairs (x, y) in which both x and y are either simultaneously above or simultaneously below their respective means, the corresponding terms in the summation that defines ρ are positive because both factors $(x - \mu_X)$ and $(y - \mu_Y)$ will be positive or both will be negative. If pairs (x, y), which yield large positive products $(x - \mu_X)(y - \mu_Y)$, contain most of the probability of the distribution, the correlation coefficient will tend to be positive. If, on the other hand, the points (x, y), in which one component is below its mean and the other above its mean, have most of the probability, then the coefficient of correlation will tend to be negative because the products $(x - \mu_X)(y - \mu_Y)$ are negative (see Exercise 4.2-6). This interpretation of the sign of the correlation coefficient will play an important role in subsequent work.

To gain additional insight into the meaning of the correlation coefficient ρ, consider the following problem. Think of the points (x, y) in the space S and their corresponding probabilities. Let us consider all possible lines in two-dimensional space, each with finite slope, that pass through the point associated with the means, namely (μ_X, μ_Y). These lines are of the form $y - \mu_Y = b(x - \mu_X)$ or, equivalently, $y = \mu_Y + b(x - \mu_X)$. For each point in S, say (x_0, y_0) so that $f(x_0, y_0) > 0$, consider the vertical distance from that point to one of these lines. Since y_0 is the height of the point above the x axis and $\mu_Y + b(x_0 - \mu_X)$ is the height of the point on the line that is directly above or below the point (x_0, y_0), then the absolute value of the difference of these two heights is the vertical distance from point (x_0, y_0) to the line $y = \mu_Y + b(x - \mu_X)$ That is, the required distance is $|y_0 - \mu_Y - b(x_0 - \mu_X)|$. Let us now square this distance and take the weighted average of all such squares; that is, let us consider the mathematical expectation

$$E\{[(Y - \mu_Y) - b(X - \mu_X)]^2\} = K(b).$$

The problem is to find that line (or that b) which minimizes this expectation of the square $\{Y - \mu_Y - b(X - \mu_X)\}^2$. This is an application of the principle of least squares, and the line is sometimes called the least squares regression line. The solution of the problem is very easy, since

$$K(b) = E\{(Y - \mu_Y)^2 - 2b(X - \mu_X)(Y - \mu_Y) + b^2(X - \mu_X)^2\}$$
$$= \sigma_Y^2 - 2b\rho\sigma_X\sigma_Y + b^2\sigma_X^2,$$

because E is a linear operator and $E[(X - \mu_X)(Y - \mu_Y)] = \rho\sigma_X\sigma_Y$. Accordingly, the derivative

$$K'(b) = -2\rho\sigma_X\sigma_Y + 2b\sigma_X^2$$

equals zero at $b = \rho\sigma_Y/\sigma_X$; and we see that $K(b)$ obtains its minimum for that b since $K''(b) = 2\sigma_X^2 > 0$. Consequently the **least squares regression line** (the line of the given form that is the best fit in the foregoing sense) is

$$y = \mu_Y + \rho\frac{\sigma_Y}{\sigma_X}(x - \mu_X).$$

Of course, if $\rho > 0$, the slope of the line is positive; but if $\rho < 0$, the slope is negative.

It is also instructive to note the value of the minimum of

$$K(b) = E\{[(Y - \mu_Y) - b(X - \mu_X)]^2\} = \sigma_Y^2 - 2b\rho\sigma_X\sigma_Y + b^2\sigma_X^2.$$

It is

$$K\left(\rho\frac{\sigma_Y}{\sigma_X}\right) = \sigma_Y^2 - 2\rho\frac{\sigma_Y}{\sigma_X}\rho\sigma_X\sigma_Y + \left(\rho\frac{\sigma_Y}{\sigma_X}\right)^2\sigma_X^2$$

$$= \sigma_Y^2 - 2\rho^2\sigma_Y^2 + \rho^2\sigma_Y^2 = \sigma_Y^2(1 - \rho^2).$$

Since $K(b)$ is the expected value of a square, it must be nonnegative for all b, and we see that $\sigma_Y^2(1 - \rho^2) \geq 0$; that is $\rho^2 \leq 1$, and hence $-1 \leq \rho \leq 1$, which is an important property of the correlation coefficient ρ. If $\rho = 0$, then $K(\rho\sigma_Y/\sigma_X) = \sigma_Y^2$; on the other hand, $K(\rho\sigma_Y/\sigma_X)$ is relatively small if ρ is close to 1 or negative 1. That is, the vertical deviations of the points with positive probability from the line $y = \mu_Y + \rho(\sigma_Y/\sigma_X)(x - \mu_X)$ are small if ρ is close to 1 or negative 1 because $K(\rho\sigma_Y/\sigma_X)$ is the expectation of the square of those deviations. Thus ρ measures, in this sense, the amount of *linearity* in the probability distribution. As a matter of fact, in the discrete case, all the points of positive probability lie on this straight line if and only if ρ is equal to 1 or -1.

Remark More generally, we could have fitted the line $y = a + bx$ by the same application of the principle of least squares. We would then have proved that the "best" line actually passes through the point (μ_X, μ_Y). Recall that in the discussion above we assumed our line to be of that form. Students will find this derivation to be an interesting exercise using partial derivatives (see Exercise 4.2-5). ∎

The following example illustrates a joint discrete distribution for which ρ is negative. In Figure 4.2-1 the line of best fit or the least squares regression line is also drawn.

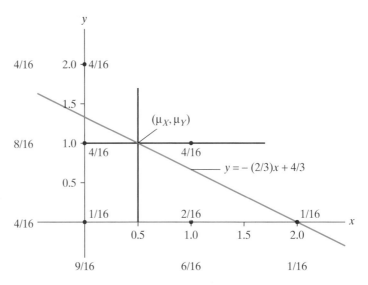

Figure 4.2-1: Trinomial distribution

EXAMPLE 4.2-2

Let X equal the number of ones and Y the number of twos and threes when a pair of fair four-sided dice are rolled. Then X and Y have a trinomial distribution with joint p.m.f.

$$f(x, y) = \frac{2!}{x!\,y!\,(2 - x - y)!}\left(\frac{1}{4}\right)^x \left(\frac{2}{4}\right)^y \left(\frac{1}{4}\right)^{2-x-y}, \qquad 0 \le x + y \le 2,$$

where x and y are nonnegative integers. Since the marginal p.m.f. of X is $b(2, 1/4)$ and the marginal p.m.f. of Y is $b(2, 1/2)$, we know that $\mu_X = 1/2$, $\text{Var}(X) = 6/16$, $\mu_Y = 1$, and $\text{Var}(Y) = 1/2$. Since $E(XY) = (1)(1)(4/16) = 4/16$, $\text{Cov}(X, Y) = 4/16 - (1/2)(1) = -4/16$; therefore, the correlation coefficient is $\rho = -1/\sqrt{3}$. Using these values for the parameters, we obtain the line of best fit, namely

$$y = 1 + \left(-\frac{1}{\sqrt{3}}\right)\sqrt{\frac{1/2}{3/8}}\left(x - \frac{1}{2}\right) = -\frac{2}{3}x + \frac{4}{3}.$$

The joint p.m.f. is displayed in Figure 4.2-1 along with the line of best fit. ◀

Suppose that X and Y are independent so that $f(x, y) \equiv f_1(x)f_2(y)$ and we want to find the expected value of the product $u(X)v(Y)$. Subject to the existence of the expectations, we know that

$$E[u(X)v(Y)] = \sum_{S_1}\sum_{S_2} u(x)v(y)f(x, y)$$

$$= \sum_{S_1}\sum_{S_2} u(x)v(y)f_1(x)f_2(y)$$

$$= \sum_{S_1} u(x)f_1(x) \sum_{S_2} v(y)f_2(y)$$

$$= E[u(X)]E[v(Y)].$$

This can be used to show that the correlation coefficient of two independent variables is zero. For, in a standard notation, we have

$$\mathrm{Cov}(X, Y) = E[(X - \mu_X)(Y - \mu_Y)]$$

$$= E(X - \mu_X)E(Y - \mu_Y) = 0.$$

The converse of this fact is not necessarily true, however; zero correlation does not in general imply independence. It is most important to keep this straight: independence implies zero correlation, but zero correlation does not necessarily imply independence. The latter is now illustrated.

EXAMPLE 4.2-3   Let X and Y have the joint p.m.f.

$$f(x, y) = \frac{1}{3}, \qquad (x, y) = (0, 1), (1, 0), (2, 1).$$

Since the support is not "rectangular," X and Y must be dependent. The means of X and Y are $\mu_X = 1$ and $\mu_Y = 2/3$, respectively. Hence

$$\mathrm{Cov}(X, Y) = E(XY) - \mu_X\mu_Y$$

$$= (0)(1)\left(\frac{1}{3}\right) + (1)(0)\left(\frac{1}{3}\right) + (2)(1)\left(\frac{1}{3}\right) - (1)\left(\frac{2}{3}\right) = 0.$$

That is, $\rho = 0$, but X and Y are dependent.   ◀

Although we have used discrete random variables to define the correlation coefficient and related concepts, these ideas carry over to the continuous case with the usual modifications, in particular integrals replacing summations. This is illustrated in Exercises 4.2-10, 4.2-11, and 4.2-12.

4.2-1 Let the random variables X and Y have the joint p.m.f

$$f(x, y) = \frac{x + y}{32}, \qquad x = 1, 2, \ y = 1, 2, 3, 4.$$

Find the means μ_X and μ_Y, the variances σ_X^2 and σ_Y^2, and the correlation coefficient ρ. Are X and Y independent or dependent?

4.2-2 Let X and Y have the joint p.m.f. defined by $f(0, 0) = f(1, 2) = 0.2$, $f(0, 1) = f(1, 1) = 0.3$.

(a) Depict the points and corresponding probabilities on a graph.
(b) Give the marginal p.m.f.s in the "margins."
(c) Compute $\mu_X, \mu_Y, \sigma_X^2, \sigma_Y^2$, Cov$(X, Y)$, and ρ.
(d) Find the equation of the least squares regression line and draw it on your graph. Does the line make sense to you intuitively?

4.2-3 Roll a fair four-sided die twice. Let X equal the outcome on the first roll and let Y equal the sum of the two rolls.

(a) Display the joint p.m.f. on a graph along with the marginal probabilities.
(b) Determine $\mu_X, \mu_Y, \sigma_X^2, \sigma_Y^2$, Cov$(X, Y)$, and ρ.
(c) Find the equation of the least squares regression line and draw it on your graph. Does the line make sense to you intuitively?

4.2-4 In the bivariate situation, show that E is a linear or distributive operator. That is, show that

$$E[a_1 u_1(X_1, X_2) + a_2 u_2(X_1, X_2)] = a_1 E[u_1(X_1, X_2)] + a_2 E[u_2(X_1, X_2)].$$

4.2-5 Let X and Y be random variables with respective means μ_X and μ_Y, respective variances σ_X^2 and σ_Y^2, and correlation coefficient ρ. Fit the line $y = a + bx$ by the method of least squares to the probability distribution by minimizing the expectation

$$K(a, b) = E[(Y - a - bX)^2]$$

with respect to a and b. HINT: Consider $\partial K/\partial a = 0$ and $\partial K/\partial b = 0$ and solve simultaneously.

4.2-6 Let X and Y have a trinomial distribution with parameters $n = 3$, $p_1 = 1/6$, and $p_2 = 1/2$. Find

(a) $E(X)$.
(b) $E(Y)$.
(c) Var(X).
(d) Var(Y).
(e) Cov(X, Y).
(f) ρ.

Note that $\rho = -\sqrt{p_1 p_2/(1 - p_1)(1 - p_2)}$.

4.2-7 Let the joint p.m.f. of X and Y be

$$f(x, y) = 1/4, \qquad (x, y) \in S = \{(0, 0), (1, 1), (1, -1), (2, 0)\}.$$

(a) Are X and Y independent?
(b) Calculate Cov(X, Y) and ρ.

This also illustrates the fact that dependent random variables can have a correlation coefficient of zero.

4.2-8 The joint p.m.f. of X and Y is $f(x, y) = 1/6$, $0 \le x + y \le 2$, where x and y are integers.

(a) Sketch the support of X and Y.
(b) Record the marginal p.m.f.s $f_1(x)$ and $f_2(y)$ in the "margins."
(c) Compute Cov(X, Y).
(d) Determine ρ, the correlation coefficient.
(e) Find the best-fitting line and draw it on your figure.

4.2-9 A certain raw material is classified as to moisture content X (in percent) and impurity Y (in percent). Let X and Y have the joint p.m.f. given by

			x	
y	1	2	3	4
2	0.10	0.20	0.30	0.05
1	0.05	0.05	0.15	0.10

(a) Find the marginal p.m.f.s and the means and the variances.
(b) Find the covariance and the correlation coefficient of X and Y.
(c) If additional heating is needed with high moisture content and additional filtering with high impurity such that the additional cost is given by the function $C = 2X + 10Y^2$ in dollars, find $E(C)$.

4.2-10 Let X and Y be random variables of the continuous type having the joint p.d.f.

$$f(x, y) = 2, \qquad 0 \le y \le x \le 1.$$

Draw a graph that illustrates the domain of this p.d.f.

(a) Find the marginal p.d.f.s of X and Y.
(b) Compute μ_X, μ_Y, σ_X^2, σ_Y^2, Cov(X, Y), and ρ.
(c) Determine the equation of the least squares regression line and draw it on your graph. Does the line make sense to you intuitively?

4.2-11 Let X and Y be random variables of the continuous type having the joint p.d.f.

$$f(x, y) = x + y, \qquad 0 < x < 1, \ 0 < y < 1.$$

Draw a graph that illustrates the domain of this p.d.f.

(a) Find the marginal p.d.f.s of X and Y.
(b) Compute μ_X, μ_Y, σ_X^2, σ_Y^2, Cov(X, Y), and ρ.
(c) Determine the equation of the least squares regression line and draw it on your graph. Does the line make sense to you intuitively?

4.2-12 Let X and Y be random variables of the continuous type having the joint p.d.f.

$$f(x, y) = 8xy, \qquad 0 \le x \le y \le 1.$$

Draw a graph that illustrates the domain of this p.d.f.

(a) Find the marginal p.d.f.s of X and Y.
(b) Compute μ_X, μ_Y, σ_X^2, σ_Y^2, Cov(X, Y), and ρ.
(c) Determine the equation of the least squares regression line and draw it on your graph. Does the line make sense to you intuitively?

4.3 CONDITIONAL DISTRIBUTIONS

Let X and Y have a joint discrete distribution with p.m.f. $f(x, y)$ on space S. Say the marginal probability mass functions are $f_1(x)$ and $f_2(y)$ with spaces S_1 and S_2, respectively. Let event $A = \{X = x\}$ and event $B = \{Y = y\}$, $(x, y) \in S$. Thus $A \cap B = \{X = x, Y = y\}$. Because

$$P(A \cap B) = P(X = x, Y = y) = f(x, y)$$

and

$$P(B) = P(Y = y) = f_2(y) > 0 \qquad \text{(since } y \in S_2\text{)},$$

we see that the conditional probability of event A given event B is

$$P(A \mid B) = \frac{P(A \cap B)}{P(B)} = \frac{f(x, y)}{f_2(y)}.$$

This leads to the following definition.

Definition 4.3-1

The *conditional probability mass function of X*, given that $Y = y$, is defined by

$$g(x \mid y) = \frac{f(x, y)}{f_2(y)}, \qquad \text{provided that } f_2(y) > 0.$$

Similarly, the *conditional probability mass function of Y*, given that $X = x$, is defined by

$$h(y \mid x) = \frac{f(x, y)}{f_1(x)}, \qquad \text{provided that } f_1(x) > 0.$$

EXAMPLE 4.3-1   Let X and Y have the joint p.m.f.

$$f(x, y) = \frac{x + y}{21}, \qquad x = 1, 2, 3, \qquad y = 1, 2.$$

In Example 4.1-4 we showed that

$$f_1(x) = \frac{2x + 3}{21}, \qquad x = 1, 2, 3,$$

and

$$f_2(y) = \frac{3y + 6}{21}, \qquad y = 1, 2.$$

Thus the conditional p.m.f. of X, given $Y = y$, is equal to

$$g(x \mid y) = \frac{(x+y)/21}{(3y+6)/21} = \frac{x+y}{3y+6}, \qquad x = 1, 2, 3, \text{ when } y = 1 \text{ or } 2.$$

For example,

$$P(X = 2 \mid Y = 2) = g(2 \mid 2) = \frac{4}{12} = \frac{1}{3}.$$

Similarly, the conditional p.m.f. of Y, given $X = x$, is equal to

$$h(y \mid x) = \frac{x+y}{2x+3}, \qquad y = 1, 2, \text{ when } x = 1, 2, \text{ or } 3.$$

The joint p.m.f. $f(x, y)$ is depicted in Figure 4.3-1(a) along with the marginal p.m.f.s. Conditionally, if $y = 2$, we would expect the outcomes of x, namely 1, 2, and 3, to occur in the ratios 3:4:5. This is precisely what $g(x \mid y)$ does:

$$g(1 \mid 2) = \frac{1+2}{12}, \qquad g(2 \mid 2) = \frac{2+2}{12}, \qquad g(3 \mid 2) = \frac{3+2}{12}.$$

(a) Joint and marginal p.m.f.s

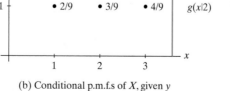

(b) Conditional p.m.f.s of X, given y (c) Conditional p.m.f.s of Y, given x

Figure 4.3-1: Joint, marginal, and conditional p.m.f.s

Figure 4.3-1(b) displays $g(x \mid 1)$ and $g(x \mid 2)$, while Figure 4.3-1(c) gives $h(y \mid 1)$, $h(y \mid 2)$, and $h(y \mid 3)$. Compare the probabilities in Figure 4.3-1(c) with those in Figure 4.3-1(a). They should agree with your intuition as well as with the formula for $h(y \mid x)$. ◄

Note that $0 \leq h(y \mid x)$. If we sum $h(y \mid x)$ over y for that fixed x, we obtain

$$\sum_y h(y \mid x) = \sum_y \frac{f(x, y)}{f_1(x)} = \frac{f_1(x)}{f_1(x)} = 1.$$

Thus $h(y \mid x)$ satisfies the conditions of a probability mass function, and so we can compute conditional probabilities such as

$$P(a < Y < b \mid X = x) = \sum_{\{y:a<y<b\}} h(y \mid x)$$

and conditional expectations such as

$$E[u(Y) \mid X = x] = \sum_y u(y)h(y \mid x)$$

in a manner similar to those associated with probabilities and expectations.

Two special conditional expectations are the **conditional mean** of Y, given $X = x$, defined by

$$\mu_{Y|x} = E(Y \mid x) = \sum_y y\, h(y \mid x),$$

and the **conditional variance** of Y, given $X = x$, defined by

$$\sigma_{Y|x}^2 = E\{[Y - E(Y \mid x)]^2 \mid x\} = \sum_y [y - E(Y \mid x)]^2 h(y \mid x),$$

which can be computed using

$$\sigma_{Y|x}^2 = E(Y^2 \mid x) - [E(Y \mid x)]^2.$$

The conditional mean $\mu_{X|y}$ and the conditional variance $\sigma_{X|y}^2$ are given by similar expressions.

EXAMPLE 4.3-2 We use the background of Example 4.3-1 and compute $\mu_{Y|x}$ and $\sigma_{Y|x}^2$, when $x = 3$:

$$\mu_{Y|3} = E(Y \mid X = 3) = \sum_{y=1}^{2} y\, h(y \mid 3)$$

$$= \sum_{y=1}^{2} y\left(\frac{3+y}{9}\right) = 1\left(\frac{4}{9}\right) + 2\left(\frac{5}{9}\right) = \frac{14}{9},$$

and

$$\sigma_{Y|3}^2 = E\left[\left(Y - \frac{14}{9}\right)^2 \bigg| X = 3\right] = \sum_{y=1}^{2}\left(y - \frac{14}{9}\right)^2\left(\frac{3+y}{9}\right)$$

$$= \frac{25}{81}\left(\frac{4}{9}\right) + \frac{16}{81}\left(\frac{5}{9}\right) = \frac{20}{81}.$$

◄

The conditional mean of X, given $Y = y$, is a function of y alone; the conditional mean of Y, given $X = x$, is a function of x alone. Suppose that the latter conditional mean is a linear function of x; that is, $E(Y \mid x) = a + bx$. Let us find the constants a and b in terms of characteristics μ_X, μ_Y, σ_X^2, σ_Y^2, and ρ. This development will shed additional light on the correlation coefficient ρ; accordingly we assume that the respective standard deviations σ_X and σ_Y are both positive so that the correlation coefficient will exist.

It is given that

$$\sum_y y\, h(y \mid x) = \sum_y y\, \frac{f(x, y)}{f_1(x)} = a + bx, \qquad \text{for } x \in S_1,$$

where S_1 is the space of X and S_2 is the space of Y. Hence

$$\sum_y y\, f(x, y) = (a + bx) f_1(x), \qquad \text{for } x \in S_1, \tag{4.3-1}$$

and

$$\sum_{x \in S_1}\sum_y y\, f(x, y) = \sum_{x \in S_1} (a + bx) f_1(x).$$

That is, with μ_X and μ_Y representing the respective means, we have

$$\mu_Y = a + b\mu_X. \tag{4.3-2}$$

In addition, if we multiply both members of Equation 4.3-1 by x and sum, we obtain

$$\sum_{x \in S_1}\sum_y xy\, f(x, y) = \sum_{x \in S_1} (ax + bx^2) f_1(x).$$

That is,

$$E(XY) = aE(X) + bE(X^2)$$

or, equivalently,

$$\mu_X\mu_Y + \rho\sigma_X\sigma_Y = a\mu_X + b(\mu_X^2 + \sigma_X^2). \tag{4.3-3}$$

The solution of Equations 4.3-2 and 4.3-3 is

$$a = \mu_Y - \rho \frac{\sigma_Y}{\sigma_X} \mu_X \qquad \text{and} \qquad b = \rho \frac{\sigma_Y}{\sigma_X},$$

which implies that if $E(Y \mid x)$ is linear, it is given by

$$E(Y \mid x) = \mu_Y + \rho \frac{\sigma_Y}{\sigma_X} (x - \mu_X).$$

So if the conditional mean of Y given $X = x$ is linear, it is exactly the same as the best-fitting line (least squares regression line) considered in Section 4.2.

By symmetry, if the conditional mean of X, given $Y = y$, is linear, it is given by

$$E(X \mid y) = \mu_X + \rho \frac{\sigma_X}{\sigma_Y} (y - \mu_Y).$$

We see that the point $[x = \mu_X, E(Y \mid x) = \mu_Y]$ satisfies the expression for $E(Y \mid x)$; and $[E(X \mid y) = \mu_X, y = \mu_Y]$ satisfies the expression for $E(X \mid y)$. That is, the point (μ_X, μ_Y) is on each of the two lines. In addition, we note that the product of the coefficient of x in $E(Y \mid x)$ and the coefficient of y in $E(X \mid y)$ equals ρ^2 and the ratio of these two coefficients equals σ_Y^2 / σ_X^2. These observations sometimes prove useful in particular problems.

EXAMPLE 4.3-3 Let X and Y have the trinomial p.m.f. with parameters n, p_1, p_2, and $1 - p_1 - p_2 = p_3$. That is,

$$f(x, y) = \frac{n!}{x! \, y! \, (n - x - y)!} p_1^x \, p_2^y \, p_3^{n-x-y},$$

where x and y are nonnegative integers such that $x + y \le n$. From the development of the trinomial distribution, we note that X and Y have marginal binomial distributions $b(n, p_1)$ and $b(n, p_2)$, respectively. Thus

$$h(y \mid x) = \frac{f(x, y)}{f_1(x)} = \frac{(n - x)!}{y! \, (n - x - y)!} \left(\frac{p_2}{1 - p_1} \right)^y \left(\frac{p_3}{1 - p_1} \right)^{n-x-y},$$

$$y = 0, 1, 2, \ldots, n - x.$$

That is, the conditional p.m.f. of Y, given $X = x$, is binomial

$$b\left[n - x, \frac{p_2}{1 - p_1} \right]$$

and thus has conditional mean

$$E(Y \mid x) = (n - x) \frac{p_2}{1 - p_1}.$$

In a similar manner, we obtain

$$E(X \mid y) = (n - y) \frac{p_1}{1 - p_2}.$$

Since each of the conditional means is linear, the product of the respective coefficients of x and y is

$$\rho^2 = \left(\frac{-p_2}{1 - p_1} \right) \left(\frac{-p_1}{1 - p_2} \right) = \frac{p_1 p_2}{(1 - p_1)(1 - p_2)}.$$

However, ρ must be negative because the coefficients of x and y are negative, thus

$$\rho = -\sqrt{\frac{p_1 p_2}{(1 - p_1)(1 - p_2)}}. \qquad \blacktriangleleft$$

In the following example we again look at conditional p.m.f.s when X and Y have a trinomial distribution.

EXAMPLE 4.3-4 Let X and Y have a trinomial distribution with $p_1 = 1/3$, $p_2 = 1/3$ and $n = 5$. Using a result from the last example, the conditional distribution of Y, given $X = x$, is $b(5 - x, (1/3)/(1 - 1/3))$ or $b(5 - x, 1/2)$. The p.m.f.s, $h(y \mid x)$ for $x = 0, 1, \ldots, 5$ are plotted in Figure 4.3-2. Note that the orientation of the axes was selected so that the shapes of these p.m.f.s can be seen. Similarly, the conditional distribution of X, given $Y = y$, is $b(n - y, 1/2)$. The p.m.f.s, $g(x \mid y)$ for $y = 0, 1, \ldots, 5$ are shown in Figure 4.3-3. $\qquad \blacktriangleleft$

Although we have used random variables of the discrete type to introduce the new definitions, they also hold for random variables of the continuous type. Let X and Y have a distribution of the continuous type with joint p.d.f. $f(x, y)$

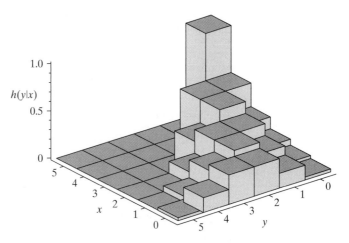

Figure 4.3-2: Conditional p.m.f.s of Y, given $X = x$

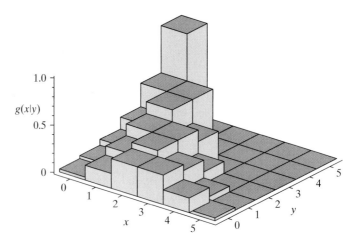

Figure 4.3-3: Conditional p.m.f.s of X, given $Y = y$

and marginal p.d.f.s $f_1(x)$ and $f_2(y)$, respectively. We have that the conditional p.d.f., mean, and variance of Y, given $X = x$, are, respectively,

$$h(y \mid x) = \frac{f(x, y)}{f_1(x)}, \qquad \text{provided that } f_1(x) > 0;$$

$$E(Y \mid x) = \int_{-\infty}^{\infty} y \, h(y \mid x) \, dy;$$

$$\mathrm{Var}(Y \mid x) = E\{[Y - E(Y \mid x)]^2 \mid x\}$$

$$= \int_{-\infty}^{\infty} [y - E(Y \mid x)]^2 \, h(y \mid x) \, dy$$

$$= E[Y^2 \mid x] - [E(Y \mid x)]^2.$$

Similar expressions are associated with the conditional distribution of X, given $Y = y$.

EXAMPLE 4.3-5 Let X and Y be the random variables of Example 4.1-9. Thus

$$\begin{array}{lll} f(x, y) & = & 2, \qquad\qquad\quad 0 \le x \le y \le 1, \\ f_1(x) & = & 2(1 - x), \quad\; 0 \le x \le 1, \\ f_2(y) & = & 2y, \qquad\qquad 0 \le y \le 1. \end{array}$$

Before we actually find the conditional p.d.f. of Y, given $X = x$, we shall give an intuitive argument. The joint p.d.f. is constant over the triangular region bounded by $y = x$, $y = 1$, and $x = 0$. If the value of X is known, say $X = x$, then the possible values of Y are between x and 1. Furthermore, we would expect Y to be uniformly distributed on the interval $[x, 1]$. That is, we would anticipate that $h(y \mid x) = 1/(1 - x)$, $x \le y \le 1$.

More formally now, we have by definition that

$$h(y \mid x) = \frac{f(x, y)}{f_1(x)} = \frac{2}{2(1 - x)} = \frac{1}{1 - x}, \qquad x \le y \le 1, \qquad 0 \le x \le 1.$$

The conditional mean of Y, given $X = x$, is

$$E(Y \mid x) = \int_x^1 y \frac{1}{1 - x} \, dy = \left[\frac{y^2}{2(1 - x)} \right]_x^1 = \frac{1 + x}{2}, \qquad 0 \le x \le 1.$$

Similarly, it can be shown that

$$E(X \mid y) = \frac{y}{2}, \qquad 0 \le y \le 1.$$

The conditional variance of Y, given $X = x$, is

$$E\{[Y - E(Y \mid x)]^2 \mid x\} = \int_x^1 \left(y - \frac{1 + x}{2} \right)^2 \frac{1}{1 - x} \, dy$$

$$= \left[\frac{1}{3(1 - x)} \left(y - \frac{1 + x}{2} \right)^3 \right]_x^1$$

$$= \frac{(1 - x)^2}{12}.$$

Recall that if a random variable W is $U(a, b)$, then $E(W) = (a + b)/2$, and $\operatorname{Var}(W) = (b - a)^2/12$. Since the conditional distribution of Y, given $X = x$, is $U(x, 1)$, we could have written down immediately that $E(Y \mid x) = (x + 1)/2$ and $\operatorname{Var}(Y \mid x) = (1 - x)^2/12$.

An illustration of a computation of a conditional probability is

$$P\left(\frac{3}{4} \le Y \le \frac{7}{8} \,\middle|\, X = \frac{1}{4} \right) = \int_{3/4}^{7/8} h\left(y \,\middle|\, \frac{1}{4} \right) dy$$

$$= \int_{3/4}^{7/8} \frac{1}{3/4} \, dy = \frac{1}{6}. \qquad \blacktriangleleft$$

In general, if $E(Y \mid x)$ is linear, it is equal to

$$E(Y \mid x) = \mu_Y + \rho\left(\frac{\sigma_Y}{\sigma_X} \right)(x - \mu_X).$$

If $E(X \mid y)$ is linear, then

$$E(X \mid y) = \mu_X + \rho\left(\frac{\sigma_X}{\sigma_Y} \right)(y - \mu_Y).$$

Thus, in Example 4.3-5, we see that the product of the coefficients of x in $E(Y \mid x)$ and y in $E(X \mid y)$ is $\rho^2 = 1/4$. Thus $\rho = 1/2$ since each coefficient is positive. Since the ratio of those coefficients is equal to $\sigma_Y^2/\sigma_X^2 = 1$, we have that $\sigma_X^2 = \sigma_Y^2$.

STATISTICAL COMMENTS **(More on Simpson's Paradox)** Let us say that a small business employs 100 men and 100 women and the salary for each person is either \$20,000, \$30,000, or \$50,000. For the men 20 make \$20,000, 20 make \$30,000, and 60 make \$50,000; these average

$$\frac{(20)(20,000) + (20)(30,000) + (60)(50,000)}{100} = \$40,000.$$

On the other hand, among the women, 60 make \$20,000, 20 make \$30,000, and 20 make \$50,000; these average

$$\frac{(60)(20,000) + (20)(30,000) + (20)(50,000)}{100} = \$28,000.$$

This seems to be a fairly large gap in these two average salaries.

It turns out that there is another variable to consider, however, namely experience. For simplicity say each employee has either 1 year or 5 years of experience. Thus we have for the men (women) a bivariate distribution with space given by Figure 4.3-4. However, we do not know the joint probabilities and hence can not compute the conditional means, given 1 year and then 5 years of experience. However, we do have the data associated with these 100 men and 100 women and estimate these conditional means. Namely, there are 20 men with 1 year of experience: 15 earn \$20,000 and 5 earn \$30,000, giving a mean of \$22,500. The other 80 men have 5 years of experience: 5 earn \$20,000;

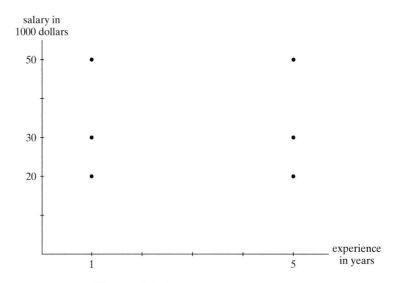

Figure 4.3-4: Salary versus experience

15 earn \$30,000; and 60 earn \$50,000 for an average of \$44,375. Considering the women, there are 80 with 1 year experience: 60 earn \$20,000, 15 earn \$30,000, and 5 earn \$50,000 averaging \$23,750. Also, 20 women have 5 years of experience: 5 earn \$30,000 and 15 earn \$50,000 for an average of \$45,000. So while there is a large gap between the \$28,000 earned on the average by women and the \$40,000 earned for the men, the women seem to have a slight advantage by comparing the conditional averages: \$23,750 to \$22,500 given 1 year of experience, and \$45,000 to \$44,375 given 5 years of experience. This information is summarized in the following table.

	Salary	Years of Experience 1-Year	Years of Experience 5-Years	Total
Men	\$20,000	15	5	20
	\$30,000	5	15	20
	\$50,000	0	60	60
Average Salary		\$22,500	\$44,375	\$40,000
Women	\$20,000	60	0	60
	\$30,000	15	5	20
	\$50,000	5	15	20
Average Salary		\$23,750	\$45,000	\$28,000

This reversal is another illustration of Simpson's paradox. We are not arguing that there has not been discrimination in the pay for different genders, for there has certainly been a great deal of that in the past. Although we know in our business (academics) most of this has been eliminated, we still insist that experience must be considered: not only years of teaching but publications, professional field, activity in professional societies, service on committees, and so on. Clearly we support the position that a woman and a man with the same credentials should receive the same rank and pay. ■

EXERCISES

4.3-1 Let X and Y have the joint p.m.f.

$$f(x, y) = \frac{x + y}{32}, \qquad x = 1, 2, \qquad y = 1, 2, 3, 4.$$

(a) Display the joint p.m.f. and the marginal p.m.f.s on a graph like Figure 4.3-1(a).
(b) Find $g(x \mid y)$ and draw a figure like Figure 4.3-1(b), depicting the conditional p.m.f.s for $y = 1, 2, 3$, and 4.
(c) Find $h(y \mid x)$ and draw a figure like Figure 4.3-1(c), depicting the conditional p.m.f.s for $x = 1$ and 2.
(d) Find $P(1 \leq Y \leq 3 \mid X = 1)$, $P(Y \leq 2 \mid X = 2)$, and $P(X = 2 \mid Y = 3)$.
(e) Find $E(Y \mid X = 1)$ and $\text{Var}(Y \mid X = 1)$.

4.3-2 Let the joint p.m.f. $f(x, y)$ of X and Y be given by the following:

(x, y)	$f(x, y)$
$(1, 1)$	3/8
$(2, 1)$	1/8
$(1, 2)$	1/8
$(2, 2)$	3/8

Find the two conditional probability mass functions and the corresponding means and variances.

4.3-3 Let W equal the weight of laundry soap in a 1-kilogram box that is distributed in Southeast Asia. Suppose that $P(W < 1) = 0.02$ and $P(W > 1.072) = 0.08$. Call a box of soap light, good, or heavy depending on whether $\{W < 1\}$, $\{1 \leq W \leq 1.072\}$, or $\{W > 1.072\}$, respectively. In $n = 50$ independent observations of these boxes, let X equal the number of light boxes and Y the number of good boxes.

(a) What is the joint p.m.f. of X and Y?

(b) Give the name of the distribution of Y along with the values of the parameters of this distribution.

(c) Given that $X = 3$, how is Y distributed conditionally?

(d) Determine $E(Y \mid X = 3)$.

(e) Find ρ, the correlation coefficient of X and Y.

4.3-4 The genes for eye color for a certain male fruit fly are (R, W). The genes for eye color for the mating female fruit fly are (R, W). Their offspring receive one gene for eye color from each parent. If an offspring ends up with either (R, R), (R, W), or (W, R), its eyes will look red. Let X equal the number of offspring having red eyes. Let Y equal the number of red-eyed offspring having (R, W) or (W, R) genes.

(a) If the total number of offspring is $n = 400$, how is X distributed?

(b) Give the values of $E(X)$ and $\text{Var}(X)$.

(c) Given that $X = 300$, how is Y distributed?

(d) Give the values of $E(Y \mid X = 300)$ and $\text{Var}(Y \mid X = 300)$.

4.3-5 Let X and Y have a trinomial distribution with $n = 2$, $p_1 = 1/4$, and $p_2 = 1/2$.

(a) Give $E(Y \mid x)$.

(b) Compare your answer in part (a) with the equation of the line of best fit in Example 4.2-2. Are they the same? Why?

4.3-6 An insurance company sells both homeowners and automobile deductible insurance. Let X be the level of deductible on the homeowners insurance and Y the level of deductible on automobile insurance. Among those who take both insurances with that company, we find the following probabilities.

	x		
y	100	500	1000
1000	0.05	0.10	0.15
500	0.10	0.20	0.05
100	0.20	0.10	0.05

(a) Compute the following probabilities:

$$P(X = 500), P(Y = 500), P(Y = 500 \mid X = 500), P(Y = 100 \mid X = 500).$$

(b) Compute the means μ_X, μ_Y, and the variances σ_X^2, σ_Y^2.
(c) Compute the conditional means $E(X \mid Y = 100)$, $E(Y \mid X = 500)$.
(d) Compute the $\mathrm{Cov}(X, Y)$.
(e) Compute the correlation coefficient $\rho = \mathrm{Cov}(X, Y)/\sigma_X \sigma_Y$.

4.3-7 Using the joint p.m.f. given in Exercise 4.2-3, find the value of $E(Y \mid x)$ for $x = 1, 2, 3, 4$. Is this linear? Do these points lie on the best-fitting line?

4.3-8 An unbiased six-sided die is cast 30 independent times. Let X be the number of ones and Y the number of twos.

(a) What is the joint p.m.f. of X and Y?
(b) Find the conditional p.m.f. of X, given $Y = y$.
(c) Compute $E(X^2 - 4XY + 3Y^2)$.

4.3-9 Let X and Y have a uniform distribution on the set of points with integer coordinates in $S = \{(x, y) : 0 \le x \le 7, x \le y \le x + 2\}$. That is, $f(x, y) = 1/24$, $(x, y) \in S$, and both x and y are integers. Find

(a) $f_1(x)$.
(b) $h(y \mid x)$.
(c) $E(Y \mid x)$.
(d) $\sigma_{Y \mid x}^2$.
(e) $f_2(y)$.

4.3-10 Let $f_1(x) = 1/10$, $x = 0, 1, 2, \ldots, 9$, and $h(y \mid x) = 1/(10 - x)$, $y = x, x + 1, \ldots, 9$. Find

(a) $f(x, y)$.
(b) $f_2(y)$.
(c) $E(Y \mid x)$.

4.3-11 An automobile repair shop makes an initial estimate X (in thousands) needed to fix a car after an accident. Say X has the p.d.f.

$$f(x) = 2 e^{-2(x - 0.2)}, \qquad 0.2 < x < \infty.$$

Given $X = x$, the final payment Y has a uniform distribution between $x - 0.1$ and $x + 0.1$. What is the expected value of Y?

4.3-12 For the random variables defined in Example 4.3-5, calculate the correlation coefficient directly from the definition

$$\rho = \frac{\mathrm{Cov}(X, Y)}{\sigma_X \sigma_Y}.$$

4.3-13 Let $f(x, y) = 1/40$, $0 \le x \le 10$, $10 - x \le y \le 14 - x$, be the joint p.d.f. of X and Y.

(a) Sketch the region for which $f(x, y) > 0$.
(b) Find $f_1(x)$, the marginal p.d.f. of X.
(c) Determine $h(y \mid x)$, the conditional p.d.f. of Y, given $X = x$.
(d) Calculate $E(Y \mid x)$, the conditional mean of Y, given $X = x$.

4.3-14 Let $f(x, y) = 1/8$, $0 \le y \le 4$, $y \le x \le y + 2$, be the joint p.d.f. of X and Y.

(a) Sketch the region for which $f(x, y) > 0$.
(b) Find $f_1(x)$, the marginal p.d.f. of X.
(c) Find $f_2(y)$, the marginal p.d.f. of Y.
(d) Determine $h(y \mid x)$, the conditional p.d.f. of Y, given $X = x$.
(e) Determine $g(x \mid y)$, the conditional p.d.f. of X, given $Y = y$.
(f) Compute $y = E(Y \mid x)$, the conditional mean of Y, given $X = x$.

     (g) Compute $x = E(X \mid y)$, the conditional mean of X, given $Y = y$.
     (h) Graph $y = E(Y \mid x)$ on your sketch in part (a). Is $y = E(Y \mid x)$ linear?
     (i) Graph $x = E(X \mid y)$ on your sketch in part (a). Is $x = E(X \mid y)$ linear?

4.3-15 Let X have a uniform distribution $U(0, 2)$, and let the conditional distribution of Y, given $X = x$, be $U(0, x^2)$.

     (a) Determine the joint p.d.f. of X and Y, $f(x, y)$.
     (b) Calculate $f_2(y)$, the marginal p.d.f. of Y.
     (c) Compute $E(X \mid y)$, the conditional mean of X, given $Y = y$.
     (d) Find $E(Y \mid x)$, the conditional mean of Y, given $X = x$.

4.3-16 Let X have a uniform distribution on the interval $(0, 1)$. Given $X = x$, let Y have a uniform distribution on the interval $(0, x)$.

     (a) Define the conditional p.d.f. of Y, given that $X = x$. Be sure to include the domain.
     (b) Find $E(Y \mid x)$.
     (c) Determine the joint p.d.f. of X and Y.
     (d) Find the marginal p.d.f. of Y.

4.3-17 The marginal distribution of X is $U(0, 1)$. The conditional distribution of Y, given $X = x$, is $U(0, e^x)$.

     (a) Determine $h(y \mid x)$, the conditional p.d.f. of Y, given $X = x$.
     (b) Find $E(Y \mid x)$.
     (c) Display the joint p.d.f. of X and Y. Sketch the region where $f(x, y) > 0$.
     (d) Find $f_2(y)$, the marginal p.d.f. of Y.

4.4 TRANSFORMATIONS OF RANDOM VARIABLES

In Section 3.5 we considered the transformation of one random variable X with p.d.f. $f(x)$. In particular, in the continuous case, if $Y = u(X)$ was an increasing or decreasing function of X, with inverse $X = v(Y)$, then the p.d.f. of Y is

$$g(y) = |v'(y)| \, f[v(y)], \qquad c < y < d,$$

where the support $c < y < d$ corresponds to the support of X, say $a < x < b$, through the transformation $x = v(y)$.

    Recall the discrete case had to be handled a little differently as illustrated by the following example.

EXAMPLE 4.4-1 Let X be binomial with parameters n and p. Since X has a discrete distribution, $Y = u(X)$ will also have a discrete distribution with the same probabilities as those in the support of X. For illustration, with $n - 3$, $p = 1/4$, and $Y = X^2$, we have

$$g(y) = \binom{3}{\sqrt{y}} \left(\frac{1}{4}\right)^{\sqrt{y}} \left(\frac{3}{4}\right)^{3-\sqrt{y}}, \qquad y = 0, 1, 4, 9. \qquad \blacktriangleleft$$

    For a more interesting problem with the binomial random variable X, suppose we were to search for a transformation $u(X/n)$ of the relative frequency X/n that would have a variance very little dependent on p itself when n is large. That is, we want the variance of $u(X/n)$ to be essentially a constant.

Consider the function $u(X/n)$ and find, using two terms of Taylor's expansion about p, that

$$u\left(\frac{X}{n}\right) \approx u(p) + u'(p)\left(\frac{X}{n} - p\right).$$

Here terms of higher powers can be disregarded if n is large enough so that X/n is close enough to p. Thus

$$\text{Var}\left[u\left(\frac{X}{n}\right)\right] \approx [u'(p)]^2 \, \text{Var}\left(\frac{X}{n} - p\right) = [u'(p)]^2 \, \frac{p(1-p)}{n}. \qquad (4.4\text{-}1)$$

However, if $\text{Var}[u(X/n)]$ is to be constant with respect to p, then

$$[u'(p)]^2 \, p(1-p) = k \quad \text{or} \quad u'(p) = \frac{c}{\sqrt{p(1-p)}},$$

where k and c are constants. We see that

$$u(p) = 2c \arcsin \sqrt{p}$$

is a solution to this differential equation. Thus with $c = 1/2$, we frequently see, in the literature, use of the arcsine transformation, namely

$$Y = \arcsin \sqrt{\frac{X}{n}},$$

which, with large n, has an approximate mean

$$\mu = \arcsin \sqrt{p}$$

and variance (using Equation 4.4-1)

$$\sigma^2 = [D_p(\arcsin \sqrt{p})]^2 \, \frac{p(1-p)}{n} = \frac{1}{4n}.$$

There should be one note of warning here: If the function $Y = u(X)$ does not have a single-valued inverse, the determination of the distribution of Y will not be as simple. We did consider two examples in Section 3.5 in which there were two inverse functions and special care was exercised. In our examples, we will not consider problems with many inverses; however, we thought that such a warning should be issued here.

When two random variables are involved, many interesting problems can result. In the case of a single-valued inverse, the rule is about the same as that in the one-variable case, with the derivative being replaced by the Jacobian. Namely, if X_1 and X_2 are two continuous-type random variables with joint p.d.f. $f(x_1, x_2)$ and if $Y_1 = u_1(X_1, X_2)$, $Y_2 = u_2(X_1, X_2)$ has the single-valued inverse $X_1 = v_1(Y_1, Y_2)$, $X_2 = v_2(Y_1, Y_2)$, then the joint p.d.f. of Y_1 and Y_2 is

$$g(y_1, y_2) = |J| \, f[v_1(y_1, y_2), v_2(y_1, y_2)], \qquad (y_1, y_2) \in S_1,$$

where the Jacobian J is the determinant

$$J = \begin{vmatrix} \dfrac{\partial x_1}{\partial y_1} & \dfrac{\partial x_1}{\partial y_2} \\[2ex] \dfrac{\partial x_2}{\partial y_1} & \dfrac{\partial x_2}{\partial y_2} \end{vmatrix}.$$

Of course, we find the support S_1 of Y_1, Y_2 by considering the mapping of the support S of X_1, X_2 under the transformation $y_1 = u_1(x_1, x_2)$, $y_2 = u_2(x_1, x_2)$. This method of finding the distribution of Y_1 and Y_2 is called the **change-of-variables technique**.

It is often the latter, namely the mapping of the support S of X_1, X_2 into that, say S_1, of Y_1, Y_2, that causes the biggest challenge. That is, in most cases, it is easy to solve for x_1 and x_2 in terms of y_1 and y_2, say

$$x_1 = v_1(y_1, y_2), \qquad x_2 = v_2(y_1, y_2),$$

and then to compute the Jacobian

$$J = \begin{vmatrix} \dfrac{\partial v_1(y_1, y_2)}{\partial y_1} & \dfrac{\partial v_1(y_1, y_2)}{\partial y_2} \\[2ex] \dfrac{\partial v_2(y_1, y_2)}{\partial y_1} & \dfrac{\partial v_2(y_1, y_2)}{\partial y_2} \end{vmatrix}.$$

However, the mapping of $(x_1, x_2) \in S$ into $(y_1, y_2) \in S_1$ can be more difficult. Let us consider two simple examples.

EXAMPLE 4.4-2 Let X_1, X_2 have the joint p.d.f.

$$f(x_1, x_2) = 2, \qquad 0 < x_1 < x_2 < 1.$$

Consider the transformation

$$Y_1 = \frac{X_1}{X_2}, \qquad Y_2 = X_2.$$

It is certainly easy enough to solve for x_1 and x_2, namely,

$$x_1 = y_1 y_2, \qquad x_2 = y_2$$

and compute

$$J = \begin{vmatrix} y_2 & y_1 \\ 0 & 1 \end{vmatrix} = y_2.$$

Let us consider S which is depicted in Figure 4.4-1(a). The boundaries of S are not part of the support, but let us see how they map. The points for which $x_1 = 0, 0 < x_2 < 1$, maps into $y_1 = 0, 0 < y_2 < 1$; $x_2 = 1, 0 \le x_1 < 1$ maps into $y_2 = 1, 0 \le y_1 < 1$; and $0 < x_1 = x_2 \le 1$ maps into $y_1 = 1, 0 < y_2 \le 1$. We depict these line segments in Figure 4.4-1(b) and mark them with the symbols corresponding to the line segments in Figure 4.4-1(a).

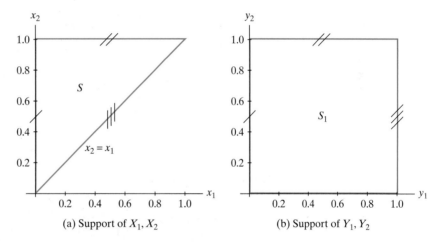

(a) Support of X_1, X_2 (b) Support of Y_1, Y_2

Figure 4.4-1: Mapping from x_1, x_2 to y_1, y_2

We note that the points for which $y_2 = 0, 0 \le y_1 < 1$, all map into the single point $x_1 = 0, x_2 = 0$. That is, this is a many-to-one mapping, and yet we are restricting ourselves to one-to-one mappings. However, the boundaries are not part of our support! Thus S_1 is as depicted in Figure 4.4-1(b), and, according to the rule, the joint p.d.f. of Y_1, Y_2 is

$$g(y_1, y_2) = |y_2| \cdot 2 = 2y_2, \qquad 0 < y_1 < 1, 0 < y_2 < 1.$$

It is interesting to note that the marginal probability density functions are

$$g_1(y_1) = \int_0^1 2y_2 \, dy_2 = 1, \qquad 0 < y_1 < 1,$$

and

$$g_2(y_2) = \int_0^1 2y_2 \, dy_1 = 2y_2, \qquad 0 < y_2 < 1.$$

Thus $Y_1 = X_1/X_2$ and $Y_2 = X_2$ are independent. Even though the computation of Y_1 depends very much on the value of Y_2, still Y_1 and Y_2 are independent in the probability sense. ◀

EXAMPLE 4.4-3 Let X_1 and X_2 be independent random variables, each with p.d.f.

$$f(x) = e^{-x}, \qquad 0 < x < \infty.$$

Hence their joint p.d.f. is

$$f(x_1)f(x_2) = e^{-x_1 - x_2}, \qquad 0 < x_1 < \infty, \ 0 < x_2 < \infty.$$

Let us consider

$$Y_1 = X_1 - X_2, \qquad Y_2 = X_1 + X_2.$$

Hence

$$x_1 = \frac{y_1 + y_2}{2}, \qquad x_2 = \frac{y_2 - y_1}{2},$$

with

$$J = \begin{vmatrix} \dfrac{1}{2} & \dfrac{1}{2} \\[2mm] -\dfrac{1}{2} & \dfrac{1}{2} \end{vmatrix} = \frac{1}{2}.$$

The region S is depicted in Figure 4.4-2(a). The line segments on the boundary, namely $x_1 = 0, 0 < x_2 < \infty$, and $x_2 = 0, 0 < x_1 < \infty$ map into the line segments $y_1 + y_2 = 0$, $y_2 > y_1$ and $y_1 = y_2$, $y_2 > -y_1$, respectively. These are shown in Figure 4.4-2(b) and the support of S_1 is depicted there. Since the region S_1 is not bounded by horizontal and vertical line segments, Y_1 and Y_2 are dependent.

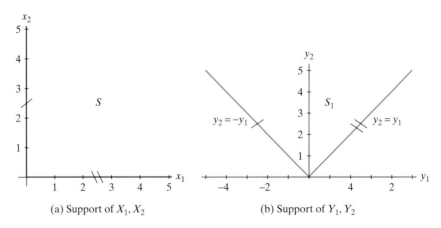

(a) Support of X_1, X_2 (b) Support of Y_1, Y_2

Figure 4.4-2: Mapping from x_1, x_2 to y_1, y_2

The joint p.d.f. of Y_1 and Y_2 is

$$g(y_1, y_2) = \frac{1}{2} e^{-y_2}, \qquad -y_2 < y_1 < y_2, \ 0 < y_2 < \infty.$$

The marginal p.d.f. of Y_2 is

$$g_2(y_2) = \int_{-y_2}^{y_2} \frac{1}{2} e^{-y_2} \, dy_1 = y_2 e^{-y_2}, \qquad 0 < y_2 < \infty.$$

That of Y_1 is

$$g_1(y_1) = \begin{cases} \displaystyle\int_{-y_1}^{\infty} \frac{1}{2} e^{-y_2} \, dy_2 = \frac{1}{2} e^{y_1}, & -\infty < y_1 < 0, \\[2ex] \displaystyle\int_{y_1}^{\infty} \frac{1}{2} e^{-y_2} \, dy_2 = \frac{1}{2} e^{-y_1}, & 0 < y_1 < \infty. \end{cases}$$

That is, the expression for $g_1(y_1)$ depends on the location of y_1, although this could be written as

$$g_1(y_1) = \frac{1}{2} e^{-|y_1|}, \qquad -\infty < y_1 < \infty,$$

which is called a **double exponential** p.d.f.   ◀

We now consider two examples that yield two important distributions. The second of these uses the distribution function technique rather than the change of variable method.

EXAMPLE 4.4-4   Let X_1 and X_2 have independent gamma distributions with parameters α, θ and β, θ, respectively. That is, the joint p.d.f. of X_1 and X_2 is

$$f(x_1, x_2) = \frac{1}{\Gamma(\alpha)\Gamma(\beta)\theta^{\alpha+\beta}} x_1^{\alpha-1} x_2^{\beta-1} \exp\left(-\frac{x_1 + x_2}{\theta}\right), \quad 0 < x_1 < \infty,$$

$$0 < x_2 < \infty.$$

Consider

$$Y_1 = \frac{X_1}{X_1 + X_2}, \qquad Y_2 = X_1 + X_2,$$

or, equivalently,

$$X_1 = Y_1 Y_2, \qquad X_2 = Y_2 - Y_1 Y_2.$$

The Jacobian is

$$J = \begin{vmatrix} y_2 & y_1 \\ -y_2 & 1 - y_1 \end{vmatrix} = y_2(1 - y_1) + y_1 y_2 = y_2.$$

Thus the joint p.d.f. $g(y_1, y_2)$ of Y_1 and Y_2 is

$$g(y_1, y_2) = |y_2| \frac{1}{\Gamma(\alpha)\Gamma(\beta)\theta^{\alpha+\beta}} (y_1 y_2)^{\alpha-1}(y_2 - y_1 y_2)^{\beta-1} e^{-y_2/\theta},$$

where the support is $0 < y_1 < 1, 0 < y_2 < \infty$, the mapping of $0 < x_i < \infty, i = 1, 2$. To see the shape of this joint p.d.f., $z = g(y_1, y_2)$ is graphed in Figure 4.4-3 with $\alpha = 4$, $\beta = 7$, and $\theta = 1$ and in Figure 4.4-3 with $\alpha = 8$, $\beta = 3$, and $\theta = 1$. To find the marginal p.d.f. of Y_1, we integrate this joint p.d.f. on y_2. We see that the marginal p.d.f. of Y_1 is

$$g_1(y_1) = \frac{y_1^{\alpha-1}(1 - y_1)^{\beta-1}}{\Gamma(\alpha)\Gamma(\beta)} \int_0^\infty \frac{y_2^{\alpha+\beta-1}}{\theta^{\alpha+\beta}} e^{-y_2/\theta} \, dy_2.$$

But the integral in this expression is that of a gamma p.d.f. with parameters $\alpha + \beta$ and θ, except for $\Gamma(\alpha + \beta)$ in the denominator; hence the integral equals $\Gamma(\alpha + \beta)$ and

$$g_1(y_1) = \frac{\Gamma(\alpha + \beta)}{\Gamma(\alpha)\Gamma(\beta)} y_1^{\alpha-1}(1 - y_1)^{\beta-1}, \qquad 0 < y_1 < 1.$$

We say that Y_1 has a **beta p.d.f.** with parameters α and β (see Figure 4.4-4). Note the relationship between Figure 4.4-3 and Figure 4.4-4. ◄

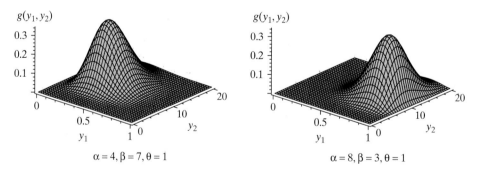

$\alpha = 4, \beta = 7, \theta = 1$ $\alpha = 8, \beta = 3, \theta = 1$

Figure 4.4-3: Joint p.d.f. of $z = g(y_1, y_2)$

The next example illustrates the distribution function technique. You calculate the same results in Exercise 4.4-2 but using the change of variables technique.

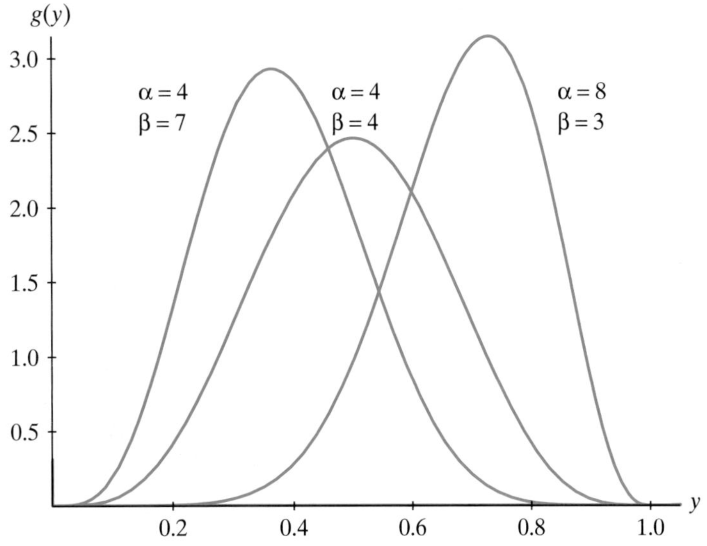

Figure 4.4-4: Beta distribution p.d.f.s

EXAMPLE 4.4-5 We let

$$F = \frac{U/r_1}{V/r_2},$$

where U and V are independent chi-square variables with r_1 and r_2 degrees of freedom, respectively. Thus the joint p.d.f. of U and V is

$$g(u, v) = \frac{u^{r_1/2-1}e^{-u/2}}{\Gamma(r_1/2)2^{r_1/2}}\frac{v^{r_2/2-1}e^{-v/2}}{\Gamma(r_2/2)2^{r_2/2}}, \qquad 0 < u < \infty, \ 0 < v < \infty.$$

In this derivation we let $W = F$ to avoid using f as a symbol for a variable. The distribution function $F(w) = P(W \le w)$ of W is

$$F(w) = P\left(\frac{U/r_1}{V/r_2} \le w\right) = P\left(U \le \frac{r_1}{r_2}w\,V\right)$$

$$= \int_0^\infty \int_0^{(r_1/r_2)wv} g(u, v)\, du\, dv.$$

That is,

$$F(w) = \frac{1}{\Gamma(r_1/2)\Gamma(r_2/2)} \int_0^\infty \left[\int_0^{(r_1/r_2)wv} \frac{u^{r_1/2-1}e^{-u/2}}{2^{(r_1+r_2)/2}}\, du\right] v^{r_2/2-1}e^{-v/2}\, dv.$$

The p.d.f. of W is the derivative of the distribution function; so applying the fundamental theorem of calculus to the inner integral, we have

$$f(w) = F'(w)$$

$$= \frac{1}{\Gamma(r_1/2)\Gamma(r_2/2)} \int_0^\infty \frac{[(r_1/r_2)vw]^{r_1/2-1}}{2^{(r_1+r_2)/2}}$$

$$\times e^{-(r_1/2r_2)(vw)} \left(\frac{r_1}{r_2} v\right) v^{r_2/2-1} e^{-v/2} \, dv$$

$$= \frac{(r_1/r_2)^{r_1/2} w^{r_1/2-1}}{\Gamma(r_1/2)\Gamma(r_2/2)} \int_0^\infty \frac{v^{(r_1+r_2)/2-1}}{2^{(r_1+r_2)/2}} e^{-(v/2)[1+(r_1/r_2)w]} \, dv.$$

In the integral, make the change of variables

$$y = \left(1 + \frac{r_1}{r_2} w\right) v \qquad \text{so that} \qquad \frac{dv}{dy} = \frac{1}{1 + (r_1/r_2)w}.$$

Thus we see that

$$f(w) = \frac{(r_1/r_2)^{r_1/2}\Gamma[(r_1+r_2)/2]w^{r_1/2-1}}{\Gamma(r_1/2)\Gamma(r_2/2)[1+(r_1w/r_2)]^{(r_1+r_2)/2}} \int_0^\infty \frac{y^{(r_1+r_2)/2-1}e^{-y/2}}{\Gamma[(r_1+r_2)/2]2^{(r_1+r_2)/2}} \, dy$$

$$= \frac{(r_1/r_2)^{r_1/2}\Gamma[(r_1+r_2)/2]w^{r_1/2-1}}{\Gamma(r_1/2)\Gamma(r_2/2)[1+(r_1w/r_2)]^{(r_1+r_2)/2}},$$

the p.d.f. of the $W = F$ distribution. Note that the integral in this last expression for $f(w)$ is equal to 1 because the integrand is like a p.d.f. of a chi-square distribution with $r_1 + r_2$ degrees of freedom. Graphs of p.d.f.s for the F distribution are given in Figure 4.4-5. ◀

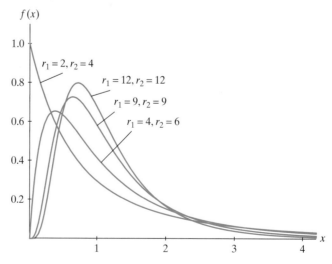

Figure 4.4-5: Graphs of F p.d.f.s

EXERCISES

4.4-1 Let X_1, X_2 denote two independent random variables, each with a $\chi^2(2)$ distribution. Find the joint p.d.f. of $Y_1 = X_1$ and $Y_2 = X_2 + X_1$. Here note that the support of Y_1, Y_2 is $0 < y_1 < y_2 < \infty$. Also find the marginal p.d.f. of each of Y_1 and Y_2. Are Y_1 and Y_2 independent?

4.4-2 Let X_1 and X_2 be independent chi-square random variables with r_1 and r_2 degrees of freedom, respectively. Let $Y_1 = (X_1/r_1)/(X_2/r_2)$ and $Y_2 = X_2$.

(a) Find the joint p.d.f. of Y_1 and Y_2.

(b) Determine the marginal p.d.f. of Y_1 and show that Y_1 has an F distribution. (This is another, but equivalent, way of finding the p.d.f. of F.)

4.4-3 Find the mean and the variance of an F random variable with r_1 and r_2 degrees of freedom by first finding $E(U)$, $E(1/V)$, $E(U^2)$, and $E(1/V^2)$.

4.4-4 Let X have a Poisson distribution with mean λ. Find a transformation $u(X)$ so that $\text{Var}[u(X)]$ is about free of λ, for large values of λ. HINT: $u(X) \approx u(\lambda) + [u'(\lambda)](X - \lambda)$, provided $[u''(\lambda)](X - \lambda)^2/2$ and higher terms can be neglected when λ is large. A solution is $u(X) = \sqrt{X}$.

4.4-5 Generalize Exercise 4.4-4 by assuming that the variance of a distribution is equal to $c\mu^p$, where c is a constant and μ is the mean (note in Exercise 4.4-4 that this is the case with $p = 1$). In particular, the transformation $Y = u(X) = X^{1-p/2}$, when $p \neq 2$, or $Y = u(X) = \ln X$, when $p = 2$, seems to produce a random variable Y whose variance is about free of μ. This is the reason transformations like $\sqrt{X}$, $\ln X$, and, more generally, X^b are so popular in applications.

4.4-6 Let X_1 and X_2 have independent gamma distributions with parameters α, θ and β, θ, respectively. Let $W = X_1/(X_1 + X_2)$. Use a method which is similar to that given in the derivation of the F distribution (Examples 4.4-5) to show that the p.d.f. of W is

$$g(w) = \frac{\Gamma(\alpha + \beta)}{\Gamma(\alpha)\Gamma(\beta)} w^{\alpha-1}(1 - w)^{\beta-1}, \qquad 0 < w < 1.$$

We say that W has a **beta distribution** with parameters α and β (see Example 4.4-4).

4.4-7 Let X_1 and X_2 be independent chi-square random variables with r_1 and r_2 degrees of freedom, respectively. Show that

(a) $U = X_1/(X_1 + X_2)$ has a beta distribution with $\alpha = r_1/2$ and $\beta = r_2/2$.

(b) $V = X_2/(X_1 + X_2)$ has a beta distribution with $\alpha = r_2/2$ and $\beta = r_1/2$.

4.4-8 Let X have a beta distribution with parameters α and β (see Example 4.4-4).

(a) Show that the mean and variance of X are

$$\mu = \frac{\alpha}{\alpha + \beta} \qquad \text{and} \qquad \sigma^2 = \frac{\alpha\beta}{(\alpha + \beta + 1)(\alpha + \beta)^2}.$$

(b) Show that the beta p.d.f. has a mode at $x = (\alpha - 1)/(\alpha + \beta - 2)$.

4.4-9 Determine the constant c such that $f(x) = cx^3(1 - x)^6$, $0 < x < 1$, is a p.d.f.

4.4-10 When α and β are integers and $0 < p < 1$, then

$$\int_0^p \frac{\Gamma(\alpha + \beta)}{\Gamma(\alpha)\Gamma(\beta)} y^{\alpha-1}(1 - y)^{\beta-1} \, dy = \sum_{y=\alpha}^n \binom{n}{y} p^y(1 - p)^{n-y},$$

where $n = \alpha + \beta - 1$. Verify this formula when $\alpha = 4$ and $\beta = 3$.

4.4-11 Evaluate

$$\int_0^{0.4} \frac{\Gamma(7)}{\Gamma(4)\Gamma(3)} y^3 (1-y)^2 \, dy$$

(a) Using integration.
(b) Using the result of Exercise 4.4-10.

4.4-12 Let W_1, W_2 be independent, each with a Cauchy distribution. In this exercise we find the p.d.f. of the sample mean, $(W_1 + W_2)/2$.

(a) Show that the p.d.f. of $X_1 = (1/2)W_1$ is

$$f(x) = \frac{2}{\pi(1+4x^2)}, \qquad -\infty < x < \infty.$$

(b) Let $Y_1 = X_1 + X_2 = \overline{W}$ and $Y_2 = X_1$, where $X_2 = (1/2)W_2$. Show that the joint p.d.f. of Y_1 and Y_2 is

$$g(y_1, y_2) = f(y_1 - y_2)f(y_2), \qquad -\infty < y_1 < \infty, -\infty < y_2 < \infty.$$

(c) Show that the p.d.f. of $Y_1 = \overline{W}$ is given by the **convolution formula**,

$$g_1(y_1) = \int_{-\infty}^{\infty} f(y_1 - y_2)f(y_2) \, dy_2.$$

(d) Show that

$$g_1(y_1) = \frac{1}{\pi(1+y_1^2)}, \qquad -\infty < y_1 < \infty.$$

That is, the p.d.f. of $\overline{W}$ is the same as that of an individual W.

4.4-13 Let X_1, X_2 be independent random variables representing lifetimes (in hours) of two key components of a device, which fails when and only when both components fail. Say each X_i has the exponential distribution with mean 1000. Let $Y_1 = \min(X_1, X_2)$ and $Y_2 = \max(X_1, X_2)$; so the space of Y_1, Y_2 is $0 < y_1 < y_2 < \infty$.

(a) Find $G(y_1, y_2) = P(Y_1 \leq y_1, Y_2 \leq y_2)$.
(b) Compute the probability that the device fails after 1200 hours; that is, compute $P(Y_2 > 1200)$.

4.4-14 A company provides earthquake insurance. The premium X is modeled by the p.d.f.

$$f(x) = \frac{x}{5^2} e^{-x/5}, \qquad 0 < x < \infty,$$

while the claims Y have the p.d.f.

$$g(y) = \frac{1}{5} e^{-y/5}, \qquad 0 < y < \infty.$$

If X and Y are independent, find the p.d.f. of $Z = X/Y$.

4.4-15 Let X be the fraction of a 24-hour period that a person is sleeping. We are trying to model the distribution of X with a beta p.d.f. Which of the following do you believe to be the best model for a person who sleeps for 8 hours on the average: (1) $\alpha = 8$, $\beta = 16$, (2) $\alpha = 16$, $\beta = 32$, (3) $\alpha = 24$, $\beta = 48$? Why? Can you construct a better model?

4.4-16 Let X and Y be the fractions of a 24-hour period that a person spends sleeping and eating, respectively. We wish to model the distribution of X and Y by the joint p.d.f.

$$f(x, y) = \frac{\Gamma(\alpha + \beta + \gamma)}{\Gamma(\alpha)\Gamma(\beta)\Gamma(\gamma)} x^{\alpha-1} y^{\beta-1} (1 - x - y)^{\gamma-1}, \qquad 0 < x, 0 < y, x + y < 1.$$

Select α, β, and γ so that this joint p.d.f. is a reasonable model for someone who sleeps for 8 hours and eats for 2 hours on the average. HINT: X has a beta distribution with parameters α, $\beta + \gamma$, and Y has a beta distribution with parameters β, $\alpha + \gamma$; thus we want $\alpha/(\alpha + \beta + \gamma) = 1/3$ and $\beta/(\alpha + \beta + \gamma) = 1/12$. Now consider the standard deviations.

4.5 SEVERAL INDEPENDENT RANDOM VARIABLES

In Section 4.1 we introduced several distributions concerning two random variables. Each of these random variables could be thought of as a measurement on some random experiment. In this section, we consider the possibility of performing several random experiments or one random experiment several times in which each trial results in one measurement which can be considered a random variable. That is, we obtain one random variable from each experiment, and thus we obtain a collection of several random variables from the several experiments. Further suppose that these experiments are performed in such a way that the events associated with any one of them are independent of the events associated with others, and hence the corresponding random variables are, in the probabilistic sense, independent.

Recall from Section 4.1 that if X_1 and X_2 are random variables of the discrete type with probability mass functions $f_1(x_1)$ and $f_2(x_2)$, respectively, and if

$$P(X_1 = x_1 \text{ and } X_2 = x_2) = P(X_1 = x_1)\, P(X_2 = x_2)$$
$$= f_1(x_1) f_2(x_2), \qquad x_1 \in S_1, \ x_2 \in S_2,$$

then X_1 and X_2 are said to be independent and the joint p.m.f. is $f_1(x_1) f_2(x_2)$.

Sometimes the two random experiments are exactly the same. For illustration, we could cast the die twice resulting in X_1 and X_2, respectively. It is reasonable to say that the p.m.f. of X_1 is $f(x_1) = 1/6, x_1 = 1, 2, \ldots, 6$ and the p.m.f. of X_2 is $f(x_2) = 1/6, x_2 = 1, 2, \ldots, 6$. Assuming independence, which would be a fair way to perform the experiments, the joint p.m.f. is then

$$f(x_1) f(x_2) = \left(\frac{1}{6}\right)\left(\frac{1}{6}\right) = \frac{1}{36}, \qquad x_1 = 1, 2, \ldots, 6; \ x_2 = 1, 2, \ldots, 6.$$

In general, if the p.m.f. $f(x)$ of the independent random variables X_1 and X_2 is the same, then the joint p.m.f. is $f(x_1) f(x_2)$. Moreover, in this case, the collection of the two random variables X_1, X_2 is called a **random sample of size $n = 2$** from a distribution with p.m.f. $f(x)$. Hence, in the two casts of the die, we say that we have a random sample of size $n = 2$ from the uniform distribution on the space $\{1, 2, 3, 4, 5, 6\}$.

EXAMPLE 4.5-1 Let X_1 and X_2 be two independent random variables resulting from two casts of an unbiased die. That is, X_1, X_2 is a random sample of size $n = 2$ from a distribution with p.m.f. $f(x) = 1/6$, $x = 1, 2, \ldots, 6$. We have that

$$E(X_1) = E(X_2) = \sum_{x=1}^{6} xf(x) = 3.5,$$

where $f(x) = 1/6$, $x = 1, 2, \ldots, 6$. Moreover,

$$\text{Var}(X_1) = \text{Var}(X_2) = \sum_{x=1}^{6} (x - 3.5)^2 f(x) = \frac{35}{12}.$$

In addition, from independence,

$$E(X_1 X_2) = E(X_1)E(X_2) = (3.5)(3.5) = 12.25$$

and

$$E[(X_1 - 3.5)(X_2 - 3.5)] = E(X_1 - 3.5)E(X_2 - 3.5) = 0.$$

If $Y = X_1 + X_2$, then

$$E(Y) = E(X_1) + E(X_2) = 3.5 + 3.5 = 7$$

and

$$\begin{aligned}
\text{Var}(Y) &= E[(X_1 + X_2 - 7)^2] = E\{[(X_1 - 3.5) + (X_2 - 3.5)]^2\} \\
&= E[(X_1 - 3.5)^2] + E[2(X_1 - 3.5)(X_2 - 3.5)] + E[(X_2 - 3.5)^2] \\
&= \text{Var}(X_1) + (2)(0) + \text{Var}(X_2) \\
&= (2)\left(\frac{35}{12}\right) = \frac{35}{6}.
\end{aligned}$$

◀

In Example 4.5-1, we can find the p.m.f. $g(y)$ of $Y = X_1 + X_2$. Since the space of Y is $\{2, 3, 4, \ldots, 12\}$, we have, by a rather straightforward calculation, that

$$g(2) = P(X_1 = 1, X_2 = 1) = f(1)f(1) = \left(\frac{1}{6}\right)\left(\frac{1}{6}\right) = \frac{1}{36},$$

$$g(3) = P(X_1 = 1, X_2 = 2 \text{ or } X_1 = 2, X_2 = 1) = \left(\frac{1}{6}\right)\left(\frac{1}{6}\right) + \left(\frac{1}{6}\right)\left(\frac{1}{6}\right) = \frac{2}{36},$$

$$g(4) = P(X_1 = 1, X_2 = 3 \text{ or } X_1 = 2, X_2 = 2 \text{ or } X_1 = 3, X_2 = 1) = \frac{3}{36},$$

and so on. This results in the p.m.f. given by

y	2	3	4	5	6	7	8	9	10	11	12
$g(y)$	$\dfrac{1}{36}$	$\dfrac{2}{36}$	$\dfrac{3}{36}$	$\dfrac{4}{36}$	$\dfrac{5}{36}$	$\dfrac{6}{36}$	$\dfrac{5}{36}$	$\dfrac{4}{36}$	$\dfrac{3}{36}$	$\dfrac{2}{36}$	$\dfrac{1}{36}$

Using this p.m.f., it is simple to calculate

$$E(Y) = \sum_{y=2}^{12} y\, g(y) = 7$$

and

$$\mathrm{Var}(Y) = \sum_{y=2}^{12} (y-7)^2\, g(y) = \frac{35}{6},$$

which agrees with the results of that example.

All of the definitions and results about two random variables of the discrete type can be carried over to two random variables of the continuous type. Moreover, the notions about two independent random variables can be extended to n independent random variables, which can be thought of as measurements on the outcomes of n random experiments. That is, if $X_1, X_2, \ldots, X_n$ are **independent**, then the joint p.d.f. is the product of the respective probability density functions, namely

$$f_1(x_1) f_2(x_2) \ldots f_n(x_n),$$

and the expected value of the product $u_1(X_1) u_2(X_2) \ldots u_n(X_n)$ is the product of the expected values of $u_1(X_1), u_2(X_2), \ldots, u_n(X_n)$.

If all n of the distributions are the same, then the collection of n independent and identically distributed random variables, $X_1, X_2, \ldots, X_n$, is said to be a **random sample of size n from that common distribution**. If $f(x)$ is the common p.m.f. or p.d.f. of these n random variables, the joint p.m.f. or p.d.f. is $f(x_1) f(x_2) \ldots f(x_n)$.

EXAMPLE 4.5-2   Let X_1, X_2, X_3 be a random sample from a distribution with p.d.f.

$$f(x) = e^{-x}, \qquad 0 < x < \infty.$$

The joint p.d.f. of these three random variables is

$$(e^{-x_1})(e^{-x_2})(e^{-x_3}) = e^{-x_1 - x_2 - x_3}, \qquad 0 < x_i < \infty, \ \ i = 1, 2, 3.$$

The probability $P(0 < X_1 < 1, 2 < X_2 < 4, 3 < X_3 < 7)$ is given by the product of three integrals,

$$\left(\int_0^1 e^{-x_1} dx_1\right)\left(\int_2^4 e^{-x_2} dx_2\right)\left(\int_3^7 e^{-x_3} dx_3\right)$$

$$= (1 - e^{-1})(e^{-2} - e^{-4})(e^{-3} - e^{-7}),$$

because of the independence of X_1, X_2, X_3. ◄

EXAMPLE 4.5-3 An electronic device runs until one of its three components fails. The lifetimes (in weeks), X_1, X_2, X_3, of these components are independent and each has the Weibull p.d.f.

$$f(x) = \frac{2x}{25} e^{-(x/5)^2}, \qquad 0 < x < \infty.$$

The probability that the device stops running in the first three weeks is equal to

$$1 - P(X_1 > 3, X_2 > 3, X_3 > 3) = 1 - P(X_1 > 3)P(X_2 > 3)P(X_3 > 3)$$

$$= 1 - \left(\int_3^\infty f(x)dx\right)^3$$

$$= 1 - \left(\left[-e^{-(x/5)^2}\right]_3^\infty\right)^3$$

$$= 1 - \left[e^{-(3/5)^2}\right]^3 = 0.660. \quad ◄$$

The special result following Example 4.5-1 with a linear function of two random variables extends to more general functions of several random variables. We accept the following without proof. (See Hogg, McKean, and Craig, 2005.)

Theorem 4.5-1 Let $X_1, X_2, \ldots, X_n$ be n independent random variables with joint p.m.f. $f_1(x_1) f_2(x_2) \ldots f_n(x_n)$. Let the random variable $Y = u(X_1, X_2, \ldots, X_n)$ have the p.m.f. $g(y)$. Then, in the discrete case,

$$E(Y) = \sum_y y\, g(y)$$

$$= \sum_{x_1} \sum_{x_2} \cdots \sum_{x_n} u(x_1, x_2, \ldots, x_n) f_1(x_1) f_2(x_2) \ldots f_n(x_n),$$

provided that these summations exist. For random variables of the continuous type, obvious changes are made; in particular, integrals replace the summations.

The next theorem proves that the expected value of the product of functions of n independent random variables is the product of their expected values.

Theorem 4.5-2   If $X_1, X_2, \ldots, X_n$ are independent random variables and, for $i = 1, 2, \ldots, n$, $E[u_i(X_i)]$ exists, then

$$E[u_1(X_1) u_2(X_2) \ldots u_n(X_n)] = E[u_1(X_1)]E[u_2(X_2)] \ldots E[u_n(X_n)].$$

Proof: In the discrete case, we have that

$$E[u_1(X_1)u_2(X_2) \cdots u_n(X_n)]$$

$$= \sum_{x_1} \sum_{x_2} \cdots \sum_{x_n} u_1(x_1) u_2(x_2) \cdots u_n(x_n) f_1(x_1) f_2(x_2) \cdots f_n(x_n)$$

$$= \sum_{x_1} u_1(x_1) f_1(x_1) \sum_{x_2} u_2(x_2) f_2(x_2) \cdots \sum_{x_n} u_n(x_n) f_n(x_n)$$

$$= E[u_1(X_1)]E[u_2(x_2)] \cdots E[u_n(X_n)].$$

In the proof for the continuous case, obvious changes are made; in particular, integrals replace summations.   ○

Remark Sometimes students recognize that $X^2 = (X)(X)$ and thus believe that $E(X^2)$ is equal to $[E(X)][E(X)] = [E(X)]^2$ because the above theorem states that the expected value of the product is the product of the expected values. However, note the hypothesis of independence in the theorem, and certainly X is not independent of itself. Incidentally, if $E(X^2)$ did equal $[E(X)]^2$, then the variance of X

$$\sigma^2 = E(X^2) - [E(X)]^2$$

would always equal zero. This really happens only in the case of degenerate (one point) distributions.   ∎

EXAMPLE 4.5-4   It is interesting to note that these two theorems allow us to determine the mean, the variance, and the moment-generating function of a function such as $Y = X_1 + X_2$, where X_1 and X_2 are independent and each has a uniform distribution on $\{1, 2, 3, 4\}$. So $\mu_1 = \mu_2 = 2.5$ and $E(X_1 + X_2) = 5$.
   The variance of Y is

$$\sigma_Y^2 = E[(Y - \mu_Y)^2] = E[(X_1 + X_2 - \mu_1 - \mu_2)^2],$$

where $\mu_i = E(X_i) = 5/2, i = 1, 2$. Thus

$$\sigma_Y^2 = E[[(X_1 - \mu_1) + (X_2 - \mu_2)]^2]$$

$$= E[(X_1 - \mu_1)^2 + 2(X_1 - \mu_1)(X_2 - \mu_2) + (X_2 - \mu_2)^2].$$

But, from an earlier result about the expected value being a linear operator, we have

$$\sigma_Y^2 = E[(X_1 - \mu_1)^2] + 2E[(X_1 - \mu_1)(X_2 - \mu_2)] + E[(X_2 - \mu_2)^2].$$

However, since X_1 and X_2 are independent, then

$$E[(X_1 - \mu_1)(X_2 - \mu_2)] = E(X_1 - \mu_1)E(X_2 - \mu_2)$$
$$= (\mu_1 - \mu_1)(\mu_2 - \mu_2) = 0.$$

Thus

$$\sigma_Y^2 = \sigma_1^2 + \sigma_2^2,$$

where

$$\sigma_i^2 = E[(X_i - \mu_i)^2], \qquad i = 1, 2.$$

In the case in which X_1 and X_2 could be considered the outcomes on two independent casts of a four-sided die, we have $\sigma_i^2 = 5/4$, $i = 1, 2$, and, hence

$$\sigma_Y^2 = \frac{5}{4} + \frac{5}{4} = \frac{5}{2}.$$

Finally, the moment-generating function is

$$M_Y(t) = E(e^{tY}) = E[e^{t(X_1 + X_2)}] = E(e^{tX_1} e^{tX_2}).$$

The independence of X_1 and X_2 implies that

$$M_Y(t) = E(e^{tX_1})E(e^{tX_2}).$$

For our example, X_1 and X_2 have the same p.m.f.

$$f(x) = \frac{1}{4}, \qquad x = 1, 2, 3, 4,$$

and thus the same moment-generating function

$$M_X(t) = \frac{1}{4}e^t + \frac{1}{4}e^{2t} + \frac{1}{4}e^{3t} + \frac{1}{4}e^{4t}.$$

So we have that $M_Y(t) = [M_X(t)]^2$ equals

$$\frac{1}{16}e^{2t} + \frac{2}{16}e^{3t} + \frac{3}{16}e^{4t} + \frac{4}{16}e^{5t} + \frac{3}{16}e^{6t} + \frac{2}{16}e^{7t} + \frac{1}{16}e^{8t}.$$

Note that the coefficient of e^{bt} is equal to the probability $P(Y = b)$; for illustration, $4/16 = P(Y = 5)$. Thus we could find the distribution of Y by determining its moment-generating function. ◀

EXERCISES

4.5-1 Let X_1 and X_2 be independent Poisson random variables with respective means $\lambda_1 = 2$ and $\lambda_2 = 3$. Find

(a) $P(X_1 = 3, X_2 = 5)$.

(b) $P(X_1 + X_2 = 1)$.

HINT. Note that this event can occur if and only if $\{X_1 = 1, X_2 = 0\}$ or $\{X_1 = 0, X_2 = 1\}$.

4.5-2 Let X_1 and X_2 be independent random variables with respective binomial distributions $b(3, 1/2)$ and $b(5, 1/2)$. Determine

(a) $P(X_1 = 2, X_2 = 4)$.

(b) $P(X_1 + X_2 = 7)$.

4.5-3 Let X_1 and X_2 be independent random variables with probability density functions $f_1(x_1) = 2x_1$, $0 < x_1 < 1$, and $f_2(x_2) = 4x_2^3$, $0 < x_2 < 1$, respectively. Compute

(a) $P(0.5 < X_1 < 1$ and $0.4 < X_2 < 0.8)$.

(b) $E(X_1^2 X_2^3)$.

4.5-4 Let X_1 and X_2 be a random sample of size $n = 2$ from the exponential distribution with p.d.f. $f(x) = 2e^{-2x}$, $0 < x < \infty$. Find

(a) $P(0.5 < X_1 < 1.0, 0.7 < X_2 < 1.2)$.

(b) $E[X_1(X_2 - 0.5)^2]$.

4.5-5 Let X_1, X_2 be a random sample of size $n = 2$ from the chi-square distribution $\chi^2(5)$.

(a) Compute the mean and the variance of $Y = X_1 + X_2$.

(b) Evaluate $E[e^{t(X_1+X_2)}] = E(e^{tX_1}e^{tX_2}) = E(e^{tX_1})E(e^{tX_2})$, the latter resulting from the independence of X_1 and X_2.

(c) Note that the expression in (b) is the m.g.f. of $\chi^2(10)$. What is its mean and variance and how do these compare with the results in (a)?

4.5-6 Suppose two independent claims are made on two insured homes, where each claim has p.d.f.

$$f(x) = \frac{4}{x^5}, \qquad 1 < x < \infty,$$

in which the unit is 1000 dollars. Find the expected value of the largest claim.

HINT: If X_1 and X_2 are the two independent claims and $Y = \max(X_1, X_2)$, then

$$G(y) = P(Y \le y) = P(X_1 \le y)P(X_2 \le y) = [P(X \le y)]^2.$$

Find $g(y) = G'(y)$ and $E(Y)$.

4.5-7 The distributions of incomes in two cities follow two Pareto type p.d.f.s:

$$f(x) = \frac{2}{x^3}, \qquad 1 < x < \infty \qquad \text{and} \qquad g(y) = \frac{3}{y^4}, \qquad 1 < y < \infty,$$

respectively. Here one unit represents 20,000 dollars. One person with income is selected at random from each city and let X and Y be their respective incomes. Compute $P(X < Y)$.

4.5-8 Let X_1, X_2, X_3 denote a random sample of size $n = 3$ from a distribution with the geometric p.m.f.

$$f(x) = \left(\frac{3}{4}\right)\left(\frac{1}{4}\right)^{x-1}, \qquad x = 1, 2, 3, \dots.$$

(a) Compute $P(X_1 = 1, X_2 = 3, X_3 = 1)$.

(b) Determine $P(X_1 + X_2 + X_3 = 5)$.

(c) If Y equals the maximum of X_1, X_2, X_3, find

$$P(Y \le 2) = P(X_1 \le 2)P(X_2 \le 2)P(X_3 \le 2).$$

4.5-9 Let X_1, X_2 be a random sample of size $n = 2$ from a distribution with p.d.f. $f(x) = 3x^2$, $0 < x < 1$. Determine

(a) $P(\max X_i < 3/4) = P(X_1 < 3/4, X_2 < 3/4)$.

(b) The mean and the variance of $Y = X_1 + X_2$.

4.5-10 Let X_1, X_2, X_3 be a random sample of size $n = 3$ from the exponential distribution with p.d.f. $f(x) = e^{-x}, 0 < x < \infty$. Find

$$P(1 < \min X_i) = P(1 < X_1, 1 < X_2, 1 < X_3).$$

4.5-11 Let X_1, X_2, X_3 be three independent random variables with binomial distributions $b(4, 1/2)$, $b(6, 1/3)$, and $b(12, 1/6)$, respectively. Find

(a) $P(X_1 = 2, X_2 = 2, X_3 = 5)$.

(b) $E(X_1 X_2 X_3)$.

(c) The mean and the variance of $Y = X_1 + X_2$.

4.5-12 Let X_1, X_2, X_3 be independent random variables that represent lifetimes (in hours) of three key components of a device. Say their respective distributions are exponential with means 1000, 1500, and 2000. Let Y be the minimum of X_1, X_2, X_3 and compute $P(Y > 1000)$.

4.5-13 Flip $n = 8$ fair coins and remove all that came up heads. Flip the other (tails) coins and remove the heads. Continue flipping the remaining coins until each has come up heads. We shall find the p.m.f. of Y, the number of trials needed. Let X_i equal the number of flips required to observe heads on coin $i, i = 1, 2, \dots, 8$. Then $Y = \max(X_1, X_2, \dots, X_8)$.

(a) Show that $P(Y \le y) = [1 - (1/2)^y]^8$.

(b) Show that $P(Y = y) = [1 - (1/2)^y]^8 - [1 - (1/2)^{y-1}]^8$, $\qquad y = 1, 2, \dots$.

(c) Use a computer algebra system such as *Maple* or *Mathematica* to show that $E(Y) = 13{,}315{,}424/3{,}011{,}805 = 4.421$.

(d) What happens to the expected value of Y as the number of coins is doubled?

4.5-14 The owner of a property that is for sale is willing to accept the maximum of four independent bids (in $100,000 units), which have a common p.d.f. $f(x) = 2x, 0 < x < 1$. What is the expected value of the highest bid?

4.5-15 A device contains three components, each of which has a lifetime in hours with the p.d.f.

$$f(x) = \frac{2x}{10^2} e^{-(x/10)^2}, \qquad 0 < x < \infty.$$

The device fails with the failure of one of the components. Assuming independent lifetimes, what is the probability the device fails in the first hour of operation?

4.5-16 Three drugs are being tested for use as the treatment of a certain disease. Let p_1, p_2, and p_3 represent the probabilities of success for the respective drugs. As three patients come in, each is given one of the drugs in a random order. After $n = 10$ "triples," assuming independence compute the probability that the maximum number of successes with one of the drugs exceeds eight if in fact $p_1 = p_2 = p_3 = 0.7$.

4.5-17 Three components are placed in series. The time in hours to failure of each has the p.d.f.

$$f(x) = \frac{x}{500^2} e^{-x/500}, \qquad 0 < x < \infty.$$

Since they are in series, we are concerned about the minimum time Y to failure of the three. Assuming independence, find the c.d.f. and the p.d.f. of Y and compute $P(Y \leq 300)$. HINT: $G(y) = P(Y \leq y) = 1 - P(Y > y) = 1 - P(\text{all three} > y)$.

4.5-18 Each of eight bearings in a bearing assembly has a diameter (in millimeters) that has the p.d.f.

$$f(x) = 10x^9, \qquad 0 < x < 1.$$

Assuming independence, find the distribution function and the p.d.f. of the maximum diameter, say Y, of these eight and compute $P(0.9999 < Y < 1)$.

4.5-19 Let X_1, X_2, X_3 represent the independent failure times in years of three components in parallel. The p.d.f.s are $f_1(x_1) = 3x_1^2$, $0 < x_1 < 1$; $f_2(x_2) = 4x_2^3$, $0 < x_2 < 1$, and $f_3(x_3) = 6x_3^5$, $0 < x_3 < 1$. Let $Y = \text{maximum}(X_1, X_2, X_3)$. Find the distribution function and p.d.f. of Y. HINT: $G(y) = P(Y \leq y) = P(\text{all three} \leq y)$.

4.6 DISTRIBUTIONS OF SUMS OF INDEPENDENT RANDOM VARIABLES

Functions of random variables are usually of interest in statistical applications; and, of course, we have already considered some functions, called **statistics**, of the observations of a random sample. Two important statistics are the sample mean $\overline{X}$ and the sample variance S^2. Of course, in a particular sample, say $x_1, x_2, \ldots, x_n$, we observed definite values of these statistics, $\bar{x}$ and s^2; however, we should recognize that each value is only one observation of the respective random variables, $\overline{X}$ and S^2. That is, each $\overline{X}$ or S^2 (or, more generally, any statistic) is also a random variable with its own distribution. In this chapter we determine the distributions of some of the important statistics and, more generally, distributions of functions of random variables. The derivations of distributions of these functions fall in that area of statistics usually referred to as **sampling distribution theory**.

In this section we restrict our attention to those functions that are linear combinations of independent random variables. We shall first prove an important theorem about the mean and the variance of such a linear combination, extending the results in the last example.

Theorem 4.6-1   If $X_1, X_2, \ldots, X_n$ are n independent random variables with respective means $\mu_1, \mu_2, \ldots, \mu_n$ and variances $\sigma_1^2, \sigma_2^2, \ldots, \sigma_n^2$, then the mean and the variance of $Y = \sum_{i=1}^{n} a_i X_i$, where $a_1, a_2, \ldots, a_n$ are real constants, are

$$\mu_Y = \sum_{i=1}^{n} a_i \mu_i \qquad \text{and} \qquad \sigma_Y^2 = \sum_{i=1}^{n} a_i^2 \sigma_i^2,$$

respectively.

Proof: We have that

$$\mu_Y = E(Y) = E\left(\sum_{i=1}^{n} a_i X_i\right) = \sum_{i=1}^{n} a_i E(X_i) = \sum_{i=1}^{n} a_i \mu_i,$$

because the expected value of the sum is the sum of the expected values (i.e., E is a linear operator). Also,

$$\sigma_Y^2 = E[(Y - \mu_Y)^2] = E\left[\left(\sum_{i=1}^{n} a_i X_i - \sum_{i=1}^{n} a_i \mu_i\right)^2\right]$$

$$= E\left\{\left[\sum_{i=1}^{n} a_i (X_i - \mu_i)\right]^2\right\} = E\left[\sum_{i=1}^{n} \sum_{j=1}^{n} a_i a_j (X_i - \mu_i)(X_j - \mu_j)\right].$$

Again using the fact that E is a linear operator, we obtain

$$\sigma_Y^2 = \sum_{i=1}^{n} \sum_{j=1}^{n} a_i a_j E[(X_i - \mu_i)(X_j - \mu_j)].$$

However, if $i \neq j$, then from the independence of X_i and X_j we have

$$E[(X_i - \mu_i)(X_j - \mu_j)] = E(X_i - \mu_i)E(X_j - \mu_j) = (\mu_i - \mu_i)(\mu_j - \mu_j) = 0.$$

Thus the variance can be written as

$$\sigma_Y^2 = \sum_{i=1}^{n} a_i^2 E[(X_i - \mu_i)^2] = \sum_{i=1}^{n} a_i^2 \sigma_i^2.$$

○

Remark While Theorem 4.6-1 gives the mean and the variance of a linear function of independent random variables, the proof can easily be modified to the case in which X_i and X_j are correlated. Then

$$E[(X_i - \mu_i)(X_j - \mu_j)] = \rho_{ij}\sigma_i\sigma_j,$$

instead of zero, where ρ_{ij} is the correlation coefficient of X_i and X_j. Thus

$$\sigma_Y^2 = \sum_{i=1}^{n} a_i^2 \sigma_i^2 + 2 \sum \sum_{i<j} a_i a_j \rho_{ij} \sigma_i \sigma_j;$$

the factor 2 appears because the sum is over $i < j$ and

$$a_i a_j \rho_{ij} \sigma_i \sigma_j = a_j a_i \rho_{ji} \sigma_j \sigma_i.$$

The mean of Y is still the same in both cases, namely

$$\mu_Y = \sum_{i=1}^{n} a_i \mu_i. \qquad \blacksquare$$

We give two illustrations of the theorem.

EXAMPLE 4.6-1   Let the independent random variables X_1 and X_2 have respectively means $\mu_1 = -4$ and $\mu_2 = 3$ and variances $\sigma_1^2 = 4$ and $\sigma_2^2 = 9$. The mean and the variance of $Y = 3X_1 - 2X_2$ are, respectively,

$$\mu_Y = (3)(-4) + (-2)(3) = -18$$

and

$$\sigma_Y^2 = (3)^2(4) + (-2)^2(9) = 72. \qquad \blacktriangleleft$$

EXAMPLE 4.6-2   Let X_1, X_2 be a random sample from a distribution with mean μ and variance σ^2. Let $Y = X_1 - X_2$; then

$$\mu_Y = \mu - \mu = 0$$

and

$$\sigma_Y^2 = (1)^2 \sigma^2 + (-1)^2 \sigma^2 = 2\sigma^2. \qquad \blacktriangleleft$$

Now consider the **mean of a random sample**, $X_1, X_2, \ldots, X_n$, from a distribution with mean μ and variance σ^2, namely

$$\overline{X} = \frac{X_1 + X_2 + \cdots + X_n}{n},$$

which is a linear function with each $a_i = 1/n$. Then

$$\mu_{\overline{X}} = \sum_{i=1}^{n} \left(\frac{1}{n}\right)\mu = \mu \qquad \text{and} \qquad \sigma_{\overline{X}}^2 = \sum_{i=1}^{n} \left(\frac{1}{n}\right)^2 \sigma^2 = \frac{\sigma^2}{n}.$$

That is, the mean of $\overline{X}$ is that of the distribution from which the sample arose, but the variance of $\overline{X}$ is that of the underlying distribution divided by n. Any function of the sample observations, $X_1, X_2, \ldots, X_n$, is called a **statistic** so here $\overline{X}$ is a statistic and also an **estimator** of the distribution mean μ.

In some applications it is sufficient to know the mean and variance of a linear combination of random variables, say Y. However, it is often helpful to know exactly how Y is distributed. The next theorem can frequently be used to find the distribution of a linear combination of independent random variables.

Theorem 4.6-2   If $X_1, X_2, \ldots, X_n$ are independent random variables with respective moment-generating functions $M_{X_i}(t)$, $i = 1, 2, 3, \ldots, n$, then the moment-generating function of $Y = \sum_{i=1}^{n} a_i X_i$ is

$$M_Y(t) = \prod_{i=1}^{n} M_{X_i}(a_i t).$$

Proof: The moment-generating function of Y is given by

$$M_Y(t) = E[e^{tY}] = E[e^{t(a_1 X_1 + a_2 X_2 + \cdots + a_n X_n)}]$$

$$= E[e^{a_1 t X_1} e^{a_2 t X_2} \ldots e^{a_n t X_n}]$$

$$= E[e^{a_1 t X_1}] E[e^{a_2 t X_2}] \ldots E[e^{a_n t X_n}]$$

using Theorem 4.5-2. However, since

$$E(e^{t X_i}) = M_{X_i}(t),$$

then

$$E(e^{a_i t X_i}) = M_{X_i}(a_i t).$$

Thus we have that

$$M_Y(t) = M_{X_1}(a_1 t) M_{X_2}(a_2 t) \ldots M_{X_n}(a_n t) = \prod_{i=1}^{n} M_{X_i}(a_i t).$$

A corollary follows immediately, and it will be used in some important examples.

Corollary 4.6-1   If $X_1, X_2, \ldots, X_n$ are observations of a random sample from a distribution with moment-generating function $M(t)$, then

(a) the moment-generating function of $Y = \sum_{i=1}^{n} X_i$ is

$$M_Y(t) = \prod_{i=1}^{n} M(t) = [M(t)]^n;$$

(b) the moment-generating function of $\overline{X} = \sum_{i=1}^{n}(1/n)X_i$ is

$$M_{\overline{X}}(t) = \prod_{i=1}^{n} M\left(\frac{t}{n}\right) = \left[M\left(\frac{t}{n}\right)\right]^n.$$ ■

Proof: For (a), let $a_i = 1$, $i = 1, 2, \ldots, n$, in Theorem 4.6-2. For (b), take $a_i = 1/n$, $i = 1, 2, \ldots, n$. ○

The following examples and the exercises give some important applications of Theorem 4.6-2 and its corollary.

EXAMPLE 4.6-3   Let $X_1, X_2, \ldots, X_n$ denote the outcomes on n Bernoulli trials. The moment-generating function of X_i, $i = 1, 2, \ldots, n$, is

$$M(t) = q + pe^t.$$

If

$$Y = \sum_{i=1}^{n} X_i,$$

then

$$M_Y(t) = \prod_{i=1}^{n} (q + pe^t) = (q + pe^t)^n.$$

Thus we again see that Y is $b(n, p)$. ◄

EXAMPLE 4.6-4   Let X_1, X_2, X_3 be the observations of a random sample of size $n = 3$ from the exponential distribution having mean θ and, of course, moment-generating function $M(t) = 1/(1 - \theta t)$, $t < 1/\theta$. The moment-generating function of $Y = X_1 + X_2 + X_3$ is

$$M_Y(t) = [(1 - \theta t)^{-1}]^3 = (1 - \theta t)^{-3}, \qquad t < 1/\theta,$$

which is that of a gamma distribution with parameters $\alpha = 3$ and θ. Thus Y has this distribution. On the other hand, the moment-generating function of $\overline{X}$ is

$$M_{\overline{X}}(t) = \left[\left(1 - \frac{\theta t}{3}\right)^{-1}\right]^3 = \left(1 - \frac{\theta t}{3}\right)^{-3}, \qquad t < 3/\theta.$$

Hence the distribution of $\overline{X}$ is gamma with the parameters $\alpha = 3$ and $\theta/3$, respectively. ◀

Theorem 4.6-3

Let $X_1, X_2, \ldots, X_n$ be independent chi-square random variables with $r_1, r_2, \ldots, r_n$ degrees of freedom, respectively. Then $Y = X_1 + X_2 + \cdots + X_n$ is $\chi^2(r_1 + r_2 + \cdots + r_n)$.

Proof: The m.g.f. of Y is, by Theorem 4.6-2 with each $a = 1$,

$$M_Y(t) = \prod_{i=1}^{n} M_{X_i}(t) = (1 - 2t)^{-r_1/2}(1 - 2t)^{-r_2/2} \cdots (1 - 2t)^{-r_n/2}$$

$$= (1 - 2t)^{-\Sigma r_i/2}, \qquad \text{with } t < 1/2,$$

which is the m.g.f. of a $\chi^2(r_1 + r_2 + \cdots + r_n)$. Thus Y is $\chi^2(r_1 + r_2 + \cdots + r_n)$. ◯

EXERCISES

4.6-1 Let X_1 and X_2 be observations of a random sample of size $n = 2$ from a distribution with p.m.f. $f(x) = x/6$, $x = 1, 2, 3$. Then find the p.m.f. of $Y = X_1 + X_2$. Determine the mean and the variance of the sum in two ways.

4.6-2 Let X_1 and X_2 be a random sample of size $n = 2$ from a distribution with p.d.f. $f(x) = 6x(1 - x), 0 < x < 1$. Find the mean and the variance of $Y = X_1 + X_2$.

4.6-3 Let X_1, X_2, X_3 be a random sample of size 3 from the distribution with p.m.f. $f(x) = 1/4$, $x = 1, 2, 3, 4$. For example, observe three independent rolls of a fair four-sided die.

  (a) Find the p.m.f. of $Y = X_1 + X_2 + X_3$.
  (b) Sketch a bar graph of the p.m.f. of Y.

4.6-4 Let X_1 and X_2 be two independent random variables with respective means μ_1 and μ_2 and variances σ_1^2 and σ_2^2. Show that the mean and the variance of $Y = X_1 X_2$ are $\mu_1 \mu_2$ and $\sigma_1^2 \sigma_2^2 + \mu_1^2 \sigma_2^2 + \mu_2^2 \sigma_1^2$, respectively. HINT: Note that $E(Y) = E(X_1)E(X_2)$ and $E(Y^2) = E(X_1^2)E(X_2^2)$.

4.6-5 Let X_1 and X_2 be two independent random variables with respective means 3 and 7 and variances 9 and 25. Compute the mean and the variance of $Y = -2X_1 + X_2$.

4.6-6 Let X_1 and X_2 have independent distributions $b(n_1, p)$ and $b(n_2, p)$. Find the moment-generating function of $Y = X_1 + X_2$. How is Y distributed?

4.6-7 Let X_1, X_2, X_3 be mutually independent random variables with Poisson distributions having means 2, 1, 4, respectively.

  (a) Find the moment-generating function of the sum $Y = X_1 + X_2 + X_3$.
  (b) How is Y distributed?
  (c) Compute $P(3 \leq Y \leq 9)$.

4.6-8 Generalize Exercise 4.6-7 by showing that the sum of n independent Poisson random variables with respective means $\mu_1, \mu_2, \ldots, \mu_n$ is Poisson with mean

$$\mu_1 + \mu_2 + \cdots + \mu_n.$$

4.6-9 Let X_1, X_2, X_3, X_4, X_5 be a random sample of size 5 from a geometric distribution with $p = 1/3$.

(a) Find the moment-generating function of $Y = X_1 + X_2 + X_3 + X_4 + X_5$.

(b) How is Y distributed?

4.6-10 Let $W = X_1 + X_2 + \cdots + X_h$, a sum of h mutually independent and identically distributed exponential random variables with mean θ. Show that W has a gamma distribution with mean $h\theta$.

4.6-11 Let X_1, X_2, X_3 denote a random sample of size 3 from a gamma distribution with $\alpha = 7$ and $\theta = 5$.

(a) Find the moment-generating function of $Y = X_1 + X_2 + X_3$.

(b) How is Y distributed?

4.6-12 The moment-generating function of X is

$$M_X(t) = \left(\frac{1}{4}\right)(e^t + e^{2t} + e^{3t} + e^{4t});$$

the moment-generating function of Y is

$$M_Y(t) = \left(\frac{1}{3}\right)(e^t + e^{2t} + e^{3t});$$

X and Y are independent random variables. Let $W = X + Y$.

(a) Find the moment-generating function of W.

(b) Give the p.m.f. of W; that is, determine $P(W = w)$, $w = 2, 3, \ldots, 7$, from the moment-generating function of W.

4.6-13 Let X equal the outcome when a four-sided die is rolled. Let Y equal the outcome when a six-sided die is rolled. Let $W = X + Y$. Assume X and Y are independent.

(a) Find the moment-generating function of W.

(b) Give the p.m.f. of W.

4.6-14 Let X and Y, with respective p.m.f.s $f(x)$ and $g(y)$, be independent discrete random variables, each of whose support is a subset of the nonnegative integers $0, 1, 2, \ldots$. Show that the p.m.f. of $W = X + Y$ is given by the **convolution formula**

$$h(w) = \sum_{x=0}^{w} f(x)g(w - x), \qquad w = 0, 1, 2, \ldots.$$

HINT: Argue that $h(w) = P(W = w)$ is the probability of the $w+1$ mutually exclusive events $(x, y = w - x)$, $x = 0, 1, \ldots, w$.

4.6-15 Let X and Y equal the outcomes when two fair four-sided dice are rolled. Let $W = X + Y$. Assuming independence, find the p.m.f. of W when

(a) The first die has two faces numbered 0 and two faces numbered 2, the second die has its faces numbered 0, 1, 4, and 5.

(b) The faces on the first die are numbered 0, 1, 2, and 3; the faces on the second die are numbered 0, 4, 8, and 12.

4.6-16 Let X and Y equal the outcomes when two fair six-sided dice are rolled. Let $W = X + Y$. Assuming independence, find the p.m.f. of W when

(a) The first die has three faces numbered 0 and three faces numbered 2, the second die has its faces numbered 0, 1, 4, 5, 8, and 9.

(b) The faces on the first die are numbered 0, 1, 2, 3, 4, and 5, the faces on the second die are numbered 0, 6, 12, 18, 24, and 30.

4.6-17 Let $X_1, X_2, \ldots, X_8$ be a random sample from a distribution having p.m.f. $f(x) = (x+1)/6, x = 0, 1, 2$.

    (a) Use Exercise 4.6-14 to find the p.m.f. of $W_1 = X_1 + X_2$.

    (b) What is the p.m.f. of $W_2 = X_3 + X_4$?

    (c) Now find the p.m.f. of $W = W_1 + W_2 = X_1 + X_2 + X_3 + X_4$.

    (d) Find the p.m.f. of $Y = X_1 + X_2 + \cdots + X_8$.

    (e) Construct probability histograms for X_1, W_1, W, and Y. Are these histograms skewed or symmetric?

4.6-18 Roll a fair four-sided die eight times and denote the outcomes by $X_1, X_2, \ldots, X_8$. Use Exercise 4.6-14 to find the p.m.f.s of

    (a) $X_1 + X_2$.   **(b)** $\displaystyle\sum_{i=1}^{4} X_i$.   **(c)** $\displaystyle\sum_{i=1}^{8} X_i$.

4.6-19 Given a fair four-sided die, let Y equal the number of rolls needed to observe each face at least once.

    (a) Argue that $Y = X_1 + X_2 + X_3 + X_4$, where X_i has a geometric distribution with $p_i = (5-i)/4, i = 1, 2, 3, 4$, and X_1, X_2, X_3, X_4 are independent.

    (b) Find the mean and variance of Y.

    (c) Find $P(Y = y)$, $y = 4, 5, 6$.

4.6-20 The number of accidents in a period of one week follows a Poisson distribution with mean two. The numbers of accidents from week to week are independent. What is the probability of exactly seven accidents in a given three weeks.

4.6-21 In a study concerning a new treatment of a certain disease, two groups of 25 participants in each were followed for five years: those in one group took the old treatment and those in the other took the new treatment. The drop out rate was 50% in both groups over that 5-year period. Let X be the number that dropped out in the first group and Y the number in the second group. Assuming independence where needed, give the sum that equals the probability that $Y \geq X + 2$. HINT: What is the distribution of $Y - X + 25$?

4.6-22 The number X of sick days taken during a year by an employee follows a Poisson distribution with mean two. Let us observe four such employees. Assuming independence, compute the probability that their total number of sick days exceeds 10.

4.6-23 Let X be the number of flaws on the outside of a new washing machine. Say $X_1, X_2, \ldots, X_{10}$ is a random sample of 10 such X values that has a Poisson distribution with mean one.

    (a) The manufacturer gives a discount to a buyer of any machine with two or more flaws. Among the 10 machines, what is the probability of giving at least 3 discounts?

    (b) Among the 10 machines, what is the probability of less than 13 flaws?

4.6-24 The number of cracks on a highway averages 0.5 per mile and follows a Poisson distribution. Assuming independence (which may not be a good assumption–why?), what is the probability that in a 40 mile stretch of that highway there are less than 15 cracks?

4.6-25 A doorman at a hotel is trying to get three taxicabs for three different couples. The arrivals of empty cabs has an exponential distribution with mean 2 minutes. Assuming independence, what is the probability that he will get all three couples taken care of within six minutes?

4.6-26 The time X in minutes of a patient visit with a cardiovascular disease specialist is modelled by a gamma p.d.f. with $\alpha = 1.5$ and $\theta = 10$. You have 4 patients ahead of you. Assuming independence, what integral gives the probability that you will wait more than 90 minutes?

4.6-27 In considering medical insurance for a certain operation, let X equal the amount (in dollars) paid for the doctor and let Y equal the amount paid to the hospital. In the past, the variances have been $\text{Var}(X) = 8100$, $\text{Var}(Y) = 10,000$, and $\text{Var}(X+Y) = 20,000$. Due to increased expenses, it was decided to increase the doctor's fee by \$500 and increase the hospital charge Y by eight percent. Calculate the variance of the new total claim, namely $X + 500 + (1.08)Y$.

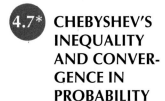

4.7* CHEBYSHEV'S INEQUALITY AND CONVERGENCE IN PROBABILITY

In this section we use Chebyshev's inequality to show, in another sense, that the sample mean, $\bar{x}$, is a good statistic to use to estimate a population mean μ; the relative frequency of success in n Bernoulli trials, y/n, is a good statistic for estimating p; and the empirical distribution function, $F_n(x)$, can be used to estimate the theoretical distribution function $F(x)$. The effect of the sample size n on these estimates is discussed.

We begin by showing that Chebyshev's inequality gives added significance to the standard deviation in terms of bounding certain probabilities. The inequality is valid for **all** distributions for which the standard deviation exists. The proof is given for the discrete case, but it holds for the continuous case with integrals replacing summations.

Theorem 4.7-1

(Chebyshev's Inequality) If the random variable X has a mean μ and variance σ^2, then for every $k \geq 1$,

$$P(|X - \mu| \geq k\sigma) \leq \frac{1}{k^2}.$$

Proof: Let $f(x)$ denote the p.m.f. of X. Then

$$\sigma^2 = E[(X - \mu)^2] = \sum_{x \in S} (x - \mu)^2 f(x)$$

$$= \sum_{x \in A} (x - \mu)^2 f(x) + \sum_{x \in A'} (x - \mu)^2 f(x), \qquad (4.7\text{-}1)$$

where

$$A = \{x : |x - \mu| \geq k\sigma\}.$$

The second term in the righthand member of Equation 4.7-1 is the sum of nonnegative numbers and thus is greater than or equal to zero. Hence

$$\sigma^2 \geq \sum_{x \in A} (x - \mu)^2 f(x).$$

However, in A, $|x - \mu| \geq k\sigma$; so

$$\sigma^2 \geq \sum_{x \in A} (k\sigma)^2 f(x) = k^2 \sigma^2 \sum_{x \in A} f(x).$$

But the latter summation equals $P(X \in A)$, and thus

$$\sigma^2 \geq k^2 \sigma^2 P(X \in A) = k^2 \sigma^2 P(|X - \mu| \geq k\sigma).$$

That is,

$$P(|X - \mu| \geq k\sigma) \leq \frac{1}{k^2}.$$

$\bigcirc$

Corollary 4.7-1   If $\varepsilon = k\sigma$, then

$$P(|X - \mu| \geq \varepsilon) \leq \frac{\sigma^2}{\varepsilon^2}.$$

$\blacksquare$

In words, Chebyshev's inequality states that the probability that X differs from its mean by at least k standard deviations is less than or equal to $1/k^2$. It follows that the probability that X differs from its mean by less than k standard deviations is at least $1 - 1/k^2$. That is,

$$P(|X - \mu| < k\sigma) \geq 1 - \frac{1}{k^2}.$$

From the corollary, it also follows that

$$P(|X - \mu| < \varepsilon) \geq 1 - \frac{\sigma^2}{\varepsilon^2}.$$

Thus Chebyshev's inequality can be used as a bound for certain probabilities. However, in many instances, the bound is not very close to the true probability.

EXAMPLE 4.7-1   If it is known that X has a mean of 25 and a variance of 16; then, since $\sigma = 4$, a lower bound for $P(17 < X < 33)$ is given by

$$P(17 < X < 33) = P(|X - 25| < 8)$$

$$= P(|X - 25| < 2\sigma) \geq 1 - \frac{1}{4} = 0.75$$

and an upper bound for $P(|X - 25| \geq 12)$ is found to be

$$P(|X - 25| \geq 12) = P(|X - \mu| \geq 3\sigma) \leq \frac{1}{9}.$$

$\blacktriangleleft$

Note that the results of the last example hold for any distribution with mean 25 and standard deviation 4. But, even stronger, the probability that any random variable X differs from its mean by 3 or more standard deviations is at most 1/9 by letting $k = 3$ in the theorem. Also the probability that any random variable X differs from its mean by less than 2 standard deviations is at least 3/4 by letting $k = 2$.

The following consideration partially indicates the value of Chebyshev's inequality in theoretical discussions. If Y is the number of successes in n Bernoulli trials with probability p of success on each trial, then Y is $b(n, p)$. Furthermore, Y/n gives the relative frequency of success and, when p is unknown, Y/n can be used as an estimate of its mean p. To gain some insight into the closeness of Y/n to p, we shall use Chebyshev's inequality. With $\varepsilon > 0$, since $\mathrm{Var}(Y/n) = pq/n$, we note from Corollary 4.7-1 that

$$P\left(\left|\frac{Y}{n} - p\right| \geq \varepsilon\right) \leq \frac{pq/n}{\varepsilon^2}$$

or, equivalently,

$$P\left(\left|\frac{Y}{n} - p\right| < \varepsilon\right) \geq 1 - \frac{pq}{n\varepsilon^2}. \qquad (4.7\text{-}2)$$

When p is completely unknown, we can use the fact that $pq = p(1 - p)$ is a maximum when $p = 1/2$ in order to find a lower bound for the probability in Equation 4.7-2. That is,

$$1 - \frac{pq}{n\varepsilon^2} \geq 1 - \frac{(1/2)(1/2)}{n\varepsilon^2}.$$

For illustration, if $\varepsilon = 0.05$ and $n = 400$,

$$P\left(\left|\frac{Y}{400} - p\right| < 0.05\right) \geq 1 - \frac{(1/2)(1/2)}{400(0.0025)} = 0.75.$$

If, on the other hand, it is known that p is close to 1/10, we would have

$$P\left(\left|\frac{Y}{400} - p\right| < 0.05\right) \geq 1 - \frac{(0.1)(0.9)}{400(0.0025)} = 0.91.$$

Note that Chebyshev's inequality is applicable to all distributions with a finite variance, and thus the bound is not always a tight one; that is, the bound is not necessarily close to the true probability.

In general, however, it should be noted that with fixed $\varepsilon > 0$ and $0 < p < 1$, we have that

$$\lim_{n \to \infty} P\left(\left|\frac{Y}{400} - p\right| < \varepsilon\right) \geq \lim_{n \to \infty} \left(1 - \frac{pq}{n\varepsilon^2}\right) = 1.$$

But since the probability of every event is less than or equal to 1, it must be that

$$\lim_{n \to \infty} P\left(\left|\frac{Y}{400} - p\right| < \varepsilon\right) = 1.$$

That is, the probability that the relative frequency Y/n is within ε of p is close to 1 when n is large enough. This is one form of the **law of large numbers**, and we say Y/n **converges in probability** to p.

This theoretical result has something to contribute to our understanding of the properties of the **empirical distribution function**, $F_n(x)$. For a fixed x, $F_n(x)$ is the proportion of sample observations that are less than or equal to x. That is, if "success" is an observed X being less than or equal to the fixed x, then $F_n(x)$ is the relative frequency of success. The *law of large numbers* states that this converges, in that probabilistic sense, to the true probability $P(X \leq x)$ of success, namely $p = F(x)$. Thus $F_n(x)$ does approach $F(x)$ in this sense and hence provides an estimate of the distribution function. We used this idea of the empirical distribution approximating the probability distribution as early as Chapter 1.

A more general form of the law of large numbers is found by considering the mean $\overline{X}$ of a random sample from a distribution with mean μ and variance σ^2. This is more general because the relative frequency Y/n can be thought of as $\overline{X}$ when the sample arises from a Bernoulli distribution. We know that

$$E(\overline{X}) = \mu \qquad \text{and} \qquad \text{Var}(\overline{X}) = \frac{\sigma^2}{n}.$$

Thus, for every $\varepsilon > 0$, using Corollary 4.7-1, we have

$$P[\,|\overline{X} - \mu| \geq \varepsilon\,] \leq \frac{\sigma^2/n}{\varepsilon^2} = \frac{\sigma^2}{n\varepsilon^2}.$$

Since probability is nonnegative, it follows that

$$\lim_{n \to \infty} P(|\overline{X} - \mu| \geq \varepsilon) \leq \lim_{n \to \infty} \frac{\sigma^2}{\varepsilon^2 n} = 0.$$

This implies that

$$\lim_{n \to \infty} P(|\overline{X} - \mu| \geq \varepsilon) = 0,$$

or, equivalently,

$$\lim_{n \to \infty} P(|\overline{X} - \mu| < \varepsilon) = 1.$$

The preceding discussion shows that the probability associated with the distribution of $\overline{X}$ becomes concentrated in an arbitrarily small interval centered at μ as n increases. This is a more general form of the law of large numbers, and we say that $\overline{X}$ converges in probability to μ.

Although Chebyshev's inequality is quite useful in theoretical discussion, it also shows that in a collection of numbers, say $x_1, x_2, \ldots, x_n$, a certain proportion of them must be within $k\sqrt{v}$ of $\overline{x}$, where $\overline{x}$ and v are the respective mean and variance of the empirical distribution defined by the numbers. That is,

$$\overline{x} = \frac{1}{n} \sum_{i=1}^{n} x_i \qquad \text{and} \qquad v = \frac{1}{n} \sum_{i=1}^{n} (x_i - \overline{x})^2.$$

To show this, think of the empirical distribution as defining the probability distribution; thus $\overline{x}$ and v are the mean and the variance of this distribution. Hence, for every $k \geq 1$, we have

$$\frac{N\{x_i : |x_i - \overline{x}| \geq k\sqrt{v}\}}{n} \leq \frac{1}{k^2}$$

or, equivalently,

$$\frac{N\{x_i : |x_i - \overline{x}| < k\sqrt{v}\}}{n} \geq 1 - \frac{1}{k^2}.$$

In words, the second inequality says, "The proportion of the numbers $x_1, x_2, \ldots, x_n$ that lie within $k\sqrt{v}$ of the mean $\overline{x}$ is at least $1 - 1/k^2$."

EXERCISES

4.7-1 If X is a random variable with mean 33 and variance 16, use Chebyshev's inequality to find

(a) A lower bound for $P(23 < X < 43)$.

(b) An upper bound for $P(|X - 33| \geq 14)$.

4.7-2 If $E(X) = 17$ and $E(X^2) = 298$, use Chebyshev's inequality to determine

(a) A lower bound for $P(10 < X < 24)$.

(b) An upper bound for $P(|X - 17| \geq 16)$.

4.7-3 Let X denote the outcome when rolling a fair die. Then $\mu = 7/2$ and $\sigma^2 = 35/12$. Note that the maximum deviation of X from μ equals 5/2. Express this deviation in terms of number of standard deviations; that is, find k where $k\sigma = 5/2$. Determine a lower bound for $P(|X - 3.5| < 2.5)$.

4.7-4 If Y is $b(n, 0.5)$, give a lower bound for $P(|Y/n - 0.5| < 0.08)$ when

(a) $n = 100$.

(b) $n = 500$.

(c) $n = 1000$.

4.7-5 If Y is $b(n, 0.25)$, give a lower bound for $P(|Y/n - 0.25| < 0.05)$ when

(a) $n = 100$.

(b) $n = 500$.

(c) $n = 1000$.

4.7-6 Let $\overline{X}$ be the mean of a random sample of size $n = 15$ from a distribution with mean $\mu = 80$ and variance $\sigma^2 = 60$. Use Chebyshev's inequality to find a lower bound for $P(75 < \overline{X} < 85)$.

4.7-7 The characteristics of the empirical distribution of test scores of 900 students are $\overline{x} = 83$ and $v = 36$, respectively. At least how many students received test scores between 71 and 95?

CHAPTER

5

THE NORMAL DISTRIBUTION

5.1 A BRIEF HISTORY OF PROBABILITY

HISTORICAL COMMENTS Most probabilists would say that the mathematics of probability began when in 1654 Chevalier de Méré, a French nobleman who liked to gamble, challenged Blaise Pascal to explain a puzzle and a problem created from his observations concerning rolls of dice. Of course, there was gambling well before this; and actually almost 200 years before this challenge, a Franciscan monk, Luca Paccioli, proposed that puzzlc.

"*A* and *B* are playing a fair game of balla. They agree to continue until one has six rounds. However, the game actually stops when *A* has won five and *B* three. How should the stakes be divided?"

And over 100 years before de Méré's challenge, a sixteenth-century doctor, Girolamo Cardano, who was also a gambler, had figured out many dice problems, but not the one that de Méré proposed. Chevalier de Méré had observed this. If a single die is tossed 4 times, the probability of obtaining at least one "six" was slightly greater that 1/2. However, keeping the same proportions, if a pair of dice is tossed 24 times, the probability of obtaining at least one double-six seemed to be slightly less than 1/2; at least he was losing money betting on it. This is when he approached Blaise Pascal with the challenge. Not wanting to work on the problems alone, Pascal formed a partnership with Pierre de Fermat, a great selection by Pascal. It was this 1654 correspondence between Pascal and Fermat that started the theory of probability.

Today an average student in probability could solve both easily. For the puzzle, note that B could only win with six rounds by winning the next three rounds, which has probability of $(1/2)^3 = 1/8$ because it was a fair game of balla. Thus A's probability of winning six rounds is $1 - 1/8 = 7/8$. Thus stakes should be divided seven units to one. For the dice problem, the probability of at least one six in four rolls of a die is

$$1 - \left(\frac{5}{6}\right)^4 = 0.518, \qquad \text{approximately,}$$

while the probability of rolling at least one double-six in 24 rolls of a pair of dice is

$$1 - \left(\frac{35}{36}\right)^{24} = 0.491, \qquad \text{approximately.}$$

It seems amazing to us that de Méré could have observed enough trials of those events to detect the slight difference in those probabilities. However, he won betting on the first but lost by betting on the second.

Incidentally, the solution to the balla puzzle led to a generalization, namely the binomial distribution, and to the famous Pascal triangle. Of course, Fermat was the great mathematician associated with "Fermat's last theorem."

Possibly the next major step in probability theory was by the Bernoullis, a remarkable Swiss family of mathematicians of the late 1600s to the late 1700s. There were eight mathematicians among them, but we shall just mention three of them, Jacob, Nicolaus II, and Daniel. While writing *Ars Conjectandi* (*The Art of Conjecture*), Jacob died in 1705; and a nephew, Nicolaus II, edited the work for publication. However, it was Jacob who discovered the important law of large numbers, which is included in our Section 4.7*.

Another nephew of Jacob, Daniel, noted in his St. Petersburg paper that "expected values are computed by multiplying each possible gain by the number of ways in which it can occur and then dividing the sum of these products by the total number of cases." His cousin, Nicolaus II, then proposed the so-called St. Petersburg Paradox. Peter continues to toss a coin until a head first appears, say on the xth trial, and he then pays Paul 2^{x-1} units (originally ducats but for convenience we use dollars). With each additional throw, the number of dollars has doubled. How much should another person pay Paul to take his place in this game? Clearly

$$E(2^{X-1}) = \sum_{x=1}^{\infty} (2^{x-1})\left(\frac{1}{2^x}\right) = \sum_{x=1}^{\infty} \frac{1}{2} = \infty.$$

However, if we consider this as a practical problem, would someone be willing to give Paul 1,000,000 dollars to take his place even though there is this unlimited expected value? We doubt it and Daniel doubted it, and it made him think about the utility of money. For example, to most of us, $3,000,000 is not worth three times that of $1,000,000. To convince you of that, suppose you had exactly $1,000,000 and a very rich man offers to bet you $2,000,000 against your $1,000,000 on the flip of a coin. You will have zero or $3,000,000 after the flip; so your expected value is

$$(\$0)\left(\frac{1}{2}\right) + (\$3,000,000)\left(\frac{1}{2}\right) = \$1,500,000,$$

much more than your $1,000,000–seemingly, a great bet and one that possibly Bill Gates would take. However, remember you have $1,000,000 for certain and you could have zero with probability 1/2. None of us with limited resources should consider taking that bet because the utility to us of that extra money is not worth the utility of that first $1,000,000. Now each of us has our own utility function. Two dollars is worth twice as much as one dollar for practically all of us. But is $200,000 worth twice as much as $100,000? It depends upon your situation; so while the utility function is a straight line for the first several dollars, it still increases but begins to bend downward someplace as the amount of money increases. This occurs at different spots for all of us. Bob Hogg (one of the authors of this text) would bet $1000 against $2000 on a flip of the coin any time, but probably not $100,000 against $200,000 so someplace between $1000 and $100,000, Hogg's utility function has started to bend downward. Daniel Bernoulli made this observation, and it is extremely useful in all kinds of businesses.

For illustration, in insurance, most of us know that the premium that we pay for all types of insurances is greater than what the company expects to pay us; that is how they make money. Seemingly insurance is a bad bet, but it really isn't always. It is true that we should self-insure less expensive items, those whose value is on that straight part of the utility function. We have even heard the "rule" that you not insure anything worth less than two months salary; this is a fairly good guide but each of us has our own utility function and must make that decision. Hogg can afford losses in the $5000 to $10,000 range (he does not like them), but he does not want to pay losses of $100,000 or more. That utility function for negative values of the argument follows that straight line for relatively small negative amounts, but again bends down for large negative amounts. If you insure expensive items, you will discover the expected utility in absolute value will now exceed the premium. This is why most people insure their life, their home, and their car (particularly on the liability side). They should not, however, insure their golf clubs, eyeglasses, furs, or jewelry (unless the latter two items are extremely valuable).

The next important person in our history is Abraham de Moivre, who was born in France in 1667, but, as a Protestant, he did not fare very well in that Catholic country. As a matter of fact, he was imprisoned for about two years for his beliefs. After his release, he went to England but there led a gloomy life as he could not find an academic position. So de Moivre supported himself tutoring or consulting with gamblers or insurance brokers. After publishing *The Doctrine of Chance*, he turned to a project that Nicolaus Bernoulli had suggested to him. Using the fact that Y is $b(n, p)$, he discovered that the relative frequency of successes, namely Y/n, which Jacob Bernoulli had proved converged in probability to p, had an interesting approximating distribution itself. De Moivre had discovered that well-known bell-shaped curve, called the normal distribution, that we study in this chapter. While de Moivre did not have computers in his day, we show that if Y is $b(100, 1/2)$, then $Y/100$ has the p.m.f. displayed in Figure 5.1-1.

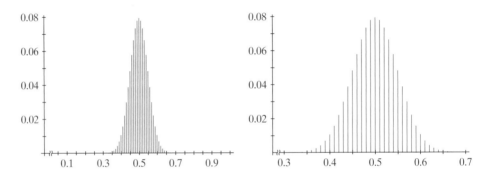

Figure 5.1-1: Bar graphs for the distribution of $Y/100$

This distribution allowed de Moivre to determine a measure of spread, which we now call a standard deviation. Also he could determine approximately the probabilities of Y/n falling in given intervals containing p. De Moivre was truly impressed with the orderliness of these random relative frequencies and he contributed that to the plan of the Almighty. Despite his great works, Abraham de Moivre died as a bitter and antisocial man, blind and in poverty, at the age of 87.

Both Jacob Bernoulli and Abraham de Moivre had clearly asked questions about the inverse probability; that is, trying to argue from the sample observations to say something about the underlying distribution or its parameters. The Reverend Thomas Bayes, who was born in 1701, was a Nonconformist and rejected most of the rituals of the Church of England. While he published nothing in mathematics when he was alive, two works were published after his death, one of which contained the essence of Bayes's theorem and a very original way of treating this inverse probability. It has had such an influence on modern statistics that many modern statisticians are associated with the Neo-Bayesian movement, and we devote Chapter 7 to some of these methods.

The final two persons who we would like to mention in this brief history of probability are Carl Friedrich Gauss and Marquis Pierre Simon de Laplace. Gauss was 29 years junior to Laplace, and Gauss was so secretive about his work it is difficult to tell who discovered the central limit theorem first. This was a generalization of de Moivre's result. In de Moivre's case he was sampling from a Bernoulli distribution where $X_i = 1$ or 0 on the ith trial, then

$$\frac{Y}{n} = \sum_{i=1}^{n} \frac{X_i}{n} = \overline{X},$$

the relative frequency of success, has that approximate normal distribution of de Moivre's. Laplace and Gauss were sampling from any distribution, provided the second moment existed, and they found that the sample mean $\overline{X}$ had an approximate normal distribution. Seemingly this central limit theorem, which we study in this chapter, was published in 1809 by Laplace just before Gauss's *Theoria Motus* in 1810. For some reason, the normal distribution is often referred to as the Gaussian distribution; people seem to forget about Laplace's contribution and (worse than that) de Moivre's original work 83 years earlier. Since then, there have been many more generalizations of the central limit theorem; in particular, most estimators of parameters in the regular case have approximate normal distributions.

There are many good histories of probability and statistics. However, the two that we find particularly interesting are Peter L. Bernstein's *Against the Gods: The Remarkable Story of Risk*, John Wiley & Sons, Inc. (1996), and Stephen M. Stigler's *The History of Statistics: The Measurement of Uncertainty Before 1900*, Harvard University Press (1986). ∎

STATISTICAL COMMENTS   (**Simulation: Central Limit Theorem**) As we think about what de Moivre, Laplace, and Gauss discovered in their times, it was truly amazing. Of course, de Moivre could compute the probabilities associated with various binomial distributions and see how they "piled up" in that bell shape, and he came up with the normal formula. Now Laplace and Gauss had an even tougher task as they could not easily find the probabilities associated with the sample mean $\overline{X}$, even with simple underlying distributions. For illustration, suppose the random sample $X_1, X_2, \ldots, X_n$ arises from a uniform distribution on the space $[0, 1)$. It is extremely difficult to compute probabilities about $\overline{X}$ unless n is very small, like $n = 2$ or $n = 3$ (unless you use a CAS). Today we can easily simulate the distribution of $\overline{X}$ for any sample size and get fairly accurate estimates of the probabilities associated with $\overline{X}$. We did this when $n = 6$ for 10,000 times, which resulted in Figure 5.1-2. Later in this

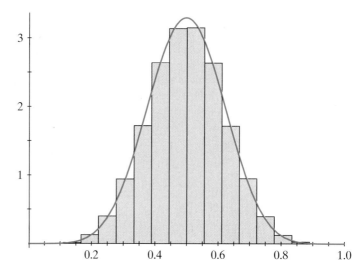

Figure 5.1-2: Simulation of 10,000 $\bar{x}$s for samples of size 6 from $U(0, 1)$

chapter we will learn that $\overline{X}$ has an approximate normal distribution with mean 1/2 and variance 1/72. We could then place this normal p.d.f. on the graph of the histogram for comparison. From this histogram we see the bell-shaped curve resulting from this simulation. None of these three outstanding mathematicians had the advantage of using the computer. A researcher today, with a new idea about a probability distribution, can check it easily with a simulation to see if the idea is worth devoting more time to it.

The simulated $\bar{x}$s were grouped into 18 classes with equal lengths of 1/18. The frequencies of the 18 classes are

$$0, \ 1, \ 9, \ 72, \ 223, \ 524, \ 957, \ 1467, \ 1744, \ 1751, \ 1464, \ 954,$$
$$528, \ 220, \ 69, \ 15, \ 2, \ 0.$$

Using these data, we can estimate certain probabilities, for example,

$$P\left(\frac{1}{6} \le \overline{X} \le \frac{2}{6}\right) \approx \frac{72 + 223 + 524}{10,000} = 0.0819$$

and

$$P\left(\frac{11}{18} \le \overline{X} \le 1\right) \approx \frac{954 + 528 + 220 + 69 + 15 + 2 + 0}{10,000} = 0.1788.$$

With a computer algebra system (CAS) it is sometimes possible to find p.d.f.s that involve rather complex calculations. For example, *Maple* was used to find the actual p.d.f. of $\overline{X}$ for this example. It is this p.d.f. that is superimposed on the histogram. Letting $u = \bar{x}$, the p.d.f. is given by

$$g(u) = \begin{cases} 6\left(\dfrac{324u^5}{5}\right), & 0 < u < 1/6, \\[2ex] 6\left(\dfrac{1}{20} - 324u^5 + 324u^4 - 108u^3 + 18u^2 - \dfrac{3u}{2}\right), & 1/6 \leq u < 2/6, \\[2ex] 6\left(\begin{array}{l} -\dfrac{79}{20} = \dfrac{117u}{2} + 648u^5 - 1296u^4 \\ + 972u^3 - 342u^2 \end{array}\right), & 2/6 \leq u < 3/6, \\[2ex] 6\left(\dfrac{731}{20} - \dfrac{693u}{2} - 648u^5 + 1944u^4 - 2268u^3\right), & 3/6 \leq u < 4/6 \\[2ex] 6\left(\begin{array}{l} -\dfrac{1829}{20} + \dfrac{1227u}{2} - 1602u^2 + 2052u^3 \\ + 324u^5 - 1296u^4 \end{array}\right), & 4/6 \leq u < 5/6 \\[2ex] 6\left(\dfrac{324}{5} - 324u + 648u^2 - 648u^3 + 324u^4 - \dfrac{324u^5}{5}\right), & 5/6 \leq u < 1. \end{cases}$$

We can also calculate

$$\int_{1/6}^{2/6} g(u)\, du = \frac{19}{240} = 0.0792$$

and

$$\int_{11/18}^{1} g(u)\, du = \frac{5,818}{32,805} = 0.17735.$$

Although these integrations are not difficult, they are tedious to do by hand. ∎

5.2 THE NORMAL DISTRIBUTION

In the previous section in the comments concerning the history of probability, mention was made that past probabilists had found that the estimators, Y/n and $\overline{X}$, of p and μ, respectively, had approximate normal distributions. The great value in the normal distribution is that many estimators of parameters have approximate normal distributions and this allows us to make statistical inferences that are explained later.

Sometimes distributions of certain measurements do have (approximate) normal distributions. This assumption is sometimes applied too frequently, however; for illustration it often bothers us when teachers "grade on the curve" because they believe that their students' grades are normally distributed and often they are not. Nevertheless, the normal distribution is extremely important and we must study its major characteristics.

We give the definition of the p.d.f. for the normal distribution, verify that it is a p.d.f., and then justify the use of μ and σ^2 in this formula. That is, we will show that μ and σ^2 are actually the mean and the variance of this distribution. The random variable X has a **normal distribution** if its p.d.f. is defined by

$$f(x) = \frac{1}{\sigma\sqrt{2\pi}} \exp\left[-\frac{(x-\mu)^2}{2\sigma^2}\right], \qquad -\infty < x < \infty,$$

where μ and σ are parameters satisfying $-\infty < \mu < \infty, 0 < \sigma < \infty$, and also where $\exp[v]$ means e^v. Briefly, we say that X is $N(\mu, \sigma^2)$.

Clearly, $f(x) > 0$. We now evaluate the integral

$$I = \int_{-\infty}^{\infty} \frac{1}{\sigma\sqrt{2\pi}} \exp\left[-\frac{(x-\mu)^2}{2\sigma^2}\right] dx,$$

showing that it is equal to 1. In I, change variables of integration by letting $z = (x - \mu)/\sigma$. Then

$$I = \int_{-\infty}^{\infty} \frac{1}{\sqrt{2\pi}} e^{-z^2/2} \, dz.$$

Since $I > 0$, if $I^2 = 1$, then $I = 1$. Now

$$I^2 = \frac{1}{2\pi} \left[\int_{-\infty}^{\infty} e^{-x^2/2} \, dx\right]\left[\int_{-\infty}^{\infty} e^{-y^2/2} \, dy\right],$$

or, equivalently,

$$I^2 = \frac{1}{2\pi} \int_{-\infty}^{\infty} \int_{-\infty}^{\infty} \exp\left(-\frac{x^2+y^2}{2}\right) dx \, dy.$$

Letting $x = r\cos\theta$, $y = r\sin\theta$ (i.e., using polar coordinates), we have

$$I^2 = \frac{1}{2\pi} \int_0^{2\pi} \int_0^{\infty} e^{-r^2/2} r \, dr \, d\theta$$

$$= \frac{1}{2\pi} \int_0^{2\pi} d\theta = \frac{1}{2\pi} 2\pi = 1.$$

Thus $I = 1$, and we have shown that $f(x)$ has the properties of a p.d.f. The moment-generating function of X is

$$M(t) = \int_{-\infty}^{\infty} \frac{e^{tx}}{\sigma\sqrt{2\pi}} \exp\left[-\frac{(x-\mu)^2}{2\sigma^2}\right] dx$$

$$= \int_{-\infty}^{\infty} \frac{1}{\sigma\sqrt{2\pi}} \exp\left\{-\frac{1}{2\sigma^2}[x^2 - 2(\mu + \sigma^2 t)x + \mu^2]\right\} dx.$$

To evaluate this integral, we complete the square in the exponent

$$x^2 - 2(\mu + \sigma^2 t)x + \mu^2 = [x - (\mu + \sigma^2 t)]^2 - 2\mu\sigma^2 t - \sigma^4 t^2.$$

Thus

$$M(t) = \exp\left(\frac{2\mu\sigma^2 t + \sigma^4 t^2}{2\sigma^2}\right) \int_{-\infty}^{\infty} \frac{1}{\sigma\sqrt{2\pi}} \exp\left\{-\frac{1}{2\sigma^2}[x - (\mu + \sigma^2 t)]^2\right\} dx.$$

Note that the integrand in the last integral is like the p.d.f. of a normal distribution with μ replaced by $\mu + \sigma^2 t$. However, the normal p.d.f. integrates to one for all real μ, in particular when μ is replaced by $\mu + \sigma^2 t$. Thus

$$M(t) = \exp\left(\frac{2\mu\sigma^2 t + \sigma^4 t^2}{2\sigma^2}\right) = \exp\left(\mu t + \frac{\sigma^2 t^2}{2}\right).$$

Now

$$M'(t) = (\mu + \sigma^2 t) \exp\left(\mu t + \frac{\sigma^2 t^2}{2}\right)$$

and

$$M''(t) = [(\mu + \sigma^2 t)^2 + \sigma^2] \exp\left(\mu t + \frac{\sigma^2 t^2}{2}\right).$$

Thus

$$E(X) = M'(0) = \mu,$$
$$\mathrm{Var}(X) = M''(0) - [M'(0)]^2 = \mu^2 + \sigma^2 - \mu^2 = \sigma^2.$$

That is, the parameters μ and σ^2 in the p.d.f. of X are the mean and the variance of X.

EXAMPLE 5.2-1   If the p.d.f. of X is

$$f(x) = \frac{1}{\sqrt{32\pi}} \exp\left[-\frac{(x + 7)^2}{32}\right], \qquad -\infty < x < \infty,$$

then X is $N(-7, 16)$. That is, X has a normal distribution with a mean $\mu = -7$, variance $\sigma^2 = 16$, and the moment-generating function

$$M(t) = \exp(-7t + 8t^2).$$

◀

EXAMPLE 5.2-2 If the moment-generating function of X is

$$M(t) = \exp(5t + 12t^2),$$

then X is $N(5, 24)$, and its p.d.f. is

$$f(x) = \frac{1}{\sqrt{48\pi}} \exp\left[-\frac{(x-5)^2}{48} \right], \qquad -\infty < x < \infty. \qquad \blacktriangleleft$$

If Z is $N(0, 1)$, we shall say that Z has a **standard normal distribution**. Moreover, the distribution function of Z is

$$\Phi(z) = P(Z \leq z) = \int_{-\infty}^{z} \frac{1}{\sqrt{2\pi}} e^{-w^2/2} \, dw.$$

It is not possible to evaluate this integral by finding an antiderivative that can be expressed as an elementary function. However, numerical approximations for integrals of this type have been tabulated and are given in Tables Va and Vb in the Appendix. The bell-shaped curved in Figure 5.2-1 represents the graph of the p.d.f. of Z, and the shaded area equals $\Phi(z_0)$.

Values of $\Phi(z)$ for $z \geq 0$ are given in Appendix Table Va. Because of the symmetry of the standard normal p.d.f., it is true that $\Phi(-z) = 1 - \Phi(z)$ for all real z. Thus Appendix Table Va is enough. However, it is sometimes convenient to be able to read $\Phi(-z)$, for $z > 0$, directly from a table. This can be done using values in Appendix Table Vb, which lists right tail probabilities.

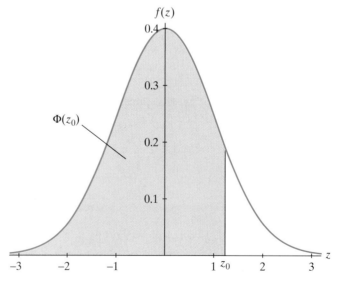

Figure 5.2-1: Standard normal p.d.f.

Again, because of the symmetry of the standard normal p.d.f., when $z > 0$, $\Phi(-z) = P(Z \leq -z) = P(Z > z)$ can be read directly from Appendix Table Vb.

EXAMPLE 5.2-3 If Z is $N(0, 1)$, then, using Appendix Table Va, we obtain

$$P(Z \leq 1.24) = \Phi(1.24) = 0.8925,$$

$$P(1.24 \leq Z \leq 2.37) = \Phi(2.37) - \Phi(1.24) = 0.9911 - 0.8925 = 0.0986,$$

$$P(-2.37 \leq Z \leq -1.24) = P(1.24 \leq Z \leq 2.37) = 0.0986.$$

Using Appendix Table Vb, we find that

$$P(Z > 1.24) = 0.1075,$$

$$P(Z \leq -2.14) = P(Z \geq 2.14) = 0.0162,$$

and using both tables, we obtain

$$P(-2.14 \leq Z \leq 0.77) = P(Z \leq 0.77) - P(Z \leq -2.14)$$
$$= 0.7794 - 0.0162 = 0.7632. \qquad \blacktriangleleft$$

There are times when we want to read the normal probability table in the opposite way, essentially finding the inverse of the standard normal distribution function. That is, given a probability p, find a constant a so that $P(Z \leq a) = p$. This is illustrated in the next example.

EXAMPLE 5.2-4 If the distribution of Z is $N(0, 1)$, to find constants a and b such that

$$P(Z \leq a) = 0.9147 \qquad \text{and} \qquad P(Z \geq b) = 0.0526,$$

find the given probabilities in Appendix Tables Va and Vb, respectively, and read off the corresponding values of z. From Appendix Table Va we see that $a = 1.37$ and from Appendix Table Vb we see that $b = 1.62$. $\qquad \blacktriangleleft$

In statistical applications we are often interested in finding a number z_α such that

$$P(Z \geq z_\alpha) = \alpha,$$

where Z is $N(0, 1)$ and α is usually less than 0.5. That is, z_α is the $100(1 - \alpha)$ percentile (sometimes called the upper 100α percent point) for the standard normal distribution (see Figure 5.2-2). The value of z_α is given in Appendix Table Va for selected values of α. For other values of α, z_α can be found in Appendix Table Vb.

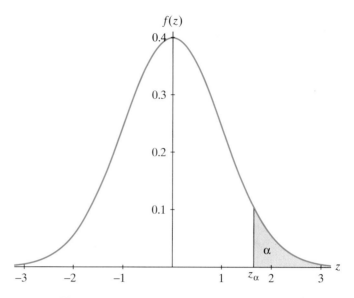

Figure 5.2-2: Upper 100α percent point, z_α

Because of the symmetry of the normal p.d.f.,

$$P(Z \le -z_\alpha) = P(Z \ge z_\alpha) = \alpha.$$

Also, since the subscript of z_α is the "right-tail" probability,

$$z_{1-\alpha} = -z_\alpha.$$

For example,

$$z_{0.95} = z_{1-0.05} = -z_{0.05}.$$

EXAMPLE 5.2-5 To find $z_{0.0125}$, note that

$$P(Z \ge z_{0.0125}) = 0.0125.$$

Thus

$$z_{0.0125} = 2.24$$

from Appendix Table Vb. Also,

$$z_{0.05} = 1.645 \qquad \text{and} \qquad z_{0.025} = 1.960$$

using the last rows of Appendix Table Va.

Remark   Recall that the $(100p)$th percentile, π_p, for a random variable X is a number such that $P(X \le \pi_p) = p$. If Z is $N(0, 1)$, since

$$P(Z \ge z_\alpha) = \alpha,$$

it follows that

$$P(Z < z_\alpha) = 1 - \alpha.$$

Thus z_α is the $100(1 - \alpha)$th percentile for the standard normal distribution, $N(0, 1)$. For example, $z_{0.05} = 1.645$ is the $100(1 - 0.05) = 95$th percentile and $z_{0.95} = -1.645$ is the $100(1 - 0.95) = 5$th percentile.   ∎

If X is $N(\mu, \sigma^2)$, the next theorem shows that the random variable $(X - \mu)/\sigma$ is $N(0, 1)$. Thus Tables Va and Vb in the Appendix can be used to find probabilities concerning X.

Theorem 5.2-1   If X is $N(\mu, \sigma^2)$, then $Z = (X - \mu)/\sigma$ is $N(0, 1)$.

Proof: The distribution function of Z is

$$P(Z \le z) = P\left(\frac{X - \mu}{\sigma} \le z\right) = P(X \le z\sigma + \mu)$$

$$= \int_{-\infty}^{z\sigma+\mu} \frac{1}{\sigma\sqrt{2\pi}} \exp\left[-\frac{(x - \mu)^2}{2\sigma^2}\right] dx.$$

In the integral representing $P(Z \le z)$, use the change of variable of integration given by $w = (x - \mu)/\sigma$ (i.e., $x = w\sigma + \mu$) to obtain

$$P(Z \le z) = \int_{-\infty}^{z} \frac{1}{\sqrt{2\pi}} e^{-w^2/2} dw.$$

But this is the expression for $\Phi(z)$, the distribution function of a standardized normal random variable. Hence Z is $N(0, 1)$.   ○

Remark   We should note that if $E(X) = \mu$ and $E[(X - \mu)^2] = \sigma^2$ exist and $Z = (X - \mu)/\sigma$, then

$$\mu_Z = E(Z) = E\left[\frac{X - \mu}{\sigma}\right] = \frac{E(X) - \mu}{\sigma} = 0$$

and

$$\sigma_Z^2 = E\left[\left(\frac{X - \mu}{\sigma}\right)^2\right] = \frac{E[(X - \mu)^2]}{\sigma^2} = \frac{\sigma^2}{\sigma^2} = 1.$$

That is, the mean and the variance of Z are zero and one, respectively, no matter what the distribution of X. The important aspect of the theorem is that if X is normally distributed, then Z is normally distributed, of course with zero mean and unit variance. ∎

This theorem can be used to find probabilities about X, which is $N(\mu, \sigma^2)$, as follows:

$$P(a \leq X \leq b) = P\left(\frac{a - \mu}{\sigma} \leq \frac{X - \mu}{\sigma} \leq \frac{b - \mu}{\sigma}\right) = \Phi\left(\frac{b - \mu}{\sigma}\right) - \Phi\left(\frac{a - \mu}{\sigma}\right),$$

since $(X - \mu)/\sigma$ is $N(0, 1)$.

EXAMPLE 5.2-6 If X is $N(3, 16)$, then

$$P(4 \leq X \leq 8) = P\left(\frac{4 - 3}{4} \leq \frac{X - 3}{4} \leq \frac{8 - 3}{4}\right)$$
$$= \Phi(1.25) - \Phi(0.25) = 0.8944 - 0.5987 = 0.2957,$$
$$P(0 \leq X \leq 5) = P\left(\frac{0 - 3}{4} \leq Z \leq \frac{5 - 3}{4}\right)$$
$$= \Phi(0.5) - \Phi(-0.75) = 0.6915 - 0.2266 = 0.4649,$$

and

$$P(-2 \leq X \leq 1) = P\left(\frac{-2 - 3}{4} \leq Z \leq \frac{1 - 3}{4}\right)$$
$$= \Phi(-0.5) - \Phi(-1.25) = 0.3085 - 0.1056 = 0.2029. \quad \triangleleft$$

EXAMPLE 5.2-7 If X is $N(25, 36)$, we find a constant c such that

$$P(|X - 25| \leq c) = 0.9544.$$

We want

$$P\left(\frac{-c}{6} \leq \frac{X - 25}{6} \leq \frac{c}{6}\right) = 0.9544.$$

Thus

$$\Phi\left(\frac{c}{6}\right) - \left[1 - \Phi\left(\frac{c}{6}\right)\right] = 0.9544$$

and

$$\Phi\left(\frac{c}{6}\right) = 0.9772.$$

Hence $c/6 = 2$ and $c = 12$. That is, the probability that X falls within two standard deviations of its mean is the same as the probability that the standard normal variable Z falls within two units (standard deviations) of zero.   ◄

In the next theorem we give a relationship between the chi-square and normal distributions.

Theorem 5.2-2   If the random variable X is $N(\mu, \sigma^2)$, $\sigma^2 > 0$, then the random variable $V = (X - \mu)^2/\sigma^2 = Z^2$ is $\chi^2(1)$.

Proof: Because $V = Z^2$, where $Z = (X - \mu)/\sigma$ is $N(0, 1)$, the distribution function $G(v)$ of V is, for $v \geq 0$,

$$G(v) = P(Z^2 \leq v) = P(-\sqrt{v} \leq Z \leq \sqrt{v}).$$

That is, with $v \geq 0$,

$$G(v) = \int_{-\sqrt{v}}^{\sqrt{v}} \frac{1}{\sqrt{2\pi}} e^{-z^2/2} \, dz = 2 \int_{0}^{\sqrt{v}} \frac{1}{\sqrt{2\pi}} e^{-z^2/2} \, dz.$$

If we change the variable of integration by writing $z = \sqrt{y}$, then, since $D_y(z) = 1/(2\sqrt{y})$, we have

$$G(v) = \int_{0}^{v} \frac{1}{\sqrt{2\pi y}} e^{-y/2} \, dy, \qquad 0 \leq v.$$

Of course, $G(v) = 0$, when $v < 0$. Hence the p.d.f. $g(v) = G'(v)$ of the continuous-type random variable V is, by one form of the fundamental theorem of calculus,

$$g(v) = \frac{1}{\sqrt{\pi}\sqrt{2}} v^{1/2-1} e^{-v/2}, \qquad 0 < v < \infty.$$

Since $g(v)$ is a p.d.f., it must be true that

$$\int_{0}^{\infty} \frac{1}{\sqrt{\pi}\sqrt{2}} v^{1/2-1} e^{-v/2} \, dv = 1.$$

The change of variables $x = v/2$ yields

$$1 = \frac{1}{\sqrt{\pi}} \int_{0}^{\infty} x^{1/2-1} e^{-x} \, dx = \frac{1}{\sqrt{\pi}} \Gamma\left(\frac{1}{2}\right).$$

Hence $\Gamma(1/2) = \sqrt{\pi}$, and thus V is $\chi^2(1)$.   ○

EXAMPLE 5.2-8 If Z is $N(0, 1)$, then

$$P(|Z| < 1.96 = \sqrt{3.841}) = 0.95$$

and, of course,

$$P(Z^2 < 3.841) = 0.95$$

from the chi-square table with $r = 1$. ◄

Given a set of observations of a random variable X, it is a challenge to determine the distribution of X. In particular, how can we decide whether or not X has an approximate normal distribution? If we have a large number of observations of X, a histogram or a stem-and-leaf diagram of the observations can often be helpful. (See Exercises 5.2-21 and 5.2-12.) For small samples a q-q plot can be used to check on whether the sample arises from a normal distribution. For example, suppose the quantiles of a sample were plotted against the corresponding quantiles of a certain normal distribution and these pairs of points were on a straight line with slope 1 and intercept 0. Of course, we would then believe that we have an ideal sample from that normal distribution with that certain mean and standard deviation. This plot, however, requires that we know the mean and the standard deviation of this normal distribution, and we usually do not. However, since the quantile, q_p, of $N(\mu, \sigma^2)$ is related to the corresponding one, z_{1-p}, of $N(0, 1)$ by $q_p = \mu + \sigma z_{1-p}$, we can always plot the quantiles of the sample against the corresponding ones of $N(0, 1)$ and get the needed information. That is, if the sample quantiles are plotted as the x coordinate of the pair and the $N(0, 1)$ quantiles as the y coordinate and if the graph is almost a straight line, then it is reasonable to assume that the sample arises from a normal distribution. Moreover, the reciprocal of the slope of that straight line is a good estimate of the standard deviation σ because $z_{1-p} = (q_p - \mu)/\sigma$.

EXAMPLE 5.2-9 When researching ground water it is often important to know the characteristics of the soil at a certain site. Many of these characteristics, such as porosity, are at least partially dependent upon the grain size. The diameter of individual grains of soil can be measured. Here are the diameters (in mm) of 30 randomly selected grains.

1.24	1.36	1.28	1.31	1.35	1.20	1.39	1.35	1.41	1.31
1.28	1.26	1.37	1.49	1.32	1.40	1.33	1.28	1.25	1.39
1.38	1.34	1.40	1.27	1.33	1.36	1.43	1.33	1.29	1.34

For these data, $\bar{x} = 1.33$ and $s^2 = 0.0040$. May we assume that these are observations of a random variable X that is $N(1.33, 0.0040)$? To help answer this question, we shall construct a q-q plot of the standard normal quantiles that correspond to $p = 1/31, 2/31, \ldots, 30/31$ versus the ordered observations. To find these quantiles, it is helpful to use the computer. Minitab could be used. We shall find the quantiles using *Maple*.

k	Diameters in mm (x)	$p = k/31$	z_{1-p}	k	Diameters in mm (x)	$p = k/31$	z_{1-p}
1	1.20	0.0323	−1.85	16	1.34	0.5161	0.04
2	1.24	0.0645	−1.52	17	1.34	0.5484	0.12
3	1.25	0.0968	−1.30	18	1.35	0.5806	0.20
4	1.26	0.1290	−1.13	19	1.35	0.6129	0.29
5	1.27	0.1613	−0.99	20	1.36	0.6452	0.37
6	1.28	0.1935	−0.86	21	1.36	0.6774	0.46
7	1.28	0.2258	−0.75	22	1.37	0.7097	0.55
8	1.28	0.2581	−0.65	23	1.38	0.7419	0.65
9	1.29	0.2903	−0.55	24	1.39	0.7742	0.75
10	1.31	0.3226	−0.46	25	1.39	0.8065	0.86
11	1.31	0.3548	−0.37	26	1.40	0.8387	0.99
12	1.32	0.3871	−0.29	27	1.40	0.8710	1.13
13	1.33	0.4194	−0.20	28	1.41	0.9032	1.30
14	1.33	0.4516	−0.12	29	1.43	0.9355	1.52
15	1.33	0.4839	−0.04	30	1.49	0.9677	1.85

A q-q plot of these data is shown in Figure 5.2-3. Note that the points do fall close to a straight line so the normal probability model seems to be appropriate based on these few data. ◄

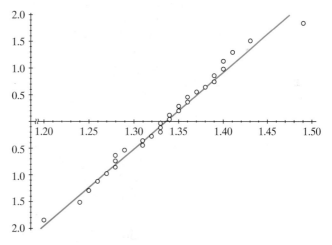

Figure 5.2-3: q-q plot, $N(0, 1)$ quantiles versus grain diameters

EXERCISES

5.2-1 If Z is $N(0, 1)$, find

    (a) $P(0.53 < Z \leq 2.06)$.
    (c) $P(Z > -1.77)$.
    (e) $P(|Z| < 1.96)$.
    (g) $P(|Z| < 2)$.

    (b) $P(-0.79 \leq Z < 1.52)$.
    (d) $P(Z > 2.89)$.
    (f) $P(|Z| < 1)$.
    (h) $P(|Z| < 3)$.

5.2-2 If Z is $N(0, 1)$, find

(a) $P(0 \leq Z \leq 0.87)$.   (b) $P(-2.64 \leq Z \leq 0)$.
(c) $P(-2.13 \leq Z \leq -0.56)$.   (d) $P(|Z| > 1.39)$.
(e) $P(Z < -1.62)$.   (f) $P(|Z| > 1)$.
(g) $P(|Z| > 2)$.   (h) $P(|Z| > 3)$.

5.2-3 Find the values of (a) $z_{0.01}$, (b) $-z_{0.005}$, (c) $z_{0.0475}$, and (d) $z_{0.985}$.

5.2-4 Find the values of (a) $z_{0.10}$, (b) $-z_{0.05}$, (c) $-z_{0.0485}$, and (d) $z_{0.9656}$.

5.2-5 If Z is $N(0, 1)$, find values of c such that

(a) $P(Z \geq c) = 0.025$.   (b) $P(|Z| \leq c) = 0.95$.
(c) $P(Z > c) = 0.05$.   (d) $P(|Z| \leq c) = 0.90$.

5.2-6 If the moment-generating function of X is $M(t) = \exp(166t + 200t^2)$, find

(a) The mean of X.   (b) The variance of X.
(c) $P(170 < X < 200)$.   (d) $P(148 \leq X \leq 172)$.

5.2-7 If X is normally distributed with a mean of 6 and a variance of 25, find

(a) $P(6 \leq X \leq 12)$.   (b) $P(0 \leq X \leq 8)$.
(c) $P(-2 < X \leq 0)$.   (d) $P(X > 21)$.
(e) $P(|X - 6| < 5)$.   (f) $P(|X - 6| < 10)$.
(g) $P(|X - 6| < 15)$.   (h) $P(|X - 6| < 12.41)$.

5.2-8 If the moment-generating function of X is $M(t) = \exp(-6t + 32t^2)$, find

(a) $P(-4 \leq X < 16)$.
(b) $P(-10 < X \leq 0)$.

5.2-9 If X is $N(650, 625)$, find

(a) $P(600 \leq X < 660)$.
(b) A constant $c > 0$ such that $P(|X - 650| \leq c) = 0.9544$.

5.2-10 If X is $N(\mu, \sigma^2)$, show that $Y = aX + b$ is $N(a\mu + b, a^2\sigma^2)$, $a \neq 0$. HINT: Find the distribution function $P(Y \leq y)$ of Y and, in the resulting integral, let $w = ax + b$ or, equivalently, $x = (w - b)/a$.

5.2-11 Find the distribution of $W = X^2$ when

(a) X is $N(0, 4)$,
(b) X is $N(0, \sigma^2)$.

5.2-12 A company manufactures windows that are then inserted into an automobile. Each window has 5 studs for attaching it. A pull-out test is used to determine the force required to pull a stud out of a window. (Note that this is an example of destructive testing.) Let X equal the force required for pulling studs out of position 4. Sixty observations of X were as follows:

159	150	147	160	155	142	143	151	154	133
151	146	140	146	137	148	154	157	142	153
135	144	135	165	118	158	126	147	123	140
125	151	153	158	144	163	150	150	137	164
137	156	139	134	171	144	160	147	155	175
162	160	149	149	158	152	165	131	150	120

(a) Construct an ordered stem-and-leaf diagram using as stems 11•, 12*, 12•, and so on.
(b) Construct a q-q plot using about every 5th observation in the ordered array and the corresponding quantiles of $N(0, 1)$.
(c) Does it look like X has a normal distribution?

5.2-13 When you fly, the usual snack is a bag of peanuts. Let X equal the weight of peanuts in a bag that has a 14-gram label weight. Sixteen observations of X, that have been ordered, are

$$
\begin{array}{cccccccc}
13.9 & 14.4 & 14.6 & 14.7 & 14.7 & 15.2 & 15.2 & 15.2 \\
15.3 & 15.4 & 15.4 & 15.5 & 15.6 & 15.6 & 15.9 & 16.4
\end{array}
$$

Determine whether X could have a normal distribution by constructing a q-q plot. What is your conclusion?

5.2-14 A candy maker produces mints that have a label weight of 20.4 grams. Assume that the distribution of the weights of these mints is $N(21.37, 0.16)$.

    (a) Let X denote the weight of a single mint selected at random from the production line. Find $P(X > 22.07)$.

    (b) Suppose that 15 mints are selected independently and weighed. Let Y equal the number of these mints that weigh less than 20.857 grams. Then find $P(Y \leq 2)$.

5.2-15 A chemistry major weighed 19 plain m&m's (in grams) on a ± 0.0001 scale. The ordered weights are

$$
\begin{array}{ccccccc}
0.7938 & 0.8032 & 0.8089 & 0.8222 & 0.8268 & 0.8383 & 0.8442 \\
0.8490 & 0.8528 & 0.8572 & 0.8674 & 0.8734 & 0.8786 & \\
0.8850 & 0.8873 & 0.8920 & 0.9069 & 0.9150 & 0.9243 &
\end{array}
$$

    (a) Construct a stem-and-leaf diagram using three-digit leaves, with stems $0.7\bullet$, $0.8*$, $0.8t$, and so on.

    (b) Construct a q-q plot of these data and the corresponding quantiles for the standard normal distribution, $N(0, 1)$.

    (c) Do these data look like observations from a normal distribution? Why?

5.2-16 Let the distribution of X be $N(\mu, \sigma^2)$ Show that the points of inflection of the graph of the p.d.f. of X occur at $x = \mu \pm \sigma$.

5.2-17 If X is $N(7, 4)$, find $P[15.364 \leq (X - 7)^2 \leq 20.096]$.

5.2-18 If the moment-generating function of X is $M(t) = e^{500t + 5000t^2}$, then find $P[27{,}060 \leq (X - 500)^2 \leq 50{,}240]$.

5.2-19 The resistance of certain resistors in an electrical circuit have a normal distribution with a mean of 10 ohms and standard deviation of 0.25 ohms. Let X be the number of ohms associated with one of these resistors. Compute

    (a) $P(X < 9.5)$,

    (b) $P(9.6 < X < 10.4)$.

5.2-20 The tensile strength (ksi) of a certain steel at $-150\,°F$ is normally distributed with a mean of 100 ksi and standard deviation of 10 ksi. If X is the ksi of a sample of this steel compute

    (a) $P(80.5 < X < 107)$,

    (b) $P(X > 115.5)$,

    (c) $P(X < 94.5)$.

5.2-21 A packaged product has a label weight of 450 grams. The company goal is to fill each package with at least 450 grams but at most 458 grams. To check these goals, a random sample of 100 packages was selected and weighed, yielding the following weights rounded to the nearest gram:

$$
\begin{array}{cccccccccc}
457 & 457 & 455 & 457 & 454 & 454 & 457 & 455 & 456 & 459 \\
457 & 458 & 456 & 456 & 461 & 457 & 458 & 452 & 457 & 460 \\
453 & 458 & 452 & 454 & 454 & 456 & 455 & 456 & 451 & 454 \\
456 & 457 & 457 & 453 & 455 & 459 & 458 & 457 & 458 & 457 \\
461 & 457 & 455 & 458 & 458 & 455 & 457 & 458 & 456 & 463 \\
455 & 455 & 455 & 456 & 456 & 456 & 455 & 456 & 460 & 456 \\
456 & 457 & 458 & 454 & 455 & 456 & 459 & 457 & 457 & 451 \\
450 & 453 & 453 & 459 & 450 & 453 & 452 & 458 & 456 & 457 \\
451 & 458 & 456 & 460 & 455 & 455 & 456 & 460 & 457 & 456 \\
457 & 456 & 460 & 459 & 457 & 455 & 461 & 455 & 457 & 457
\end{array}
$$

(a) Construct a frequency table and a corresponding relative frequency histogram.

(b) Do these data seem to come from a normal distribution with mean μ about equal to $\bar{x} = 456.2$ and variance about equal to $s^2 = 5.96$? Sketch the p.d.f. for the normal distribution $N(456.2, 5.96)$ on your histogram.

5.2-22 The strength X of a certain material is such that its distribution is found by $X = e^Y$, where Y is $N(10, 1)$. Find the distribution function and p.d.f. of X and compute $P(10,000 < X < 20,000)$. NOTE: The random variable X is said to have a **lognormal distribution**.

5.2-23 The "fill" problem is important in many industries, like those making cereal, toothpaste, beer, and so on. If such an industry claims it is selling 12 ounces of its product in a container, it must have a mean greater that 12 ounces or else the FDA will crack down, although the FDA will allow a very small percentage of the containers to have less than 12 ounces.

(a) If the contents X of a container has a $N(\mu_1 = 12.1, \sigma^2)$ distribution, find σ so that $P(X < 12) = 0.01$.

(b) If $\sigma = 0.05$, find μ so that $P(X < 12) = 0.01$.

5.2-24 Nine measurements are taken on the strength of a certain metal. In order they are 7.2, 8.9, 9.7, 10.5, 10.9, 11.7, 12.9, 13.9, 15.3, and these values correspond to the 10th, 20th,..., 90th percentiles of this sample. Plotting these against the same percentiles of $N(0, 1)$, does it seem reasonable that the underlying distribution of strengths could be normal?

5.2-25 Assume that the fill X of a filling machine for a beverage has a normal distribution with $\mu = 12.2$ and $\sigma = 0.1$, measured in fluid ounces.

(a) Compute $P(X < 12)$.

(b) In 50 independent such measurements, compute the probability that at least one is under 12 ounces.

5.2-26 The serum zinc level X in micrograms per deciliter for males between ages 15 and 17 has a distribution which is approximately normal with $\mu = 90$ and $\sigma = 15$. Compute the conditional probability $P(X > 120 \mid X > 105)$.

5.2-27 The graphs of the moment-generating functions of three normal distributions, $N(0, 1)$, $N(-1, 1)$, and $N(2, 1)$ are given in Figure 5.2-4. Identify them.

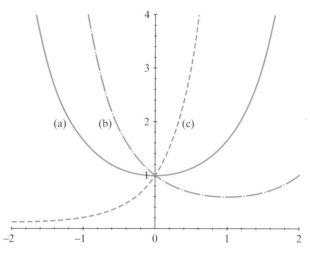

Figure 5.2-4: Moment-generating functions for some normal distributions

5.3   RANDOM FUNCTIONS ASSOCIATED WITH NORMAL DISTRIBUTIONS

In statistical applications, it is often assumed that the population from which a sample is taken is normally distributed, $N(\mu, \sigma^2)$. There is then interest in estimating the parameters μ and σ^2 or in testing conjectures about these parameters. The usual statistics that are used in these activities are the sample mean $\overline{X}$ and the sample variance S^2; thus we need to know something about the distribution of these statistics or functions of these statistics.

Theorem 5.3-1

If $X_1, X_2, \ldots, X_n$ are observations of a random sample of size n from the normal distribution $N(\mu, \sigma^2)$, then the distribution of the sample mean $\overline{X} = (1/n) \sum_{i=1}^{n} X_i$ is $N(\mu, \sigma^2/n)$.

Proof: Since the moment-generating function of each X is

$$M_X(t) = \exp\left(\mu t + \frac{\sigma^2 t^2}{2}\right),$$

the moment-generating function of

$$\overline{X} = \frac{1}{n} \sum_{i=1}^{n} X_i$$

is, from the corollary of Theorem 4.6-2, equal to

$$M_{\overline{X}}(t) = \left\{\exp\left[\mu\left(\frac{t}{n}\right) + \frac{\sigma^2 (t/n)^2}{2}\right]\right\}^n$$

$$= \exp\left[\mu t + \frac{(\sigma^2/n)t^2}{2}\right].$$

However, the moment-generating function uniquely determines the distribution of the random variable. Since this one is that associated with the normal distribution $N(\mu, \sigma^2/n)$, the sample mean $\overline{X}$ is $N(\mu, \sigma^2/n)$.   ◯

Theorem 5.3-1 shows that if $X_1, X_2, \ldots, X_n$ is a random sample from the normal distribution, $N(\mu, \sigma^2)$, then the probability distribution of $\overline{X}$ is also normal with the same mean μ but a variance σ^2/n. This means that $\overline{X}$ has a greater probability of falling in an interval containing μ than does a single observation, say X_1. For example, if $\mu = 50$, $\sigma^2 = 16$, $n = 64$, then $P(49 < \overline{X} < 51) = 0.9544$, whereas $P(49 < X_1 < 51) = 0.1974$. This is illustrated again in the next example.

EXAMPLE 5.3-1

Let $X_1, X_2, \ldots, X_n$ be a random sample from the $N(50, 16)$ distribution. We know that the distribution of $\overline{X}$ is $N(50, 16/n)$. To illustrate the effect of n, the graph of the p.d.f. of $\overline{X}$ is given in Figure 5.3-1 for $n = 1, 4, 16$, and 64. When $n = 64$, compare the areas that represent $P(49 < \overline{X} < 51)$ and $P(49 < X_1 < 51)$.   ◀

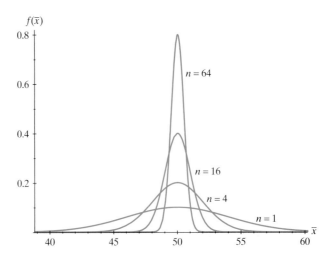

Figure 5.3-1: p.d.f.s of means of samples from $N(50, 16)$

Before we can find the distribution of S^2, or $(n-1)S^2/\sigma^2$, we give two preliminary results that are important in their own right. The first is a direct result of the material in Section 4.6 and gives an additive property for independent chi-square random variables, namely if $X_1, X_2, \ldots, X_n$ are independent and have chi-square distributions with $r_1, r_2, \ldots, r_n$ degrees of freedom, respectively, then

$$Y = \sum_{i=1}^{n} X_i \quad \text{is} \quad \chi^2(r_1 + r_2 + \cdots + r_n).$$

This is Theorem 4.6-3.

The next theorem combines and extends the results of Theorems 5.2-2 and 4.6-3 and gives one interpretation of degrees of freedom.

Theorem 5.3-2   Let $Z_1, Z_2, \ldots, Z_n$ have standard normal distributions, $N(0, 1)$. If these random variables are independent, then $W = Z_1^2 + Z_2^2 + \cdots + Z_n^2$ has a distribution that is $\chi^2(n)$.

Proof: By Theorem 5.2-2, Z_i^2 is $\chi^2(1)$ for $i = 1, 2, \ldots, n$. From Theorem 4.6-3, with $k = n$, $Y = W$, and $r_i = 1$, we see that W is $\chi^2(n)$.   ○

Corollary 5.3-1   If $X_1, X_2, \ldots, X_n$ are independent and have normal distributions $N(\mu_i, \sigma_i^2)$, $i = 1, 2, \ldots, n$, respectively, then the distribution of

$$W = \sum_{i=1}^{n} \frac{(X_i - \mu_i)^2}{\sigma_i^2}$$

is $\chi^2(n)$.   ■

Proof: This follows from Theorem 5.3-2 since $Z_i = (X_i - \mu_i)/\sigma_i$ is $N(0, 1)$, $i = 1, 2, \ldots, n$.   ○

Note that the number of terms in the summation and the number of degrees of freedom are equal in Theorem 5.3-2 and the corollary.

The following theorem gives an important result that will be used in statistical applications. In connection with those, we will use the sample variance s^2 to estimate the variance, σ^2, when sampling from the normal distribution, $N(\mu, \sigma^2)$. More will be said about s^2 at the time of its use.

Theorem 5.3-3   Let $X_1, X_2, \ldots, X_n$ be observations of a random sample of size n from the normal distribution $N(\mu, \sigma^2)$. The statistics, the sample mean,

$$\overline{X} = \frac{1}{n} \sum_{i=1}^{n} X_i,$$

and the sample variance,

$$S^2 = \frac{1}{n-1} \sum_{i=1}^{n} (X_i - \overline{X})^2,$$

are independent and

$$\frac{(n-1)S^2}{\sigma^2} = \frac{\displaystyle\sum_{i=1}^{n} (X_i - \overline{X})^2}{\sigma^2} \quad \text{is } \chi^2(n-1).$$

Proof: We are not prepared to prove the independence of $\overline{X}$ and S^2 at this time (see Section 6.3* for a proof) so we accept it without proof here. To prove the second part, note that

$$W = \sum_{i=1}^{n} \left(\frac{X_i - \mu}{\sigma}\right)^2 = \sum_{i=1}^{n} \left[\frac{(X_i - \overline{X}) + (\overline{X} - \mu)}{\sigma}\right]^2$$

$$= \sum_{i=1}^{n} \left(\frac{X_i - \overline{X}}{\sigma}\right)^2 + \frac{n(\overline{X} - \mu)^2}{\sigma^2} \qquad (5.3\text{-}1)$$

because the cross-product term is equal to

$$2 \sum_{i=1}^{n} \frac{(\overline{X} - \mu)(X_i - \overline{X})}{\sigma^2} = \frac{2(\overline{X} - \mu)}{\sigma^2} \sum_{i=1}^{n} (X_i - \overline{X}) = 0.$$

But $Y_i = (X_i - \mu)/\sigma$, $i = 1, 2, \ldots n$, are standardized normal variables that are independent. Hence $W = \sum_{i=1}^{n} Y_i^2$ is $\chi^2(n)$ by Corollary 5.3-1. Moreover, since $\overline{X}$ is $N(\mu, \sigma^2/n)$, then

$$Z^2 = \left(\frac{\overline{X} - \mu}{\sigma/\sqrt{n}}\right)^2 = \frac{n(\overline{X} - \mu)^2}{\sigma^2}$$

is $\chi^2(1)$ by Theorem 5.2-2. In this notation, Equation 5.3-1 becomes

$$W = \frac{(n-1)S^2}{\sigma^2} + Z^2.$$

However, from the fact that $\overline{X}$ and S^2 are independent, we have that Z^2 and S^2 are also independent. In the moment-generating function of W, this independence permits us to write

$$E[e^{tW}] = E[e^{t\{(n-1)S^2/\sigma^2 + Z^2\}}] = E[e^{t(n-1)S^2/\sigma^2} e^{tZ^2}]$$
$$= E[e^{t(n-1)S^2/\sigma^2}]E[e^{tZ^2}].$$

Since W and Z^2 have chi-square distributions, we can substitute their moment-generating functions to obtain

$$(1 - 2t)^{-n/2} = E[e^{t(n-1)S^2/\sigma^2}](1 - 2t)^{-1/2}.$$

Equivalently, we have

$$E[e^{t(n-1)S^2/\sigma^2}] = (1 - 2t)^{-(n-1)/2}, \qquad t < \frac{1}{2}.$$

This, of course, is the moment-generating function of a $\chi^2(n-1)$ variable, and accordingly $(n-1)S^2/\sigma^2$ has this distribution. ○

Combining the results of Corollary 5.3-1 and Theorem 5.3-3, we see that when sampling from a normal distribution,

$$U = \sum_{i=1}^{n} \frac{(X_i - \mu)^2}{\sigma^2}$$

is $\chi^2(n)$, and

$$W = \sum_{i=1}^{n} \frac{(X_i - \overline{X})^2}{\sigma^2}$$

is $\chi^2(n-1)$. That is, when the population mean, μ, in $\sum_{i=1}^{n}(X_i - \mu)^2$ is replaced by the sample mean, $\overline{X}$, one degree of freedom is lost. There are more general situations in which a degree of freedom is lost for each parameter estimated in certain chi-square random variables.

EXAMPLE 5.3-2   Let X_1, X_2, X_3, X_4 be a random sample of size 4 from the normal distribution, $N(76.4, 383)$. Then

$$U = \sum_{i=1}^{4} \frac{(X_i - 76.4)^2}{383} \quad \text{is} \quad \chi^2(4),$$

$$W = \sum_{i=1}^{4} \frac{(X_i - \overline{X})^2}{383} \quad \text{is} \quad \chi^2(3),$$

and, for example,

$$P(0.711 \le U \le 7.779) = 0.90 - 0.05 = 0.85,$$

$$P(0.352 \le W \le 6.251) = 0.90 - 0.05 = 0.85. \qquad \blacktriangleleft$$

In later sections we shall illustrate the importance of the chi-square distribution in applications.

We now prove another theorem which deals with linear functions of independent normally distributed random variables.

Theorem 5.3-4   If $X_1, X_2, \ldots, X_n$ are n mutually independent normal variables with means $\mu_1, \mu_2, \ldots, \mu_n$ and variances $\sigma_1^2, \sigma_2^2, \ldots, \sigma_n^2$, respectively, then the linear function

$$Y = \sum_{i=1}^{n} c_i X_i$$

has the normal distribution

$$N\left(\sum_{i=1}^{n} c_i \mu_i, \sum_{i=1}^{n} c_i^2 \sigma_i^2\right).$$

Proof: By Theorem 4.6-2, we have

$$M_Y(t) = \prod_{i=1}^{n} M_{X_i}(c_i t) = \prod_{i=1}^{n} \exp(\mu_i c_i t + \sigma_i^2 c_i^2 t^2 / 2)$$

because $M_{X_i}(t) = \exp(\mu_i t + \sigma_i^2 t^2 / 2), i = 1, 2, \ldots, n$. Thus

$$M_Y(t) = \exp\left[\left(\sum_{i=1}^{n} c_i \mu_i\right) t + \left(\sum_{i=1}^{n} c_i^2 \sigma_i^2\right)\left(\frac{t^2}{2}\right)\right].$$

This is the moment-generating function of a distribution which is

$$N\left(\sum_{i=1}^{n} c_i \mu_i, \sum_{i=1}^{n} c_i^2 \sigma_i^2\right)$$

and thus Y has this normal distribution. ○

From Theorem 5.3-4 we can make the observation that the difference of two independent normally distributed random variables, say $Y = X_1 - X_2$, has the normal distribution $N(\mu_1 - \mu_2, \sigma_1^2 + \sigma_2^2)$.

EXAMPLE 5.3-3 Let X_1 and X_2 equal the number of pounds of butterfat produced by two Holstein cows (one selected at random from those on the Koopman farm and one selected at random from those on the Vliestra farm, respectively) during the 305-day lactation period following the births of calves. Assume that the distribution of X_1 is $N(693.2, 22820)$ and the distribution of X_2 is $N(631.7, 19205)$. Moreover, if X_1 and X_2 are independent, we shall find $P(X_1 > X_2)$. That is, we shall find the probability that the butterfat produced by the Koopman farm cow exceeds that produced by the Vliestra farm cow. (Sketch p.d.f.s on the same graph for these two normal distributions.) If we let $Y = X_1 - X_2$, then the distribution of Y is $N(693.2 - 631.7, 22820 + 19205)$. Thus,

$$P(X_1 > X_2) = P(Y > 0) = P\left(\frac{Y - 61.5}{\sqrt{42025}} > \frac{0 - 61.5}{205}\right)$$

$$= P(Z > -0.30) = 0.6179.$$ ◄

EXAMPLE 5.3-4 (**Box-Muller Transformation**) Consider the following transformation, where X_1 and X_2 are independent, each with the uniform distribution $U(0, 1)$. Let

$$Z_1 = \sqrt{-2 \ln X_1} \cos(2\pi X_2), \qquad Z_2 = \sqrt{-2 \ln X_1} \sin(2\pi X_2)$$

or, equivalently,

$$X_1 = \exp\left(-\frac{Z_1^2 + Z_2^2}{2}\right) = e^{-q/2}, \qquad X_2 = \frac{1}{2\pi}\arctan\left(\frac{Z_2}{Z_1}\right),$$

which has Jacobian

$$J = \begin{vmatrix} -z_1 e^{-q/2} & -z_2 e^{-q/2} \\ \dfrac{-z_2}{2\pi(z_1^2 + z_2^2)} & \dfrac{z_1}{2\pi(z_1^2 + z_2^2)} \end{vmatrix} = \frac{-1}{2\pi} e^{-q/2}.$$

Since the joint p.d.f. of X_1 and X_2 is

$$f(x_1, x_2) = 1, \qquad 0 < x_1 < 1, \, 0 < x_2 < 1,$$

we have that the joint p.d.f. of Z_1 and Z_2 is

$$g(z_1, z_2) = \left| \frac{-1}{2\pi} e^{-q/2} \right| \quad (1)$$

$$= \frac{1}{2\pi} \exp\left(-\frac{z_1^2 + z_2^2}{2} \right), \qquad -\infty < z_1 < \infty, -\infty < z_2 < \infty.$$

Note that there is some difficulty with the definition of this transformation, particularly when $z_1 = 0$. However, these difficulties occur at events with probability zero and hence cause no problems. (See Exercise 5.3-21.) Summarizing, from two independent $U(0, 1)$ random variables we have generated two independent $N(0, 1)$ random variables through this **Box-Muller transformation**. ◀

The next example illustrates the distribution function technique. We do the same example in Exercise 5.3-16 using the change of variable technique.

EXAMPLE 5.3-5   (**Student's t distribution**) We let

$$T = \frac{Z}{\sqrt{U/r}},$$

where Z is a random variable that is $N(0, 1)$, U is a random variable that is $\chi^2(r)$, and Z and U are independent. Thus the joint p.d.f. of Z and U is

$$g(z, u) = \frac{1}{\sqrt{2\pi}} e^{-z^2/2} \frac{1}{\Gamma(r/2)2^{r/2}} u^{r/2-1} e^{-u/2}, \qquad -\infty < z < \infty, 0 < u < \infty.$$

The distribution function $F(t) = P(T \le t)$ of T is given by

$$F(t) = P(Z/\sqrt{U/r} \le t)$$

$$= P(Z \le \sqrt{U/r}\, t)$$

$$= \int_0^{\infty} \int_{-\infty}^{\sqrt{(u/r)}\, t} g(z, u)\, dz\, du.$$

That is,

$$F(t) = \frac{1}{\sqrt{\pi}\, \Gamma(r/2)} \int_0^{\infty} \left[\int_{-\infty}^{\sqrt{(u/r)}\, t} \frac{e^{-z^2/2}}{2^{(r+1)/2}}\, dz \right] u^{r/2-1} e^{-u/2}\, du.$$

The p.d.f. of T is the derivative of the distribution function; so, applying the fundamental theorem of calculus to the inner integral, we see that

$$f(t) = F'(t) = \frac{1}{\sqrt{\pi}\, \Gamma(r/2)} \int_0^{\infty} \frac{e^{-(u/2)(t^2/r)}}{2^{(r+1)/2}} \sqrt{\frac{u}{r}}\, u^{r/2-1} e^{-u/2}\, du$$

$$= \frac{1}{\sqrt{\pi r}\, \Gamma(r/2)} \int_0^{\infty} \frac{u^{(r+1)/2-1}}{2^{(r+1)/2}} e^{-(u/2)(1+t^2/r)}\, du.$$

In the integral make the change of variables

$$y = (1 + t^2/r)u \qquad \text{so that} \qquad \frac{du}{dy} = \frac{1}{1 + t^2/r}.$$

Thus we find that

$$f(t) = \frac{\Gamma((r+1)/2)}{\sqrt{\pi r}\,\Gamma(r/2)} \left[\frac{1}{(1 + t^2/r)^{(r+1)/2}} \right] \int_0^\infty \frac{y^{(r+1)/2 - 1}}{\Gamma((r+1)/2)\,2^{(r+1)/2}} e^{-y/2}\, dy.$$

The integral in this last expression for $f(t)$ is equal to 1 because the integrand is like the p.d.f. of a chi-square distribution with $r + 1$ degrees of freedom. Thus the p.d.f. is

$$f(t) = \frac{\Gamma((r+1)/2)}{\sqrt{\pi r}\,\Gamma(r/2)} \frac{1}{(1 + t^2/r)^{(r+1)/2}}, \qquad -\infty < t < \infty.$$

Graphs of the p.d.f. of T when $r = 1, 3$, and 7 along with the $N(0, 1)$ p.d.f. are given in Figure 5.3-2. In this figure we see that the tails of the t distribution are heavier than those of a normal one; that is, there is more extreme probability in the t distribution than in the standardized normal one. ◄

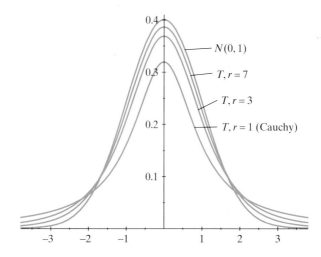

Figure 5.3-2: t distribution p.d.f.s

Remark This distribution was first discovered by W. S. Gosset when he was working for an Irish brewery. Because Gosset published under the pseudonym Student, this distribution is sometimes known as Student's t distribution. ■

EXERCISES

5.3-1 Let $X_1, X_2, \ldots, X_{16}$ be a random sample from a normal distribution $N(77, 25)$. Compute

(a) $P(77 < \overline{X} < 79.5)$. (b) $P(74.2 < \overline{X} < 78.4)$.

5.3-2 Let X be $N(50, 36)$. Using the same set of axes, sketch the graphs of the probability density functions of

(a) X.
(b) $\overline{X}$, the mean of a random sample of size 9 from this distribution.
(c) $\overline{X}$, the mean of a random sample of size 36 from this distribution.

5.3-3 Let X equal the widest diameter (in millimeters) of the fetal head measured between the 16th and 25th weeks of pregnancy. Assume that the distribution of X is $N(46.58, 40.96)$. Let $\overline{X}$ be the sample mean of a random sample of $n = 16$ observations of X.

(a) Give the values of $E(\overline{X})$ and $\mathrm{Var}(\overline{X})$.
(b) Find $P(44.42 \leq \overline{X} \leq 48.98)$.

5.3-4 Let X equal the weight of the soap in a "6-pound" box. Assume that the distribution of X is $N(6.05, 0.0004)$.

(a) Find $P(X < 6.0171)$.
(b) If nine boxes of soap are selected at random from the production line, find the probability that at most two boxes weigh less than 6.0171 each. HINT: Let Y equal the number of boxes that weigh less than 6.0171.
(c) Let $\overline{X}$ be the sample mean of the nine boxes. Find $P(\overline{X} \leq 6.035)$.

5.3-5 Let $Z_1, Z_2, \ldots, Z_7$ be a random sample from the standard normal distribution $N(0, 1)$. Let $W = Z_1^2 + Z_2^2 + \cdots + Z_7^2$. Find $P(1.69 < W < 14.07)$.

5.3-6 If $X_1, X_2, \ldots, X_{16}$ is a random sample of size $n = 16$ from the normal distribution $N(50, 100)$, determine

(a) $P\left(796.2 \leq \sum_{i=1}^{16} (X_i - 50)^2 \leq 2630 \right)$.

(b) $P\left(726.1 \leq \sum_{i=1}^{16} (X_i - \overline{X})^2 \leq 2500 \right)$.

5.3-7 Let X equal the weight (in grams) of a nail of the type that is used for making decks. Assume that the distribution of X is $N(8.78, 0.16)$. Let $\overline{X}$ be the mean of a random sample of the weights of $n = 9$ nails.

(a) Sketch, on the same set of axes, the graphs of the p.d.f.s of X and of $\overline{X}$.
(b) Let S^2 be the sample variance of the 9 weights. Find constants a and b so that $P(a \leq S^2 \leq b) = 0.90$.

HINT: That probability is equivalent to $P(8a/0.16 \leq 8S^2/0.16 \leq 8b/0.16)$ and $8S^2/0.16$ is $\chi^2(8)$. Find $8a/0.16$ and $8b/0.16$ from the tables.

5.3-8 At a heat-treating company, iron castings and steel forgings are heat-treated to achieve desired mechanical properties and machinability. One steel forging is annealed to soften the part for each machining. Two lots of this part, made of 1020 steel, are heat-treated in two different furnaces. The specification for this part is 36-66 on the Rockwell G scale. Let X_1 and X_2 equal the respective hardness measurements for parts selected randomly from furnaces 1 and 2. Assume that the distributions of X_1 and X_2 are $N(47.88, 2.19)$ and $N(43.04, 14.89)$, respectively.

(a) Sketch the p.d.f.s of X_1 and X_2 on the same graph.
(b) Compute $P(X_1 > X_2)$, assuming independence of X_1 and X_2.

5.3-9 Let X equal the force required to pull a stud out of a window that is to be inserted into an automobile. Assume that the distribution of X is $N(147.8, 12.3^2)$.

(a) Find $P(X < 163.3)$.

(b) If $\overline{X}$ is the mean and S^2 is the variance of a random sample of size $n = 25$ from this distribution of X, determine $P(\overline{X} \leq 150.9)$.

(c) Find constants a and b so that $P(a \leq S^2 \leq b) = 0.90$. (See the Hint in Exercise 5.3-7.)

5.3-10 Suppose that the distribution of the weight of a prepackaged "1-pound bag" of carrots is $N(1.18, 0.07^2)$ and the distribution of the weight of a prepackaged "3-pound bag" of carrots is $N(3.22, 0.09^2)$. Selecting bags at random, find the probability that the sum of three 1-pound bags exceeds the weight of one 3-pound bag. HINT: First determine the distribution of Y, the sum of the three, and then compute $P(Y > W)$, where W is the weight of the 3-pound bag.

5.3-11 Let X denote the wing length in millimeters of a male gallinule and Y the wing length in millimeters of a female gallinule. Assume that X is $N(184.09, 39.37)$ and Y is $N(171.93, 50.88)$ and that X and Y are independent. If a male and a female gallinule are captured, what is the probability that X is greater than Y?

5.3-12 Suppose that for a particular population of students SAT mathematics scores are $N(529, 5732)$ and SAT verbal scores are $N(474, 6368)$. Select two students at random, and let X equal the first student's math score and Y the second student's verbal score. Find $P(X > Y)$.

5.3-13 Suppose that the length of life in hours, say X, of a light bulb manufactured by company A is $N(800, 14400)$ and the length of life in hours, say Y, of a light bulb manufactured by company B is $N(850, 2500)$. One bulb is selected from each company and is burned until "death."

(a) Find the probability that the length of life of the bulb from company A exceeds the length of life of the bulb from company B by at least 15 hours.

(b) Find the probability that at least one of the bulbs "lives" for at least 920 hours.

5.3-14 Let X and Y equal the number of miles per gallon for compact cars and mid-sized cars, respectively, as reported in fuel economy ratings. Assume that $\mu_X = 24.5$, $\sigma_X = 3.8$, $\mu_Y = 21.3$, and $\sigma_Y = 2.7$. Let $\overline{X}$ and $\overline{Y}$ be the sample means of independent random samples of eight observations of X and Y, respectively.

(a) What are the values of the means and variances of $\overline{X}$ and $\overline{Y}$?

(b) Assuming that $\overline{X}$ and $\overline{Y}$ are each (approximately) normally distributed, how is $\overline{X} - \overline{Y}$ distributed?

(c) Find the (approximate) probability, $P(\overline{X} > \overline{Y})$.

5.3-15 Let X equal the weight of a fat free Fig Newton cookie. Assume that the distribution of X is $N(14.22, 0.0854)$. These cookies are sold in packages that have a label weight of 340 grams. The number of cookies in a package is usually 24, 25, or 26. Assuming that a package is filled with a random sample of cookies, how many cookies should be put into a package to be quite certain (say with a probability of at least 0.95) that the total weight of the cookies exceeds 340 grams? Also keep in mind that extra cookies in a package decreases profit.

5.3-16 Let the independent random variables X_1 and X_2 be $N(0, 1)$ and $\chi^2(r)$, respectively. Let $Y_1 = X_1/\sqrt{X_2/r}$ and $Y_2 = X_2$.

(a) Find the joint p.d.f. of Y_1 and Y_2.

(b) Determine the marginal p.d.f. of Y_1 and show that Y_1 has a t distribution. (This is another, but equivalent, way of finding the p.d.f. of T.)

5.3-17 Let T have a t distribution with r degrees of freedom. Show that $E(T) = 0, r \geq 2$, and $\text{Var}(T) = r/(r-2)$, provided that $r \geq 3$, by first finding $E(Z)$, $E(1/\sqrt{U})$, $E(Z^2)$, and $E(1/U)$.

5.3-18 A consumer buys n light bulbs, each of which has a lifetime that has a mean of 800 hours, a standard deviation of 100 hours, and a normal distribution. A light bulb is replaced by another as soon as it burns out. Assuming independence of the lifetimes, find the smallest n so that the succession of light bulbs produces light for at least 10,000 hours with probability of 0.90.

5.3-19 A marketing research firm suggests to a company that two possible competing products can generate incomes X and Y (in millions) that are $N(3, 1)$ and $N(3.5, 4)$, respectively. Clearly $P(X < Y) > 1/2$. However, the company would prefer the one with the smaller variance if in fact $P(X > 2) > P(Y > 2)$. Which product does the company select?

5.3-20 For Company A, there is an 80% chance that no claims are made; but if one or more claims are made, the total claims has a normal distribution with mean 20,000 and standard deviation 5000. For Company B, there is a 90% chance of no claims; but if one or more claims are made, the total claims has a normal distribution with mean 30,000 and standard deviation 6000. Assuming independence, what is the probability that B's total claims exceed those of A?

5.3-21 In Example 5.3-4 verify that the given transformation maps $\{(x_1, x_2) : 0 < x_1 < 1, 0 < x_2 < 1\}$ onto $\{(z_1, z_2) : -\infty < z_1 < \infty, -\infty < z_2 < \infty\}$ except for a set of points that has probability 0. HINT: What is the image of vertical line segments? What is the image of horizontal line segments?

5.3-22 Let X be $N(0, 1)$. Find the p.d.f. of $Y = |X|$, a distribution that is often called the **half normal**. HINT: Here $y \in S_y = \{y : 0 < y < \infty\}$. Consider the two transformations $x_1 = -y, -\infty < x_1 < 0$, and $x_2 = y, 0 < y < \infty$.

5.4 THE CENTRAL LIMIT THEOREM

In Section 4.6 we found that the mean $\overline{X}$ of a random sample of size n from a distribution with mean μ and variance $\sigma^2 > 0$ is a random variable with the properties that

$$E(\overline{X}) = \mu \quad \text{and} \quad \text{Var}(\overline{X}) = \frac{\sigma^2}{n}.$$

Thus, as n increases, the variance of $\overline{X}$ decreases. Consequently, the distribution of $\overline{X}$ clearly depends on n, and we see that we are dealing with sequences of distributions.

In Theorem 5.3-1 we considered the p.d.f. of $\overline{X}$ when sampling from the normal distribution $N(\mu, \sigma^2)$. We showed that the distribution of $\overline{X}$ is $N(\mu, \sigma^2/n)$, and in Figure 5.3-1, by graphing the p.d.f.s for several values of n, we illustrated that as n increases, the probability becomes concentrated in a small interval centered at μ. That is, as n increases, $\overline{X}$ tends to converge to μ, or $(\overline{X} - \mu)$ tends to converge to 0 in a probability sense.

In general, if we let

$$W = \frac{\sqrt{n}}{\sigma}(\overline{X} - \mu) = \frac{\overline{X} - \mu}{\sigma/\sqrt{n}} = \frac{Y - n\mu}{\sqrt{n}\,\sigma},$$

where Y is the sum of a random sample of size n from some distribution with mean μ and variance σ^2, then for each positive integer n,

$$E(W) = E\left[\frac{\overline{X} - \mu}{\sigma/\sqrt{n}}\right] = \frac{E(\overline{X}) - \mu}{\sigma/\sqrt{n}} = \frac{\mu - \mu}{\sigma/\sqrt{n}} = 0$$

and

$$\text{Var}(W) = E(W^2) = E\left[\frac{(\overline{X} - \mu)^2}{\sigma^2/n}\right] = \frac{E[(\overline{X} - \mu)^2]}{\sigma^2/n} = \frac{\sigma^2/n}{\sigma^2/n} = 1.$$

Thus, while $\overline{X} - \mu$ "degenerates" to zero, the factor $\sqrt{n}/\sigma$ in $\sqrt{n}(\overline{X} - \mu)/\sigma$ "spreads out" the probability enough to prevent this degeneration. What then is the distribution of W as n increases? One observation that might shed some light on the answer to this question can be made immediately. If the sample arises from a normal distribution then, from Theorem 5.3-1, we know that $\overline{X}$ is $N(\mu, \sigma^2/n)$ and hence W is $N(0, 1)$ for each positive n. Thus, in the limit, the distribution of W must be $N(0, 1)$. So if the solution of the question does not depend on the underlying distribution (i.e., it is unique), the answer must be $N(0, 1)$. As we will see, this is exactly the case, and this result is so important it is called the central limit theorem, the proof of which is given in Section 5.7*.

Theorem 5.4-1

(Central Limit Theorem) If $\overline{X}$ is the mean of a random sample $X_1, X_2, \ldots,$ X_n of size n from a distribution with a finite mean μ and a finite positive variance σ^2, then the distribution of

$$W = \frac{\overline{X} - \mu}{\sigma/\sqrt{n}} = \frac{\sum\limits_{i=1}^{n} X_i - n\mu}{\sqrt{n}\,\sigma}$$

is $N(0, 1)$ in the limit as $n \to \infty$.

A practical use of the central limit theorem is approximating, when n is "sufficiently large," the distribution function of W, namely

$$P(W \leq w) \approx \int_{-\infty}^{w} \frac{1}{\sqrt{2\pi}} e^{-z^2/2}\, dz = \Phi(w).$$

We present some illustrations of this application, discuss "sufficiently large," and try to give an intuitive feeling for the central limit theorem.

EXAMPLE 5.4-1

Let $\overline{X}$ be the mean of a random sample of $n = 25$ currents (in milliamperes) in a strip of wire in which each measurement has a mean of 15 and a variance of 4. Then $\overline{X}$ has an approximate $N(15, 4/25 = 0.16)$ distribution. For illustration,

$$P(14.4 < \overline{X} < 15.6) = P\left(\frac{14.5 - 15}{0.4} < \frac{\overline{X} - 15}{0.4} < \frac{15.6 - 15}{0.4}\right)$$

$$\approx \Phi(1.5) - \Phi(-1.5) = 0.9332 - 0.0668 = 0.8664. \quad \blacktriangleleft$$

EXAMPLE 5.4-2 Let $X_1, X_2, \ldots, X_{20}$ denote a random sample of size 20 from the uniform distribution $U(0, 1)$. Here $E(X_i) = 1/2$ and $\mathrm{Var}(X_i) = 1/12$, for $i = 1, 2, \ldots, 20$. If $Y = X_1 + X_2 + \cdots + X_{20}$, then

$$P(Y \le 9.1) = P\left(\frac{Y - 20(1/2)}{\sqrt{20/12}} \le \frac{9.1 - 10}{\sqrt{20/12}}\right) = P(W \le -0.697)$$

$$\approx \Phi(-0.697)$$

$$= 0.2423.$$

Also,

$$P(8.5 \le Y \le 11.7) = P\left(\frac{8.5 - 10}{\sqrt{5/3}} \le \frac{Y - 10}{\sqrt{5/3}} \le \frac{11.7 - 10}{\sqrt{5/3}}\right)$$

$$= P(-1.162 \le W \le 1.317)$$

$$\approx \Phi(1.317) - \Phi(-1.162)$$

$$= 0.9061 - 0.1226 = 0.7835. \qquad \blacktriangleleft$$

EXAMPLE 5.4-3 Let $\overline{X}$ denote the mean of a random sample of size 25 from the distribution whose p.d.f. is $f(x) = x^3/4$, $0 < x < 2$. It is easy to show that $\mu = 8/5 = 1.6$ and $\sigma^2 = 8/75$. Thus

$$P(1.5 \le \overline{X} \le 1.65) = P\left(\frac{1.5 - 1.6}{\sqrt{8/75}/\sqrt{25}} \le \frac{\overline{X} - 1.6}{\sqrt{8/75}/\sqrt{25}} \le \frac{1.65 - 1.6}{\sqrt{8/75}/\sqrt{25}}\right)$$

$$= P(-1.531 \le W \le 0.765)$$

$$\approx \Phi(0.765) - \Phi(-1.531)$$

$$= 0.7779 - 0.0629 = 0.7150. \qquad \blacktriangleleft$$

These examples have shown how the central limit theorem can be used for approximating certain probabilities concerning the mean $\overline{X}$ or the sum $Y = \sum_{i=1}^{n} X_i$ of a random sample. That is, $\overline{X}$ is approximately $N(\mu, \sigma^2/n)$, and Y is approximately $N(n\mu, n\sigma^2)$ when n is "sufficiently large," where μ and σ^2 are the mean and the variance of the underlying distribution from which the sample arose. Generally, if n is greater than 25 or 30, these approximations will be good. However, if the underlying distribution is symmetric, unimodal, and of the continuous type, a value of n as small as 4 or 5 can yield a very adequate approximation. Moreover, if the original distribution is approximately normal, $\overline{X}$ would have a distribution very close to normal when n equals 2 or 3. In fact, we know that if the sample is taken from $N(\mu, \sigma^2)$, $\overline{X}$ is exactly $N(\mu, \sigma^2/n)$ for every $n = 1, 2, 3, \ldots$.

The following examples will help to illustrate the previous remarks and will give the reader a better intuitive feeling about the central limit theorem. In particular, we shall see how the size of n affects the distribution of $\overline{X}$ and $Y = \sum_{i=1}^{n} X_i$ for samples from several underlying distributions.

EXAMPLE 5.4-4 Let X_1, X_2, X_3, X_4 be a random sample of size 4 from the uniform distribution $U(0, 1)$ with p.d.f. $f(x) = 1, 0 < x < 1$. Then $\mu = 1/2$ and $\sigma^2 = 1/12$. We shall compare the graph of the p.d.f. of

$$Y = \sum_{i=1}^{n} X_i$$

with the graph of the $N[n(1/2), n(1/12)]$ p.d.f. for $n = 2$ and 4, respectively.

By methods given in Section 4.4, we can determine that the p.d.f. of $Y = X_1 + X_2$ is

$$g(y) = \begin{cases} y, & 0 < y \leq 1, \\ 2 - y, & 1 < y < 2. \end{cases}$$

This is the triangular p.d.f. that is graphed in Figure 5.4-1(a). In this figure the $N[2(1/2), 2(1/12)]$ p.d.f. is also graphed.

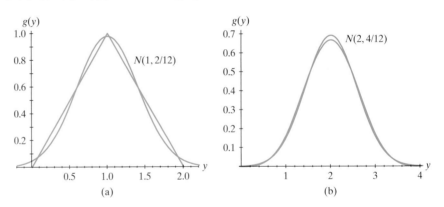

Figure 5.4-1: p.d.f.s of sums of uniform random variables

Moreover, the p.d.f. of $Y = X_1 + X_2 + X_3 + X_4$ is

$$g(y) = \begin{cases} \dfrac{y^3}{6}, & 0 \leq y < 1, \\[2mm] \dfrac{-3y^3 + 12y^2 - 12y + 4}{6}, & 1 \leq y < 2, \\[2mm] \dfrac{3y^3 - 24y^2 + 60y - 44}{6}, & 2 \leq y < 3, \\[2mm] \dfrac{-y^3 + 12y^2 - 48y + 64}{6}, & 3 \leq y \leq 4. \end{cases}$$

This p.d.f. is graphed in Figure 5.4-1(b) along with the $N[4(1/2), 4(1/12)]$ p.d.f. If we are interested in finding $P(1.7 \leq Y \leq 3.2)$, this could be done by evaluating

$$\int_{1.7}^{3.2} g(y)\,dy,$$

which is tedious (see Exercise 5.4-10). It is much easier to use a normal approximation, which results in a number very close to the exact value.   ◀

In Example 5.4-4 and Exercise 5.4-10 we show that even for a small value of n, like $n = 4$, the sum of the sample items has an approximate normal distribution. The following example illustrates that for some underlying distributions (particularly skewed ones) n must be quite large to obtain a satisfactory approximation. In order to keep the scale on the horizontal axis the same for each value of n, we will use the following result.

Let $f(x)$ and $F(x)$ be the p.d.f. and distribution function of a random variable, X, of the continuous type having mean μ and variance σ^2. Let $W = (X - \mu)/\sigma$. The distribution function of W is given by

$$G(w) = P(W \leq w) = P\left(\frac{X - \mu}{\sigma} \leq w\right)$$

$$= P(X \leq \sigma w + \mu) = F(\sigma w + \mu).$$

Thus the p.d.f. of W is given by

$$g(w) = F'(\sigma w + \mu) = \sigma f(\sigma w + \mu).$$

EXAMPLE 5.4-5   Let $X_1, X_2, \ldots, X_n$ be a random sample of size n from a chi-square distribution with one degree of freedom. If

$$Y = \sum_{i=1}^{n} X_i,$$

then Y is $\chi^2(n)$, and $E(Y) = n$, $\text{Var}(Y) = 2n$. Let

$$W = \frac{Y - n}{\sqrt{2n}}.$$

The p.d.f. of W is given by

$$g(w) = \sqrt{2n}\,\frac{(\sqrt{2n}\,w + n)^{n/2-1}}{\Gamma\left(\frac{n}{2}\right)2^{n/2}}\,e^{-(\sqrt{2n}\,w+n)/2}, \qquad -n/\sqrt{2n} < w < \infty.$$

Note that $w > -n/\sqrt{2n}$ corresponds to $y > 0$. In Figure 5.4-2(a) and (b), the graph of W is given along with the $N(0, 1)$ p.d.f. for $n = 20$ and 100, respectively. ◄

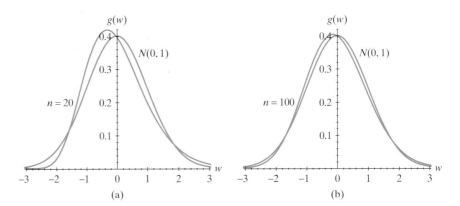

Figure 5.4-2: p.d.f.s of sums of chi-square random variables

In order to have an intuitive feeling about how the sample size n affects the distribution of $W = (\overline{X} - \mu)/(\sigma/\sqrt{n})$, it is helpful to simulate values of W on a computer using different values of n and different underlying distributions. The next example illustrates this.

Remark We simulate observations from a distribution of X having a continuous-type distribution function $F(x)$ as follows. Suppose $F(a) = 0$, $F(b) = 1$, and $F(x)$ is strictly increasing for $a < x < b$. Let $Y = F(x)$ and let the distribution of Y be $U(0, 1)$. If y is an observed value of Y, then $x = F^{-1}(y)$ is an observed value of X (see Section 3.5). Thus, if y is the value of a computer-generated random number, then $x = F^{-1}(y)$ is the simulated value of X. ■

EXAMPLE 5.4-6

It is often difficult to find the exact distribution of the random variable $W = (\overline{X} - \mu)/(\sigma/\sqrt{n})$ unless you use a computer algebra system such as *Maple*. In this example we give some empirical evidence about the distribution of W by simulating random samples on the computer. We also superimpose the theoretical p.d.f. of W which was found using *Maple*. Let $X_1, X_2, \ldots, X_n$ denote a random sample of size n from the distribution with p.d.f. $f(x)$, distribution function $F(x)$, mean μ, and variance σ^2. We simulated 1000 random samples of size $n = 2$ and $n = 7$ from each of two distributions. The value of W was computed for each sample, thus obtaining 1000 observed values of W. A histogram of these 1000 values was constructed using 21 intervals of equal length. A relative frequency histogram of the observations of W, the p.d.f. for the standard normal distribution, and the theoretical p.d.f. of W are given.

(a) In Figure 5.4-3, $f(x) = (x+1)/2$ and $F(x) = (x+1)^2/4$ for $-1 < x < 1$; $\mu = 1/3$, $\sigma^2 = 2/9$; and $n = 2$ and 7.

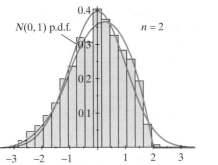
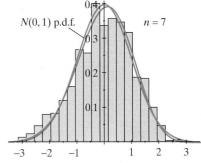

Figure 5.4-3: p.d.f.s of $(\overline{X} - \mu)/(\sigma/\sqrt{n})$, underlying distribution triangular

(b) In Figure 5.4-4, $f(x) = (3/2)x^2$ and $F(x) = (x^3 + 1)/2$ for $-1 < x < 1$; $\mu = 0$, $\sigma^2 = 3/5$; and $n = 2$ and 7. (Sketch the graph of $y = f(x)$. Give an argument as to why the histogram for $n = 2$ looks the way it does.) ◀

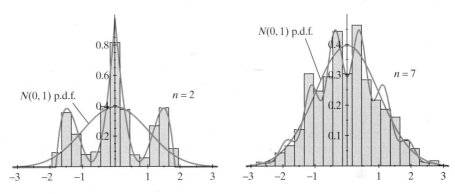

Figure 5.4-4: p.d.f.s of $(\overline{X} - \mu)/(\sigma/\sqrt{n})$, underlying distribution U-shaped

Note very clearly that these examples have not proved anything. They are presented to give evidence of the truth of the central limit theorem and they do give a nice feeling for what is happening.

So far all the illustrations have concerned distributions of the continuous type. However, the hypotheses for the central limit theorem do not require the distribution to be continuous. We shall consider applications of the central limit theorem for discrete-type distributions in the next section.

EXERCISES

5.4-1 Let $\overline{X}$ be the mean of a random sample of size 12 from the uniform distribution on the interval $(0, 1)$. Approximate $P(1/2 \leq \overline{X} \leq 2/3)$.

5.4-2 Let $Y = X_1 + X_2 + \cdots + X_{15}$ be the sum of a random sample of size 15 from the distribution whose p.d.f. is $f(x) = (3/2)x^2$, $-1 < x < 1$. Approximate

$$P(-0.3 \leq Y \leq 1.5).$$

5.4-3 Let $\overline{X}$ be the mean of a random sample of size 36 from an exponential distribution with mean 3. Approximate $P(2.5 \le \overline{X} \le 4)$.

5.4-4 Approximate $P(39.75 \le \overline{X} \le 41.25)$, where $\overline{X}$ is the mean of a random sample of size 32 from a distribution with mean $\mu = 40$ and variance $\sigma^2 = 8$.

5.4-5 Let $X_1, X_2, \ldots, X_{18}$ be a random sample of size 18 from a chi-square distribution with $r = 1$. Recall that $\mu = 1, \sigma^2 = 2$.

(a) How is $Y = \sum_{i=1}^{18} X_i$ distributed?
(b) Using the result of part (a), we see from Table IV in the Appendix that

$$P(Y \le 9.390) = 0.05 \text{ and } P(Y \le 34.80) = 0.99.$$

Compare these two probabilities with the approximations found using the central limit theorem.

5.4-6 A random sample of size $n = 18$ is taken from the distribution with p.d.f $f(x) = 1 - x/2, 0 \le x \le 2$.

(a) Find μ and σ^2.   **(b)** Find, approximately, $P(2/3 \le \overline{X} \le 5/6)$.

5.4-7 Let X equal the maximal oxygen intake of a human on a treadmill, where the measurements are in milliliters of oxygen per minute per kilogram of weight. Assume that for a particular population the mean of X is $\mu = 54.030$ and the standard deviation is $\sigma = 5.8$. Let $\overline{X}$ be the sample mean of a random sample of size $n = 47$. Find $P(52.761 \le \overline{X} \le 54.453)$, approximately.

5.4-8 Let X equal the weight in grams of a miniature candy bar. Assume that $\mu = E(X) = 24.43$ and $\sigma^2 = \text{Var}(X) = 2.20$. Let $\overline{X}$ be the sample mean of a random sample of $n = 30$ candy bars. Find

(a) $E(\overline{X})$.   **(b)** $\text{Var}(\overline{X})$.   **(c)** $P(24.17 \le \overline{X} \le 24.82)$, approximately.

5.4-9 Let X equal the birth weight in grams of a baby born in the Sudan. Assume that $E(X) = 3320$ and $\text{Var}(X) = 660^2$. Let $\overline{X}$ be the sample mean of a random sample of size $n = 225$. Find $P(3233.76 \le \overline{X} \le 3406.24)$, approximately.

5.4-10 In Example 5.4-4, with $n = 4$, compute $P(1.7 \le Y \le 3.2)$ and compare this answer with the normal approximation of this probability.

5.4-11 Five measurements in "ohms per square" of the electronic conductive coating (that allows light to pass through it) on a thin, clear piece of glass were made and the average was calculated. This was repeated 150 times, yielding the following averages:

83.86	75.86	79.65	90.57	95.37	97.97	77.00	80.80	83.53	83.17
80.20	81.42	89.14	88.68	85.15	90.11	89.03	85.00	82.57	79.46
81.20	82.80	81.28	74.64	69.85	75.60	76.60	78.30	88.51	86.32
84.03	95.27	90.05	77.50	75.14	81.33	86.12	78.30	80.30	84.96
79.30	88.96	82.76	83.13	79.60	86.20	85.16	87.86	91.00	91.10
84.04	92.10	79.20	83.76	87.87	86.71	81.89	85.72	75.84	74.27
93.08	81.75	75.66	75.35	76.55	84.86	90.68	91.02	90.97	98.30
91.84	97.41	73.60	90.65	80.20	74.75	90.35	79.66	86.88	83.00
86.24	80.50	74.25	91.20	70.16	78.40	85.60	80.82	75.95	80.75
81.86	82.18	82.98	84.00	76.85	85.00	79.50	86.56	83.30	72.40
79.20	86.20	82.36	84.08	86.11	88.25	88.93	93.12	78.30	77.24
82.52	81.37	83.72	86.90	84.37	92.60	95.01	78.95	81.40	88.40
76.10	85.33	82.95	80.20	88.21	83.49	81.00	82.72	81.12	83.62
91.18	85.90	79.01	77.56	81.13	80.60	81.65	70.70	69.36	79.09
71.35	67.20	67.43	69.95	66.76	76.35	69.45	80.13	84.26	88.13

(a) Construct a frequency table for these 150 observations using 10 intervals of equal length, and using (66.495, 69.695) for the first class interval. Which is the modal class?

(b) Construct a relative frequency histogram for the grouped data.

(c) Superimpose a normal p.d.f. using the mean 82.661 and variance 42.2279 of these 150 values as the true mean and variance. Do these 150 means of five measurements seem to be normally distributed?

5.4-12 A church has pledges (in dollars) with a mean of 2000 and a standard deviation of 500. A random sample of size $n = 25$ is taken from all of this church's pledges and the sample mean $\overline{X}$ is considered. Approximate $P(\overline{X} > 2050)$.

5.4-13 A company has a one-year group life policy with an insurance company that divides its employees into two classes as follows.

Class	Probability of Death	Benefit	Number in Class
A	0.01	$20,000	1000
B	0.03	$10,000	500

The insurance company wants to collect a premium from that company that equals the 90th percentile of the distribution of the total claims. What should that premium be?

5.4-14 Let X and Y equal the respective numbers of hours a randomly selected child watches movies or cartoons on TV during a month. From experience, it is known that $E(X) = 30$, $E(Y) = 50$, $\text{Var}(X) = 52$, $\text{Var}(Y) = 64$, and $\text{Cov}(X, Y) = 14$. Twenty-five children are selected at random. Let Z equal the total number of hours these 25 children watch TV movies or cartoons in the next month. Approximate $P(1970 < Z < 2090)$. HINT: Use remark after Theorem 4.6-1.

5.4-15 The tensile strength X of paper has $\mu = 30$ and $\sigma = 3$ (pounds per square inch). A random sample of size $n = 100$ is taken from the distribution of tensile strengths. Compute the probability that the sample mean $\overline{X}$ is greater than 29.5 pounds per square inch.

5.4-16 At certain times during the year, a bus company runs a special van holding ten passengers from Iowa City to Chicago. After the opening of sales of the tickets, the time (in minutes) between sales of tickets for the trip has a gamma distribution with $\alpha = 3$ and $\theta = 2$.

(a) Assuming independence, record an integral that gives the probability of being sold out within one hour.

(b) Approximate the answer in part (a).

5.4-17 Let X_1, X_2, X_3, X_4 represent the random times in days needed to complete four steps of a project. These times are independent and have gamma distributions with common $\theta = 2$ and $\alpha_1 = 3, \alpha_2 = 2, \alpha_3 = 5, \alpha_4 = 3$, respectively. One step must be completed before the next can be started. Let Y equal the total time needed to complete the project.

(a) Find an integral that represents $P(Y \le 25)$.

(b) Approximate part (a) using a normal distribution. Is this justified?

5.4-18 Assume that the sick leave taken by the typical worker per year has $\mu = 10$, $\sigma = 2$, measured in days. A firm has $n = 20$ employees. Assuming independence, how many sick days should the firm budget if the financial officer wants the probability of exceeding the budgeted days to be less than 20%?

5.5 **APPROXI-MATIONS FOR DISCRETE DISTRIBUTIONS**

In this section we illustrate how the normal distribution can be used to approximate probabilities for certain discrete-type distributions. One of the more important discrete distributions is the binomial distribution. To see how the central limit theorem can be applied, recall that a binomial random variable can be described as the sum of Bernoulli random variables. That is, let $X_1, X_2, \ldots, X_n$ be a random sample from a Bernoulli distribution with a mean $\mu = p$ and a variance $\sigma^2 = p(1 - p)$, where $0 < p < 1$. Then $Y = \sum_{i=1}^{n} X_i$ is $b(n, p)$. The central limit theorem states that the distribution of

$$W = \frac{Y - np}{\sqrt{np(1 - p)}} = \frac{\overline{X} - p}{\sqrt{p(1 - p)/n}}$$

is $N(0, 1)$ in the limit as $n \to \infty$. Thus, if n is sufficiently large, the distribution of Y is approximately $N[np, np(1 - p)]$, and probabilities for the binomial distribution $b(n, p)$ can be approximated using this normal distribution. A rule often stated is that n is "sufficiently large" if $np \geq 5$ and $n(1 - p) \geq 5$. This can be used as a rough guide although as p deviates more and more from 0.5, we need larger and larger sample sizes, because the underlying Bernoulli distribution becomes more skewed.

Note that we shall be approximating probabilities for a discrete distribution with probabilities for a continuous distribution. Let us discuss a reasonable procedure in this situation. If V is $N(\mu, \sigma^2)$, $P(a < V < b)$ is equivalent to the area bounded by the p.d.f. of V, the v axis, $v = a$, and $v = b$. If Y is $b(n, p)$, recall that the probability histogram for Y was defined as follows. For each y such that $k - 1/2 < y = k < k + 1/2$, let

$$f(k) = \frac{n!}{k!(n - k)!} \, p^k (1 - p)^{n-k}, \qquad k = 0, 1, 2, \ldots, n.$$

Then $P(Y = k)$ can be represented by the area of the rectangle with a height of $P(Y = k)$ and a base of length 1 centered at k. Figure 5.5-1 shows the graph

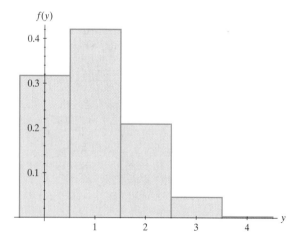

Figure 5.5-1: Probability histogram for $b(4, 1/4)$

of the probability histogram for the binomial distribution $b(4, 1/4)$. When using the normal distribution to approximate probabilities for the binomial distribution, areas under the p.d.f. for the normal distribution will be used to approximate areas of rectangles in the probability histogram for the binomial distribution.

EXAMPLE 5.5-1 Let Y be $b(10, 1/2)$. Then, using the central limit theorem, $P(a < Y < b)$ can be approximated using the normal distribution with mean $10(1/2) = 5$ and variance $10(1/2)(1/2) = 5/2$. Figure 5.5-2 shows the graph of the probability histogram for $b(10, 1/2)$ and the graph of the p.d.f. of the normal distribution $N(5, 5/2)$. Note that the area of the rectangle whose base is

$$\left(k - \frac{1}{2}, k + \frac{1}{2}\right)$$

and the area under the normal curve between $k - 1/2$ and $k + 1/2$ are approximately equal for each integer k.   ◄

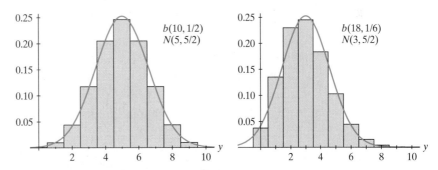

Figure 5.5-2: Normal approximation for the binomial distribution

EXAMPLE 5.5-2 Let Y be $b(18, 1/6)$. Because $np = 18(1/6) = 3 < 5$, the normal approximation is not as good here. Figure 5.5-2 illustrates this by depicting the skewed probability histogram for $b(18, 1/6)$ and the symmetric p.d.f. of the normal distribution $N(3, 5/2)$.   ◄

EXAMPLE 5.5-3 Let Y have the binomial distribution of Example 5.5-1 and Figure 5.5-2, namely $b(10, 1/2)$. Then

$$P(3 \le Y < 6) = P(2.5 \le Y \le 5.5)$$

because $P(Y = 6)$ is not in the desired answer. But the latter equals

$$P\left(\frac{2.5 - 5}{\sqrt{10/4}} \le \frac{Y - 5}{\sqrt{10/4}} \le \frac{5.5 - 5}{\sqrt{10/4}}\right) \approx \Phi(0.316) - \Phi(-1.581)$$

$$= 0.6240 - 0.0570 = 0.5670.$$

Using Appendix Table II, we find that $P(3 \le Y < 6) = 0.5683$.   ◄

EXAMPLE 5.5-4 Let Y be $b(36, 1/2)$. Then, since

$$\mu = (36)(1/2) = 18 \quad \text{and} \quad \sigma^2 = (36)(1/2)(1/2) = 9,$$

$$
\begin{aligned}
P(12 < Y \le 18) &= P(12.5 \le Y \le 18.5) \\
&= P\left(\frac{12.5 - 18}{\sqrt{9}} \le \frac{Y - 18}{\sqrt{9}} \le \frac{18.5 - 18}{\sqrt{9}}\right) \\
&\approx \Phi(0.167) - \Phi(-1.833) \\
&= 0.5329.
\end{aligned}
$$

Note that 12 was increased to 12.5 because $P(Y = 12)$ is not included in the desired probability. Using the binomial formula, we find that (you may verify this answer using your calculator)

$$P(12 < Y \le 18) = P(13 \le Y \le 18) = 0.5334.$$

Also,

$$
\begin{aligned}
P(Y = 20) &= P(19.5 \le Y \le 20.5) \\
&= P\left(\frac{19.5 - 18}{\sqrt{9}} \le \frac{Y - 18}{\sqrt{9}} \le \frac{20.5 - 18}{\sqrt{9}}\right) \\
&\approx \Phi(0.833) - \Phi(0.5) \\
&= 0.1060.
\end{aligned}
$$

Using the binomial formula, we have $P(Y = 20) = 0.1063$. So, in this situation, the approximation is extremely good. ◀

Note that, in general, if Y is $b(n, p)$, n is large, and $k = 0, 1, \ldots, n$,

$$P(Y \le k) \approx \Phi\left(\frac{k + 1/2 - np}{\sqrt{npq}}\right)$$

and

$$P(Y < k) \approx \Phi\left(\frac{k - 1/2 - np}{\sqrt{npq}}\right),$$

because in the first case k is included and in the second it is not.

We now show how the Poisson distribution with large enough mean can be approximated using a normal distribution.

EXAMPLE 5.5-5 A random variable having a Poisson distribution with mean 20 can be thought of as the sum Y of the observations of a random sample of size 20 from a Poisson distribution with mean 1. Thus

$$W = \frac{Y - 20}{\sqrt{20}}$$

has a distribution that is approximately $N(0, 1)$, and the distribution of Y is approximately $N(20, 20)$ (see Figure 5.5-3). So, for illustration,

$$P(16 < Y \le 21) = P(16.5 \le Y \le 21.5)$$

$$= P\left(\frac{16.5 - 20}{\sqrt{20}} \le \frac{Y - 20}{\sqrt{20}} \le \frac{21.5 - 20}{\sqrt{20}}\right)$$

$$\approx \Phi(0.335) - \Phi(-0.783)$$

$$= 0.4142.$$

Note that 16 is increased to 16.5 because $Y = 16$ is not included in the event $\{16 < Y \le 21\}$. The answer using the Poisson formula is 0.4226. ◄

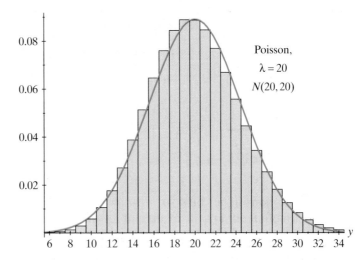

Figure 5.5-3: Normal approximation of Poisson, $\lambda = 20$

In general, if Y has a Poisson distribution with mean λ, then the distribution of

$$W = \frac{Y - \lambda}{\sqrt{\lambda}}$$

is approximately $N(0, 1)$ when λ is sufficiently large.

EXERCISES

5.5-1 Let the distribution of Y be $b(25, 1/2)$. Find the following probabilities in two ways: exactly using Appendix Table II and approximately using the central limit theorem. Compare the two results in each of the three cases.

(a) $P(10 < Y \le 12)$.   (b) $P(12 \le Y < 15)$.   (c) $P(Y = 12)$.

5.5-2 Among the gifted 7th-graders who score very high on a mathematics exam, approximately 20% are left-handed or ambidextrous. Let X equal the number of left-handed or ambidextrous students among a random sample of $n = 25$ gifted 7th-graders. Find $P(2 < X < 9)$.

(a) Using Appendix Table II.

(b) Approximately, using the central limit theorem.

Remark Since X has a skewed distribution, the approximation is not as good as for the symmetrical distribution where $p = 0.50$ even though $np = 5$. ∎

5.5-3 A public opinion poll in Southern California was conducted to determine whether Southern Californians are prepared for the big earthquake that experts predict will devastate the region sometime in the next 50 years. It was learned that "60% have not secured objects in their homes that might fall and cause injury and damage during a temblor." In a random sample of $n = 864$ Southern Californians, let X equal the number who "have not secured objects in their homes." Find $P(496 \leq X \leq 548)$, approximately.

5.5-4 Let X equal the number out of $n = 48$ mature aster seeds that will germinate when $p = 0.75$ is the probability that a particular seed germinates. Approximate $P(35 \leq X \leq 40)$.

5.5-5 Let p equal the proportion of all college students who would say yes to the question, "Would you drink from the same glass as your friend if you suspected that this friend were an AIDS virus carrier?" Assume that $p = 0.10$. Let X equal the number of students out of a random sample of size $n = 100$ who would say yes to this question. Approximate $P(X \leq 11)$.

Remark The value used for p was based on a poll conducted in a class at San Diego State University. It would be interesting for you to conduct a survey at your college to estimate the value of p. ∎

5.5-6 In adults the pneumococcus bacterium cause 70% of the pneumonia cases. In a random sample of $n = 84$ adults who have pneumonia, let X equal the number whose pneumonia was caused by the pneumococcus bacterium. Use the normal distribution to find, approximately, $P(X \leq 52)$.

5.5-7 Let $X_1, X_2, \ldots, X_{48}$ be a random sample of size 48 from the distribution with p.d.f. $f(x) = 1/x^2$, $1 < x < \infty$. Approximate the probability that at most 10 of these random variables have values greater than 4. HINT: Let the ith trial be a success if $X_i > 4, i = 1, 2, \ldots, 48$ and let Y equal the number of successes.

5.5-8 A candy maker produces mints that have a label weight of 20.4 grams. Assume that the distribution of the weights of these mints is $N(21.37, 0.16)$.

(a) Let X denote the weight of a single mint selected at random from the production line. Find $P(X < 20.857)$.

(b) During a particular shift 100 mints are selected at random and weighed. Let Y equal the number of these mints that weigh less than 20.857 grams. Approximate $P(Y \leq 5)$.

(c) Let $\overline{X}$ equal the sample mean of the 100 mints selected and weighed on a particular shift. Find $P(21.31 \leq \overline{X} \leq 21.39)$.

5.5-9 Let X equal the number of alpha particles emitted by barium-133 per second and counted by a Geiger counter. Assume that X has a Poisson distribution with $\lambda = 49$. Approximate $P(45 < X < 60)$.

5.5-10 Let X equal the number of alpha particles counted by a Geiger counter during 30 seconds. Assume that the distribution of X is Poisson with a mean of 4829. Determine (approximately) $P(4776 \leq X \leq 4857)$.

5.5-11 Let $X_1, X_2, \ldots, X_{30}$ be a random sample of size 30 from a Poisson distribution with a mean of 2/3. Approximate

(a) $P\left(15 < \sum_{i=1}^{30} X_i \leq 22\right)$.   (b) $P\left(21 \leq \sum_{i=1}^{30} X_i < 27\right)$.

5.5-12 In the casino game roulette, the probability of winning with a bet on red is $p = 18/38$. Let Y equal the number of winning bets out of 1000 independent bets that are placed. Find $P(Y > 500)$ approximately.

5.5-13 About 60% of all Americans have a sedentary life style. Select $n = 96$ Americans at random (assume independence). What is the probability that between 50 and 60, inclusive, do not exercise regularly?

5.5-14 If X is $b(100, 0.1)$, find the approximate value of $P(12 \leq X \leq 14)$ using

(a) The normal approximation.
(b) The Poisson approximation.
(c) The binomial.

5.5-15 Let $X_1, X_2, \ldots, X_{36}$ be a random sample of size 36 from the geometric distribution with p.m.f. $f(x) = (1/4)^{x-1}(3/4), x = 1, 2, 3, \ldots$. Approximate

(a) $P\left(46 \leq \sum_{i=1}^{36} X_i \leq 49\right)$.   (b) $P(1.25 \leq \overline{X} \leq 1.50)$.

HINT: Observe that the distribution of the sum is of the discrete type.

5.5-16 A die is rolled 24 independent times. Let Y be the sum of the 24 resulting values. Recalling that Y is a random variable of the discrete type, approximate

(a) $P(Y \geq 86)$.   (b) $P(Y < 86)$.   (c) $P(70 < Y \leq 86)$.

5.5-17 In the United States the probability that a child dies in his or her first year of life is about $p = 0.01$ (it is slightly less than this in truth). Consider a group of 5000 such infants. What is the probability that between 45 and 53 (including 45 and 53) die in the first year of life?

5.5-18 Let Y equal the sum of $n = 100$ Bernoulli trials. That is, Y is $b(100, p)$. For each of (i) $p = 0.1$, (ii) $p = 0.5$, and (iii) $p = 0.8$,

(a) Draw on the same graph the approximating normal p.d.f.s.
(b) Find, approximately, $P(|Y/100 - p| \leq 0.015)$.

5.5-19 The number of trees in one acre has a Poisson distribution with mean 60. Assuming independence, compute approximately $P(5950 \leq X \leq 6100)$, where X is the number of trees in 100 acres.

5.5-20 A communication system for a company has 35 outside lines. If the number X of lines in use at a given time follows a Poisson distribution with mean 25, compute the probability that an incoming call cannot find an open line.

5.5-21 The number of flaws in a certain unit of material has a Poisson distribution with the mean of two. Fifty of these units are taken at random.

(a) What is the probability that the total number of flaws in the 50 is less that 110?
(b) What is the probability that at least two of the 50 have more than 6 flaws?

5.5-22 The number X of flaws on a certain tape of length one yard follows a Poisson distribution with mean 0.3. We examine $n = 100$ such tapes and count the total number Y of flaws.

(a) Assuming independence, what is the distribution of Y?
(b) Approximate $P(Y \leq 25)$.

5.6   THE BIVARIATE NORMAL DISTRIBUTION

Let X and Y be random variables with joint p.d.f. $f(x, y)$ of the continuous type. Many applications are concerned with the conditional distribution of one of the random variables, say Y, given that $X = x$. For example, X and Y might be a student's grade point averages from high school and from the first year in college, respectively. Persons in the field of educational testing and measurement are extremely interested in the conditional distribution of Y, given $X = x$, in such situations.

Suppose that we have an application in which we can make the following three assumptions about the conditional distribution of Y, given $X = x$:

(a) It is normal for each real x.
(b) Its mean $E(Y \mid x)$ is a linear function of x.
(c) Its variance is constant; that is, it does not depend upon the given value of x.

Of course, assumption (b), along with a result given in Section 4.3 implies that

$$E(Y \mid x) = \mu_Y + \rho \frac{\sigma_Y}{\sigma_X} (x - \mu_X).$$

Let us consider the implication of assumption (c). The conditional variance is given by

$$\sigma_{Y|x}^2 = \int_{-\infty}^{\infty} \left[y - \mu_Y - \rho \frac{\sigma_Y}{\sigma_X} (x - \mu_X) \right]^2 h(y \mid x) \, dy,$$

where $h(y \mid x)$ is the conditional p.d.f. of Y given $X = x$. Multiply each member of this equation by $f_1(x)$ and integrate on x. Since $\sigma_{Y|x}^2$ is a constant, the lefthand member is equal to $\sigma_{Y|x}^2$. Thus we have

$$\sigma_{Y|x}^2 = \int_{-\infty}^{\infty} \int_{-\infty}^{\infty} \left[y - \mu_Y - \rho \frac{\sigma_Y}{\sigma_X} (x - \mu_X) \right]^2 h(y \mid x) f_1(x) \, dy \, dx. \quad (5.6\text{-}1)$$

However, $h(y \mid x) f_1(x) = f(x, y)$; hence the righthand member is just an expectation and Equation 5.6-1 can be written as

$$\sigma_{Y|x}^2 = E\left\{ (Y - \mu_Y)^2 - 2\rho \frac{\sigma_Y}{\sigma_X} (X - \mu_X)(Y - \mu_Y) + \rho^2 \frac{\sigma_Y^2}{\sigma_X^2} (X - \mu_X)^2 \right\}.$$

But using the fact that the expectation E is a linear operator, we have, recalling $E[(X - \mu_X)(Y - \mu_Y)] = \rho \sigma_X \sigma_Y$, that

$$\sigma_{Y|x}^2 = \sigma_Y^2 - 2\rho \frac{\sigma_Y}{\sigma_X} \rho \sigma_X \sigma_Y + \rho^2 \frac{\sigma_Y^2}{\sigma_X^2} \sigma_X^2$$

$$= \sigma_Y^2 - 2\rho^2 \sigma_Y^2 + \rho^2 \sigma_Y^2 = \sigma_Y^2 (1 - \rho^2).$$

That is, the conditional variance of Y, for each given x, is $\sigma_Y^2(1-\rho^2)$. These facts about the conditional mean and variance, along with assumption (a), require that the conditional p.d.f. of Y, given $X = x$, be

$$h(y \mid x) = \frac{1}{\sigma_Y \sqrt{2\pi} \sqrt{1 - \rho^2}} \exp\left[-\frac{[y - \mu_Y - \rho(\sigma_Y/\sigma_X)(x - \mu_X)]^2}{2\sigma_Y^2(1 - \rho^2)} \right],$$

$$-\infty < y < \infty, \text{ for every real } x.$$

Before we make any assumptions about the distribution of X, we give an example and figure to illustrate the implications of our current assumptions.

EXAMPLE 5.6-1   Let $\mu_X = 10$, $\sigma_X^2 = 9$, $\mu_Y = 12$, $\sigma_Y^2 = 16$, and $\rho = 0.6$. We have seen that assumptions (a), (b), and (c) imply that the conditional distribution of Y, given $X = x$, is

$$N\left[12 + (0.6)\left(\frac{4}{3}\right)(x - 10), 16(1 - 0.6^2) \right].$$

In Figure 5.6-1 the conditional mean line

$$E(Y \mid x) = 12 + (0.6)\left(\frac{4}{3}\right)(x - 10) = 0.8x + 4$$

has been graphed. For each of $x = 5, 10,$ and 15, the conditional p.d.f. of Y, given $X = x$, is displayed. ◄

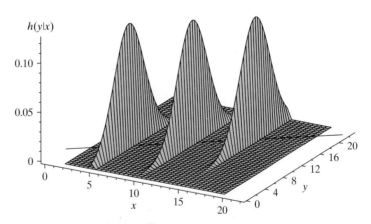

Figure 5.6-1: Conditional p.d.f. of Y, given $x = 5, 10, 15$

Up to this point, nothing has been said about the distribution of X other than that it has mean μ_X and positive variance σ_X^2. Suppose, in addition, we assume that this distribution is also normal; that is, the marginal p.d.f. of X is

$$f_1(x) = \frac{1}{\sigma_X \sqrt{2\pi}} \exp\left[-\frac{(x - \mu_X)^2}{2\sigma_X^2} \right], \qquad -\infty < x < \infty.$$

Hence the joint p.d.f. of X and Y is given by the product

$$f(x, y) = h(y \mid x) f_1(x) = \frac{1}{2\pi \sigma_X \sigma_Y \sqrt{1 - \rho^2}} \exp\left[-\frac{q(x, y)}{2}\right], \qquad (5.6\text{-}2)$$

where it can be shown (see Exercise 5.6-2) that

$$q(x, y) = \frac{1}{1 - \rho^2}\left[\left(\frac{x - \mu_X}{\sigma_X}\right)^2 - 2\rho\left(\frac{x - \mu_X}{\sigma_X}\right)\left(\frac{y - \mu_Y}{\sigma_Y}\right) + \left(\frac{y - \mu_Y}{\sigma_Y}\right)^2\right].$$

A joint p.d.f. of this form is called a **bivariate normal p.d.f.**

EXAMPLE 5.6-2 Let us assume that in a certain population of college students, the respective grade point averages, say X and Y, in high school and the first year in college have an approximate bivariate normal distribution with parameters $\mu_X = 2.9$, $\mu_Y = 2.4$, $\sigma_X = 0.4$, $\sigma_Y = 0.5$, and $\rho = 0.8$.
Then, for illustration,

$$P(2.1 < Y < 3.3) = P\left(\frac{2.1 - 2.4}{0.5} < \frac{Y - 2.4}{0.5} < \frac{3.3 - 2.4}{0.5}\right)$$

$$= \Phi(1.8) - \Phi(-0.6) = 0.6898.$$

Since the conditional p.d.f. of Y, given $X = 3.2$, is normal with mean

$$24 + (0.8)\left(\frac{0.5}{0.4}\right)(3.2 - 2.9) = 2.7$$

and standard deviation $(0.5)\sqrt{1 - 0.64} = 0.3$, we have that

$$P(2.1 < Y < 3.3 \mid X = 3.2)$$

$$= P\left(\frac{2.1 - 2.7}{0.3} < \frac{Y - 2.7}{0.3} < \frac{3.3 - 2.7}{0.3} \,\middle|\, X = 3.2\right)$$

$$= \Phi(2) - \Phi(-2) = 0.9544.$$

From a practical point of view, however, the reader should be warned that the correlation coefficient of these grade point averages is, in many instances, much smaller than 0.8. ◀

Since x and y enter the bivariate normal p.d.f. in a similar manner, the roles of X and Y could have been interchanged. That is, Y could have been assigned the marginal normal p.d.f. $N(\mu_Y, \sigma_Y^2)$, and the conditional p.d.f. of X, given $Y = y$, would have then been normal, with mean $\mu_X + \rho(\sigma_X/\sigma_Y)(y - \mu_Y)$ and variance $\sigma_X^2(1 - \rho^2)$. Although this is fairly obvious, we do want to make special note of it.

In order to have a better understanding of the geometry of the bivariate normal distribution, consider the graph of $z = f(x, y)$, where $f(x, y)$ is given by Equation 5.6-2. If we intersect this surface with planes parallel to the yz plane, that is, with $x = x_0$, we have

$$f(x_0, y) = f_1(x_0)h(y \mid x_0).$$

In this equation $f_1(x_0)$ is a constant, and $h(y \mid x_0)$ is a normal p.d.f. Thus $z = f(x_0, y)$ is bell shaped; that is, has the shape of a normal p.d.f. However, note that it is not necessarily a p.d.f. because of the factor $f_1(x_0)$. Similarly, intersections of the surface $z = f(x, y)$ with planes $y = y_0$, parallel to the xz plane will be bell shaped.

If

$$0 < z_0 < \frac{1}{2\pi \sigma_X \sigma_Y \sqrt{1 - \rho^2}},$$

then

$$0 < z_0 2\pi \sigma_X \sigma_Y \sqrt{1 - \rho^2} < 1.$$

If we intersect $z = f(x, y)$ with the plane $z = z_0$, which is parallel to the xy plane, we have

$$z_0 2\pi \sigma_X \sigma_Y \sqrt{1 - \rho^2} = \exp\left[\frac{-q(x, y)}{2}\right].$$

Taking the natural logarithm of each side, we obtain

$$\left(\frac{x - \mu_X}{\sigma_X}\right)^2 - 2\rho\left(\frac{x - \mu_X}{\sigma_X}\right)\left(\frac{y - \mu_Y}{\sigma_Y}\right) + \left(\frac{y - \mu_Y}{\sigma_Y}\right)^2$$

$$= -2(1 - \rho^2)\ln(z_0 2\pi \sigma_X \sigma_Y \sqrt{1 - \rho^2}).$$

Thus we see that these intersections are ellipses.

EXAMPLE 5.6-3 With $\mu_X = 10$, $\sigma_X^2 = 9$, $\mu_Y = 12$, $\sigma_Y^2 = 16$, and $\rho = 0.6$, the bivariate normal p.d.f. has been graphed in Figure 5.6-2. For $\rho = 0.6$, level curves or contours are given in Figure 5.6-3. The conditional mean line,

$$E(Y \mid x) = 12 + (0.6)\left(\frac{4}{3}\right)(x - 10) = 0.8x + 4,$$

is also drawn on Figure 5.6-3. Note that this line intersects the level curves at points through which vertical tangents can be drawn to the ellipses. ◄

We close this section by observing another important property of the correlation coefficient ρ if X and Y have a bivariate normal distribution. In Equation 5.6-2 of the product $h(y \mid x)f_1(x)$, let us consider the factor $h(y \mid x)$ if

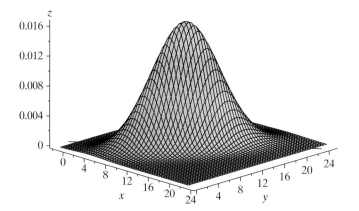

Figure 5.6-2: Bivariate normal p.d.f.

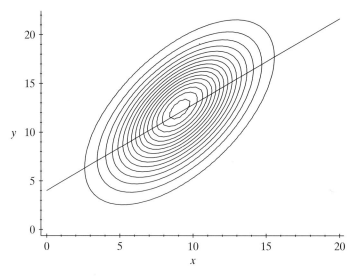

Figure 5.6-3: Contours for the bivariate normal distribution

$\rho = 0$. We see that this product, which is the joint p.d.f. of X and Y, equals $f_1(x)f_2(y)$ because $h(y \mid x)$ is, when $\rho = 0$, a normal p.d.f. with mean μ_Y and variance σ_Y^2. That is, if $\rho = 0$, the joint p.d.f. factors into the product of the two marginal probability density functions and hence, X and Y are independent random variables. Of course, if X and Y are any independent random variables (not necessarily normal), we know that ρ, if it exists, is always equal to zero. Thus we have proved the following.

Theorem 5.6-1   If X and Y have a bivariate normal distribution with correlation coefficient ρ, then X and Y are independent if and only if $\rho = 0$.

Thus, in the bivariate normal case, $\rho = 0$ does imply independence of X and Y.

It should be mentioned here than these characteristics of the bivariate normal distribution can be extended to the trivariate normal distribution or, more generally, the multivariate normal distribution. This is done in more advanced texts assuming some knowledge of matrices; for illustration, see Hogg, McKean, and Craig (2005).

EXERCISES

5.6-1 Let X and Y have a bivariate normal distribution with parameters $\mu_X = -3, \mu_Y = 10$, $\sigma_X^2 = 25, \sigma_Y^2 = 9$, and $\rho = 3/5$. Compute

  (a) $P(-5 < X < 5)$.
  (b) $P(-5 < X < 5 \mid Y = 13)$.
  (c) $P(7 < Y < 16)$.
  (d) $P(7 < Y < 16 \mid X = 2)$.

5.6-2 Show that the expression in the exponent of Equation 5.6-2 is equal to the function $q(x, y)$ given in the text.

5.6-3 Let X and Y have a bivariate normal distribution with parameters $\mu_X = 2.8$, $\mu_Y = 110, \sigma_X^2 = 0.16, \sigma_Y^2 = 100$, and $\rho = 0.6$. Compute

  (a) $P(106 < Y < 124)$.
  (b) $P(106 < Y < 124 \mid X = 3.2)$.

5.6-4 Let X and Y have a bivariate normal distribution with $\mu_X = 70, \sigma_X^2 = 100, \mu_Y = 80$, $\sigma_Y^2 = 169$, and $\rho = 5/13$. Find

  (a) $E(Y \mid X = 72)$.
  (b) $\mathrm{Var}(Y \mid X = 72)$.
  (c) $P(Y \le 84, \mid X = 72)$.

5.6-5 Let X denote the height in centimeters and Y the weight in kilograms of male college students. Assume that X and Y have a bivariate normal distribution with parameters $\mu_X = 185, \sigma_X^2 = 100, \mu_Y = 84, \sigma_Y^2 = 64$, and $\rho = 3/5$.

  (a) Determine the conditional distribution of Y, given that $X = 190$.
  (b) Find $P(86.4 < Y < 95.36 \mid X = 190)$.

5.6-6 For a freshman taking introductory statistics and majoring in psychology, let X equal the student's ACT mathematics score and Y the student's ACT verbal score. Assume that X and Y have a bivariate normal distribution with $\mu_X = 22.7$, $\sigma_X^2 = 17.64, \mu_Y = 22.7, \sigma_Y^2 = 12.25$, and $\rho = 0.78$. Find

  (a) $P(18.5 < Y < 25.5)$.
  (b) $E(Y \mid x)$.
  (c) $\mathrm{Var}(Y \mid x)$.
  (d) $P(18.5 < Y < 25.5 \mid X = 23)$.
  (e) $P(18.5 < Y < 25.5 \mid X = 25)$.
  (f) For $x = 21, 23$, and 25, draw a graph of $z = h(y \mid x)$ similar to Figure 5.6-1.

5.6-7 For a pair of gallinules, let X equal the weight in grams of the male and Y the weight in grams of the female. Assume that X and Y have a bivariate normal distribution with $\mu_X = 415, \sigma_X^2 = 611, \mu_Y = 347, \sigma_Y^2 = 689$, and $\rho = -0.25$. Find

  (a) $P(309.2 < Y < 380.6)$.
  (b) $E(Y \mid x)$.
  (c) $\mathrm{Var}(Y \mid x)$.
  (d) $P(309.2 < Y < 380.6 \mid X = 385.1)$.

5.6-8 Let X and Y have a bivariate normal distribution with parameters $\mu_X = 10, \sigma_X^2 = 9$, $\mu_Y = 15, \sigma_Y^2 = 16$, and $\rho = 0$. Find

   (a) $P(13.6 < Y < 17.2)$.
   (b) $E(Y \mid x)$.
   (c) $\text{Var}(Y \mid x)$.
   (d) $P(13.6 < Y < 17.2 \mid X = 9.1)$.

5.6-9 Let X and Y have a bivariate normal distribution. Find two different lines, $a(x)$ and $b(x)$, parallel to and equidistant from $E(Y \mid x)$, such that

$$P[a(x) < Y < b(x) \mid X = x] = 0.9544$$

for all real x. Plot $a(x)$, $b(x)$, and $E(Y \mid x)$ when $\mu_X = 2$, $\mu_Y = -1$, $\sigma_X = 3$, $\sigma_Y = 5$, and $\rho = 3/5$.

5.6-10 In a college health fitness program, let X denote the weight in kilograms of a male freshman at the beginning of the program and let Y denote his weight change during a semester. Assume that X and Y have a bivariate normal distribution with $\mu_X = 72.30$, $\sigma_X^2 = 110.25$, $\mu_Y = 2.80$, $\sigma_Y^2 = 2.89$, and $\rho = -0.57$. (The lighter students tend to gain weight, while the heavier students tend to lose weight.) Find

   (a) $P(2.80 \leq Y \leq 5.35)$.
   (b) $P(2.76 \leq y \leq 5.34 \mid X = 82.3)$.

5.6-11 For a female freshman in a health fitness program, let X equal her percentage of body fat at the beginning of the program and let Y equal the change in her percentage of body fat measured at the end of the program. Assume that X and Y have a bivariate normal distribution with $\mu_X = 24.5$, $\sigma_X^2 = 4.8^2 = 23.04$, $\mu_Y = -0.2$, $\sigma_Y^2 = 3.0^2 = 9.0$, and $\rho = -0.32$. Find

   (a) $P(1.3 \leq Y \leq 5.8)$.
   (b) $\mu_{Y \mid x}$, the conditional mean of Y, given $X = x$.
   (c) $\sigma_{Y \mid x}^2$, the conditional variance of Y, given $X = x$.
   (d) $P(1.3 \leq Y \leq 5.8 \mid X = 18)$.

5.6-12 For a male freshman in a health fitness program, let X equal his percentage of body fat at the beginning of the program and let Y equal the change in his percentage of body fat measured at the end of the program. Assume that X and Y have a bivariate normal distribution with $\mu_X = 15.00$, $\sigma_X^2 = 4.5^2$, $\mu_Y = -1.55$, $\sigma_Y^2 = 1.5^2$, and $\rho = -0.60$. Find

   (a) $P(0.205 \leq Y \leq 0.805)$.
   (b) $P(0.21 \leq Y \leq 0.81 \mid X = 20)$.

5.7* LIMITING MOMENT-GENERATING FUNCTIONS

We would like to begin this section by showing that the binomial distribution can be approximated by the Poisson distribution when n is sufficiently large and p fairly small. Of course, this was proved in Section 3.7 by showing that, under these conditions, the binomial p.m.f. is close to that of the Poisson. Here, however, we show that the binomial moment-generating function is close to that of the Poisson distribution. We do this by taking the limit of a moment-generating function.

Consider the moment-generating function of Y, which is $b(n, p)$. We shall take the limit of this as $n \to \infty$ such that $np = \lambda$ is a constant; thus $p \to 0$. The moment-generating function of Y is

$$M(t) = (1 - p + pe^t)^n.$$

Because $p = \lambda/n$, we have that

$$M(t) = \left[1 - \frac{\lambda}{n} + \frac{\lambda}{n}e^t \right]^n$$

$$= \left[1 + \frac{\lambda(e^t - 1)}{n} \right]^n.$$

Since

$$\lim_{n \to \infty} \left(1 + \frac{b}{n} \right)^n = e^b,$$

we have

$$\lim_{n \to \infty} M(t) = e^{\lambda(e^t - 1)},$$

which exists for all real t. But this is the moment-generating function of a Poisson random variable with mean λ. Hence this Poisson distribution seems like a reasonable approximation to the binomial distribution when n is large and p is small. This approximation is usually found to be fairly successful if $n \geq 20$ and $p \leq 0.05$ and very successful if $n \geq 100$ and $p \leq 0.10$, but it is not bad if these bounds are violated somewhat. That is, it could be used in other situations too; we only want to stress that the approximation becomes better with larger n and smaller p.

The preceding result illustrates the theorem we now state without proof.

Theorem 5.7-1   If a sequence of moment-generating functions approaches a certain m.g.f., say $M(t)$, then the limit of the corresponding distributions must be the distribution corresponding to $M(t)$.

Remark This theorem certainly appeals to one's intuition! In a more advanced course, the proof of this theorem is given and there the existence of the moment-generating function is not even needed, for we would use the characteristic function $\phi(t) = E(e^{itX})$ instead.   ■

The following example illustrates graphically the convergence of the binomial moment-generating functions to that of a Poisson distribution.

EXAMPLE 5.7-1   Consider the moment-generating function for the Poisson distribution with $\lambda = 5$ and those for three binomial distributions for which $np = 5$, namely, $b(50, 1/10)$, $b(100, 1/20)$, and $b(200, 1/40)$. These four moment-generating functions are, respectively:

$$M(t) = e^{5(e^t - 1)}$$

$$M(t) = (0.9 + 0.1e^t)^{50}$$

$$M(t) = (0.95 + 0.05e^t)^{100}$$

$$M(t) = (0.975 + 0.025e^t)^{200}$$

The graphs of these moment-generating functions are shown in Figure 5.7-1. Although the proof and the figure show the convergence of the binomial moment-generating functions to that of the Poisson distribution, the other graphs in Figure 5.7-1 show more clearly how the Poisson distribution can be used to approximate binomial probabilities with large n and small p. ◄

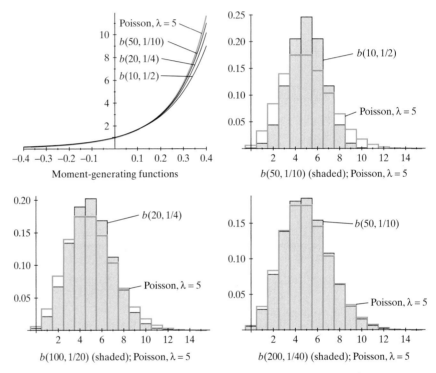

Figure 5.7-1: Poisson approximation of the binomial distribution

The next example gives a numerical approximation.

EXAMPLE 5.7-2   Let Y be $b(50, 1/25)$. Then

$$P(Y \leq 1) = \left(\frac{24}{25}\right)^{50} + 50\left(\frac{1}{25}\right)\left(\frac{24}{25}\right)^{49} = 0.400.$$

Since $\lambda = np = 2$, the Poisson approximation is

$$P(Y \leq 1) \approx 0.406,$$

from Table III in the Appendix. ◄

Theorem 5.7-1 is used to prove the central limit theorem. To help in the understanding of this proof, let us first consider a different problem, that of the limiting distribution of the mean $\overline{X}$ of a random sample $X_1, X_2, \ldots, X_n$, from a distribution with mean μ. If the distribution has moment-generating function $M(t)$, the moment-generating function of $\overline{X}$ is $[M(t/n)]^n$. But, by Taylor's expansion, there exists a number t_1 between 0 and t/n such that

$$M\left(\frac{t}{n}\right) = M(0) + M'(t_1)\frac{t}{n}$$

$$= 1 + \frac{\mu t}{n} + \frac{[M'(t_1) - M'(0)]t}{n}$$

because $M(0) = 1$ and $M'(0) = \mu$. Since $M'(t)$ is continuous at $t = 0$ and since $t_1 \to 0$ as $n \to \infty$, we know that

$$\lim_{n\to\infty} [M'(t_1) - M'(0)] = 0.$$

Thus, using a result from advanced calculus, we obtain

$$\lim_{n\to\infty} \left[M\left(\frac{t}{n}\right) \right]^n = \lim_{n\to\infty} \left\{ 1 + \frac{\mu t}{n} + \frac{[M'(t_1) - M'(0)]t}{n} \right\}^n$$

$$= e^{\mu t}$$

for all real t. But this limit is the moment-generating function of a degenerate distribution with all of the probability on μ. Accordingly, $\overline{X}$ has this limiting distribution, indicating that $\overline{X}$ converges to μ in a certain sense. This is one form of the law of large numbers.

We have seen that, in some probability sense, $\overline{X}$ converges to μ in the limit, or, equivalently, $\overline{X} - \mu$ converges to zero. Let us multiply the difference $\overline{X} - \mu$ by some function of n so that the result will not converge to zero. In our search for such a function, it is natural to consider

$$W = \frac{\overline{X} - \mu}{\sigma/\sqrt{n}} = \frac{\sqrt{n}(\overline{X} - \mu)}{\sigma} = \frac{Y - n\mu}{\sqrt{n}\,\sigma},$$

where Y is the sum of the observations of the random sample. The reason for this is that W is a standardized random variable by the remark after the proof of Theorem 5.2-1; that is W has mean 0 and variance 1 for each positive integer n. We are now ready to prove the central limit theorem, which is stated in Section 5.4.

Proof: (of the Central Limit Theorem)
We first consider

$$E[\exp(tW)] = E\left\{\exp\left[\left(\frac{t}{\sqrt{n}\sigma}\right)\left(\sum_{i=1}^{n} X_i - n\mu\right)\right]\right\}$$

$$= E\left\{\exp\left[\left(\frac{t}{\sqrt{n}}\right)\left(\frac{X_1 - \mu}{\sigma}\right)\right]\cdots\exp\left[\left(\frac{t}{\sqrt{n}}\right)\left(\frac{X_n - \mu}{\sigma}\right)\right]\right\}$$

$$= E\left\{\exp\left[\left(\frac{t}{\sqrt{n}}\right)\left(\frac{X_1 - \mu}{\sigma}\right)\right]\right\}\cdots E\left\{\exp\left[\left(\frac{t}{\sqrt{n}}\right)\left(\frac{X_n - \mu}{\sigma}\right)\right]\right\},$$

which follows from the independence of $X_1, X_2, \ldots, X_n$. Then

$$E[\exp(tW)] = \left[m\left(\frac{t}{\sqrt{n}}\right)\right]^n, \qquad -h < \frac{t}{\sqrt{n}} < h,$$

where

$$m(t) = E\left\{\exp\left[t\left(\frac{X_i - \mu}{\sigma}\right)\right]\right\}, \qquad -h < t < h,$$

is the common moment-generating function of each

$$Y_i = \frac{X_i - \mu}{\sigma}, \qquad i = 1, 2, \ldots, n.$$

Since $E(Y_i) = 0$ and $E(Y_i^2) = 1$, it must be that

$$m(0) = 1, \qquad m'(0) = E\left(\frac{X_i - \mu}{\sigma}\right) = 0, \qquad m''(0) = E\left[\left(\frac{X_i - \mu}{\sigma}\right)^2\right] = 1.$$

Hence, using Taylor's formula with a remainder, there exists a number t_1 between 0 and t such that

$$m(t) = m(0) + m'(0)t + \frac{m''(t_1)t^2}{2} = 1 + \frac{m''(t_1)t^2}{2}.$$

By adding and subtracting $t^2/2$, we have that

$$m(t) = 1 + \frac{t^2}{2} + \frac{[m''(t_1) - 1]t^2}{2}.$$

Using this expression of $m(t)$ in $E[\exp(tW)]$, we can represent the moment-generating function of W by

$$E[\exp(tW)] = \left\{1 + \frac{1}{2}\left(\frac{t}{\sqrt{n}}\right)^2 + \frac{1}{2}[m''(t_1) - 1]\left(\frac{t}{\sqrt{n}}\right)^2\right\}^n$$

$$= \left\{1 + \frac{t^2}{2n} + \frac{[m''(t_1) - 1]t^2}{2n}\right\}^n, \qquad -\sqrt{n}\,h < t < \sqrt{n}\,h,$$

where now t_1 is between 0 and $t/\sqrt{n}$. Since $m''(t)$ is continuous at $t = 0$ and $t_1 \to 0$ as $n \to \infty$, we have that

$$\lim_{n\to\infty} [m''(t_1) - 1] = 1 - 1 = 0.$$

Thus, using a result from advanced calculus, we have that

$$\lim_{n\to\infty} E[\exp(tW)] = \lim_{n\to\infty}\left\{1 + \frac{t^2}{2n} + \frac{[m''(t_1) - 1]t^2}{2n}\right\}^n$$

$$= \lim_{n\to\infty}\left\{1 + \frac{t^2}{2n}\right\}^n = e^{t^2/2}$$

for all real t, which is the m.g.f. of the standard normal distribution, $N(0, 1)$. This means that the limiting distribution of

$$W = \frac{\overline{X} - \mu}{\sigma/\sqrt{n}} = \frac{\displaystyle\sum_{i=1}^{n} X_i - n\mu}{\sqrt{n}\,\sigma}$$

is $N(0, 1)$. This completes the proof of the central limit theorem.   ◯

Examples of the use of the central limit theorem as an approximating distribution have been given in Sections 5.4 and 5.5.

To help appreciate the proof of the central limit theorem, the following example graphically illustrates the convergence of the moment-generating functions for two distributions.

EXAMPLE 5.7-3   Let $X_1, X_2, \ldots, X_n$ be a random sample of size n from an exponential distribution with $\theta = 2$. The moment-generating function of $(\overline{X} - \theta)/(\theta/\sqrt{n})$ is

$$M_n(t) = \frac{e^{-t\sqrt{n}}}{(1 - t/\sqrt{n}\,)^n}.$$

The central limit theorem says that, as n increases, this moment-generating function approaches that of the standard normal distribution, namely,

$$M(t) = e^{t^2/2}.$$

The moment-generating functions for $M(t)$ and $M_n(t)$, $n = 5, 15, 50$, are shown in Figure 5.7-2(a). Also see Figure 5.4-2, in which samples were taken from a $\chi^2(1)$ distribution and recall that the exponential distribution with $\theta = 2$ is $\chi^2(2)$.

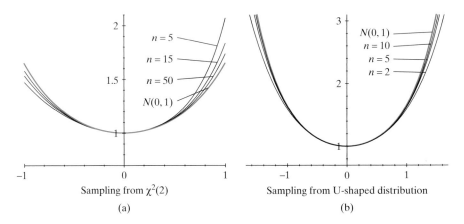

(a) Sampling from $\chi^2(2)$
(b) Sampling from U-shaped distribution

Figure 5.7-2: Convergence of moment-generating functions

In Example 5.4-6 a U-shaped distribution was considered for which the p.d.f. is $f(x) = (3/2)x^2$, $-1 < x < 1$. For this distribution, $\mu = 0$ and $\sigma^2 = 3/5$. Its moment-generating function, for $t \neq 0$, is

$$M(t) = \left(\frac{3}{2}\right) \frac{e^t t^2 - 2e^t t + 2e^t - e^{-t}t^2 - 2e^{-t}t - 2e^{-t}}{t^3}.$$

Of course, $M(0) = 1$. The moment-generating function of

$$W_n = \frac{\overline{X} - 0}{\sqrt{3/(5n)}}$$

is

$$E[e^{tW_n}] = \left\{E\left[\exp\left(\sqrt{\frac{5}{3n}}\,t\right)\right]\right\}^n = \left[M\left(\sqrt{\frac{5}{3n}}\,t\right)\right]^n.$$

The graphs of these moment-generating functions when $n = 2, 5, 10$ and the graph of the moment-generating function for the standard normal distribution are shown in Figure 5.7-2(b). Note how much more quickly these moment-generating functions converge compared to those for the exponential distribution. ◄

EXERCISES

5.7-1 Let Y be the number of defectives in a box of 50 articles taken from the output of a machine. Each article is defective with probability 0.01. What is the probability that $Y = 0, 1, 2,$ or 3?

  (a) Using the binomial distribution.
  (b) Using the Poisson approximation.

5.7-2 The probability that a certain type of inoculation takes effect is 0.995. Use the Poisson distribution to approximate the probability that at most 2 out of 400 people given the inoculation find that it has not taken effect.

  HINT: Let $p = 1 - 0.995 = 0.005$.

5.7-3 Let S^2 be the sample variance of a random sample of size n from $N(\mu, \sigma)$. Show that the limit, as $n \to \infty$, of the moment-generating function of S^2 is $e^{\sigma^2 t}$. Thus, in the limit, the distribution of S^2 is degenerate with probability 1 at σ^2.

5.7-4 Let Y be $\chi^2(n)$. Use the central limit theorem to demonstrate that $W = (Y - n)/\sqrt{2n}$ has a limiting distribution function that is $N(0, 1)$.

  HINT: Think of Y as being the sum of a random sample from a certain distribution.

5.7-5 Let Y have a Poisson distribution with mean $3n$. Use the central limit theorem to show that $W = (Y - 3n)/\sqrt{3n}$ has a limiting distribution that is $N(0, 1)$.

IMPORTANCE OF UNDERSTANDING VARIABILITY

This centerpiece is written as a transition as we move from probability to statistical inference. Much of this discussion involves improving situations in a manufacturing context, but clearly it can be applied to the services industries as well. As a matter of fact, really understanding variation, which is the key element of probability and statistics, is extremely important in so many phases of life. Yet as frequently happens in studying a textbook like this, we are so involved in the mathematics of probability, distributions, sampling distributions, and even descriptive statistics that we miss the real meanings of some of these concepts, particularly as far as the role they play in day-to-day activities.

Variability explains what is commonly called the "sophomore jinx": Some freshmen (rookies) have great first years (above their averages) and then most of them do worse the second year. The same explanation can be applied to movies that have sequels; often the original was outstanding, but even with the same cast and same director, the sequel is usually worse than the original. This phenomenon is sometimes called *regression to the mean*.

We also note that workers who do very well (above their averages) and who thus might be given some reward usually do worse during the next work period. On the other hand, those who have bad periods (below average) and are reprimanded frequently do better the next. Yet it is wrong to think that a policy of reprimand is better than that of a pat on the back. Both situations can be explained by understanding the worker's pattern of variation. If the employer really wants to improve the outcome of the process, he or she must consider ways of improving the level (average). The workers can sometimes make small adjustments, but they need road maps to make the major ones. Thus it is really the responsibility of management to improve the process substantially. That is, management must realize that working harder with techniques that have failed will not improve the situation much. Changes must be made, and often data and the resulting statistical analysis can suggest changes for the better.

It is also disturbing to see the following situation: Suppose that the manager has created a team (say of 10) of outstanding workers, and it is time to give raises. The workers are ranked from one to ten. (It is always true that 10 workers, no matter how good or how bad, will always get the ranks 1, 2, 3, ... , 10.) Then say they get raises according to their ranks, which might have been determined simply by a random process for this particular period. That is, in the next period, their ranks might be entirely different. How does the small raise make the one with lowest rank feel if, in fact, all 10 are members of an outstanding team and there might be little or no difference among the persons ranked 1 and 10? Not good, and he or she is less likely to help the ones with higher ranks in a future period. That is, this does not promote teamwork. As a matter of fact, there is always this danger in any kind of reward, such as the "worker of the month." Others are not likely to help that person the next month, because they want to be the worker of the month and receive the corresponding bonus.

A better way of rewarding a good team would be to give them essentially the same raises. Clearly, a good manager will continuously monitor performance and supply appropriate feedback to the workers. Of course, if one worker is consistently on the high side, then he or she should be considered for promotion or a substantial raise. On the other hand, if some worker is on the low end most of the time, some help for this worker is in order: maybe additional training or even a transfer to a different department for which the worker's talents are more suitable. It is important not to demean this worker because each of us needs to take pride in our accomplishments. If at all possible, firing the worker should be avoided, possibly by finding some other suitable job in the company.

At one state university, the legislators appropriated enough money for 2% raises one year. The administrators thought this was not adequate; so they had the "brilliant" idea to give one-third of the faculty 6% raises and zero to the other two-thirds. Thus, in a department of 18, 6 received raises and 12 did not. Who ranked those 18? Was there much difference between the sixth and seventh? Or between the fifth and tenth for that matter? So the person who received a raise might have people on either side of him or her that got nothing. It created a terrible morale problem. It certainly would have been better to give every faculty member an approximate 2% raise and blame it on the legislators. Of course, we must recognize differences among people and accordingly they must be treated differently, but certainly not just due to chance variation.

Some teachers will announce to a class of 30 students that there will be only so many A's, so many B's, and so on. This is wrong! If those students are competing for those given number of A's, why would anyone want to help anyone else? It completely destroys any sort of teamwork, and yet that is what students must learn to do when they are on the job later in life. Clearly, it is better to say that all of them can earn an A (they probably won't) and encourage them to work together. In that way, the teacher has a better opportunity to improve the level of the entire class which, after all, is the real objective. Too frequently mathematics and science courses discourage interaction among students, and teachers must do everything possible to break down those barriers.

Many of these preceding observations are those of W. Edwards Deming, an esteemed statistician who went to Japan after World War II and taught the Japanese how to make quality products. For his work there, he was awarded the Emperor's Medal and the Japanese established the Deming Prize to be awarded each year to the company or individual contributing the most to quality improvement. One of the things Deming stressed is the need for "profound knowledge," and a major item in that is understanding variation and statistical theory.

In making quality products, you want to reduce the variation as much as possible and move the level closer to the target. Deming believed that barriers between departments, between management and workers, and among workers must be broken down to improve communications and the ability to work as a team. The lines of communication, all the way from suppliers to customers, must be open to help reduce variation and improve products and services. For example, he argued that a company should not buy only on price tag but have a few reliable suppliers (possibly one) for a single part because that will reduce variation. That is, many different suppliers would obviously increase variation. Moreover, he argued that you should become partners, friends if you like, with your suppliers. You learn to trust the other; in that way, you can use methods like "just in time," in which your inventory can be kept reasonably small, to keep costs down.

If each of us thinks about these ideas, we might become obsessed by understanding variation, and that might make a big difference in our everyday lives. For illustration, once we have selected good suppliers, we continue to go to the same barber, the same service station, the same clothier, the same bank, and on and on. We like to buy items, even if a little more expensive, from places that will give us good service if something goes wrong. If this does not happen, then of course we must consider changing suppliers.

Clearly, listening to the customer can help improve the quality of our products and services. We can then meet–or even exceed–the expectations of our customers. We should continually try to improve by reducing the variation and moving the process to a better level (high is not always good, as in golf). Although the customer–as well as the supplier–must be part of the total team, more often management must continue to look for better ways to do things, ways that the customer never would have imagined. For example, in the early days of automobiles not many owners would have thought of driving on pneumatic tires. Harvey Firestone did, and in this way exceeded the expectations of those early customers.

Deming also preached constancy of purpose. If management ideas tend to change too much, employees really do not know what to do and cannot do their best. That is, they are mixed up, increasing variation. It is better for them to receive a constant signal from their employer, a signal that changes only if research dictates ways of improving.

More training and education for everyone associated with a company also decreases variation by teaching how to make the product more uniform.

Workers must know that they do not have to be afraid to make suggestions to improve the process. Often, being team members will make it easier for workers to speak up without fear of reprisal.

Many of us remember playing a game called "telephone": One whispers a message to the next person, who whispers the message to the next, and so on. The message at the end is compared to the original message, and it is usually much different from the original. This is like trying to hit a target (say at zero) with a random variable X_1. Then starting with X_1 as the center and adding on another error X_2, and so on, creating the sum of the errors, $X_1 + X_2 + \cdots + X_n$, which has an ever-increasing variance with independent errors. Deming would say, "We are off to the Milky Way." Yet we actually do this in business and industry by letting worker train worker. Once errors are introduced, they will stay there, and others will add on and on. (Incidentally, we might say the same thing about too many layers of management.) To decrease variation, wouldn't it be better to have a master instructor train each worker (or have fewer layers of management)?

Deming also noted that requiring quotas does not help the quality. A foreman who has a quota of 100 per day will often ship out 90 good and 10 bad items just to make the quota. Clearly, it would reduce the variation and satisfy the customer better if only the 90 good ones were shipped.

This leads to the point that a final inspection of products often does not really improve the quality. With such a procedure, you can only eliminate the bad ones and send on the good ones. Improvements in the design of the products and manufacturing processes are needed. If these are done well, often with the help of statistical methods, that final inspection can be eliminated. That is, improvements should be made early in the manufacturing process rather than try to correct things at an end inspection by weeding out the bad items.

Once you recognize that there is variation in almost every process, you begin to think like a statistician. Often you want to reduce that variation and center what variation that remains at the right level, because then "the doors will fit better." But while we want to reduce undesirable variation, we do not want to reduce all variation: Most of us do not want to eat Chinese food every night even though we enjoy it from time to time. Certain variation is "the spice of life."

This desire to understand variation is the key to statistical thinking. As a matter of fact, in the quality improvement area, Ron Snee, another quality guru, defines statistical thinking as "thought processes, which recognize that variation is all around us and present in everything we do, all work is a series of interconnected processes, and identifying, characterizing, quantifying, controlling, and reducing variation provide opportunities for improvement."

In a sense, statistical thinking can affect everything we do, including our relationships with others. All of us have our "ups and downs," and we must seriously think about how to deal with others, recognizing this variation.

It is the authors' opinion that this quality improvement must begin with the individual by creating a personal vision, goals, and principles by which to live. Hopefully one of those individuals is the chief of the organization for then things go much smoother. Bob Galvin, who was the CEO of Motorola during the 1980s, believes that "quality improvement is a daily, personal, priority obligation," and his company made great improvements in the quality of their products during that period. Much of this progress was due to understanding variation and statistical methods. Even Motorola's program was called Six-Sigma, in which *sigma* referred to a standard deviation, a name which is a delight to a statistician. But the description of that program is another story for another day.

Each of us should use as much statistical thinking as we know to help us build trust with the persons with whom we deal directly or indirectly. Let us not do what those administrators of that university did when they did not understand the variability associated with ranking people. That action destroyed any trust that was there and created a mean competition among the members of that faculty. Why should anyone help others when their gain only hurts that individual? While statistical thinking cannot solve all problems, it most certainly can help in many situations. We hope you use what statistics you know to benefit you the rest of your life.

Remark Even the definitions and distributions of the t and F random variables say something about understanding variability. Both t and F are defined as the ratio of two independent random variables. The resulting distributions have heavier tails than the respective random variables in the numerators. For example, the t random variable has a standardized normal random variable in the numerator and the p.d.f. of t is much heavier in the tails than that of $N(0, 1)$. That is, there is a tendency to get more outliers in sampling from the t distribution than from $N(0, 1)$. Thus ratios that measure something per something else often exhibit more variability than occurs in the normal distribution, particularly if that quantity in the denominator is a random variable. And we often observe the situation in which the denominator is a variable, creating a ratio that has more variability than we think it should have. For illustration, suppose we observe successive sample averages, say $\bar{x}_1, \bar{x}_2, \ldots, \bar{x}_k$. However, the sample sizes change from one sample to the next, say $n_1, n_2, \ldots, n_k$. If these sample sizes vary enough, the $\bar{x}$ values could look as if they arose from an approximate t distribution due to the fact that the variance of the averages is changing due to the different sample sizes. That is why in practice we often see more variation in a plot of sample means than we might expect. For illustration, the Dow-Jones index in the stock market is a type of average and those deviations tend to be more like observations from a t distribution than normal observations due to the fact that the volume of sales varies each day. ∎

CHAPTER

6

ESTIMATION

6.1 SAMPLE CHARACTERISTICS

In the first five chapters we mainly studied probability and probability distributions. In the following chapters we consider situations in which we take a random sample from a distribution that is not completely known and try to say something about the unknown distribution from the observed sample observations. This procedure is called making a **statistical inference**. The **estimation** of unknown characteristics of the distribution is an extremely important part of statistical inference. Estimating these characteristics is usually done by constructing functions of the sample observations, and these functions are called **statistics**. We already have mentioned as statistics the mean $\overline{X}$ and the variance S^2 of a random sample and made a few observations about their distributional properties, particularly when sampling from a normal distribution. Let us begin by reviewing these statistics and percentiles of the sample.

If the sample size n is large and if the corresponding histogram is bell shaped, then the percentages associated with the normal distribution would be approximately true and we could use the following empirical rule.

 **Empirical Rule:** Let $x_1, x_2, \ldots, x_n$ have a sample mean $\overline{x}$ and sample standard deviation s. If the histogram of these data is bell shaped, then

- approximately 68% of the data are in the interval $(\overline{x} - s, \overline{x} + s)$,

- approximately 95% of the data are in the interval $(\bar{x} - 2s, \bar{x} + 2s)$,
- approximately 99.7% of the data are in the interval $(\bar{x} - 3s, \bar{x} + 3s)$.   ◄

When you draw a histogram, it is useful to indicate on the histogram the location of $\bar{x}$ as well as the points $\bar{x} \pm s$ and $\bar{x} \pm 2s$.

EXAMPLE 6.1-1   A manufacturer of fluoride toothpaste regularly measures the concentration of fluoride in the toothpaste to make sure that it is within the specifications of $0.85 - 1.10$ mg/g. Table 6.1-1 lists 100 such measurements.

Table 6.1-1: Concentrations of Fluoride in *mg/g* in Toothpaste

0.98	0.92	0.89	0.90	0.94	0.99	0.86	0.85	1.06	1.01
1.03	0.85	0.95	0.90	1.03	0.87	1.02	0.88	0.92	0.88
0.88	0.90	0.98	0.96	0.98	0.93	0.98	0.92	1.00	0.95
0.88	0.90	1.01	0.98	0.85	0.91	0.95	1.01	0.88	0.89
0.99	0.95	0.90	0.88	0.92	0.89	0.90	0.95	0.93	0.96
0.93	0.91	0.92	0.86	0.87	0.91	0.89	0.93	0.93	0.95
0.92	0.88	0.87	0.98	0.98	0.91	0.93	1.00	0.90	0.93
0.89	0.97	0.98	0.91	0.88	0.89	1.00	0.93	0.92	0.97
0.97	0.91	0.85	0.92	0.87	0.86	0.91	0.92	0.95	0.97
0.88	1.05	0.91	0.89	0.92	0.94	0.90	1.00	0.90	0.93

The minimum of these measurements is 0.85 and the maximum is 1.06. The range is $1.06 - 0.85 = 0.21$. We shall use $k = 8$ classes of length 0.03. Note that $8(0.03) = 0.24 > 0.21$. We start at 0.835 and end at 1.075. These boundaries are the same distance below the minimum and above the maximum. In Table 6.1-2 we also give the values of the heights of each rectangle in the relative frequency histogram so that the total area of the histogram is 1. These heights are given by

$$h(x) = \frac{f_i}{(0.03)(100)} = \frac{f_i}{3}.$$

The plots of the relative frequency histogram and polygon are given in Figure 6.1-1.

If you are using a computer program to analyze a set of data, it is very easy to find the sample mean, the sample variance, and the sample standard deviation. However, if you have only grouped data or if you are not using a computer, you can obtain close approximations of these values by computing the mean $\bar{u}$ and variance s_u^2 of the grouped data using the class marks weighted with their respective frequencies. We have

$$\bar{u} = \frac{1}{n} \sum_{i=1}^{k} f_i u_i$$

$$= \frac{1}{100} \sum_{i=1}^{8} f_i u_i = \frac{92.83}{100} = 0.9283,$$

Table 6.1-2: Frequency Table of Fluoride Concentrations				
Class Interval	Class Mark (u_i)	Tabulation	Frequency (f_i)	$h(x) = f_i/3$
(0.835, 0.865)	0.85	$\cancel{||||}\ ||$	7	7/3
(0.865, 0.895)	0.88	$\cancel{||||}\ \cancel{||||}\ \cancel{||||}\ \cancel{||||}$	20	20/3
(0.895, 0.925)	0.91	$\cancel{||||}\ \cancel{||||}\ \cancel{||||}\ \cancel{||||}\ \cancel{||||}\ ||$	27	27/3
(0.925, 0.955)	0.94	$\cancel{||||}\ \cancel{||||}\ \cancel{||||}\ |||$	18	18/3
(0.955, 0.985)	0.97	$\cancel{||||}\ \cancel{||||}\ ||||$	14	14/3
(0.985, 1.015)	1.00	$\cancel{||||}\ ||||$	9	9/3
(1.015, 1.045)	1.03	$|||$	3	3/3
(1.045, 1.075)	1.06	$||$	2	2/3

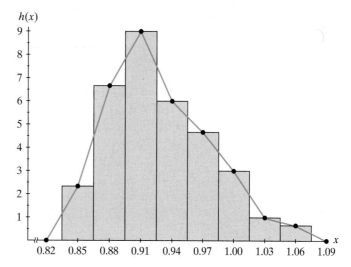

Figure 6.1-1: Concentrations of fluoride in toothpaste

$$s_u^2 = \frac{1}{n-1}\sum_{i=1}^{k} f_i(u_i - \bar{u})^2 = \frac{\displaystyle\sum_{i=1}^{k} f_i u_i^2 - \frac{1}{n}\left(\sum_{i=1}^{k} f_i u_i\right)^2}{n-1}$$

$$= \frac{0.237411}{99} = 0.002398.$$

Thus

$$s_u = \sqrt{0.002398} = 0.04897.$$

These compare rather favorably with $\bar{x} = 0.9293$ and $s_x = 0.04895$ of the original data.   ◄

We can also get an idea of the shape of the underlying distribution using a stem-and-leaf diagram.

EXAMPLE 6.1-2 We continue to use the data given in Table 6.1-1. For convenience, we use 0.02 as the length of a class interval. The ordered stem-and-leaf display is given in Table 6.1-3.

Table 6.1-3: Ordered Stem-and-Leaf Diagram of Fluoride Concentrations			
Stems	**Leaves**	**Frequency**	**Depths**
$0.8f$	5 5 5 5	4	4
$0.8s$	6 6 6 7 7 7 7	7	11
$0.8\bullet$	8 8 8 8 8 8 8 8 9 9 9 9 9 9 9 9	16	27
$0.9*$	0 0 0 0 0 0 0 0 0 1 1 1 1 1 1 1 1	17	44
$0.9t$	2 2 2 2 2 2 2 2 2 2 2 3 3 3 3 3 3 3 3	19	(19)
$0.9f$	4 4 5 5 5 5 5 5 5	9	37
$0.9s$	6 6 7 7 7 7	6	28
$0.9\bullet$	8 8 8 8 8 8 8 8 9 9	10	22
$1.0*$	0 0 0 0 1 1 1	7	12
$1.0t$	2 3 3	3	5
$1.0f$	5	1	2
$1.0s$	6	1	1

This ordered stem-and-leaf diagram is useful when finding sample percentiles of the data. ◄

We now find some of the sample percentiles associated with the fluoride data. Since $n = 100$, $(n + 1)(0.25) = 25.25$, $(n + 1)(0.50) = 50.5$, and $(n + 1)(0.75) = 75.75$ so that the 25th, 50th, and 75th percentiles are

$$\tilde{\pi}_{0.25} = (0.75)y_{25} + (0.25)y_{26} = (0.75)(0.89) + (0.25)(0.89) = 0.89,$$

$$\tilde{\pi}_{0.50} = (0.50)y_{50} + (0.50)y_{51} = (0.50)(0.92) + (0.50)(0.92) = 0.92,$$

$$\tilde{\pi}_{0.75} = (0.25)y_{75} + (0.75)y_{76} = (0.25)(0.97) + (0.75)(0.97) = 0.97.$$

These three percentiles are often called the **first quartile**, the **median** or **second quartile**, and the **third quartile**, respectively. Along with the smallest (the **minimum**) and largest (the **maximum**) values, they give the **five-number summary** of a set of data. Furthermore, the difference between the third and first quartiles is called the **interquartile range**, **IQR**. Here it is equal to

$$\tilde{q}_3 - \tilde{q}_1 = \tilde{\pi}_{0.75} - \tilde{\pi}_{0.25} = 0.97 - 0.89 = 0.08.$$

There is a graphical means for displaying the five-number summary of a set of data that is called a **box-and-whisker diagram**. To construct a horizontal box-and-whisker diagram, or more simply, a **box plot**, draw a horizontal axis that is scaled to the data. Above the axis draw a rectangular box with the left and right sides drawn at $\tilde{q}_1$ and $\tilde{q}_3$ with a vertical line segment drawn at the median, $\tilde{q}_2 = \tilde{m}$. A left whisker is drawn as a horizontal line segment from the minimum to the midpoint of the left side of the box and a right whisker is drawn as a horizontal line segment from the midpoint of the right side of the box to the maximum. Note that the length of the box is equal to the IQR. The left and right whiskers represent the first and fourth quarters of the data while the two middle quarters of the data are represented, respectively, by the two sections of the box, one to the left and one to the right of the median line.

EXAMPLE 6.1-3   Using the fluoride data shown in Table 6.1-3, we found that the five-number summary is given by

$$y_1 = 0.85, \ \tilde{q}_1 = 0.89, \ \tilde{q}_2 = \tilde{m} = 0.92, \ \tilde{q}_3 = 0.97, \ y_{100} = 1.06.$$

The box-and-whisker diagram or box plot of these data is given in Figure 6.1-2. The fact that the long whisker is to the right and the right half of the box is larger than the left half of the box lead us to say that these data are slightly *skewed to the right*. Note that this can also be seen in the histogram and in the stem-and-leaf diagram.   ◄

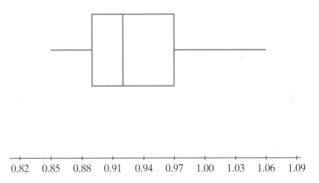

0.82   0.85   0.88   0.91   0.94   0.97   1.00   1.03   1.06   1.09

Figure 6.1-2: Box plot of fluoride concentrations

The next example illustrates how the box plot depicts data that are *skewed to the left*.

EXAMPLE 6.1-4   The following data give the ordered weights (in grams) of 39 gold coins that were produced during the reign of Verica, a pre-Roman British King. These coins display the vine-leaf as an emblem.

4.90	5.06	5.07	5.08	5.15	5.17	5.18	5.19	5.24	5.25
5.25	5.25	5.25	5.27	5.27	5.27	5.27	5.28	5.28	5.28
5.29	5.30	5.30	5.30	5.30	5.31	5.31	5.31	5.31	5.31
5.32	5.32	5.33	5.34	5.35	5.35	5.35	5.36	5.37	

For these data the minimum and maximum are 4.90 and 5.37. Since

$$(39 + 1)(1/4) = 10, \qquad (39 + 1)(2/4) = 20, \qquad (39 + 1)(3/4) = 30,$$

we have that

$$\tilde{q}_1 = y_{10} = 5.25,$$
$$\tilde{m} = y_{20} = 5.28,$$
$$\tilde{q}_3 = y_{30} = 5.31.$$

Thus the five-number summary is given by

$$y_1 = 4.90, \ \ \tilde{q}_1 = 5.25, \ \ \tilde{q}_2 = \tilde{m} = 5.28, \ \ \tilde{q}_3 = 5.31, \ \ y_{39} = 5.37.$$

The box plot associated with these data is given in Figure 6.1-3. Note that the box plot indicates that the data are skewed to the left. ◀

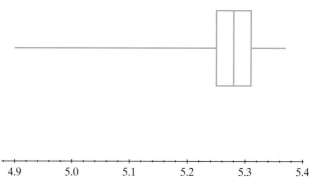

Figure 6.1-3: Box plot for weights of 39 gold coins

Sometimes we are interested in picking out observations that seem to be much larger or much smaller than most of the other observations. That is, we are looking for outliers. Tukey suggested a method for defining outliers that is resistant to the effect of one or two extreme values and makes use of the IQR. In a box-and-whisker diagram, construct **inner fences** to the left and right of the box at a distance of 1.5 times the IQR. **Outer fences** are constructed in the same way at a distance of 3 times the IQR. Observations that lie between the inner and outer fences are called **suspected outliers**. Observations that lie beyond the outer fences are called **outliers**. The observations beyond the inner fences are denoted with a circle (•) and the whiskers are drawn only to the extreme values within or on the inner fences. When you are analyzing a set of data, suspected outliers deserve a closer look and outliers should be looked at very carefully. It does not follow that suspected outliers should be removed from the data unless some error (such as a recording error) has been made. Moreover, it is sometimes important to determine the cause of extreme values, because outliers can often provide useful insights to the situation under consideration (such as a better way of doing things).

EXAMPLE 6.1-5   Continuing with Example 6.1-4, we find that the interquartile range is IQR = 5.31 − 5.25 = 0.06. Thus the inner fences would be constructed at a distance of 1.5(0.06) = 0.09 to the left and right of the box, and the outer fences would be constructed at a distance of 3(0.06) = 0.18 to the left and right of the box. Figure 6.1-4 shows a box plot with the fences. Of course, since the maximum is 0.06 greater than $\tilde{q}_3$, there are no fences to the right. From this box plot we see that there are three suspected outliers and two outliers. (You may speculate as to why there are outliers with these data and why they fall to the left. That is, they are lighter than expected.) Note that many computer programs simply plot outliers and suspected outliers using an asterisk and do not print fences. ◄

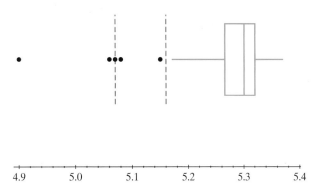

Figure 6.1-4: Box plot for weights of 39 gold coins with fences and outliers

Some functions of two or more order statistics are quite important in modern statistics. We mention and illustrate two more, along with the range and the IQR, using the 100 fluoride concentrations shown in Table 6.1-3.

(a) **Midrange** = average of the extremes

$$= \frac{y_1 + y_n}{2} = \frac{0.85 + 1.06}{2} = 0.955,$$

(b) **Trimean** = $\dfrac{(\text{1st quartile}) + 2(\text{2nd quartile}) + (\text{3rd quartile})}{4}$

$$= \frac{\tilde{q}_1 + 2\tilde{q}_2 + \tilde{q}_3}{4} = \frac{0.89 + 2(0.92) + 0.97}{4} = 0.925,$$

(c) **Range** = difference of the extremes

$$= y_n - y_1 = 1.06 - 0.85 = 0.21,$$

(d) **Interquartile range** = difference of third and first quartiles

$$= \tilde{q}_3 - \tilde{q}_1$$
$$= 0.97 - 0.89 = 0.08.$$

Thus we see that the mean, the median, the midrange, and the trimean are measures of the middle of the sample. In some sense, the standard deviation, the range, and the interquartile range provide measures of spread of the sample.

EXERCISES

6.1-1 For determining half-lives of radioactive isotopes, it is important to know what the background radiation is in a given detector over a period of time. Data were collected in a γ-ray detection experiment over 98 ten-second intervals (observations of a random variable X), yielding the following data:

58	50	57	58	64	63	54	64	59	41	43	56	60	50
46	59	54	60	59	60	67	52	65	63	55	61	68	58
63	36	42	54	58	54	40	60	64	56	61	51	48	50
60	42	62	67	58	49	66	58	57	59	52	54	53	53
57	43	73	65	45	43	57	55	73	62	68	55	51	55
53	68	58	53	51	73	44	50	53	62	58	47	63	59
59	56	60	59	50	52	62	51	66	51	56	53	59	57

(a) Find the sample mean.
(b) Find the sample variance.

6.1-2 Let X equal the number of calls per hour received by 911 between midnight and noon and reported in the *Holland Sentinel*. On October 29 and October 30, the following numbers of calls were reported:

$$0 \ 1 \ 1 \ 1 \ 0 \ 1 \ 2 \ 1 \ 4 \ 1 \ 2 \ 3 \ 0 \ 3 \ 0 \ 1 \ 0 \ 1 \ 1 \ 2 \ 3 \ 0 \ 2 \ 2$$

(a) Find the sample mean.
(b) Find the sample variance.

6.1-3 Two different saws are used to cut columns for a gazebo that has a diameter of 16 feet. The company was interested in comparing the two saws. They measured 9 columns cut by each of the saws, say X and Y, yielding the following lengths in feet.

Saw X:	8.02	8.10	8.04	8.04	8.00	8.11	8.07	8.02	8.04
Saw Y:	8.04	8.04	8.10	8.06	8.08	8.10	8.07	8.08	8.06

(a) Find the sample means and sample standard deviations of each set of measurements.
(b) Find the five-number summary for each set of measurements and then construct box plots of the two sets of measurements on the same graph.
(c) How would you compare the two saws?

6.1-4 In Exercise 3.1-7 the underwater weights are given 82 male students. Here are the underwater weights for 100 female students.

2.0	2.0	2.1	1.6	1.9	2.0	2.0	1.3	1.3	1.2	2.3	1.9
2.1	1.2	2.0	1.6	1.1	2.2	2.2	1.4	1.7	2.4	1.8	1.7
2.0	2.1	1.6	1.7	1.8	0.7	1.9	1.7	1.7	1.1	2.0	2.3
0.5	1.3	2.7	1.8	2.0	1.7	1.2	0.7	1.1	1.1	1.7	1.7
1.2	1.2	0.7	2.3	1.7	2.4	1.0	2.4	1.4	1.9	2.5	2.2
2.1	1.4	2.4	1.8	2.5	1.3	0.5	1.7	1.9	1.8	1.3	2.0
2.2	1.7	2.0	2.5	1.2	1.4	1.4	1.2	2.2	2.0	1.8	1.4
1.9	1.4	1.3	2.5	1.2	1.5	0.8	2.0	2.2	1.8	2.0	1.6
1.5	1.6	1.5	2.6								

(a) Calculate the sample mean and sample standard deviation of these data.
(b) Construct a histogram of these data. Locate on the histogram the points $\bar{x}, \bar{x} \pm s$ and $\bar{x} \pm 2s$.

   (c) Do these look like data from a normal distribution? Why?
   (d) Construct a box-and-whisker diagram of these data and the data in Exercise 3.1-7 on the same graph. Describe what this graph shows.

6.1-5 When you purchase "one-pound bags" of carrots, you can either buy "baby" carrots or regular carrots. We shall compare the weights of 75 bags of each of these types of carrots. The following table gives the weights of the bags of baby carrots.

1.03	1.03	1.06	1.02	1.03	1.03	1.03	1.02	1.03	1.03
1.06	1.04	1.05	1.03	1.04	1.03	1.05	1.06	1.04	1.04
1.03	1.04	1.04	1.06	1.03	1.04	1.05	1.04	1.04	1.02
1.03	1.05	1.05	1.03	1.04	1.03	1.04	1.04	1.03	1.04
1.03	1.04	1.04	1.04	1.05	1.04	1.04	1.03	1.03	1.05
1.04	1.04	1.05	1.04	1.03	1.03	1.05	1.03	1.04	1.05
1.04	1.04	1.04	1.05	1.03	1.04	1.04	1.04	1.04	1.03
1.05	1.05	1.05	1.03	1.04					

The following table gives the weights of the regular-sized carrots.

1.29	1.10	1.28	1.29	1.23	1.20	1.31	1.25	1.13	1.26
1.19	1.33	1.24	1.20	1.26	1.24	1.11	1.14	1.15	1.15
1.19	1.26	1.14	1.20	1.20	1.20	1.24	1.25	1.28	1.24
1.26	1.20	1.30	1.23	1.26	1.16	1.34	1.10	1.22	1.27
1.21	1.09	1.23	1.03	1.32	1.21	1.23	1.34	1.19	1.18
1.20	1.20	1.13	1.43	1.19	1.05	1.16	1.19	1.07	1.21
1.36	1.21	1.00	1.23	1.22	1.13	1.24	1.10	1.18	1.26
1.12	1.10	1.19	1.10	1.24					

   (a) Calculate the five-number summary of each set of weights.
   (b) Construct box plots for each set of weights on the same graph.
   (c) If the carrots are the same price per package, which is the better buy? Which type of carrots would you select?

6.1-6 The distances in feet of Barry Bonds's $n = 45$ home runs in 2003 are the following.

463	390	310	370	410	400	420	400	425
430	420	390	420	400	370	475	415	415
415	390	460	325	381	392	402	470	383
402	430	440	410	390	420	370	475	405
390	450	410	420	480	400	370	400	390

   (a) Construct an ordered stem-and-leaf diagram of these distances.
   (b) Calculate the five number summary of these data.
   (c) Construct a box plot of these distances.
   (d) Determine if there are any outliers and, if there are, modify your box plot to include inner and outer fences.

6.1-7 Let X denote the concentration of $CaCO_3$ in milligrams per liter. Twenty observations of X are

130.8	129.9	131.5	131.2	129.5
132.7	131.5	127.8	133.7	132.2
134.8	131.7	133.9	129.8	131.4
128.8	132.7	132.8	131.4	131.3

   (a) Construct an ordered stem-and-leaf display, using stems of 127, 128, ... , 134.
   (b) Find the midrange, range, interquartile range, median, sample mean, and sample variance.
   (c) Draw a box-and-whisker diagram.

6.1-8 For the roulette data in Exercise 3.1-9,

   (a) Determine the order statistics.
   (b) Find the five-number summary of the data.

(c) Draw a box-and-whisker diagram.
(d) Find the locations of the inner and outer fences and draw a box plot that shows the fences, the suspected outliers, and the outliers.
(e) Find $\bar{x}$, the sample mean.
(f) In your opinion does the median or sample mean give a better measure of the center of these data?

6.1-9 For the insurance data in Exercise 3.1-10,

(a) Find the five-number summary of the data and draw a box-and-whisker diagram.
(b) Calculate the IQR and the locations of the inner and outer fences.
(c) Draw a box plot that shows the fences, suspected outliers, and outliers.

6.1-10 In Exercise 3.1-6 25K race times are given for 125 male runners, 50–54 years old. In this exercise we give similar data for 76 female runners, 50–54 years old, who competed in the 25K Fifth Third River Bank Run. The data are given in minutes.

118.05	180.45	123.80	125.23	125.27	125.90	126.90	128.37	129.30
130.08	131.43	131.72	132.17	133.00	133.25	133.42	133.43	133.50
134.25	136.98	137.10	137.28	138.52	139.80	140.87	142.85	143.75
145.05	145.07	145.48	146.55	146.67	148.07	148.75	148.93	149.77
150.10	150.22	150.62	150.82	150.85	151.10	151.43	152.25	152.60
152.90	155.55	157.02	159.58	159.70	159.80	159.92	159.97	161.33
162.17	163.60	165.45	166.23	168.10	168.60	169.17	169.93	170.88
171.20	173.23	174.22	178.57	178.83	183.08	186.07	186.48	189.77
192.52	197.88	199.80	203.92					

(a) Construct an ordered stem-and-leaf display.
(b) Find the five-number summary of the data and draw a box-and-whisker diagram.
(c) Draw, on the same figure, the box plot for these data and the box plot for the data in Exercise 3.1-6. How would you compare these two sets of data?
(d) Find the 90th percentile.

6.1-11 Using the costs of the "300 level" textbooks in Exercise 3.1-3,

(a) Construct an ordered stem-and-leaf display using integer stems.
(b) Calculate the five-number summary of the data and draw a box plot.
(c) Are there any suspected outliers?

6.1-12 The weights (in grams) of 25 indicator housings used on gauges are as follows:

102.0	106.3	106.6	108.8	107.7
106.1	105.9	106.7	106.8	110.2
101.7	106.6	106.3	110.2	109.9
102.0	105.8	109.1	106.7	107.3
102.0	106.8	110.0	107.9	109.3

(a) Construct an ordered stem-and-leaf display using integers as the stems and tenths as the leaves.
(b) Find the five-number summary of the data and draw a box plot.
(c) Are there any suspected outliers or outliers?

6.2 POINT ESTIMATION

In earlier chapters we have alluded to estimating characteristics of the distribution from the corresponding ones of the sample, hoping that the latter would be reasonably close to the former. For example, the sample mean $\bar{x}$ can be thought of as an estimate of the distribution mean μ, and the sample variance s^2 can be used as an estimate of the distribution variance σ^2. Even the relative frequency histogram associated with a sample can be taken as an estimate of the p.d.f. of the underlying distribution. But how good are these estimates? What makes an estimate good? Can we say anything about the closeness of an estimate to an unknown parameter?

In this section we consider random variables for which the functional form of the p.m.f. or p.d.f. is known, but the distribution depends on an unknown parameter, say θ, that may have any value in a set, say Ω, which is called the **parameter space**. For example, perhaps it is known that $f(x; \theta) = (1/\theta)e^{-x/\theta}$, $0 < x < \infty$ and that $\theta \in \Omega = \{\theta : 0 < \theta < \infty\}$. In certain instances, it might be necessary for the experimenter to select precisely one member of the family $\{f(x, \theta), \theta \in \Omega\}$ as the most likely p.d.f. of the random variable. That is, the experimenter needs a point estimate of the parameter θ, namely the value of the parameter that corresponds to the selected p.d.f.

In estimation we take a random sample from the distribution to elicit some information about the unknown parameter θ. That is, we repeat the experiment n independent times, observe the sample, $X_1, X_2, \ldots, X_n$, and try to estimate the value of θ using the observations $x_1, x_2, \ldots, x_n$. The function of $X_1, X_2, \ldots, X_n$ used to estimate θ, say the statistic $u(X_1, X_2, \ldots, X_n)$, is called an **estimator** of θ. We want it to be such that the computed **estimate** $u(x_1, x_2, \ldots, x_n)$ is usually close to θ. Since we are estimating one member of $\theta \in \Omega$, such as estimator is often called a **point estimator**.

The following illustration should help motivate one principle that is often used in finding point estimates. Suppose that X is $b(1, p)$ so that the p.m.f. of X is

$$f(x; p) = p^x(1 - p)^{1-x}, \qquad x = 0, 1, \qquad 0 \le p \le 1.$$

We note that $p \in \Omega = \{p : 0 \le p \le 1\}$, where Ω represents the parameter space, that is, the space of all possible values of the parameter, p. Given a random sample $X_1, X_2, \ldots, X_n$, the problem is to find an estimator $u(X_1, X_2, \ldots, X_n)$ such that $u(x_1, x_2, \ldots, x_n)$ is a good point estimate of p, where $x_1, x_2, \ldots, x_n$ are the observed values of the random sample. Now the probability that $X_1, X_2, \ldots, X_n$ takes these particular values is (with $\Sigma\, x_i$ denoting $\sum_{i=1}^{n} x_i$)

$$P(X_1 = x_1, \ldots, X_n = x_n) = \prod_{i=1}^{n} p^{x_i}(1 - p)^{1-x_i} = p^{\Sigma\, x_i}(1 - p)^{n - \Sigma\, x_i},$$

which is the joint p.m.f. of $X_1, X_2, \ldots, X_n$ evaluated at the observed values. One reasonable way to proceed toward finding a good estimate of p is to regard this probability (or joint p.m.f.) as a function of p and find the value of p that maximizes it. That is, find the p value most likely to have produced these sample values. The joint p.m.f., when regarded as a function of p, is frequently called the **likelihood function**. Thus here the likelihood function is

$$L(p) = L(p; x_1, x_2, \ldots, x_n)$$

$$= f(x_1; p)f(x_2; p) \cdots f(x_n; p)$$

$$= p^{\Sigma\, x_i}(1 - p)^{n - \Sigma\, x_i}, \qquad 0 \le p \le 1.$$

To find the value of p that maximizes $L(p)$, we first take its derivative when $0 < p < 1$:

$$\frac{dL(p)}{dp} = (\Sigma\, x_i)\, p^{\Sigma\, x_i - 1}(1-p)^{n-\Sigma\, x_i} - (n - \Sigma\, x_i)\, p^{\Sigma\, x_i}(1-p)^{n-\Sigma\, x_i - 1}.$$

Setting this first derivative equal to zero gives us, with the restriction $0 < p < 1$,

$$p^{\Sigma\, x_i}(1-p)^{n-\Sigma\, x_i}\left[\frac{\Sigma\, x_i}{p} - \frac{n - \Sigma\, x_i}{1-p}\right] = 0.$$

Since $0 < p < 1$, this equals zero when

$$\frac{\Sigma\, x_i}{p} - \frac{n - \Sigma\, x_i}{1-p} = 0. \tag{6.2-1}$$

Multiplying each member of Equation 6.2-1 by $p(1-p)$ and simplifying, we obtain

$$\sum_{i=1}^{n} x_i - np = 0$$

or, equivalently,

$$p = \frac{\sum_{i=1}^{n} x_i}{n} = \overline{x}.$$

The corresponding statistic, namely $(\sum_{i=1}^{n} X_i)/n = \overline{X}$, is called the **maximum likelihood estimator** and is denoted by $\widehat{p}$; that is,

$$\widehat{p} = \frac{1}{n}\sum_{i=1}^{n} X_i = \overline{X}.$$

When finding a maximum likelihood estimator, it is often easier to find the value of the parameter that maximizes the natural logarithm of the likelihood function rather than the value of the parameter that maximizes the likelihood function itself. Because the natural logarithm function is an increasing function, the solutions will be the same. To see this, the example we have been considering gives us, for $0 < p < 1$,

$$\ln L(p) = \left(\sum_{i=1}^{n} x_i\right)\ln p + \left(n - \sum_{i=1}^{n} x_i\right)\ln(1-p).$$

To find the maximum, we set the first derivative equal to zero to obtain

$$\frac{d\,[\ln L(p)]}{dp} = \left(\sum_{i=1}^{n} x_i\right)\left(\frac{1}{p}\right) + \left(n - \sum_{i=1}^{n} x_i\right)\left(\frac{-1}{1-p}\right) = 0,$$

which is the same as Equation 6.2-1. Thus the solution is $p = \bar{x}$ and the maximum likelihood estimator for p is $\widehat{p} = \bar{X}$.

Motivated by the preceding illustration, we present the formal definition of maximum likelihood estimators. This definition is used in both the discrete and continuous cases.

Let $X_1, X_2, \ldots, X_n$ be a random sample from a distribution that depends on one or more unknown parameters $\theta_1, \theta_2, \ldots, \theta_m$ with p.m.f. or p.d.f. denoted by $f(x; \theta_1, \theta_2, \ldots, \theta_m)$. Suppose that $(\theta_1, \theta_2, \ldots, \theta_m)$ is restricted to a given parameter space Ω. Then the joint p.m.f. or p.d.f. of $X_1, X_2, \ldots, X_n$, namely

$$L(\theta_1, \theta_2, \ldots, \theta_m) = f(x_1; \theta_1, \ldots, \theta_m) f(x_2; \theta_1, \ldots, \theta_m)$$
$$\cdots f(x_n; \theta_1, \ldots, \theta_m), \quad (\theta_1, \theta_2, \ldots, \theta_m) \in \Omega,$$

when regarded as a function of $\theta_1, \theta_2, \ldots, \theta_m$, is called the **likelihood function**. Say

$$[u_1(x_1, \ldots, x_n), u_2(x_1, \ldots, x_n), \ldots, u_m(x_1, \ldots, x_n)]$$

is that m-tuple in Ω that maximizes $L(\theta_1, \theta_2, \ldots, \theta_m)$. Then

$$\widehat{\theta_1} = u_1(X_1, \ldots, X_n),$$
$$\widehat{\theta_2} = u_2(X_1, \ldots, X_n),$$
$$\vdots$$
$$\widehat{\theta_m} = u_m(X_1, \ldots, X_n)$$

are maximum likelihood estimators of $\theta_1, \theta_2, \ldots, \theta_m$, respectively; and the corresponding observed values of these statistics, namely

$$u_1(x_1, \ldots, x_n), u_2(x_1, \ldots, x_n), \ldots, u_m(x_1, \ldots, x_n),$$

are called **maximum likelihood estimates**. In many practical cases, these estimators (and estimates) are unique.

For many applications there is just one unknown parameter. In these cases the likelihood function is given by

$$L(\theta) = \prod_{i=1}^{n} f(x_i; \theta).$$

Some additional examples will help clarify these definitions.

EXAMPLE 6.2-1   Let $X_1, X_2, \ldots, X_n$ be a random sample from the exponential distribution with p.d.f.

$$f(x; \theta) = \frac{1}{\theta} e^{-x/\theta}, \qquad 0 < x < \infty, \qquad \theta \in \Omega = \{\theta : 0 < \theta < \infty\}.$$

The likelihood function is given by

$$L(\theta) = L(\theta; x_1, x_2, \ldots, x_n)$$

$$= \left(\frac{1}{\theta} e^{-x_1/\theta}\right)\left(\frac{1}{\theta} e^{-x_2/\theta}\right) \cdots \left(\frac{1}{\theta} e^{-x_n/\theta}\right)$$

$$= \frac{1}{\theta^n} \exp\left(-\frac{\displaystyle\sum_{i=1}^{n} x_i}{\theta}\right), \qquad 0 < \theta < \infty.$$

The natural logarithm of $L(\theta)$ is

$$\ln L(\theta) = -(n)\ln(\theta) - \frac{1}{\theta}\sum_{i=1}^{n} x_i, \qquad 0 < \theta < \infty.$$

Thus

$$\frac{d\,[\ln L(\theta)]}{d\theta} = \frac{-n}{\theta} + \frac{\displaystyle\sum_{i=1}^{n} x_i}{\theta^2} = 0.$$

The solution of this equation for θ is

$$\theta = \frac{1}{n}\sum_{i=1}^{n} x_i = \overline{x}.$$

Note that

$$\frac{d\,[\ln L(\theta)]}{d\theta} = \frac{1}{\theta}\left(-n + \frac{n\overline{x}}{\theta}\right) \begin{array}{ll} > 0, & \theta < \overline{x}, \\ = 0, & \theta = \overline{x}, \\ < 0, & \theta > \overline{x}. \end{array}$$

Hence $\ln L(\theta)$ does have a maximum at $\overline{x}$, and thus the maximum likelihood estimator for θ is

$$\widehat{\theta} = \overline{X} = \frac{1}{n}\sum_{i=1}^{n} X_i.$$

◀

EXAMPLE 6.2-2 Let $X_1, X_2, \ldots, X_n$ be a random sample from the geometric distribution with p.m.f. $f(x; p) = (1-p)^{x-1}p$, $x = 1, 2, 3, \ldots$. The likelihood function is given by

$$L(p) = (1-p)^{x_1-1}p(1-p)^{x_2-1}p \cdots (1-p)^{x_n-1}p$$

$$= p^n(1-p)^{\sum x_i - n}, \qquad 0 \le p \le 1.$$

The natural logarithm of $L(p)$ is

$$\ln L(p) = n \ln p + \left(\sum_{i=1}^{n} x_i - n \right) \ln(1 - p), \qquad 0 < p < 1.$$

Thus, restricting p to $0 < p < 1$ so as to be able to take the derivative, we have

$$\frac{d \ln L(p)}{dp} = \frac{n}{p} - \frac{\displaystyle\sum_{i=1}^{n} x_i - n}{1 - p} = 0.$$

Solving for p, we obtain

$$p = \frac{n}{\displaystyle\sum_{i=1}^{n} x_i} = \frac{1}{\bar{x}},$$

and this solution provides a maximum. So the maximum likelihood estimator of p is

$$\widehat{p} = \frac{n}{\displaystyle\sum_{i=1}^{n} X_i} = \frac{1}{\bar{X}}.$$

This estimator agrees with our intuition because, in n observations of a geometric random variable, there are n successes in the $\sum_{i=1}^{n} x_i$ trials. Thus the estimate of p is the number of successes divided by the total number of trials. ◄

In the following important example we find the maximum likelihood estimators of the parameters associated with the normal distribution.

EXAMPLE 6.2-3   Let $X_1, X_2, \ldots, X_n$ be a random sample from $N(\theta_1, \theta_2)$, where

$$\Omega = \{(\theta_1, \theta_2) : -\infty < \theta_1 < \infty, 0 < \theta_2 < \infty\}.$$

That is, here we let $\theta_1 = \mu$ and $\theta_2 = \sigma^2$. Then

$$L(\theta_1, \theta_2) = \prod_{i=1}^{n} \frac{1}{\sqrt{2\pi\theta_2}} \exp\left[-\frac{(x_i - \theta_1)^2}{2\theta_2} \right]$$

or, equivalently,

$$L(\theta_1, \theta_2) = \left(\frac{1}{\sqrt{2\pi\theta_2}} \right)^n \exp\left[-\frac{\displaystyle\sum_{i=1}^{n} (x_i - \theta_1)^2}{2\theta_2} \right], \qquad (\theta_1, \theta_2) \in \Omega.$$

The natural logarithm of the likelihood function is

$$\ln L(\theta_1, \theta_2) = -\frac{n}{2} \ln(2\pi\theta_2) - \frac{\sum_{i=1}^{n} (x_i - \theta_1)^2}{2\theta_2}.$$

The partial derivatives with respect to θ_1 and θ_2 are

$$\frac{\partial (\ln L)}{\partial \theta_1} = \frac{1}{\theta_2} \sum_{i=1}^{n} (x_i - \theta_1)$$

and

$$\frac{\partial (\ln L)}{\partial \theta_2} = \frac{-n}{2\theta_2} + \frac{1}{2\theta_2^2} \sum_{i=1}^{n} (x_i - \theta_1)^2.$$

The equation $\partial (\ln L)/\partial \theta_1 = 0$ has the solution $\theta_1 = \bar{x}$. Setting $\partial (\ln L)/\partial \theta_2 = 0$ and replacing θ_1 by $\bar{x}$ yields

$$\theta_2 = \frac{1}{n} \sum_{i=1}^{n} (x_i - \bar{x})^2.$$

By considering the usual condition on the second partial derivatives, these solutions do provide a maximum. Thus the maximum likelihood estimators of $\mu = \theta_1$ and $\sigma^2 = \theta_2$ are

$$\widehat{\theta}_1 = \overline{X} \quad \text{and} \quad \widehat{\theta}_2 = \frac{1}{n} \sum_{i=1}^{n} (X_i - \overline{X})^2 = V. \qquad \blacktriangleleft$$

It is interesting to note that in each of our first illustration, where $\widehat{p} = \overline{X}$, and Example 6.2-1, where $\widehat{\theta} = \overline{X}$, the expected value of the estimator is equal to the corresponding parameter. This observation leads to the following definition.

Definition 6.2-1   If $E[u(X_1, X_2, \ldots, X_n)] = \theta$, the statistic $u(X_1, X_2, \ldots X_n)$ is called an *unbiased estimator* of θ. Otherwise, it is said to be *biased*.

EXAMPLE 6.2-4   We have shown that when sampling from $N(\theta_1 = \mu, \theta_2 = \sigma^2)$, the maximum likelihood estimators of μ and σ^2 are

$$\widehat{\theta}_1 = \widehat{\mu} = \overline{X} \quad \text{and} \quad \widehat{\theta}_2 = \widehat{\sigma^2} = \frac{(n-1)S^2}{n}.$$

Recalling that the distribution of $\overline{X}$ is $N(\mu, \sigma^2/n)$, we see that $E(\overline{X}) = \mu$ and thus $\overline{X}$ is an unbiased estimator of μ.

In Theorem 5.3-3 we showed that the distribution of $(n-1)S^2/\sigma^2$ is $\chi^2(n-1)$. Thus

$$E(S^2) = E\left[\frac{\sigma^2}{n-1}\frac{(n-1)S^2}{\sigma^2}\right] = \frac{\sigma^2}{n-1}(n-1) = \sigma^2.$$

That is, the sample variance

$$S^2 = \frac{1}{n-1}\sum_{i=1}^{n}(X_i - \overline{X})^2$$

is an unbiased estimator of σ^2. Thus since

$$E(\widehat{\theta_2}) = \frac{n-1}{n}E(S^2) = \frac{n-1}{n}\sigma^2,$$

$\widehat{\theta_2}$ is a biased estimator of $\theta_2 = \sigma^2$. ◀

Sometimes it is impossible to find maximum likelihood estimators in a convenient closed form and numerical methods must be used to maximize the likelihood function. For illustration, suppose that $X_1, X_2, \ldots, X_n$ is a random sample from a gamma distribution with parameters $\alpha = \theta_1$ and $\beta = \theta_2$, where $\theta_1 > 0, \theta_2 > 0$. It is difficult to maximize

$$L(\theta_1, \theta_2; x_1, \ldots, x_n) = \left[\frac{1}{\Gamma(\theta_1)\theta_2^{\theta_1}}\right]^n (x_1 x_2 \ldots x_n)^{\theta_1 - 1}\exp\left(-\sum_{i=1}^{n} x_i/\theta_2\right)$$

with respect to θ_1 and θ_2, owing to the presence of the gamma function $\Gamma(\theta_1)$. Thus numerical methods must be used to maximize L once $x_1, x_2, \ldots, x_n$ are observed.

There are other ways, however, to easily obtain point estimates of θ_1 and θ_2. One of the early methods was to simply equate the first sample moment to the first theoretical moment. If needed, the two second moments are equated; then the third moments and so on until we have enough equations to solve for the parameters. For illustration, in the gamma distribution situation, let us simply equate the first two moments of the distribution to the corresponding moments of the empirical distribution. This seems like a reasonable way in which to find estimators, since the empirical distribution converges in some sense to the probability distribution, and hence corresponding moments should be about equal. Here in this illustration we have

$$\theta_1\theta_2 = \overline{X}, \qquad \theta_1\theta_2^2 = V,$$

the solutions of which are

$$\widetilde{\theta_1} = \frac{\overline{X}^2}{V} \qquad \text{and} \qquad \widetilde{\theta_2} = \frac{V}{\overline{X}}.$$

We say that these latter two statistics, $\widetilde{\theta}_1$ and $\widetilde{\theta}_2$, are respective estimators of θ_1 and θ_2 found by the **method of moments**.

To generalize this discussion, let $X_1, X_2, \ldots, X_n$ be a random sample of size n from a distribution with p.d.f. $f(x; \theta_1, \theta_2, \ldots, \theta_r), (\theta_1, \ldots, \theta_r) \in \Omega$. The expectation $E(X^k)$ is frequently called the kth moment of the distribution, $k = 1, 2, 3, \ldots$. The sum $M_k = \sum_{i=1}^{n} X_i^k / n$ is the kth moment of the sample, $k = 1, 2, 3, \ldots$. The method of moments can be described as follows. Equate $E(X^k)$ to M_k, beginning with $k = 1$ and continuing until there are enough equations to provide unique solutions for $\theta_1, \theta_2, \ldots, \theta_r$, say $h_i(M_1, M_2, \ldots), i = 1, 2, \ldots, r$, respectively. It should be noted that this could be done in an equivalent manner by equating $\mu = E(X)$ to $\overline{X}$ and $E[(X - \mu)^k]$ to $\sum_{i=1}^{n} (X_i - \overline{X})^k / n, k = 2, 3$, and so on until unique solutions for $\theta_1, \theta_2, \ldots, \theta_r$ are obtained. This alternative procedure was used in the preceding illustration. In most practical cases, the estimator $\widetilde{\theta}_i = h_i(M_1, M_2, \ldots)$ of θ_i, found by the method of moments, is an estimator of θ_i that in some sense gets close to that parameter when n is large, $i = 1, 2, \ldots, r$.

The following two examples illustrate the method of moments technique for finding estimators, the first for a one-parameter family and the second for a two-parameter family.

EXAMPLE 6.2-5 Let $X_1, X_2, \ldots, X_n$, be a random sample of size n from the distribution with p.d.f. $f(x; \theta) = \theta x^{\theta-1}, 0 < x < 1, 0 < \theta < \infty$. Sketch the graphs of this p.d.f. for $\theta = 1/4, 1$, and 4. Note that sets of observations for these three values of θ would look very different. How do we estimate the value of θ? The mean of this distribution is given by

$$E(X) = \int_0^1 x \theta x^{\theta-1} \, dx = \frac{\theta}{\theta + 1}.$$

We shall set the population mean equal to the sample mean and solve for θ. We have

$$\overline{x} = \frac{\theta}{\theta + 1}$$

or, solving for θ, we obtain the method of moments estimator,

$$\widetilde{\theta} = \frac{\overline{X}}{1 - \overline{X}}.$$

Thus an estimate of θ by the method of moments is $\overline{x}/(1 - \overline{x})$. ◀

Recall that in the method of moments, if there are two parameters that have to be estimated, the first two sample moments are set equal to the first two population moments that are given in terms of the unknown parameters. These two equations are then solved simultaneously for the unknown parameters.

EXAMPLE 6.2-6 Let the distribution of X be $N(\mu, \sigma^2)$. Then

$$E(X) = \mu \qquad \text{and} \qquad E(X^2) = \sigma^2 + \mu^2.$$

Given a random sample of size n, the first two moments are given by

$$m_1 = \frac{1}{n}\sum_{i=1}^{n} x_i \text{ and } m_2 = \frac{1}{n}\sum_{i=1}^{n} x_i^2.$$

We set $m_1 = E(X)$ and $m_2 = E(X^2)$ and solve for μ and σ^2. That is,

$$\frac{1}{n}\sum_{i=1}^{n} x_i = \mu \qquad \text{and} \qquad \frac{1}{n}\sum_{i=1}^{n} x_i^2 = \sigma^2 + \mu^2.$$

The first equation yields $\overline{x}$ as the estimate of μ. Replacing μ^2 with $\overline{x}^2$ in the second equation and solving for σ^2, we obtain

$$\frac{1}{n}\sum_{i=1}^{n} x_i^2 - \overline{x}^2 = \sum_{i=1}^{n} \frac{(x_i - \overline{x})^2}{n} = v$$

for the solution of σ^2. Thus the method of moment estimators for μ and σ^2 are $\widetilde{\mu} = \overline{X}$ and $\widetilde{\sigma^2} = V$. Of course, $\widetilde{\mu} = \overline{X}$ is unbiased whereas $\widetilde{\sigma^2} = V$ is biased. ◄

In Example 6.2-4 we showed that $\overline{X}$ and S^2 are unbiased estimators of μ and σ^2, respectively, when sampling from a normal distribution. This is also true when sampling from any population with finite variance σ^2. That is, $E(\overline{X}) = \mu$ and $E(S^2) = \sigma^2$, provided that the sample arises from a distribution with variance $\sigma^2 < \infty$ (see Exercise 6.2-13). Although S^2 is an unbiased estimator of σ^2, S is a biased estimator of σ. In Exercise 6.2-14, you are asked to show that, when sampling from a normal distribution, cS is an unbiased estimator for σ, where

$$c = \frac{\sqrt{n-1}\,\Gamma\left(\dfrac{n-1}{2}\right)}{\sqrt{2}\,\Gamma\left(\dfrac{n}{2}\right)}.$$

EXERCISES

6.2-1 Let $X_1, X_2, \ldots, X_n$ be a random sample from $N(\mu, \sigma^2)$, where the mean $\theta = \mu$ is such that $-\infty < \theta < \infty$ and σ^2 is a known positive number. Show that the maximum likelihood estimator for θ is $\widehat{\theta} = \overline{X}$.

6.2-2 A random sample $X_1, X_2, \ldots, X_n$ of size n is taken from $N(\mu, \sigma^2)$, where the variance $\theta = \sigma^2$ is such that $0 < \theta < \infty$ and μ is a known real number. Show that the maximum likelihood estimator for θ is $\widehat{\theta} = (1/n)\sum_{i=1}^{n}(X_i - \mu)^2$ and that this estimator is an unbiased estimator of θ.

6.2-3 A random sample $X_1, X_2, \ldots, X_n$ of size n is taken from a Poisson distribution with a mean of $\lambda, 0 < \lambda < \infty$.

(a) Show that the maximum likelihood estimator for λ is $\widehat{\lambda} = \overline{X}$.

(b) Let X equal the number of flaws per 100 feet of a used computer tape. Assume that X has a Poisson distribution with a mean of λ. If 40 observations of X yielded 5 zeros, 7 ones, 12 twos, 9 threes, 5 fours, 1 five, and 1 six, find the maximum likelihood estimate of λ.

6.2-4 For determining half-lives of radioactive isotopes, it is important to know what the background radiation is in a given detector over a period of time. The following data were taken in a γ-ray detection experiment over 98 ten-second intervals:

58	50	57	58	64	63	54	64	59	41	43	56	60	50
46	59	54	60	59	60	67	52	65	63	55	61	68	58
63	36	42	54	58	54	40	60	64	56	61	51	48	50
60	42	62	67	58	49	66	58	57	59	52	54	53	53
57	43	73	65	45	43	57	55	73	62	68	55	51	55
53	68	58	53	51	73	44	50	53	62	58	47	63	59
59	56	60	59	50	52	62	51	66	51	56	53	59	57

Assume that these are observations of a Poisson random variable with mean λ.

(a) Find the values of $\overline{x}$ and s^2 if you did not do this in Exercise 2.3-17.

(b) What is the value of the maximum likelihood estimator of λ?

(c) Is S^2 an unbiased estimator of λ?

(d) Which of $\overline{x}$ and s^2 would you recommend for estimating λ? Why? You could compare the variance of $\overline{X}$ with the variance of S^2, which is

$$\mathrm{Var}(S^2) = \frac{\lambda(2\lambda n + n - 1)}{n(n-1)}.$$

6.2-5 Let $X_1, X_2, \ldots, X_n$ be a random sample from distributions with the following probability density functions. In each case find the maximum likelihood estimator $\widehat{\theta}$.

(a) $f(x; \theta) = (1/\theta^2) x\, e^{-x/\theta}$, $\qquad 0 < x < \infty, \ 0 < \theta < \infty$.

(b) $f(x; \theta) = (1/2\theta^3) x^2\, e^{-x/\theta}$, $\qquad 0 < x < \infty, \ 0 < \theta < \infty$.

(c) $f(x; \theta) = (1/2)\, e^{-|x-\theta|}$, $\qquad -\infty < x < \infty, \ -\infty < \theta < \infty$.

HINT: This involves minimizing $\sum |x_i - \theta|$, which is a difficult problem. When $n = 5$, do it for $x_1 = 6.1, x_2 = -1.1, x_3 = 3.2, x_4 = 0.7, x_5 = 1.7$ and you will see the answer. Also see Exercise 2.2-10.

6.2-6 Find the maximum likelihood estimates for $\theta_1 = \mu$ and $\theta_2 = \sigma^2$ if a random sample of size 15 from $N(\mu, \sigma^2)$ yielded the following values:

31.5	36.9	33.8	30.1	33.9
35.2	29.6	34.4	30.5	34.2
31.6	36.7	35.8	34.5	32.7

6.2-7 Let $f(x; \theta) = \theta x^{\theta-1}, 0 < x < 1, \theta \in \Omega = \{\theta : 0 < \theta < \infty\}$. Let $X_1, X_2, \ldots, X_n$ denote a random sample of size n from this distribution.

(a) Sketch the p.d.f. of X for (i) $\theta = 1/2$, (ii) $\theta = 1$, and (iii) $\theta = 2$.

(b) Show that $\widehat{\theta} = -n/\ln\left(\prod_{i=1}^{n} X_i\right)$ is the maximum likelihood estimator of θ.

(c) For each of the following three sets of 10 observations from this distribution, calculate the values of the maximum likelihood estimate and the method of moments estimate for θ.

(i)	0.0256	0.3051	0.0278	0.8971	0.0739
	0.3191	0.7379	0.3671	0.9763	0.0102
(ii)	0.9960	0.3125	0.4374	0.7464	0.8278
	0.9518	0.9924	0.7112	0.2228	0.8609
(iii)	0.4698	0.3675	0.5991	0.9513	0.6049
	0.9917	0.1551	0.0710	0.2110	0.2154

6.2-8 Let $f(x; \theta) = (1/\theta)x^{(1-\theta)/\theta}$, $0 < x < 1$, $0 < \theta < \infty$.

(a) Show that the maximum likelihood estimator of θ is $\widehat{\theta} = -(1/n)\sum_{i=1}^{n} \ln X_i$.

(b) Show that $E(\widehat{\theta}) = \theta$ and thus $\widehat{\theta}$ is an unbiased estimator of θ.

6.2-9 Let $X_1, X_2, \ldots, X_n$ be a random sample of size n from the exponential distribution whose p.d.f. is $f(x; \theta) = (1/\theta)e^{-x/\theta}$, $0 < x < \infty$, $0 < \theta < \infty$.

(a) Show that $\overline{X}$ is an unbiased estimator of θ.

(b) Show that the variance of $\overline{X}$ is θ^2/n.

(c) What is a good estimate of θ if a random sample of size 5 yielded the sample values 3.5. 8.1. 0.9, 4.4, and 0.5?

6.2-10 Let $X_1, X_2, \ldots, X_n$ be a random sample of size n from a geometric distribution for which p is the probability of success.

(a) Use the method of moments to find a point estimate for p.

(b) Explain intuitively why your estimate makes good sense.

(c) Use the following data to give a point estimate of p:

3	34	7	4	19	2	1	19	43	2
22	4	19	11	7	1	2	21	15	16

6.2-11 Out of 50,000,000 instant winner lottery tickets, the proportion of winning tickets is p. Each day, for 20 consecutive days, a bettor purchased tickets, one at a time, until a winning ticket was purchased. The numbers of tickets that were purchased each day to obtain the winning ticket were

1	26	19	6	6	1	2	3	1	23
19	3	6	8	4	1	18	34	1	8

By making reasonable assumptions, find the maximum likelihood estimate of p based on these data.

6.2-12 Let $X_1, X_2, \ldots, X_n$ be a random sample from $b(1, p)$ (i.e., n Bernoulli trials). Thus

$$Y = \sum_{i=1}^{n} X_i \text{ is } b(n, p).$$

(a) Show that $\overline{X} = Y/n$ is an unbiased estimator of p.

(b) Show that $\text{Var}(\overline{X}) = p(1 - p)/n$.

(c) Show that $E[\overline{X}(1 - \overline{X})/n] = (n - 1)[p(1 - p)/n^2]$.

(d) Find the value of c so that $c\overline{X}(1 - \overline{X})$ is an unbiased estimator of $\text{Var}(\overline{X}) = p(1 - p)/n$.

6.2-13 Let $X_1, X_2, \ldots, X_n$ be a random sample from a distribution having finite variance σ^2. Show that

$$S^2 = \sum_{i=1}^{n} \frac{(X_i - \overline{X})^2}{n - 1}$$

is an unbiased estimator of σ^2. HINT: Write

$$S^2 = \frac{1}{n-1}\left(\sum_{i=1}^{n} X_i^2 - n\overline{X}^2\right)$$

and compute $E(S^2)$.

6.2-14 Let $X_1, X_2, \ldots, X_n$ be a random sample of size n from a normal distribution.

(a) Show that an unbiased estimator for σ is cS, where

$$c = \frac{\sqrt{n-1}\, \Gamma\left(\dfrac{n-1}{2}\right)}{\sqrt{2}\, \Gamma\left(\dfrac{n}{2}\right)}.$$

HINT: Recall that the distribution of $(n-1)S^2/\sigma^2$ is $\chi^2(n-1)$.

(b) Find the value of c when $n = 5, n = 6$.

(c) Graph c as a function of n. What is the limit of c as n increases without bound?

6.2-15 Let $X_1, X_2, \ldots, X_n$ be a random sample from a uniform distribution on the interval $(\theta - 1, \theta + 1)$.

(a) Find the method of moments estimator for θ.

(b) Is your estimator in part (a) an unbiased estimator for θ?

(c) Given the following $n = 5$ observations of X, give a point estimate of θ:

$$6.61 \quad 7.70 \quad 6.98 \quad 8.36 \quad 7.26$$

(d) The method of moments estimator actually has greater variance than the estimator $[\min(X_i) + \max(X_i)]/2$, which is a maximum likelihood estimator of θ. Compute the value of this estimator for the $n = 5$ observations in (c).

6.2-16 Given the following 25 observations from a gamma distribution with mean $\mu = \alpha\theta$ and variance $\sigma^2 = \alpha\theta^2$, use the method of moments estimators to find point estimates of α and θ.

6.9	7.3	6.7	6.4	6.3	5.9	7.0	7.1	6.5	7.6	7.2	7.1	6.1
7.3	7.6	7.6	6.7	6.3	5.7	6.7	7.5	5.3	5.4	7.4	6.9	

6.2-17 Let X equal the number of telephone calls per "five minutes" that are received at Great Lakes Pizza Co. in the evening. Assume that the distribution of X is Poisson with mean λ.

(a) Given n observations of X, find the method of moments estimate of λ.

(b) Give a point estimate of λ using the following 12 observations of X:

$$1 \quad 2 \quad 1 \quad 1 \quad 2 \quad 4 \quad 0 \quad 1 \quad 0 \quad 1 \quad 1 \quad 0$$

(c) Compare the values of $\bar{x}$ and s^2. Does this information support the assumption that X has a Poisson distribution?

6.2-18 An urn contains 64 balls of which N_1 are orange and N_2 are blue. A random sample of $n = 8$ balls is selected from the urn without replacement and X is equal to the number of orange balls in the sample. This experiment was repeated 30 times (the 8 balls being returned to the urn before each repetition), yielding the following data:

3	0	0	1	1	1	1	3	1	1	2	0	1	3	1
0	1	0	2	1	1	2	3	2	2	4	3	1	1	2

Using these data, guess the value of N_1 and give a reason for your guess.

6.2-19 A Geiger counter was set up in the physics laboratory to record the number of alpha particle emissions of carbon-14 in half a second. Ten observations yielded the following data:

$$4 \quad 6 \quad 9 \quad 6 \quad 10 \quad 11 \quad 6 \quad 3 \quad 7 \quad 10$$

Give the value of an unbiased estimate for λ, the mean number of counts per half second, assuming that these are observations of a Poisson random variable.

6.3* SUFFICIENT STATISTICS

We first define a sufficient statistic $Y = u(X_1, X_2, \ldots, X_n)$ for a parameter using a statement that, in most books, is given as a necessary and sufficient condition for sufficiency, namely the well-known Fisher-Neyman factorization theorem. We do this because we find readers at this level can apply this easily. However, using this as a definition, we shall note, by examples, the implications of this definition, one of which is sometimes used as the definition. The understanding of this example (6.3-3) is most important in the appreciation of the value of sufficient statistics.

Definition 6.3-1

(Factorization Theorem) Let $X_1, X_2, \ldots, X_n$ denote random variables with joint p.d.f. or p.m.f. $f(x_1, x_2, \ldots, x_n; \theta)$, which depends on the parameter θ. The statistic $Y = u(X_1, X_2, \ldots, X_n)$ is sufficient for θ if and only if

$$f(x_1, x_2, \ldots, x_n; \theta) = \phi[u(x_1, x_2, \ldots, x_n); \theta]h(x_1, x_2, \ldots, x_n),$$

where ϕ depends on $x_1, x_2, \ldots, x_n$ only through $u(x_1, \ldots, x_n)$ and $h(x_1, \ldots, x_n)$ does not depend on θ.

Let us consider several important examples and consequences of this definition. We first note, however, that in all instances in this book the random variables $X_1, X_2, \ldots, X_n$ will be of a random sample and hence their joint p.d.f. or p.m.f. will be of the form

$$f(x_1; \theta)f(x_2; \theta)\cdots f(x_n; \theta).$$

EXAMPLE 6.3-1

Let $X_1, X_2, \ldots, X_n$ denote a random sample from a Poisson distribution with parameter $\lambda > 0$. Then

$$f(x_1; \lambda)f(x_2; \lambda)\cdots f(x_n; \lambda) = \frac{\lambda^{\Sigma x_i}e^{-n\lambda}}{x_1!x_2!\ldots x_n!} = (\lambda^{n\bar{x}}e^{-n\lambda})\left(\frac{1}{x_1!x_2!\ldots x_n!}\right),$$

where $\bar{x} = (1/n)\sum x_i$. Thus, from the factorization theorem (Definition 6.3-1), it is clear that the sample mean $\overline{X}$ is a sufficient statistic for λ. It can easily be shown that the maximum likelihood estimator for λ is also $\overline{X}$; so here the maximum likelihood estimator is a function of the sufficient statistic. ◄

In Example 6.3-1, if we replace $n\bar{x}$ by $\sum x_i$, it is quite obvious that the sum $\sum X_i$ is also a sufficient statistic for λ. This certainly agrees with our intuition because if we know one of the statistics, $\overline{X}$ or $\sum X_i$, we can easily find the other. If we generalize this, we see that if Y is sufficient for a parameter θ, then every single-valued function of Y, not involving θ but with a single-valued inverse, is also a sufficient statistic for θ. Again the reason is that knowing either Y or that function of Y, we know the other. More formally, if $W = v(Y) = v[u(X_1, X_2, \ldots, X_n)]$ is that function and $Y = v^{-1}(W)$ is the single-valued inverse, then the display of the factorization theorem can be written as

$$f(x_1, x_2, \ldots, x_n; \theta) = \phi[v^{-1}\{v[u(x_1, x_2, \ldots, x_n)]\}; \theta] h(x_1, x_2, \ldots, x_n).$$

The first factor of the righthand member of this equation depends on $x_1, x_2, \ldots, x_n$ through $v[u(x_1, x_2, \ldots, x_n)]$, so $W = v[u(X_1, X_2, \ldots, X_n)]$ is a sufficient statistic for θ. We illustrate this fact and the factorization theorem with an underlying distribution of the continuous type.

EXAMPLE 6.3-2 Let $X_1, X_2, \ldots, X_n$ be a random sample from $N(\mu, 1)$, $-\infty < \mu < \infty$. The joint p.d.f. of these random variables is

$$\frac{1}{(2\pi)^{n/2}} \exp\left[-\frac{1}{2}\sum_{i=1}^{n}(x_i - \mu)^2\right]$$

$$= \frac{1}{(2\pi)^{n/2}} \exp\left[-\frac{1}{2}\sum_{i=1}^{n}[(x_i - \bar{x}) + (\bar{x} - \mu)]^2\right]$$

$$= \left\{\exp\left[-\frac{n}{2}(\bar{x} - \mu)^2\right]\right\} \left\{\frac{1}{(2\pi)^{n/2}} \exp\left[-\frac{1}{2}\sum_{i=1}^{n}(x_i - \bar{x})^2\right]\right\}.$$

From the factorization theorem we see that $\bar{X}$ is sufficient for μ. Now $\bar{X}^3$ is also sufficient for μ because knowing $\bar{X}^3$ is equivalent to having knowledge of the value of $\bar{X}$. However, $\bar{X}^2$ does not have this property, and it is not sufficient for μ. ◄

One extremely important consequence of the sufficiency of a statistic Y is that the conditional probability of any given event A in the support of $X_1, X_2, \ldots, X_n$, given $Y = y$, does not depend on θ. This is sometimes used as the definition of sufficiency and is illustrated by the following example.

EXAMPLE 6.3-3 Let $X_1, X_2, \ldots, X_n$ be a random sample from a distribution with p.m.f.

$$f(x; p) = p^x(1 - p)^{1-x}, \qquad x = 0, 1,$$

where the parameter p is between zero and one. We know that

$$Y = X_1 + X_2 + \cdots + X_n$$

is $b(n, p)$ and Y is sufficient for p because the joint p.m.f. of $X_1, X_2, \ldots, X_n$ is

$$p^{x_1}(1 - p)^{1-x_1} \cdots p^{x_n}(1 - p)^{1-x_n} = [p^{\Sigma x_i}(1 - p)^{n-\Sigma x_i}](1),$$

where $\phi(y; p) = p^y(1 - p)^{n-y}$ and $h(x_1, x_2, \ldots, x_n) = 1$. What then is the conditional probability $P(X_1 = x_1, \ldots, X_n = x_n \mid Y = y)$, where $y = 0, 1, \ldots, n-1$, or n? Unless the sum of the nonnegative integers $x_1, x_2, \ldots, x_n$ equals y, this conditional probability is obviously equal to zero, which does not depend on p. Hence it is only interesting to consider the solution when $y = x_1 + \cdots + x_n$. From the definition of conditional probability we have

$$P(X_1 = x_1, \ldots, X_n = x_n \mid Y = y) = \frac{P(X_1 = x_1, \ldots, X_n = x_n)}{P(Y = y)}$$

$$= \frac{p^{x_1}(1-p)^{1-x_1} \cdots p^{x_n}(1-p)^{1-x_n}}{\binom{n}{y} p^y (1-p)^{n-y}}$$

$$= \frac{1}{\binom{n}{y}},$$

where $y = x_1 + \cdots + x_n$. Since y equals the number of ones in the collection $x_1, x_2, \ldots, x_n$, this answer is only the probability of selecting a particular arrangement, namely $x_1, x_2, \ldots, x_n$ of y ones and $n - y$ zeros, and does not depend on the parameter p. That is, given that the sufficient statistic $Y = y$, the conditional probability of $X_1 = x_1, X_2 = x_2, \ldots, X_n = x_n$ does not depend on the parameter p. ◀

It is interesting to observe that the underlying p.d.f. or p.m.f. in Examples 6.3-1, 6.3-2, and 6.3-3 can be written in the exponential form

$$f(x; \theta) = \exp[K(x)p(\theta) + S(x) + q(\theta)],$$

where the support is free of θ. That is, we have, respectively,

$$\frac{e^{-\lambda}\lambda^x}{x!} = \exp\{x \ln \lambda - \ln x! - \lambda\}, \qquad x = 0, 1, 2, \ldots,$$

$$\frac{1}{\sqrt{2\pi}} e^{-(x-\mu)^2/2} = \exp\left\{ x\mu - \frac{x^2}{2} - \frac{\mu^2}{2} - \frac{1}{2}\ln(2\pi) \right\}, \qquad -\infty < x < \infty,$$

and

$$p^x(1-p)^{1-x} = \exp\left\{ x \ln\left(\frac{p}{1-p} \right) + \ln(1-p) \right\}, \qquad x = 0, 1.$$

In each of these examples, the sum $\sum X_i$ of the observations of the random sample was a sufficient statistic for the parameter. This is generalized by Theorem 6.3-1.

Theorem 6.3-1

Let $X_1, X_2, \ldots, X_n$ be a random sample from a distribution with a p.d.f. or p.m.f. of the exponential form

$$f(x; \theta) = \exp[K(x)p(\theta) + S(x) + q(\theta)]$$

on a support free of θ. The statistic $\sum_{i=1}^{n} K(X_i)$ is sufficient for θ.

Proof: The joint p.d.f. (p.m.f.) of $X_1, X_2, \ldots, X_n$ is

$$\exp\left[p(\theta) \sum_{i=1}^{n} K(x_i) + \sum_{i=1}^{n} S(x_i) + nq(\theta) \right]$$

$$= \left\{ \exp\left[p(\theta) \sum_{i=1}^{n} K(x_i) + nq(\theta) \right] \right\} \left\{ \exp\left[\sum_{i=1}^{n} S(x_i) \right] \right\}.$$

In accordance with the factorization theorem, the statistic $\sum_{i=1}^{n} K(X_i)$ is sufficient for θ. ○

In many cases Theorem 6.3-1 permits the student to find a sufficient statistic for the parameter with very little effort, as shown by the following example.

EXAMPLE 6.3-4 Let $X_1, X_2, \ldots, X_n$ be a random sample from an exponential distribution with p.d.f.

$$f(x; \theta) = \frac{1}{\theta} e^{-x/\theta} = \exp\left[x\left(-\frac{1}{\theta}\right) - \ln\theta \right], \qquad 0 < x < \infty,$$

provided $0 < \theta < \infty$. Here $K(x) = x$. Thus $\sum_{i=1}^{n} X_i$ is sufficient for θ; of course, $\overline{X} = \sum_{i=1}^{n} X_i/n$ is also sufficient. ◀

We should also note that if there is a sufficient statistic for the parameter under consideration and if the maximum likelihood estimator of this parameter is unique, then the maximum likelihood estimator is a function of the sufficient statistic. To see this heuristically, consider the following. If a sufficient statistic exists, then the likelihood function is

$$L(\theta) = f(x_1, x_2, \ldots, x_n; \theta) = \phi[u(x_1, x_2, \ldots, x_n); \theta] h(x_1, x_2, \ldots, x_n).$$

Since $h(x_1, x_2, \ldots, x_n)$ does not depend on θ, we maximize $L(\theta)$ by maximizing $\phi[u(x_1, x_2, \ldots, x_n); \theta]$. But ϕ is a function of $x_1, x_2, \ldots, x_n$ only through the statistic $u(x_1, x_2, \ldots, x_n)$. Thus, if there is a unique value of θ that maximizes ϕ, then it must be a function of $u(x_1, x_2, \ldots, x_n)$. That is, $\widehat{\theta}$ is a function of the sufficient statistic $u(X_1, X_2, \ldots, X_n)$. This fact was alluded to in Example 6.3-1, but it could be checked using other examples and exercises.

In many cases, we have two (or more) parameters, say θ_1 and θ_2. All of the preceding concepts can be extended to these situations. For illustration, the definition (factorization theorem) in the case of two parameters becomes the following: If

$$f(x_1, \ldots, x_n; \theta_1, \theta_2) = \phi[u_1(x_1, \ldots, x_n), u_2(x_1, \ldots, x_n); \theta_1, \theta_2] h(x_1, \ldots, x_n),$$

where ϕ depends on $x_1, x_2, \ldots, x_n$ only through $u_1(x_1, \ldots, x_n), u_2(x_1, \ldots, x_n)$, and $h(x_1, x_2, \ldots, x_n)$ does not depend upon θ_1 or θ_2, then $Y_1 = u_1(X_1, X_2, \ldots, X_n)$ and $Y_2 = u_2(X_1, X_2, \ldots, X_n)$ are **joint sufficient statistics** for θ_1 and θ_2.

EXAMPLE 6.3-5   Let $X_1, X_2, \ldots, X_n$ denote a random sample from a normal distribution $N(\theta_1 = \mu, \theta_2 = \sigma^2)$. Then

$$\prod_{i=1}^{n} f(x_i; \theta_1, \theta_2) = \left(\frac{1}{\sqrt{2\pi\theta_2}} \right)^n \exp\left[-\sum_{i=1}^{n} (x_i - \theta_1)^2 \bigg/ 2\theta_2 \right]$$

$$= \exp\left[\left(-\frac{1}{2\theta_2} \right) \sum_{i=1}^{n} x_i^2 + \left(\frac{\theta_1}{\theta_2} \right) \sum_{i=1}^{n} x_i - \frac{n\theta_1^2}{2\theta_2} - n \ln \sqrt{2\pi\theta_2} \right] \cdot (1).$$

Thus

$$Y_1 = \sum_{i=1}^{n} X_i^2 \qquad \text{and} \qquad Y_2 = \sum_{i=1}^{n} X_i$$

are joint sufficient statistics for θ_1 and θ_2. Of course, the single-valued functions of Y_1 and Y_2, namely

$$\overline{X} = \frac{Y_2}{n} \qquad \text{and} \qquad S^2 = \frac{Y_1 - Y_2^2/n}{n-1},$$

are also joint sufficient statistics for θ_1 and θ_2.   ◄

Actually we can see from Definition 6.3-1 and Example 6.3-5 that if we can write the p.d.f. in the exponential form, it is easy to find the joint sufficient statistics. In that example,

$$f(x; \theta_1, \theta_2) = \exp\left(\frac{-1}{2\theta_2} x^2 + \frac{\theta_1}{\theta_2} x - \frac{\theta_1^2}{2\theta_2} - \ln \sqrt{2\pi\theta_2} \right);$$

so

$$Y_1 = \sum_{i=1}^{n} X_i^2 \qquad \text{and} \qquad Y_2 = \sum_{i=1}^{n} X_i$$

are joint sufficient statistics for θ_1 and θ_2. A much more complicated illustration is given if we take a random sample $(X_1, Y_1), (X_2, Y_2), \ldots, (X_n, Y_n)$ from a bivariate normal distribution with parameters $\theta_1 = \mu_X$, $\theta_2 = \mu_Y$, $\theta_3 = \sigma_X^2$, $\theta_4 = \sigma_Y^2$, and $\theta_5 = \rho$. In Exercise 6.3-5 we write the bivariate normal p.d.f. $f(x, y; \theta_1, \theta_2, \theta_3, \theta_4, \theta_5)$ in exponential form and see that $Z_1 = \sum_{i=1}^{n} X_i^2$, $Z_2 = \sum_{i=1}^{n} Y_i^2$, $Z_3 = \sum_{i=1}^{n} X_i Y_i$, $Z_4 = \sum_{i=1}^{n} X_i$, and $Z_5 = \sum_{i=1}^{n} Y_i$ are joint sufficient statistics for $\theta_1, \theta_2, \theta_3, \theta_4$, and θ_5. Of course, the single-valued functions

$$\overline{X} = \frac{Z_4}{n}, \qquad \overline{Y} = \frac{Z_5}{n}, \qquad S_X^2 = \frac{Z_1 - Z_4^2/n}{n-1},$$

$$S_Y^2 = \frac{Z_2 - Z_5^2/n}{n-1}, \qquad R = \frac{(Z_3 - Z_4 Z_5/n)/(n-1)}{S_X S_Y},$$

are also joint sufficient statistics for those parameters.

The important point to stress for cases in which sufficient statistics exist is that once the sufficient statistics are given there is no additional information about the parameters left in the remaining (conditional) distribution. That is, all statistical inferences should be based upon these sufficient statistics. To help convince the reader of this in point estimation, we state and prove the well known **Rao-Blackwell theorem**.

Theorem 6.3-2　Let $X_1, X_2, \ldots, X_n$ be a random sample from a distribution with p.d.f. or p.m.f. $f(x; \theta)$, $\theta \in \Omega$. Let $Y_1 = u_1(X_1, X_2, \ldots, X_n)$ be a sufficient statistic for θ and let $Y_2 = u_2(X_1, X_2, \ldots, X_n)$ be an unbiased estimator of θ, where Y_2 is not a function of Y_1 alone. Then $E(Y_2 \mid y_1) = u(y_1)$ defines a statistic $u(Y_1)$, a function of the sufficient statistic Y_1, which is an unbiased estimator of θ and its variance is less than that of Y_2.

Proof:　Let $g(y_1, y_2; \theta)$ be the joint p.d.f. or p.m.f. of Y_1 and Y_2. Let $g_1(y_1; \theta)$ be the marginal of Y_1 and thus

$$\frac{g(y_1, y_2; \theta)}{g_1(y_1; \theta)} = h(y_2 \mid y_1)$$

is the conditional of Y_2, given $Y_1 = y_1$. This does not depend upon θ since Y_1 is a sufficient statistic for θ. Of course, in the continuous case,

$$u(y_1) = \int_{y_2} y_2 h(y_2 \mid y_1)\, dy_2 = \int_{y_2} y_2 \frac{g(y_1, y_2; \theta)}{g_1(y_1; \theta)}\, dy_2$$

$$E[u(Y_1)] = \int_{y_1} \left(\int_{y_2} y_2 \frac{g(y_1, y_2; \theta)}{g_1(y_1; \theta)}\, dy_2 \right) g_1(y_1; \theta)\, dy_1$$

$$= \int_{y_1} \int_{y_2} y_2\, g(y_1, y_2; \theta)\, dy_2\, dy_1 = \theta$$

because Y_2 is an unbiased estimator of θ. Thus $u(Y_1)$ is also an unbiased estimator of θ.

Consider

$$\mathrm{Var}(Y_2) = E[(Y_2 - \theta)^2] = E[\{Y_2 - u(Y_1) + u(Y_1) - \theta\}^2]$$

$$= E[\{Y_2 - u(Y_1)\}^2] + E[\{u(Y_1) - \theta\}^2] + 2E[\{Y_2 - u(Y_1)\}\{u(Y_1) - \theta\}].$$

However, the latter expression is equal to

$$\int_{y_1} [u(y_1) - \theta] \left\{ \int_{y_2} [y_2 - u(y_1)]\, h(y_2 \mid y_1)\, dy_2 \right\} g(y_1; \theta)\, dy_1 = 0$$

because $u(y_1)$ is the mean $E(Y_2 \mid y_1)$ of Y_2 in the conditional distribution given by $h(y_2 \mid y_1)$. Thus

$$\mathrm{Var}(Y_2) = E[\{Y_2 - u(Y_1)\}^2] + \mathrm{Var}[u(Y_1)].$$

However, $E[\{(Y_2 - u(Y_1)\}^2] \geq 0$ as it is the expected value of a positive expression, and so

$$\text{Var}(Y_2) \geq \text{Var}[u(Y_1)].$$

The importance of this theorem is that it shows that for every other unbiased estimator of θ we can always find an unbiased estimator based on the sufficient statistic which has a smaller variance than the first unbiased estimator. Hence, in that sense, the one based upon the sufficient statistic is better than the first one. More important, we might as well begin our search for an unbiased estimator with smallest variance by considering only those based upon the sufficient statistics. Moreover, in an advanced course we show that if the underlying distribution is described by a p.d.f. or p.m.f. of the exponential form, then, if an unbiased estimator exists, there is only one function of the sufficient statistic that is unbiased. That is, it is unique. (See Hogg, McKean, and Craig, 2005.)

There is one other useful result involving a sufficient statistic Y for a parameter θ, particularly with a p.d.f. of the exponential form. It is that if another statistic Z has a distribution free of θ, then Y and Z are independent. This is the reason $Z = (n-1)S^2$ is independent of $Y = \overline{X}$ when the sample arises from a distribution that is $N(\theta, \sigma^2)$. The sample mean is the sufficient statistic for θ and

$$Z = (n-1)S^2 = \sum_{i=1}^{n}(X_i - \overline{X})^2$$

has a distribution free of θ. To see this we note that its m.g.f., $E(e^{tZ})$, is

$$\int_{-\infty}^{\infty}\int_{-\infty}^{\infty}\cdots\int_{-\infty}^{\infty} \exp\left[t\sum_{i=1}^{n}(x_i - \overline{x})^2\right]$$

$$\times \left(\frac{1}{\sqrt{2\pi}\sigma}\right)^n \exp\left[-\frac{\sum(x_i - \theta)^2}{2\sigma^2}\right]dx_1 dx_2 \ldots dx_n.$$

Changing variables, $x_i - \theta = w_i$, $i = 1, 2, \ldots, n$, this becomes

$$\int_{-\infty}^{\infty}\int_{-\infty}^{\infty}\cdots\int_{-\infty}^{\infty} \exp\left[t\sum_{i=1}^{n}(w_i - \overline{w})^2\right]$$

$$\times \left(\frac{1}{\sqrt{2\pi}\sigma}\right)^n \exp\left[-\frac{\sum w_i^2}{2\sigma^2}\right]dw_1 dw_2 \ldots dw_n,$$

which is free of θ.

An outline of the proof of this result is given by noting

$$\int_{y}[h(z \mid y) - g_2(z)]\, g_1(y; \theta)\, dy = g_2(z) - g_2(z) = 0$$

for all $\theta \in \Omega$. However, because $h(z \mid y)$ is free of θ due to the hypothesis of sufficiency, $h(z \mid y) - g_2(z)$ is free of θ since Z has a distribution free of θ. Due to the fact that $N(\theta, \sigma^2)$ is of the exponential form and thus $Y = \overline{X}$ has a p.d.f. $g_1(y \mid \theta)$ that is called a **complete kernel**, $h(z \mid y) - g_2(z)$ must be equal to zero. That is,

$$h(z \mid y) = g_2(z),$$

which means Z and Y are independent. This proves the independence of $\overline{X}$ and S^2, which was stated in Theorem 5.3-3.

EXAMPLE 6.3-6 Let $X_1, X_2, \ldots, X_n$ be a random sample from a gamma distribution with α (given) and $\theta > 0$. We know that $Y = \Sigma X_i$ is a sufficient statistic for θ since the gamma p.d.f. is of the exponential form. Clearly

$$Z = \frac{\sum_{i=1}^{n} a_i X_i}{\sum_{i=1}^{n} X_i},$$

where not all constants $a_1, a_2, \ldots, a_n$ are equal, has a distribution free of the spread parameter θ because the moment-generating function of Z, namely

$$E(e^{tZ}) = \int_0^\infty \int_0^\infty \cdots \int_0^\infty \frac{1}{[\Gamma(\alpha)]^n \theta^{n\alpha}} (x_1 x_2 \cdots x_n)^{\alpha-1} e^{-\Sigma x_i/\theta} dx_1 dx_2 \ldots dx_n,$$

does not depend upon θ as seen by the transformation $w_i = x_i/\theta$, $i = 1, 2, \ldots, n$. So Y and Z are independent statistics. ◀

This special case of independence of Y and Z concerning one sufficient statistic Y and one parameter θ was first observed by Hogg (1953) and then generalized to several sufficient statistics for more than one parameter by Basu (1955) and is usually called **Basu's theorem**.

Due to these results, sufficient statistics are extremely important and thus future estimation problems are based upon the sufficient statistics, when they exist.

EXERCISES

6.3-1 Let $X_1, X_2, \ldots, X_n$ be a random sample from the distribution with p.d.f. or p.m.m. $f(x; p) = p(1-p)^{x-1}$, $x = 1, 2, 3, \ldots$, where $0 < p < 1$.

(a) Show that $Y = \sum_{i=1}^{n} X_i$ is a sufficient statistic for p.

(b) Find a function of $Y = \sum_{i=1}^{n} X_i$ that is an unbiased estimator of $\theta = 1/p$.

6.3-2 Let $X_1, X_2, \ldots, X_n$ be a random sample from a Poisson distribution with mean $\lambda > 0$. Find the conditional probability $P(X_1 = x_1, \ldots, X_n = x_n | Y = y)$, where $Y = X_1 + \cdots + X_n$ and the nonnegative integers $x_1, x_2, \ldots, x_n$ sum to y, showing that this probability does not depend on λ.

6.3-3 Let $X_1, X_2, \ldots, X_n$ be a random sample from $N(0, \sigma^2)$.

(a) Find a sufficient statistic Y for σ^2.

(b) Show that the maximum likelihood estimator for σ^2 is a function of Y.

(c) Is the maximum likelihood estimator for σ^2 unbiased?

6.3-4 Let $X_1, X_2, \ldots, X_n$ be a random sample from a distribution with p.d.f. $f(x; \theta) = \theta x^{\theta-1}$, $0 < x < 1$, where $0 < \theta$.

(a) Find the sufficient statistic Y for θ.

(b) Show that the maximum likelihood estimator $\widehat{\theta}$ is a function of Y.

(c) Argue that $\widehat{\theta}$ is also sufficient for θ.

6.3-5 Write the bivariate normal p.d.f. $f(x, y; \theta_1, \theta_2, \theta_3, \theta_4, \theta_5)$ in exponential form and show that $Z_1 = \sum_{i=1}^{n} X_i^2$, $Z_2 = \sum_{i=1}^{n} Y_i^2$, $Z_3 = \sum_{i=1}^{n} X_i Y_i$, $Z_4 = \sum_{i=1}^{n} X_i$, and $Z_5 = \sum_{i=1}^{n} Y_i$ are joint sufficient statistics for $\theta_1, \theta_2, \theta_3, \theta_4$, and θ_5.

6.3-6 Let $X_1, X_2, \ldots, X_n$ be a random sample from a gamma distribution with known parameter α and unknown parameter $\theta > 0$.

(a) Show that $Y = \Sigma X_i$ is a sufficient statistic for θ.

(b) Show that the maximum likelihood estimator of θ is a function of the sufficient statistics and it is an unbiased estimator of θ.

6.3-7 Let $X_1, X_2, \ldots, X_n$ be a random sample from a gamma distribution with $\alpha = 1$ and $1/\theta > 0$. Show that $Y = \Sigma X_i$ is a sufficient statistic, Y has a gamma distribution with parameters n and $1/\theta$, and $(n-1)/Y$ is an unbiased estimator of θ.

6.3-8 Let $X_1, X_2, \ldots, X_n$ be a random sample from $N(0, \theta)$, where $\sigma^2 = \theta > 0$ is unknown. Argue that the sufficient statistic $Y = \Sigma X_i^2$ for θ and $Z = \Sigma a_i X_i / \Sigma X_i$ are independent.

HINT: Let $x_i = \theta w_i$, $i = 1, 2, \ldots, n$ in the multivariate integral representing $E[e^{tZ}]$.

6.3-9 Let $X_1, X_2, \ldots, X_n$ be a random sample from $N(\theta_1, \theta_2)$. Show that the sufficient statistics $Y_1 = \overline{X}$ and $Y_2 = S^2$ are independent of the statistic

$$Z = \sum_{i=1}^{n-1} \frac{(X_{i+1} - X_i)^2}{S^2}$$

because Z has a distribution free of θ_1 and θ_2.

HINT: Let $w_i = (x_i - \theta_1)/\sqrt{\theta_2}$, $i = 1, 2, \ldots, n$ in the multivariate integral representing $E[e^{tZ}]$.

6.4 CONFIDENCE INTERVALS FOR MEANS

Given a random sample $X_1, X_2, \ldots, X_n$ from a normal distribution $N(\mu, \sigma^2)$, we shall now consider the closeness of $\overline{X}$, the unbiased estimator of μ, to the unknown mean μ. To do this, we use the error structure (distribution) of $\overline{X}$, namely that $\overline{X}$ is $N(\mu, \sigma^2/n)$, to construct what is called a confidence interval for the unknown parameter μ, when the variance σ^2 is known. For the probability $1 - \alpha$, we can find a number $z_{\alpha/2}$ from Table V in the Appendix such that

$$P\left(-z_{\alpha/2} \leq \frac{\overline{X} - \mu}{\sigma/\sqrt{n}} \leq z_{\alpha/2}\right) = 1 - \alpha.$$

For example, if $1 - \alpha = 0.95$, then $z_{\alpha/2} = z_{0.025} = 1.96$ and if $1 - \alpha = 0.90$, then $z_{\alpha/2} = z_{0.05} = 1.645$. Now recalling that $\sigma > 0$, we see that the following inequalities are equivalent:

$$-z_{\alpha/2} \leq \frac{\overline{X} - \mu}{\sigma/\sqrt{n}} \leq z_{\alpha/2},$$

$$-z_{\alpha/2}\left(\frac{\sigma}{\sqrt{n}}\right) \leq \overline{X} - \mu \leq z_{\alpha/2}\left(\frac{\sigma}{\sqrt{n}}\right),$$

$$-\overline{X} - z_{\alpha/2}\left(\frac{\sigma}{\sqrt{n}}\right) \leq -\mu \leq -\overline{X} + z_{\alpha/2}\left(\frac{\sigma}{\sqrt{n}}\right),$$

$$\overline{X} + z_{\alpha/2}\left(\frac{\sigma}{\sqrt{n}}\right) \geq \mu \geq \overline{X} - z_{\alpha/2}\left(\frac{\sigma}{\sqrt{n}}\right).$$

Thus, since the probability of the first of these is $1 - \alpha$, the probability of the last must also be $1 - \alpha$ because the latter is true if and only if the former is true. That is, we have

$$P\left[\overline{X} - z_{\alpha/2}\left(\frac{\sigma}{\sqrt{n}}\right) \leq \mu \leq \overline{X} + z_{\alpha/2}\left(\frac{\sigma}{\sqrt{n}}\right)\right] = 1 - \alpha.$$

So the probability that the random interval

$$\left[\overline{X} - z_{\alpha/2}\left(\frac{\sigma}{\sqrt{n}}\right), \overline{X} + z_{\alpha/2}\left(\frac{\sigma}{\sqrt{n}}\right)\right]$$

includes the unknown mean μ is $1 - \alpha$.

Once the sample is observed and the sample mean computed to equal $\overline{x}$, the interval $[\overline{x} - z_{\alpha/2}(\sigma/\sqrt{n}), \overline{x} + z_{\alpha/2}(\sigma/\sqrt{n})]$ is a known interval. Since the probability that the random interval covers μ before the sample is drawn is equal to $1 - \alpha$, we now call the computed interval, $\overline{x} \pm z_{\alpha/2}(\sigma/\sqrt{n})$ (for brevity), a $100(1 - \alpha)\%$ **confidence interval** for the unknown mean μ. For illustration, $\overline{x} \pm 1.96(\sigma/\sqrt{n})$ is a 95% confidence interval for μ. The number $100(1 - \alpha)\%$, or equivalently, $1 - \alpha$, is called the **confidence coefficient**.

We see that the confidence interval for μ is centered at the point estimate $\overline{x}$ and is completed by subtracting and adding the quantity $z_{\alpha/2}(\sigma/\sqrt{n})$. Note that as n increases, $z_{\alpha/2}(\sigma/\sqrt{n})$ decreases, resulting in a shorter confidence interval with the same confidence coefficient $1 - \alpha$. A shorter confidence interval indicates that we have more reliance in $\overline{x}$ as an estimate of μ. Statisticians who are not restricted by time, money, effort, or availability of observations can obviously make the confidence interval as short as they like by increasing the sample size n. For a fixed sample size n, the length of the confidence interval can also be shortened by decreasing the confidence coefficient $1 - \alpha$. But if this is done, we achieve a shorter confidence interval by losing some confidence.

EXAMPLE 6.4-1 Let X equal the length of life of a 60-watt light bulb marketed by a certain manufacturer of light bulbs. Assume that the distribution of X is $N(\mu, 1296)$. If a random sample of $n = 27$ bulbs were tested until they burned out, yielding a sample mean of $\overline{x} = 1478$ hours, then a 95% confidence interval for μ is

$$\left[\overline{x} - z_{0.025}\left(\frac{\sigma}{\sqrt{n}}\right), \overline{x} + z_{0.025}\left(\frac{\sigma}{\sqrt{n}}\right)\right]$$

$$= \left[1478 - 1.96\left(\frac{36}{\sqrt{27}}\right), 1478 + 1.96\left(\frac{36}{\sqrt{27}}\right)\right]$$

$$= [1478 - 13.58, 1478 + 13.58]$$
$$= [1464.42, 1491.58].$$ ◄

The next example will help to give a better intuitive feeling for the interpretation of a confidence interval.

EXAMPLE 6.4-2 Let $\bar{x}$ be the observed sample mean of five items of a random sample from the normal distribution $N(\mu, 16)$. A 90% confidence interval for the unknown mean μ is

$$\left[\bar{x} - 1.645\sqrt{\frac{16}{5}}, \bar{x} + 1.645\sqrt{\frac{16}{5}} \right].$$

For a particular sample this interval either does or does not contain the mean μ. However, if many such intervals were calculated, it should be true that about 90% of them contain the mean μ. Fifty random samples of size five from the normal distribution $N(50, 16)$ were simulated on a computer. A 90% confidence interval was calculated for each random sample, as if the mean were unknown. Figure 6.4-1(a) depicts each of these 50 intervals as a line segment. For these 50 intervals, 45 (or 90%) of them contain the mean, $\mu = 50$. In other simulations of 50 confidence intervals, the number of 90% confidence intervals containing the mean could be larger or smaller. (In fact, if W is the random variable that counts the number of 90% confidence intervals that contain the mean, then the distribution of W is $b(50, 0.90)$.) ◄

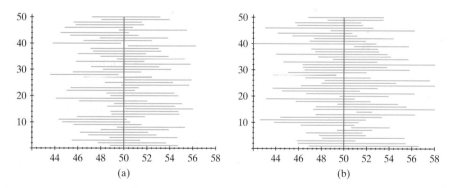

Figure 6.4-1: Confidence intervals using z and t

If we cannot assume that the distribution from which the sample arose is normal, we can still obtain an approximate confidence interval for μ. By the central limit theorem the ratio $(\overline{X} - \mu)/(\sigma/\sqrt{n})$ has, provided that n is large enough, the approximate normal distribution $N(0, 1)$ when the underlying distribution is not normal. In this case

$$P\left(-z_{\alpha/2} \leq \frac{\overline{X} - \mu}{\sigma/\sqrt{n}} \leq z_{\alpha/2} \right) \approx 1 - \alpha,$$

and

$$\left[\overline{x} - z_{\alpha/2}\left(\frac{\sigma}{\sqrt{n}}\right), \overline{x} + z_{\alpha/2}\left(\frac{\sigma}{\sqrt{n}}\right) \right]$$

is an approximate $100(1 - \alpha)\%$ confidence interval for μ.

The closeness of the approximate probability $1 - \alpha$ to the exact probability depends on both the underlying distribution and the sample size. When the underlying distribution is unimodal (has only one mode) and continuous, the approximation is usually quite good for even small n, such as $n = 5$. As the underlying distribution becomes "less normal" (i.e., badly skewed or discrete), a larger sample size might be required to keep a reasonably accurate approximation. But, in almost all cases, an n of at least 30 is usually quite adequate.

EXAMPLE 6.4-3 Let X equal the amount of orange juice (in grams per day) consumed by an American. Suppose it is known that the standard deviation of X is $\sigma = 96$. To estimate the mean μ of X, an orange growers' association took a random sample of $n = 576$ Americans and found that they consumed, on the average, $\overline{x} = 133$ grams of orange juice per day. Thus an approximate 90% confidence interval for μ is

$$133 \pm 1.645\left(\frac{96}{\sqrt{576}}\right) \qquad \text{or} \qquad [133 - 6.58, 133 + 6.58] = [126.42, 139.58].$$

◀

If σ^2 is unknown and the sample size n is 30 or greater, we shall use the fact that the ratio $(\overline{X} - \mu)/(S/\sqrt{n})$ has an approximate normal distribution $N(0, 1)$. This statement is true whether or not the underlying distribution is normal. However, if the underlying distribution is badly skewed or contaminated with occasional outliers, most statisticians would prefer to have a larger sample size, say 50 or more, and even that might not produce good results. After this next example, we consider what to do when n is small.

EXAMPLE 6.4-4 Lake Macatawa, an inlet lake on the east side of Lake Michigan, is divided into an east basin and a west basin. To measure the effect on the lake of salting city streets in the winter, students took 32 samples of water from the west basin and measured the amount of sodium in parts per million in order to make a statistical inference about the unknown mean μ. They obtained the following data:

13.0	18.5	16.4	14.8	19.4	17.3	23.2	24.9
20.8	19.3	18.8	23.1	15.2	19.9	19.1	18.1
25.1	16.8	20.4	17.4	25.2	23.1	15.3	19.4
16.0	21.7	15.2	21.3	21.5	16.8	15.6	17.6

For these data $\bar{x} = 19.07$ and $s^2 = 10.60$. Thus an approximate 95% confidence interval for μ is

$$\bar{x} \pm 1.96\left(\frac{s}{\sqrt{n}}\right) \qquad \text{or} \qquad 19.07 \pm 1.96\sqrt{\frac{10.60}{32}} \qquad \text{or} \qquad [17.94, 20.20].$$

◀

So we have found a confidence interval for the mean μ of a normal distribution, assuming that the value of the standard deviation σ is known and when σ is unknown but the sample size is large. However, in many applications, the sample sizes are small and we do not know the value of the standard deviation, although in some cases we might have a very good idea about its value. For illustration, a manufacturer of light bulbs probably has a good notion from past experience of the value of the standard deviation of the length of life of different types of light bulbs. But certainly, most of the time, the investigator will not have any more idea about the standard deviation than about the mean–and frequently less. Let us consider how to proceed under these circumstances.

If the random sample arises from a normal distribution, we use the fact that

$$T = \frac{\dfrac{\overline{X} - \mu}{\sigma/\sqrt{n}}}{\sqrt{\dfrac{(n-1)S^2}{\sigma^2}\Bigg/(n-1)}} = \frac{\overline{X} - \mu}{S/\sqrt{n}}$$

has a t distribution with $r = n - 1$ degrees of freedom (see Theorem 5.3-3 and Example 5.3-5), where S^2 is the usual unbiased estimator of σ^2. Select $t_{\alpha/2}(n{-}1)$ so that $P[T \geq t_{\alpha/2}(n{-}1)] = \alpha/2$ (see Table VI in the Appendix). Then

$$1 - \alpha = P\left[-t_{\alpha/2}(n{-}1) \leq \frac{\overline{X} - \mu}{S/\sqrt{n}} \leq t_{\alpha/2}(n{-}1)\right]$$

$$= P\left[-t_{\alpha/2}(n{-}1)\left(\frac{S}{\sqrt{n}}\right) \leq \overline{X} - \mu \leq t_{\alpha/2}(n{-}1)\left(\frac{S}{\sqrt{n}}\right)\right]$$

$$= P\left[-\overline{X} - t_{\alpha/2}(n{-}1)\left(\frac{S}{\sqrt{n}}\right) \leq -\mu \leq -\overline{X} + t_{\alpha/2}(n{-}1)\left(\frac{S}{\sqrt{n}}\right)\right]$$

$$= P\left[\overline{X} - t_{\alpha/2}(n{-}1)\left(\frac{S}{\sqrt{n}}\right) \leq \mu \leq \overline{X} + t_{\alpha/2}(n{-}1)\left(\frac{S}{\sqrt{n}}\right)\right].$$

Thus the observations of a random sample provide $\bar{x}$ and s^2 and

$$\left[\bar{x} - t_{\alpha/2}(n{-}1)\left(\frac{s}{\sqrt{n}}\right), \bar{x} + t_{\alpha/2}(n{-}1)\left(\frac{s}{\sqrt{n}}\right)\right]$$

is a $100(1 - \alpha)\%$ confidence interval for μ.

EXAMPLE 6.4-5 Let X equal the amount of butterfat in pounds produced by a typical cow during a 305-day milk production period between her first and second calves. Assume that the distribution of X is $N(\mu, \sigma^2)$. To estimate μ a farmer measured the butterfat production for $n = 20$ cows yielding the following data:

481	537	513	583	453	510	570	500	457	555
618	327	350	643	499	421	505	637	599	392

For these data, $\bar{x} = 507.50$ and $s = 89.75$. Thus a point estimate of μ is $\bar{x} = 507.50$. Since $t_{0.05}(19) = 1.729$, a 90% confidence interval for μ is

$$507.50 \pm 1.729 \left(\frac{89.75}{\sqrt{20}} \right)$$

$$507.50 \pm 34.70, \qquad \text{or, equivalently,} \qquad [472.80, 542.20]. \qquad \blacktriangleleft$$

Let T have a t distribution with $n - 1$ degrees of freedom. Then $t_{\alpha/2}(n-1) > z_{\alpha/2}$. Consequently, we would expect the interval $\bar{x} \pm z_{\alpha/2}(\sigma/\sqrt{n})$ to be shorter than the interval $\bar{x} \pm t_{\alpha/2}(n-1)(s/\sqrt{n})$. After all, we have more information, namely the value of σ, in constructing the first interval. However, the length of the second interval is very much dependent on the value of s. If the observed s is smaller than σ, a shorter confidence interval could result by the second procedure. But on the average, $\bar{x} \pm z_{\alpha/2}(\sigma/\sqrt{n})$ is the shorter of the two confidence intervals (Exercise 6.4-16).

EXAMPLE 6.4-6 In Example 6.4-2, 50 confidence intervals were simulated for the mean of a normal distribution assuming that the variance was known. For those same data a 90% confidence interval was calculated for μ using $\bar{x} \pm 2.132(s/\sqrt{5})$. For those particular 50 intervals, 46 contained the mean $\mu = 50$. These are depicted in Figure 6.4-1(b). Note the different lengths of these intervals. Some are longer and some are shorter than the corresponding z intervals. The average length of these intervals is 7.137, which is quite close to the expected length of such an interval which is 7.169. (See Exercise 6.4-16.) The length of the intervals that use z and $\sigma = 4$ is 5.885. $\qquad \blacktriangleleft$

If we are not able to assume that the underlying distribution is normal but μ and σ are both unknown, approximate confidence intervals for μ can still be constructed using

$$T = \frac{\bar{X} - \mu}{S/\sqrt{n}},$$

which now only has an approximate t distribution. Generally, this approximation is quite good for many nonnormal distributions (i.e., it is robust), in particular, if the underlying distribution is symmetric, unimodal, and of the continuous type. However, if the distribution is highly skewed, there is great danger using this approximation. In such a situation, it would be safer to use certain nonparametric methods for finding a confidence interval for the median of the distribution, one of which is given in Section 6.10*.

There is one other aspect of confidence intervals that should be mentioned. So far we have created only what are called **two-sided confidence intervals** for the mean μ. Sometimes it happens that you might want only a lower (or upper) bound on μ. We proceed as follows.

Say $\overline{X}$ is the mean of a random sample of size n from the normal distribution $N(\mu, \sigma^2)$, where say for the moment that σ^2 is known. Then

$$P\left(\frac{\overline{X} - \mu}{\sigma/\sqrt{n}} \le z_\alpha\right) = 1 - \alpha,$$

or equivalently,

$$P\left[\overline{X} - z_\alpha\left(\frac{\sigma}{\sqrt{n}}\right) \le \mu\right] = 1 - \alpha.$$

Once $\overline{X}$ is observed to be equal to $\overline{x}$, then $[\overline{x} - z_\alpha(\sigma/\sqrt{n}), \infty)$ is a $100(1 - \alpha)\%$ **one-sided confidence interval** for μ. That is, with the confidence coefficient of $1 - \alpha$, $\overline{x} - z_\alpha(\sigma/\sqrt{n})$ is a lower bound for μ. Similarly, $(-\infty, \overline{x} + z_\alpha(\sigma/\sqrt{n}))$ is a one-sided confidence interval for μ and $\overline{x} + z_\alpha(\sigma/\sqrt{n})$ provides an upper bound for μ with confidence coefficient $1 - \alpha$.

When σ is unknown, we would use $T = (\overline{X} - \mu)/(S/\sqrt{n})$ to find the corresponding lower or upper bounds for μ, namely $\overline{x} - t_\alpha(n-1)(s/\sqrt{n})$ and $\overline{x} + t_\alpha(n-1)(s/\sqrt{n})$.

EXERCISES

6.4-1 A random sample of size 16 from the normal distribution $N(\mu, 25)$ yielded $\overline{x} = 73.8$. Find a 95% confidence interval for μ.

6.4-2 A random sample of size 8 from $N(\mu, 72)$ yielded $\overline{x} = 85$. Find the following confidence intervals for μ.

(a) 99%.   **(b)** 95%.   **(c)** 90%.   **(d)** 80%.

6.4-3 A pet store sells gerbil food in "2-pound" bags that are weighed on the platform of an old 25-pound scale. Suppose it is known that the standard deviation of weights is $\sigma = 0.12$ pound. If a sample of $n = 16$ bags of gerbil food were weighed carefully in a laboratory and the average weight was $\overline{x} = 2.09$ pounds, find an approximate 95% confidence interval for μ, the mean weight of gerbil food in the "2-pound" bags sold by the pet store.

6.4-4 Let X equal the weight in grams of a "52-gram" snack pack of candies. Assume that the distribution of X is $N(\mu, 4)$. A random sample of $n = 10$ observations of X yielded the following data:

$$\begin{array}{ccccc}
55.95 & 56.54 & 57.58 & 55.13 & 57.48 \\
56.06 & 59.93 & 58.30 & 52.57 & 58.46
\end{array}$$

(a) Give a point estimate for μ.
(b) Find the endpoints for a 95% confidence interval for μ.
(c) Based on these very limited data, what is the probability that an individual snack pack selected at random is filled with less than 52 grams of candy?

6.4-5 As a clue to the amount of organic waste in Lake Macatawa (see Example 6.4-4), a count was made of the number of bacteria colonies in 100 milliliters of water. The number of colonies, in hundreds, for $n = 30$ samples of water from the east basin yielded

93	140	8	120	3	120	33	70	91	61
7	100	19	98	110	23	14	94	57	9
66	53	28	76	58	9	73	49	37	92

Find an approximate 90% confidence interval for the mean number, say μ_E, of colonies in 100 milliliters of water in the east basin.

6.4-6 To determine whether bacteria count was lower in the west basin of Lake Macatawa than in the east basin, $n = 37$ samples of water were taken from the west basin, and the number of bacteria colonies in 100 milliliters of water was counted. The sample characteristics were $\bar{x} = 11.95$ and $s = 11.80$, measured in hundreds of colonies. Find the approximate 95% confidence interval for the mean number of colonies, say μ_W, in 100 milliliters of water in the west basin.

6.4-7 Thirteen tons of cheese is stored in some old gypsum mines, including "22-pound" wheels (label weight). A random sample of $n = 9$ of these wheels yielded the following weights in pounds:

$$21.50 \quad 18.95 \quad 18.55 \quad 19.40 \quad 19.15$$
$$22.35 \quad 22.90 \quad 22.20 \quad 23.10$$

Assuming that the distribution of the weights of the wheels of cheese is $N(\mu, \sigma^2)$, find a 95% confidence interval for μ.

6.4-8 Assume that the yield per acre for a particular variety of soybeans is $N(\mu, \sigma^2)$. For a random sample of $n = 5$ plots, the yields in bushels per acre were 37.4, 48.8, 46.9, 55.0, and 44.0.

(a) Give a point estimate for μ.

(b) Find a 90% confidence interval for μ.

6.4-9 In a study of maximal aerobic capacity (*Journal of Applied Physiology 65, 6 (December 1988), p. 2696*), 12 women were used as subjects, and one measurement that was made was blood plasma volume. The following data give their blood plasma volumes in liters:

$$3.15 \quad 2.99 \quad 2.77 \quad 3.12 \quad 2.45 \quad 3.85$$
$$2.99 \quad 3.87 \quad 4.06 \quad 2.94 \quad 3.53 \quad 3.20$$

Assume that these are observations of a normally distributed random variable X that has mean μ and standard deviation σ.

(a) Give the value of a point estimate of μ.

(b) Determine a point estimates of σ^2 and σ.

(c) Find a 90% confidence interval for μ.

6.4-10 During the Friday night shift, $n = 28$ mints were selected at random from a production line and weighed. They had an average weight of $\bar{x} = 21.45$ grams and $s = 0.31$ grams. Give the lower endpoint of a 90% one-sided confidence interval for μ, the mean weight of all the mints.

6.4-11 A farm grows grapes for jelly. The following data are measurements of sugar in the grapes of a sample taken from each of 30 truckloads:

16.0	15.2	12.0	16.9	14.4	16.3	15.6	12.9	15.3	15.1
15.8	15.5	12.5	14.5	14.9	15.1	16.0	12.5	14.3	15.4
15.4	13.0	12.6	14.9	15.1	15.3	12.4	17.2	14.7	14.8

Assume that these are observations of a random variable X that has mean μ and standard deviation σ.

(a) Give point estimates of μ and σ.

(b) Find an approximate 90% confidence interval for μ.

6.4-12 A leakage test was conducted to determine the effectiveness of a seal designed to keep the inside of a plug air tight. An air needle was inserted in the plug and this was placed under water. The pressure was then increased until leakage was observed. Let X equal the pressure in pounds per square inch. Assume that the distribution of X is $N(\mu, \sigma^2)$. Using the following $n = 10$ observations of X,

$$3.1 \quad 3.3 \quad 4.5 \quad 2.8 \quad 3.5 \quad 3.5 \quad 3.7 \quad 4.2 \quad 3.9 \quad 3.3$$

(a) Find a point estimate of μ.

(b) Find a point estimate of σ.

(c) Find a 95% one-sided confidence interval for μ that provides an upper bound for μ.

6.4-13 When researching ground water it is often important to know the characteristics of the soil at a certain site. For example, let X equal the diameter of an individual grain of soil and assume that the distribution of X is $N(\mu, \sigma^2)$. (See Example 3.1-14, where this assumption is discussed.) We again give the diameters of $n = 30$ individual grains.

1.24	1.36	1.28	1.31	1.35	1.20	1.39	1.35	1.41	1.31
1.28	1.26	1.37	1.49	1.32	1.40	1.33	1.28	1.25	1.39
1.38	1.34	1.40	1.27	1.33	1.36	1.43	1.33	1.29	1.34

Find a 90% confidence interval for μ.

6.4-14 In nuclear physics detectors are often used to measure the energy of a particle. To calibrate a detector, particles of known energy are directed into a detector. The values of signals from 15 different detectors, for the same energy, are

260	216	259	206	265	284	291	229
232	250	225	242	240	252	236	

(a) Find a 95% confidence interval for μ, assuming that these are observations of a $N(\mu, \sigma^2)$ distribution.

(b) Construct a box-and-whisker diagram of these data.

(c) Are these detectors doing a good job or a poor job of putting out the same signal for the same input energy?

6.4-15 Students took $n = 35$ samples of water from the east basin of Lake Macatawa (see Example 6.4-4) and measured the amount of sodium in parts per million. For their data they calculated $\bar{x} = 24.11$ and $s^2 = 24.44$. Find an approximate 90% confidence interval for μ, the mean of the amount of sodium in parts per million.

6.4-16 Let $X_1, X_2, \ldots, X_n$ be a random sample of size n from the normal distribution $N(\mu, \sigma^2)$. Calculate the expected length of a 95% confidence interval for μ assuming that $n = 5$ and the variance is

(a) known,

(b) unknown.

HINT: To find $E(S)$, determine $E[\sqrt{(n-1)S^2/\sigma^2}]$, recalling $(n-1)S^2/\sigma^2$ is $\chi^2(n-1)$. Also see Exercise 6.2-14.

6.4-17 Let X equal the weight (in pounds) of a "12-ounce" can of buttermilk biscuits. Assume that the distribution of X is $N(\mu, \sigma^2)$. Use the following 18 net weights to find a 95% one-sided confidence interval for μ that provides a lower bound for μ:

0.75	0.78	0.77	0.75	0.77	0.76	0.78	0.76	0.78
0.78	0.75	0.74	0.78	0.75	0.74	0.79	0.75	0.77

6.4-18 A manufacturer of soap powder packages the soap in "6-pound" boxes. To check the filling machine, they took a sample of $n = 1219$ boxes and weighed them. Given that $\bar{x} = 6.05$ pounds and $s = 0.02$ pounds, give the endpoints for a 99% confidence interval for μ, the mean weight of the boxes of soap filled by this machine.

6.4-19 A study was conducted to measure the amount of cervical spine movement induced by different methods of gaining access to the mouth and nose to begin resuscitation of a football player who is wearing a helmet and the time it takes to complete each method. One method is using a manual screw driver to remove the side clips holding the face mask in place and then flipping the mask up. Twelve measured times in seconds for the manual screw driver are

$$33.8 \quad 31.6 \quad 28.5 \quad 29.9 \quad 29.8 \quad 26.0 \quad 35.7 \quad 27.2 \quad 29.1 \quad 32.1 \quad 26.1 \quad 24.1$$

Assume that these are independent observations of a normally distributed random variable that is $N(\mu, \sigma^2)$.

(a) Find point estimates of μ and σ.

(b) Find a 95% one-sided confidence interval for μ that provides an upper bound for μ.

(c) Does the assumption of normality seem to be justified? Why?

6.4-20 Let X equal the weight of an unbreaded $2.00 fish fry at a local restaurant in East Greenbush, New York. Assume that the distribution of X is $N(\mu, \sigma^2)$. A random sample of $n = 24$ weights (in ounces) were

4.4	3.8	5.1	4.6	4.5	4.5	4.8	4.1
3.9	4.2	4.4	4.9	5.0	4.3	4.4	3.6
5.2	4.8	4.4	4.6	4.6	5.0	4.0	4.5

(a) Find point estimates of μ, σ^2, and σ.

(b) Find a 95% one-sided confidence interval for μ that gives a lower bound for μ.

(c) Does the assumption that these weights are normal seem to be justified? Why?

6.4-21 An interior automotive supplier places several electrical wires in a harness. A pull test measures the force required to pull spliced wires apart. A customer requires that each wire that is spliced into the harness must withstand a pull force of 20 pounds. Let X equal the pull force required to pull 20 gauge wires apart. Assume that the distribution of X is $N(\mu, \sigma^2)$. The following data give 20 observations of X.

28.8	24.4	30.1	25.6	26.4	23.9	22.1	22.5	27.6	28.1
20.8	27.7	24.4	25.1	24.6	26.3	28.2	22.2	26.3	24.4

(a) Find point estimates for μ and σ.

(b) Find a 99% one-sided confidence interval for μ that provides a lower bound for μ.

6.5 CONFIDENCE INTERVALS FOR DIFFERENCE OF TWO MEANS

Suppose that we are interested in comparing the means of two normal distributions. Let $X_1, X_2, \ldots, X_n$ and $Y_1, Y_2, \ldots, Y_m$ be, respectively, two independent random samples of sizes n and m from the two normal distributions $N(\mu_X, \sigma_X^2)$ and $N(\mu_Y, \sigma_Y^2)$. Suppose, for now, that σ_X^2 and σ_Y^2 are known. The random samples are independent and thus the respective sample means $\overline{X}$ and $\overline{Y}$ are also with distributions $N(\mu_X, \sigma_X^2/n)$ and $N(\mu_Y, \sigma_Y^2/m)$. Thus we know that the distribution of $W = \overline{X} - \overline{Y}$ is $N(\mu_X - \mu_Y, \sigma_X^2/n + \sigma_Y^2/m)$ and

$$P\left(-z_{\alpha/2} \leq \frac{(\overline{X} - \overline{Y}) - (\mu_X - \mu_Y)}{\sqrt{\sigma_X^2/n + \sigma_Y^2/m}} \leq z_{\alpha/2}\right) = 1 - \alpha,$$

which can be rewritten as

$$P[(\overline{X} - \overline{Y}) - z_{\alpha/2}\sigma_W \leq \mu_X - \mu_Y \leq (\overline{X} - \overline{Y}) + z_{\alpha/2}\sigma_W] = 1 - \alpha,$$

where $\sigma_W = \sqrt{\sigma_X^2/n + \sigma_Y^2/m}$ is the standard deviation of $\overline{X} - \overline{Y}$. Once the experiments have been performed and the means $\overline{x}$ and $\overline{y}$ computed, then

$$[\overline{x} - \overline{y} - z_{\alpha/2}\sigma_W, \; \overline{x} - \overline{y} + z_{\alpha/2}\sigma_W]$$

or, equivalently, $\overline{x} - \overline{y} \pm z_{\alpha/2}\sigma_W$ provides a $100(1 - \alpha)\%$ confidence interval for $\mu_X - \mu_Y$. Note that this interval is centered at the point estimate $\overline{x} - \overline{y}$ of $\mu_X - \mu_Y$ and completed by subtracting and adding the product of $z_{\alpha/2}$ and the standard deviation of the point estimator.

EXAMPLE 6.5-1   In the preceding discussion, let $n = 15$, $m = 8$, $\overline{x} = 70.1$, $\overline{y} = 75.3$, $\sigma_X^2 = 60$, $\sigma_Y^2 = 40$, and $1 - \alpha = 0.90$. Thus $1 - \alpha/2 = 0.95 = \Phi(1.645)$. Hence

$$1.645\sigma_W = 1.645\sqrt{\frac{60}{15} + \frac{40}{8}} = 4.935,$$

and, since $\overline{x} - \overline{y} = -5.2$, we have that

$$[-5.2 - 4.935, -5.2 + 4.935] = [-10.135, -0.265]$$

is a 90% confidence interval for $\mu_X - \mu_Y$. Since the confidence interval does not include zero, we suspect that μ_Y is greater than μ_X. ◀

If the sample sizes are large and σ_X and σ_Y are unknown, we can replace σ_X^2 and σ_Y^2 with s_X^2 and s_Y^2, where s_X^2 and s_Y^2 are the values of the respective unbiased estimates of the variances. This means that

$$\overline{x} - \overline{y} \pm z_{\alpha/2}\sqrt{\frac{s_X^2}{n} + \frac{s_Y^2}{m}}$$

serves as an approximate $100(1 - \alpha)\%$ confidence interval for $\mu_X - \mu_Y$.

Now consider the problem of constructing confidence intervals for the difference of the means of two normal distributions when the variances are unknown but the sample sizes are small. Let $X_1, X_2, \ldots, X_n$ and $Y_1, Y_2, \ldots, Y_m$ be two independent random samples from the distributions $N(\mu_X, \sigma_X^2)$ and $N(\mu_Y, \sigma_Y^2)$, respectively. If the sample sizes are not large (say considerably smaller than 30), this problem can be a difficult one. However, even in these cases, if we can assume common, but unknown, variances, say $\sigma_X^2 = \sigma_Y^2 = \sigma^2$, there is a way out of our difficulty.

We know that

$$Z = \frac{\overline{X} - \overline{Y} - (\mu_X - \mu_Y)}{\sqrt{\sigma^2/n + \sigma^2/m}}$$

is $N(0, 1)$. Moreover, since the random samples are independent,

$$U = \frac{(n - 1)S_X^2}{\sigma^2} + \frac{(m - 1)S_Y^2}{\sigma^2}$$

is the sum of two independent chi-square random variables; and thus the distribution of U is $\chi^2(n+m-2)$. In addition, the independence of the sample means and sample variances implies that Z and U are independent. According to the definition of a T random variable,

$$T = \frac{Z}{\sqrt{U/(n+m-2)}}$$

has a t distribution with $n+m-2$ degrees of freedom. That is,

$$T = \frac{\dfrac{\overline{X} - \overline{Y} - (\mu_X - \mu_Y)}{\sqrt{\sigma^2/n + \sigma^2/m}}}{\sqrt{\left[\dfrac{(n-1)S_X^2}{\sigma^2} + \dfrac{(m-1)S_Y^2}{\sigma^2}\right] / (n+m-2)}}$$

$$= \frac{\overline{X} - \overline{Y} - (\mu_X - \mu_Y)}{\sqrt{\left[\dfrac{(n-1)S_X^2 + (m-1)S_Y^2}{n+m-2}\right]\left[\dfrac{1}{n} + \dfrac{1}{m}\right]}}$$

has a t distribution with $r = n+m-2$ degrees of freedom. Thus, with $t_0 = t_{\alpha/2}(n+m-2)$,

$$P(-t_0 \le T \le t_0) = 1 - \alpha.$$

Solving the inequality for $\mu_X - \mu_Y$ yields

$$P\left(\overline{X} - \overline{Y} - t_0 S_P \sqrt{\frac{1}{n} + \frac{1}{m}} \le \mu_X - \mu_Y \le \overline{X} - \overline{Y} + t_0 S_P \sqrt{\frac{1}{n} + \frac{1}{m}}\right),$$

where the pooled estimator of the common standard deviation is

$$S_P = \sqrt{\frac{(n-1)S_X^2 + (m-1)S_Y^2}{n+m-2}}.$$

If $\overline{x}$, $\overline{y}$, and s_p are the observed values of $\overline{X}$, $\overline{Y}$, and S_P, then

$$\left[\overline{x} - \overline{y} - t_0 s_P \sqrt{\frac{1}{n} + \frac{1}{m}}, \overline{x} - \overline{y} + t_0 s_P \sqrt{\frac{1}{n} + \frac{1}{m}}\right]$$

is a $100(1-\alpha)\%$ confidence interval for $\mu_X - \mu_Y$.

EXAMPLE 6.5-2 Suppose that scores on a standardized test in mathematics taken by students from large and small high schools are $N(\mu_X, \sigma^2)$ and $N(\mu_Y, \sigma^2)$, respectively, where σ^2 is unknown. If a random sample of $n = 9$ students from large high schools yielded $\bar{x} = 81.31$, $s_x^2 = 60.76$ and a random sample of $m = 15$ students from small high schools yielded $\bar{y} = 78.61$, $s_y^2 = 48.24$, the endpoints for a 95% confidence interval for $\mu_X - \mu_Y$ are given by

$$81.31 - 78.61 \pm 2.074 \sqrt{\frac{8(60.76) + 14(48.24)}{22}} \sqrt{\frac{1}{9} + \frac{1}{15}}$$

because $t_{0.025}(22) = 2.074$. The 95% confidence interval is $[-3.65, 9.05]$. ◄

Remark The assumption of equal variances, namely $\sigma_X^2 = \sigma_Y^2$, can be modified somewhat so that we are still able to find a confidence interval for $\mu_X - \mu_Y$. That is, if we know the ratio σ_X^2/σ_Y^2 of the variances, we can still make this type of statistical inference using a random variable with a t distribution (see Exercise 6.5-10). However, if we do not know the ratio of the variances and yet suspect that the unknown σ_X^2 and σ_Y^2 differ by a great deal, what do we do? It is safest to return to

$$\frac{\overline{X} - \overline{Y} - (\mu_X - \mu_Y)}{\sqrt{\sigma_X^2/n + \sigma_Y^2/m}}$$

for the inference about $\mu_X - \mu_Y$ but replacing σ_X^2 and σ_Y^2 by their respective estimators S_X^2 and S_Y^2. That is, consider

$$W = \frac{\overline{X} - \overline{Y} - (\mu_X - \mu_Y)}{\sqrt{S_X^2/n + S_Y^2/m}}.$$

What is the distribution of W? As before, we note that if n and m are large enough and the underlying distributions are close to normal (or at least not badly skewed), W has an approximate normal distribution and a confidence interval for $\mu_X - \mu_Y$ can be found by considering

$$P(-z_{\alpha/2} \leq W \leq z_{\alpha/2}) \approx 1 - \alpha.$$

However, if n and m are smaller, then Welch has proposed a Student's t distribution as the approximating one for W. Welch's proposal was later modified by Aspin. [See A. A. Aspin, "Tables for Use in Comparisons Whose Accuracy Involves Two Variances, Separately Estimated," *Biometrika*, **36** (1949), pp. 290–296, with an appendix by B. L. Welch in which he makes the suggestion used here.] The approximating Student's t distribution has r degrees of freedom, where

$$\frac{1}{r} = \frac{c^2}{n-1} + \frac{(1-c)^2}{m-1} \qquad \text{and} \qquad c = \frac{s_x^2/n}{s_x^2/n + s_y^2/m}.$$

An equivalent formula for r is

$$r = \frac{\left(\dfrac{s_x^2}{n} + \dfrac{s_y^2}{m}\right)^2}{\dfrac{1}{n-1}\left(\dfrac{s_x^2}{n}\right)^2 + \dfrac{1}{m-1}\left(\dfrac{s_y^2}{m}\right)^2}.$$

In particular, the assignment of r by this rule provides protection in the case in which the smaller sample size is associated with the larger variance by greatly reducing the number of degrees of freedom from the usual $n+m-2$. Of course, this increases the value of $t_{\alpha/2}$. If r is not an integer, then use the greatest integer in r; that is, $\lfloor r \rfloor$, the "floor" or greatest integer in r is the number of degrees of freedom associated with the approximating Student's t distribution. An approximate $100(1-\alpha)$ percent confidence interval for $\mu_X - \mu_Y$ is given by

$$\bar{x} - \bar{y} \pm t_{\alpha/2}(r)\sqrt{\frac{s_x^2}{n} + \frac{s_y^2}{m}}. \qquad \blacksquare$$

In some applications, two measurements, say X and Y, are taken on the same subject. In these cases, X and Y are dependent random variables. Many times these are "before" and "after" measurements, such as weight before and after participating in a diet and exercise program. To compare the means of X and Y, it is not permissible to use the t statistics and confidence intervals that we just developed, because there X and Y are independent. Instead we proceed as follows.

Let $(X_1, Y_1), (X_2, Y_2), \ldots, (X_n, Y_n)$ be n pairs of dependent measurements. Let $D_i = X_i - Y_i$, $i = 1, 2, \ldots, n$. Suppose that $D_1, D_2, \ldots, D_n$ could be thought of as a random sample from $N(\mu_D, \sigma_D^2)$, where μ_D and σ_D are the mean and standard deviation of each difference. To form a confidence interval for $\mu_X - \mu_Y$, use

$$T = \frac{\bar{D} - \mu_D}{S_D/\sqrt{n}},$$

where $\bar{D}$ and S_D are the sample mean and sample standard deviation of the n differences. Thus T is a t statistic with $n-1$ degrees of freedom. The endpoints for a $100(1-\alpha)\%$ confidence interval for $\mu_D = \mu_X - \mu_Y$ are then

$$\bar{d} \pm t_{\alpha/2}(n-1)\frac{s_d}{\sqrt{n}},$$

where $\bar{d}$ and s_d are the observed mean and standard deviation of the sample. Of course, this is the same as the confidence interval for a single mean in the last section.

EXAMPLE 6.5-3   An experiment was conducted to compare people's reaction times to a red light versus a green light. When signaled with either the red or the green light, the subject was asked to hit a switch to turn off the light. When the switch was hit, a clock was turned off and the reaction time in seconds was recorded. The following results give the reaction times for eight subjects.

Subject	Red (X)	Green (Y)	$D = X - Y$
1	0.30	0.43	−0.13
2	0.23	0.32	−0.09
3	0.41	0.58	−0.17
4	0.53	0.46	0.07
5	0.24	0.27	−0.03
6	0.36	0.41	−0.05
7	0.38	0.38	0.00
8	0.51	0.61	−0.10

For these data, $\bar{d} = -0.0625$ and $s_d = 0.0765$. To form a 95% confidence interval for $\mu_D = \mu_X - \mu_Y$, we find from Appendix Table VI that $t_{0.025}(7) = 2.365$. Thus the endpoints for the confidence interval are

$$-0.0625 \pm 2.365 \frac{0.07675}{\sqrt{8}} \qquad \text{or} \qquad [-0.1265, 0.0015].$$

In this very limited data set, zero is included in the confidence interval but is close to the endpoint 0.0015. We suspect that if more data were taken, zero might not be included in the confidence interval. Accordingly, if this actually happens, it would seem that people react faster to a red light.   ◀

Of course, we can find one-sided confidence intervals for the difference of the means, $\mu_X - \mu_Y$. Suppose that we believe that we have changed some characteristic of the X distribution and created a Y distribution such that we think $\mu_X > \mu_Y$. Let us find a one-sided 95% confidence interval which is a lower bound for $\mu_X - \mu_Y$. Say this lower bound is greater than zero. Then we would feel 95% confident that the mean μ_X is larger than the mean μ_Y. That is, the change that was made seemed to decrease the mean; this would be good in some cases, such as in golf or racing. In other cases, in which we hope the change would be such that $\mu_X < \mu_Y$ we would find a one-sided confidence interval which is an upper bound for $\mu_X - \mu_Y$, and we would hope that it would be less than zero. These ideas are illustrated in Exercises 6.5-5, 6.5-12, and 6.5-13.

EXERCISES

6.5-1 The length of life of brand X light bulbs is assumed to be $N(\mu_X, 784)$. The length of life of brand Y light bulbs is assumed to be $N(\mu_Y, 627)$ and independent of that of X. If a random sample of $n = 56$ brand X light bulbs yielded a mean of $\bar{x} = 937.4$ hours and a random sample of size $m = 57$ brand Y light bulbs yielded a mean of $\bar{y} = 988.9$ hours, find a 90% confidence interval for $\mu_X - \mu_Y$.

6.5-2 Let $X_1, X_2, \ldots, X_5$ be a random sample of SAT mathematics scores, assumed to be $N(\mu_X, \sigma^2)$, and let $Y_1, Y_2, \ldots, Y_8$ be an independent random sample of SAT verbal scores, assumed to be $N(\mu_Y, \sigma^2)$ If the following data are observed, find a 90% confidence interval for $\mu_X - \mu_Y$:

$$
\begin{array}{lllll}
x_1 = 644 & x_2 = 493 & x_3 = 532 & x_4 = 462 & x_5 = 565 \\
y_1 = 623 & y_2 = 472 & y_3 = 492 & y_4 = 661 & y_5 = 540 \\
y_6 = 502 & y_7 = 549 & y_8 = 518 &&
\end{array}
$$

6.5-3 Independent random samples of the heights of adult males living in two countries yielded the following results: $n = 12$, $\bar{x} = 65.7$ inches, $s_x = 4$ inches and $m = 15$, $\bar{y} = 68.2$ inches, $s_y = 3$ inches. Find an approximate 98% confidence interval for the difference $\mu_X - \mu_Y$ of the means of the populations of heights. Assume that $\sigma_X^2 = \sigma_Y^2$.

6.5-4 [*Medicine and Science in Sports and Exercise* (January 1990)] Let X and Y equal, respectively, the blood volumes in milliliters for a male who is a paraplegic and participates in vigorous physical activities and a male who is able-bodied and participates in normal activities. Assume that X is $N(\mu_X, \sigma_X^2)$ and Y is $N(\mu_Y, \sigma_Y^2)$, respectively. Using the following $n = 7$ observations of X:

$$1612 \quad 1352 \quad 1456 \quad 1222 \quad 1560 \quad 1456 \quad 1924$$

and $m = 10$ observations of Y:

$$
\begin{array}{ccccc}
1082 & 1300 & 1092 & 1040 & 910 \\
1248 & 1092 & 1040 & 1092 & 1288
\end{array}
$$

(a) Give a point estimate for $\mu_X - \mu_Y$.

(b) Find a 95% confidence interval for $\mu_X - \mu_Y$. Since the variances σ_X^2 and σ_Y^2 might not be equal, use Welch's T.

6.5-5 A biologist who studies spiders was interested in comparing the lengths of female and male green lynx spiders. Assume that the length X of the male spider is approximately $N(\mu_X, \sigma_X^2)$ and the length Y of the female spider is approximately $N(\mu_Y, \sigma_Y^2)$. Find an approximate one-sided 95% confidence interval which is an upper bound for $\mu_X - \mu_Y$ using $n = 30$ observations of X:

$$
\begin{array}{cccccc}
5.20 & 4.70 & 5.75 & 7.50 & 6.45 & 6.55 \\
4.70 & 4.80 & 5.95 & 5.20 & 6.35 & 6.95 \\
5.70 & 6.20 & 5.40 & 6.20 & 5.85 & 6.80 \\
5.65 & 5.50 & 5.65 & 5.85 & 5.75 & 6.35 \\
5.75 & 5.95 & 5.90 & 7.00 & 6.10 & 5.80
\end{array}
$$

and $m = 30$ observations of Y:

$$
\begin{array}{cccccc}
8.25 & 9.95 & 5.90 & 7.05 & 8.45 & 7.55 \\
9.80 & 10.80 & 6.60 & 7.55 & 8.10 & 9.10 \\
6.10 & 9.30 & 8.75 & 7.00 & 7.80 & 8.00 \\
9.00 & 6.30 & 8.35 & 8.70 & 8.00 & 7.50 \\
9.50 & 8.30 & 7.05 & 8.30 & 7.95 & 9.60
\end{array}
$$

where the measurements are in millimeters.

6.5-6 Consider the butterfat production (in pounds) for a cow during a 305-day milk production period following the birth of a calf. Let X and Y equal the butterfat production for such cows on a farm in Wisconsin and a farm in Michigan. Twelve observations of X are

$$
\begin{array}{cccccc}
649 & 657 & 714 & 877 & 975 & 468 \\
567 & 849 & 721 & 791 & 874 & 405
\end{array}
$$

Sixteen observations of Y are

$$
\begin{array}{cccccccc}
699 & 891 & 632 & 815 & 589 & 764 & 524 & 727 \\
597 & 868 & 652 & 978 & 479 & 733 & 549 & 790
\end{array}
$$

(a) Assuming that X is $N(\mu_X, \sigma^2)$ and Y is $N(\mu_Y, \sigma^2)$, find a 95% confidence interval for $\mu_X - \mu_Y$.
(b) Construct box-and-whisker diagrams for these two sets of data on the same graph.
(c) Does there seem to be a significant difference in butterfat production for cows on these two farms?

6.5-7 An interior automotive supplier is considering changing its electrical wire harness to save money. The idea is to replace a current 20-gauge wire with a 22-gauge wire. Since not all wires in the harness can be changed, the new wire must work with the current wire splice process. To determine if the new wire is compatible, random samples were selected and measured with a pull test. A pull test measures the force required to pull the spliced wires apart. The minimum pull force required by the customer is 20 pounds. Twenty observations of the forces needed for the current wire are

28.8	24.4	30.1	25.6	26.4	23.9	22.1	22.5	27.6	28.1
20.8	27.7	24.4	25.1	24.6	26.3	28.2	22.2	26.3	24.4

Twenty observations of the forces needed for the new wire are

14.1	12.2	14.0	14.6	8.5	12.6	13.7	14.8	14.1	13.2
12.1	11.4	10.1	14.2	13.6	13.1	11.9	14.8	11.1	13.5

(a) Does the current wire meet the customer's specifications?
(b) Find a 90% confidence interval for the difference of the means for these two sets of wire.
(c) Construct box-and-whisker diagrams of the two sets of data on the same figure.
(d) What is your recommendation for this company?

6.5-8 A test was conducted to determine if a wedge on the end of a plug fitting designed to hold a seal onto that plug was doing its job. The data taken were in the form of measurements of the force required to remove a seal from the plug first with the wedge in place, say X, and the force required without the plug, say Y. Assume that the distributions of X and Y are $N(\mu_X, \sigma^2)$ and $N(\mu_Y, \sigma^2)$. Ten independent observations of X are

$$3.26 \quad 2.26 \quad 2.62 \quad 2.62 \quad 2.36 \quad 3.00 \quad 2.62 \quad 2.40 \quad 2.30 \quad 2.40$$

Ten independent observations of Y are

$$1.80 \quad 1.46 \quad 1.54 \quad 1.42 \quad 1.32 \quad 1.56 \quad 1.36 \quad 1.64 \quad 2.00 \quad 1.54$$

(a) Find a 95% confidence interval for $\mu_X - \mu_Y$.
(b) Construct box-and-whisker diagrams of these data on the same figure.
(c) Is the wedge necessary?

6.5-9 Students are weighed (in kilograms) at the beginning and the end of a semester-long health-fitness program. Let the random variable D equal the weight change for a student, postweight minus preweight. Assume that the distribution of D is $N(\mu_D, \sigma_D^2)$. A random sample of $n = 12$ female students yielded the following observations of D:

2.0	−0.5	1.4	−2.2	0.3	−0.8
3.7	−0.1	0.6	0.2	0.9	−0.1

(a) Give a point estimate of μ_D.
(b) Find a 95% confidence interval for μ_D.

6.5-10 Let $\overline{X}, \overline{Y}, S_X^2$, and S_Y^2 be the respective sample means and unbiased estimates of the variances using independent samples of sizes n and m from the normal distributions $N(\mu_X, \sigma_X^2)$ and $N(\mu_Y, \sigma_Y^2)$, where μ_X, μ_Y, σ_X^2, and σ_Y^2 are unknown. If, however, $\sigma_X^2/\sigma_Y^2 = d$, a known constant, argue that

(a) $\dfrac{(\overline{X} - \overline{Y}) - (\mu_X - \mu_Y)}{\sqrt{d\sigma_Y^2/n + \sigma_Y^2/m}}$ is $N(0, 1)$.

(b) $\dfrac{(n - 1)S_X^2}{d\sigma_Y^2} + \dfrac{(m - 1)S_Y^2}{\sigma_Y^2}$ is $\chi^2(n+m-2)$.

(c) The two random variables in (a) and (b) are independent.

(d) With these results, construct a random variable (not depending upon σ_Y^2) that has a t distribution and can be used to construct a confidence interval for $\mu_X - \mu_Y$.

6.5-11 Students in a semester-long health-fitness program have their percentage of body fat measured at the beginning of the semester and at the end of the semester. The following measurements give these percentages for 10 men and for 10 women.

Males		Females	
Pre %	Post %	Pre %	Post %
11.10	9.97	22.90	22.89
19.50	15.80	31.60	33.47
14.00	13.02	27.70	25.75
8.30	9.28	21.70	19.80
12.40	11.51	19.36	18.00
7.89	7.40	25.03	22.33
12.10	10.70	26.90	25.26
8.30	10.40	25.75	24.90
12.31	11.40	23.63	21.80
10.00	11.95	25.06	24.28

(a) Find a 90% confidence interval for the mean of the difference in the percentages for the males.

(b) Find a 90% confidence interval for the mean of the difference in the percentages for the females.

(c) Based on these data, have these percentages decreased?

(d) If possible, check whether each set of differences come from a normal distribution.

6.5-12 Twenty-four 9th- and 10th-grade high school girls were put on an ultraheavy rope-jumping program. The following data give the time difference for each girl–"before program time" minus "after program time"–for the 40-yard dash.

0.28	0.01	0.13	0.33	−0.03	0.07	−0.18	−0.14
−0.33	0.01	0.22	0.29	−0.08	0.23	0.08	0.04
−0.30	−0.08	0.09	0.70	0.33	−0.34	0.50	0.06

(a) Give a point estimate of μ_D, the mean of the difference in race times.

(b) Find a one-sided 95% confidence interval which is a lower bound for μ_D.

(c) Does it look like the rope-jumping program was effective?

6.5-13 The Biomechanics Lab at Hope College tested healthy old women and healthy young women to discover whether or not lower extremity response time to a stimulus is a function of age. Let X and Y, respectively, equal the independent response times for these two groups when taking steps in the anterior direction. Find a one-sided 95% confidence interval which is a lower bound for $\mu_X - \mu_Y$ given that $n = 60$ observations of X yielded $\overline{x} = 671$ and $s_x = 129$ while $m = 60$ observations of Y yielded $\overline{y} = 480$ and $s_y = 93$.

6.6* CONFIDENCE INTERVALS FOR VARIANCES

In this section we find confidence intervals for the variance of a normal distribution and for the ratio of the variances of two normal distributions. The confidence interval for the variance σ^2 is based on the sample variance

$$S^2 = \frac{1}{n-1} \sum_{i=1}^{n} (X_i - \overline{X})^2.$$

We use the fact that the distribution of $(n-1)S^2/\sigma^2$ is $\chi^2(n-1)$ to find a confidence interval for σ^2. Select constants a and b from Table IV in the Appendix with $n-1$ degrees of freedom such that

$$P\left(a \leq \frac{(n-1)S^2}{\sigma^2} \leq b\right) = 1 - \alpha.$$

One way to do this is by selecting a and b so that $a = \chi^2_{1-\alpha/2}(n-1)$ and $b = \chi^2_{\alpha/2}(n-1)$. That is, select a and b so that the probabilities in the two tails are equal. Then, solving the inequalities, we have

$$1 - \alpha = P\left(\frac{a}{(n-1)S^2} \leq \frac{1}{\sigma^2} \leq \frac{b}{(n-1)S^2}\right)$$

$$= P\left(\frac{(n-1)S^2}{b} \leq \sigma^2 \leq \frac{(n-1)S^2}{a}\right).$$

Thus the probability that the random interval $[(n-1)S^2/b, (n-1)S^2/a]$ contains the unknown σ^2 is $1 - \alpha$. Once the values of $X_1, X_2, \ldots, X_n$ are observed to be $x_1, x_2, \ldots, x_n$ and s^2 is computed, then the interval $[(n-1)s^2/b, (n-1)s^2/a]$ is a $100(1-\alpha)\%$ confidence interval for σ^2. It follows that a $100(1-\alpha)\%$ confidence interval for σ, the standard deviation, is given by

$$\left[\sqrt{\frac{(n-1)s^2}{b}}, \sqrt{\frac{(n-1)s^2}{a}}\right] = \left[\sqrt{\frac{n-1}{b}}\, s, \sqrt{\frac{n-1}{a}}\, s\right].$$

EXAMPLE 6.6-1   Assume that the time in days required for maturation of seeds of a species of Guardiola, a flowering plant found in Mexico, is $N(\mu, \sigma^2)$. A random sample of $n = 13$ seeds, both parents having narrow leaves, yielded $\overline{x} = 18.97$ days and

$$12s^2 = \sum_{i=1}^{13} (x_i - \overline{x})^2 = 128.41.$$

A 90% confidence interval for σ^2 is

$$\left[\frac{128.41}{21.03}, \frac{128.41}{5.226}\right] = [6.11, 24.57]$$

because $5.226 = \chi^2_{0.95}(12)$ and $21.03 = \chi^2_{0.05}(12)$ from Table IV in the Appendix. The corresponding 90% confidence interval for σ is

$$[\sqrt{6.11}, \sqrt{24.57}] = [2.47, 4.96].$$ ◄

Although a and b are generally selected so that the probabilities in the two tails are equal, the resulting $100(1 - \alpha)\%$ confidence interval is not the shortest that can be formed using the available data. Table X in the Appendix gives solutions for a and b that yield confidence intervals of minimum length for the standard deviation. (See Exercise 6.6-14.)

EXAMPLE 6.6-2 Using the data in Example 6.6-1, a 90% confidence interval for σ of minimum length is (since $a = 5.940$ and $b = 24.202$ from Table X)

$$\left[\sqrt{\frac{128.41}{24.202}}, \sqrt{\frac{128.41}{5.940}}\right] = [2.30, 4.65].$$

The length of this interval is 2.35, whereas the length of the interval given in Example 6.6-1 is 2.49. To see why this new interval is shorter, carefully sketch a graph of the $\chi^2(12)$ p.d.f. and compare $\chi^2_{0.95}(12)$ to a and $\chi^2_{0.05}(12)$ to b on the graph. ◄

There are occasions when it is of interest to compare the variances of two normal distributions. We do this by finding a confidence interval for σ^2_X/σ^2_Y using the ratio of S^2_X/σ^2_X and S^2_Y/σ^2_Y, where S^2_X and S^2_Y are the two sample variances based on two independent samples of sizes n and m from $N(\mu_X, \sigma^2_X)$ and $N(\mu_Y, \sigma^2_Y)$, respectively. However the reciprocal of that ratio can be rewritten as follows:

$$\frac{\dfrac{S^2_Y}{\sigma^2_Y}}{\dfrac{S^2_X}{\sigma^2_X}} = \frac{\left[\dfrac{(m-1)S^2_Y}{\sigma^2_Y}\right] \Big/ (m-1)}{\left[\dfrac{(n-1)S^2_X}{\sigma^2_X}\right] \Big/ (n-1)}.$$

Since $(m-1)S^2_Y/\sigma^2_Y$ and $(n-1)S^2_X/\sigma^2_X$ are independent chi-square variables with $(m-1)$ and $(n-1)$ degrees of freedom, respectively, we know from Example 4.4-5 that the distribution of this ratio is $F(m-1, n-1)$. That is,

$$F = \frac{\dfrac{(m-1)S^2_Y}{\sigma^2_Y(m-1)}}{\dfrac{(n-1)S^2_X}{\sigma^2_X(n-1)}} = \frac{\dfrac{S^2_Y}{\sigma^2_Y}}{\dfrac{S^2_X}{\sigma^2_X}}$$

has an F distribution with $r_1 = m-1$ and $r_2 = n-1$ degrees of freedom. This is the ratio that we want to use to find a confidence interval for σ^2_X/σ^2_Y.

To form the confidence interval, select constants c and d from Table VII in the Appendix so that

$$1 - \alpha = P\left(c \le \frac{S_Y^2/\sigma_Y^2}{S_X^2/\sigma_X^2} \le d\right)$$

$$= P\left(c\,\frac{S_X^2}{S_Y^2} \le \frac{\sigma_X^2}{\sigma_Y^2} \le d\,\frac{S_X^2}{S_Y^2}\right).$$

Because of the limitations of Table VII, we generally let $c = F_{1-\alpha/2}(m\!-\!1, n\!-\!1) = 1/F_{\alpha/2}(n-1, m-1)$ and $d = F_{\alpha/2}(m-1, n-1)$. If s_x^2 and s_y^2 are the observed values of S_X^2 and S_Y^2, respectively, then

$$\left[\frac{1}{F_{\alpha/2}(n-1, m-1)}\,\frac{s_x^2}{s_y^2},\; F_{\alpha/2}(m-1, n-1)\,\frac{s_x^2}{s_y^2}\right]$$

is a $100(1 - \alpha)\%$ confidence interval for σ_X^2/σ_Y^2. By taking square roots of both endpoints, we would obtain a $100(1 - \alpha)\%$ confidence interval for σ_X/σ_Y.

EXAMPLE 6.6-3   In Example 6.6-1, denote σ^2 by σ_X^2. There $(n - 1)s_x^2 = 12s^2 = 128.41$. Assume that the time in days required for maturation of seeds of a species of Guardiola, both parents having broad leaves, is $N(\mu_Y, \sigma_Y^2)$. A random sample of size $m = 9$ seeds yielded $\overline{y} = 23.20$ and

$$8s_y^2 = \sum_{i=1}^{8} (y_i - \overline{y})^2 = 36.72.$$

A 98% confidence interval for σ_X^2/σ_Y^2 is given by

$$\left[\left(\frac{1}{5.67}\right)\frac{(128.41)/12}{(36.72)/8},\; (4.50)\,\frac{(128.41)/12}{(36.72)/8}\right] = [0.41, 10.49]$$

because $F_{0.01}(12, 8) = 5.67$ and $F_{0.01}(8, 12) = 4.50$. It follows that a 98% confidence interval for σ_X/σ_Y is

$$[\sqrt{0.41},\, \sqrt{10.49}] = [0.64, 3.24]. \qquad \blacktriangleleft$$

Although we are able to formally find a confidence interval for the ratio of two distribution variances and/or standard deviations, we should point out that these intervals are generally not too useful because they are often very wide. Moreover, these intervals are not very robust. That is, the confidence coefficients are not very accurate if we deviate much from underlying normal distributions, because, in those instances, the distribution of $(n - 1)S^2/\sigma^2$

could deviate greatly from $\chi^2(n-1)$. This same statement is not true for confidence intervals for μ or $\mu_X - \mu_Y$ based upon the sample means, because the sample means have approximate normal distributions. Hence the confidence coefficients are reasonably accurate in the situations associated with means. Having said that, we do warn the reader that in cases that are highly skewed or have very heavy tails, there are better ways of finding confidence intervals for the "middles" than using the sample means. We consider a few of these later in the text.

EXERCISES

6.6-1 Let X equal the length (in centimeters) of a certain species of fish when caught in the spring. A random sample of $n = 13$ observations of X is

13.1	5.1	18.0	8.7	16.5	9.8	6.8
12.0	17.8	25.4	19.2	15.8	23.0	

(a) Give a point estimate of the standard deviation σ of this species of fish.
(b) Find a 95% confidence interval for σ.

6.6-2 A random sample of $n = 9$ wheels of cheese yielded the following weights in pounds, assumed to be $N(\mu, \sigma^2)$:

21.50	18.95	18.55	19.40	19.15
22.35	22.90	22.20	23.10	

(a) Give a point estimate of σ.
(b) Find a 95% confidence interval for σ.
(c) Find a 90% confidence interval for σ.

6.6-3 A student who works in a blood lab tested 25 men for cholesterol levels and found the following values:

164	272	261	248	235	192	203	278	268
230	242	305	286	310	345	289	326	
335	297	328	400	228	194	338	252	

Assume that these values represent observations of a random sample taken from $N(\mu, \sigma^2)$.

(a) Calculate the sample mean and sample variance for these data.
(b) Find a 90% confidence interval for σ^2.
(c) Find a 90% confidence interval for σ.
(d) Find a 90% confidence interval for σ that has minimum length.
(e) Does the assumption of normality seem to be valid?

6.6-4 For the data in Exercise 6.4-9, find a 95% confidence interval for

(a) σ^2.
(b) σ.
(c) σ having minimum length.

6.6-5 Let $X_1, X_2, X_3, \ldots, X_n$ be a random sample from $N(\mu, \sigma^2)$, with known mean μ. Describe how you would construct a confidence interval for the unknown variance σ^2.

HINT: Use the fact that $\sum_{i=1}^{n}(X_i - \mu)^2/\sigma^2$ is $\chi^2(n)$.

6.6-6 Let $X_1, X_2, \ldots, X_n$ be a random sample of size n from an exponential distribution with unknown mean of $\mu = \theta$.

(a) Show that the distribution of $W = (2/\theta) \sum_{i=1}^{n} X_i$ is $\chi^2(2n)$.
HINT: Find the moment-generating function of W.

(b) Use W to construct a $100(1 - \alpha)\%$ confidence interval for θ.

(c) If $n = 7$ and $\bar{x} = 93.6$, give the endpoints for a 90% confidence interval for the mean θ.

6.6-7 Let X equal the time in seconds between phone calls at Great Lakes Pizza Company. Sixteen observations of X are

$$
\begin{array}{cccccccc}
82 & 42 & 185 & 66 & 384 & 27 & 334 & 545 \\
650 & 127 & 35 & 45 & 285 & 133 & 120 & 471
\end{array}
$$

(a) Assuming that X has an exponential distribution with mean θ, use Exercise 6.6-6 to construct a 95% confidence interval for θ.

(b) Calculate the values of $\bar{x}$ and s. Do they support the assumption that X has an exponential distribution?

(c) Construct a q-q plot or a box-and-whiskers display to confirm or deny the assumption that X has an exponential distribution.

6.6-8 Let X and Y equal the weights of a phosphorus-free laundry detergent in a "6-pound" box and a "12-pound" box, respectively. Assume that the distributions of X and Y are $N(\mu_X, \sigma_X^2)$ and $N(\mu_Y, \sigma_Y^2)$, respectively. A random sample of $n = 10$ observations of X yielded a sample mean of $\bar{x} = 6.10$ pounds with a sample variance of $s_x^2 = 0.0040$, while an independent random sample of $m = 9$ observations of Y yielded a sample mean of $\bar{y} = 12.10$ pounds with a sample variance of $s_y^2 = 0.0076$.

(a) Give a point estimate of σ_X^2/σ_Y^2.

(b) Find a 95% confidence interval for σ_X^2/σ_Y^2.

6.6-9 Let X and Y equal the number of milligrams of tar in filtered and nonfiltered cigarettes, respectively. Assume that the distributions of X and Y are $N(\mu_X, \sigma_X^2)$ and $N(\mu_Y, \sigma_Y^2)$, respectively. A random sample of $n = 9$ observations of X was

$$0.9 \quad 1.1 \quad 0.1 \quad 0.7 \quad 0.3 \quad 0.9 \quad 0.8 \quad 1.0 \quad 0.4$$

and an independent random sample of $m = 11$ observations of Y was

$$1.5 \quad 0.9 \quad 1.6 \quad 0.5 \quad 1.4 \quad 1.9 \quad 1.0 \quad 1.2 \quad 1.3 \quad 1.6 \quad 2.1$$

(a) Give a point estimate of σ_X^2/σ_Y^2.

(b) Find a one-sided 95% confidence interval which is an upper bound for σ_X^2/σ_Y^2. Does this confidence interval include the number one?

6.6-10 A candy maker produces mints that have a label weight of 20.4 grams. For quality assurance, $n = 16$ mints were selected at random from the Wednesday morning shift, resulting in the statistics $\bar{x} = 21.95$ grams and $s_x = 0.197$. On Wednesday afternoon $m = 13$ mints were selected at random, giving $\bar{y} = 21.88$ grams and $s_y = 0.318$. Find a 90% confidence interval for σ_X/σ_Y, the ratio of the standard deviations of the mints produced by the morning and by the afternoon shifts, respectively.

6.6-11 Let X and Y equal the concentration in parts per billion of chromium in the blood for healthy persons and for persons with a suspected disease, respectively. Assume that the distributions of X and Y are $N(\mu_X, \sigma_X^2)$ and $N(\mu_Y, \sigma_Y^2)$, respectively. Using $n = 8$ observations of X,

$$15 \quad 23 \quad 12 \quad 18 \quad 9 \quad 28 \quad 11 \quad 10$$

and $m = 10$ observations of Y,

$$25 \quad 20 \quad 35 \quad 15 \quad 40 \quad 16 \quad 10 \quad 22 \quad 18 \quad 32$$

(a) Give a point estimate of σ_X^2/σ_Y^2.

(b) Find a one-sided 95% confidence interval which is an upper bound for σ_X^2/σ_Y^2.

6.6-12 Some nurses were interested in the effect of prenatal care on the birthweight of babies. Mothers were divided into two groups, and their babies' weights were compared. The birthweight in ounces of babies of mothers who had received 5 or fewer prenatal visits were

$$
\begin{array}{ccccccc}
49 & 108 & 110 & 82 & 93 & 114 & 134 \\
114 & 96 & 52 & 101 & 114 & 120 & 116
\end{array}
$$

and the birthweights of babies of mothers who had received 6 or more prenatal visits were

$$
\begin{array}{ccccccc}
133 & 108 & 93 & 119 & 119 & 98 & 106 \\
87 & 153 & 116 & 129 & 97 & 110 & 131
\end{array}
$$

Assuming that these are respectively independent observations of X and Y, which are $N(\mu_X, \sigma_X^2)$ and $N(\mu_Y, \sigma_Y^2)$, find a 95% confidence interval for

(a) σ_X^2/σ_Y^2.

(b) σ_X/σ_Y.

6.6-13 Using a random sample of size n from the normal distribution $N(\mu, \sigma^2)$, find the $100(1 - \alpha)\%$ confidence interval for μ of minimum length based on the statistic $T = (\overline{X} - \mu)/(S/\sqrt{n})$, where $\overline{X}$ and S^2 are the unbiased estimators of μ and σ^2, respectively.

HINT: (i) Show that $1 - \alpha = P(a \le T \le b) = P[\overline{X} - b(S/\sqrt{n}) \le \mu \le \overline{X} - a(S/\sqrt{n})]$, and (ii) the length of the resulting confidence interval is $L = (s/\sqrt{n})(b - a)$. (iii) Minimize L subject to the condition $\int_a^b g(t)\,dt = 1 - \alpha$, where $g(t)$ is the p.d.f. of a t random variable with $r = n - 1$ degrees of freedom.

6.6-14 Let $X_1, X_2, \ldots, X_n$ be a random sample of size n from a normal distribution, $N(\mu, \sigma^2)$. Select a and b so that

$$
P\left(a \le \frac{(n-1)S^2}{\sigma^2} \le b\right) = 1 - \alpha.
$$

So a $100(1 - \alpha)\%$ confidence interval for σ is $[\sqrt{(n-1)/b}\,s, \sqrt{(n-1)/a}\,s]$. Find values of a and b that minimize the length of this confidence interval. That is, minimize

$$
k = s\sqrt{n-1}\left(\frac{1}{\sqrt{a}} - \frac{1}{\sqrt{b}}\right)
$$

under the restriction

$$
G(b) - G(a) = \int_a^b g(u)\,du = 1 - \alpha,
$$

where $G(u)$ and $g(u)$ are the distribution function and p.d.f. of a $\chi^2(n-1)$ distribution, respectively.

HINT: Due to the restriction, b is a function of a. In particular, by taking derivatives of the restricting equation with respect to a, show that $\dfrac{db}{da} = \dfrac{g(a)}{g(b)}$. Determine $\dfrac{dk}{da}$.

By setting $\dfrac{dk}{da} = 0$, show that a and b must satisfy

$$
a^{n/2}e^{-a/2} - b^{n/2}e^{-b/2} = 0.
$$

This condition, along with the restriction, was used to calculate the values in Appendix Table X.

6.6-15 Let X and Y equal the playing times for a $33\frac{1}{3}$ rpm long-playing record album and a compact disc, respectively. Observe that $n = 9$ observations of X were

$$40.83 \quad 43.18 \quad 35.72 \quad 38.68 \quad 37.17 \quad 39.75 \quad 24.76 \quad 34.58 \quad 33.98$$

and $m = 13$ independent observations of Y were

42.82	64.42	56.92	39.92	72.38	47.26	64.58
38.20	72.75	39.09	39.07	33.70	62.02	

(a) Determine a 95% confidence interval for σ_X^2/σ_Y^2, assuming that the distributions of X and Y are $N(\mu_X, \sigma_X^2)$ and $N(\mu_Y, \sigma_Y^2)$, respectively.

(b) Does the assumption of normality seem to be valid for these data? Why?

6.6-16 Find a 95% confidence interval for the ratio of the variances in

(a) Exercise 6.5-6,
(b) Exercise 6.5-7,
(c) Exercise 6.5-8.

6.7   CONFIDENCE INTERVALS FOR PROPORTIONS

We have suggested that the histogram is a good description of how the observations of a random sample are distributed. We might naturally inquire about the accuracy of those relative frequencies (or percentages) associated with the various classes. To illustrate: in Example 3.1-1 concerning $n = 40$ candy bar weights, we found that the relative frequency of the class interval $(22.25, 23.15)$ was $8/40 = 0.20$, that is, 20%. If we think of this collection of 40 weights as a random sample observed from a larger population of candy bar weights, how close is 20% to the true percentage (or 0.20 to the true proportion) of weights in that class interval for the entire population of weights for this type of candy bar?

In considering this problem, we generalize it somewhat by treating the class interval $(22.25, 23.15)$ as "success." That is, there is some true probability of success p, namely the proportion of the population in that interval. Let Y equal the frequency of measurements in the interval out of the n observations so that (under the assumptions of independence and constant probability p) Y has the binomial distribution $b(n, p)$. Thus the problem is to determine the accuracy of the relative frequency Y/n as an estimator of p. We solve this by finding for the unknown p a confidence interval based on Y/n.

In general, when observing n Bernoulli trials with probability p of success on each trial, we shall find a confidence interval for p based on Y/n, where Y is the number of successes and Y/n is an unbiased point estimator for p.

In Section 5.5 we noted that

$$\frac{Y - np}{\sqrt{np(1 - p)}} = \frac{(Y/n) - p}{\sqrt{p(1 - p)/n}}$$

has an approximate normal distribution $N(0, 1)$, provided that n is large enough. This means that, for a given probability $1 - \alpha$, we can find a $z_{\alpha/2}$ in

Table V in the Appendix such that

$$P\left[-z_{\alpha/2} \le \frac{(Y/n) - p}{\sqrt{p(1-p)/n}} \le z_{\alpha/2}\right] \approx 1 - \alpha. \tag{6.7-1}$$

If we proceed as we did when we found a confidence interval for μ in Section 6.4, we would obtain

$$P\left[\frac{Y}{n} - z_{\alpha/2}\sqrt{\frac{p(1-p)}{n}} \le p \le \frac{Y}{n} + z_{\alpha/2}\sqrt{\frac{p(1-p)}{n}}\right] \approx 1 - \alpha. \quad .$$

Unfortunately, the unknown parameter p appears in the endpoints of this inequality. There are two ways out of this dilemma. We could make an additional approximation, namely replacing p with Y/n in $p(1-p)/n$ in the endpoints. That is, it is still true, if n is large enough, that

$$P\left[\frac{Y}{n} - z_{\alpha/2}\sqrt{\frac{(Y/n)(1-Y/n)}{n}} \le p \le \frac{Y}{n} + z_{\alpha/2}\sqrt{\frac{(Y/n)(1-Y/n)}{n}}\right] \approx 1 - \alpha.$$

Thus, for large n, if the observed Y equals y, the interval

$$\left[\frac{y}{n} - z_{\alpha/2}\sqrt{\frac{(y/n)(1-y/n)}{n}}, \frac{y}{n} + z_{\alpha/2}\sqrt{\frac{(y/n)(1-y/n)}{n}}\right]$$

serves as an approximate $100(1-\alpha)\%$ confidence interval for p. Frequently, this is written as

$$\frac{y}{n} \pm z_{\alpha/2}\sqrt{\frac{(y/n)(1-y/n)}{n}} \tag{6.7-2}$$

for brevity. This clearly notes, as does $\bar{x} \pm z_{\alpha/2}(\sigma/\sqrt{n})$ in Section 6.4, the reliability of the estimate y/n; namely, we are $100(1-\alpha)\%$ confident that p is within $z_{\alpha/2}\sqrt{(y/n)(1-y/n)/n}$ of $\widehat{p} = y/n$.

A second way to solve for p, in the inequality in Equation 6.7-1, is to note that

$$\frac{|Y/n - p|}{\sqrt{p(1-p)/n}} \le z_{\alpha/2}$$

is equivalent to

$$H(p) = \left(\frac{Y}{n} - p\right)^2 - \frac{z_{\alpha/2}^2\, p(1-p)}{n} \le 0. \tag{6.7-3}$$

But $H(p)$ is a quadratic expression in p. Thus we can find those values of p for which $H(p) \leq 0$ by finding the two zeros of $H(p)$. Letting $\widehat{p} = Y/n$ and $z_0 = z_{\alpha/2}$ in Equation 6.7-3, we have

$$H(p) = \left(1 + \frac{z_0^2}{n}\right)p^2 - \left(2\widehat{p} + \frac{z_0^2}{n}\right)p + \widehat{p}^2.$$

By the quadratic formula, the zeros of $H(p)$ are, after simplifying,

$$\frac{\widehat{p} + z_0^2/(2n) \pm z_0\sqrt{\widehat{p}(1 - \widehat{p})/n + z_0^2/(4n^2)}}{1 + z_0^2/n}, \tag{6.7-4}$$

and these zeros give the endpoints for an approximate $100(1 - \alpha)\%$ confidence interval for p. If n is large, $z_0^2/(2n)$, $z_0^2/(4n^2)$, and z_0^2/n are small. Thus the confidence intervals given by Equation 6.7-2 and Equation 6.7-4 are approximately equal when n is large.

EXAMPLE 6.7-1   Let us return to the example of the histogram of the candy bar weights, Example 3.1-1 with $n = 36$ and $y/n = 0.222$. If $1 - \alpha = 0.90$ so that $z_{\alpha/2} = 1.645$, then, using Equation 6.7-2, the endpoints

$$0.222 \pm 1.645\sqrt{\frac{(0.222)(0.778)}{36}}$$

serve as an approximate 90% confidence interval for the true fraction p. That is, $[0.108, 0.336]$, which is the same as $[10.8\%, 33.6\%]$, is an approximate 90% confidence interval for the percentage of weights of the entire population in the interval $(22.25, 23.15)$. If we had used the endpoints given by Equation 6.7-4, the confidence interval would be $[0.130, 0.353]$. Because of the small sample size, there is a difference in the lengths of these intervals. If the sample size had been $n = 360$, the two 90% confidence intervals would have been $(0.186, 0.258)$ and $(0.188, 0.260)$, respectively, which differ very little.   ◀

EXAMPLE 6.7-2   In a certain political campaign, one candidate has a poll taken at random among the voting population. The results are $y = 185$ out of $n = 351$ voters favor this candidate. Even though $y/n = 185/351 = 0.527$, should the candidate feel very confident of winning? An approximate 95% confidence interval for the fraction p of the voting population who favor this candidate is, using Equation 6.7-2,

$$0.527 \pm 1.96\sqrt{\frac{(0.527)(0.473)}{351}}$$

or, equivalently, $[0.475, 0.579]$. Thus there is a good possibility that p is less than 50%, and the candidate should certainly take this into account in campaigning.   ◀

One-sided confidence intervals are sometimes appropriate for p. For example, we may be interested in an upper bound on the proportion of defectives when manufacturing some item. Or we may be interested in a lower bound on the proportion of voters who favor a particular candidate. The one-sided confidence interval for p given by

$$\left[0, \frac{y}{n} + z_\alpha \sqrt{\frac{y/n(1 - y/n)}{n}} \right]$$

provides an upper bound for p while

$$\left[\frac{y}{n} - z_\alpha \sqrt{\frac{y/n(1 - y/n)}{n}}, 1 \right]$$

provides a lower bound for p.

Frequently, there are two (or more) possible independent ways of performing an experiment; suppose these have probabilities of success p_1 and p_2, respectively. Let n_1 and n_2 be the number of independent trials associated with these two methods, and let us say they result in Y_1 and Y_2 successes, respectively. In order to make a statistical inference about the difference $p_1 - p_2$, we proceed as follows.

Since the independent random variables Y_1/n_1 and Y_2/n_2 have respective means p_1 and p_2 and variances $p_1(1 - p_1)/n_1$ and $p_2(1 - p_2)/n_2$, we know from Section 4.6 that the difference $Y_1/n_1 - Y_2/n_2$ must have mean $p_1 - p_2$ and variance

$$\frac{p_1(1 - p_1)}{n_1} + \frac{p_2(1 - p_2)}{n_2}.$$

(Recall that the variances are added to get the variance of a difference of two independent random variables.) Moreover, the fact that Y_1/n_1 and Y_2/n_2 have approximate normal distributions would suggest that the difference

$$\frac{Y_1}{n_1} - \frac{Y_2}{n_2}$$

would have an approximate normal distribution with the above mean and variance (see Theorem 5.3-4). That is,

$$\frac{(Y_1/n_1) - (Y_2/n_2) - (p_1 - p_2)}{\sqrt{p_1(1 - p_1)/n_1 + p_2(1 - p_2)/n_2}}$$

has an approximate normal distribution $N(0, 1)$. If we now replace p_1 and p_2 in the denominator of this ratio by Y_1/n_1 and Y_2/n_2, respectively, it is still true for large enough n_1 and n_2, that the new ratio will be approximately $N(0, 1)$. Thus, for a given $1 - \alpha$, we can find $z_{\alpha/2}$ from Table V in the Appendix so that

$$P\left[-z_{\alpha/2} \leq \frac{(Y_1/n_1) - (Y_2/n_2) - (p_1 - p_2)}{\sqrt{\dfrac{(Y_1/n_1)(1 - Y_1/n_1)}{n_1} + \dfrac{(Y_2/n_2)(1 - Y_2/n_2)}{n_2}}} \leq z_{\alpha/2} \right] \approx 1 - \alpha.$$

This can be solved to obtain, once Y_1 and Y_2 are observed to be y_1 and y_2, an approximate $100(1 - \alpha)\%$ confidence interval

$$\frac{y_1}{n_1} - \frac{y_2}{n_2} \pm z_{\alpha/2} \sqrt{\frac{(y_1/n_1)(1 - y_1/n_1)}{n_1} + \frac{(y_2/n_2)(1 - y_2/n_2)}{n_2}}$$

for the unknown difference $p_1 - p_2$. Note again how this form indicates the reliability of the estimate $y_1/n_1 - y_2/n_2$ of the difference $p_1 - p_2$.

EXAMPLE 6.7-3   Two detergents were tested for their ability to remove stains of a certain type. An inspector judged the first one to be successful on 63 out of 91 independent trials and the second one to be successful on 42 out of 79 independent trials. The respective relative frequencies of success are 0.692 and 0.532. An approximate 90% confidence interval for the difference $p_1 - p_2$ of the two detergents is

$$0.692 - 0.532 \pm 1.645 \sqrt{\frac{(0.692)(0.308)}{91} + \frac{(0.532)(0.468)}{79}},$$

or, equivalently, $[0.038, 0.282]$. Accordingly, it seems that the first detergent is definitely better than the second one for removing these stains.   ◀

EXERCISES

6.7-1 A machine shop manufactures toggle levers. A lever is flawed if a standard nut cannot be screwed onto the threads. Let p equal the proportion of flawed toggle levers that the shop manufactures. If there were 24 flawed levers out of a sample of 642 that were selected randomly from the production line,

    (a) Give a point estimate of p.
    (b) Find an approximate 95% confidence interval for p using Equation 6.7-2.
    (c) Find an approximate 95% confidence interval for p using Equation 6.7-4.
    (d) Find a one-sided 95% confidence interval for p that provides an upper bound for p.

6.7-2 Let p equal the proportion of letters mailed in the Netherlands that are delivered the next day. If $y = 142$ out of a random sample of $n = 200$ letters were delivered the day after they were mailed, find an approximate 90% confidence interval for p.

6.7-3 Let p equal the proportion of adult Americans who favor a law requiring a teenager to have her parents' consent before having an abortion. In a survey of 1000 adult Americans (conducted by *Time*/CNN and reported in *Time* on July 9, 1990), 690 said they favored such a law.

    (a) Give a point estimate of p.
    (b) Find an approximate 95% confidence interval for p.

6.7-4 Let p equal the proportion of Americans who favor the death penalty. If a random sample of $n = 1234$ Americans yielded $y = 864$ who favored the death penalty, find an approximate 95% confidence interval for p.

6.7-5 Let p equal the proportion of triathletes who suffered a training-related overuse injury during the past year. Out of 330 triathletes who responded to a survey, 167 indicated that they had suffered a training-related injury during the past year. Using these data,

(a) Give a point estimate of p.

(b) Find an approximate 90% confidence interval for p.

(c) Do you think that the 330 triathletes who responded to the survey may be considered as a random sample from the population of triathletes?

6.7-6 Let p equal the proportion of Americans who select jogging as one of their recreational activities. If 1497 out of a random sample of 5757 selected jogging, find an approximate 98% confidence interval for p.

6.7-7 In order to estimate the percentage of a large class of college freshmen that had high school GPAs from 3.2 to 3.6 inclusive, a sample of $n = 50$ students was taken, and $y = 9$ students fell in this class. Give a 95% confidence interval for the percentage of this freshman class having a high school GPA of 3.2 to 3.6.

6.7-8 A proportion, p, that many public opinion polls estimate is the number of Americans who would say yes to the question, "If something were to happen to the President of the United States, do you think that the Vice President would be qualified to take over as President?" In one such random sample of 1022 adults, 388 said yes.

(a) Based on the given data, find a point estimate of p.

(b) Find an approximate 90% confidence interval for p.

(c) Give updated answers to this question if new poll results are available.

6.7-9 To obtain an estimate of the proportion, p, of New York City residents who feel that the quality of life in New York City has become worse in the past few years, a telephone poll by *Time*/CNN on August 2–5, 1990, revealed that 686 out of 1009 residents said that life has become worse.

(a) Give a point estimate of p.

(b) Find an approximate 98% confidence interval for p.

6.7-10 *Time*, January 17, 1994, reported that 58% of adult Americans said yes to the question "If you or your spouse were pregnant, would you want the unborn child tested for genetic defects?" This was based on a telephone poll of 500 adult Americans. Let p equal the proportion of all adult Americans who would say yes to this question.

(a) Give the endpoints for a 90% confidence interval for p.

(b) The poll takers claimed that their sampling error was 4.5%. What is their confidence level?

6.7-11 *Time*, January 29, 1996, reported that 48% of adult Americans "like the principle of a flat tax." It further claimed that this estimate had a sampling error of $\pm 3\%$. Let p equal the proportion of all adult Americans who like the principle of a flat tax.

(a) Give the endpoints for a 95% confidence interval for p given that the sample size was $n = 800$.

(b) Given that the reported sampling error was $\pm 3\%$ with $n = 800$, what is their confidence level?

6.7-12 *Time*, March 29, 1993, reported the proportions of adult Americans who favor "stricter gun-control laws." A telephone poll of 800 adult Americans, of whom 374 were gun owners and 426 did not own guns, showed that 206 gun owners and 338 non-gun owners favor stricter gun-control laws. Let p_1 and p_2 be the respective proportions of gun owners and non-gun owners who favor stricter gun-control laws.

(a) Give point estimates of p_1 and p_2.

(b) Find a 95% confidence interval for $p_1 - p_2$.

6.7-13 In developing countries in Africa and the Americas, let p_1 and p_2 be the respective proportions of women with nutritional anemia. Find an approximate 90% confidence interval for $p_1 - p_2$ given that a random sample of $n_1 = 2100$ African women yielded $y_1 = 840$ with nutritional anemia and a random sample of $n_2 = 1900$ women from the Americas yielded $y_2 = 323$ women with nutritional anemia.

6.7-14 A candy manufacturer selects mints at random from the production line and weighs them. For one week, the day shift weighed $n_1 = 194$ mints, and the night shift weighed $n_2 = 162$ mints. The numbers of these mints that weighed at most 21 grams was $y_1 = 28$ for the day shift and $y_2 = 11$ for the night shift. Let p_1 and p_2 denote the proportions of mints that weigh at most 21 grams for the day and night shifts, respectively.

   (a) Give a point estimate of p_1.
   (b) Give the endpoints for a 95% confidence interval for p_1.
   (c) Give a point estimate of $p_1 - p_2$.
   (d) Find a one-sided 95% confidence interval that gives a lower bound for $p_1 - p_2$.

6.7-15 Consider the following two groups of women: Group 1–Women who spend less than $500 annually on clothes; Group 2–Women who spend over $1000 annually on clothes. Let p_1 and p_2 equal the proportions of women in these two groups, respectively, who believe that clothes are too expensive. If 1009 out of a random sample of 1230 women from group 1 and 207 out of a random sample 340 from group 2 believe that clothes are too expensive,

   (a) Give a point estimate of $p_1 - p_2$.
   (b) Find an approximate 95% confidence interval for $p_1 - p_2$.

6.7-16 For developing countries in Asia (excluding China) and Africa, let p_1 and p_2 be the respective proportions of preschool children with chronic malnutrition (stunting). If respective random samples of $n_1 = 1300$ and $n_2 = 1100$ yielded $y_1 = 520$ and $y_2 = 385$ children with chronic malnutrition, find an approximate 95% confidence interval for $p_1 - p_2$.

6.7-17 The following question was asked in a *Newsweek* poll: "Would you prefer to live in a neighborhood with mostly whites, with mostly blacks, or in a neighborhood mixed half and half?" Let p_1 and p_2 equal the proportion of black and white adult respondents, respectively, who prefer "half and half." If 207 out of 305 black adults and 291 out of 632 white adults prefer "half and half,"

   (a) Give a point estimate of $p_1 - p_2$.
   (b) Find an approximate one-sided 90% confidence interval that gives a lower bound for $p_1 - p_2$.

6.7-18 An environmental survey contained a question asking what the respondent thought was the major cause of air pollution in this country, giving the choices "automobiles," "factories," and "incinerators." Two versions of the test, A and B, were used. Let p_A and p_B be the respective proportions of people using forms A and B who select "factories." If 170 out of 460 people who used version A chose "factories" and 141 out of 440 people who used version B chose "factories,"

   (a) Find a 95% confidence interval for $p_A - p_B$.
   (b) Do the forms seem to be consistent concerning this answer? Why?

6.8   **SAMPLE SIZE**

In statistical consulting, the first question frequently asked is: "How large should the sample size be to estimate a mean?" In order to convince the inquirer that the answer will depend on the variation associated with the random variable under observation, the statistician could correctly respond, "Only one observation is needed, provided that the standard deviation of the distribution is zero." That is, if σ equals zero, then the value of that one observation would necessarily equal the unknown mean of the distribution. This, of course, is an extreme case and one that is not met in practice; however, it should help convince persons that the smaller the variance, the smaller the sample size needed to achieve a given degree of accuracy. This will become clearer as we consider several examples. Let us begin with a problem that involves a statistical inference about the unknown mean of a distribution.

EXAMPLE 6.8-1   A mathematics department wishes to evaluate a new method of teaching calculus that does mathematics using a computer. At the end of the course, the evaluation will be made on the basis of scores of the participating students on a standard test. There is particular interest in estimating μ, the mean score for students taking calculus using the computer. Thus there is a desire to determine the number of students, n, who are to be selected at random from a larger group of students to take the course taught using the computer. Since new computing equipment must be purchased, they cannot afford to let all of the students take calculus the new way. In addition, some of the staff question the value of this approach and hence do not want to expose every student to this new procedure. So, let us find the sample size n such that we are fairly confident that $\bar{x} \pm 1$ contains the unknown test mean μ. From past experience it is believed that the standard deviation associated with this type of test is about 15. (The mean is also known when students take the standard calculus course.) Accordingly, using the fact that the sample mean of the test scores, $\overline{X}$, is approximately $N(\mu, \sigma^2/n)$, we see that the interval given by $\bar{x} \pm 1.96(15/\sqrt{n})$ will serve as an approximate 95% confidence interval for μ. That is, we want

$$1.96\left(\frac{15}{\sqrt{n}}\right) = 1$$

or, equivalently,

$$\sqrt{n} = 29.4 \qquad \text{and thus} \qquad n \approx 864.36$$

or $n = 865$ because n must be an integer.   ◄

It is quite likely that, in the preceding example, it had not been anticipated that as many as 865 students would be needed in this study. If that is the case, the statistician must discuss with those involved in the experiment whether or not the accuracy and the confidence level could be relaxed some. For illustration, rather than requiring $\bar{x} \pm 1$ to be a 95% confidence interval for μ, possibly $\bar{x} \pm 2$ would be a satisfactory 80% one. If this modification is acceptable, we now have

$$1.282\left(\frac{15}{\sqrt{n}}\right) = 2$$

or, equivalently,

$$\sqrt{n} = 9.615 \qquad \text{and} \qquad n \approx 92.4.$$

Since n must be an integer, we would probably use 93 in practice. Most likely, the persons involved in this project would find this a more reasonable sample size. Of course, any sample size greater than 93 could be used. Then either the length of the confidence interval could be decreased from that of $\bar{x} \pm 2$ or the confidence coefficient could be increased from 80% or a combination of both. Also, since there might be some question of whether the standard deviation σ actually equals 15, the sample standard deviation s would no doubt be used in the construction of the interval. For example, suppose that the sample characteristics observed are

$$n = 145, \qquad \bar{x} = 77.2, \qquad s = 13.2;$$

then

$$\bar{x} \pm \frac{1.282s}{\sqrt{n}} \qquad \text{or} \qquad 77.2 \pm 1.41$$

provides an approximate 80% confidence interval for μ.

In general, if we want the $100(1 - \alpha)\%$ confidence interval for μ, $\bar{x} \pm z_{\alpha/2}(\sigma/\sqrt{n})$, to be no longer than that given by $\bar{x} \pm \varepsilon$, the sample size n is the solution of

$$\varepsilon = \frac{z_{\alpha/2}\sigma}{\sqrt{n}}, \qquad \text{where} \quad \Phi(z_{\alpha/2}) = 1 - \frac{\alpha}{2}.$$

That is,

$$n = \frac{z_{\alpha/2}^2 \sigma^2}{\varepsilon^2}, \tag{6.8-1}$$

where it is assumed that σ^2 is known. We sometimes call $\varepsilon = z_{\alpha/2}(\sigma/\sqrt{n})$ the **maximum error of the estimate**. If the experimenter has no idea about the value of σ^2, it may be necessary to first take a preliminary sample to estimate σ^2.

The type of statistic we see most often in newspapers and magazines is an estimate of a proportion p. We might, for example, want to know the percentage of the labor force that is unemployed or the percentage of voters favoring a certain candidate. Sometimes extremely important decisions are made on the basis of these estimates. If this is the case, we would most certainly desire short confidence intervals for p with large confidence coefficients. We recognize that these conditions will require a large sample size. On the other hand, if the fraction p being estimated is not too important, an estimate associated with a longer confidence interval with a smaller confidence coefficient is satisfactory; and thus a smaller sample size can be used.

EXAMPLE 6.8-2 Suppose we know that the unemployment rate has been about 8% (0.08). However, we wish to update our estimate in order to make an important decision about the national economic policy. Accordingly, let us say we wish to be 99% confident that the new estimate of p is within 0.001 of the true p. If we assume Bernoulli trials (an assumption that might be questioned), the relative frequency y/n, based upon a large sample size n, provides the approximate 99% confidence interval

$$\frac{y}{n} \pm 2.576\sqrt{\frac{(y/n)(1 - y/n)}{n}}.$$

Although we do not know y/n exactly before sampling, we do know, since y/n will be near 0.08, that

$$2.576\sqrt{\frac{(y/n)(1 - y/n)}{n}} \approx 2.576\sqrt{\frac{(0.08)(0.92)}{n}}$$

and we want this number to equal 0.001. That is,

$$2.576\sqrt{\frac{(0.08)(0.92)}{n}} = 0.001$$

or, equivalently,

$$\sqrt{n} = 2576\sqrt{0.0736} \qquad \text{and} \qquad n \approx 488{,}394.$$

That is, under our assumptions, such a sample size is needed in order to achieve the reliability and the accuracy desired. Because n is so large, we would probably be willing to increase the error, say to 0.01 and perhaps reduce the confidence level to 98%. The required sample size required would then be

$$\sqrt{n} = (2.326/0.01)\sqrt{0.0736} \qquad \text{and} \qquad n \approx 3{,}982,$$

which is a more reasonable sample size.   ◄

From the preceding example we hope that the student will recognize how important it is to know the sample size (or length of the confidence interval and confidence coefficient) before he or she can place much weight on a statement such as 51% of the voters seem to favor candidate A, 46% favor candidate B, and 3% are undecided. Is this statement based on a sample of 100 or 2000 or 10,000 voters? If we assume Bernoulli trials, the approximate 95% confidence intervals for the fraction of voters favoring candidate A in these cases are, respectively, $[0.41, 0.61]$, $[0.49, 0.53]$, and $[0.50, 0.52]$. Quite obviously, the first interval, with $n = 100$, does not assure candidate A of the support of at least half the voters, whereas the interval with $n = 10{,}000$ is more convincing.

In general, to find the required sample size to estimate p, recall that the point estimate of p is $\widehat{p} = y/n$ and an approximate $1 - \alpha$ confidence interval for p is

$$\widehat{p} \pm z_{\alpha/2}\sqrt{\frac{\widehat{p}(1 - \widehat{p})}{n}}.$$

Suppose we want an estimate of p that is within ε of the unknown p with $100(1 - \alpha)\%$ confidence, where $\varepsilon = z_{\alpha/2}\sqrt{\widehat{p}(1 - \widehat{p})/n}$ is the **maximum error of the point estimate** $\widehat{p} = y/n$. Since $\widehat{p}$ is unknown before the experiment is run, we cannot use the value of $\widehat{p}$ in our determination of n. However, if it is known that p is about equal to p^*, the necessary sample size n is the solution of

$$\varepsilon = \frac{z_{\alpha/2}\sqrt{p^*(1 - p^*)}}{\sqrt{n}}.$$

That is,

$$n = \frac{z_{\alpha/2}^2 p^*(1 - p^*)}{\varepsilon^2}. \qquad (6.8\text{-}2)$$

It is often true, however, that we do not have a strong prior idea about p, as we did in Example 6.8-2 about the rate of unemployment. It is interesting to observe that no matter what value p takes between zero and one, it is always true that $p^*(1 - p^*) \leq 1/4$. Hence,

$$n = \frac{z_{\alpha/2}^2 p^*(1 - p^*)}{\varepsilon^2} \leq \frac{z_{\alpha/2}^2}{4\varepsilon^2}.$$

Thus, if we want the $100(1 - \alpha)\%$ confidence interval for p to be no longer than that given by $y/n \pm \varepsilon$, a solution for n that provides this protection is

$$n = \frac{z_{\alpha/2}^2}{4\varepsilon^2}. \qquad (6.8\text{-}3)$$

EXAMPLE 6.8-3 A possible gubernatorial candidate wants to assess initial support among the voters before making a candidacy announcement. If the fraction p of voters who are favorable, without any advance publicity, is around 0.15, the candidate will enter the race. From a poll of n voters selected at random, the candidate would like the estimate y/n to be within 0.03 of p. That is, the decision will be based on a 95% confidence interval of the form $y/n \pm 0.03$. Since the candidate has no idea about the magnitude of p, a consulting statistician formulates the equation

$$n = \frac{(1.96)^2}{4(0.03)^2} = 1067.11.$$

Thus the sample size should be around 1068 to achieve the desired reliability and accuracy. Suppose that 1068 voters around the state were selected at random and interviewed and $y = 214$ express support for this candidate. Then $\widehat{p} = 214/1068 = 0.20$ is a point estimate of p and an approximate 95% confidence interval for p is

$$0.20 \pm 1.96\sqrt{(0.20)(0.80)/n} \qquad \text{or} \qquad 0.20 \pm 0.024.$$

That is, we are 95% confident that p belongs to the interval $[0.176, 0.224]$. On the basis of this sample, the candidate decided to run for office. Note that for a confidence coefficient of 95%, we found a sample size so that the maximum error of the estimate would be 0.03. From the data that were collected, the maximum error of the estimate is only 0.024. We ended up with a smaller error because we found the sample size assuming that $p = 0.50$ while, in fact, p is closer to 0.20.   ◄

Suppose that you want to estimate the proportion p of a student body that favors a new policy. How large should the sample be? If p is close to 1/2 and you want to be 95% confident that the maximum error of the estimate is $\varepsilon = 0.02$, then

$$n = \frac{(1.96)^2}{4(0.02)^2} = 2401.$$

Such a sample size makes sense at a large university. However, if you are a student at a small college, the entire enrollment could be less than 2401. Thus we now give a procedure that can be used to determine the sample size when the population is small relative to the desired sample size.

Let N equal the size of a population and assume that N_1 of them have a certain characteristic C (e.g., favor a new policy). Let $p = N_1/N$. Then $1 - p = 1 - N_1/N$. If we take a sample of size n without replacement, then X, the number of observations with the characteristic C, has a hypergeometric distribution. The mean and variance of X are

$$\mu = n\left(\frac{N_1}{N}\right) = np$$

and

$$\sigma^2 = n\left(\frac{N_1}{N}\right)\left(1 - \frac{N_1}{N}\right)\left(\frac{N-n}{N-1}\right) = np(1-p)\left(\frac{N-n}{N-1}\right).$$

The mean and variance of X/n are

$$E\left(\frac{X}{n}\right) = \frac{\mu}{n} = p$$

and

$$\mathrm{Var}\left(\frac{X}{n}\right) = \frac{\sigma^2}{n^2} = \frac{p(1-p)}{n}\left(\frac{N-n}{N-1}\right).$$

To find an approximate confidence interval for p, we can use the normal approximation:

$$P\left[-z_{\alpha/2} \leq \frac{(X/n) - p}{\sqrt{\dfrac{p(1-p)}{n}\left(\dfrac{N-n}{N-1}\right)}} \leq z_{\alpha/2}\right] \approx 1 - \alpha.$$

Thus $1 - \alpha \approx$

$$P\left[\frac{X}{n} - z_{\alpha/2}\sqrt{\frac{p(1-p)}{n}\left(\frac{N-n}{N-1}\right)} \le p \le \frac{X}{n} + z_{\alpha/2}\sqrt{\frac{p(1-p)}{n}\left(\frac{N-n}{N-1}\right)}\right].$$

An approximate $1 - \alpha$ confidence interval for p is, replacing p under the radical with $\widehat{p} = x/n$,

$$\widehat{p} \pm z_{\alpha/2}\sqrt{\frac{\widehat{p}(1-\widehat{p})}{n}\left(\frac{N-n}{N-1}\right)}.$$

This is similar to the confidence interval for p when the distribution of X is $b(n, p)$. If N is large relative to n, then

$$\frac{N-n}{N-1} = \frac{1 - n/N}{1 - 1/N} \approx 1,$$

so in this case the two intervals are essentially equal.

Suppose now that we are interested in determining the sample size n that is required to have $1 - \alpha$ confidence that the maximum error of the estimate of p is ε. We let

$$\varepsilon = z_{\alpha/2}\sqrt{\frac{p(1-p)}{n}\left(\frac{N-n}{N-1}\right)}$$

and solve for n. After some simplification, we obtain

$$n = \frac{Nz_{\alpha/2}^2\, p(1-p)}{(N-1)\varepsilon^2 + z_{\alpha/2}^2\, p(1-p)}$$

$$= \frac{z_{\alpha/2}^2 p(1-p)/\varepsilon^2}{\dfrac{N-1}{N} + \dfrac{z_{\alpha/2}^2\, p(1-p)/\varepsilon}{N}}.$$

If we let

$$m = \frac{z_{\alpha/2}^2 p^*(1-p^*)}{\varepsilon^2},$$

which is the n value given by Equation 6.8-2, we then choose for our sample size n

$$n = \frac{m}{1 + \dfrac{m-1}{N}}.$$

If we know nothing about p, set $p^* = 1/2$ to determine m. For example, if the size of the student body is $N = 4000$ and $1 - \alpha = 0.95$, $\varepsilon = 0.02$, and we let $p^* = 1/2$, then $m = 2401$ and

$$n = \frac{2401}{1 + 2400/4000} = 1501$$

rounded up to the nearest integer. Thus we would sample approximately 37.5% of the student body.

EXAMPLE 6.8-4 Suppose that a college of $N = 3000$ students is interested in assessing student support for a new form for teacher evaluation. To estimate the proportion p in favor of the new form, how large a sample is required so that with 95% confidence the maximum error of the estimate of p is $\varepsilon = 0.03$? If we assume that p is completely unknown, we use $p^* = 1/2$ to obtain

$$m = \frac{(1.96)^2}{4(0.03)^2} = 1068,$$

rounding up to the nearest integer. Thus the desired sample size is

$$n = \frac{1068}{1 + 1067/3000} = 788,$$

rounding up to the nearest integer. ◀

EXERCISES

6.8-1 Let X equal the tarsus length for a male grackle. Assume that the distribution of X is $N(\mu, 4.84)$. Find the sample size n that is needed so that we are 95% confident that the maximum error of the estimate of μ is 0.4.

6.8-2 Let X equal the excess weight of soap in a "1000-gram" bottle. Assume that the distribution of X is $N(\mu, 169)$. What sample size is required so that we have 95% confidence that the maximum error of the estimate of μ is 1.5?

6.8-3 A company packages powdered soap in "6-pound" boxes. The sample mean and standard deviation of the soap in these boxes are currently 6.09 and 0.02 pounds. If the mean fill can be lowered by 0.01 pounds, $14,000 would be saved per year. Adjustments were made in the filling equipment, but it can be assumed that the standard deviation remains unchanged.

    (a) How large a sample is needed so that the maximum error of the estimate of the new μ is $\varepsilon = 0.001$ with 90% confidence?

    (b) A random sample of size $n = 1219$ yielded $\bar{x} = 6.048$ and $s = 0.022$. Calculate a 90% confidence interval for μ.

    (c) Estimate the savings per year with these new adjustments.

    (d) Estimate the proportion of boxes that will now weigh less than 6 pounds.

6.8-4 The length in centimeters of $n = 29$ fish (species nezumia) yielded an average length of $\bar{x} = 16.82$ and $s^2 = 34.9$. Determine the size of a new sample so that $\bar{x} \pm 0.5$ is an approximate 95% confidence interval for μ.

6.8-5 A quality engineer wanted to be 98% confident that the maximum error of the estimate of the mean strength, μ, of the left hinge on a vanity cover molded by a machine is 0.25. A preliminary sample of size $n = 32$ parts yielded a sample mean of $\bar{x} = 35.68$ and a standard deviation of $s = 1.723$.

   (a) How large a sample is required?
   (b) Does this seem to be a reasonable sample size, noting that destructive testing is needed to obtain the data?

6.8-6 A light bulb manufacturer sells a light bulb that has a mean life of 1450 hours with a standard deviation of 33.7 hours. A new manufacturing process is being tested and there is interest in knowing the mean life μ of the new bulbs. How large a sample is required so that $\bar{x} \pm 5$ is a 95% confidence interval for μ? You may assume that the change in the standard deviation is minimal.

6.8-7 For a public opinion poll for a close presidential election, let p denote the proportion of voters who favor candidate A. How large a sample should be taken if we want the maximum error of the estimate of p to be equal to

   (a) 0.03 with 95% confidence?
   (b) 0.02 with 95% confidence?
   (c) 0.03 with 90% confidence?

6.8-8 Let p equal the proportion of all college and university students who would say yes to the question, "Would you drink from the same glass as your friend if you suspected that this friend were an AIDS virus carrier?" Find the sample size required to be 95% confident that the maximum error of the estimate of p is 0.025. (For your information, it was reported in *Sociology and Social Research* **72**, 2 [January 1988] that 30 out of 375 San Diego State University students answered yes to this question.)

6.8-9 A die has been loaded to change the probability of rolling a 6. In order to estimate p, the new probability of rolling a 6, how many times must the die be rolled so that we are 99% confident that the maximum error of the estimate of p is $\varepsilon = 0.02$?

6.8-10 Some college professors and students examined 137 Canadian geese for patent schistosome in the year they hatched. Of these 137 birds, 54 were infected. They were interested in estimating p, the proportion of infected birds of this type. For future studies determine the sample size n so that the estimate of p is within $\varepsilon = 0.04$ of the unknown p with 90% confidence.

6.8-11 According to the Department of Health Care of the Elderly, Sherwood Hospital, Nottingham, England, only 25% of the patients who were using canes had canes of the correct length. Suppose that you were interested in estimating the proportion p of Americans who used canes of the correct length. How large a sample is required so that, with 95% confidence, the maximum error of the estimate of p is 0.04?

6.8-12 A seed distributor claims that 80% of its beet seeds will germinate. How many seeds must be tested for germination in order to estimate p, the true proportion that will germinate, so that the maximum error of the estimate is $\varepsilon = 0.03$ with 90% confidence?

6.8-13 Some dentists were interested in studying the fusion of embryonic rat palates by using a standard transplantation technique. When no treatment is used, the probability of fusion approximately equals 0.89. They would like to estimate p, the probability of fusion, when vitamin A is lacking.

   (a) How large a sample n of rat embryos is needed for $y/n \pm 0.10$ to be a 95% confidence interval for p?
   (b) If $y = 44$ out of $n = 60$ palates showed fusion, give a 95% confidence interval for p.

6.8-14 Let p equal the proportion of New York City residents who feel that the quality of life in New York City has become worse in the past few years. To update the estimate given in Exercise 6.7-9, how large a sample is required to be 98% confident that the maximum error of the estimate of p is 0.025?

6.8-15 Let p equal the proportion of triathletes who suffered a training-related overuse injury during the past year. (See Exercise 6.7-5.) How large a sample would be required to estimate p so that with 95% confidence the maximum error of the estimate of p is 0.04?

6.8-16 Let p equal the proportion of college students who favor a new policy for alcohol consumption on campus. How large a sample is required to estimate p so that with 95% confidence the maximum error of the estimate of p is 0.04 when the size of the student body is

   (a) $N = 1500$?
   (b) $N = 15,000$?
   (c) $N = 25,000$?

6.8-17 Out of 1000 welds that have been made on a tower, it is suspected that 15% of the welds are defective. To estimate p, the proportion of defective welds, how many welds must be inspected to have 95% confidence, approximately, that the maximum error of the estimate of p is 0.04?

6.8-18 If Y_1/n and Y_2/n are the respective independent relative frequencies of successes associated with the two binomial distributions $b(n, p_1)$ and $b(n, p_2)$, compute n such that the approximate probability that the random interval $Y_1/n - Y_2/n \pm 0.05$ covers $p_1 - p_2$ is at least 0.80. HINT: Take $p_1^* = p_2^* = 1/2$ to provide an upper bound for n.

6.9* ORDER STATISTICS

The **order statistics** are the observations of the random sample arranged, or ordered, in magnitude from the smallest to the largest. In recent years, the importance of order statistics has increased owing to the more frequent use of nonparametric inferences and robust procedures. However, order statistics have always been prominent because, among other things, they are needed to determine rather simple statistics such as the sample median, the sample range, and the empirical distribution function. Earlier we discussed observed order statistics in connection with descriptive and exploratory statistical methods. We will consider certain interesting aspects about their distributions in this section.

In most of our discussions about order statistics, we will assume that the n independent observations come from a continuous-type distribution. This means, among other things, that the probability of any two observations being equal is zero. That is, the probability is *one* that the observations can be ordered from smallest to largest without having two equal values. Of course, in practice, we do frequently observe *ties*; but if the probability of this is small, the following distribution theory will hold approximately. Thus, in the discussion here, we are assuming that the probability of ties is zero.

EXAMPLE 6.9-1 The values $x_1 = 0.62$, $x_2 = 0.98$, $x_3 = 0.31$, $x_4 = 0.81$, and $x_5 = 0.53$ are the $n = 5$ observed values of five independent trials of an experiment with p.d.f. $f(x) = 2x, 0 < x < 1$. The observed order statistics are

$$y_1 = 0.31 < y_2 = 0.53 < y_3 = 0.62 < y_4 = 0.81 < y_5 = 0.98.$$

Recall that the middle observation in the ordered arrangement, here $y_3 = 0.62$, is called the sample median and the difference of the largest and the smallest, here

$$y_5 - y_1 = 0.98 - 0.31 = 0.67,$$

is called the sample range.   ◄

If $X_1, X_2, \ldots, X_n$ are observations of a random sample of size n from a continuous-type distribution, we let the random variables

$$Y_1 < Y_2 < \cdots < Y_n$$

denote the order statistics of that sample. That is,

$$Y_1 = \text{smallest of } X_1, X_2, \ldots, X_n,$$
$$Y_2 = \text{second smallest of } X_1, X_2, \ldots, X_n,$$
$$\vdots$$
$$Y_n = \text{largest of } X_1, X_2, \ldots, X_n.$$

There is a very simple method for determining the distribution function of the rth order statistic, Y_r. This procedure depends on the binomial distribution and is illustrated in Example 6.9-2.

EXAMPLE 6.9-2   Let $Y_1 < Y_2 < Y_3 < Y_4 < Y_5$ be the order statistics associated with n independent observations X_1, X_2, X_3, X_4, X_5, each from the distribution with p.d.f. $f(x) = 2x$, $0 < x < 1$. Consider $P(Y_4 < 1/2)$. For the event $\{Y_4 < 1/2\}$ to occur, at least four of the random variables X_1, X_2, X_3, X_4, X_5 must be less than $1/2$ because Y_4 is the fourth smallest among the five observations. Thus if the event $\{X_i < 1/2\}$, $i = 1, 2, \ldots, 5$, is called "success," we must have at least four successes in the five mutually independent trials, each of which has probability of success

$$P\left(X_i \leq \frac{1}{2}\right) = \int_0^{1/2} 2x\, dx = \left(\frac{1}{2}\right)^2 = \frac{1}{4}.$$

Thus

$$P\left(Y_4 \leq \frac{1}{2}\right) = \binom{5}{4}\left(\frac{1}{4}\right)^4\left(\frac{3}{4}\right) + \left(\frac{1}{4}\right)^5 = 0.0156.$$

In general, if $0 < y < 1$, then the distribution function of Y_4 is

$$G(y) = P(Y_4 < y) = \binom{5}{4}(y^2)^4(1 - y^2) + (y^2)^5,$$

since this represents the probability of at least four "successes" in five independent trials, each of which has probability of success

$$P(X_i < y) = \int_0^y 2x \, dx = y^2.$$

The p.d.f. of Y_4 is therefore, for $0 < y < 1$,

$$g(y) = G'(y) = \binom{5}{4} 4(y^2)^3 (2y)(1 - y^2) + \binom{5}{4}(y^2)^4(-2y) + 5(y^2)^4(2y)$$

$$= \frac{5!}{3! \, 1!} (y^2)^3 (1 - y^2)(2y), \qquad 0 < y < 1.$$

Note that in this example, the distribution function of each X is $F(x) = x^2$ when $0 < x < 1$. Thus

$$g(y) = \frac{5!}{3! \, 1!} [F(y)]^3 [1 - F(y)] f(y), \qquad 0 < y < 1. \qquad \blacktriangleleft$$

The preceding example should make the following generalization easier to read. Let $Y_1 < Y_2 < \cdots < Y_n$ be the order statistics of n independent observations from a distribution of the continuous type with distribution function $F(x)$ and p.d.f. $F'(x) = f(x)$, where $0 < F(x) < 1$ for $a < x < b$ and $F(a) = 0$, $F(b) = 1$. (It is possible that $a = -\infty$ and/or $b = +\infty$.) The event that the rth order statistic Y_r is at most y, $\{Y_r \leq y\}$, can occur if and only if at least r of the n observations are less than or equal to y. That is, here the probability of "success" on each trial is $F(y)$ and we must have at least r successes. Thus

$$G_r(y) = P(Y_r \leq y) = \sum_{k=r}^{n} \binom{n}{k}[F(y)]^k[1 - F(y)]^{n-k}.$$

Rewriting this slightly, we have

$$G_r(y) = \sum_{k=r}^{n-1} \binom{n}{k}[F(y)]^k[1 - F(y)]^{n-k} + [F(y)]^n.$$

Thus the p.d.f. of Y_r is

$$g_r(y) = G'_r(y) = \sum_{k=r}^{n-1} \binom{n}{k}(k)[F(y)]^{k-1} f(y)[1 - F(y)]^{n-k}$$

$$+ \sum_{k=r}^{n-1} \binom{n}{k}[F(y)]^k(n - k)[1 - F(y)]^{n-k-1}[-f(y)]$$

$$+ n[F(y)]^{n-1} f(y). \qquad (6.9\text{-}1)$$

However, since

$$\binom{n}{k} k = \frac{n!}{(k-1)!\,(n-k)!} \qquad \text{and} \qquad \binom{n}{k}(n-k) = \frac{n!}{k!\,(n-k-1)!},$$

we have that the p.d.f. of Y_r is

$$g_r(y) = \frac{n!}{(r-1)!\,(n-r)!} [F(y)]^{r-1}[1 - F(y)]^{n-r} f(y), \qquad a < y < b,$$

which is the first term of the first summation in $g_r(y) = G_r'(y)$, Equation 6.9-1. The remaining terms in $g_r(y) = G_r'(y)$ sum to zero because the second term of the first summation (when $k = r + 1$) equals the negative of the first term in the second summation (when $k = r$), and so on. Finally, the last term of the second summation equals the negative of $n[F(y)]^{n-1} f(y)$. To see this clearly, the student is urged to write out a number of terms in these summations (see Exercise 6.9-4).

It is worth noting that the p.d.f. of the smallest order statistic is

$$g_1(y) = n[1 - F(y)]^{n-1} f(y), \qquad a < y < b,$$

and the p.d.f. of the largest order statistic is

$$g_n(y) = n[F(y)]^{n-1} f(y), \qquad a < y < b.$$

Remark   There is one very satisfactory way to construct heuristically the expression for the p.d.f. of Y_r. To do this, we must recall the multinomial probability and then consider the probability element $g_r(y)(\Delta y)$ of Y_r. If the length Δy is *very* small, $g_r(y)(\Delta y)$ represents approximately the probability

$$P(y < Y_r \le y + \Delta y).$$

Thus we want the probability, $g_r(y)(\Delta y)$, that $(r - 1)$ items fall less than y, $(n - r)$ items are greater than $y + \Delta y$, and one item falls between y and $y + \Delta y$. Recall that the probabilities on a single trial are

$$P(X \le y) = F(y)$$
$$P(X > y + \Delta y) = 1 - F(y + \Delta y) \approx 1 - F(y)$$
$$P(y < X \le y + \Delta y) \approx f(y)(\Delta y)$$

Thus the multinomial probability is approximately

$$g_r(y)(\Delta y) = \frac{n!}{(r-1)!\,1!\,(n-r)!} [F(y)]^{r-1}[1 - F(y)]^{n-r}[f(y)(\Delta y)].$$

If we divide both sides by the length Δy, the formula for $g_r(y)$ results.   ■

EXAMPLE 6.9-3 Returning to Example 6.9-2, we shall now graph the p.d.f.s of the order statistics $Y_1 < Y_2 < Y_3 < Y_4 < Y_5$ when sampling from a distribution with p.d.f. $f(x) = 2x, 0 < x < 1$. These graphs are given in Figure 6.9-1. The respective p.d.f.s and their means are as follows:

$$g_1(y) = 10y(1 - y^2)^4; \qquad \mu_1 = \frac{256}{693},$$

$$g_2(y) = 40y^3(1 - y^2)^3; \qquad \mu_2 = \frac{128}{231},$$

$$g_3(y) = 60y^5(1 - y^2)^2; \qquad \mu_3 = \frac{160}{231},$$

$$g_4(y) = 40y^7(1 - y^2); \qquad \mu_4 = \frac{80}{99},$$

$$g_5(y) = 10y^9; \qquad \mu_5 = \frac{10}{11}.$$

◀

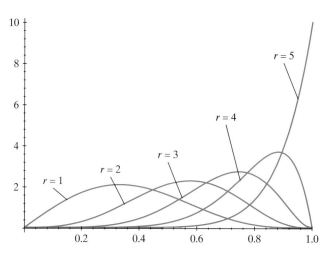

Figure 6.9-1: p.d.f.s of order statistics, $f(x) = 2x, 0 < x < 1$

EXAMPLE 6.9-4 Let $Y_1 < Y_2 < Y_3 < Y_4$ be the order statistics of a random sample X_1, X_2, X_3, X_4 from a uniform distribution with p.d.f. $f(x; \theta) = 1/\theta, 0 < x \leq \theta$. The likelihood function is

$$f(x; \theta) = \left(\frac{1}{\theta}\right)^4, \qquad 0 < x_i \leq \theta, \ i = 1, 2, 3, 4,$$

and equals zero if $\theta < x_i$. To maximize $L(\theta)$ we must make θ as small as possible; hence

$$\widehat{\theta} = \max(X_i) = Y_4$$

because θ cannot be less than any X_i. Since $F(x; \theta) = x/\theta$, $0 < x \le \theta$, then the p.d.f. of Y_4 is

$$g_4(y_4) = \frac{4!}{3!1!} \left(\frac{y_4}{\theta}\right)^3 \left(\frac{1}{\theta}\right) = 4\frac{y_4^3}{\theta^4}, \qquad 0 < y_4 \le \theta.$$

Accordingly,

$$E(Y_4) = \int_0^\theta y_4 \cdot 4\frac{y_4^3}{\theta^4}\, dy_4 = \frac{4}{5}\theta$$

and Y_4 is a biased estimator of θ. However, $5Y_4/4$ is unbiased. Since, with $0 < c < 1$,

$$\int_{c\theta}^\theta 4\, y_4^3/\theta^4\, dy_4 = 1 - c^4;$$

thus

$$1 - c^4 = P(c\theta < Y_4 < \theta) = P(Y_4 < \theta < Y_4/c).$$

Then, with an observed Y_4 of y_4, y_4 to y_4/c would be a $100(1 - c^4)$ percent confidence interval for θ. Say $c = (0.05)^{1/4} = 0.47$, then y_4 to $2.11y_4$ is a 95% confidence interval for θ.   ◀

Recall that, in Theorem 3.5-2, we proved that if X has a distribution function $F(x)$ of the continuous type, then $F(X)$ has a uniform distribution on the interval zero to one. If $Y_1 < Y_2 < \cdots < Y_n$ are the order statistics of n independent observations $X_1, X_2, \ldots, X_n$, then

$$F(Y_1) < F(Y_2) < \cdots < F(Y_n)$$

because F is a nondecreasing function and the probability of an equality is again zero. Note that this last display could be looked upon as an ordering of the mutually independent random variables $F(X_1), F(X_2), \ldots, F(X_n)$, each of which is $U(0, 1)$. That is,

$$W_1 = F(Y_1) < W_2 = F(Y_2) < \cdots < W_n = F(Y_n)$$

can be thought of as the order statistics of n independent observations from that uniform distribution. Since the distribution function of $U(0, 1)$ is $G(w) = w, 0 < w < 1$, the p.d.f. of the rth order statistic $W_r = F(Y_r)$ is

$$h_r(w) = \frac{n!}{(r-1)!\,(n-r)!}\, w^{r-1}(1 - w)^{n-r}, \qquad 0 < w < 1.$$

Of course, the mean, $E(W_r) = E[F(Y_r)]$ of $W_r = F(Y_r)$, is given by the integral

$$E(W_r) = \int_0^1 w \frac{n!}{(r-1)!\,(n-r)!}\, w^{r-1}(1 - w)^{n-r}\, dw.$$

This can be evaluated by integrating by parts several times, but it is easier to obtain the answer if we rewrite it as follows:

$$E(W_r) = \left(\frac{r}{n+1}\right) \int_0^1 \frac{(n+1)!}{r!\,(n-r)!} w^r (1-w)^{n-r}\, dw.$$

The integrand in this last expression can be thought of as the p.d.f. of the $(r+1)$st order statistic of $n+1$ independent observations from a $U(0,1)$ distribution. This is a beta p.d.f. with $\alpha = r+1$ and $\beta = n-r+1$ and hence the integral must equal one and

$$E(W_r) = \frac{r}{n+1}, \qquad r = 1, 2, \ldots, n.$$

There is an extremely interesting interpretation of $W_r = F(Y_r)$. Note that $F(Y_r)$ is the cumulated probability up to and including Y_r or, equivalently, the area under $f(x) = F'(x)$ but less than Y_r. Hence $F(Y_r)$ can be treated as a random area. Since $F(Y_{r-1})$ is also a random area, $F(Y_r) - F(Y_{r-1})$ is the random area under $f(x)$ between Y_{r-1} and Y_r. The expected value of the random area between any two adjacent order statistics is then

$$E[F(Y_r) - F(Y_{r-1})] = E[F(Y_r)] - E[F(Y_{r-1})]$$
$$= \frac{r}{n+1} - \frac{r-1}{n+1} = \frac{1}{n+1}.$$

Also, it is easy to show (see Exercise 6.9-6)

$$E[F(Y_1)] = \frac{1}{n+1} \qquad \text{and} \qquad E[1 - F(Y_n)] = \frac{1}{n+1}.$$

That is, the order statistics $Y_1 < Y_2 < \cdots < Y_n$ partition the support of X into $n+1$ parts and thus create $n+1$ areas under $f(x)$ and above the x axis. "On the average," each of the $n+1$ areas equals $1/(n+1)$.

If we recall that the $(100p)$th percentile π_p is such that the area under $f(x)$ to the left of π_p is p, the preceding discussion suggests that we let Y_r be an estimator of π_p, where $p = r/(n+1)$. For this reason, we define the $(100p)$**th percentile of the sample** as Y_r, where $r = (n+1)p$. In case $(n+1)p$ is not an integer, we use a weighted average (or an average) of the two adjacent order statistics Y_r and Y_{r+1}, where r is the greatest integer $\lfloor (n+1)p \rfloor$ in $(n+1)p$. In particular, the sample median is

$$\widetilde{m} = \begin{cases} Y_{(n+1)/2}, & \text{when } n \text{ is odd,} \\[2mm] \dfrac{Y_{n/2} + Y_{(n/2)+1}}{2}, & \text{when } n \text{ is even.} \end{cases}$$

EXAMPLE 6.9-5 Let X equal the weight of soap in a "1000-gram" bottle. A random sample of $n = 12$ observations of X yielded the following weights that have been ordered:

1013	1019	1021	1024	1026	1028
1033	1035	1039	1040	1043	1047

Since $n = 12$ is even, the sample median is

$$\tilde{m} = \frac{y_6 + y_7}{2} = \frac{1028 + 1033}{2} = 1030.5.$$

The location of the 25th percentile (or first quartile) is

$$(n + 1)(0.25) = (12 + 1)(0.25) = 3.25.$$

Thus the first quartile, using a weighted average, is

$$\tilde{q}_1 = y_3 + (0.25)(y_4 - y_3) = (0.75)y_3 + (0.25)y_4$$
$$= (0.75)(1021) + (0.25)(1024) = 1021.75.$$

Similarly, the 75th percentile (or third quartile) is

$$\tilde{q}_3 = y_9 + (0.75)(y_{10} - y_9) = (0.25)y_9 + (0.75)y_{10}$$
$$= (0.25)(1039) + (0.75)(1040) = 1039.75$$

because $(12 + 1)(0.75) = 9.75$. Since $(12 + 1)(0.60) = 7.8$, the 60th percentile is

$$\tilde{\pi}_{0.60} = (0.2)y_7 + (0.8)y_8 = (0.2)(1033) + (0.8)(1035) = 1034.6. \quad \blacktriangleleft$$

EXERCISES

6.9-1 One of the tasks performed by a computer operator is that of testing tapes in order to detect bad records (i.e., records that could not store information in the form of positive or negative charges). Let X equal the distance in feet between bad records. Ten observations of X are

$$67 \quad 6 \quad 7 \quad 35 \quad 78 \quad 28 \quad 74 \quad 5 \quad 9 \quad 37$$

(a) Find the order statistics.
(b) Find the median and 80th percentile of the sample.
(c) Determine the first and third quartiles (i.e., 25th and 75th percentiles) of the sample.

6.9-2 Let X equal the forced vital capacity (the volume of air a person can expel from their lungs) for a male freshman. Seventeen observations of X, which have been ordered, are

$$3.7 \quad 3.8 \quad 4.0 \quad 4.3 \quad 4.7 \quad 4.8 \quad 4.9 \quad 5.0$$
$$5.2 \quad 5.4 \quad 5.6 \quad 5.6 \quad 5.6 \quad 5.7 \quad 6.2 \quad 6.8 \quad 7.6$$

(a) Find the median, the first quartile, and the third quartile.
(b) Find the 35th and 65th percentiles.

6.9-3 Let $Y_1 < Y_2 < Y_3 < Y_4 < Y_5$ be the order statistics of 5 independent observations from an exponential distribution that has a mean of $\theta = 3$.

(a) Find the p.d.f. of the sample median Y_3.
(b) Compute the probability that Y_4 is less than 5.
(c) Determine $P(1 < Y_1)$.

6.9-4 In the expression for $g_r(y) = G'_r(y)$ in Equation 6.9-1, let $n = 6, r = 3$, and write out the summations, showing that the "telescoping" suggested in the text is achieved.

6.9-5 Let $Y_1 < Y_2 < \cdots < Y_8$ be the order statistics of 8 independent observations from a continuous-type distribution with 70th percentile $\pi_{0.7} = 27.3$.

(a) Determine $P(Y_7 < 27.3)$.

(b) Find $P(Y_5 < 27.3 < Y_8)$.

6.9-6 Let $W_1 < W_2 < \cdots < W_n$ be the order statistics of n independent observations from a $U(0, 1)$ distribution.

(a) Find the p.d.f. of W_1 and that of W_n.

(b) Use the results of (a) to verify that $E(W_1) = 1/(n + 1)$ and $E(W_n) = n/(n + 1)$.

6.9-7 Let $Y_1 < Y_2 < \cdots < Y_{19}$ be the order statistics of $n = 19$ independent observations from the exponential distribution with mean θ.

(a) What is the p.d.f. of Y_1?

(b) Find, using integration, the value of $E[F(Y_1)]$, where F is the distribution function of the exponential distribution.

6.9-8 Let $W_1 < W_2 < \cdots < W_n$ be the order statistics of n independent observations from a $U(0, 1)$ distribution.

(a) Show that $E(W_r^2) = r(r+1)/(n+1)(n+2)$ using a technique similar to that used in determining that $E(W_r) = r/(n+1)$.

(b) Find the variance of W_r.

6.9-9 Let $X_1, X_2, \ldots, X_{10}$ be a random sample of size $n = 10$ from a distribution with p.d.f. $f(x; \theta) = e^{-(x-\theta)}$, $\theta \leq x < \infty$.

(a) Show that $Y_1 = \min(X_i)$ is the maximum likelihood estimator of θ.

(b) Find the p.d.f. of Y_1 and show that $E(Y_1) = \theta + 1/10$ so that $Y_1 - 1/10$ is an unbiased estimator of θ.

(c) Compute $P(\theta \leq Y_1 \leq \theta + c)$ and use this to construct a 95% confidence interval for θ.

6.10* DISTRIBUTION-FREE CONFIDENCE INTERVALS FOR PERCENTILES

In Section 6.9* we defined the sample percentiles in terms of the order statistics and noted that the sample percentiles can be used to estimate the corresponding distribution percentiles. In this section we use the order statistics to construct confidence intervals for the unknown distribution percentiles. Since little is assumed about the underlying distribution (except it is of the continuous type) in the construction of these confidence intervals, they are often called **distribution-free confidence intervals**.

If $Y_1 < Y_2 < Y_3 < Y_4 < Y_5$ are the order statistics of a random sample of size $n = 5$ from a continuous-type distribution, then the sample median Y_3 could be thought of as an estimator of the distribution median $\pi_{0.5}$. We shall let $m = \pi_{0.5}$. We could simply use the sample median Y_3 as an estimator of the distribution median m. However, we are certain that all of us recognize that, with only a sample of size 5, we would be quite lucky if the observed $Y_3 = y_3$ were very close to m. Thus we now describe how a confidence interval can be constructed for m.

Instead of simply using Y_3 as an estimator of m, let us also compute the probability that the random interval (Y_1, Y_5) includes m. That is, let us determine $P(Y_1 < m < Y_5)$. This is easy if we say that we have success if an individual item, say X, is less than m; thus the probability of success on one of the independent trials is $P(X < m) = 0.5$. In order for the first order statistic Y_1 to be less than m and the last order statistic Y_5 to be greater than m, we must have at least one success but not five successes. That is,

$$P(Y_1 < m < Y_5) = \sum_{k=1}^{4} \binom{5}{k} \left(\frac{1}{2}\right)^k \left(\frac{1}{2}\right)^{5-k}$$

$$= 1 - \left(\frac{1}{2}\right)^5 - \left(\frac{1}{2}\right)^5 = \frac{15}{16}.$$

So the probability that the random interval (Y_1, Y_5) includes m is $15/16 \approx 0.94$. Suppose that this random sample is actually taken and the order statistics are observed to equal $y_1 < y_2 < y_3 < y_4 < y_5$, respectively. Then (y_1, y_5) is a 94% confidence interval for m.

It is interesting to note what happens as the sample size increases. Let $Y_1 < Y_2 < \cdots < Y_n$ be the order statistics of a random sample of size n from a distribution of the continuous type. Thus $P(Y_1 < m < Y_n)$ is the probability that there is at least one "success" but not n successes, where the probability of success on each trial is $P(X < m) = 0.5$. Consequently,

$$P(Y_1 < m < Y_n) = \sum_{k=1}^{n-1} \binom{n}{k} \left(\frac{1}{2}\right)^k \left(\frac{1}{2}\right)^{n-k}$$

$$= 1 - \left(\frac{1}{2}\right)^n - \left(\frac{1}{2}\right)^n = 1 - \left(\frac{1}{2}\right)^{n-1}.$$

This probability increases as n increases so that the corresponding confidence interval (y_1, y_n) would have a very large confidence coefficient, $1 - (1/2)^{n-1}$. Unfortunately, the interval (y_1, y_n) tends to get wider as n increases, and thus we are not "pinning down" m very well. However, if we used the interval (y_2, y_{n-1}) or (y_3, y_{n-2}), we would obtain shorter intervals but also smaller confidence coefficients. Let us investigate this possibility further.

With the order statistics $Y_1 < Y_2 < \cdots < Y_n$ associated with a random sample of size n from a continuous-type distribution, consider $P(Y_i < m < Y_j)$, where $i < j$. For example, we might want

$$P(Y_2 < m < Y_{n-1}) \quad \text{or} \quad P(Y_3 < m < Y_{n-2}).$$

On each of the n independent trials we say that we have success if that X is less than m; thus the probability of success on each trial is $P(X < m) = 0.5$. Consequently, to have the ith order statistic Y_i less than m and the jth order statistic greater than m, we must have at least i successes but fewer than j successes (or else $Y_j < m$). That is,

$$P(Y_i < m < Y_j) = \sum_{k=i}^{j-1} \binom{n}{k} \left(\frac{1}{2}\right)^k \left(\frac{1}{2}\right)^{n-k} = 1 - \alpha.$$

For particular values of n, i, and j, this probability, say $1 - \alpha$, which is the sum of probabilities from a binomial distribution, can be calculated directly or approximated by an area under the normal p.d.f. provided n is large enough. The observed interval (y_i, y_j) could then serve as a $100(1 - a)\%$ confidence interval for the unknown distribution median.

EXAMPLE 6.10-1 The lengths in centimeters of $n = 9$ fish of a particular species (*nezumia*) captured off the New England coast were 32.5, 27.6, 29.3, 30.1, 15.5, 21.7, 22.8, 21.2, 19.0. Thus the observed order statistics are

$$15.5 < 19.0 < 21.2 < 21.7 < 22.8 < 27.6 < 29.3 < 30.1 < 32.5.$$

Before the sample is drawn, we know that

$$P(Y_2 < m < Y_8) = \sum_{k=2}^{7} \binom{9}{k}\left(\frac{1}{2}\right)^k \left(\frac{1}{2}\right)^{9-k} = 0.9805 - 0.0195 = 0.9610,$$

from Table II in the Appendix. Thus the confidence interval ($y_2 = 19.0$, $y_8 = 30.1$) for m, the median of the lengths of all fish of this species, has a 96.1% confidence coefficient. ◀

So that the student need not compute many of these probabilities, we give in Table 6.10-1 the necessary information for constructing confidence intervals of the form (y_i, y_{n+1-i}) for the unknown m for sample sizes $n = 5, 6, \ldots, 20$. The subscript i is selected so that the confidence coefficient $P(Y_i < m < Y_{n+1-i})$ is greater than 90% and as close to 95% as possible.

For sample sizes larger than 20, we approximate those binomial probabilities with areas under the normal curve. To illustrate how good these approximations are, we compute the probability corresponding to $n = 16$ in Table 6.10-1. Here, using Table II, we have

$$1 - \alpha = P(Y_5 < m < Y_{12}) = \sum_{k=5}^{11} \binom{16}{k}\left(\frac{1}{2}\right)^k \left(\frac{1}{2}\right)^{16-k}$$

$$= P(W = 5, 6, \ldots, 11) = 0.9616 - 0.0384 = 0.9232,$$

where W is $b(16, 1/2)$. The normal approximation gives

$$1 - \alpha = P(4.5 < W < 11.5) = P\left(\frac{4.5 - 8}{2} < \frac{W - 8}{2} < \frac{11.5 - 8}{2}\right)$$

because W has mean $np = 8$ and variance $np(1 - p) = 4$. The standardized variable $Z = (W - 8)/2$ has an approximate normal distribution. Thus

$$1 - \alpha \approx \Phi\left(\frac{3.5}{2}\right) - \Phi\left(\frac{-3.5}{2}\right) = \Phi(1.75) - \Phi(-1.75)$$

$$= 0.9599 - 0.0401 = 0.9198.$$

This compares very favorably with the probability 0.9232 recorded in Table 6.10-1.

Table 6.10-1: Information for Confidence Intervals for m		
n	$(i, n+1-i)$	$P(Y_i < m < Y_{n+1-i})$
5	$(1, 5)$	0.9376
6	$(1, 6)$	0.9688
7	$(1, 7)$	0.9844
8	$(2, 7)$	0.9296
9	$(2, 8)$	0.9610
10	$(2, 9)$	0.9786
11	$(3, 9)$	0.9346
12	$(3, 10)$	0.9614
13	$(3, 11)$	0.9776
14	$(4, 11)$	0.9426
15	$(4, 12)$	0.9648
16	$(5, 12)$	0.9232
17	$(5, 13)$	0.9510
18	$(5, 14)$	0.9692
19	$(6, 14)$	0.9364
20	$(6, 15)$	0.9586

The argument used to find a confidence interval for the median m of a distribution of the continuous type can be applied to any percentile π_p. In this case we say that we have success on a single trial if that X is less than π_p. Thus the probability of success on each of the independent trials is $P(X < \pi_p) = p$. Accordingly, with $i < j$, $1 - \alpha = P(Y_i < \pi_p < Y_j)$ is the probability that we have at least i successes but fewer than j successes. Thus

$$1 - \alpha = P(Y_i < \pi_p < Y_j) = \sum_{k=i}^{j-1} \binom{n}{k} p^k (1-p)^{n-k}.$$

Once the sample is observed and the order statistics determined, the known interval (y_i, y_j) could serve as a $100(1 - \alpha)\%$ confidence interval for the unknown distribution percentile π_p.

EXAMPLE 6.10-2 Let the following numbers represent the order statistics of the $n = 27$ observations obtained in a random sample from a certain population of incomes (measured in hundreds of dollars).

161	180	192	205	229	264
169	183	193	213	241	291
171	184	196	221	243	317
174	186	200	222	256	376
179	187	204			

Say we are interested in estimating the 25th percentile $\pi_{0.25}$ of the population. Since $(n + 1)p = 28(1/4) = 7$, the 7th order statistic, namely $y_7 = 183$, would be a point estimate of $\pi_{0.25}$. To find a confidence interval for $\pi_{0.25}$, let us move down and up a few order statistics from y_7, say to y_4 and y_{10}. What is the confidence coefficient associated with the interval (y_4, y_{10})? Before the sample was drawn, we had

$$1 - \alpha = P(Y_4 < \pi_{0.25} < Y_{10}) = \sum_{k=4}^{9} \binom{27}{k}(0.25)^k(0.75)^{27-k}$$

$$= P(3.5 < W < 9.5),$$

where W is $b(27, 1/4)$ with mean $27/4 = 6.75$ and variance $81/16$. Hence

$$1 - \alpha \approx \Phi\left(\frac{9.5 - 6.75}{9/4}\right) - \Phi\left(\frac{3.5 - 6.75}{9/4}\right)$$

$$= \Phi\left(\frac{11}{9}\right) - \Phi\left(-\frac{13}{9}\right) = 0.8149.$$

Thus $(y_4 = 174, \ y_{10} = 187)$ is an 81.49% confidence interval for $\pi_{0.25}$. It should be noted that we could choose other intervals, such as $(y_3 = 171, y_{11} = 192)$, and these would have different confidence coefficients. The persons involved in the study must select the desired confidence coefficient, and then the appropriate order statistics are taken, usually quite symmetrically about the $(n + 1)p$th order statistic. ◀

When the number of observations is large, it is important to be able to determine rather easily the order statistics. A stem-and-leaf diagram, as introduced in Section 3.1, can be helpful in determining the needed order statistics. This is illustrated in the next example.

EXAMPLE 6.10-3 The measurements of butterfat produced by $n = 90$ cows during a 305-day milk production period following their first calf are summarized in Table 6.10-2, in which each leaf consists of two digits.

From this display it is quite easy to see that $y_8 = 392$. It takes a little more work to show that $y_{38} = 494$ and $y_{53} = 526$. The interval $(494, 526)$ serves as a confidence interval for the unknown median m of all butterfat production for the given breed of cows. Its confidence coefficient is

$$P(Y_{38} < m < Y_{53}) = \sum_{k=38}^{52} \binom{90}{k}\left(\frac{1}{2}\right)^k\left(\frac{1}{2}\right)^{90-k}$$

$$\approx \Phi\left(\frac{52.5 - 45}{\sqrt{22.5}}\right) - \Phi\left(\frac{37.5 - 45}{\sqrt{22.5}}\right)$$

$$= \Phi(1.58) - \Phi(-1.58) = 0.8858.$$

Table 6.10-2: Ordered Stem-and-Leaf Diagram of Butterfat Production

Stems	Leaves										Depths
2s	74										1
2•											1
3*											1
3t	27	39									3
3f	45	50									5
3s											5
3•	80	88	92	94	95						10
4*	17	18									12
4t	21	22	27	34	37	39					18
4f	44	52	53	53	57	58					24
4s	60	64	66	70	70	72	75	78			32
4•	81	86	89	91	92	94	96	97	99		41
5*	00	00	01	02	05	09	10	13	13	16	(10)
5t	24	26	31	32	32	37	37	39			39
5f	40	41	44	55							31
5s	61	70	73	74							27
5•	83	83	86	93	99						23
6*	07	08	11	12	13	17	18	19			18
6t	27	28	35	37							10
6f	43	43	45								6
6s	72										3
6•	91	96									2

Similarly, $(y_{17} = 437, y_{29} = 470)$ is a confidence interval for the first quartile $\pi_{0.25}$ with confidence coefficient

$$P(Y_{17} < \pi_{0.25} < Y_{29}) \approx \Phi\left(\frac{28.5 - 22.5}{\sqrt{16.875}}\right) - \Phi\left(\frac{16.5 - 22.5}{\sqrt{16.875}}\right)$$

$$= \Phi(1.46) - \Phi(-1.46) = 0.8558. \qquad \blacktriangleleft$$

It is interesting to compare the lengths of a confidence interval for the mean μ using $\bar{x} \pm t_{\alpha/2}(n-1)(s/\sqrt{n})$ and a $100(1-\alpha)\%$ one for the median m using the distribution-free techniques of this section. Usually, if the sample arises from a distribution that does not deviate too much from the normal, the confidence interval based upon $\bar{x}$ is much shorter. After all, we assume much more creating that confidence interval. With the distribution-free method, all we assume is that the distribution be of the continuous type. So if the distribution is highly skewed or heavy tailed so that outliers could exist, the distribution-free techniques is safer and much more robust. Moreover, the distribution-free technique provides a way to get confidence intervals for various percentiles, and investigators are often interested in these.

6.10-1 Let $Y_1 < Y_2 < Y_3 < Y_4 < Y_5 < Y_6$ be the order statistics of a random sample of size $n = 6$ from a distribution of the continuous type having $(100p)$th percentile π_p. Compute

(a) $P(Y_2 < \pi_{0.5} < Y_5)$.

(b) $P(Y_1 < \pi_{0.25} < Y_4)$.

(c) $P(Y_4 < \pi_{0.9} < Y_6)$.

6.10-2 For $n = 12$ 1995 model automobiles whose horse power is between 290 and 390, the following measurements give the time in seconds for the car to go from 0 to 60 mph.

$$6.2 \quad 7.9 \quad 6.6 \quad 6.4 \quad 5.2 \quad 4.9 \quad 5.5 \quad 7.0 \quad 6.6 \quad 5.4 \quad 5.3 \quad 5.1$$

(a) Find a 96.14% confidence interval for the median, m.

(b) The interval (y_1, y_7) could serve as a confidence interval for $\pi_{0.3}$. Find it and give its confidence coefficient?

6.10-3 A sample of $n = 9$ electrochromic mirrors was used to measure the following low end reflectivity percentages.

$$7.12 \quad 7.22 \quad 6.78 \quad 6.31 \quad 5.99 \quad 6.58 \quad 7.80 \quad 7.40 \quad 7.05$$

(a) Find the endpoints for an approximate 95% confidence interval for the median, m.

(b) The interval (y_3, y_7) could serve as a confidence interval for m. Find it and give its confidence coefficient.

6.10-4 Let m denote the median weight of "80 pound" bags of water softener pellets. Use the following random sample of $n = 14$ weights to find an approximate 95% confidence interval for m.

$$80.51 \quad 80.28 \quad 80.40 \quad 80.35 \quad 80.38 \quad 80.28 \quad 80.27$$
$$80.16 \quad 80.59 \quad 80.56 \quad 80.32 \quad 80.27 \quad 80.53 \quad 80.32$$

(a) Find a 96.14% confidence interval for m.

(b) The interval (y_6, y_{12}) could serve as a confidence interval for $\pi_{0.6}$. What is its confidence coefficient?

6.10-5 Use the following weights of an observed random sample of $n = 11$ pieces of candy to find an approximate 95% confidence interval for the median weight, m. Also give the exact confidence level.

$$2.76 \quad 2.96 \quad 2.67 \quad 2.97 \quad 2.77 \quad 2.72$$
$$2.74 \quad 2.84 \quad 2.61 \quad 2.82 \quad 2.72$$

6.10-6 Use the 81 weights of mints listed in Exercise 3.1-16 to find an approximate 95% confidence interval for

(a) $\pi_{0.25}$.

(b) $\pi_{0.5}$.

(c) $\pi_{0.75}$.

6.10-7 A biologist who studies spiders selected a random sample of 20 male green lynx spiders (a spider that does not weave a web but chases and leaps on its prey) and measured in millimeters the lengths of one of the front legs of the 20 spiders. Use the following measurements to construct a confidence interval for m that has a confidence coefficient about equal to 0.95.

$$15.10 \quad 13.55 \quad 15.75 \quad 20.00 \quad 15.45$$
$$13.60 \quad 16.45 \quad 14.05 \quad 16.95 \quad 19.05$$
$$16.40 \quad 17.05 \quad 15.25 \quad 16.65 \quad 16.25$$
$$17.75 \quad 15.40 \quad 16.80 \quad 17.55 \quad 19.05$$

6.10-8 The biologist (Exercise 6.10-7) also selected a random sample of 20 female green lynx spiders and measured in millimeters the length of one of their front legs. Use the following data to construct a confidence interval for m that has a confidence coefficient about equal to 0.95.

15.85	18.00	11.45	15.60	16.10
18.80	12.85	15.15	13.30	16.65
16.25	16.15	15.25	12.10	16.20
14.80	14.60	17.05	14.15	15.85

6.10-9 The following 25 observations give the time in seconds between submissions of computer programs to a printer queue.

79	315	445	350	136	723	198	75	161
13	215	24	57	152	238	288	272	
9	315	11	51	98	620	244	34	

  (a) List the observations in order of magnitude.

  (b) Give point estimates of $\pi_{0.25}$, m, and $\pi_{0.75}$.

  (c) Find the following confidence intervals and give the confidence level (using Table II) for

    (i) (y_3, y_{10}), a confidence interval for $\pi_{0.25}$.

    (ii) (y_9, y_{17}), a confidence interval for the median m.

    (iii) (y_{16}, y_{23}), a confidence interval for $\pi_{0.75}$.

  (d) Find a confidence interval for μ, whose confidence coefficient corresponds to that of (d), part (ii), using $\bar{x} \pm t_{\alpha/2}(24)(s/\sqrt{25})$. Compare these two confidence intervals of the middles. Is this surprising? Why?

6.10-10 Let X equal the weight in grams of a miniature candy bar. A random sample of candy bar weights yielded the following observations of X.

22.4	23.7	22.4	23.6	24.0	24.5	23.4	23.9
22.8	22.5	23.4	23.0	22.8	22.6	24.0	23.3

  (a) Give a point estimate of the median, $m = \pi_{0.50}$.

  (b) Find a confidence interval for m with confidence level between 90% and 95%. Give the value of the exact confidence coefficient.

6.10-11 Let X equal the amount of fluoride in a certain brand of toothpaste. The specifications are 0.85–1.10 mg/g. Table 6.1-1 lists 100 such measurements.

  (a) Give a point estimate of the median, $m = \pi_{0.50}$.

  (b) Find an approximate 95% confidence interval for the median m. Give the exact confidence level if you have computing capability.

  (c) Give a point estimate for the first quartile.

  (d) Find an approximate 95% confidence interval for the first quartile and, if possible, give the exact confidence level.

  (e) Give a point estimate for the third quartile.

  (f) Find an approximate 95% confidence interval for the third quartile and, if possible, give the exact confidence level.

6.10-12 When placed in solutions of varying ionic strength, paramecia grow blisters in order to counteract the flow of water. The following 60 measurements in microns are blister lengths.

7.42	5.73	3.80	5.20	11.66	8.51	6.31	8.49
10.31	6.92	7.36	5.92	6.74	8.93	9.61	11.38
12.78	11.43	6.57	13.50	10.58	8.03	10.07	8.71
10.09	11.16	7.22	10.10	6.32	10.30	10.75	11.51
11.55	11.41	9.40	4.74	6.52	12.10	6.01	5.73
7.57	7.80	6.84	6.95	8.93	8.92	5.51	6.71
10.40	13.44	9.33	8.57	7.08	8.11	13.34	6.58
8.82	7.70	12.22	7.46				

(a) Construct an ordered stem-and-leaf diagram.

(b) Give a point estimate of the median, $m = \pi_{0.50}$.

(c) Find an approximate 95% confidence interval for m.

(d) Give a point estimate for the 40th percentile, $\pi_{0.40}$.

(e) Find an approximate 90% confidence interval for $\pi_{0.40}$.

6.10-13 With the stem-and-leaf display of the distances of McGwire's homers given in Example 3.1-5, find approximate 95% confidence intervals for $\pi_{0.25}$, $\pi_{0.5}$, and $\pi_{0.75}$.

6.10-14 Let $Y_1 < Y_2 < \cdots < Y_8$ be the order statistics of 8 independent observations from a continuous-type distribution with 70th percentile $\pi_{0.7} = 27.3$.

(a) Determine $P(Y_7 < 27.3)$.

(b) Find $P(Y_5 < 27.3 < Y_8)$.

6.10-15 Let X equal the weight in grams of peanuts in a "14-gram" snack package. The order statistics of a sample of size $n = 16$ are

13.9	14.4	14.6	14.7	14.7	15.2	15.2	15.2
15.3	15.4	15.4	15.5	15.6	15.6	15.9	16.4

(a) Find the 30th percentile of this sample.

(b) Find $P(Y_2 < \pi_{0.30} < Y_8)$, where Y_2 and Y_8 are the respective second and eighth order statistics of 16 independent observations.

6.11 A SIMPLE REGRESSION PROBLEM

There is often interest in the relation between two variables, for example, a student's scholastic aptitude test score in mathematics and this same student's grade in calculus. Frequently, one of these variables, say x, is known in advance of the other, and hence there is interest in predicting a future random variable Y. Since Y is a random variable, we cannot predict its future observed value $Y = y$ with certainty. Let us first concentrate on the problem of estimating the mean of Y, that is, $E(Y)$. Now $E(Y)$ is usually a function of x. For example, in our illustration with the calculus grade, say Y, we would expect $E(Y)$ to increase with increasing mathematics aptitude score x. Sometimes $E(Y) = \mu(x)$ is assumed to be of a given form, such as linear or quadratic or exponential; that is, $\mu(x)$ could be assumed to be equal to $\alpha + \beta x$ or $\alpha + \beta x + \gamma x^2$ or $\alpha e^{\beta x}$. To estimate $E(Y) = \mu(x)$, or, equivalently, the parameters α, β, and γ, we observe the random variable Y for each of n different values of x, say $x_1, x_2, \ldots, x_n$. Once the n independent experiments have been performed, we have n pairs of known numbers $(x_1, y_1), (x_2, y_2), \ldots, (x_n, y_n)$. These pairs are then used to estimate the mean $E(Y)$. Problems like this are often classified under **regression** because $E(Y) = \mu(x)$ is frequently called a regression curve.

Remark A model for the mean like $\alpha + \beta x + \gamma x^2$, is called a linear model because it is linear in the parameters, α, β, and γ. Thus $\alpha e^{\beta x}$ is not a linear model because it is not linear in α and β. ∎

Let us begin with the case in which $E(Y) = \mu(x)$ is a linear function. The data points are $(x_1, y_1), (x_2, y_2), \ldots, (x_n, y_n)$; so the first problem is that of fitting a straight line to the set of data (see Figure 6.11-1). In addition to assuming that the mean of Y is a linear function, we assume that for a particular value of x,

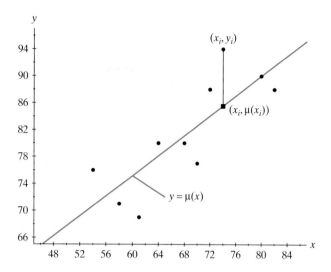

Figure 6.11-1: Scatter plot and least squares regression line

the value of Y will differ from its mean by a random amount ε. We further assume that the distribution of ε is $N(0, \sigma^2)$. So we have for our linear model

$$Y_i = \alpha_1 + \beta x_i + \varepsilon_i,$$

where ε_i, for $i = 1, 2, \ldots, n$, are independent and $N(0, \sigma^2)$.

We shall now find point estimates for α_1, β, and σ^2. For convenience we let $\alpha_1 = \alpha - \beta \bar{x}$ so that

$$Y_i = \alpha + \beta(x_i - \bar{x}) + \varepsilon_i, \quad \text{where } \bar{x} = \frac{1}{n} \sum_{i=1}^{n} x_i.$$

Then Y_i is equal to a constant, $\alpha + \beta(x_i - \bar{x})$, plus a normal random variable ε_i. Hence $Y_1, Y_2, \ldots, Y_n$ are mutually independent normal variables with respective means $\alpha + \beta(x_i - \bar{x})$, $i = 1, 2, \ldots, n$, and unknown variance σ^2. Their joint p.d.f. is therefore the product of the individual probability density functions; that is, the likelihood function equals

$$L(\alpha, \beta, \sigma^2) = \prod_{i=1}^{n} \frac{1}{\sqrt{2\pi\sigma^2}} \exp\left\{ -\frac{[y_i - \alpha - \beta(x_i - \bar{x})]^2}{2\sigma^2} \right\}$$

$$= \left(\frac{1}{2\pi\sigma^2} \right)^{n/2} \exp\left\{ -\frac{\sum_{i=1}^{n} [y_i - \alpha - \beta(x_i - \bar{x})]^2}{2\sigma^2} \right\}.$$

To maximize $L(\alpha, \beta, \sigma^2)$ or, equivalently, to minimize

$$-\ln L(\alpha, \beta, \sigma^2) = \frac{n}{2}\ln(2\pi\sigma^2) + \frac{\sum_{i=1}^{n}[y_i - \alpha - \beta(x_i - \overline{x})]^2}{2\sigma^2},$$

we must select α and β to minimize

$$H(\alpha, \beta) = \sum_{i=1}^{n}[y_i - \alpha - \beta(x_i - \overline{x})]^2.$$

Since $|y_i - \alpha - \beta(x_i - \overline{x})| = |y_i - \mu(x_i)|$ is the vertical distance from the point (x_i, y_i) to the line $y = \mu(x)$, we note that $H(\alpha, \beta)$ represents the sum of the squares of those distances. Thus selecting α and β so that the sum of the squares is minimized means that we are fitting the straight line to the data by the **method of least squares**.

To minimize $H(\alpha, \beta)$, we find the two first partial derivatives

$$\frac{\partial H(\alpha, \beta)}{\partial\alpha} = 2\sum_{i=1}^{n}[y_i - \alpha - \beta(x_i - \overline{x})](-1)$$

and

$$\frac{\partial H(\alpha, \beta)}{\partial\beta} = 2\sum_{i=1}^{n}[y_i - \alpha - \beta(x_i - \overline{x})][-(x_i - \overline{x})].$$

Setting $\partial H(\alpha, \beta)/\partial\alpha = 0$, we obtain

$$\sum_{i=1}^{n} y_i - n\alpha - \beta\sum_{i=1}^{n}(x_i - \overline{x}) = 0.$$

Since

$$\sum_{i=1}^{n}(x_i - \overline{x}) = 0,$$

we have that

$$\sum_{i=1}^{n} y_i - n\alpha = 0$$

and thus

$$\widehat{\alpha} = \overline{Y}.$$

The equation $\partial H(\alpha, \beta)/\partial\beta = 0$ yields, with α replaced by $\bar{y}$,

$$\sum_{i=1}^{n}(y_i - \bar{y})(x_i - \bar{x}) - \beta\sum_{i=1}^{n}(x_i - \bar{x})^2 = 0$$

or, equivalently,

$$\widehat{\beta} = \frac{\displaystyle\sum_{i=1}^{n}(Y_i - \bar{Y})(x_i - \bar{x})}{\displaystyle\sum_{i=1}^{n}(x_i - \bar{x})^2} = \frac{\displaystyle\sum_{i=1}^{n}Y_i(x_i - \bar{x})}{\displaystyle\sum_{i=1}^{n}(x_i - \bar{x})^2}.$$

Thus, to find the mean line of best fit, $\mu(x) = \alpha + \beta(x_i - \bar{x})$, we use

$$\widehat{\alpha} = \bar{y} \tag{6.11-1}$$

and

$$\widehat{\beta} = \frac{\displaystyle\sum_{i=1}^{n}y_i(x_i - \bar{x})}{\displaystyle\sum_{i=1}^{n}(x_i - \bar{x})^2} = \frac{\displaystyle\sum_{i=1}^{n}x_iy_i - \left(\frac{1}{n}\right)\left(\sum_{i=1}^{n}x_i\right)\left(\sum_{i=1}^{n}y_i\right)}{\displaystyle\sum_{i=1}^{n}x_i^2 - \left(\frac{1}{n}\right)\left(\sum_{i=1}^{n}x_i\right)^2}. \tag{6.11-2}$$

To find the maximum likelihood estimator of σ^2, consider the partial derivative

$$\frac{\partial[-\ln L(\alpha, \beta, \sigma^2)]}{\partial(\sigma^2)} = \frac{n}{2\sigma^2} - \frac{\displaystyle\sum_{i=1}^{n}[y_i - \alpha - \beta(x_i - \bar{x})]^2}{2(\sigma^2)^2}.$$

Setting this equal to zero and replacing α and β by their solutions $\widehat{\alpha}$ and $\widehat{\beta}$, we obtain

$$\widehat{\sigma^2} = \frac{1}{n}\sum_{i=1}^{n}[Y_i - \widehat{\alpha} - \widehat{\beta}(x_i - \bar{x})]^2. \tag{6.11-3}$$

A formula for $n\widehat{\sigma^2}$ useful in its calculation is

$$n\widehat{\sigma^2} = \sum_{i=1}^{n}y_i^2 - \frac{1}{n}\left(\sum_{i=1}^{n}y_i\right)^2 - \widehat{\beta}\sum_{i=1}^{n}x_iy_i + \widehat{\beta}\left(\frac{1}{n}\right)\left(\sum_{i=1}^{n}x_i\right)\left(\sum_{i=1}^{n}y_i\right). \tag{6.11-4}$$

Note that the summand in Equation 6.11-3 for $\widehat{\sigma^2}$ is the square of the difference between the value of Y_i and the predicted mean of Y_i. Let $\widehat{Y}_i = \widehat{\alpha} + \widehat{\beta}(x_i - \overline{x})$, the predicted mean value of Y_i. The difference

$$Y_i - \widehat{Y}_i = Y_i - \widehat{\alpha} - \widehat{\beta}(x_i - \overline{x})$$

is called the ith **residual**, $i = 1, 2, \ldots, n$. The maximum likelihood estimate of σ^2 is then the sum of the squares of the residuals divided by n. It should always be true that the sum of the residuals is equal to zero. However, in practice, due to round off, the sum of the observed residuals, $y_i - \widehat{y}_i$, sometimes differs slightly from zero. A graph of the residuals plotted as a scatter plot of the points x_i, $y_i - \widehat{y}_i$, $i = 1, 2, \ldots, n$, can graphically show whether or not linear regression provides the best fit.

Table 6.11-1: Calculations for Test Score Data

x	y	x^2	xy	y^2	$\widehat{y}$	$y - \widehat{y}$	$(y - \widehat{y})^2$
70	77	4,900	5,390	5,929	82.561566	−5.561566	30.931016
74	94	5,476	6,956	8,836	85.529956	8.470044	71.741645
72	88	5,184	6,336	7,744	84.045761	3.954239	15.636006
68	80	4,624	5,440	6,400	81.077371	−1.077371	1.160728
58	71	3,364	4,118	5,041	73.656395	−2.656395	7.056434
54	76	2,916	4,104	5,776	70.688004	5.311996	28.217302
82	88	6,724	7,216	7,744	91.466737	−3.466737	12.018265
64	80	4,096	5,120	6,400	78.108980	1.891020	3.575957
80	90	6,400	7,200	8,100	89.982542	0.017458	0.000305
61	69	3,721	4,209	4,761	75.882687	−6.882687	47.371380
683	813	47,405	56,089	66,731		0.000001	217.709038

EXAMPLE 6.11-1   The data plotted in Figure 6.11-1 are 10 pairs of test scores of 10 students in a psychology class, x being the score on a preliminary test and y the score on the final examination. The values of x and y are shown in Table 6.11-1. The sums that are needed to calculate estimates of the parameters are also given. Of course, the estimates of α and β have to be found before the residuals can be calculated.

Thus $\widehat{\alpha} = 813/10 = 81.3$,

$$\widehat{\beta} = \frac{56{,}089 - (683)(813)/10}{47{,}405 - (683)(683)/10} = \frac{561.1}{756.1} = 0.742.$$

Since $\overline{x} = 683/10 = 68.3$, the least squares regression line is

$$\widehat{y} = 81.3 + (0.742)(x - 68.3).$$

The maximum likelihood estimate of σ^2 is

$$\widehat{\sigma^2} = \frac{217.709038}{10} = 21.7709.$$

A residual plot for these data is shown in Figure 6.11-2.   ◄

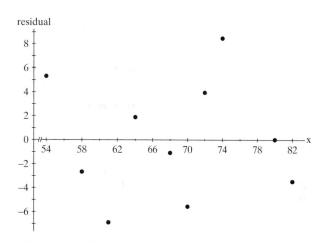

Figure 6.11-2: Residuals plot for data in Table 6.11-1

We shall now consider the problem of finding the distributions of $\widehat{\alpha}$, $\widehat{\beta}$, and $\widehat{\sigma^2}$ (or distributions of functions of these estimators). We would like to be able to say something about the error of the estimates to find confidence intervals for the parameters.

During the preceding discussion we have treated $x_1, x_2, \ldots, x_n$ as constants. Of course, many times they can be set by the experimenter; for example, a chemist in experimentation might produce a compound at many different temperatures. But it is also true that these numbers might be observations on an earlier random variable, such as an SAT score or preliminary test grade (as in Example 6.11-1), but we consider the problem on the condition that these x values are given in either case. Thus, in finding the distributions of $\widehat{\alpha}$, $\widehat{\beta}$, and $\widehat{\sigma^2}$, the only random variables are $Y_1, Y_2, \ldots, Y_n$.

Since $\widehat{\alpha}$ is a linear function of independent and normally distributed random variables, $\widehat{\alpha}$ has a normal distribution with mean

$$E(\widehat{\alpha}) = E\left(\frac{1}{n}\sum_{i=1}^{n} Y_i\right) = \frac{1}{n}\sum_{i=1}^{n} E(Y_i)$$

$$= \frac{1}{n}\sum_{i=1}^{n} [\alpha + \beta(x_i - \overline{x})] = \alpha,$$

and variance

$$\mathrm{Var}(\widehat{\alpha}) = \sum_{i=1}^{n} \left(\frac{1}{n}\right)^2 \mathrm{Var}(Y_i) = \frac{\sigma^2}{n}.$$

The estimator $\widehat{\beta}$ is also a linear function of $Y_1, Y_2, \ldots, Y_n$ and hence has a normal distribution with mean

$$E(\widehat{\beta}) = \frac{\displaystyle\sum_{i=1}^{n} (x_i - \overline{x})E(Y_i)}{\displaystyle\sum_{i=1}^{n} (x_i - \overline{x})^2}$$

$$= \frac{\displaystyle\sum_{i=1}^{n} (x_i - \overline{x})[\alpha + \beta(x_i - \overline{x})]}{\displaystyle\sum_{i=1}^{n} (x_i - \overline{x})^2}$$

$$= \frac{\alpha \displaystyle\sum_{i=1}^{n} (x_i - \overline{x}) + \beta \displaystyle\sum_{i=1}^{n} (x_i - \overline{x})^2}{\displaystyle\sum_{i=1}^{n} (x_i - \overline{x})^2} = \beta$$

and variance

$$\mathrm{Var}(\widehat{\beta}) = \sum_{i=1}^{n} \left[\frac{x_i - \overline{x}}{\sum_{j=1}^{n} (x_j - \overline{x})^2} \right]^2 \mathrm{Var}(Y_i)$$

$$= \frac{\displaystyle\sum_{i=1}^{n} (x_i - \overline{x})^2}{\left[\displaystyle\sum_{i=1}^{n} (x_i - \overline{x})^2 \right]^2} \sigma^2 = \frac{\sigma^2}{\displaystyle\sum_{i=1}^{n} (x_i - \overline{x})^2}.$$

It can be shown (Exercise 6.11-6) that

$$\sum_{i=1}^{n} [Y_i - \alpha - \beta(x_i - \overline{x})]^2 = \sum_{i=1}^{n} \{(\widehat{\alpha} - \alpha) + (\widehat{\beta} - \beta)(x_i - \overline{x})$$

$$+ [Y_i - \widehat{\alpha} - \widehat{\beta}(x_i - \overline{x})]\}^2$$

$$= n(\widehat{\alpha} - \alpha)^2 + (\widehat{\beta} - \beta)^2 \sum_{i=1}^{n} (x_i - \overline{x})^2$$

$$+ \sum_{i=1}^{n} [Y_i - \widehat{\alpha} - \widehat{\beta}(x_i - \overline{x})]^2. \qquad (6.11\text{-}5)$$

From the fact that $Y_i, \widehat{\alpha}$, and $\widehat{\beta}$ have normal distributions, we know that each of

$$\frac{[Y_i - \alpha - \beta(x_i - \overline{x})]^2}{\sigma^2}, \quad \frac{(\widehat{\alpha} - \alpha)^2}{\left[\dfrac{\sigma^2}{n}\right]}, \quad \frac{(\widehat{\beta} - \beta)^2}{\left[\dfrac{\sigma^2}{\displaystyle\sum_{i=1}^{n}(x_i - \overline{x})^2}\right]}$$

has a chi-square distribution with one degree of freedom. Since $Y_1, Y_2, \ldots, Y_n$ are mutually independent, then

$$\frac{\displaystyle\sum_{i=1}^{n}[Y_i - \alpha - \beta(x_i - \overline{x})]^2}{\sigma^2}$$

is $\chi^2(n)$. That is, the left-hand member of Equation 6.11-5 divided by σ^2 is $\chi^2(n)$ and is equal to the sum of two $\chi^2(1)$ variables and

$$\frac{\displaystyle\sum_{i=1}^{n}[Y_i - \widehat{\alpha} - \widehat{\beta}(x_i - \overline{x})]^2}{\sigma^2} = \frac{n\widehat{\sigma^2}}{\sigma^2} \geq 0.$$

Thus we might then guess that $n\widehat{\sigma^2}/\sigma^2$ is $\chi^2(n-2)$. This is true, and, moreover, $\widehat{\alpha}, \widehat{\beta}$, and $\widehat{\sigma^2}$ are mutually independent. [For a proof, see Hogg, McKean, and Craig, *Introduction to Mathematical Statistics*, 6th ed. (Upper Saddle River, NJ: Prentice Hall, 2005).]

Suppose now that we are interested in forming a confidence interval for β, the slope of the line. We can use the fact that

$$T_1 = \frac{\sqrt{\displaystyle\sum_{i=1}^{n}(x_i - \overline{x})^2}\left(\dfrac{\widehat{\beta} - \beta}{\sigma}\right)}{\sqrt{\dfrac{n\widehat{\sigma^2}}{\sigma^2(n-2)}}} = \frac{\widehat{\beta} - \beta}{\sqrt{\dfrac{n\widehat{\sigma^2}}{(n-2)\displaystyle\sum_{i=1}^{n}(x_i - \overline{x})^2}}}$$

has a t distribution with $n - 2$ degrees of freedom. Thus

$$P\left[-t_{\gamma/2}(n-2) \leq \frac{\widehat{\beta} - \beta}{\sqrt{\dfrac{n\widehat{\sigma^2}}{(n-2)\displaystyle\sum_{i=1}^{n}(x_i - \overline{x})^2}}} \leq t_{\gamma/2}(n-2)\right] = 1 - \gamma.$$

It follows that

$$\left[\widehat{\beta} - t_{\gamma/2}(n-2) \sqrt{\frac{n\widehat{\sigma}^2}{(n-2)\sum\limits_{i=1}^{n}(x_i - \overline{x})^2}} , \right.$$

$$\left. \widehat{\beta} + t_{\gamma/2}(n-2) \sqrt{\frac{n\widehat{\sigma}^2}{(n-2)\sum\limits_{i=1}^{n}(x_i - \overline{x})^2}} \right]$$

is a $100(1 - \gamma)\%$ confidence interval for β.

Similarly,

$$T_2 = \frac{\dfrac{\sqrt{n}(\widehat{\alpha} - \alpha)}{\sigma}}{\sqrt{\dfrac{n\widehat{\sigma}^2}{\sigma^2(n-2)}}} = \frac{\widehat{\alpha} - \alpha}{\sqrt{\dfrac{\widehat{\sigma}^2}{n-2}}}$$

has a t distribution with $n - 2$ degrees of freedom. Thus T_2 can be used to make inferences about α (see Exercise 6.11-7). The fact that $n\widehat{\sigma}^2/\sigma^2$ has a chi-square distribution with $n - 2$ degrees of freedom can be used to make inferences about the variance σ^2 (see Exercise 6.11-8).

STATISTICAL COMMENTS We now give an illustration using data from the *Challenger* explosion on January 28, 1986. You may recall that the *Challenger* space shuttle was launched from Cape Kennedy in Florida on a very cold January morning. Meteorologists had forecasted temperatures (as of January 27) in the range of $26°$–$29°$ Fahrenheit. The night before the launch there was much debate among engineers and NASA officials whether a launch under such low-temperature conditions would be advisable. Several engineers advised against a launch because they thought that O-ring failures were related to temperature. Data on O-ring failures experienced in previous launches were available and were studied the night before the launch. There were seven previous incidents of known distressed O-rings. Figure 6.11-3 displays this information; it is a simple scatter plot of the number of distressed rings per launch against temperature at launch.

From this plot alone there does not seem to be a strong relationship between the number of O-ring failures and temperature. Based on this information, along with many other technical and political considerations, it was decided to launch the *Challenger* space shuttle. As you all know, the launch resulted in disaster: the loss of seven lives and billions of dollars, and a serious setback to the space program.

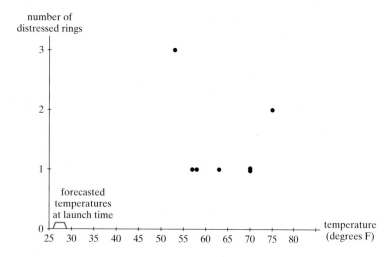

Figure 6.11-3: Number of distressed rings per launch versus temperature

One may argue that engineers looked at the scatter plot of the number of failures against temperature but could not see a relationship. However, this argument misses the fact that engineers did not display *all the data that were relevant to this question.* They looked only at instances where there were failures; they ignored the cases where there were no failures. In fact, there were 17 prior launches in which no failures occurred. A scatter plot of the number of distressed O-rings per launch against temperature using data from all previous shuttle launches is given in Figure 6.11-4.

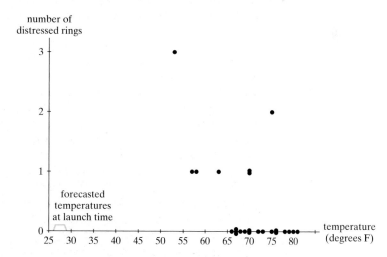

Figure 6.11-4: Number of distressed rings per launch versus temperature (all data)

It is difficult to look at these data and not see a relationship between failures and temperature. Moreover, one recognizes that an extrapolation is required and that an inference about the number of failures outside the observed range of temperature is needed. The actual temperature at launch was 31°F, while the lowest temperature recorded at a previous launch was 53°F. It is always very dangerous to extrapolate inferences to a region for which one does not have data. If NASA officials had looked at this plot, certainly the launch would have been delayed.

This example shows why it is important to have statistically minded engineers involved in important decisions. Applied statisticians like R. Snee have told us time and time again, "In God We Trust; Others Must Have Data."

This comment raises two important points. It illustrates the importance of scatter plots where we plot one variable against another. The other point that is made so well in the *Challenger* example is the importance of plotting *relevant data*. Yes, it is true that some data were used in making the decision. But not all the relevant data were utilized. It takes knowledge of statistics to make good decisions, as well as subject knowledge, common sense, and an ability to question the relevance of information.   ■

EXERCISES

6.11-1 The midterm and final exam scores of 10 students in a statistics course are tabulated as shown.

(a) Calculate the least squares regression line for these data.
(b) Plot the points and the least squares regression line on the same graph.
(c) Find the value of $\widehat{\sigma^2}$.

Midterm	Final	Midterm	Final
70	87	67	73
74	79	70	83
80	88	64	79
84	98	74	91
80	96	82	94

6.11-2 The final course grade in calculus was predicted on the basis of the student's high school grade point average in mathematics, Scholastic Aptitude Test (SAT) score in mathematics, and score on a mathematics entrance examination. The predicted grades X and the earned grades Y for 10 students are given (2.0 represents a C, 2.3 a C+, 2.7 a B−, etc.).

(a) Calculate the least squares regression line for these data.
(b) Plot the points and the least squares regression line on the same graph.
(c) Find the value of $\widehat{\sigma^2}$.

x	y	x	y
2.0	1.3	2.7	3.0
3.3	3.3	4.0	4.0
3.7	3.3	3.7	3.0
2.0	2.0	3.0	2.7
2.3	1.7	2.3	3.0

6.11-3 Students' scores on the mathematics portion of the ACT examination, X, and on the final examination in first semester calculus (200 points possible), Y, are given.

(a) Calculate the least squares regression line for these data.
(b) Plot the points and the least squares regression line on the same graph.

x	y	x	y
25	138	20	100
20	84	25	143
26	104	26	141
26	112	28	161
28	88	25	124
28	132	31	118
29	90	30	168
32	183		

6.11-4 Chemists often use ion-sensitive electrodes (ISEs) to measure the ion concentration of aqueous solutions. These devices measure the migration of the charge of these ions and give a reading in millivolts (mV). A standard curve is produced by measuring known concentrations (in ppm) and fitting a line to the millivolt data. The following table gives the concentrations in ppm and the voltage in mV for calcium ISE.

ppm	mV	ppm	mV
0	1.72	100	2.91
0	1.68	100	3.00
0	1.74	100	2.89
50	2.04	150	4.47
50	2.11	150	4.51
50	2.17	150	4.43
75	2.40	200	6.67
75	2.32	200	6.66
75	2.33	200	6.57

(a) Calculate the least squares regression line for mV versus ppm.
(b) Plot the points and the least squares regression line on the same graph.
(c) Calculate and plot the residuals. Does linear regression seem to be appropriate? (See Exercise 6.12-15.)

6.11-5 A student who considered himself to be a "car guy" was interested in how horse power and weight of a car affected the time that it takes a car to go from 0 to 60. The following table gives, for each of 14 cars, the horse power, time in seconds to go from 0 to 60 mph, and the weight in pounds.

Horse Power	0–60	Weight	Horse Power	0–60	Weight
230	8.1	3516	282	6.2	3627
225	7.8	3690	300	6.4	3892
375	4.7	2976	220	7.7	3377
322	6.6	4215	250	7.0	3625
190	8.4	3761	315	5.3	3230
150	8.4	2940	200	6.2	2657
178	7.2	2818	300	5.5	3518

    (a) Calculate the least squares regression line for "0–60" versus horse power.
    (b) Plot the points and the least squares regression line on the same graph.
    (c) Calculate the least squares regression line for "0–60" versus weight.
    (d) Plot the points and the least squares regression line on the same graph.
    (e) Which of the two variables, horse power or weight, has the most effect on the "0–60" time?

6.11-6 Show that

$$\sum_{i=1}^{n}[Y_i - \alpha - \beta(x_i - \overline{x})]^2 = n(\widehat{\alpha} - \alpha)^2 + (\widehat{\beta} - \beta)^2 \sum_{i=1}^{n}(x_i - \overline{x})^2$$

$$+ \sum_{i=1}^{n}[Y_i - \widehat{\alpha} - \widehat{\beta}(x_i - \overline{x})]^2.$$

Remark In Exercises 6.11-7 to 6.11-14, to find the confidence intervals, we need the distributional assumptions made in this section. ∎

6.11-7 Show that the endpoints for a $100(1 - \gamma)\%$ confidence interval for α are

$$\widehat{\alpha} \pm t_{\gamma/2}(n-2)\sqrt{\frac{\widehat{\sigma^2}}{n - 2}}.$$

6.11-8 Show that a $100(1 - \gamma)\%$ confidence interval for σ^2 is

$$\left[\frac{n\widehat{\sigma^2}}{\chi^2_{\gamma/2}(n-2)}, \frac{n\widehat{\sigma^2}}{\chi^2_{1-\gamma/2}(n-2)}\right].$$

6.11-9 Find 95% confidence intervals for α, β, and σ^2 for the data in Exercise 6.11-1.

6.11-10 Find 95% confidence intervals for α, β, and σ^2 for the data in Exercise 6.11-2.

6.11-11 In a "48.1-gram" package of candies, let x equal the number of pieces of candy and let Y equal the total weight of the candies. For each of 5 different values of x, 4 observations of Y are given.

x	y	x	y
54	48.8	56	50.8
54	49.4	56	50.9
54	49.2	57	50.7
54	50.4	57	51.6
55	49.5	57	51.3
55	49.0	57	50.8
55	50.2	58	52.1
55	48.9	58	51.3
56	49.9	58	51.4
56	50.1	58	52.0

(a) Calculate the least squares regression line for these data.
(b) Plot the points and the least squares regression line on the same graph.
(c) Find the value of $\widehat{\sigma^2}$.
(d) Find a 95% confidence interval for β under the usual assumptions.

6.11-12 The Federal Trade Commission measured the number of milligrams of tar and carbon monoxide (CO) per cigarette for all domestic cigarettes. Let x and y equal the measurements of tar and CO, respectively, for 100-millimeter filtered and mentholated cigarettes. A sample of 12 brands yielded the following data:

Brand	x	y	Brand	x	y
Capri	9	6	Now	3	4
Carlton	4	6	Salem	17	18
Kent	14	14	Triumph	6	8
Kool Milds	12	12	True	7	8
Marlboro Lights	10	12	Vantage	8	13
Merit Ultras	5	7	Virginia Slims	15	13

(a) Calculate the least squares regression line for these data.
(b) Plot the points and the least squares regression line on the same graph.
(c) Find point estimates for α, β, and σ^2.
(d) Find 95% confidence intervals for α, β, and σ^2 under the usual assumptions.

6.11-13 Let X and Y equal the lengths in inches of a foot and a hand, respectively. The following measurements were made on 15 women.

x	y	x	y	x	y
9.00	6.50	10.00	7.00	9.25	7.00
8.50	6.25	9.50	6.50	10.00	7.50
9.25	7.25	9.00	7.00	10.00	7.25
9.75	7.00	9.25	7.00	9.75	7.25
9.00	6.75	9.50	7.00	9.50	7.25

(a) Calculate the least squares regression line for these data.
(b) Plot the points and the least squares regression line on the same graph.
(c) Find point estimates for α, β, and σ^2.
(d) Find 95% confidence intervals for α, β, and σ^2 under the usual assumptions.

6.11-14 Let x and y equal the ACT scores in social science and natural science for a student who is applying for admission to a small liberal arts college. A sample of $n = 15$ such students yielded the following data.

x	y	x	y	x	y
32	28	30	27	26	32
23	25	17	23	16	22
23	24	20	30	21	28
23	32	17	18	24	31
26	31	18	18	30	26

(a) Calculate the least squares regression line for these data.
(b) Plot the points and the least squares regression line on the same graph.
(c) Find point estimates for α, β, and σ^2.
(d) Find 95% confidence intervals for α, β, and σ^2 under the usual assumptions.

6.11-15 The ACT composite score and the first year college gpa are given in the following table for 18 students.

ACT	GPA	ACT	GPA
23	2.40	25	3.67
29	3.10	23	3.03
15	2.51	30	3.07
23	2.98	16	2.28
28	3.93	24	2.45
27	1.91	26	3.83
23	1.69	27	3.78
26	2.88	21	2.57
20	2.60	19	3.42

(a) Calculate the least squares regression line for these data.
(b) Plot the points and the least squares regression line on the same graph.
(c) Find 95% confidence intervals for α, β, and σ^2.

6.12* MORE REGRESSION

In this section, we use the notation and assumptions of Section 6.11. We have noted that $\widehat{Y} = \widehat{\alpha} + \widehat{\beta}(x - \overline{x})$ is a point estimate for the mean of Y for some given x, or we could think of this as a prediction of the value of Y for this given x. But how close is $\widehat{Y}$ to the mean of Y or to Y itself? We shall now find a confidence interval for $\alpha + \beta(x - \bar{x})$ and a prediction interval for Y, given a particular value of x.

To find a confidence interval for

$$E(Y) = \mu(x) = \alpha + \beta(x - \bar{x}),$$

let

$$\widehat{Y} = \widehat{\alpha} + \widehat{\beta}(x - \bar{x}).$$

Recall that $\widehat{Y}$ is a linear combination of normally and independently distributed random variables, $\widehat{\alpha}$ and $\widehat{\beta}$, so that $\widehat{Y}$ has a normal distribution. Furthermore,

$$E(\widehat{Y}) = E[\widehat{\alpha} + \widehat{\beta}(x - \bar{x})]$$
$$= \alpha + \beta(x - \bar{x})$$

and

$$\mathrm{Var}(\widehat{Y}) = \mathrm{Var}[\widehat{\alpha} + \widehat{\beta}(x - \bar{x})]$$
$$= \frac{\sigma^2}{n} + \frac{\sigma^2}{\displaystyle\sum_{i=1}^{n}(x_i - \bar{x})^2}(x - \bar{x})^2$$
$$= \sigma^2 \left[\frac{1}{n} + \frac{(x - \bar{x})^2}{\displaystyle\sum_{i=1}^{n}(x_i - \bar{x})^2} \right].$$

Recall that the distribution of $n\widehat{\sigma^2}/\sigma^2$ is $\chi^2(n-2)$. Since $\widehat{\alpha}$ and $\widehat{\beta}$ are independent of $\widehat{\sigma^2}$, we can form the t statistic

$$T = \frac{\widehat{\alpha} + \widehat{\beta}(x - \bar{x}) - [\alpha + \beta(x - \bar{x})]}{\sigma\sqrt{\dfrac{1}{n} + \dfrac{(x - \bar{x})^2}{\displaystyle\sum_{i=1}^{n}(x_i - \bar{x})^2}} \Bigg/ \sqrt{\dfrac{n\widehat{\sigma^2}}{(n-2)\sigma^2}}},$$

which has a t distribution with $r = n - 2$ degrees of freedom. Select $t_{\gamma/2}(n-2)$ from Table VI in the Appendix so that

$$P[-t_{\gamma/2}(n-2) \leq T \leq t_{\gamma/2}(n-2)] = 1 - \gamma.$$

This becomes

$$P[\widehat{\alpha} + \widehat{\beta}(x - \overline{x}) - ct_{\gamma/2}(n-2) \le \alpha + \beta(x - \overline{x}) \le$$
$$\widehat{\alpha} + \widehat{\beta}(x - \overline{x}) + ct_{\gamma/2}(n-2)] = 1 - \gamma,$$

where

$$c = \sqrt{\frac{n\widehat{\sigma^2}}{n-2}} \sqrt{\frac{1}{n} + \frac{(x - \overline{x})^2}{\sum_{i=1}^{n}(x_i - \overline{x})^2}}.$$

Thus the endpoints for a $100(1-\gamma)\%$ confidence interval for $\mu(x) = \alpha + \beta(x - \overline{x})$ are

$$\widehat{\alpha} + \widehat{\beta}(x - \overline{x}) \pm ct_{\gamma/2}(n-2).$$

Note that the width of this interval depends on the particular value of x because c depends on x (see Example 6.12-1).

We have used $(x_1, y_1), (x_2, y_2), \ldots, (x_n, y_n)$ to estimate α and β. Suppose that we are given a value of x, say x_{n+1}. A point estimate of the corresponding value of Y is

$$\widehat{y}_{n+1} = \widehat{\alpha} + \widehat{\beta}(x_{n+1} - \overline{x}).$$

However, $\widehat{y}_{n+1}$ is just one possible value of the random variable

$$Y_{n+1} = \alpha + \beta(x_{n+1} - \overline{x}) + \varepsilon_{n+1},$$

What can we say about possible values for Y_{n+1}? We shall now obtain a **prediction interval** for Y_{n+1} when $x = x_{n+1}$ that is similar to the confidence interval for the mean of Y when $x = x_{n+1}$.

We have that

$$Y_{n+1} = \alpha + \beta(x_{n+1} - \overline{x}) + \varepsilon_{n+1},$$

where ε_{n+1} is $N(0, \sigma^2)$ and $\overline{x} = (1/n)\sum_{i=1}^{n} x_i$. Now

$$W = Y_{n+1} - \widehat{\alpha} - \widehat{\beta}(x_{n+1} - \overline{x})$$

is a linear combination of normally and independently distributed random variables, so W has a normal distribution. The mean of W is

$$E(W) = E[Y_{n+1} - \widehat{\alpha} - \widehat{\beta}(x_{n+1} - \overline{x})]$$
$$= \alpha + \beta(x_{n+1} - \overline{x}) - \alpha - \beta(x_{n+1} - \overline{x}) = 0.$$

Since $Y_{n+1}, \widehat{\alpha}$ and $\widehat{\beta}$ are independent, the variance of W is

$$\text{Var}(W) = \sigma^2 + \frac{\sigma^2}{n} + \frac{\sigma^2}{\sum_{i=1}^{n}(x_i - \overline{x})^2}(x_{n+1} - \overline{x})^2$$

$$= \sigma^2 \left[1 + \frac{1}{n} + \frac{(x_{n+1} - \overline{x})^2}{\sum_{i=1}^{n}(x_i - \overline{x})^2} \right].$$

Recall that $n\widehat{\sigma^2}/[(n-2)\sigma^2]$ is $\chi^2(n-2)$. Since $Y_{n+1}, \widehat{\alpha}$, and $\widehat{\beta}$ are independent of $\widehat{\sigma^2}$, we can form the t statistic

$$T = \frac{\dfrac{Y_{n+1} - \widehat{\alpha} - \widehat{\beta}(x_{n+1} - \overline{x})}{\sigma \sqrt{1 + \dfrac{1}{n} + \dfrac{(x_{n+1} - \overline{x})^2}{\sum_{i=1}^{n}(x_i - \overline{x})^2}}}}{\sqrt{\dfrac{n\widehat{\sigma^2}}{(n-2)\sigma^2}}},$$

which has a t distribution with $r = n - 2$ degrees of freedom. Select a constant $t_{\gamma/2}(n-2)$ from Table VI in the Appendix so that

$$P[-t_{\gamma/2}(n-2) \leq T \leq t_{\gamma/2}(n-2)] = 1 - \gamma.$$

Solving this inequality for Y_{n+1}, we have

$$P[\widehat{\alpha} + \widehat{\beta}(x_{n+1} - \overline{x}) - d\, t_{\gamma/2}(n-2) \leq Y_{n+1} \leq$$
$$\widehat{\alpha} + \widehat{\beta}(x_{n+1} - \overline{x}) + dt_{\gamma/2}(n-2)] = 1 - \gamma,$$

where

$$d = \sqrt{\frac{n\widehat{\sigma^2}}{n-2}} \sqrt{1 + \frac{1}{n} + \frac{(x_{n+1} - \overline{x})^2}{\sum_{i=1}^{n}(x_i - \overline{x})^2}}.$$

Thus the endpoints for a $100(1-\gamma)\%$ prediction interval for Y_{n+1} are

$$\widehat{\alpha} + \widehat{\beta}(x_{n+1} - \overline{x}) \pm dt_{\gamma/2}(n-2).$$

We shall now illustrate a 95% confidence interval for $\mu(x)$ and a 95% prediction interval for Y for a given value of x using the data in Example 6.11-1. To find such intervals, the calculating formulas that are given in equations 6.11-1, 6.11-2, and 6.11-4 can be used.

EXAMPLE 6.12-1   To find a 95% confidence interval for $\mu(x)$ using the data in Example 6.11-1, note that we have already found that $\bar{x} = 68.3$, $\hat{\alpha} = 81.3$, $\hat{\beta} = 561.1/756.1 = 0.7421$, and $\widehat{\sigma^2} = 21.7709$. We also need

$$\sum_{i=1}^{n}(x_i - \bar{x})^2 = \sum_{i=1}^{n} x_i^2 - \left(\frac{1}{n}\right)\left(\sum_{i=1}^{n} x_i\right)^2$$

$$= 47{,}405 - \frac{683^2}{10} = 756.1.$$

For 95% confidence, $t_{0.025}(8) = 2.306$. When $x = 60$, the endpoints for a 95% confidence interval for $\mu(60)$ are

$$81.3 + 0.7421(60 - 68.3) \pm \left[\sqrt{\frac{10(21.7709)}{8}}\sqrt{\frac{1}{10} + \frac{(60 - 68.3)^2}{756.1}}\right](2.306)$$

or

$$75.1406 \pm 5.2589.$$

Similarly, when $x = 70$ the endpoints for a 95% confidence interval for $\mu(70)$ are

$$82.5616 \pm 3.8761.$$

Note that the lengths of these intervals depend on the particular value of x. A 95% confidence band for $\mu(x)$ is graphed in Figure 6.12-1 along with the scatter diagram and $\hat{y} = \hat{\alpha} + \hat{\beta}(x - \bar{x})$.

The endpoints for a 95% prediction interval for Y when $x = 60$ are

$$81.3 + 0.7421(60 - 68.3) \pm \left[\sqrt{\frac{10(21.7709)}{8}}\sqrt{1.1 + \frac{(60 - 68.3)^2}{756.1}}\right](2.306)$$

or

$$75.1406 \pm 13.1289.$$

Note that this interval is much wider than the confidence interval for $\mu(60)$. In Figure 6.12-2 the 95% prediction band for Y is graphed along with the scatter diagram and the least squares regression line.   ◄

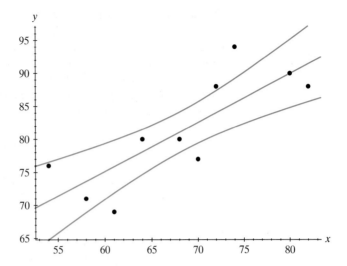

Figure 6.12-1: A 95% confidence band for $\mu(x)$

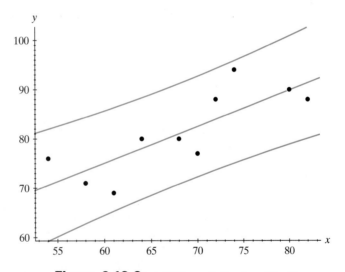

Figure 6.12-2: A 95% prediction band for Y

We now generalize the simple regression model to the **multiple regression** case. Say we observe several x values, say $x_1, x_2, \ldots, x_k$, along with the y value. For example, suppose that x_1 equals the student's ACT composite score, x_2 equals the student's high school class rank, and y equals the student's first-year GPA in college. We want to estimate a regression function $E(Y) = \mu(x_1, x_2, \ldots, x_k)$ from some observed data. If

$$\mu(x_1, x_2, \ldots, x_k) = \beta_1 x_1 + \beta_2 x_2 + \cdots + \beta_k x_k,$$

then we say that we have a **linear model** because this expression is linear in the coefficients $\beta_1, \beta_2, \ldots, \beta_k$.

Note, for illustration, that the model in Section 6.11 is linear in $\alpha = \beta_1$ and $\beta = \beta_2$, with $x_1 = 1$ and $x_2 = x$, giving the mean $\alpha + \beta x$. (For convenience, there the mean of the x values was subtracted from x.) Suppose, however, that we had wished to use the cubic function $\beta_1 + \beta_2 x + \beta_3 x^2 + \beta_4 x^3$ as the mean. This cubic expression still provides a linear model (i.e., linear in the β-values), and we would take $x_1 = 1$, $x_2 = x$, $x_3 = x^2$, and $x_4 = x^3$.

Say our n observation points are

$$(x_{1j}, x_{2j}, \ldots, x_{kj}, y_j), \qquad j = 1, 2, \ldots, n.$$

To fit the linear model $\beta_1 x_1 + \beta_2 x_2 + \cdots + \beta_k x_k$ by the method of least squares, we minimize

$$G = \sum_{j=1}^{n} (y_j - \beta_1 x_{1j} - \beta_2 x_{2j} - \cdots - \beta_k x_{kj})^2.$$

If we equate to zero the k first partial derivatives

$$\frac{\partial G}{\partial \beta_i} = \sum_{j=1}^{n} (-2)(y_j - \beta_1 x_{1j} - \beta_2 x_{2j} - \cdots - \beta_k x_{kj})(x_{ij}), \qquad i = 1, 2, \ldots, k,$$

we obtain the k **normal equations**

$$\beta_1 \sum_{j=1}^{n} x_{1j}^2 + \beta_2 \sum_{j=1}^{n} x_{1j} x_{2j} + \cdots + \beta_k \sum_{j=1}^{n} x_{1j} x_{kj} = \sum_{j=1}^{n} x_{1j} y_j,$$

$$\beta_1 \sum_{j=1}^{n} x_{2j} x_{1j} + \beta_2 \sum_{j=1}^{n} x_{2j}^2 + \cdots + \beta_k \sum_{j=1}^{n} x_{2j} x_{kj} = \sum_{j=1}^{n} x_{2j} y_j,$$

$$\vdots \qquad\qquad \vdots \qquad\qquad \ddots \qquad\qquad \vdots \qquad\qquad \vdots$$

$$\beta_1 \sum_{j=1}^{n} x_{kj} x_{1j} + \beta_2 \sum_{j=1}^{n} x_{kj} x_{2j} + \cdots + \beta_k \sum_{j=1}^{n} x_{kj}^2 = \sum_{j=1}^{n} x_{kj} y_j.$$

The solution of these k equations provides the least squares estimates of $\beta_1, \beta_2, \ldots, \beta_k$. These estimates are also maximum likelihood estimates of $\beta_1, \beta_2, \ldots, \beta_k$, provided that the random variables $Y_1, Y_2, \ldots, Y_n$ are mutually independent and Y_j is $N(\beta_1 x_{1j} + \beta_2 x_{2j} + \cdots + \beta_k x_{kj}, \sigma^2)$, $j = 1, 2, \ldots, k$.

By the method of least squares, fit $y = \beta_1 x_1 + \beta_2 x_2 + \beta_3 x_3$ to the five observed points (x_1, x_2, x_3, y):

$$(1, 1, 0, 4), \quad (1, 0, 1, 3), \quad (1, 2, 3, 2), \quad (1, 3, 0, 6), \quad (1, 0, 0, 1).$$

Note that $x_1 = 1$ in each point; so we are really fitting $y = \beta_1 + \beta_2 x_2 + \beta_3 x_3$. Since

$$\sum_{j=1}^{5} x_{1j}^2 = 5, \quad \sum_{j=1}^{5} x_{1j}x_{2j} = 6, \quad \sum_{j=1}^{5} x_{1j}x_{3j} = 4, \quad \sum_{j=1}^{5} x_{1j}y_j = 16,$$

$$\sum_{j=1}^{5} x_{2j}x_{1j} = 6, \quad \sum_{j=1}^{5} x_{2j}^2 = 14, \quad \sum_{j=1}^{5} x_{2j}x_{3j} = 6, \quad \sum_{j=1}^{5} x_{2j}y_j = 26,$$

$$\sum_{j=1}^{5} x_{3j}x_{1j} = 4, \quad \sum_{j=1}^{5} x_{3j}x_{2j} = 6, \quad \sum_{j=1}^{5} x_{3j}^2 = 10, \quad \sum_{j=1}^{5} x_{3j}y_j = 9,$$

the normal equations are

$$5\beta_1 + 6\beta_2 + 4\beta_3 = 16,$$
$$6\beta_1 + 14\beta_2 + 6\beta_3 = 26,$$
$$4\beta_1 + 6\beta_2 + 10\beta_3 = 9.$$

Solving these three linear equations in three unknowns, we obtain

$$\widehat{\beta}_1 = \frac{274}{112}, \qquad \widehat{\beta}_2 = \frac{127}{112}, \qquad \widehat{\beta}_3 = -\frac{85}{112}.$$

Thus the least squares fit is

$$y = \frac{274x_1 + 127x_2 - 85x_3}{112}.$$

If x_1 always equals 1, then the equation reads

$$y = \frac{274 + 127x_2 - 85x_3}{112}.$$

It is interesting to observe that the usual two-sample problem is actually a linear model. Let $\beta_1 = \mu_1$ and $\beta_2 = \mu_2$ and consider n pairs of (x_1, x_2) that equal $(1, 0)$ and m pairs that equal $(0, 1)$. This would require each of the first n Y-variables, namely $Y_1, Y_2, \ldots, Y_n$, to have the mean

$$\beta_1 \cdot 1 + \beta_2 \cdot 0 = \beta_1 = \mu_1$$

and the next m Y variables, namely $Y_{n+1}, Y_{n+2}, \ldots, Y_{n+m}$, to have the mean

$$\beta_1 \cdot 0 + \beta_2 \cdot 1 = \beta_2 = \mu_2.$$

This is the background of the two-sample problem with the usual $X_1, X_2, \ldots, X_n$ and $Y_1, Y_2, \ldots, Y_m$ replaced by $Y_1, Y_2, \ldots, Y_n$ and $Y_{n+1}, Y_{n+2}, \ldots, Y_{n+m}$.

EXERCISES

6.12-1 For the data given in Exercise 6.11-1, with the usual assumptions,

    (a) Find a 95% confidence interval for $\mu(x)$ when $x = 68, 75$, and 82.
    (b) Find a 95% prediction interval for Y when $x = 68, 75$, and 82.

6.12-2 For the data given in Exercise 6.11-2, with the usual assumptions,

    (a) Find a 95% confidence interval for $\mu(x)$ when $x = 2, 3$, and 4.
    (b) Find a 95% prediction interval for Y when $x = 2, 3$, and 4.

6.12-3 For the data given in Exercise 6.11-3, with the usual assumptions,

    (a) Find a 95% confidence interval for $\mu(x)$ when $x = 20, 25$, and 30.
    (b) Find a 95% prediction interval for Y when $x = 20, 25$, and 30.

6.12-4 For the candy data given in Exercise 6.11-11, with the usual assumptions,

    (a) Find a 90% confidence interval for $\mu(x)$ when $x = 54, 56$, and 58.
    (b) Determine a 90% prediction interval for Y when $x = 54, 56$, and 58.

6.12-5 For the cigarette data in Exercise 6.11-12, with the usual assumptions,

    (a) Find a 95% confidence interval for $\mu(x)$ when $x = 5, 10$, and 15.
    (b) Determine a 95% prediction interval for Y when $x = 5, 10$, and 15.

6.12-6 A computer center recorded the number of programs it maintained during each of 10 consecutive years.

    (a) Calculate the least squares regression line for these data.
    (b) Plot the points and the line on the same graph.
    (c) Find a 95% prediction interval for the number of programs in year 11 under the usual assumptions.

Year	Number of Programs
1	430
2	480
3	565
4	790
5	885
6	960
7	1200
8	1380
9	1530
10	1591

6.12-7 For the ACT scores in Exercise 6.11-14, with the usual assumptions,

    (a) Find a 95% confidence interval for $\mu(x)$ when $x = 17, 20, 23, 26$, and 29.
    (b) Determine a 90% prediction interval for Y when $x = 17, 20, 23, 26$, and 29.

6.12-8 By the method of least squares, fit the regression plane $y = \beta_1 + \beta_2 x_1 + \beta_3 x_2$ to the 12 observations of (x_1, x_2, y): $(1, 1, 6), (0, 2, 3), (3, 0, 10), (-2, 0, -4), (-1, 2, 0), (0, 0, 1), (2, 1, 8), (-1, -1, -2), (0, -3, -3), (2, 1, 5), (1, 1, 1), (-1, 0, -2)$.

6.12-9 By the method of least squares, fit the cubic $y = \beta_1 + \beta_2 x + \beta_3 x^2 + \beta_4 x^3$ to the 10 observed data points (x, y): $(0, 1)$, $(-1, -3)$, $(0, 3)$, $(1, 3)$, $(-1, -1)$, $(2, 10)$, $(0, 0)$, $(-2, -9)$, $(-1, -2)$, $(2, 8)$.

6.12-10 The following data give the times of 22 male swimmers who entered the 50 yard freestyle race in the conference championship meet. For these data, x is the swimmer's best time for the season and y is the swimmer's time in the meet.

x	y	x	y	x	y	x	y
24.97	23.98	24.76	23.63	23.80	23.61	22.84	23.58
24.07	23.57	22.98	23.17	22.93	22.83	23.41	22.75
22.10	22.74	23.08	22.70	23.59	22.62	23.80	22.53
22.38	22.39	22.91	22.34	22.08	21.95	22.46	21.73
21.34	21.68	22.76	21.66	21.82	21.68	21.80	21.58
22.26	21.57	21.36	21.35				

(a) Graph the points and the regression line for these data.
(b) Calculate and plot the residuals. Does linear regression seem to be appropriate?
(c) Find a 90% confidence band for the regression line. Interpret.
(d) Find a 90% prediction band for these data. Interpret.
(e) Find 95% confidence intervals for α, β, and σ^2.

6.12-11 Explain why the model $\mu(x) = \beta_1 e^{\beta_2 x}$ is not a linear model. Would the logarithm, $\ln \mu(x)$, be a linear model?

6.12-12 We would like to fit the quadratic curve $y = \beta_1 + \beta_2 x + \beta_3 x^2$ to a set of points (x_1, y_1), (x_2, y_2), ..., (x_n, y_n) by the method of least squares. To do this, let

$$h(\beta_1, \beta_2, \beta_3) = \sum_{i=1}^{n} (y_i - \beta_1 - \beta_2 x_i - \beta_3 x_i^2)^2.$$

(a) By setting the three first partial derivatives of h with respect to β_1, β_2, and β_3 equal to 0, show that β_1, β_2, and β_3 satisfy the following set of equations (called normal equations), all sums going from 1 to n:

$$\beta_1 n + \beta_2 \sum x_i + \beta_3 \sum x_i^2 = \sum y_i;$$

$$\beta_1 \sum x_i + \beta_2 \sum x_i^2 + \beta_3 \sum x_i^3 = \sum x_i y_i;$$

$$\beta_1 \sum x_i^2 + \beta_2 \sum x_i^3 + \beta_3 \sum x_i^4 = \sum x_i^2 y_i.$$

(b) For the following data:

(6.91, 17.52)	(4.32, 22.69)	(2.38, 17.61)	(7.98, 14.29)
(8.26, 10.77)	(2.00, 12.87)	(3.10, 18.63)	(7.69, 16.77)
(2.21, 14.97)	(3.42, 19.16)	(8.18, 11.15)	(5.39, 22.41)
(1.19, 7.50)	(3.21, 19.06)	(5.47, 23.89)	(7.35, 16.63)
(2.32, 15.09)	(7.54, 14.75)	(1.27, 10.75)	(7.33, 17.42)
(8.41, 9.40)	(8.72, 9.83)	(6.09, 22.33)	(5.30, 21.37)
(7.30, 17.36)			

$n = 25$, $\sum x_i = 133.34$, $\sum x_i^2 = 867.75$, $\sum x_i^3 = 6197.21$, $\sum x_i^4 = 46,318.88$, $\sum y_i = 404.22$, $\sum x_i y_i = 2138.38$, $\sum x_i^2 y_i = 13,380.30$. Show that $\beta_1 = -1.88$, $\beta_2 = 9.86$, and $\beta_3 = -0.995$.

  (c) Plot the points and the linear regression line for these data.

  (d) Calculate and plot the residuals. Does linear regression seem to be appropriate?

  (e) Show that the least squares quadratic regression line is $\hat{y} = -1.88 + 9.86x - 0.995x^2$.

  (f) Plot the points and this least squares quadratic regression curve on the same graph.

  (g) Plot the residuals for quadratic regression and compare this plot with that in part (b).

6.12-13 (The following information comes from the Westview Blueberry Farm and National Oceanic and Atmospheric Administration Reports [NOAA].) For the paired data, (x, y), x gives the Holland rainfall for June and y gives the blueberry production in thousands of pounds from the Westview Blueberry Farm. The data come from the years 1971 to 1989 for those years in which the last frost occurred May 10 or earlier.

$$
\begin{array}{llll}
(4.11, 56.2) & (5.49, 45.3) & (5.35, 31.0) & (6.53, 30.1) \\
(5.18, 40.0) & (4.89, 38.5) & (2.09, 50.0) & (1.40, 45.8) \\
(4.52, 45.9) & (1.11, 32.4) & (0.60, 18.2) & (3.80, 56.1)
\end{array}
$$

  (a) Find the correlation coefficient for these data.

  (b) Find the least squares regression line.

  (c) Make a scatter plot of the data with the least squares regression line on the plot.

  (d) Calculate and plot the residuals. Does linear regression seem to be appropriate?

  (e) Find the least squares quadratic regression curve.

  (f) Calculate and plot the residuals. Does quadratic regression seem to be appropriate?

  (g) Give a short interpretation.

6.12-14 *Parade*, March 13, 1994, stated that "Heart surgeries, including bypass and angioplasty, totaled 196,000 in 1980 and reached 839,000 in 1991." (*Source*: National Hospital Discharge Survey, 1991.) The data for the years 1980 to 1991 in thousands are

$$
\begin{array}{llllll}
(1980, 196) & (1981, 217) & (1982, 243) & (1983, 275) & (1984, 314) & (1985, 379) \\
(1986, 490) & (1987, 588) & (1988, 674) & (1989, 719) & (1990, 781) & (1991, 839)
\end{array}
$$

  (a) Make a scatter plot (here a time-sequence) of the data with the least squares regression line superimposed.

  (b) If possible, find a polynomial that provides a better fit of these data. Construct a scatter plot of the data with this polynomial superimposed.

  (c) Calculate and plot the residuals for the polynomial that you select.

6.12-15 If, in Exercise 6.11-4, a linear regression fit does not seem appropriate, fit the quadratic curve $\beta_1 + \beta_2 x + \beta_3 x^2$. Does this seem to provide a better fit?

6.13* RESAMPLING METHODS

Sampling and resampling methods have become more useful in recent years due to the power of computers. These methods are even used in very beginning courses to convince students that statistics have distributions; that is, statistics are random variables with distributions. At this stage in the book, the reader should be convinced that this is true, although we did use some sampling in Chapter 5 to help sell the idea that the sample mean has an approximate normal distribution.

  Resampling methods, however, are used for more than showing that statistics have certain distributions. Rather, they are needed in finding approximate distributions of certain statistics, which are used to make statistical inferences.

We already know a great deal about the distribution of $\overline{X}$ and resampling methods are not needed for $\overline{X}$. In particular, $\overline{X}$ has an approximate normal distribution with mean μ and standard deviation $\sigma/\sqrt{n}$. Of course, if the latter is unknown, we can estimate it by $s/\sqrt{n}$ and note that $(\overline{X} - \mu)/(s/\sqrt{n})$ has an approximate $N(0, 1)$ distribution, provided the sample size is large enough and the underlying distribution is not too badly skewed with a long heavy tail.

We even know something about the distribution of S^2 *if the random sample arises from a normal distribution* or one fairly close to it. However, the statistic S^2 is not very robust in that its distribution changes a great deal as the underlying distribution deviates much from the normal. It is not like $\overline{X}$, which always has an approximate normal distribution provided the mean μ and variance σ^2 of the underlying distribution exist. So what do we do about distributions of statistics like the sample variance S^2 whose given distribution depends so much on having underlying normal-type assumptions? We use resampling methods that essentially substitute computation for theory. We need to have some idea about the distributions of these various estimators to make statistical inference about the corresponding parameter.

Let us explain resampling by considering very carefully a simple illustration. In this, the sample values are selected to be integers just to keep the computations easy. Say the observed random sample of size $n = 10$ resulted in the following order statistics:

$$2, \ 4, \ 6, \ 10, \ 14, \ 20, \ 26, \ 34, \ 42, \ 52.$$

Clearly this sample arose from a distribution that is skewed to the right as the median is $(14 + 20)/2 = 17$, and the left tail is much shorter than the right tail. A simple box-and-whisker display would support this observation.

Suppose we would like to get some idea about the distribution of the median, m, a statistic which is the average of two middle order statistics in this case. Without any additional assumptions, our best guess at the underlying distribution is the empirical distribution, which is formed by placing the "probability" $1/n = 1/10$ on each of the order statistics. We shall now take a random sample of size $n = 10$ from this empirical distribution. Note this is the **resampling** aspect, and this new sample has the same size as the first sample. Note, however, we do not necessarily obtain the same 10 values because each of the 10 original observations has a probability of 1/10 of being selected on each of the 10 trials. As we are going to continue this resampling many times, we would normally do this on the computer; but here we shall use the table of random numbers so that the reader can understand fully what is happening.

Let us turn to Table IX and start with the fifth column (in blocks of four digits) and the third block down (here in blocks of five four-digit numbers). That is, the first four-digit number we find is 5418. Now we want to select 10 values from the 10 order statistics, some of which can be repeated as we want each to have probability of 1/10 of being selected on each trial. Accordingly, let us decide to read down, selecting the first integer of each of the four digits. We obtain

$$5, \ 6, \ 0, \ 9, \ 4, \ 2, \ 7, \ 5, \ 9, \ 0.$$

Let 5 represent the selection of the 5th order statistic, the 6 selects the 6th, 0 selects the 10th, and so on. Thus, referring back to the order statistics, our sample is

$$14, \ 20, \ 52, \ 42, \ 10, \ 4, \ 26, \ 14, \ 42, \ 52,$$

the median of which is $(20 + 26)/2 = 23$.

We do this again; let us say by continuing down that same column of the first digits of four; obtaining

$$0, \ 5, \ 6, \ 2, \ 9, \ 3, \ 6, \ 4, \ 2, \ 5.$$

This yields the sample of

$$52, \ 14, \ 20, \ 4, \ 42, \ 6, \ 26, \ 10, \ 4, \ 14,$$

which has the median $(14 + 14)/2 = 14$.

This is repeated, continuing down that same column:

$$0, \ 2, \ 7, \ 6, \ 7, \ 2, \ 4, \ 8, \ 7, \ 5.$$

This yields the sample of

$$52, \ 4, \ 26, \ 20, \ 26, \ 4, \ 10, \ 34, \ 26, \ 14,$$

which has the median $(20 + 26)/2 = 23$.

Let us continue, agreeing that we jump to the top of the second of the four digits in the fifth column after we reach the bottom:

$$6, \ 9, \ 8, \ 1, \ 1, \ 6, \ 9, \ 5, \ 1, \ 6,$$

yielding

$$20, \ 42, \ 34, \ 2, \ 2, \ 20, \ 42, \ 14, \ 2, \ 20$$

with a median of $(20 + 20)/2 = 20$.

This is continued a large number of times, maybe more than $N = 1000$, yielding N values of the median which can be used to construct a histogram or stem-and-leaf display. These displays give us some idea about the distribution of the median when the random sample of size $n = 10$ arises from this empirical distribution. However, the empirical distribution is our best guess of the underlying distribution and thus, hopefully, the distribution of the median will be something like that given by our histogram or stem-and-leaf display.

In most instances the sample size is larger than $n = 10$ so that the empirical distribution is a better estimate of the underlying distribution and the resampling is repeated a large number of times. However, so that the reader is certain to understand the process, Exercise 8.10-1 suggests that the illustration be continued until $N = 20$ resamples have been taken and an ordered stem-and-leaf display of the 20 medians is constructed. In the next illustration, a more realistic position is taken and the sample size is larger, $n = 40$, and the resampling is done $N = 1000$ times. Needless to say, this is done with the computer.

While we usually do not know the underlying distribution, we will state that, in the next illustration, it is of the Cauchy type because there are certain basic ideas we want to review or introduce for the first time. The p.d.f. of the Cauchy is

$$f(x) = \frac{1}{\pi(1+x^2)}, \qquad -\infty < x < \infty.$$

The distribution function is

$$F(x) = \int_{-\infty}^{x} \frac{1}{\pi(1+w^2)}\, dw = \frac{1}{\pi}\arctan x + \frac{1}{2}, \qquad -\infty < x < \infty.$$

If we want to generate some X values that have this distribution, we let Y have the uniform distribution $U(0, 1)$ and define X by

$$Y = F(X) = \frac{1}{\pi}\arctan X + \frac{1}{2}$$

or, equivalently,

$$X = \tan\left[\pi\left(Y - \frac{1}{2}\right)\right].$$

We can find some Y values in Table IX of random numbers, but instead we generate 40 values of Y on the computer and then calculate the 40 values of X. Let us now add $\theta = 5$ to each X value to create a sample from a Cauchy distribution with a median of 5. That is, we have a random sample of 40 W values, where $W = X + 5$. We will consider some statistics used to estimate the median, θ, of this distribution. Of course, usually the value of the median is unknown, but here we know that it is equal to $\theta = 5$ and our statistics are estimates of this known number. These 40 values of W are the following, after ordering:

-7.34	-5.92	-2.98	0.19	0.77	0.95	2.86	3.17	3.76	4.20
4.20	4.27	4.31	4.42	4.60	4.73	4.84	4.87	4.90	4.96
4.98	5.00	5.09	5.09	5.14	5.22	5.23	5.42	5.50	5.83
5.94	5.95	6.00	6.01	6.24	6.82	9.62	10.03	18.27	93.62

It is interesting to observe that many of these 40 values are close to the $\theta = 5$, being between 3 and 7; they are almost as if they had arisen from a normal distribution with mean $\mu = 5$ and $\sigma^2 = 1$. But then we note the outliers; these very large or small values occur because of the heavy and long tails of the Cauchy distribution. This suggests that the sample mean $\overline{X}$ is not a very good estimator of the middle. And it is not as $\overline{w} = 6.67$ in this sample. In a more theoretical course, it can be shown that due to the fact that the mean μ and the variance σ^2 do not exist for a Cauchy distribution, $\overline{X}$ is not any better than a single observation X_i in estimating the median θ. The sample median $\widetilde{m}$ is a much better estimate of θ as it is not influenced by the outliers. Here the median equals 4.97, which is fairly close to 5. Actually, the maximum likelihood estimator found by maximizing

$$L(\theta) = \prod_{i=1}^{40} \frac{1}{\pi[1 + (x_i - \theta)^2]}$$

is extremely good but requires difficult numerical methods to compute. Then advanced theory shows that, in the case of a Cauchy distribution, a trimmed mean found by ordering the sample and discarding the smallest and largest $3/8 = 37.5\%$ of the sample and averaging the middle 25% is almost as good as the maximum likelihood estimator, but it is much easier to compute. This trimmed mean is usually denoted by $\overline{X}_{0.375}$; but we use $\overline{X}_t$ for brevity and $\overline{x}_t = 4.96$. For this sample it is not quite as good as the median; but, for most samples, it is better.

It is interesting to note that in our first illustration, the median was the trimmed mean $\overline{X}_{0.40}$ in which 40% of the smallest and largest ordered statistics were discarded and we averaged the middle 20%. Trimmed means are often very useful and many times are used with a smaller trimming percentage. For example, in sporting events like skating and diving, often the smallest and largest of the judges' scores are discarded.

For this Cauchy example, let us resample from the empirical distribution created by placing the "probability" 1/40 on each of our 40 observations. With each of these samples, we find our trimmed mean $\overline{X}_t$. That is, we order the observations of each resample and average the middle 25% of the order statistics, namely the middle 10 order statistics. We do this $N = 1000$ times; thus obtaining $N = 1000$ values of $\overline{X}_t$. These are summarized with the histogram in Figure 6.13-1.

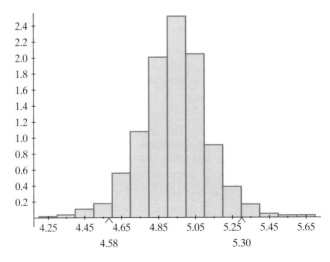

Figure 6.13-1: Histogram of the Trimmed Means

From this resampling procedure, which is called **bootstrapping**, we have some idea about the distribution if the sample arises from the empirical distribution and, hopefully, from the underlying distribution, which is approximated by the empirical distribution. While the distribution of the sample mean, $\overline{X}$, is not normal if the sample arises from a Cauchy-type distribution, the approximate distribution of $\overline{X}_t$ is normal. From the histogram in Figure 6.13-1 of trimmed mean values, this looks to be the case. This observation is supported by the q-q plot in Figure 6.13-2 of the quantities of a standard normal distribution versus those of the 1000 $\overline{x}_t$ values which is very close to being a straight line.

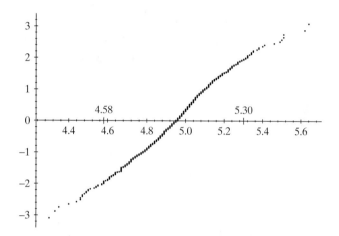

Figure 6.13-2: Quantiles of $N(0, 1)$ versus quantiles of trimmed means

Remark The expression "to pull oneself up by his or her own bootstraps" seems to come from *The Surprising Adventures of Baron Munchausen* by Rudolph Erich Raspe. The Baron had fallen from the sky and found himself in a hole nine fathoms deep and had no idea how to get out. He comments as follows: "Looking down I observed that I had on a pair of boots with exceptionally sturdy straps. Grasping them firmly, I pulled with all my might. Soon I had hoisted myself to the top and stepped out on terra firma without further ado."

Of course, here in *bootstrapping*, statisticians pull themselves up by their bootstraps (the empirical distributions) by recognizing the empirical distribution is the best estimate of the underlying distribution without a lot of other assumptions. So they use the empirical distribution as if it is the underlying distribution to find approximately the distribution of statistics of interest.   ■

How do we find a confidence interval for θ? Recall that the middle of the distribution of $\overline{X}_t - \theta$ is zero. So a guess at θ would be the amount needed to move the histogram of $\overline{X}_t$ values over so that zero is more or less in the middle of this translated histogram. We recognize that this histogram was just generated using the original sample $X_1, X_2, \ldots, X_{40}$ and thus is really only an estimate of the distribution of $\overline{X}_t$.

We could get a point estimate of θ by moving it over until its median (or mean) is at zero. Clearly, however, there is some error in doing so; and we really want some bounds for θ as given by a confidence interval.

To find that confidence interval, let us proceed as follows: In the $N = 1000$ bootstrapped values of $\overline{X}_t$, we find two points, say c and d, such that about 25 values are less than c and about 25 are greater than d. That is, c and d are about on the respective 2.5 and 97.5 percentiles of the empirical distribution of these $N = 1000$ bootstrapped $\overline{X}_t$ values. Thus, θ should be big enough so that over 2.5% of the $\overline{X}_t$ values are less than c and small enough so that over 2.5% of the $\overline{X}_t$ values are greater than d. This requires that $c < \theta$ and $\theta < d$; thus $c < \theta < d$ serves as an approximate 95% confidence interval for θ as found by the **percentile method**. With our bootstrapped distribution of $N = 1000$ $\overline{X}_t$ values, this 95% confidence interval for θ is 4.58 to 5.30 and these two points are marked on the histogram and the q-q plot. Clearly, we could change this percentage to other values, such as 90%. Also, we can test hypotheses about θ by constructing two-sided or one-side confidence intervals for θ.

This **percentile method**, associated with bootstrap method, is a nonparametric procedure as we make no assumptions about the underlying distribution. It is interesting to compare the answer to that obtained by using the order statistics $Y_1 < Y_2 < \cdots < Y_{40}$. If the sample arises from a continuous-type distribution, we have, when θ is the median, that

$$P(Y_{14} < \theta < Y_{27}) = \sum_{k=14}^{26} \binom{40}{k} \left(\frac{1}{2}\right)^{40}$$

$$\approx \Phi\left(\frac{26.5 - 20}{\sqrt{10}}\right) - \Phi\left(\frac{13.5 - 20}{\sqrt{10}}\right) \approx 0.96.$$

Since in our illustration $Y_{14} = 4.42$ and $Y_{27} = 5.23$, then 4.42 to 5.23 is an approximate 96% confidence interval for θ. Of course, $\theta = 5$ is included in each of the two confidence intervals. In this case, the bootstrap confidence interval is a little more symmetric about $\theta = 5$ and somewhat shorter, but required much more work.

We have now illustrated bootstrapping, which allows us to substitute computation for theory to make statistical inferences about characteristics of the underlying distribution. This method is becoming more important as we encounter complicated data sets which clearly do not satisfy certain underlying assumptions. When sampling from a skewed, very long heavy tailed distribution, we know that $T = (\overline{X} - \mu)/(S/\sqrt{n})$ has a distribution that is much different from $N(0, 1)$ as $\overline{X}$ and S are highly correlated. Thus, we would need to use a bootstrap method to help estimate the distribution of T. Bootstrapping and other resampling methods are proving to be extremely useful tools and are worth mentioning in a first course in mathematical statistics.

EXERCISES

6.13-1 Continue the first illustration of this section that concerns the median until the resampling has been done $N = 20$ times. Construct a stem-and-leaf display for the 20 resulting values of the bootstrapped medians.

6.13-2 If time and computing facilities are available, consider the $n = 40$ wind-related losses of Example 3.1-3. To illustrate bootstrapping, take resamples of size $n = 40$ as many as $N = 100$ times computing $T = (\overline{X} - 5)/(S/\sqrt{40})$ each time. Here the value 5 is the median of the original sample. Construct a histogram of the bootstrapped values of T.

6.14* ASYMPTOTIC DISTRIBUTIONS OF MAXIMUM LIKELIHOOD ESTIMATORS

Let us consider a distribution of the continuous type with p.d.f. $f(x; \theta)$ such that the parameter θ is not involved in the support of the distribution. Moreover, we want $f(x; \theta)$ to enjoy a number of mathematical properties that we do not list here. However, in particular, we want to be able to find the maximum likelihood estimator $\widehat{\theta}$ by solving

$$\frac{\partial[\ln L(\theta)]}{\partial \theta} = 0,$$

where here we use a partial derivative sign because $L(\theta)$ involves $x_1, x_2, \ldots, x_n$ too.

That is,

$$\frac{\partial[\ln L(\widehat{\theta})]}{\partial \theta} = 0,$$

where now, with $\widehat{\theta}$ in this expression, $L(\widehat{\theta}) = f(X_1; \widehat{\theta})f(X_2; \widehat{\theta}) \cdots f(X_n; \widehat{\theta})$. We can approximate the left-hand member of this latter equation by a linear function found from the first two terms of a Taylor's series expanded about θ, namely

$$\frac{\partial[\ln L(\theta)]}{\partial \theta} + (\widehat{\theta} - \theta)\frac{\partial^2[\ln L(\theta)]}{\partial \theta^2} \approx 0,$$

when $L(\theta) = f(X_1; \theta)f(X_2; \theta) \cdots f(X_n; \theta)$.

Obviously, this approximation is good enough only if $\widehat{\theta}$ is close to θ, and an adequate mathematical proof involves those conditions that we have not given here. (See Hogg, McKean, and Craig, 2005.) But a heuristic argument can be made by solving for $\widehat{\theta} - \theta$ to obtain

$$\widehat{\theta} - \theta = \frac{\dfrac{\partial[\ln L(\theta)]}{\partial \theta}}{-\dfrac{\partial^2[\ln L(\theta)]}{\partial \theta^2}}. \tag{6.14-1}$$

Recall that

$$\ln L(\theta) = \ln f(X_1; \theta) + \ln f(X_2; \theta) + \cdots + \ln f(X_n; \theta)$$

and

$$\frac{\partial \ln L(\theta)}{\partial \theta} = \sum_{i=1}^{n} \frac{\partial[\ln f(X_i; \theta)]}{\partial \theta}, \qquad (6.14\text{-}2)$$

which is the numerator in Equation 6.14-1. However, this Expression 6.14-2 is the sum of the n independent and identically distributed random variables

$$Y_i = \frac{\partial[\ln f(X_i; \theta)]}{\partial \theta}, \qquad i = 1, 2, \ldots, n;$$

and thus, by the central limit theorem, has an approximate normal distribution with mean (in the continuous case) equal to

$$\int_{-\infty}^{\infty} \frac{\partial[\ln f(x; \theta)]}{\partial \theta} f(x; \theta)\, dx = \int_{-\infty}^{\infty} \frac{\partial[f(x; \theta)]}{\partial \theta} \frac{f(x; \theta)}{f(x; \theta)}\, dx$$

$$= \int_{-\infty}^{\infty} \frac{\partial[f(x; \theta)]}{\partial \theta}\, dx$$

$$= \frac{\partial}{\partial \theta} \left[\int_{-\infty}^{\infty} f(x; \theta)\, dx \right]$$

$$= \frac{\partial}{\partial \theta}[1]$$

$$= 0.$$

Clearly, we need a certain mathematical condition so that it is permissible to interchange the operations of integration and differentiation in those last steps. Of course, the integral of $f(x; \theta)$ is equal to one because it is a p.d.f.

Since we now know that the mean of each Y is

$$\int_{-\infty}^{\infty} \frac{\partial[\ln f(x; \theta)]}{\partial \theta} f(x; \theta)\, dx = 0,$$

let us take derivatives of each member of this equation with respect to θ, obtaining

$$\int_{-\infty}^{\infty} \left\{ \frac{\partial^2[\ln f(x; \theta)]}{\partial \theta^2} f(x; \theta) + \frac{\partial[\ln f(x; \theta)]}{\partial \theta} \frac{\partial[f(x; \theta)]}{\partial \theta} \right\} dx = 0.$$

However,

$$\frac{\partial[f(x;\theta)]}{\partial\theta} = \frac{\partial[\ln f(x;\theta)]}{\partial\theta} f(x;\theta);$$

so

$$\int_{-\infty}^{\infty} \left\{ \frac{\partial[\ln f(x;\theta)]}{\partial\theta} \right\}^2 f(x;\theta)\, dx = -\int_{-\infty}^{\infty} \frac{\partial^2[\ln f(x;\theta)]}{\partial\theta^2} f(x;\theta)\, dx.$$

Since $E(Y) = 0$, this last expression gives the variance of $Y = \partial[\ln f(X;\theta)]/\partial\theta$. Then the variance of Expression 6.14-2 is n times this value, namely

$$-nE\left\{ \frac{\partial^2[\ln f(X;\theta)]}{\partial\theta^2} \right\}.$$

Let us rewrite Equation 6.14-1 as

$$\frac{\sqrt{n}\,(\widehat{\theta} - \theta)}{\dfrac{1}{\sqrt{-E\{\partial^2[\ln f(X;\theta)]/\partial\theta^2\}}}} = \frac{\dfrac{\partial[\ln L(\theta)]/\partial\theta}{\sqrt{-nE\{\partial^2[\ln f(X;\theta)]/\partial\theta^2\}}}}{\dfrac{-\dfrac{1}{n}\dfrac{\partial^2[\ln L(\theta)]}{\partial\theta^2}}{E\{-\partial^2[\ln f(X;\theta)]/\partial\theta^2\}}}. \qquad (6.14\text{-}3)$$

The numerator of the right-hand member of Equation 6.14-3 has an approximate $N(0, 1)$ distribution; and those unstated mathematical conditions require, in some sense, for

$$-\frac{1}{n}\frac{\partial^2[\ln L(\theta)]}{\partial\theta^2} \qquad \text{to converge to} \qquad E\{-\partial^2[\ln f(X;\theta)]/\partial\theta^2\}.$$

Accordingly, the ratios given in Equation 6.14-3 must be approximately $N(0, 1)$. That is, $\widehat{\theta}$ has an approximate normal distribution with mean θ and standard deviation

$$\frac{1}{\sqrt{-nE\{\partial^2[\ln f(X;\theta)]/\partial\theta^2\}}}.$$

EXAMPLE 6.14-1 (continuation of Example 6.2-1). With the underlying exponential p.d.f.

$$f(x;\theta) = \frac{1}{\theta} e^{-x/\theta}, \qquad 0 < x < \infty, \qquad \theta \in \Omega = \{\theta : 0 < \theta < \infty\},$$

$\overline{X}$ is the maximum likelihood estimator. Since

$$\ln f(x;\theta) = -\ln\theta - \frac{x}{\theta}$$

and

$$\frac{\partial[\ln f(x;\theta)]}{\partial\theta} = -\frac{1}{\theta} + \frac{x}{\theta^2} \qquad \text{and} \qquad \frac{\partial^2[\ln f(x;\theta)]}{\partial\theta} = \frac{1}{\theta^2} - \frac{2x}{\theta^3},$$

we have

$$-E\left[\frac{1}{\theta^2} - \frac{2X}{\theta^3}\right] = -\frac{1}{\theta^2} + \frac{2\theta}{\theta^3} = \frac{1}{\theta^2}$$

because $E(X) = \theta$. That is, $\overline{X}$ has an approximate normal distribution with mean θ and standard deviation $\theta/\sqrt{n}$. Thus the random interval $\overline{X} \pm 1.96(\theta/\sqrt{n})$ has an approximate probability of 0.95 for covering θ. Substituting the observed $\overline{x}$ for θ, as well as for $\overline{X}$, we say that $\overline{x} \pm 1.96\overline{x}/\sqrt{n}$ is an approximate 95% confidence interval for θ. ◀

While development of this result used a continuous-type distribution, the result holds for the discrete type also as long as the support does not involve the parameter. This is illustrated in the next example.

EXAMPLE 6.14-2 (continuation of Exercise 6.2-3). The maximum likelihood estimator for λ if the random sample arises from a Poisson distribution with p.m.f.

$$f(x;\lambda) = \frac{\lambda^x e^{-\lambda}}{x!}, \qquad x = 0, 1, 2, \ldots; \qquad \lambda \in \Omega = \{\lambda : 0 < \lambda < \infty\},$$

is $\widehat{\lambda} = \overline{X}$. Now

$$\ln f(x;\lambda) = x \ln \lambda - \lambda - \ln x!$$

and

$$\frac{\partial[\ln f(x;\lambda)]}{\partial\lambda} = \frac{x}{\lambda} - 1 \qquad \text{and} \qquad \frac{\partial^2[\ln f(x;\lambda)]}{\partial\lambda^2} = -\frac{x}{\lambda^2}.$$

Thus

$$-E\left(-\frac{X}{\lambda^2}\right) = \frac{\lambda}{\lambda^2} = \frac{1}{\lambda}$$

and $\widehat{\lambda} = \overline{X}$ has an approximate normal distribution with mean λ and standard deviation $\sqrt{\lambda/n}$. Finally, $\overline{x} \pm 1.645\sqrt{\overline{x}/n}$ serves as an approximate 90% confidence interval for λ. With the data in Exercise 6.2-3, $\overline{x} = 2.225$ and hence this interval is from 1.887 to 2.563. ◀

It is interesting that there is another theorem which is somewhat related to the preceding result in that the variance of $\widehat{\theta}$ serves as a lower bound for the variance of every unbiased estimator of θ. Thus we know that if a certain unbiased estimator has a variance equal to that lower bound, we cannot find a better one and hence it is the best in the sense of being the minimum variance unbiased estimator. So, in the limit, the maximum likelihood estimator is this type of best estimator.

We describe this **Rao-Cramér inequality** here without proof. Let $X_1, X_2, \ldots, X_n$ be a random sample from a distribution of the continuous type with p.d.f. $f(x; \theta), \theta \in \Omega = \{\theta: c < \theta < d\}$, where the support of X does not depend upon θ so that we can differentiate, with respect to θ, under integral signs like that in the following integral:

$$\int_{-\infty}^{\infty} f(x; \theta)\, dx = 1.$$

If $Y = u(X_1, X_2, \ldots, X_n)$ is an unbiased estimator of θ, then

$$\mathrm{Var}(Y) \geq \frac{1}{n \int_{-\infty}^{\infty} \{[\partial \ln f(x; \theta)/\partial\theta]\}^2 f(x; \theta)\, dx}$$

$$= \frac{-1}{n \int_{-\infty}^{\infty} [\partial^2 \ln f(x; \theta)/\partial\theta^2] f(x; \theta)\, dx}.$$

Note that the two integrals in the respective denominators are the expectations

$$E\left\{\left[\frac{\partial \ln f(X; \theta)}{\partial\theta}\right]^2\right\} \quad \text{and} \quad E\left[\frac{\partial^2 \ln f(X; \theta)}{\partial\theta^2}\right];$$

sometimes one is easier to compute than the other. While the Rao-Cramér lower bound has been stated for a continuous-type distribution, it is also true for a discrete-type distribution with summations replacing integrals.

We have computed this lower bound for each of two distributions: exponential and Poisson. Those respective lower bounds were θ^2/n and λ/n; see Examples 6.14-1 and 6.14-2. Since, in each case, the variance of $\overline{X}$ equals the lower hound, then $\overline{X}$ is the minimum variance unbiased estimator.

Let us consider another example.

EXAMPLE 6.14-3 (continuation of Exercise 6.2-7). Let the p.d.f. of X be given by

$$f(x; \theta) = \theta x^{\theta-1}, \qquad 0 < x < 1, \qquad \theta \in \Omega = \{\theta: 0 < \theta < \infty\}.$$

We then have

$$\ln f(x; \theta) = \ln \theta + (\theta - 1)\ln x,$$

$$\frac{\partial \ln f(x; \theta)}{\partial\theta} = \frac{1}{\theta} + \ln x,$$

and

$$\frac{\partial^2 \ln f(x; \theta)}{\partial \theta^2} = -\frac{1}{\theta^2}.$$

Since $E(-1/\theta^2) = -1/\theta^2$, the lower bound of the variance of every unbiased estimator of θ is θ^2/n. Moreover, the maximum likelihood estimator $\widehat{\theta} = -n/\ln \prod_{i=1}^{n} X_i$ has an approximate normal distribution with mean θ and variance θ^2/n. Thus, in a limiting sense, $\widehat{\theta}$ is the minimum variance unbiased estimator of θ. ◀

To measure the value of estimators, their variances are compared to the Rao-Cramér lower bound. The ratio of the Rao-Cramér lower bound to the actual variance of any unbiased estimator is called the **efficiency** of that estimator. An estimator with efficiency of 50%, say, requires that $1/0.5 = 2$ times as many sample observations are needed to do as well in estimation as can be done with the minimum variance unbiased estimator (the 100% efficient estimator).

EXERCISES

6.14-1 Let $X_1, X_2, \ldots, X_n$ be a random sample from $N(\theta, \sigma^2)$, where σ^2 is known.
  (a) Show that $Y = (X_1 + X_2)/2$ is an unbiased estimator of θ.
  (b) Find the Rao-Cramér lower bound for the variance of an unbiased estimator of θ for a general n.
  (c) What is the efficiency of Y in part (a)?

6.14-2 Let $X_1, X_2, \ldots, X_n$ denote a random sample from $b(1, p)$. We know that $\overline{X}$ is an unbiased estimator of p and that $\text{Var}(\overline{X}) = p(1 - p)/n$. (See Exercise 6.2-12.)
  (a) Find the Rao-Cramér lower bound for the variance of every unbiased estimator of p.
  (b) What is the efficiency of $\overline{X}$ as an estimator of p?

6.14-3 (Continuation of Exercise 6.2-2.) When sampling from a normal distribution with known mean μ, $\widehat{\theta} = \sum_{i=1}^{n}(X_i - \mu)^2/n$ is the maximum likelihood estimator of $\theta = \sigma^2$.
  (a) Determine the Rao-Cramér lower bound.
  (b) What is the approximate distribution of $\widehat{\theta}$?
  (c) What is the exact distribution of $n\widehat{\theta}/\theta$, where $\theta = \sigma^2$?

6.14-4 Find the Rao-Cramér lower bound and thus the asymptotic variance of the maximum likelihood estimator $\widehat{\theta}$ if the random sample $X_1, X_2, \ldots, X_n$ is taken from each of the distributions having the following p.d.f.s:
  (a) $f(x; \theta) = (1/\theta^2) x \, e^{-x/\theta},$ $\quad 0 < x < \infty,$ $\quad 0 < \theta < \infty.$
  (b) $f(x; \theta) = (1/2\theta^3) x^2 \, e^{-x/\theta},$ $\quad 0 < x < \infty,$ $\quad 0 < \theta < \infty.$
  (c) $f(x; \theta) = (1/\theta) x^{(1-\theta)/\theta},$ $\quad 0 < x < 1,$ $\quad 0 < \theta < \infty.$

CHAPTER

BAYESIAN METHODS

7.1* SUBJECTIVE PROBABILITY

Subjective probability is the foundation for Bayesian methods. Thus we devote this short section to understanding how a Bayesian would assign probabilities to certain events. In a thorough study of Bayesian methods, there would be greater discussion about how a Bayesian would assess the prior probabilities.

Consider an event A and let us say that we believe that $P(A) = 2/5$. This means that the odds *against* the event A are

$$O(A) = \frac{1 - P(A)}{P(A)} = \frac{1 - 2/5}{2/5} = \frac{3}{2}$$

or 3 to 2. Likewise the odds *for* A are

$$\frac{P(A)}{1 - P(A)} = \frac{2/5}{1 - 2/5} = \frac{2}{3}$$

or 2 to 3. Moreover, if a person really believes that $P(A) = 2/5$ and is willing to bet, he or she would take either side of this bet:

1. Win 3 dollars (say) if A occurs and lose 2 dollars if A does not occur or
2. Win 2 dollars if A does not occur and lose 3 dollars if A does occur.

If this is not the case, then that person should review his or her subjective probability $P(A)$.

Another way of thinking about this subjective probability, $P(A)$, is to say it is equal to the amount you are willing to pay to receive one dollar [the bet of $P(A)$ plus the winnings of $\{1 - P(A)\}$] if A actually occurs. That is, you lose $P(A)$ if A does not occur and you win $1 - P(A)$ if A occurs. So if you believe $P(A) = 2/5$, you would be willing to pay 2/5 of a dollar to play "a game" in which you win 3/5 (plus your bet of 2/5) if A occurs. Note that $(2/5)/(3/5) = 2/3$ or 2 to 3 are the odds in favor of A. Or $(3/5)/(2/5) = 3/2$ or 3 to 2 are the odds against A. Again you must be willing to take either side of the bet.

This is really much like two children dividing a candy bar as equally as possible: One divides it and the other gets to choose which of the two parts seems most desirable; that is, the larger. Accordingly, the child dividing the candy bar tries extremely hard to cut it as equally as possible. Clearly, this is exactly what the person selecting the subjective probability does as he or she must be willing to take either side of the bet with the odds established.

Now every person has his or her own personal probability of almost any statement A, which could be something like "There must be some type of life on the planet Mars." Since we often think of A as something on which we might bet, let us consider the statement A that Iowa will beat Michigan in their football game this year. A man who studies the strengths of those two teams might offer the subjective probability $P(A) = 1/3$ or odds against Iowa of

$$O(A) = \frac{1 - 1/3}{1/3} = 2 = \frac{2}{1},$$

that is, 2 to 1 and be willing to take either side of the bet. Suppose his opponent in this gamble was a strong Michigan fan, and he might think that he will try to offer 3 to 1 odds to that individual hoping that she will bet on Michigan, and he would have a very satisfactory bet: he wins three units if Iowa wins and loses one if Michigan wins. Does he dare do that if he really believes $P(A) = 1/3$? His opponent also knows a lot about these two teams and actually thinks $P(A) = 1/3$ is about right; so she takes the 3 to 1 bet, but chooses Iowa as she likes those 3 to 1 odds even though she is a Michigan fan. So the first person has a bad bet according to his subjective probability $P(A) = 1/3$.

Recall, however, that each person has her own subjective probability and that the Michigan fan might really believe $P(A) = 1/5$ or odds of $(1 - 1/5)/(1/5)$; that is, 4 to 1 against Iowa. Thus she would be glad to bet 3 units to 1 unit; that is, give three units if Iowa wins and accept one unit if Michigan wins. In this case, each player would have a satisfactory bet, consistent with their subjective probabilities. We are certain that this is why bookmakers (bookies) tend to stay in business and often make a great deal of money. They offer odds, and others think that they are wrong and thus bet. Later we shall point out how bookies (and racetracks, casinos, and even state lotteries) can "cover themselves." Thus it is a good rule, in our opinion, never to bet against people who are professional gamblers. How can casinos afford to have such fancy buildings? The persons that bet against them pay for those nice facilities.

Incidentally we have no objection to a friendly poker game, say, in which none of the participants could "be hurt" by the stakes involved. Even one of the authors has participated in friendly games and wagers; and the readers might be warned about playing against statisticians, as well as professional gamblers.

Let us now explain how a racetrack can always make money; but, of course, the owners have certain fixed costs that must be covered so that they must make some. To keep it simple, suppose there are only three horses, A, B, and C, and we will only worry about bets to "win," not "place" or "show." The following table shows the amount of money bet on each horse to win and the "subjective probability" determined by all the bettors.

Horse	Amount	Probability
A	$50,000	0.50
B	$40,000	0.40
C	$10,000	0.10
Total	$100,000	1.00

In a sense, these are not really subjective probabilities as we have been describing them, but some sort of degrees of belief in the various horses as determined by all the bettors.

If we used those probabilities and the associated odds to determine fair payoffs (track bets are made in "two dollar" units), a two dollar bet on A would yield $4.00 if A wins: the two dollar bet plus two more. A two dollar bet on B would yield $5.00 if B wins: Recall that the odds are $(1 - 0.4)/0.4 = 1.5$ against B so the track puts up $3 against the bettor's $2. Finally, a two dollar bet on C would yield $20 if C wins: the $2 bet plus $18 more as the odds are $(1 - 0.1)/0.1 = 9$ to one against C.

If the track would make these payoffs, what would be the total payoff in case A wins, in case B wins, or in case C wins? There are 25,000 $2 bets in that $50,000 bet on A; so if A wins

$$25{,}000 \times \$4 = \$100{,}000$$

would be the total payoff. There are 20,000 $2 bets in that $40,000 bet on B; so if B wins

$$20{,}000 \times \$5 = \$100{,}000$$

would be the total payoff. Similarly, for C winning, the total payoff is

$$5000 \times \$20 = \$100{,}000.$$

That is, with these probabilities, the track would pay off as much as the total amount bet. They could never keep the track in business under that situation; so they must take something off the top to stay in business and this is now explained.

A typical percentage for the track's take is 17%; so we will use that. In our example this amounts to $17,000, and thus there is only $83,000 to distribute. Accordingly the payoffs are adjusted by the factor $83/100 = 0.83$. That is, instead of payoffs of $4, $5, and $20, they are, respectively,

$$4(0.83) = 3.32$$

$$5(0.83) = 4.15$$

$$6(0.83) = 16.60$$

and the three possible total payoffs are

$$25,000(3.32) = 83,000$$

$$20,000(4.15) = 83,000$$

$$5000(16.60) = 83,000,$$

respectively. Hence this track is always assured to making the money necessary to cover all expenses, provided they have enough bettors.

Other gambling establishments cover themselves by having the probabilities slightly favor the "house." Sometimes the take is nothing like 17%, but the probabilities of the bettor winning are slightly under 0.50. For example, in American roulette, the probability of winning by betting on "red" is $18/38 = 0.474$ because there are 18 out of the 38 equally likely places that favor red. Actually, the game that provides a probability closest to one-half is "craps," if you bet on the caster. There, the probability of the caster winning is 0.493, which is extremely close to 0.5. Sometimes the side bets in craps are not so favorable; be careful making those. With such slim odds favoring the house, how does a casino make money? Mainly because they have lots of betting. And, of course, some games are not quite as favorable as roulette or craps; for example, the slots. (As a side note we mention that, in state lotteries, the expected payoff for a one dollar bet is usually about 45 or 50 cents; so these are bad bets although states claim they do good things with their winnings.)

In addition, it should be noted that many bookies have two sets of odd, like possibly 2 to 1 if you bet on Iowa and 3 to 1 if you bet on Michigan. That is, they will pay three dollars $(2 + 1 = 3)$ for a one dollar bet on Iowa, if Iowa wins. On the other hand, they will pay four dollars $(1 + 3 = 4)$ for a three dollar bet on Michigan, if Michigan wins. Clearly, if they can get the right number of bettors on each side they have created a position that they cannot lose. Suppose 10,000 people bet on Iowa (you might not need 10,000 people as some will bet more than the minimum) and 7500 bet on Michigan. The bookies take in

$$(10,000)(\$1) + (7500)(\$3) = \$32,500.$$

Suppose Iowa wins, they pay off $2 + 1 = 3$ for each dollar bet or a total of

$$(10,000)(\$3) = \$30,000.$$

If Michigan wins, they pay off $1 + 3 = 4$ for each three dollar bet or a total of

$$(7500)(\$4) = \$30,000.$$

In either case, they make money, namely,

$$\$32,500 - \$30,000 = \$2500.$$

Clearly, they must balance the bettors or change the odds (which they do) to have a sure win situation. That is, just like the racetrack, they change the odds depending upon the amount of money bet on each side. So if the bookies can get enough bettors, they can make lots of money, and they generally do! Hence, do not bet against professional gamblers unless you achieve a certain amount of enjoyment from betting; but, even then, make certain that stakes are such that you can afford possible losses. (For further discussion of betting, see Berry (1996)).

Now, not all betting is bad. For example, most people are betting when they buy insurance. The insurance companies do exactly the same thing that casinos do as they have the probabilities and the corresponding premiums favor them somewhat. Is this bad? Certainly not, and one of the authors is associated with a Department of Statistics and Actuarial Science and has helped teach many actuaries over the years and even has a daughter who is an actuary. However, to understand the value of buying insurance, the reader must understand the utility of money.

Most of us would say that four dollars are worth twice as much as two dollars. Also, $100 are worth 20 times as much as five dollars. But to most individuals, is $2,000,000 worth twice as much as $1,000,000? For very rich people, it is. However, suppose a person has retired and has exactly $1,000,000. A very rich person offers that person the following bet: Toss a coin fairly and if a head occurs, the rich person gives the retiree $2,000,000; but if a tail results, the retiree gives the rich person that $1,000,000. That is, the retiree gains $2,000,000 with probability 0.5 and loses $1,000,000 with probability 0.5. The expected gain is

$$(\$2,000,000)(0.5) + (-\$1,000,000)(0.5) = \$500,000;$$

seemingly this is a great bet, but any reasonable person with only $1,000,000 would turn it down. Why? That first $1,000,000 is worth a lot more that the second $1,000,000 or a third $1,000,000. Clearly with the bet he could have zero or $3,000,000, each with probability 0.5, but the retiree would rather have $1,000,000 with certainty. While the utility function $u(x)$ of money for this individual is just about $u(x) = x$ for relatively small values of x, it starts to bend downward for larger values of x. Where this bending occurs is also subjective; Bill Gates would immediately accept the proposed bet because his utility function is $u(x) = x$ into the millions, but Bob Hogg would not take the bet because his utility function is bending downward long before $1,000,000.

So why do we buy insurance? To keep it simple, suppose a man wishes to buy a term life insurance valued at $100,000 for one year, possibly to protcct a young family, say a wife and two children, in case he dies. Say his probability of dying in that year is 0.002. What is a fair premium for this insurance if it will cost the company $100,000 with probability 0.002 and zero with probability

0.998? The answer is

$$(\$100{,}000)(0.002) + (\$0)(0.998) = \$200.$$

However, the company asks for a loading factor of 15%, say, or $30. Thus the charge is $230, which seemingly is not a fair bet. However, if he does die, the widow with two children have nothing without the insurance and $100,000 with the insurance. Most men under those circumstances would buy the insurance because the value of the policy is worth much more that the $230 premium in his mind and hence the bet is more than fair to him *personally*. You will note that insurance companies, like casinos, have nice facilities, but we believe in that type of betting.

So it is smart to insure valuable items, like life, liability associated with car accidents, homes, and so on. Things of smaller value should not be insured; that is, you carry your own insurance. The cutoff value differs with each individual, although we find the rule of not insuring anything with value under one or two months salary is a good one; most of our utility functions are $u(x) = x$ in that range and there is no sense paying a "load." Bill Gates does not need to insure for collision on a new car of value $40,000 (he carries liability, however, as most states require it); Bob Hogg does. However, once the value of the car drops to under $10,000, Bob Hogg even drops the collision insurance. But note that these are personal decisions and depend upon the subjective probabilities and a person's utility of money.

Let us now say the reader is willing to accept that the subjective probability $P(A)$ as the fair price for event A, given that you will receive one unit [win $1 - P(A)$] in case A occurs and, of course, lose $P(A)$ if it does not occur. Then, it turns out, all rules (definitions and theorems) in Section 1.2 follow for subjective probabilities. We do not give proofs of them all, but only one of them and some of the others are left as exercises. This proof was given to us in a personal communication from George Woodworth, University of Iowa.

Theorem 7.1-1   If A_1 and A_2 are mutually exclusive, then

$$P(A_1 \cup A_2) = P(A_1) + P(A_2).$$

Proof: Suppose a person thinks a fair price for A_1 is $p_1 = P(A_1)$ and that for A_2 is $p_2 = P(A_2)$. However, that person believes the fair price for $A_1 \cup A_2$ is p_3 which differs from $p_1 + p_2$. Say, $p_3 < p_1 + p_2$ and let the difference be $d = (p_1 + p_2) - p_3$. A gambler offers this person the price $p_3 + d/4$ for $A_1 \cup A_2$. That person takes the offer because it is better than p_3. The gambler sells A_1 at a discount price of $p_1 - d/4$ and sells A_2 at a discount price of $p_2 - d/4$ to that person. Being a rational person with those given prices of p_1, p_2, and p_3, all three of these deals seem very satisfactory. However, that person received $p_3 + d/4$ and that person paid $p_1 + p_2 - d/2$. Thus before any bets are paid off, that person has

$$p_3 + \frac{d}{4} - \left(p_1 + p_2 - \frac{d}{2} \right) = p_3 - p_1 - p_2 + \frac{3d}{4} = -\frac{d}{4}.$$

That is, the person is down $d/4$ before any bets are settled.

1. Suppose A_1 happens: The gambler has $A_1 \cup A_2$ and the person has A_1; so they exchange units and the person is still down $d/4$. The same thing occurs if A_2 happens.
2. Suppose neither A_1 nor A_2 happens; then the gambler and that person receive zero, and the person is still down $d/4$.
3. Of course, A_1 and A_2 cannot occur together since they are mutually exclusive.

Thus we see that it is bad for that person to assign

$$p_3 = P(A_1 \cup A_2) < p_1 + p_2 = P(A_1) + P(A_2),$$

because the gambler can put that person in a position to lose $(p_1 + p_2 - p_3)/4$ no matter what happens. This is sometimes referred to as a *Dutch book*.

The argument when $p_3 > p_1 + p_2$ is similar and also can lead to a Dutch book; it is left as an exercise. Thus p_3 must equal $p_1 + p_2$ to avoid a Dutch book; that is, $P(A_1 \cup A_2) = P(A_1) + P(A_2)$.   ○

EXERCISES

7.1-1 Think of some event A (it could or could not be a sporting outcome) that might or might not happen on your college campus. Get several students to assign their personal probabilities to A. If they differ and the students are not opposed to betting, could you find a friendly (small stakes) bet or bets among them? For example, if two students, S_1 and S_2, assigned $P_1(A) = 0.5$ and $P_2(A) = 0.3$, respectively, they might agree that their odds of $(1 - 0.4)/0.4 = 1.5$ to one would be reasonable as they are between respective odds. That is, S_1 would place one unit on A to 1.5 units on A' by S_2. According to their subjective probabilities, this would be a satisfactory bet for both.

7.1-2 As in Exercise 7.1-1, consider some A around your campus. Let one student, S_1, determine $P_1(A)$, the subjective probability. Based upon this $P_1(A)$, let a second student, S_2, take either side of a friendly bet established by these odds. Try this several times with students exchanging roles.

7.1-3 In American roulette, suppose that 3800 consecutive five dollar bets were placed on red. Show that the casino's expected gain is $1000. Determine the standard deviation associated with the casino's gain.

7.1-4 On a certain fight involving boxer A and boxer B, A is the favorite over B. A bookie finds that $30,000 is bet on A and only $5000 is bet on B. Determine two sets of odds, one for those betting on A and one for those betting on B.

HINT: Note that the bets suggest a "subjective probability" of $5000/35,000 = 1/7$ for B or odds of 6 to 1 against B. Find odds on either side of the 6 to 1 for the two possible odds. Make certain that your choice (it is not unique) will be such that the bookie always makes money.

7.1-5 The following amounts are bet on horses A, B, C, D, E to win.

Horse	Amount
A	$600,000
B	$200,000
C	$100,000
D	$75,000
E	$25,000
Total	$1,000,000

Suppose the track wants to take 20% off the top, namely $200,000. Determine the payoff for winning with a two dollar bet on each of the five horses.

7.1-6 In the proof of Theorem 7.1-1, we considered the case in which $p_3 < p_1 + p_2$. Now, say the person believes that $p_3 > p_1 + p_2$ and create a Dutch book for him.

HINT: Let $d = p_3 - (p_1 + p_2)$. The gambler buys from the person A_1 at a premium price of $p_1 + d/4$ and A_2 for $p_2 + d/4$. Then the gambler sells $A_1 \cup A_2$ to that person at a discount of $p_3 - d/4$. All those are good deals for that person who believes that p_1, p_2, p_3 are correct with $p_3 > p_1 + p_2$. Show that the person has a Dutch book.

7.1-7 Show that $P(S) = 1$.

HINT: Suppose a person thinks $P(S) = p \neq 1$. Consider two cases: $p > 1$ and $p < 1$. In the first case, say $d = p - 1$ and the gambler sells the person S at a discount price $1 + d/2$. Of course, S happens and the gambler pays the person one unit, but he is down $1 + d/2 - 1 = d/2$; so he has a Dutch book. Do the other case.

7.1-8 Show that $P(A') = 1 - P(A)$.

HINT: $A' \cup A = S$ and use the result of Theorem 7.1-1 and Exercise 7.1-7.

7.2* BAYESIAN ESTIMATION

We now describe another approach to estimation that is used by a group of statisticians who call themselves Bayesians. To understand fully their approach would require more text than we can allocate to this topic, but let us begin this brief introduction by considering a simple application of the theorem of the Reverend Thomas Bayes (see Section 1.6).

EXAMPLE 7.2-1

Suppose we know that we are going to select an observation from a Poisson distribution with mean λ equal to 2 or 4. Moreover, we believe prior to performing the experiment that $\lambda = 2$ has about four times as much chance of being the parameter as does $\lambda = 4$; that is, the prior probabilities are $P(\lambda = 2) = 0.8$ and $P(\lambda = 4) = 0.2$. The experiment is now performed and we observe $x = 6$. At this point, our intuition now tells us that $\lambda = 2$ seems less likely than before as the observation $x = 6$ is much more probable with $\lambda = 4$ than with $\lambda = 2$, because, in an obvious notation,

$$P(X = 6 \mid \lambda = 2) = 0.995 - 0.983 = 0.012$$

and

$$P(X = 6 \mid \lambda = 4) = 0.889 - 0.785 = 0.104,$$

from Table III in the Appendix. Our intuition can be supported by computing the conditional probability of $\lambda = 2$, given $X = 6$;

$$P(\lambda = 2 \mid X = 6) = \frac{P(\lambda = 2, X = 6)}{P(X = 6)}$$

$$= \frac{P(\lambda = 2)P(X = 6 \mid \lambda = 2)}{P(\lambda = 2)P(X = 6 \mid \lambda = 2) + P(\lambda = 4)P(X = 6 \mid \lambda = 4)}$$

$$= \frac{(0.8)(0.012)}{(0.8)(0.012) + (0.2)(0.104)} = 0.316.$$

This conditional probability is called the posterior probability of $\lambda = 2$, given the data (here $x = 6$). In a similar fashion, the posterior probability of $\lambda = 4$ is found to be 0.684. Thus we see that the probability of $\lambda = 2$ has decreased from 0.8 (the prior probability) to 0.316 (the posterior probability) with the observation of $x = 6$. ◀

In a more practical application the parameter, say θ, can possibly take many more than two values as in Example 7.2-1. Somehow Bayesians must assign prior probabilities to this total parameter space through a prior p.d.f. $h(\theta)$. They have developed procedures for assessing these prior probabilities, and we clearly cannot do justice to these methods here. Somehow $h(\theta)$ reflects the prior weights that the Bayesian wants to assign to the various possible values of θ. In some instances, if $h(\theta)$ is a constant and thus θ has the uniform prior distribution, we say that the Bayesian has a *noninformative* prior. If, in fact, some knowledge of θ exists in advance of experimentation, noninformative priors should be avoided, if at all possible.

Also, in more practical examples, we usually take several observations, not just one. That is, we take a random sample and there is frequently a good statistic, say Y, for the parameter θ. Say we are considering a continuous case and the p.d.f. of Y, say $g(y; \theta)$, can be thought of as the conditional p.d.f. of Y, given θ; and henceforth in this section we write $g(y; \theta) = g(y \mid \theta)$. Thus we can treat

$$g(y \mid \theta)h(\theta) = k(y, \theta)$$

as the joint p.d.f. of the statistic Y and the parameter. Of course, the marginal p.d.f. of Y is

$$k_1(y) = \int_{-\infty}^{\infty} h(\theta)g(y \mid \theta)\, d\theta.$$

Thus

$$\frac{k(y, \theta)}{k_1(y)} = \frac{g(y \mid \theta)h(\theta)}{k_1(y)} = k(\theta \mid y)$$

would serve as the conditional p.d.f. of the parameter, given $Y = y$. This is essentially Bayes's theorem, and $k(\theta \mid y)$ is called the *posterior p.d.f. of* θ, given $Y = y$.

Bayesians believe that everything that needs to be known about the parameter is summarized in this posterior p.d.f. $k(\theta \mid y)$. Suppose, for illustration, they were pressed into making a point estimate of the parameter θ. They would note that they would be guessing the value of a random variable, here θ, given its p.d.f. $k(\theta \mid y)$. There are many ways that this could be done: For illustration, the mean, the median, or the mode of that distribution would be reasonable guesses. However, in the final analysis, the best guess would clearly depend upon the penalties for various errors created by incorrect guesses. For illustration, if we were penalized by taking the square of the error between the guess, say $w(y)$, and real value of the parameter θ, clearly we would use the conditional mean

$$w(y) = \int_{-\infty}^{\infty} \theta k(\theta \mid y)\, d\theta$$

as our Bayes estimate of θ. The reason for this is that, in general, if Z is a random variable, then the function of b, $E[(Z - b)^2]$, is minimized by $b = E(Z)$. (See Example 2.2-4.) Likewise, if the penalty (loss) function is the absolute value of the error, $|\theta - w(y)|$, then we use the median of the distribution as, with any random variable Z, $E[|Z - b|]$ is minimized when b equals the median of the distribution of Z. (See Exercise 2.2-10.)

EXAMPLE 7.2-2 Suppose that Y has a binomial distribution with parameters n and $p = \theta$. Thus the p.m.f. of Y, given θ, is

$$g(y \mid \theta) = \binom{n}{y} \theta^y (1 - \theta)^{n-y}, \qquad y = 0, 1, 2, \ldots, n.$$

Let us take the prior p.d.f. of the parameter to be the beta p.d.f.

$$h(\theta) = \frac{\Gamma(\alpha + \beta)}{\Gamma(\alpha)\Gamma(\beta)} \theta^{\alpha - 1} (1 - \theta)^{\beta - 1}, \qquad 0 < \theta < 1.$$

Such a prior p.d.f. provides a Bayesian a great deal of flexibility through the selection of the parameters α and β. Thus the joint probabilities can be described by a product of a binomial p.m.f. with parameters n and θ and this beta p.d.f., namely,

$$k(y, \theta) = \binom{n}{y} \frac{\Gamma(\alpha + \beta)}{\Gamma(\alpha)\Gamma(\beta)} \theta^{y + \alpha - 1} (1 - \theta)^{n - y + \beta - 1},$$

on the support given by $y = 0, 1, 2, \ldots, n$ and $0 < \theta < 1$. We find

$$k_1(y) = \int_0^1 k(y, \theta)\, d\theta$$

$$= \binom{n}{y} \frac{\Gamma(\alpha + \beta)}{\Gamma(\alpha)\Gamma(\beta)} \frac{\Gamma(\alpha + y)\Gamma(n + \beta - y)}{\Gamma(n + \alpha + \beta)}$$

on the support $y = 0, 1, 2, \ldots, n$ by comparing the integral to one involving a beta p.d.f. with parameters $y + \alpha$ and $n - y + \beta$. Therefore,

$$k(\theta \mid y) = \frac{k(y, \theta)}{k_1(y)}$$

$$= \frac{\Gamma(n + \alpha + \beta)}{\Gamma(\alpha + y)\Gamma(n + \beta - y)} \theta^{y+\alpha-1}(1 - \theta)^{n-y+\beta-1}, \qquad 0 < \theta < 1,$$

which is a beta p.d.f. with parameters $y + \alpha$ and $n - y + \beta$. With the square error loss function we must minimize, with respect to $w(y)$, the integral

$$\int_0^1 [\theta - w(y)]^2 \, k(\theta \mid y) \, d\theta$$

to obtain the Bayes solution. But, as noted earlier, if Z is a random variable with a second moment, $E[(Z - b)^2]$ is minimized by $b = E(Z)$. In the preceding display, θ is like the Z with p.d.f. $k(\theta \mid y)$, and $w(y)$ is like the b, so the minimization is accomplished by taking

$$w(y) = E(\theta \mid y) = \frac{\alpha + y}{\alpha + \beta + n}$$

which is the mean of the beta distribution with parameters $y + \alpha$ and $n - y + \beta$. (See Exercise 4.4-8.) It is very instructive to note that this Bayes solution can be written as

$$w(y) = \left(\frac{n}{\alpha + \beta + n}\right)\left(\frac{y}{n}\right) + \left(\frac{\alpha + \beta}{\alpha + \beta + n}\right)\left(\frac{\alpha}{\alpha + \beta}\right),$$

which is a weighted average of the maximum likelihood estimate y/n of θ and the mean $\alpha/(\alpha + \beta)$ of the prior p.d.f. of the parameter. Moreover, the respective weights are $n/(\alpha + \beta + n)$ and $(\alpha + \beta)/(\alpha + \beta + n)$. Thus we see that α and β should be selected so that not only is $\alpha/(\alpha + \beta)$ the desired prior mean, but the sum $\alpha + \beta$ also plays a role corresponding to a sample size. That is, if we want our prior opinion to have as much weight as a sample size of 20, we would take $\alpha + \beta = 20$. So if our prior mean is 3/4, we have that α and β are selected such that $\alpha = 15$ and $\beta = 5$. That is, the prior p.d.f. of θ is beta$(15, 5)$. If we observe $n = 40$ and $y = 28$, then the posterior p.d.f. is beta$(28 + 15 = 43, 12 + 5 = 17)$. The prior and posterior p.d.f.s are shown in Figure 7.2-1.   ◀

   In Example 7.2-2 it is quite convenient to note that it is not really necessary to determine $k_1(y)$ to find $k(\theta \mid y)$. If we divide $k(y, \theta)$ by $k_1(y)$, we get the product of a factor, which depends on y but does *not* depend on θ, say $c(y)$, and

$$\theta^{y+\alpha-1}(1 - \theta)^{n-y+\beta-1}.$$

That is,
$$k(\theta \mid y) = c(y)\,\theta^{y+\alpha-1}(1 - \theta)^{n-y+\beta-1}, \qquad 0 < \theta < 1.$$

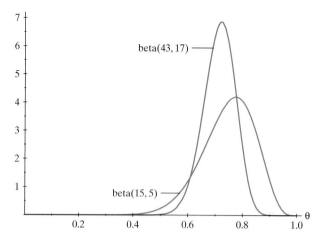

Figure 7.2-1: Beta prior and posterior p.d.f.s

However, $c(y)$ must be that "constant" needed to make $k(\theta \mid y)$ a p.d.f., namely

$$c(y) = \frac{\Gamma(n + \alpha + \beta)}{\Gamma(y + \alpha)\Gamma(n - y + \beta)}.$$

Accordingly, Bayesians frequently write that $k(\theta \mid y)$ is proportional to $k(y, \theta) = g(y \mid \theta)h(\theta)$; that is,

$$k(\theta \mid y) \propto g(y \mid \theta)\, h(\theta).$$

Then, to actually form the p.d.f. $k(\theta \mid y)$, they simply find a "constant," which is some function of y, so that the expression integrates to one.

EXAMPLE 7.2-3 Suppose that $Y = \overline{X}$ is the mean of a random sample of size n that arises from the normal distribution $N(\theta, \sigma^2)$, where σ^2 is known. Then $g(y \mid \theta)$ is $N(\theta, \sigma^2/n)$. Further suppose that we are able to assign prior weights to θ through a prior p.d.f. $h(\theta)$, which is $N(\theta_0, \sigma_0^2)$. Then we have that

$$k(\theta \mid y) \propto \frac{1}{\sqrt{2\pi}\,(\sigma/\sqrt{n})}\frac{1}{\sqrt{2\pi}\,\sigma_0} \exp\left[-\frac{(y - \theta)^2}{2(\sigma^2/n)} - \frac{(\theta - \theta_0)^2}{2\sigma_0^2}\right].$$

If we eliminate all constant factors (including factors involving y only), then we have

$$k(\theta \mid y) \propto \exp\left[-\frac{(\sigma_0^2 + \sigma^2/n)\theta^2 - 2(y\sigma_0^2 + \theta_0\sigma^2/n)\theta}{2(\sigma^2/n)\sigma_0^2}\right].$$

This can be simplified, by completing the square, to read (after eliminating factors not involving θ)

$$k(\theta \mid y) \propto \exp\left\{-\frac{[\theta - (y\sigma_0^2 + \theta_0\sigma^2/n)/(\sigma_0^2 + \sigma^2/n)]^2}{[2(\sigma^2/n)\sigma_0^2]/[\sigma_0^2 + (\sigma^2/n)]}\right\}.$$

That is, the posterior p.d.f. of the parameter is obviously normal with mean

$$\frac{y\sigma_0^2 + \theta_0\sigma^2/n}{\sigma_0^2 + \sigma^2/n} = \left(\frac{\sigma_0^2}{\sigma_0^2 + \sigma^2/n}\right) y + \left(\frac{\sigma^2/n}{\sigma_0^2 + \sigma^2/n}\right)\theta_0$$

and variance $(\sigma^2/n)\sigma_0^2/(\sigma_0^2 + \sigma^2/n)$. If the square error loss function is used, then this posterior mean is the Bayes solution. Again note that it is a weighted average of the maximum likelihood estimate $y = \bar{x}$ and the prior mean θ_0. The Bayes solution $w(y)$ will always be a value between the prior judgment and the usual estimate. Also, note here and in Example 7.2-2 that the Bayes solution gets closer to the maximum likelihood estimate as n increases. Thus the Bayesian procedures permit the decision maker to enter his or her prior opinions into the solution in a very formal way so that the influence of those prior notions will be less and less as n increases. ◄

In Bayesian statistics, all the information is contained in the posterior p.d.f. $k(\theta \mid y)$. In Examples 7.2-2 and 7.2-3 we found Bayesian point estimates using the square error loss function. It should be noted that if the loss function is the absolute value of the error, $|w(y) - \theta|$, then the Bayes solution would be the median of the posterior distribution of the parameter, which is given by $k(\theta \mid y)$. Hence the Bayes solution changes, as it *should*, with different loss functions.

Finally, if an interval estimate of θ is desired, we would find two functions of y, say $u(y)$ and $v(y)$, such that

$$\int_{u(y)}^{v(y)} k(\theta \mid y)\, d\theta = 1 - \alpha,$$

where α is small, say $\alpha = 0.05$. Then the observed interval $u(y)$ to $v(y)$ would serve as an interval estimate for the parameter in the sense that the posterior probability of the parameter being in that interval is $1 - \alpha$. In Example 7.2-3, where the posterior p.d.f. of the parameter was normal, the interval

$$\frac{y\sigma_0^2 + \theta_0\sigma^2/n}{\sigma_0^2 + \sigma^2/n} \pm 1.96 \sqrt{\frac{(\sigma^2/n)\sigma_0^2}{\sigma_0^2 + \sigma^2/n}}$$

serves as an interval estimate for θ with posterior probability of 0.95.

In closing this short section on Bayesian estimation, it should be noted that we could begin with the sample observations, $X_1, X_2, \ldots, X_n$, rather than some statistic Y. Then, in our discussion, we would replace $g(y \mid \theta)$ by the likelihood function

$$L(\theta) = f(x_1 \mid \theta) f(x_2 \mid \theta) \cdots f(x_n \mid \theta),$$

which is the joint p.d.f. of $X_1, X_2, \ldots, X_n$, given θ. Thus we find that

$$k(\theta \mid x_1, x_2, \ldots, x_n) \propto h(\theta) f(x_1 \mid \theta) f(x_2 \mid \theta) \cdots f(x_n \mid \theta) = h(\theta) L(\theta).$$

Now $k(\theta \mid x_1, x_2, \ldots, x_n)$ contains all the information about θ, given the data. Depending on the loss function we would choose our Bayesian estimate of

θ as some characteristic of this posterior distribution, like the mean or the median. It is interesting to observe that if the loss function is zero for some small neighborhood about the true parameter θ and is some large positive constant otherwise, then the Bayesian estimate, $w(x_1, x_2, \ldots, x_n)$, is essentially the mode of this conditional p.d.f., $k(\theta \mid x_1, x_2, \ldots, x_n)$. The reason for this is that we want to take the estimate, $w(x_1, x_2, \ldots, x_n)$, so that it has as much posterior probability as possible in a small neighborhood around it. Finally, note that if $h(\theta)$ if a constant (noninformative prior), then this Bayesian estimate using the mode is exactly the same as the maximum likelihood estimate. More generally, if $h(\theta)$ is not a constant, the Bayesian estimate using the mode can be thought of as a weighted maximum likelihood estimate, in which the weights reflect prior opinion about θ. That is, that value of θ which maximizes $h(\theta)L(\theta)$ is the mode of the posterior distribution of the parameter given the data and can be used as the Bayesian estimate associated with the appropriate loss function.

EXAMPLE 7.2-4 Let us reconsider Example 7.2-2, but now say $X_1, X_2, \ldots, X_n$ is a random sample from the Bernoulli distribution with p.m.f.

$$f(x \mid \theta) = \theta^x (1 - \theta)^{1-x}, \qquad x = 0, 1.$$

With the same prior p.d.f. of θ, the joint distribution of $X_1, X_2, \ldots, X_n$ and θ is given by

$$\frac{\Gamma(\alpha + \beta)}{\Gamma(\alpha)\Gamma(\beta)} \theta^{\alpha-1}(1 - \theta)^{\beta-1} \theta^{\sum_{i=1}^{n} x_i}(1 - \theta)^{n - \sum_{i=1}^{n} x_i}, \qquad 0 < \theta < 1, \ x_i = 0, 1.$$

Of course, the posterior p.d.f. of θ, given $X_1 = x_1, X_2 = x_2, \ldots, X_n = x_n$ is such that

$$k(\theta \mid x_1, x_2, \ldots, x_n) \propto \theta^{\sum_{i=1}^{n} x_i + \alpha - 1}(1 - \theta)^{n - \sum_{i=1}^{n} x_i + \beta - 1}, \qquad 0 < \theta < 1,$$

which is beta with $\alpha^* = \Sigma x_i + \alpha$, $\beta^* = n - \Sigma x_i + \beta$. The conditional mean of θ is

$$\frac{\displaystyle\sum_{i=1}^{n} x_i + \alpha}{n + \alpha + \beta} = \left(\frac{n}{n + \alpha + \beta}\right)\frac{\displaystyle\sum_{i=1}^{n} x_i}{n} + \left(\frac{\alpha + \beta}{n + \alpha + \beta}\right)\left(\frac{\alpha}{\alpha + \beta}\right),$$

which with $y = \Sigma x_i$ is exactly the same result as that of Example 7.2-2. ◀

EXERCISES

7.2-1 Let Y be the sum of the observations of a random sample from a Poisson distribution with mean θ. Let the prior p.d.f. of θ be a gamma one with parameters α and β.

(a) Find the posterior p.d.f. of θ, given $Y = y$.

(b) If the loss function is $[w(y) - \theta]^2$, find the Bayesian point estimate $w(y)$.

(c) Show that this $w(y)$ is a weighted average of the maximum likelihood estimate y/n and the prior mean $\alpha\beta$, with respective weights of $n/(n+1/\beta)$ and $(1/\beta)/(n+1/\beta)$.

7.2-2 Let $X_1, X_2, \ldots, X_n$ be a random sample from a gamma distribution with known α and $\theta = 1/\tau$. Say τ has a prior p.d.f. which is gamma with parameters α_0 and θ_0 so that the prior mean is $\alpha_0 \theta_0$.

   (a) Find the posterior p.d.f. of τ, given $X_1 = x_1, X_2 = x_2, \ldots, X_n = x_n$.
   (b) Find the mean of this posterior distribution and write it as a function of the sample mean $\overline{X}$ and $\alpha_0 \theta_0$.
   (c) Explain how you would find a 95% interval estimate of τ if $n = 10, \alpha = 3, \alpha_0 = 10$, and $\theta_0 = 2$.

7.2-3 In Example 7.2-2 take $n = 30, \alpha = 15$, and $\beta = 5$.

   (a) Using the square error loss, compute the expected loss (risk function) associated with the Bayes solution $w(Y)$.
   (b) The risk function associated with the usual estimator Y/n is, of course, $\theta(1-\theta)/30$. Find those values of θ for which the risk function in part (a) is less than $\theta(1-\theta)/30$. In particular, if the prior mean $\alpha/(\alpha + \beta) = 3/4$ is a reasonable guess, then the risk function in part (a) is the better of the two for what values of θ (i.e., it is smaller in a neighborhood of $\theta = 3/4$)?

7.2-4 Consider a random sample $X_1, X_2, \ldots, X_n$ from a distribution with p.d.f.

$$f(x \mid \theta) = 3\theta x^2 e^{-\theta x^3}, \qquad 0 < x < \infty.$$

Let θ have a prior p.d.f. which is gamma with $\alpha = 4$ and the usual $\theta = 1/4$. Find the conditional mean of θ, given $X_1 = x_1, X_2 = x_2, \ldots, X_n = x_n$.

**7.3*   MORE
BAYESIAN
CONCEPTS**

Let $X_1, X_2, \ldots, X_n$ be a random sample from a distribution with p.d.f. (p.m.f.) $f(x \mid \theta)$ and let $h(\theta)$ be the prior p.d.f. Then the distribution associated with the marginal p.d.f. of $X_1, X_2, \ldots, X_n$, namely

$$k_1(x_1, x_2, \ldots, x_n) = \int_{-\infty}^{\infty} f(x_1 \mid \theta) \, f(x_2 \mid \theta) \cdots f(x_n \mid \theta) \, h(\theta) \, d\theta,$$

is called the **predictive distribution** because it provides the best description of the probabilities on $X_1, X_2, \ldots, X_n$. Often this creates some interesting distributions. For illustration, suppose there is only one X with the normal p.d.f.

$$f(x \mid \theta) = \frac{\sqrt{\theta}}{\sqrt{2\pi}} e^{-(\theta x^2)/2}, \qquad -\infty < x < \infty.$$

Here $\theta = 1/\sigma^2$, the inverse of the variance, is called the **precision** of X. Say this precision has the gamma p.d.f.

$$h(\theta) = \frac{1}{\Gamma(\alpha)\beta^\alpha} \theta^{\alpha-1} e^{-\theta/\beta}, \qquad 0 < \theta < \infty.$$

Then the predictive p.d.f. is

$$k_1(x) = \int_0^\infty \frac{\theta^{\alpha + \frac{1}{2} - 1} e^{-\left(\frac{x^2}{2} + \frac{1}{\beta}\right)\theta}}{\Gamma(\alpha)\beta^\alpha \sqrt{2\pi}} \, d\theta$$

$$= \frac{\Gamma(\alpha + 1/2)}{\Gamma(\alpha)\beta^\alpha \sqrt{2\pi}} \frac{1}{(1/\beta + x^2/2)^{\alpha+1/2}}, \qquad -\infty < x < \infty.$$

Note that if $\alpha = r/2$ and $\beta = 2/r$, where r is a positive integer, then

$$k_1(x) \propto \frac{1}{(1 + x^2/r)^{(r+1)/2}}, \qquad -\infty < x < \infty,$$

which is a t p.d.f. with r degrees of freedom. So if the inverse of the variance, or precision θ, of a normal distribution varies as a gamma random variable, a generalization of a t distribution has been created which has heavier tails than the normal distribution. This **mixture** of normals (different from a mixed distribution), by weighing with the gamma distribution, is by a process often called **compounding**.

Another illustration of compounding is given in the next example.

EXAMPLE 7.3-1 Suppose X has a gamma distribution with the two parameters k and θ^{-1} (that is, the usual α is replaced by k and θ by its reciprocal). Say $h(\theta)$ is gamma with parameters α and β; so that

$$
\begin{aligned}
k_1(x) &= \int_0^\infty \frac{\theta^k x^{k-1} e^{-\theta x}}{\Gamma(k)} \frac{1}{\Gamma(\alpha)\beta^\alpha} \theta^{\alpha-1} e^{-\theta/\beta} \, d\theta \\
&= \int_0^\infty \frac{x^{k-1} \theta^{k+\alpha-1} e^{-\theta(x+1/\beta)}}{\Gamma(k)\Gamma(\alpha)\beta^\alpha} \, d\theta \\
&= \frac{\Gamma(k+\alpha)x^{k-1}}{\Gamma(k)\Gamma(\alpha)\beta^\alpha} \frac{1}{(x+1/\beta)^{k+\alpha}} \\
&= \frac{\Gamma(k+x)\beta^k x^{k-1}}{\Gamma(k)\Gamma(\alpha)(1+\beta x)^{k+\alpha}}, \qquad 0 < x < \infty.
\end{aligned}
$$

Of course, this is a generalization of the F distribution, which we obtain by letting $\alpha = r_2/2$, $k = r_1/2$, and $\beta = r_1/r_2$. ◀

Note how well the prior $h(\theta)$ "fits" with $f(x \mid \theta)$ or $f(x_1 \mid \theta)f(x_2 \mid \theta) \cdots f(x_n \mid \theta)$ in all of our examples and the posterior distribution is of exactly the same form as the prior. In Example 7.2-2, the prior was beta and the posterior was beta. In Example 7.2-3, both the prior and posterior were normal. In Example 7.3-1, both the prior and the posterior (if we had found it) were gamma. When this occurs, we say that class of prior p.d.f.s (p.m.f.s) is a **conjugate family of priors**. Obviously, this makes the mathematics easier and usually the parameters in the prior distribution give us enough flexibility to obtain good fits.

EXAMPLE 7.3-2 (Berry, 1996) This example deals with *predictive probabilities*, and it concerns breakage of glass panels in high-rise buildings. This case involved 39 such panels, and of the 39 panels that broke, it was known that three broke due to nickel sulfide (NiS) stones found in them. Due to loss of evidence, the causes of breakage of the other 36 were unknown. So the court wanted to know whether the manufacturer of the panels or the builder was at fault for the breakage of these 36 panels.

From expert testimony it was thought that usually about 5% breakage is caused by NiS stones. That is, if this p is selected from a beta distribution, we have

$$\frac{\alpha}{\alpha + \beta} = 0.05. \tag{7.3-1}$$

Moreover, the expert thought that if two panels from the same lot break and one breakage was caused by NiS stones, then, due to the pervasive nature of the manufacturing process, the probability of the second breaking due to NiS stones increases to about 95%. Thus the posterior estimate of the p (see Example 7.2-2) with one "success" after one trial is

$$\frac{\alpha + 1}{\alpha + \beta + 1} = 0.95. \tag{7.3-2}$$

Solving Equations 7.3-1 and 7.3-2 for α and β, we obtain

$$\alpha = \frac{1}{360} \quad \text{and} \quad \beta = \frac{19}{360}.$$

Now updating the posterior probability with 3 "successes" out of 3 trials, we obtain the posterior estimate of p to be

$$\frac{\alpha + 3}{\alpha + \beta + 3} = \frac{1/360 + 3}{20/360 + 3} = \frac{1081}{1100} = 0.983.$$

Of course, the court hearing the case wanted to know the expert's probability that all of the remaining 36 panels broke because of NiS stones. Using updated probabilities after the 3rd break, then the 4th, and so on, we obtain the product

$$\left(\frac{1/360 + 3}{20/360 + 3} \right) \left(\frac{1/360 + 4}{20/360 + 4} \right) \left(\frac{1/360 + 5}{20/360 + 5} \right) \cdots \left(\frac{1/360 + 38}{20/360 + 38} \right) = 0.8664.$$

That is, the expert's probability that all other 36 breakages were caused by NiS stones is about 87%, which is the needed value in the court's decision.   ◀

We now look at a situation in which we have two unknown parameters, and we will use, for convenience, what is called a **noninformative prior**. These noninformative priors usually put uniform distributions on the parameters. Let us begin with a random sample $X_1, X_2, \ldots, X_n$ from the normal distribution $N(\theta_1, \theta_2)$ and suppose we have little prior knowledge about θ_1 and θ_2. We then use the noninformative prior that θ_1 and $\ln \theta_2$ are uniform and independent, namely

$$h_1(\theta_1)h_2(\theta_2) \propto \frac{1}{\theta_2}, \quad -\infty < \theta_1 < \infty, \quad 0 < \theta_2 < \infty.$$

Of course, we immediately note that we cannot find a constant c so that c/θ_2 is a joint p.d.f. on that support. That is, this noninformative prior p.d.f. is not a p.d.f. at all and it is called an **improper** prior. However, we use it anyway because it will be satisfactory when multiplied by the joint p.d.f. of $X_1, X_2, \ldots, X_n$. We have the product

$$\left(\frac{1}{\theta_2}\right)\left(\frac{1}{\sqrt{2\pi}\theta_2}\right)^n e^{-\sum_{i=1}^{n}\left(\frac{x_i-\theta_1}{\theta_2}\right)^2/2}.$$

Thus

$$k_{12}(\theta_1, \theta_2 \mid x_1, x_2, \ldots, x_n) \propto \left(\frac{1}{\theta_2}\right)^{\frac{n}{2}+1} \exp\left[-\frac{1}{2}\left\{(n-1)s^2 + n(\bar{x}-\theta_1)^2\right\}/\theta_2\right]$$

since $\sum_{i=1}^{n}(x_i-\theta_1)^2 = (n-1)s^2 + n(\bar{x}-\theta_1)^2 = D$. Then

$$k_1(\theta_1 \mid x_1, x_2, \ldots, x_n) \propto \int_0^\infty k_{12}(\theta_1, \theta_2 \mid x_1, x_2, \ldots, x_n)\, d\theta_2.$$

Changing variables $z = 1/\theta_2$, we obtain

$$k_1(\theta_1 \mid x_1, x_2, \ldots, x_n) \propto \int_0^\infty \frac{z^{n/2+1}}{z^2} e^{-\frac{1}{2}Dz}\, dz$$

$$\propto D^{-n/2} = \left[(n-1)s^2 + n(\bar{x}-\theta_1)^2\right]^{-n/2}.$$

To get this p.d.f. in a more familiar form, let $t = (\theta_1 - \bar{x})/(s/\sqrt{n})$ with Jacobian $s/\sqrt{n}$ to yield

$$k(t \mid x_1, x_2, \ldots, x_n) \propto \frac{1}{[1+t^2/(n-1)]^{[(n-1)+1]/2}}, \qquad -\infty < t < \infty.$$

That is, the conditional p.d.f. of t, given $x_1, x_2, \ldots, x_n$, is Student's t with $n-1$ degrees of freedom. Thus a $(1-\alpha)$ probability interval for θ_1 is given by

$$-t_{\alpha/2} < \frac{\theta_1 - \bar{x}}{s/\sqrt{n}} < t_{\alpha/2}$$

or

$$\bar{x} - t_{\alpha/2}\, s/\sqrt{n} < \theta_1 < \bar{x} + t_{\alpha/2}\, s/\sqrt{n},$$

which is the same as the standard $100(1-\alpha)\%$ confidence interval for θ_1.

The reason that we get the same answer in this case is that we use a noninformative prior. Bayesians do not like to use a noninformative prior if they really know something about the parameters. For example, say they believe that the precision $1/\theta_2$ has a gamma distribution with parameters α and β instead of the noninformative prior. This is a much more difficult integration to find the conditional p.d.f. of θ_1. However, it can be done, but we leave it to a more advanced course. (See Hogg, McKean, and Craig, 2005.)

EXAMPLE 7.3-3 (Johnson and Albert, 1999) The data in this example are a sample of $n = 13$ measurements of the NOAA/EPA ultraviolet (UV) index taken in Los Angeles. The 13 values were collected from archival data of every Sunday in October during the years 1995-1997; this data base is maintained by the National Oceanic and Atmospheric Administration. The 13 UV readings are

$$7, \; 6, \; 5, \; 5, \; 3, \; 6, \; 5, \; 5, \; 3, \; 5, \; 5, \; 4, \; 4;$$

and, while integer values, we assume that they are taken from a $N(\mu, \sigma^2)$ distribution.

The Bayesian analysis, using a noninformative prior in the preceding discussion, implies that, with $\mu = \theta_1$,

$$\frac{\mu - 4.846}{0.317}, \qquad \text{where} \qquad \overline{x} = 4.846 \qquad \text{and} \qquad \frac{s}{\sqrt{n}} = 0.317,$$

has a posterior t distribution with $n - 1 = 12$ degrees of freedom. For illustration, a posterior 95% probability interval for μ is

$$(4.846 - [t_{0.025}(12)][0.317], \; 4.846 + [t_{0.025}(12)][0.317]) = (4.155, \; 5.537). \quad \blacktriangleleft$$

EXAMPLE 7.3-4 Tsutakawa et. al. (1985) discuss the rates of death from stomach cancer for males in the age bracket 45–61 for the 20 largest cities in Missouri. Suppose p_i, $i = 1, 2, \ldots, 20$, are the 20 respective probabilities of death due to stomach cancer. For one model we can assume $p_1, p_2, \ldots, p_{20}$ are taken independently from a beta distribution with parameters α and β. If y_i is the number of "successes" out of n_i observations in city i, then the posterior mean of p_i is

$$\widehat{p_i} \left(\frac{n_i}{n_i + \alpha + \beta} \right) + \left(\frac{\alpha}{\alpha + \beta} \right) \left(\frac{\alpha + \beta}{n_i + \alpha + \beta} \right), \quad i = 1, 2, \ldots, 20,$$

where $\widehat{p_i} = y_i / n_i$. Of course, the parameters α and β are unknown, but we have assumed that $p_1, p_2, \ldots, p_{20}$ arose from a similar distribution for these cities in Missouri, that is, our prior knowledge concerning the proportions is *exchangeable*. So it would be reasonable to estimate $\alpha/(\alpha + \beta)$, the prior mean of a proportion, using

$$\overline{y} = \frac{y_1 + y_2 + \cdots + y_{20}}{n_1 + n_2 + \cdots + n_{20}} = \frac{71}{71,478} = 0.000993$$

for the data given in Table 7.3-1. Thus the posterior estimate of p_i is found by *shrinking* $\hat{p}_i$ towards the pooled estimate of the mean $\alpha/(\alpha + \beta)$, namely $\bar{y}$. That is, the posterior estimate is

$$\hat{p}_i \left(\frac{n_i}{n_i + \alpha + \beta} \right) + \bar{y} \left(\frac{\alpha + \beta}{n_i + \alpha + \beta} \right).$$

The only question remaining is how much weight should be given to the prior, represented by $\alpha + \beta$, relative to $n_1, n_2, \ldots, n_{20}$. Considering the sizes of the samples from the various cities, we selected $\alpha + \beta = 3000$ (which means the prior is worth about a sample of 3000) which resulted in the posterior probabilities given in Table 7.3-1. Note how this type of shrinkage tends to pull the posterior estimates much closer to the average, particularly for those associated with small sample sizes. Baseball fans might try this type of shrinkage in predicting some of the final batting averages of the better batters about a quarter of the way through the season. ◀

Table 7.3-1: Cancer Mortality Rates

y_i	n_i	$\hat{p}_i$	posterior estimate	y_i	n_i	$\hat{p}_i$	posterior estimate
0	1083	0	0.00073	0	855	0	0.00077
2	3461	0.00058	0.00077	0	657	0	0.00081
1	1208	0.00083	0.00095	1	1025	0.00098	0.00099
0	527	0	0.00084	2	1668	0.00120	0.00107
1	583	0.00172	0.00111	3	582	0.00515	0.00167
0	917	0	0.00076	1	857	0.00117	0.00103
1	680	0.00147	0.00108	1	917	0.00109	0.00102
54	53637	0.00101	0.00101	0	874	0	0.00077
0	395	0	0.00088	1	581	0.00172	0.00111
3	588	0.00510	0.00167	0	383	0	0.00088

It is clear that difficult integration caused Bayesians great problems until very recent times in which advances in computer methods "solved" many of these problems. As a simple illustration, suppose the p.d.f. of a good statistic Y is $f(y \mid \theta)$ and the prior p.d.f. $h(\theta)$ is such that

$$k(\theta \mid y) = \frac{f(y \mid \theta) h(\theta)}{\displaystyle\int_{-\infty}^{\infty} f(y \mid \theta) h(\theta) \, d\theta}$$

is not a nice p.d.f. with which to deal. In particular, say that we have a square error loss and we wish to determine $E(\theta \mid y)$, namely

$$\delta(y) = \frac{\displaystyle\int_{-\infty}^{\infty} \theta f(y \mid \theta) h(\theta) \, d\theta}{\displaystyle\int_{-\infty}^{\infty} f(y \mid \theta) h(\theta) \, d\theta},$$

but cannot do it easily. Call $f(y \mid \theta) = w(\theta)$ and thus we wish to evaluate the ratio

$$\frac{E[\theta \, w(\theta)]}{E[w(\theta)]},$$

where y is given and the expected values are taken with respect to θ. Simply generate a number of θ values, say $\theta_1, \theta_2, \ldots, \theta_m$ (where m is large), from the distribution given by $h(\theta)$. Then estimate the numerator and denominator of the desired ratio by

$$\sum_{i=1}^{m} \frac{\theta_i \, w(\theta_i)}{m} \qquad \text{and} \qquad \sum_{i=1}^{m} \frac{w(\theta_i)}{m},$$

respectively, to obtain

$$\tau = \frac{\displaystyle\sum_{i=1}^{m} \theta_i \, w(\theta_i)/m}{\displaystyle\sum_{i=1}^{m} w(\theta_i)/m}.$$

In addition to this simple Monte Carlo procedure, there are additional ones extremely useful in Bayesian inferences. Two of these are the *Gibbs sampler* and *Markov chain Monte Carlo (MCMC)*. The latter is used in *hierarchical Bayes models* in which the prior has another parameter which has its own prior. That is, we have

$$f(y \mid \theta), \qquad h(\theta \mid \tau), \qquad \text{and} \qquad g(\tau).$$

Hence

$$k(\theta, \tau \mid y) = \frac{f(y \mid \theta) \, h(\theta \mid \tau) \, g(\tau)}{\displaystyle\int_{-\infty}^{\infty} \int_{-\infty}^{\infty} f(y \mid \theta) \, h(\theta \mid \tau) \, g(\tau) \, d\theta \, d\tau}$$

and

$$k_1(\theta \mid y) = \int_{-\infty}^{\infty} k(\theta, \tau \mid y) \, d\tau.$$

Thus a Bayes estimator, for square error loss, is

$$\int_{-\infty}^{\infty} \theta \, k_1(\theta \mid y) \, d\theta.$$

Using the Gibbs sampler, we can generate a stream of values $(\theta_1, \tau_1), (\theta_2, \tau_2), \ldots$ that allows us to estimate $k(\theta, \tau \mid y)$ and $\int_{-\infty}^{\infty} \theta \, k_1(\theta \mid y) \, d\theta$. These procedures are the MCMC procedures. For additional references see Hogg, McKean, and Craig (2005).

EXERCISES

7.3-1 Let X have a Poisson distribution with parameter θ. Let θ be $\Gamma(\alpha, \beta)$. Show that the marginal p.m.f. of X (the compound distribution) is

$$k_1(x) = \frac{\Gamma(\alpha + x)\,\beta^x}{\Gamma(\alpha)\,x!\,(1 + \beta)^{\alpha+x}}, \qquad x = 0, 1, 2, 3, \dots,$$

which is a generalization of the negative binomial distribution.

7.3-2 Suppose X is $b(n, \theta)$ and θ is beta(α, β). Show that the marginal p.d.f. of X (the compound distribution) is

$$k_1(x) = \frac{n!\,\Gamma(\alpha + \beta)\,\Gamma(x + \alpha)\,\Gamma(n - x + \beta)}{x!\,(n - x)!\,\Gamma(\alpha)\,\Gamma(\beta)\,\Gamma(n + \alpha + \beta)}, \qquad x = 0, 1, 2, \dots, n.$$

7.3-3 Let X have the geometric p.m.f. $\theta(1 - \theta)^{x-1}$, $x = 1, 2, 3, \dots$, where θ is beta with parameters α and β. Show that the compound p.m.f. is

$$\frac{\Gamma(\alpha + \beta)\,\Gamma(\alpha + 1)\,\Gamma(\beta + x - 1)}{\Gamma(\alpha)\,\Gamma(\beta)\,\Gamma(\alpha + \beta + x)}, \qquad x = 1, 2, 3, \dots.$$

With $\alpha = 1$, this is one form of **Zipf's law**

$$\frac{\beta}{(\beta + x)(\beta + x - 1)}, \qquad x = 1, 2, 3, \dots.$$

7.3-4 Let X have the p.d.f.

$$f(x \mid \theta) = \theta\tau x^{\tau-1}e^{-\theta x^\tau}, \qquad 0 < x < \infty,$$

where the distribution of θ is $\Gamma(\alpha, \beta)$. Find the compound distribution of X, which is called the **Burr distribution**.

7.3-5 Let $X_1, X_2, \dots, X_n$ be a random sample from a gamma distribution with $\alpha = 1, \theta$. Let $h(\theta) \propto 1/\theta$, $0 < \theta < \infty$ be an improper noninformative prior.

    (a) Find the posterior p.d.f. of θ.
    (b) Change variables $z = 1/\theta$ and show that the posterior distribution of Z is $\Gamma(n, 1/y)$, where $y = \sum_{i=1}^{n} x_i$.
    (c) Use $2yz$ to obtain a $(1 - \alpha)$ probability interval for z and, of course, for θ.

7.3-6 Let X_1, X_2 be a random sample from the Cauchy distribution with p.d.f.

$$f(x \mid \theta_1, \theta_2) = \frac{1}{\pi}\,\frac{\theta_2}{\theta_2^2 + (x - \theta_1)^2}, \qquad -\infty < x < \infty,$$

and $-\infty < \theta_1 < \infty$, $0 < \theta_2 < \infty$. Consider the noninformative prior $h(\theta_1, \theta_2) \propto 1$ on that support. Obtain the posterior p.d.f. (except for constants) of θ_1, θ_2 if $x_1 = 3$, $x_2 = 7$. For estimates find θ_1, θ_2 which maximizes this posterior p.d.f.; that is, find the mode of that posterior. (This might require some reasonable "trial and error" or an advanced method of maximizing a function of two variables.)

CHAPTER

8

TESTS OF STATISTICAL HYPOTHESES

TESTS ABOUT PROPORTIONS

A first major area of statistical inference, namely estimation of parameters, was introduced in Chapter 6. In this chapter we consider a second major one, **tests of statistical hypotheses**. This very important topic is introduced through an illustration.

Suppose a manufacturer of a certain printed circuit observes that about $p = 0.06$ of the circuits fail. An engineer and statistician working together suggest some changes that might improve the design of the product. To test this new procedure, it was agreed that $n = 200$ circuits would be produced using the proposed method and then checked. Let Y equal the number of these 200 circuits that fail. Clearly, if the number of failures, Y, is such that $Y/200$ is about equal to 0.06, then it seems that the new procedure has not resulted in an improvement. On the other hand, if Y is small so that $Y/200$ is about 0.02 or 0.03, we might believe that the new method is better than the old. On the other hand, if $Y/200$ is 0.09 or 0.10, the proposed method has perhaps caused a greater proportion of failures.

What we need to establish is a formal rule that tells us when to accept the new procedure as an improvement. In addition, we must know the consequences of this rule. For an example of a rule, we could accept the new procedure as an improvement if $Y \leq 7$ or $Y/n \leq 0.035$.

We do note, however, that the probability of failure could still be about $p = 0.06$ even with the new procedure, and yet we could observe 7 or fewer failures in $n = 200$ trials. That is, we would accept the new method as being an improvement when, in fact, it was not. This decision is a mistake which we call a **Type I error**. On the other hand, the new procedure might actually improve the product so that p is much smaller, say $p = 0.03$, and yet we could observe $y = 9$ failures so that $y/200 = 0.045$. Thus we would not accept the new method as resulting in an improvement when in fact it had. This decision would also be a mistake which we call a **Type II error**. We must study the probabilities of these two types of errors to understand fully the consequences of our rule.

Let us begin by modeling the situation. If we believe these trials, using the new procedure, are independent and have about the same probability of failure on each trial, then Y is binomial $b(200, p)$. We wish to make a statistical inference about p using the unbiased estimator $\widehat{p} = Y/200$. Of course, we could construct a one-sided confidence interval, say one that has 95% confidence of providing an upper bound for p, obtaining

$$\left[0, \ \widehat{p} + 1.645 \sqrt{\frac{\widehat{p}(1 - \widehat{p})}{200}} \ \right].$$

This inference is very appropriate and many statisticians simply do this. If the limits of this confidence interval contain 0.06, they would not say the new procedure is necessarily better, at least until more data are taken. If, on the other hand, the upper limit of this confidence interval is less than 0.06, then they feel 95% confident that the true p is now less than 0.06. Hence they would support the fact that the new procedure has improved the manufacturing of these printed circuits.

While this use of confidence intervals is highly appropriate and later we indicate the relationship of confidence intervals and tests of hypotheses, every student of statistics should also have some understanding of the basic concepts in the latter area. Here, in our illustration, we are testing whether or not the probability of failure has or has not decreased from 0.06 when the new manufacturing procedure is used. The *no change* hypothesis, $H_0: p = 0.06$, is called the **null hypothesis**. Since $H_0: p = 0.06$ completely specifies the distribution it is called a **simple hypothesis**; thus $H_0: p = 0.06$ is a **simple null hypothesis**. The *research worker's* (here the engineer and/or the statistician) hypothesis $H_1: p < 0.06$ is called the **alternative hypothesis**. Since $H_1: p < 0.06$ does not completely specify the distribution, it is a **composite hypothesis** because it is composed of many simple hypotheses. Our rule of rejecting H_0 and accepting H_1 if $Y \leq 7$, and otherwise accepting H_0 is called a **test of a statistical hypothesis**. We now see that the two types of errors can be recorded as follows.

Type I error: Rejecting H_0 and accepting H_1 when H_0 is true;

Type II error: Failing to reject H_0 when H_1 is true, that is, when H_0 is false.

Since, in our illustration, we make a Type I error if $Y \leq 7$ when in fact $p = 0.06$, we can calculate the probability of this error, which we denote by α and call the **significance level of the test**. Under our assumptions, it is

$$\alpha = P(Y \leq 7; \; p = 0.06) = \sum_{y=0}^{7} \binom{200}{y} (0.06)^y (0.94)^{200-y}.$$

Since n is rather large and p is small, these binomial probabilities can be approximated very well by Poisson probabilities with $\lambda = 200(0.06) = 12$. That is, from the Poisson table, the probability of the Type I error is

$$\alpha \approx \sum_{y=0}^{7} \frac{12^y e^{-12}}{y!} = 0.090.$$

Thus, the approximate significance level of this test is $\alpha = 0.090$. (The exact value of α using the binomial distribution is 0.083.)

  This value of α is reasonably small. However, what about the probability of Type II error in case p has been improved to 0.03, say? This error occurs if $Y > 7$ when, in fact, $p = 0.03$; hence its probability, denoted by β, is

$$\beta = P(Y > 7; \; p = 0.03) = \sum_{y=8}^{200} \binom{200}{y} (0.03)^y (0.97)^{200-y}.$$

Again we use the Poisson approximation, here with $\lambda = 200(0.03) = 6$, to obtain

$$\beta \approx 1 - \sum_{y=0}^{7} \frac{6^y e^{-6}}{y!} = 1 - 0.744 = 0.256.$$

(The exact probability using the binomial distribution is 0.254 so the approximation is very good.) The engineer and the statistician who created this new procedure probably are not too pleased with this answer. That is, they note that if their new procedure of manufacturing circuits has actually decreased the probability of failure to 0.03 from 0.06 (*a big improvement*), there is still a good chance, 0.256, that H_0: $p = 0.06$ is accepted and their improvement rejected. Thus, in their eyes, this test of H_0: $p = 0.06$ against H_1: $p = 0.03$ is unsatisfactory. More will be said about modifying tests in Section 9.1* so that satisfactory values of the probabilities of the two types of errors, namely α and β, can be obtained.

  Without worrying more about the probability of the Type II error here, we present a frequently used procedure for testing H_0': $p = p_0$, where p_0 is some specified probability of success. This test is based upon the fact that the number of successes Y in n independent Bernoulli trials is such that Y/n has an approximate normal distribution $N[p_0, p_0(1 - p_0)/n]$, provided H_0: $p = p_0$

is true and n is large. Suppose the alternative hypothesis is $H_1: p > p_0$; that is, it has been hypothesized by a research worker that something has been done to increase the probability of success. Consider the test of $H_0: p = p_0$ against $H_1: p > p_0$ that rejects H_0 and accepts H_1 if and only if

$$Z = \frac{Y/n - p_0}{\sqrt{p_0(1 - p_0)/n}} \geq z_\alpha.$$

That is, if Y/n exceeds p_0 by z_α standard deviations of Y/n, we reject H_0 and accept the hypothesis $H_1: p > p_0$. Since, under H_0, Z is approximately $N(0, 1)$, the approximate probability of this occurring when $H_0: p = p_0$ is true is α. So the significance level of this test is approximately α.

If the alternative is $H_1: p < p_0$ instead of $H_1: p > p_0$, then the appropriate α-level test is given by $Z \leq -z_\alpha$. Hence if Y/n is smaller than p_0 by z_α standard deviations of Y/n, we accept $H_1: p < p_0$.

EXAMPLE 8.1-1 It was claimed that many commercially manufactured dice are not fair because the "spots" are really indentations so that, for example, the 6-side is lighter than the 1-side. Let p equal the probability of rolling a 6 with one of these dice. To test $H_0: p = 1/6$ against the alternative hypothesis $H_1: p > 1/6$, several of these dice will be rolled to yield a total of $n = 8000$ observations. Let Y equal the number of times that six resulted in the 8000 trials. The test statistic is

$$Z = \frac{Y/n - 1/6}{\sqrt{(1/6)(5/6)/n}} = \frac{Y/8000 - 1/6}{\sqrt{(1/6)(5/6)/8000}}.$$

If we use a significance level of $\alpha = 0.05$, the critical region is

$$z \geq z_{0.05} = 1.645.$$

The results of the experiment yielded $y = 1389$, so that the calculated value of the test statistic is

$$z = \frac{1389/8000 - 1/6}{\sqrt{(1/6)(5/6)/8000}} = 1.670.$$

Since

$$z = 1.670 > 1.645,$$

the null hypothesis is rejected and these experimental results indicate that these dice favor a 6 more than a fair die would. (You could perform your own experiment to check out other dice. Also see Exercise 8.1-6.) ◀

Tests of $H_0: p = p_0$ against $H_1: p < p_0$ or $H_0: p = p_0$ against $H_1: p > p_0$ are called **one-sided** tests because the alternative hypotheses are **one-sided**. There are times when **two-sided** alternatives and tests are appropriate as in $H_1: p \neq p_0$. For illustration, suppose that the pass rate in the usual beginning statistics course is p_0. There has been an intervention (say some new teaching method) and it is not known whether the pass rate will increase, decrease, or stay about the same. Thus we test the null (no change) hypothesis H_0: $p = p_0$ against the two-sided alternative H_1: $p \neq p_0$. A test with the approximate significance level α for doing this is to reject H_0: $p = p_0$ if

$$|Z| = \frac{|Y/n - p_0|}{\sqrt{p_0(1 - p_0)/n}} \geq z_{\alpha/2},$$

since, under H_0, $P(|Z| \geq z_{\alpha/2}) \approx \alpha$. These tests of approximate significance level α are summarized in Table 8.1-1. The rejection region for H_0 is often called the **critical region** of the test, and we use that terminology in the table.

Table 8.1-1: Tests of Hypotheses for One Proportion

H_0	H_1	Critical Region
$p = p_0$	$p > p_0$	$z = \dfrac{y/n - p_0}{\sqrt{p_0(1 - p_0)/n}} \geq z_\alpha$
$p = p_0$	$p < p_0$	$z = \dfrac{y/n - p_0}{\sqrt{p_0(1 - p_0)/n}} \leq -z_\alpha$
$p = p_0$	$p \neq p_0$	$|z| = \dfrac{|y/n - p_0|}{\sqrt{p_0(1 - p_0)/n}} \geq z_{\alpha/2}$

For a number of years there has been another value associated with a statistical test, and most statistical computer programs automatically print this out; it is called the **probability value** or, for brevity, *p*-value. The *p*-**value** associated with a test is the probability, under the null hypothesis H_0, that the test statistic (a random variable) is equal to or exceeds the observed value (a constant) of the test statistic in the direction of the alternative hypothesis. Rather than select the critical region ahead of time, the *p*-value of a test can be reported and the reader then makes a decision.

In Example 8.1-1, the value of the test statistic was $z = 1.67$. Because the alternative hypothesis was H_1: $p > 1/6$, the *p*-value is

$$p\text{-value} = P(Z \geq 1.67) = 0.0475.$$

Note that this *p*-value is less than $\alpha = 0.05$ and this would lead to rejection of H_0 at an $\alpha = 0.05$ significance level. If the alternative hypothesis was two sided, H_1: $p \neq 1/6$, then the *p*-value would be $P(|Z| \geq 1.67) = 0.095$ and would not lead to rejection of H_0 at $\alpha = 0.05$.

Often there is interest in tests about p_1 and p_2, the probabilities of success for two different distributions or the proportions of two different populations having a certain characteristic. For example, if p_1 and p_2 denote the respective proportions of homeowners and renters who vote in favor of a proposal to reduce property tax, a politician might be interested in testing H_0: $p_1 = p_2$ against the one-sided alternative hypothesis H_1: $p_1 > p_2$.

Let Y_1 and Y_2 represent, respectively, the numbers of observed successes in n_1 and n_2 independent trials with probabilities of success p_1 and p_2. Recall that the distribution of $\widehat{p}_1 = Y_1/n_1$ is approximately $N[p_1, p_1(1 - p_1)/n_1]$ and the distribution of $\widehat{p}_2 = Y_2/n_2$ is approximately $N[p_2, p_2(1 - p_2)/n_2]$. Thus the distribution of $\widehat{p}_1 - \widehat{p}_2$ is $N[p_1 - p_2, p_1(1 - p_1)/n_1 + p_2(1 - p_2)/n_2]$, approximately. It follows that the distribution of

$$Z = \frac{Y_1/n_1 - Y_2/n_2 - (p_1 - p_2)}{\sqrt{p_1(1 - p_1)/n_1 + p_2(1 - p_2)/n_2}} \tag{8.1-1}$$

is approximately $N(0, 1)$. To test H_0: $p_1 - p_2 = 0$ or, equivalently, H_0: $p_1 = p_2$, let $p = p_1 = p_2$ be the common value under H_0. We shall estimate p with $\widehat{p} = (Y_1 + Y_2)/(n_1 + n_2)$. Replacing p_1 and p_2 in the denominator of Equation 8.1-1 with this estimate, we obtain the test statistic

$$Z = \frac{\widehat{p}_1 - \widehat{p}_2 - 0}{\sqrt{\widehat{p}(1 - \widehat{p})(1/n_1 + 1/n_2)}},$$

which has an approximate $N(0, 1)$ distribution when the null hypothesis is true.

The three possible alternative hypotheses and their critical regions are summarized in Table 8.1-2.

	Table 8.1-2: Tests of Hypotheses for Two Proportions	
H_0	H_1	Critical Region
$p_1 = p_2$	$p_1 > p_2$	$z = \dfrac{\widehat{p}_1 - \widehat{p}_2}{\sqrt{\widehat{p}(1 - \widehat{p})(1/n_1 + 1/n_2)}} \geq z_\alpha$
$p_1 = p_2$	$p_1 < p_2$	$z = \dfrac{\widehat{p}_1 - \widehat{p}_2}{\sqrt{\widehat{p}(1 - \widehat{p})(1/n_1 + 1/n_2)}} \leq -z_\alpha$
$p_1 = p_2$	$p_1 \neq p_2$	$|z| = \dfrac{|\widehat{p}_1 - \widehat{p}_2|}{\sqrt{\widehat{p}(1 - \widehat{p})(1/n_1 + 1/n_2)}} \geq z_{\alpha/2}$

Remark In testing both H_0: $p = p_0$ and H_0: $p_1 = p_2$, statisticians sometimes use different denominators for z. For tests of a single proportions, $\sqrt{p_0(1 - p_0)/n}$ can be replaced by $\sqrt{(y/n)(1 - y/n)/n}$; and, for testing equality of two proportions, the following denominator can be used:

$$\sqrt{\frac{\widehat{p}_1(1 - \widehat{p}_1)}{n_1} + \frac{\widehat{p}_2(1 - \widehat{p}_2)}{n_2}}.$$

We do not have a strong preference one way or the other since the two methods provide about the same numerical result. The substitutions do provide better estimates of the standard deviations of the numerators when the null hypotheses are clearly false. There is some advantage to this if the null hypothesis is likely to be false. In addition, the substitutions also tie together the use of confidence intervals and tests of hypotheses. For example, if the null hypothesis is H_0: $p = p_0$, then the alternative hypothesis H_1: $p < p_0$ is accepted if

$$z = \frac{\widehat{p} - p_0}{\sqrt{\dfrac{\widehat{p}(1 - \widehat{p})}{n}}} \leq -z_\alpha.$$

This is equivalent to the statement that

$$p_0 \notin \left[0, \widehat{p} + z_\alpha \sqrt{\frac{\widehat{p}(1 - \widehat{p})}{n}} \right),$$

where the latter is a one-sided confidence interval providing an upper bound for p. Or if the alternative hypothesis is H_1: $p \neq p_0$, then H_0 is rejected if

$$\frac{|\widehat{p} - p_0|}{\sqrt{\dfrac{\widehat{p}(1 - \widehat{p})}{n}}} \geq z_{\alpha/2}.$$

This is equivalent to

$$p_0 \notin \left(\widehat{p} - z_{\alpha/2} \sqrt{\frac{\widehat{p}(1 - \widehat{p})}{n}}, \, \widehat{p} + z_{\alpha/2} \sqrt{\frac{\widehat{p}(1 - \widehat{p})}{n}} \right),$$

where the latter is a confidence interval for p. However, using the forms given in Tables 8.1-1 and 8.1-2, we do get better approximations to α-level significance tests. Thus there are trade-offs and it is difficult to say one is better than the other. Fortunately, the numerical answers are about the same.

   In the second situation in which the estimates of p_1 and p_2 are the observed $\widehat{p_1} = y_1/n_1$ and $\widehat{p_2} = y_2/n_2$, we have, with large values of n_1 and n_2, an approximate 95% confidence interval for $p_1 - p_2$ given by

$$\frac{y_1}{n_1} - \frac{y_2}{n_2} \pm 1.96 \sqrt{\frac{(y_1/n_1)(1 - y_1/n_1)}{n_1} + \frac{(y_2/n_2)(1 - y_2/n_2)}{n_2}}.$$

If $p_1 - p_2 = 0$ is not in this interval, we reject H_0: $p_1 - p_2 = 0$ at the $\alpha = 0.05$ significance level. This is equivalent to saying we reject H_0: $p_1 - p_2 = 0$ if

$$\frac{\left| \dfrac{y_1}{n_1} - \dfrac{y_2}{n_2} \right|}{\sqrt{\dfrac{(y_1/n_1)(1 - y_1/n_1)}{n_1} + \dfrac{(y_2/n_2)(1 - y_2/n_2)}{n_2}}} \geq 1.96.$$

In general, if the estimator $\widehat{\theta}$ (often maximum likelihood) of θ has an approximate (sometimes exact) normal distribution $N(\theta, \sigma_{\widehat{\theta}}^2)$, then $H_0: \theta = \theta_0$ is rejected in favor of $H_1: \theta \neq \theta_0$ at the α significance level if

$$\theta_0 \notin (\widehat{\theta} - z_{\alpha/2}\, \sigma_{\widehat{\theta}}, \ \widehat{\theta} + z_{\alpha/2}\, \sigma_{\widehat{\theta}})$$

or, equivalently,

$$\frac{|\widehat{\theta} - \theta_0|}{\sigma_{\widehat{\theta}}} \geq z_{\alpha/2}.$$

It should be mentioned that often $\sigma_{\widehat{\theta}}$ depends upon some unknown parameter that must be estimated and substituted in $\sigma_{\widehat{\theta}}$ to obtain $\widehat{\sigma}_{\widehat{\theta}}$. Sometimes $\sigma_{\widehat{\theta}}$ or its estimate is called the **standard error** of $\widehat{\theta}$. This was the case in our last illustration when, with $\theta = p_1 - p_2$ and $\widehat{\theta} = \widehat{p}_1 - \widehat{p}_2$, we substituted y_1/n_1 for p_1 and y_2/n_2 for p_2 in

$$\sqrt{\frac{p_1(1 - p_1)}{n_1} + \frac{p_2(1 - p_2)}{n_2}}$$

to obtain the standard error of $\widehat{p}_1 - \widehat{p}_2 = \widehat{\theta}$.   ■

EXERCISES

8.1-1 Bowl A contains 100 red balls and 200 white balls; bowl B contains 200 red balls and 100 white balls. Let p denote the probability of drawing a red ball from a bowl, but say p is unknown, since it is unknown whether bowl A or bowl B is being used. We shall test the simple null hypothesis $H_0: p = 1/3$ against the simple alternative hypothesis $H_1: p = 2/3$. Draw three balls at random, one at a time and with replacement from the selected bowl. Let X equal the number of red balls drawn. Then let the critical region be $C = \{x : x = 2, 3\}$. What are the values of α and β, the probabilities of Type I and Type II errors, respectively?

8.1-2 A bowl contains two red balls, two white balls, and fifth ball that is either red or white. Let p denote the probability of drawing a red ball from the bowl. We shall test the simple null hypothesis $H_0: p = 3/5$ against the simple alternative hypothesis $H_1: p = 2/5$. Draw four balls at random from the bowl, one at a time and with replacement. Let X equal the number of red balls drawn.

(a) Define a critical region C for this test in terms of X.
(b) For the critical region C defined in part (a), find the values of α and β.

8.1-3 Let Y be $b(100, p)$. To test $H_0: p = 0.08$ against $H_1: p < 0.08$, we reject H_0 and accept H_1 if and only if $Y \leq 6$.

(a) Determine the significance level α of the test.
(b) Find the probability of the Type II error if in fact $p = 0.04$.

8.1-4 Let p denote the probability that, for a particular tennis player, the first serve is good. Since $p = 0.40$, this player decided to take lessons in order to increase p. When the lessons are completed, the hypothesis $H_0: p = 0.40$ will be tested against $H_1: p > 0.40$ based on $n = 25$ trials. Let y equal the number of first serves that are good, and let the critical region be defined by $C = \{y : y \geq 13\}$.

(a) Determine $\alpha = P(Y \geq 13; p = 0.40)$. Use Table II in the Appendix.
(b) Find $\beta = P(Y < 13)$ when $p = 0.60$; that is, $\beta = P(Y \leq 12; p = 0.60)$. Use Table II.

8.1-5 Let Y be $b(192, p)$. We reject H_0: $p = 0.75$ and accept H_1: $p > 0.75$ if and only if $Y \geq 152$. Use the normal approximation to determine

(a) $\alpha = P(Y \geq 152; p = 0.75)$.
(b) $\beta = P(Y < 152)$ when $p = 0.80$.

8.1-6 To determine whether the 1-side on a commercially manufactured die is heavy and the 6-side is light (see Example 8.1-1), some students kept track of the number of observed 1s when observing $n = 8000$ rolls of the dice. Let p equal the probability of rolling a one with such a die. We shall test the null hypothesis H_0: $p = 1/6$ against the alternative hypothesis H_1: $p < 1/6$.

(a) Define the test statistic and an $\alpha = 0.05$ critical region.
(b) If $y = 1265$ ones were observed in 8000 rolls, calculate the value of the test statistic and state your conclusion.
(c) Is 1/6 in the 95% one-sided confidence interval providing an upper bound for p?

8.1-7 If a newborn baby has a birth weight that is less than 2500 grams (5.5 pounds), we say that the baby has a low birth weight. The proportion of babies with a low birth weight is an indicator of nutrition (or lack of nutrition) for the mothers. For the United States, approximately 7% of babies have a low birth weight. Let p equal the proportion of babies born in the Sudan who weigh less than 2500 grams. We shall test the null hypothesis H_0: $p = 0.07$ against the alternative hypothesis H_1: $p > 0.07$. If $y = 23$ babies out of a random sample of $n = 209$ babies weighed less than 2500 grams, what is your conclusion at a significance level of

(a) $\alpha = 0.05$?
(b) $\alpha = 0.01$?
(c) Find the p-value for this test.

8.1-8 It was claimed that 75% of all dentists recommend a certain brand of gum for their gum-chewing patients. A consumer group doubted this claim and decided to test H_0: $p = 0.75$ against the alternative hypothesis H_1: $p < 0.75$, where p is the proportion of dentists who recommend this brand of gum. A survey of 390 dentists found that 273 recommended this brand of gum. Which hypothesis would you accept if the significance level is

(a) $\alpha = 0.05$?
(b) $\alpha = 0.01$?
(c) Find the p-value for this test.

8.1-9 It was claimed that the proportion of Americans who select jogging as one of their recreational activities is $p = 0.25$. A shoe manufacturer thought that p was larger than 0.25. They decided to test the null hypothesis H_0: $p = 0.25$ against the alternative hypothesis H_1: $p > 0.25$. If $y = 1497$ out of a random sample of $n = 5757$ selected jogging, what is your conclusion at a significance level of

(a) $\alpha = 0.05$?
(b) $\alpha = 0.025$?
(c) Find the p-value for this test.

8.1-10 Let p equal the proportion of drivers who use a seat belt in a state that does not have a mandatory seat belt law. It was claimed that $p = 0.14$. An advertising campaign was conducted to increase this proportion. Two months after the campaign, $y = 104$ out of a random sample of $n = 590$ drivers were wearing their seat belts. Was the campaign successful?

(a) Define the null and alternative hypotheses.
(b) Define a critical region with an $\alpha = 0.01$ significance level.
(c) What is your conclusion?

8.1-11 The management of the Tiger baseball team decided to sell only low-alcohol beer in their ballpark to help combat rowdy fan conduct. They claimed that more than 40% of the fans would approve of this decision. Let p equal the proportion of Tiger fans on opening day who approved of this decision. We shall test the null hypothesis H_0: $p = 0.40$ against the alternative hypothesis H_1: $p > 0.40$.

(a) Define a critical region that has an $\alpha = 0.05$ significance level.
(b) If out of a random sample of $n = 1278$ fans, $y = 550$ said that they approved of this new policy, what is your conclusion?

8.1-12 Because of tourism in the State, it was proposed that public schools in Michigan begin after Labor Day. To determine whether support for this change was greater than 65%, a public poll was taken. Let p equal the proportion of Michigan adults who favor a post Labor Day start. We shall test H_0: $p = 0.65$ against H_1: $p > 0.65$.

(a) Define a test statistic and an $\alpha = 0.025$ critical region.
(b) Given that 414 out of a sample of 600 favor a post Labor Day start, calculate the value of the test statistic.
(c) Find the p-value and state your conclusion.
(d) Find a 95% one-sided confidence interval that gives a lower bound for p.

8.1-13 Let p equal the proportion of women who agree that "men are basically selfish and self-centered." Suppose that in the past, it was believed that $p = 0.40$. It is now claimed that p has increased.

(a) Define the null and alternative hypotheses, a test statistic and critical region that has an approximate $\alpha = 0.01$ significance level. Sketch a standard normal p.d.f. and illustrate this critical region.
(b) The *Detroit Free Press* (April 26, 1990) reported that $y = 1260$ out of a random sample of $n = 3000$ women agree with the statement. What is the conclusion of your test? Locate the calculated value of your test statistic on the p.d.f. in part (a).

8.1-14 A physician at a local hospital wrote that "In medicine we are now asked to perform quality assurance studies by various regulatory governmental or quasi–governmental agencies." He wrote, "Dr. X was notified that for a particular procedure, his complication rate of 20% last year ($n = 20$) was unacceptable for continued privileges. So he took a one month remedial course for improving his technique. His complication rate is now 15% ($n = 15$). Has he really improved?" (You may assume that some rounding off was done in the letter and that the 15% is really 2/15.)

8.1-15 According to a population census in 1986, the percentage of males who are 18 or 19 years old that are married was 3.7%. We shall test whether this percentage increased from 1986 to 1988.

(a) Define the null and alternative hypotheses.
(b) Define a critical region that has an approximate significance level of $\alpha = 0.01$. Sketch a standard normal p.d.f. to illustrate this critical region.
(c) If $y = 20$ out of a random sample of $n = 300$ males, each 18 or 19 years old, were married (*U.S. Bureau of the Census, Statistical Abstract of the United States: 1988*), what is your conclusion? Show the calculated value of the test statistic on your figure in part (b).

8.1-16 Let p equal the proportion of yellow candies in a package of mixed colors. It is claimed that $p = 0.20$.

(a) Define the test statistic and critical region with a significance level of $\alpha = 0.05$ for testing H_0: $p = 0.20$ against a two-sided alternative hypothesis.
(b) To perform the test, each of 20 students counted the number of yellow candies, y, and the total number, n, in a 48.1-gram package, yielding the following ratios, y/n: 8/56, 13/55, 12/58, 13/56, 14/57, 5/54, 14/56, 15/57, 11/54, 13/55, 10/57, 8/59, 10/54, 11/55, 12/56, 11/57, 6/54, 7/58, 12/58, 14/58. If each individual makes a test of H_0: $p = 0.20$, what proportion of the students rejected the null hypothesis?
(c) If we may assume that the null hypothesis is true, what proportion of the students would you have expected to reject the null hypothesis?

(d) For each of the 20 ratios in part (b), a 95% confidence interval for p can be calculated. What proportion of these 95% confidence intervals contain $p = 0.20$?

(e) If the 20 results are pooled so that $\sum_{i=1}^{20} y_i$ equals the number of yellow candies and $\sum_{i=1}^{20} n_i$ equals the total sample size, do we reject H_0: $p = 0.20$?

8.1-17 A machine shop that manufactures toggle levers has both a day and a night shift. A toggle lever is defective if a standard nut cannot be screwed onto the threads. Let p_1 and p_2 be the proportion of defective levers among those manufactured by the day and night shifts, respectively. We shall test the null hypothesis, H_0: $p_1 = p_2$, against a two-sided alternative hypothesis based on two random samples, each of 1000 levers taken from the production of the respective shifts.

(a) Define the test statistic and a critical region that has an $\alpha = 0.05$ significance level. Sketch a standard normal p.d.f. illustrating this critical region.

(b) If $y_1 = 37$ and $y_2 = 53$ defectives were observed for the day and night shifts, respectively, calculate the value of the test statistic. Locate the calculated test statistic on your figure in part (a) and state your conclusion.

8.1-18 *Time, April 18, 1994*, reported the results of a telephone poll of 800 adult Americans, 605 of them nonsmokers, who were asked the following question: "Should the federal tax on cigarettes be raised by $1.25 to pay for health care reform?" Let p_1 and p_2 equal the proportions of nonsmokers and smokers, respectively, who would say yes to this question. Given that $y_1 = 351$ nonsmokers and $y_2 = 41$ smokers said yes,

(a) With $\alpha = 0.05$, test H_0: $p_1 = p_2$ against H_1: $p_1 \neq p_2$.

(b) Find a 95% confidence interval for $p_1 - p_2$. Is this in agreement with the conclusion of part (a)?

(c) Find a 95% confidence interval for p, the proportion of adult Americans who would say yes.

8.1-19 Let p_m and p_f be the respective proportions of male and female white-crowned sparrows that return to their hatching site. Give the endpoints for a 95% confidence interval for $p_m - p_f$, given that 124 out of 894 males and 70 out of 700 females returned. (*The Condor*, 1992, pp. 117–133.) Does this agree with the conclusion of a test of H_0: $p_1 = p_2$ against H_1: $p_1 \neq p_2$ with $\alpha = 0.05$?

8.1-20 For developing countries in Africa and the Americas, let p_1 and p_2 be the respective proportions of babies with a low birth weight (below 2500 grams). We shall test H_0: $p_1 = p_2$ against the alternative hypothesis H_1: $p_1 > p_2$.

(a) Define a critical region that has an $\alpha = 0.05$ significance level.

(b) If respective random samples of sizes $n_1 = 900$ and $n_2 = 700$ yielded $y_1 = 135$ and $y_2 = 77$ babies with a low birth weight, what is your conclusion?

(c) What would your decision be with a significance level of $\alpha = 0.01$?

(d) What is the p-value of your test?

8.2 TESTS ABOUT ONE MEAN AND ONE VARIANCE

Many applications assume that we are sampling from a normal distribution. We begin this section with such an application for which we also assume that the variance is known.

EXAMPLE 8.2-1

Let X equal the breaking strength of a steel bar. If the steel bar is manufactured by process I, X is $N(50, 36)$. It is hoped that if process II (a new process) is used, X will be $N(55, 36)$. Given a large number of steel bars manufactured by process II, how could we test whether the increase in the mean breaking strength was realized?

That is, we are assuming X is $N(\mu, 36)$ and μ is equal to 50 or 55. We want to test the null hypothesis H_0: $\mu = 50$ against alternative hypothesis H_1: $\mu = 55$. Note that each of these hypotheses completely specifies the distribution of X. That is, H_0 states that X is $N(50, 36)$, and H_1 states that X is $N(55, 36)$. Recall that an hypothesis that completely specifies the distribution of X is called a **simple hypothesis**; otherwise it is called a **composite hypothesis** (composed of at least two simple hypotheses). For example, H_1: $\mu > 50$ would be a composite hypothesis because it is composed of all normal distributions with $\sigma^2 = 36$ and means greater than 50. In order to test which of the two hypotheses, H_0 or H_1, is true, we shall set up a rule based on the breaking strengths $x_1, x_2, \ldots, x_n$ of n bars (the observed values of a random sample of size n from this new normal distribution). The rule leads to a decision to accept or reject H_0; so it is necessary to partition the sample space into two parts, say C and C', so that if $(x_1, x_2, \ldots, x_n) \in C$, H_0 is rejected, and if $(x_1, x_2, \ldots, x_n) \in C'$, H_0 is accepted (not rejected). The rejection region C for H_0 is called the critical region for the test. Often the partitioning of the sample space is specified in terms of the values of a statistic called the test statistic. In this illustration we could let $\overline{X}$ be the test statistic and, for example, take $C = \{(x_1, x_2, \ldots, x_n): \overline{x} \geq 53\}$. We could then define the critical region as those values of the test statistic for which H_0 is rejected. That is, the given critical region is equivalent to defining $C = \{\overline{x}: \overline{x} \geq 53\}$ in the $\overline{x}$ space. If $(x_1, x_2, \ldots, x_n) \in C$ when H_0 is true, H_0 would be rejected when it is true, a Type I error. If $(x_1, x_2, \ldots, x_n) \in C'$ when H_1 is true, H_0 would be accepted, that is, not rejected, when in fact H_1 is true, a Type II error. Recall that the probability of a Type I error is called the significance level of the test and is denoted by α. That is,

$$\alpha = P[(X_1, X_2, \ldots, X_n) \in C; H_0]$$

is the probability that $(X_1, X_2, \ldots, X_n)$ falls in C when H_0 is true. The probability of a Type II error is denoted by β; that is,

$$\beta = P[(X_1, X_2, \ldots, X_n) \in C'; H_1]$$

is the probability of accepting (failing to reject) H_0 when it is false. For illustration, suppose $n = 16$ bars were tested and $C = \{\overline{x}: \overline{x} \geq 53\}$. Then $\overline{X}$ is $N(50, 36/16)$ when H_0 is true and is $N(55, 36/16)$ when H_1 is true. Thus

$$\alpha = P(\overline{X} \geq 53; H_0) = P\left(\frac{\overline{X} - 50}{6/4} \geq \frac{53 - 50}{6/4}; H_0\right)$$

$$= 1 - \Phi(2) = 0.0228$$

and

$$\beta = P(\overline{X} < 53; H_1) = P\left(\frac{\overline{X} - 55}{6/4} < \frac{53 - 55}{6/4}; H_1\right)$$

$$= \Phi\left(-\frac{4}{3}\right) = 1 - 0.9087 = 0.0913.$$

See Figure 8.2-1 for the graphs of the probability density functions of $\overline{X}$ when H_0 and H_1 are true, respectively. Note that a decrease in the size of α leads to an increase in the size of β, and vice versa. Both α and β can be decreased if the sample size n is increased. ◄

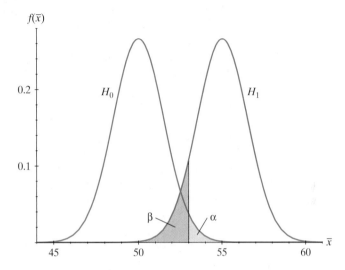

Figure 8.2-1: p.d.f. of $\overline{X}$ under H_0 and H_1

The definition for p-value is given in Section 8.1. We now give an example of p-value when testing an hypothesis about a mean.

EXAMPLE 8.2-2   Assume that the underlying distribution is normal with unknown mean μ but known variance $\sigma^2 = 100$. Say we are testing $H_0: \mu = 60$ against $H_1: \mu > 60$ with a sample mean $\overline{X}$ based on $n = 52$ observations. Suppose that we obtain the observed sample mean of $\overline{x} = 62.75$. If we compute the probability of obtaining an $\overline{X}$ of that value of 62.75 or greater when $\mu = 60$, then we obtain the p-value associated with $\overline{x} = 62.75$. That is,

$$p\text{-value} = P(\overline{X} \geq 62.75; \ \mu = 60)$$

$$= P\left(\frac{\overline{X} - 60}{10/\sqrt{52}} \geq \frac{62.75 - 60}{10/\sqrt{52}}; \ \mu = 60\right)$$

$$= 1 - \Phi\left(\frac{62.75 - 60}{10/\sqrt{52}}\right) = 1 - \Phi(1.983) = 0.0237.$$

If this p-value is small, we tend to reject the hypothesis $H_0: \mu = 60$. For example, rejection of $H_0: \mu = 60$ if the p-value is less than or equal to $\alpha = 0.05$ is exactly the same as rejection if

$$\overline{x} \geq 60 + (1.645)\left(\frac{10}{\sqrt{52}}\right) = 62.718.$$

Here

$$p\text{-value} = 0.0237 < \alpha = 0.05 \qquad \text{and} \qquad \overline{x} = 62.75 > 62.718.$$

To help the reader keep the definition of p-value in mind, we note that it can be thought of as that **tail-end probability**, under H_0, of the distribution of the statistic, here $\overline{X}$, beyond the observed value of the statistic. See Figure 8.2-2 for the p-value associated with $\overline{x} = 62.75$.

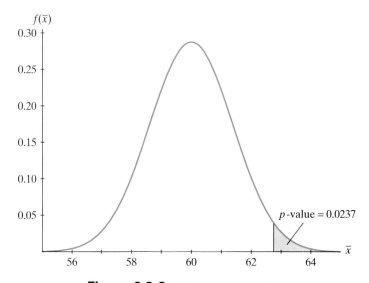

Figure 8.2-2: Illustration of p-value

If the alternative was the two-sided H_1: $\mu \neq 60$, then the p-value would have been double the 0.0237; that is, then the p-value $= 2(0.0237) = 0.0474$ because we include both tails. ◀

When we sample from a normal distribution the null hypothesis is generally of the form H_0: $\mu = \mu_0$. There are essentially three possibilities for the alternative hypothesis, namely that μ has increased, (i) H_1: $\mu > \mu_0$; that μ has decreased, (ii) H_1: $\mu < \mu_0$; or that μ has changed, but it is not known if it has increased or decreased, which leads to a two-sided alternative hypothesis (iii) H_1: $\mu \neq \mu_0$.

To test H_0: $\mu = \mu_0$ against one of these three alternative hypotheses, a random sample is taken from the distribution, and an observed sample mean, $\overline{x}$, that is close to μ_0 supports H_0. The closeness of $\overline{x}$ to μ_0 is measured in terms of standard deviations of $\overline{X}$, $\sigma/\sqrt{n}$ when σ is known, which is sometimes called the **standard error of the mean**. Thus the test statistic could be defined by

$$Z = \frac{\overline{X} - \mu_0}{\sqrt{\sigma^2/n}} = \frac{\overline{X} - \mu_0}{\sigma/\sqrt{n}}, \tag{8.2-1}$$

and the critical regions, at a significance level α, for the three respective alternative hypotheses would be (i) $z \geq z_\alpha$, (ii) $z \leq -z_\alpha$, and (iii) $|z| \geq z_{\alpha/2}$. In terms of $\bar{x}$ these three critical regions become (i) $\bar{x} \geq \mu_0 + z_\alpha(\sigma/\sqrt{n})$, (ii) $\bar{x} \leq \mu_0 - z_\alpha(\sigma/\sqrt{n})$, and (iii) $|\bar{x} - \mu_0| \geq z_{\alpha/2}(\sigma/\sqrt{n})$.

These tests and critical regions are summarized in Table 8.2-1. The underlying assumption is that the distribution is $N(\mu, \sigma^2)$ and σ^2 is known.

Table 8.2-1: Tests of Hypotheses About One Mean, Variance Known

H_0	H_1	Critical Region
$\mu = \mu_0$	$\mu > \mu_0$	$z \geq z_\alpha$ or $\bar{x} \geq \mu_0 + z_\alpha \sigma/\sqrt{n}$
$\mu = \mu_0$	$\mu < \mu_0$	$z \leq -z_\alpha$ or $\bar{x} \leq \mu_0 - z_\alpha \sigma/\sqrt{n}$
$\mu = \mu_0$	$\mu \neq \mu_0$	$|z| \geq z_{\alpha/2}$ or $|\bar{x} - \mu_0| \geq z_{\alpha/2} \sigma/\sqrt{n}$

It is usually not true that the variance σ^2 is known. We now take a more realistic position and assume that the variance is unknown. Suppose our null hypothesis is H_0: $\mu = \mu_0$ and the two-sided alternative hypothesis is H_1: $\mu \neq \mu_0$. If a random sample $X_1, X_2, \ldots, X_n$ is taken from a normal distribution $N(\mu, \sigma^2)$, we recall from Section 6.4 that a confidence interval for μ was based on

$$T = \frac{\overline{X} - \mu}{\sqrt{S^2/n}} = \frac{\overline{X} - \mu}{S/\sqrt{n}}.$$

This suggests that T might be a good statistic to use for the test of H_0: $\mu = \mu_0$ with μ replaced by μ_0. In addition, it is the natural statistic to use if we replace σ^2/n by its unbiased estimator S^2/n in $(\overline{X} - \mu_0)/\sqrt{\sigma^2/n}$ in Equation 8.2-1. If $\mu = \mu_0$, we know that T has a t distribution with $n - 1$ degrees of freedom. Thus, with $\mu = \mu_0$,

$$P[\,|T| \geq t_{\alpha/2}(n-1)] = P\left[\frac{|\overline{X} - \mu_0|}{S/\sqrt{n}} \geq t_{\alpha/2}(n-1)\right] = \alpha.$$

Accordingly, if $\bar{x}$ and s are the sample mean and sample standard deviation, the rule that rejects H_0: $\mu = \mu_0$ and accepts H_1: $\mu \neq \mu_0$ if and only if

$$|t| = \frac{|\bar{x} - \mu_0|}{s/\sqrt{n}} \geq t_{\alpha/2}(n-1)$$

provides a test of this hypothesis with significance level α. It should be noted that this rule is equivalent to rejecting H_0: $\mu = \mu_0$ if μ_0 is not in the open $100(1 - \alpha)\%$ confidence interval

$$(\bar{x} - t_{\alpha/2}(n-1)(s/\sqrt{n}), \ \bar{x} + t_{\alpha/2}(n-1)(s/\sqrt{n})).$$

Table 8.2-2 summarizes tests of hypotheses for a single mean, along with the three possible alternative hypotheses, when the underlying distribution is $N(\mu, \sigma^2)$, σ^2 is unknown, $t = (\bar{x} - \mu_0)/(s/\sqrt{n})$, and $n \leq 30$. If $n > 30$, use Table 8.2-1 for approximate tests with σ replaced by s.

Table 8.2-2: Tests of Hypotheses for One Mean, Variance Unknown

H_0	H_1	Critical Region
$\mu = \mu_0$	$\mu > \mu_0$	$t \geq t_\alpha(n-1)$ or $\bar{x} \geq \mu_0 + t_\alpha(n-1)s/\sqrt{n}$
$\mu = \mu_0$	$\mu < \mu_0$	$t \leq -t_\alpha(n-1)$ or $\bar{x} \leq \mu_0 - t_\alpha(n-1)s/\sqrt{n}$
$\mu = \mu_0$	$\mu \neq \mu_0$	$|t| \geq t_{\alpha/2}(n-1)$ or $|\bar{x} - \mu_0| \geq t_{\alpha/2}(n-1)s/\sqrt{n}$

EXAMPLE 8.2-3 Let X (in millimeters) equal the growth in 15 days of a tumor induced in a mouse. Assume that the distribution of X is $N(\mu, \sigma^2)$. We shall test the null hypothesis H_0: $\mu = \mu_0 = 4.0$ millimeters against the two-sided alternative hypothesis H_1: $\mu \neq 4.0$. If we use $n = 9$ observations and a significance level of $\alpha = 0.10$, the critical region is

$$|t| = \frac{|\bar{x} - 4.0|}{s/\sqrt{9}} \geq t_{\alpha/2}(8) = 1.860.$$

If we are given that $n = 9$, $\bar{x} = 4.3$, and $s = 1.2$, we see that

$$t = \frac{4.3 - 4.0}{1.2/\sqrt{9}} = \frac{0.3}{0.4} = 0.75.$$

Thus

$$|t| = |0.75| < 1.860$$

and we accept (do not reject) H_0: $\mu = 4.0$ at the $\alpha = 10\%$ significance level. See Figure 8.2-3. The p-value is

$$p\text{-value} = P(|T| \geq 0.75) = 2P(T \geq 0.75).$$

With our t tables with 8 degrees of freedom, we cannot find this p-value exactly. It is about 0.50 because

$$P(|T| \geq 0.706) = 2P(T \geq 0.706) = 0.50,$$

and the computer gives the p-value to be 0.475. See Figure 8.2-3. ◄

Remark In discussing the test of a statistical hypothesis, the word *accept H_0* might better be replaced by *do not reject H_0*. That is, if, in Example 8.2-3, $\bar{x}$ is close enough to 4.0 so that we accept $\mu = 4.0$, we do not want that acceptance to imply that μ is actually equal to 4.0. We want to say that the data do not deviate enough from $\mu = 4.0$ for us to reject that hypothesis; that is, we

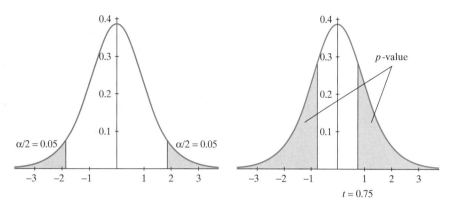

Figure 8.2-3: Test about mean of tumor growths

do not reject $\mu = 4.0$ with these observed data. With this understanding, we sometimes use *accept*, and sometimes *fail to reject* or *do not reject*, the null hypothesis. ■

The next example illustrates the use of the t statistic with a one-sided alternative hypothesis.

EXAMPLE 8.2-4   In attempting to control the strength of the wastes discharged into a nearby river, a paper firm has taken a number of measures. Members of the firm believe that they have reduced the oxygen-consuming power of their wastes from a previous mean μ of 500 (measured in permanganate in parts per million). They plan to test $H_0: \mu = 500$ against $H_1: \mu < 500$, using readings taken on $n = 25$ consecutive days. If these 25 values can be treated as a random sample, then the critical region, for a significance level of $\alpha = 0.01$, is

$$t = \frac{\bar{x} - 500}{s/\sqrt{25}} \leq -t_{0.01}(24) = -2.492.$$

The observed values of the sample mean and sample standard deviation were $\bar{x} = 308.8$ and $s = 115.15$. Since

$$t = \frac{308.8 - 500}{115.15/\sqrt{25}} = -8.30 < -2.492,$$

we clearly reject the null hypothesis and accept $H_1: \mu < 500$. It should be noted, however, that although an improvement has been made, there still might exist the question of whether the improvement is adequate. The one-sided 99% confidence interval for μ,

$$[0, 308.8 + 2.492(115.25/\sqrt{25})] = [0, 366.191],$$

provides an upper bound for μ and may help the company answer this question. ◄

There is, oftentimes, interest in comparing the means of two different distributions or populations. We must consider two situations, that in which X and Y are dependent, and that in which X and Y are independent. We consider the independent case in the next section.

If X and Y are dependent, let $W = X - Y$, and the hypothesis that $\mu_X = \mu_Y$ would be replaced with the hypothesis H_0: $\mu_W = 0$. For example, suppose that X and Y equal the resting pulse rate for a person before and after taking an 8-week program in aerobic dance. We would be interested in testing H_0: $\mu_W = 0$ (no change) against H_1: $\mu_W > 0$ (the aerobic dance program decreased resting pulse rate). Because X and Y are measurements on the same person, X and Y are clearly dependent. If we can assume that the distribution of W is (approximately) $N(\mu_W, \sigma^2)$, then the appropriate t-test for a single mean could be used, selecting from Table 8.2-2. This is often called a **paired t test**.

EXAMPLE 8.2-5 Twenty-four girls in the 9th and 10th grades were put on an ultraheavy rope-jumping program. Someone thought that such a program would increase their speed when running the 40-yard dash. Let W equal the difference in time to run the 40-yard dash-the "before program time" minus the "after program time." Assume that the distribution of W is (approximately) $N(\mu_W, \sigma_W^2)$. We shall test the null hypothesis H_0: $\mu_W = 0$ against the alternative hypothesis H_1: $\mu_W > 0$. The test statistic and critical region that has an $\alpha = 0.05$ significance level are given by

$$t = \frac{\overline{w} - 0}{s_w/\sqrt{24}} \geq t_{0.05}(23) = 1.714.$$

The following data give the difference in time that it took each girl to run the 40-yard dash, with positive numbers indicating a faster time after the program:

0.28	0.01	0.13	0.33	−0.03	0.07	−0.18	−0.14
−0.33	0.01	0.22	0.29	−0.08	0.23	0.08	0.04
−0.30	−0.08	0.09	0.70	0.33	−0.34	0.50	0.06

For these data, $\overline{w} = 0.079$ and $s_w = 0.255$. Thus the observed value of the test statistic is

$$t = \frac{0.079 - 0}{0.255/\sqrt{24}} = 1.518.$$

Since $1.518 < 1.714$, the null hypothesis is not rejected. Note, however, that $t_{0.10}(23) = 1.319$ and $t = 1.518 > 1.319$. Thus, the null hypothesis would be rejected at an $\alpha = 0.10$ significance level. Another way of saying this is that

$$0.05 < p\text{-value} < 0.10.$$

It would be instructive for you to draw a figure illustrating this. ◀

Since there has been no previous example of a test of a hypothesis concerning variances, an example follows a brief discussion and then rules are given formally.

A psychology professor claims that the variance of IQ scores for college students is equal to $\sigma^2 = 100$. To test this claim, it is decided to test the hypothesis $H_0: \sigma^2 = 100$ against $H_1: \sigma^2 \neq 100$, a two-sided alternative hypothesis. A random sample of n students will be selected, and the test will be based on the observed unbiased estimate s^2 of the variance σ^2 of their IQ scores. The hypothesis H_0 will be rejected if s^2 differs "too much" from $\sigma^2 = 100$. That is, H_0 will be rejected if $s^2 \leq c_1$ or $s^2 \geq c_2$ for some constants $c_1 < 100$ and $c_2 > 100$. Generally, c_1 and c_2 are selected so that

$$P(S^2 \leq c_1) = \alpha/2 \quad \text{and} \quad P(S^2 \geq c_2) = \alpha/2.$$

If we may assume that IQ scores are normally distributed, then the distribution of $(n-1)S^2/100$ is $\chi^2(n-1)$ when H_0 is true. Thus the chi-square probability table can be used to give us

$$P\left[\frac{(n-1)S^2}{100} \leq \frac{(n-1)c_1}{100}\right] = \alpha/2 \quad \text{and} \quad P\left[\frac{(n-1)S^2}{100} \geq \frac{(n-1)c_2}{100}\right] = \alpha/2.$$

Accordingly,

$$\frac{(n-1)c_1}{100} = \chi^2_{1-\alpha/2}(n-1) \quad \text{and} \quad \frac{(n-1)c_2}{100} = \chi^2_{\alpha/2}(n-1).$$

Hence

$$c_1 = [100/(n-1)]\chi^2_{1-\alpha/2}(n-1) \quad \text{and} \quad c_2 = [100/(n-1)]\chi^2_{\alpha/2}(n-1).$$

This test can also be based on the value of the chi-square statistic

$$\chi^2 = \frac{(n-1)S^2}{100}.$$

We would reject the null hypothesis if

$$\chi^2 \leq \chi^2_{1-\alpha/2}(n-1) \quad \text{or} \quad \chi^2 \geq \chi^2_{\alpha/2}(n-1).$$

EXAMPLE 8.2-6   Suppose that in the discussion above, $n = 23$ and $\alpha = 0.05$. To test $H_0: \sigma^2 = 100$ against $H_1: \sigma^2 \neq 100$, we let $\chi^2 = (n-1)S^2/\sigma_0^2 = 22S^2/100$. Because

$$\chi^2_{0.975}(22) = 10.98 \quad \text{and} \quad \chi^2_{0.025}(22) = 36.78,$$

H_0 will be rejected if

$$\chi^2 = \frac{22s^2}{100} \leq 10.98 \quad \text{or} \quad \chi^2 = \frac{22s^2}{100} \geq 36.78$$

or, equivalently, if

$$s^2 \le c_1 = \frac{100(10.98)}{22} = 49.91 \qquad \text{or} \qquad s^2 \ge c_2 = \frac{100(36.78)}{22} = 167.18.$$

(Exercise 8.2-14 gives some insight as to why these critical values are so far apart.) Given that the observed value of the sample variance was $s^2 = 147.82$, the hypothesis $H_0: \sigma^2 = 100$ was not rejected. Note that the 95% confidence interval for σ^2,

$$\left[\frac{(22)(147.82)}{36.78}, \ \frac{(22)(147.82)}{10.98} \right] = [88.42, \ 296.18],$$

contains $\sigma^2 = 100$. Also, the observed value of the chi-square test statistic is

$$\chi^2 = \frac{22(147.82)}{100} = 32.52.$$

Because

$$10.98 < 32.52 < 36.78,$$

we would again accept H_0 as expected. See Figure 8.2-4.

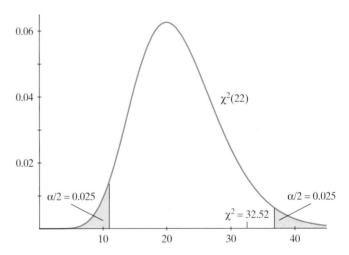

Figure 8.2-4: Test about variance of IQ scores

If $H_1: \sigma^2 > 100$ had been the alternative hypothesis, $H_0: \sigma^2 = 100$ would have been rejected if

$$\chi^2 = \frac{22s^2}{100} \ge \chi_{0.05}^2(22) = 33.92$$

or, equivalently,

$$s^2 \geq \frac{100\chi^2_{0.05}(22)}{22} = \frac{(100)(33.92)}{22} = 154.18.$$

Because

$$\chi^2 = 32.52 < 33.92 \qquad \text{and} \qquad s^2 = 147.82 < 154.18,$$

H_0 would not be rejected in favor of this one-sided alternative hypothesis for $\alpha = 0.05$, although we suspect a slightly larger data set might lead to rejection. ◄

Table 8.2-3 summarizes tests of hypotheses for a single variance. The critical region is given in terms of the sample variance. The critical region is also given in terms of the chi-square test statistic

$$\chi^2 = \frac{(n-1)s^2}{\sigma_0^2}.$$

Table 8.2-3: Tests About the Variance		
H_0	**H_1**	**Critical Region**
$\sigma^2 = \sigma_0^2$	$\sigma^2 > \sigma_0^2$	$s^2 \geq \dfrac{\sigma_0^2\chi^2_\alpha(n-1)}{n-1}$ or $\chi^2 \geq \chi^2_\alpha(n-1)$
$\sigma^2 = \sigma_0^2$	$\sigma^2 < \sigma_0^2$	$s^2 \leq \dfrac{\sigma_0^2\chi^2_{1-\alpha}(n-1)}{n-1}$ or $\chi^2 \leq \chi^2_{1-\alpha}(n-1)$
$\sigma^2 = \sigma_0^2$	$\sigma^2 \neq \sigma_0^2$	$s^2 \leq \dfrac{\sigma_0^2\chi^2_{1-\alpha/2}(n-1)}{n-1}$ or $s^2 \geq \dfrac{\sigma_0^2\chi^2_{\alpha/2}(n-1)}{n-1}$ or $\chi^2 \leq \chi^2_{1-\alpha/2}(n-1)$ or $\chi^2 \geq \chi^2_{\alpha/2}(n-1)$

There are two ways of viewing a statistical test. One of these is through the p-value of the test; this is becoming more popular and is included in most computer printouts. (This was discussed in Section 8.1.) After observing the test statistic, the p-value is the probability, under the hypothesis H_0, of the test statistic being as extreme (in the direction of rejection of H_0) as the observed one. That is, the p-value is the tail-end probability. For illustration, say a golfer averages about 90, with a standard deviation of 3, and she takes some lessons to improve. To test that possible improvement, namely $H_0: \mu = 90$ against $H_1: \mu < 90$, she plays $n = 16$ rounds of golf. Assume a normal distribution with $\sigma = 3$, if the golfer averaged $\bar{x} = 87.94$, then

$$p\text{-value} = P(\overline{X} \leq 87.94) = P\left(\frac{\overline{X}-90}{3/4} \leq \frac{87.94-90}{3/4}\right) = 0.0027.$$

The fact that the p-value < 0.05 is equivalent to the fact that $\bar{x} < 88.77$ because $P(\overline{X} \leq 88.77; \mu = 90) = 0.05$. Since $\bar{x} = 87.94$ is an observed value of a random variable, namely $\overline{X}$, then the p-value, a function of $\bar{x}$, is also an observed value of a random variable. That is, before the random experiment is performed, the probability that the p-value is less than α is equal to α when the null hypothesis is true. Many statisticians believe that the observed p-value provides an understandable measure of the truth of H_0: The smaller the p-value, the less they believe in H_0.

Three additional examples of the p-value are given by referring to Examples 8.2-3, 8.2-4, and 8.2-6. In two-sided tests for means and proportions, the p-value is the probability of the extreme values in both directions. With the mouse data (Example 8.2-3), the p-value is

$$p\text{-value} = P(|T| \geq 0.75).$$

In Table VI we see that if T has a t distribution with 8 degrees of freedom, then $P(T \geq 0.706) = 0.25$. Thus $P(|T| \geq 0.706) = 0.50$ and the p-value will be a little smaller than 0.50. In fact, $P(|T| \geq 0.75) = 0.47$ (a probability that can be found with certain computer programs), which is not less than $\alpha = 0.10$, and thus we do not reject H_0 at that significance level. In the example concerned with waste (Example 8.2-4), the p-value is essentially zero, since $P(Z \leq -8.30) \approx 0$, where Z is a standardized normal variable, and we reject H_0. In the example about the variance of IQ scores (Example 8.2-6), with the one-sided alternative, $H_1: \sigma^2 > 100$, the p-value is

$$p\text{-value} = P(W \geq 32.52),$$

where $W = 22S^2/100$ has a chi-square distribution with 22 degrees of freedom when H_0 is true. From Table IV we see that

$$P(W \geq 30.81) = 0.10 \qquad \text{and} \qquad P(W \geq 33.92) = 0.05.$$

Thus

$$0.05 < p\text{-value} = P(W \geq 32.52) < 0.10$$

and the null hypothesis would not be rejected at an $\alpha = 0.05$ significance level. To find the p-value for a two-sided alternative for tests about the variance, double the tail probability beyond the chi-square test statistic if the sample variance is larger than σ_0^2, or double the tail probability below the chi-square test statistic if the sample variance is below σ_0^2. So in Example 8.2-6, since $s^2 = 147.82 > 100$, the p-value equals $2[P(W \geq 32.52)]$ and thus

$$0.10 < p\text{-value} < 0.20.$$

The other way of looking at tests of hypotheses is through the consideration of confidence intervals. particularly for two-sided alternatives and the corresponding tests. For example, with the mouse data (Example 8.2-3), a 90% confidence interval for the unknown mean is

$$4.3 \pm (1.86)(1.2)/\sqrt{9} \quad \text{or} \quad [3.56, 5.04],$$

since $t_{0.05}(8) = 1.86$. Note that this confidence interval covers the hypothesized value $\mu = 4.0$ and we do not reject H_0: $\mu = 4.0$. If the confidence interval did not cover $\mu = 4.0$, then we would have rejected H_0: $\mu = 4.0$. Many statisticians believe that estimation is much more important than tests of hypotheses and accordingly approach statistical tests through confidence intervals. For one-sided tests, we use one-sided confidence intervals.

EXERCISES

8.2-1 Assume that IQ scores for a certain population are approximately $N(\mu, 100)$. To test H_0: $\mu = 110$ against the one-sided alternative hypothesis H_1: $\mu > 110$, we take a random sample of size $n = 16$ from this population and observe $\bar{x} = 113.5$. Do we accept or reject H_0 at the

(a) 5% significance level?
(b) 10% significance level?
(c) What is the p-value of this test?

8.2-2 Assume that the weight of cereal in a "10-ounce box" is $N(\mu, \sigma^2)$. To test H_0: $\mu = 10.1$ against H_1: $\mu > 10.1$, we take a random sample of size $n = 16$ and observe that $\bar{x} = 10.4$ and $s = 0.4$.

(a) Do we accept or reject H_0 at the 5% significance level?
(b) What is the approximate p-value of this test?

8.2-3 Let X equal the Brinell hardness measurement of ductile iron subcritically annealed. Assume that the distribution of X is $N(\mu, 100)$. We shall test the null hypothesis H_0: $\mu = 170$ against the alternative hypothesis H_1: $\mu > 170$ using $n = 25$ observations of X.

(a) Define the test statistic and a critical region that has a significance level of $\alpha = 0.05$. Sketch a figure showing this critical region.
(b) A random sample of $n = 25$ observations of X yielded the following measurements.

170	167	174	179	179	156	163	156	187
156	183	179	174	179	170	156	187	
179	183	174	187	167	159	170	179	

Calculate the value of the test statistic and clearly give your conclusion.
(c) Give the approximate p-value of this test.

8.2-4 Let X equal the thickness of spearmint gum manufactured for vending machines. Assume that the distribution of X is $N(\mu, \sigma^2)$. The target thickness is 7.5 hundredths of an inch. We shall test the null hypothesis H_0: $\mu = 7.5$ against a two-sided alternative hypothesis using 10 observations.

(a) Define the test statistic and critical region for an $\alpha = 0.05$ significance level. Sketch a figure illustrating this critical region.
(b) Calculate the value of the test statistic and clearly give your decision using the following $n = 10$ thicknesses in hundredths of an inch for pieces of gum that were selected randomly from the production line.

| 7.65 | 7.60 | 7.65 | 7.70 | 7.55 |
| 7.55 | 7.40 | 7.40 | 7.50 | 7.50 |

(c) Is $\mu = 7.50$ contained in a 95% confidence interval for μ?

8.2-5 The mean birth weight in the United States is $\mu = 3315$ grams with a standard deviation of $\sigma = 575$. Let X equal the birth weight in grams in Jerusalem. Assume that the distribution of X is $N(\mu, \sigma^2)$, We shall test the null hypothesis $H_0: \mu = 3315$ against the alternative hypothesis $H_1: \mu < 3315$ using a random sample of size $n = 30$.

(a) Define a critical region that has a significance level of $\alpha = 0.05$.

(b) If the random sample of $n = 30$ yielded $\bar{x} = 3189$ and $s = 488$, what is your conclusion?

(c) What is the approximate p-value of your test?

8.2-6 Let X equal the forced vital capacity (FVC) in liters for a female college student. (This is the amount of air that a student can force out of her lungs.) Assume that the distribution of X is (approximately) $N(\mu, \sigma^2)$. Suppose it is known that $\mu = 3.4$ liters. A volleyball coach claims that the FVC of volleyball players is greater than 3.4. She plans to test her claim using a random sample of size $n = 9$.

(a) Define the null hypothesis.

(b) Define the alternative (coach's) hypothesis.

(c) Define the test statistic.

(d) Define a critical region for which $\alpha = 0.05$. Draw a figure illustrating your critical region.

(e) Calculate the value of the test statistic given that the random sample yielded the following forced vital capacities.

$$3.4 \quad 3.6 \quad 3.8 \quad 3.3 \quad 3.4 \quad 3.5 \quad 3.7 \quad 3.6 \quad 3.7$$

(f) What is your conclusion?

(g) What is the approximate p-value of this test?

8.2-7 Vitamin B_6 is one of the vitamins in a multiple vitamin pill manufactured by a pharmaceutical company. The pills are produced with a mean of 50 milligrams of vitamin B_6 per pill. The company believes that there is a deterioration of 1 milligram per month, so that after 3 months they expect that $\mu = 47$. A consumer group suspects that $\mu < 47$ after 3 months.

(a) Define a critical region to test $H_0: \mu = 47$ against $H_1: \mu < 47$ at an $\alpha = 0.05$ significance level based on a random sample of size $n = 20$.

(b) If the 20 pills yielded a mean of $\bar{x} = 46.94$ with a standard deviation of $s = 0.15$, what is your conclusion?

(c) What is the approximate p-value of this test?

8.2-8 Assume that the birth weight in grams of a baby born in the United States is $N(3315, 525^2)$, boys and girls combined. Let X equal the weight of a baby girl who is born at home in Ottawa County and assume that the distribution of X is $N(\mu_X, \sigma_X^2)$.

(a) Using 11 observations of X, give the test statistic and critical region for testing $H_0: \mu_X = 3315$ against the alternative hypothesis $H_1: \mu_X > 3315$ (home-born babies are heavier) if $\alpha = 0.01$.

(b) Calculate the value of the test statistic and give your conclusion using the following weights.

$$3119 \quad 2657 \quad 3459 \quad 3629 \quad 3345 \quad 3629$$
$$3515 \quad 3856 \quad 3629 \quad 3345 \quad 3062$$

(c) What is the approximate p-value of the test?

(d) Give the test statistic and critical region for testing $H_0: \sigma_X^2 = 525^2$ against the alternative hypothesis $H_1: \sigma_X^2 < 525^2$ (less variation of weights of home-born babies) if $\alpha = 0.05$.

(e) Calculate the value of your test statistic and state your conclusion.

(f) What is the approximate p-value of this second test?

8.2-9 Let Y equal the weight in grams of a baby boy who is born at home in Ottawa County and assume that the distribution of Y is $N(\mu_Y, \sigma_Y^2)$. Use the following weights

$$4082 \quad 3686 \quad 4111 \quad 3686 \quad 3175 \quad 4139$$
$$3686 \quad 3430 \quad 3289 \quad 3657 \quad 4082$$

to answer the questions in Exercise 8.2-8. What other hypothesis is suggested by these two exercises?

8.2-10 A company that manufactures brackets for an auto maker regularly selects brackets from the production line and performs a torque test. The goal is for mean torque to equal 125. Let X equal the torque and assume that X is $N(\mu, \sigma^2)$. We shall use a sample of size $n = 15$ to test $H_0: \mu = 125$ against a two-sided alternative hypothesis.

(a) Give the test statistic and a critical region with significance level $\alpha = 0.05$. Sketch a figure illustrating the critical region.

(b) Use the following observations to calculate the value of the test statistic and state your conclusion.

$$128 \quad 149 \quad 136 \quad 114 \quad 126 \quad 142 \quad 124 \quad 136$$
$$122 \quad 118 \quad 122 \quad 129 \quad 118 \quad 122 \quad 129$$

8.2-11 Let X equal the number of pounds of butterfat produced by a Holstein cow during the 305-day milking period following the birth of a calf. We shall test the null hypothesis $H_0: \sigma^2 = 140^2$ against the alternative hypothesis $H_1: \sigma^2 > 140^2$.

(a) Give the test statistic and a critical region that has a significance level of $\alpha = 0.05$, assuming that there are $n = 25$ observations.

(b) Calculate the value of the test statistic and give your conclusion using the following 25 observations of X.

$$425 \quad 710 \quad 661 \quad 664 \quad 732 \quad 714 \quad 934 \quad 761 \quad 744$$
$$653 \quad 725 \quad 657 \quad 421 \quad 573 \quad 535 \quad 602 \quad 537 \quad 405$$
$$874 \quad 791 \quad 721 \quad 849 \quad 567 \quad 468 \quad 975$$

(c) Find a 98% one-sided confidence interval that gives an upper bound for μ.

8.2-12 In May the fill weights of 6-pound boxes of laundry soap had a mean of 6.13 pounds with a standard deviation of 0.095. The goal was to decrease the standard deviation. The company decided to adjust the filling machines and then test $H_0: \sigma = 0.095$ against $H_1: \sigma < 0.095$. In June a random sample of size $n = 20$ yielded $\bar{x} = 6.10$ and $s = 0.065$.

(a) At an $\alpha = 0.05$ significance level, was the company successful?

(b) What is the approximate p-value of your test?

8.2-13 The mean birth weight in the United States is $\mu = 3315$ grams with a standard deviation of $\sigma = 575$. Let X equal the birth weight in Rwanda. Assume that the distribution of X is $N(\mu, \sigma^2)$. We shall test the hypothesis $H_0: \sigma = 575$ against the alternative hypothesis $H_1: \sigma < 575$ at an $\alpha = 0.10$ significance level.

(a) What is your decision if a random sample of size $n = 81$ yielded $\bar{x} = 2819$ and $s = 496$?

(b) What is the approximate p-value of this test?

8.2-14 Let $X_1, X_2, \ldots, X_{23}$ be a random sample from a normal distribution that has variance $\sigma^2 = 100$. Let $S^2 = (1/22) \sum_{i=1}^{23} (X_i - \bar{X})^2$ be the sample variance. Of course, $22S^2/100$ is $\chi^2(22)$. Show that $\text{Var}(S^2) = 10,000/11$. Thus the standard deviation of S^2 is 30.15, and this helps explain the critical region $s^2 \leq 49.91$ or $s^2 \geq 167.18$ in Example 8.2-6.

8.2-15 The ornamental ground cover *Vinca minor* is rapidly spreading through the Hope College Biology Field Station because it can outcompete the small, native woody vegetation. In an attempt to discover if *Vinca minor* utilized natural chemical weapons to inhibit growth of the native vegetation, Hope biology students recently conducted an experiment in which they treated 33 sunflower seedlings with extracts taken from *Vinca minor* roots for several weeks and then measured the height of the seedlings. Let X equal the height of one of these seedlings and assume that the distribution of X is $N(\mu, \sigma^2)$. The observed growths (in cm) were

11.5	11.8	15.7	16.1	14.1	10.5	15.2	19.0	12.8	12.4	19.2
13.5	16.5	13.5	14.4	16.7	10.9	13.0	15.1	17.1	13.3	12.4
8.5	14.3	12.9	11.1	15.0	13.3	15.8	13.5	9.3	12.2	10.3

The students also planted some control sunflower seedlings that had a mean height of 15.7 cm. We shall test the null hypothesis $H_0: \mu = 15.7$ against the alternative hypothesis $H_1: \mu < 15.7$.

(a) Calculate the value of the test statistic and give limits for the p-value of this test.
(b) What is your conclusion?
(c) Find an approximate 98% one-sided confidence interval that gives an upper bound for μ.

8.2-16 Let $X_1, X_2, \ldots, X_{19}$ be a random sample of size $n = 19$ from the normal distribution $N(\mu, \sigma^2)$.

(a) Find a critical region, C, of size $\alpha = 0.05$ for testing $H_0: \sigma^2 = 30$ against H_1: $\sigma^2 = 80$.
(b) Find the approximate value of β, the probability of Type II error, for the critical region C of part (a).

8.2-17 Each of 51 golfers hit three golf balls of brand X and three golf balls of brand Y in a random order. Let X_i and Y_i equal the averages of the distances traveled by the brand X and brand Y golf balls hit by the ith golfer, $i = 1, 2, \ldots, 51$. Let $W_i = X_i - Y_i$, $i = 1, 2, \ldots, 51$. Test $H_0: \mu_W = 0$ against $H_1: \mu_W > 0$, where μ_W is the mean of the differences. If $\overline{w} = 2.07$ and $s_w^2 = 84.63$, would H_0 be accepted or rejected at an $\alpha = 0.05$ significance level?

8.2-18 To test whether a golf ball of brand A can be hit a greater distance off the tee than a golf ball of brand B, each of 17 golfers hit a ball of each brand, eight hitting ball A before ball B and nine hitting ball B before ball A. Assume that the differences of the paired A distance and B distance are approximately normally distributed and test the null hypothesis $H_0: \mu_D = 0$ against the alternative hypothesis $H_1: \mu_D > 0$ using a paired t test with the 17 differences. Let $\alpha = 0.05$.

Golfer	Distance for Ball A	Distance for Ball B	Golfer	Distance for Ball A	Distance for Ball B
1	265	252	10	274	260
2	272	276	11	274	267
3	246	243	12	269	267
4	260	246	13	244	251
5	274	275	14	212	222
6	263	246	15	235	235
7	255	244	16	254	255
8	258	245	17	224	231
9	276	259			

8.2-19 A vendor of milk products produces and sells low-fat dry milk to a company that uses it to produce baby formula. In order to determine the fat content of the milk, both the company and the vendor take a sample from each lot and test it for fat content in percent. Ten sets of paired test results are

Lot Number	Company Test Results (X)	Vendor Test Results (Y)
1	0.50	0.79
2	0.58	0.71
3	0.90	0.82
4	1.17	0.82
5	1.14	0.73
6	1.25	0.77
7	0.75	0.72
8	1.22	0.79
9	0.74	0.72
10	0.80	0.91

Let μ_D denote the mean of the difference $X - Y$. Test H_0: $\mu_D = 0$ against H_1: $\mu_D > 0$ using a paired t test with the differences. Let $\alpha = 0.05$.

8.2-20 A company that manufactures motors receives reels of 10,000 terminals per reel. Before using a reel of terminals, 20 terminals are randomly selected to be tested. The test is the amount of pressure needed to pull the terminal apart from its mate. This amount of pressure should continue to increase from test to test as the terminal is "roughed up." (Since this is destructive testing, a terminal that is tested cannot be used in a motor.) Let W equal the difference of the pressures: "test No. 1 pressure" minus "test No. 2 pressure." Assume that the distribution of W is $N(\mu_W, \sigma_W^2)$. We shall test the null hypothesis H_0: $\mu_W = 0$ against the alternative hypothesis H_1: $\mu_W < 0$ using 20 pairs of observations.

(a) Give the test statistic and a critical region that has a significance level of $\alpha = 0.05$. Sketch a figure illustrating this critical region.

(b) Use the following data to calculate the value of the test statistic and clearly state your conclusion.

Terminal	Test No. 1	Test No. 2	Terminal	Test No. 1	Test No. 2
1	2.5	3.8	11	7.3	8.2
2	4.0	3.9	12	7.2	6.6
3	5.2	4.7	13	5.9	6.8
4	4.9	6.0	14	7.5	6.6
5	5.2	5.7	15	7.1	7.5
6	6.0	5.7	16	7.2	7.5
7	5.2	5.0	17	6.1	7.3
8	6.6	6.2	18	6.3	7.1
9	6.7	7.3	19	6.5	7.2
10	6.6	6.5	20	6.5	6.7

(c) What would the conclusion be if $\alpha = 0.01$?

(d) What is the approximate p-value of this test?

8.2-21 The following data give the times of 22 male swimmers who entered the 50 yard freestyle race in the Conference Championship Meet. For these data, x is the swimmer's best time for the season and y is the swimmer's time in the meet.

$$
\begin{array}{lllll}
(24.97, 23.98) & (24.76, 23.63) & (23.80, 23.61) & (22.84, 23.58) & (24.07, 23.57) \\
(22.93, 22.83) & (23.41, 22.75) & (22.10, 22.74) & (23.08, 22.70) & (23.59, 22.62) \\
(22.38, 22.39) & (22.91, 22.34) & (22.08, 21.95) & (22.46, 21.73) & (21.34, 21.68) \\
(22.76, 21.66) & (21.82, 21.68) & (21.80, 21.58) & (22.26, 21.57) & (21.36, 21.35) \\
(22.98, 23.17) & (23.80, 22.53) & & &
\end{array}
$$

Let $d = x - y$, a swimmer's best time minus the swimmer's meet time. Assume that the distribution of D is $N(\mu_D, \sigma_D^2)$. Test the null hypothesis H_0: $\mu_D = 0$ against the alternative hypothesis H_1: $\mu_D > 0$. You may select the significance level. Interpret your conclusion.

8.3 TESTS OF THE EQUALITY OF TWO NORMAL DISTRIBUTIONS

Let independent random variables X and Y have normal distributions $N(\mu_X, \sigma_X^2)$ and $N(\mu_Y, \sigma_Y^2)$, respectively. There are times when we are interested in testing whether the distributions of X and Y are the same. So if the assumption of normality is valid, we would be interested in testing whether the two variances are equal and whether the two means are equal.

We first consider a test of the equality of the two means. When X and Y are independent and normally distributed, we can test hypotheses about their means using the same t statistic that was used for constructing a confidence interval for $\mu_X - \mu_Y$ in Section 6.5. Recall that the t statistic used for constructing the confidence interval assumed that the variances of X and Y are equal. That is why we shall later consider a test for the equality of two variances.

We begin with an example and then give a table that lists some hypotheses and critical regions. A botanist is interested in comparing the growth response of dwarf pea stems to two different levels of the hormone indoleacetic acid (IAA). Using 16-day-old pea plants, the botanist obtains 5-millimeter sections and floats these sections on solutions with different hormone concentrations to observe the effect of the hormone on the growth of the pea stem. Let X and Y denote, respectively, the independent growths that can be attributed to the hormone during the first 26 hours after sectioning for $(0.5)(10)^{-4}$ and 10^{-4} levels of concentration of IAA. The botanist would like to test the null hypothesis H_0: $\mu_X - \mu_Y = 0$ against the alternative hypothesis H_1: $\mu_X - \mu_Y < 0$. If we can assume X and Y are independent and normally distributed with common variance, respective random samples of sizes n and m give a test based on the statistic

$$
T = \frac{\overline{X} - \overline{Y}}{\sqrt{\{[(n-1)S_X^2 + (m-1)S_Y^2]/(n+m-2)\}(1/n+1/m)}} \tag{8.3-1}
$$

$$
= \frac{\overline{X} - \overline{Y}}{S_P\sqrt{1/n+1/m}}
$$

where

$$
S_P = \sqrt{\frac{(n-1)S_X^2 + (m-1)S_Y^2}{n+m-2}}. \tag{8.3-2}
$$

T has a t distribution with $r = n + m - 2$ degrees of freedom when H_0 is true and the variances are (approximately) equal. The hypothesis H_0 will be rejected in favor of H_1 if the observed value of T is less than $-t_\alpha(n+m-2)$.

EXAMPLE 8.3-1 In the preceding discussion, the botanist measured the growths of pea stem segments, in millimeters, for $n = 11$ observations of X:

$$0.8 \quad 1.8 \quad 1.0 \quad 0.1 \quad 0.9 \quad 1.7 \quad 1.0 \quad 1.4 \quad 0.9 \quad 1.2 \quad 0.5$$

and $m = 13$ observations of Y:

$$1.0 \quad 0.8 \quad 1.6 \quad 2.6 \quad 1.3 \quad 1.1 \quad 2.4$$
$$1.8 \quad 2.5 \quad 1.4 \quad 1.9 \quad 2.0 \quad 1.2$$

For these data, $\bar{x} = 1.03$, $s_x^2 = 0.24$, $\bar{y} = 1.66$, and $s_y^2 = 0.35$. The critical region for testing $H_0: \mu_X - \mu_Y = 0$ against $H_1: \mu_X - \mu_Y < 0$ is $t \leq -t_{0.05}(22) = -1.717$, where t is the two-sample t found in Equation 8.3-1. Since

$$t = \frac{1.03 - 1.66}{\sqrt{\{[10(0.24) + 12(0.35)]/(11 + 13 - 2)\}(1/11 + 1/13)}}$$

$$= -2.81 < -1.717,$$

H_0 is clearly rejected at an $\alpha = 0.05$ significance level. Notice that the approximate p-value of this test is 0.005 because $-t_{0.005}(22) = -2.819$. See Figure 8.3-1. Also, the sample variances do not differ too much; thus most statisticians would use this two-sample t test.

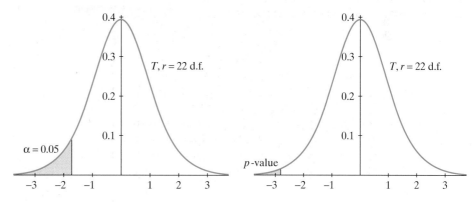

Figure 8.3-1: Critical region and p-value for pea stem growths

It is also instructive to construct box-and-whisker diagrams to gain a visual comparison of the two samples. For these two sets of data, the five-number summaries (minimum, three quartiles, maximum) are

$$0.1 \quad 0.8 \quad 1.0 \quad 1.4 \quad 1.8$$

for the X sample and

$$0.8 \quad 1.15 \quad 1.6 \quad 2.2 \quad 2.6$$

for the Y sample. The two box plots are shown in Figure 8.3-2.   ◄

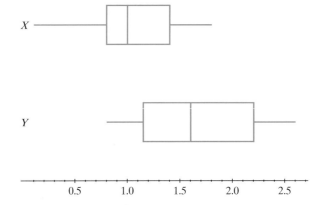

Figure 8.3-2: Box plots for pea stem growths

Based on independent random samples of sizes n and m, let $\bar{x}$, $\bar{y}$, and s_p^2 represent the observed unbiased estimates of the respective parameters μ_X, μ_Y, and $\sigma_X^2 = \sigma_Y^2$ of two normal distributions with common variance. Then α-level tests of certain hypotheses are given in Table 8.3-1, when $\sigma_X^2 = \sigma_Y^2$. If the common variance assumption is violated, but not too badly, the test is satisfactory but the significance levels are only approximate. The t statistic along with s_p are given in Equations 8.3-1 and 8.3-2, respectively.

Table 8.3-1: Tests of Hypotheses for Equality of Two Means		
H_0	H_1	**Critical Region**
$\mu_X = \mu_Y$	$\mu_X > \mu_Y$	$t \geq t_\alpha(n+m-2)$ or
		$\bar{x} - \bar{y} \geq t_\alpha(n+m-2)s_p\sqrt{1/n + 1/m}$
$\mu_X = \mu_Y$	$\mu_X < \mu_Y$	$t \leq -t_\alpha(n+m-2)$ or
		$\bar{x} - \bar{y} \leq -t_\alpha(n+m-2)s_p\sqrt{1/n + 1/m}$
$\mu_X = \mu_Y$	$\mu_X \neq \mu_Y$	$|t| \geq t_{\alpha/2}(n+m-2)$ or
		$|\bar{x} - \bar{y}| \geq t_{\alpha/2}(n+m-2)s_p\sqrt{1/n + 1/m}$

Remark Again to emphasize the relationship between confidence intervals and tests of hypotheses, we note that each of the tests in Table 8.3-1 has a corresponding confidence interval. For illustration, the first one-sided test is equivalent to saying we reject H_0: $\mu_X - \mu_Y = 0$ if zero is not in the one-sided confidence interval with lower bound

$$\bar{x} - \bar{y} - t_\alpha(n+m-2)s_p\sqrt{1/n + 1/m}.$$ ∎

EXAMPLE 8.3-2   A product is packaged using a machine with 24 filler heads numbered 1 to 24, with the odd numbered heads on one side of the machine and the even on the other side. Let X and Y equal the fill weights in grams when a package is filled by an odd-numbered head and an even-numbered head, respectively. Assume that the distributions of X and Y are $N(\mu_X, \sigma^2)$ and $N(\mu_Y, \sigma^2)$, respectively, and that X and Y are independent. We would like to test the null hypothesis $H_0: \mu_X - \mu_Y = 0$ against the alternative hypothesis $H_1: \mu_X - \mu_Y \neq 0$. To perform the test, after the machine has been set up and is running, we shall select one package at random from each filler head and weigh it. The test statistic is that given by Equation 8.3-1 with $n = m = 12$. At an $\alpha = 0.10$ significance level, the critical region is $|t| \geq t_{0.05}(22) = 1.717$.

For the $n = 12$ observations of X,

$$\begin{array}{cccccc}
1071 & 1076 & 1070 & 1083 & 1082 & 1067 \\
1078 & 1080 & 1075 & 1084 & 1075 & 1080
\end{array}$$

$\overline{x} = 1076.75$ and $s_x^2 = 29.30$. For the $m = 12$ observations of Y,

$$\begin{array}{cccccc}
1074 & 1069 & 1075 & 1067 & 1068 & 1079 \\
1082 & 1064 & 1070 & 1073 & 1072 & 1075
\end{array}$$

$\overline{y} = 1072.33$ and $s_y^2 = 26.24$. The calculated value of the test statistic is

$$t = \frac{1076.75 - 1072.33}{\sqrt{\dfrac{11(29.30) + 11(26.24)}{22}\left(\dfrac{1}{12} + \dfrac{1}{12}\right)}} = 2.05.$$

Since

$$|t| = |2.05| = 2.05 > 1.717,$$

the null hypothesis is rejected at an $\alpha = 0.10$ significance level. Note, however, that

$$|t| = 2.05 < 2.074 = t_{0.025}(22)$$

so that the null hypothesis would not be rejected at an $\alpha = 0.05$ significance level. That is, the p-value is between 0.05 and 0.10.

Again it is instructive to construct box plots on the same graph for these two sets of data. The box plots in Figure 8.3-3 were constructed using the five-number summary for the observations of X: 1067, 1072, 1077, 1081.5, 1084, and the five-number summary for the observations of Y: 1064, 1068.25, 1072.5, 1075, 1082. It looks like additional sampling would be advisable to test that the filler heads on the two sides of the machine are filling in a similar manner. If not, some corrective action needs to be taken.   ◄

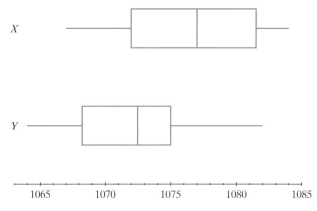

Figure 8.3-3: Box plots for fill weights

We would like to give two modifications of tests about two means. If we are able to assume that we know the variances of X and Y, then the appropriate test statistic to use for testing $H_0: \mu_X = \mu_Y$ is

$$Z = \frac{\overline{X} - \overline{Y}}{\sqrt{\dfrac{\sigma_X^2}{n} + \dfrac{\sigma_Y^2}{m}}}, \qquad (8.3\text{-}3)$$

which has a standard normal distribution when the null hypothesis is true and, of course, the populations are normally distributed. If the variances are unknown and the sample sizes are large, replace σ_X^2 with S_X^2 and σ_Y^2 with S_Y^2 in Equation 8.3-3. The resulting statistic will have an approximate $N(0, 1)$ distribution.

EXAMPLE 8.3-3 The target thickness for Fruit Flavored Gum and for Fruit Flavored Bubble Gum is 6.7 hundredths of an inch. Let the independent random variables X and Y equal the respective thicknesses of these gums in hundredths of an inch and assume that their distributions are $N(\mu_X, \sigma_X^2)$ and $N(\mu_Y, \sigma_Y^2)$, respectively. Because bubble gum has more elasticity than regular gum, it seems as if it would be harder to roll it out to the correct thickness. Thus we shall test the null hypothesis $H_0: \mu_X = \mu_Y$ against the alternative hypothesis $H_1: \mu_X < \mu_Y$ using samples of sizes $n = 50$ and $m = 40$.

Because the variances are unknown and the sample sizes are large, the test statistic that is used is

$$Z = \frac{\overline{X} - \overline{Y}}{\sqrt{\dfrac{S_X^2}{50} + \dfrac{S_Y^2}{40}}}.$$

At an approximate significance level of $\alpha = 0.01$, the critical region is

$$z \le -z_{0.01} = -2.326.$$

The observed values of X were

6.85	6.60	6.70	6.75	6.75	6.90	6.85	6.90	6.70	6.85
6.60	6.70	6.75	6.70	6.70	6.70	6.55	6.60	6.95	6.95
6.80	6.80	6.70	6.75	6.60	6.70	6.65	6.55	6.55	6.60
6.60	6.70	6.80	6.75	6.60	6.75	6.50	6.75	6.70	6.65
6.70	6.70	6.55	6.65	6.60	6.65	6.60	6.65	6.80	6.60

for which $\bar{x} = 6.701$ and $s_x = 0.108$. The observed values of Y were

7.10	7.05	6.70	6.75	6.90	6.90	6.65	6.60	6.55	6.55
6.85	6.90	6.60	6.85	6.95	7.10	6.95	6.90	7.15	7.05
6.70	6.90	6.85	6.95	7.05	6.75	6.90	6.80	6.70	6.75
6.90	6.90	6.70	6.70	6.90	6.90	6.70	6.70	6.90	6.95

for which $\bar{y} = 6.841$ and $s_y = 0.155$. Since the calculated value of the test statistic is

$$z = \frac{6.701 - 6.841}{\sqrt{0.108^2/50 + 0.155^2/40}} = -4.848 < -2.326,$$

the null hypothesis is clearly rejected.

The box-and-whisker diagrams in Figure 8.3-4 were constructed using the five-number summary of the observations of X: 6.50, 6.60, 6.70, 6.75, 6.95, and the five-number summary of the observations of Y: 6.55, 6.70, 6.90, 6.94, 7.15. This graphical display also confirms our conclusion. ◄

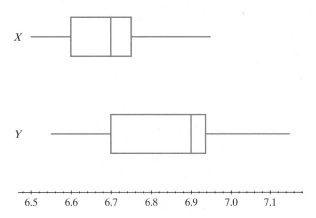

Figure 8.3-4: Box plots for gum thicknesses

Remark   To have satisfactory tests, our assumptions must be satisfied reasonably well. As long as the underlying distributions are not highly skewed, the normal assumptions are not too critical as $\overline{X}$ and $\overline{Y}$ have approximate normal distributions by the central limit theorem. As distributions become nonnormal and highly skewed, the sample mean and sample variance become more dependent and that causes problems using the Student's t as an approximating distribution for T. In these cases, nonparametric methods described later could be used.

When the distributions are close to normal but the variances seem to differ by a great deal, the t statistic should again be avoided, particularly if the sample sizes are also different. In that case, use Z or the modification by substituting the sample variances for the distribution variances. In the latter situation, if n and m are large enough, there is no problem. With small n and m, most statisticians would use Welch's suggestion (or other modifications of it); that is, use an approximating Student's t distribution with $[r]$ degrees of freedom, where

$$\frac{1}{r} = \frac{c^2}{n-1} + \frac{(1-c)^2}{m-1} \quad \text{and} \quad c = \frac{s_x^2/n}{s_x^2/n + s_y^2/m}.$$

An equivalent formula for r is

$$r = \frac{\left(\dfrac{s_x^2}{n} + \dfrac{s_y^2}{m}\right)^2}{\dfrac{1}{n-1}\left(\dfrac{s_x^2}{n}\right)^2 + \dfrac{1}{m-1}\left(\dfrac{s_y^2}{m}\right)^2}.$$

We actually give a test for the equality of variances. This could be used to decide whether to use T or a modification of Z. However, most statisticians do not place much confidence in this test of $\sigma_X^2 = \sigma_Y^2$ and would use a modification of Z (possibly Welch's) if they suspected that the variances differ greatly. ∎

We now give a test for the equality of two variances when sampling from normal populations. Let the independent random variables X and Y have respective distributions that are $N(\mu_X, \sigma_X^2)$ and $N(\mu_Y, \sigma_Y^2)$. To test the null hypothesis $H_0: \sigma_X^2/\sigma_Y^2 = 1$ (or, equivalently, $\sigma_X^2 = \sigma_Y^2$), take random samples of n observations of X and m observations of Y. Recall that $(n-1)S_X^2/\sigma_X^2$ and $(m-1)S_Y^2/\sigma_Y^2$ have independent chi-square distributions $\chi^2(n-1)$ and $\chi^2(m-1)$, respectively. Thus, when H_0 is true,

$$F = \frac{\dfrac{(n-1)S_X^2}{\sigma_X^2(n-1)}}{\dfrac{(m-1)S_Y^2}{\sigma_Y^2(m-1)}} = \frac{S_X^2}{S_Y^2}$$

has an F distribution with $r_1 = n - 1$ and $r_2 = m - 1$ degrees of freedom. This F statistic is our test statistic. When H_0 is true, we would expect the observed value of F to be close to 1.

Three possible alternative hypotheses along with critical regions of size α are summarized in Table 8.3-2. Recall that $1/F$, the reciprocal of F, has an F

distribution with $m - 1$ and $n - 1$ degrees of freedom so all critical regions may be written in terms of right-tail rejection regions so that the critical values can be selected easily from Appendix Table VII.

Table 8.3-2: Tests of Hypotheses of the Equality of Variances

H_0	H_1	Critical Region
$\sigma_X^2 = \sigma_Y^2$	$\sigma_X^2 > \sigma_Y^2$	$\dfrac{s_x^2}{s_y^2} \geq F_\alpha(n-1, m-1)$
$\sigma_X^2 = \sigma_Y^2$	$\sigma_X^2 < \sigma_Y^2$	$\dfrac{s_y^2}{s_x^2} \geq F_\alpha(m-1, n-1)$
$\sigma_X^2 = \sigma_Y^2$	$\sigma_X^2 \neq \sigma_Y^2$	$\dfrac{s_x^2}{s_y^2} \geq F_{\alpha/2}(n-1, m-1)$ or $\dfrac{s_y^2}{s_x^2} \geq F_{\alpha/2}(m-1, n-1)$

EXAMPLE 8.3-4   A biologist who studies spiders believes that not only do female green lynx spiders tend to be longer than their male counterparts but also that the lengths of the female spiders seem to vary more than those of the male spiders. We shall test whether this latter belief is true. Suppose that the distribution of the length X of male spiders is $N(\mu_X, \sigma_X^2)$ and the length Y of female spiders is $N(\mu_Y, \sigma_Y^2)$, and X and Y are independent. We shall test $H_0: \sigma_X^2/\sigma_Y^2 = 1$ (i.e., $\sigma_X^2 = \sigma_Y^2$) against the alternative hypothesis $H_1: \sigma_X^2/\sigma_Y^2 < 1$ (i.e., $\sigma_X^2 < \sigma_Y^2$.) If we use $n = 30$ and $m = 30$ observations of X and Y, respectively, a critical region that has a significance level of $\alpha = 0.01$ is

$$\frac{s_y^2}{s_x^2} \geq F_{0.01}(29, 29) = 2.42,$$

approximately, using interpolation in Appendix Table VII. In Exercise 6.5-5, $n = 30$ observations of X yielded $\bar{x} = 5.917$ and $s_x^2 = 0.4399$, while $m = 30$ observations of Y yielded $\bar{y} = 8.153$ and $s_y^2 = 1.4100$. Since

$$\frac{s_y^2}{s_x^2} = \frac{1.4100}{0.4399} = 3.2053 > 2.42,$$

the null hypothesis is rejected in favor of the biologist's belief. This is illustrated in Figure 8.3-5. (See Exercise 8.3-16.)   ◄

In Example 8.3-1 we used a t statistic for testing the equality of means that assumed the variances were equal. In the next example we shall test whether that assumption is valid.

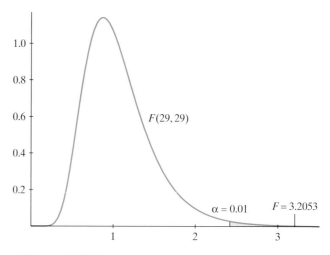

Figure 8.3-5: Test of variances for green lynx spiders

EXAMPLE 8.3-5 For Example 8.3-1, given $n = 11$ observations of X and $m = 13$ observations of Y, where X is $N(\mu_X, \sigma_X^2)$ and Y is $N(\mu_Y, \sigma_Y^2)$, we shall test the null hypothesis $H_0: \sigma_X^2/\sigma_Y^2 = 1$ against a two-sided alternative hypothesis. At an $\alpha = 0.05$ significance level, H_0 is rejected if

$$s_x^2/s_y^2 \geq F_{0.025}(10, 12) = 3.37$$

or

$$s_y^2/s_x^2 \geq F_{0.025}(12, 10) = 3.62.$$

Using the data in Example 8.3-1, we obtain

$$s_x^2/s_y^2 = 0.24/0.35 = 0.686 \qquad \text{and} \qquad s_y^2/s_x^2 = 1.458,$$

so we do not reject H_0. Thus the assumption of equal variances for the t-statistic that was used in Example 8.3-1 seems to be valid. See Figure 8.3-6 noting that $F_{0.975}(10, 12) = 1/F_{0.025}(12, 10) = 1/3.62 = 0.276.$ ◀

EXERCISES

8.3-1 The botanist in Example 8.3-1 is really interested in testing for synergistic interaction. That is, given two hormones, gibberellin (GA$_3$) and indoleacetic acid (IAA), let X_1 and X_2 equal the growth responses (in millimeters) of dwarf pea stem segments to GA$_3$ and IAA, respectively and separately. Let $X = X_1 + X_2$. Let Y equal the growth response when both hormones are present. Assuming that X is $N(\mu_X, \sigma^2)$ and Y is $N(\mu_Y, \sigma^2)$, the botanist is interested in testing the hypothesis $H_0: \mu_X = \mu_Y$ against the alternative hypothesis of synergistic interaction $H_1: \mu_X < \mu_Y$.

(a) Using $n = m = 10$ observations of X and Y, define the test statistic and critical region. Sketch a figure of the t p.d.f. and show the critical region on your figure. Let $\alpha = 0.05$.

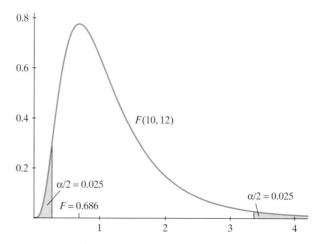

Figure 8.3-6: Test of variances for pea stem growths

(b) Given $n = 10$ observations of X,

$$2.1 \quad 2.6 \quad 2.6 \quad 3.4 \quad 2.1 \quad 1.7 \quad 2.6 \quad 2.6 \quad 2.2 \quad 1.2$$

and $m = 10$ observations of Y,

$$3.5 \quad 3.9 \quad 3.0 \quad 2.3 \quad 2.1 \quad 3.1 \quad 3.6 \quad 1.8 \quad 2.9 \quad 3.3$$

calculate the value of the test statistic and state your conclusion. Locate the test statistic on your figure.

(c) Construct two box plots on the same figure. Does this confirm your conclusion?

8.3-2 Let X and Y denote the weights in grams of male and female gallinules, respectively. Assume that X is $N(\mu_X, \sigma_X^2)$ and Y is $N(\mu_Y, \sigma_Y^2)$.

(a) Given $n = 16$ observations of X and $m = 13$ observations of Y, define a test statistic and critical region for testing the null hypothesis H_0: $\mu_X = \mu_Y$ against the one-sided alternative hypothesis H_1: $\mu_X > \mu_Y$. Let $\alpha = 0.01$. (Assume the variances are equal.)

(b) Given that $\bar{x} = 415.16$, $s_x^2 = 1356.75$, $\bar{y} = 347.40$, and $s_y^2 = 692.21$, calculate the value of the test statistic and state your conclusion.

(c) Test whether the assumption of equal variances is valid. Let $\alpha = 0.05$.

(d) Despite the fact that $\sigma_X^2 = \sigma_Y^2$ is accepted in part (c), let us say we suspect the equality is not valid. Thus use the test proposed by Welch.

8.3-3 Let X equal the weight in grams of a Low Fat Strawberry Kudo and Y the weight of a Low Fat Blueberry Kudo. Assume that the distributions of X and Y are $N(\mu_X, \sigma_X^2)$ and $N(\mu_Y, \sigma_Y^2)$, respectively. Use the following $n = 9$ observations of X,

$$21.7 \quad 21.0 \quad 21.2 \quad 20.7 \quad 20.4 \quad 21.9 \quad 20.2 \quad 21.6 \quad 20.6$$

and $m = 13$ observations of Y,

$$21.5 \quad 20.5 \quad 20.3 \quad 21.6 \quad 21.7 \quad 21.3 \quad 23.0 \quad 21.3 \quad 18.9 \quad 20.0 \quad 20.4 \quad 20.8 \quad 20.3$$

to answer the following questions.

(a) Test the null hypothesis H_0: $\sigma_X^2/\sigma_Y^2 = 1$ against a two-sided alternative hypothesis. You may select the significance level.

(b) Test the null hypothesis H_0: $\mu_X = \mu_Y$ against a two-sided alternative hypothesis. You may select the significance level. Also select the appropriate test statistic, depending on your answer to part (a).

(c) Construct and interpret box-and-whisker diagrams to support your conclusions.

8.3-4 Among the data collected for the World Health Organization air quality monitoring project is a measure of suspended particles in $\mu g/m^3$. Let X and Y equal the concentration of suspended particles in $\mu g/m^3$ in the city center (commercial district), for Melbourne and Houston, respectively. Using $n = 13$ observations of X and $m = 16$ observations of Y, we shall test $H_0: \mu_X = \mu_Y$ against $H_1: \mu_X < \mu_Y$.

    (a) Define the test statistic and critical region, assuming that the variances are equal. Let $\alpha = 0.05$.

    (b) If $\bar{x} = 72.9$, $s_x = 25.6$, $\bar{y} = 81.7$, and $s_y = 28.3$, calculate the value of the test statistic and state your conclusion.

    (c) Give limits for the p-value of this test.

    (d) Test whether the assumption of equal variances is valid. Let $\alpha = 0.05$.

8.3-5 Some nurses in County Public Health conducted a survey of women who had received inadequate prenatal care. They used information from birth certificates to select mothers for the survey. The mothers that were selected were divided into two groups: 14 mothers who said they had 5 or fewer prenatal visits and 14 mothers who said they had 6 or more prenatal visits. Let X and Y equal the respective birthweights of the babies from these two sets of mothers and assume that the distribution of X is $N(\mu_X, \sigma^2)$ and the distribution of Y is $N(\mu_Y, \sigma^2)$,

    (a) Define the test statistic and critical region for testing $H_0: \mu_X - \mu_Y = 0$ against $H_1: \mu_X - \mu_Y < 0$. Let $\alpha = 0.05$.

    (b) Given that the observations of X were

49	108	110	82	93	114	134
114	96	52	101	114	120	116

and the observations of Y were

133	108	93	119	119	98	106
131	87	153	116	129	97	110

calculate the value of the test statistic and state your conclusion.

    (c) Approximate the p-value.

    (d) Construct box plots on the same figure for these two sets of data. Do the box plots support your conclusion?

    (e) Test whether the assumption of equal variances is valid. Let $\alpha = 0.05$.

8.3-6 Let X and Y equal the forces required to pull stud No. 3 and stud No. 4 out of a window that has been manufactured for an automobile. Assume that the distributions of X and Y are $N(\mu_X, \sigma_X^2)$ and $N(\mu_Y, \sigma_Y^2)$, respectively.

    (a) If $m = n = 10$ observations are selected randomly, define a test statistic and critical region for testing $H_0: \mu_X - \mu_Y = 0$ against a two-sided alternative hypothesis. Let $\alpha = 0.05$. Assume that the variances are equal.

    (b) Given n = 10 observations of X,

$$111 \quad 120 \quad 139 \quad 136 \quad 138 \quad 149 \quad 143 \quad 145 \quad 111 \quad 123$$

and $m = 10$ observations of Y,

$$152 \quad 155 \quad 133 \quad 134 \quad 119 \quad 155 \quad 142 \quad 146 \quad 157 \quad 149$$

calculate the value of the test statistic and clearly state your conclusion.

    (c) What is the approximate p-value of this test?

    (d) Construct box plots on the same figure for these two sets of data. Do the box plots confirm your decision in part (b)?

    (e) Test whether the assumption of equal variances is valid.

8.3-7 Let X and Y equal the number of milligrams of tar in filtered and nonfiltered cigarettes, respectively. Assume that the distributions of X and Y are $N(\mu_X, \sigma_X^2)$ and $N(\mu_Y, \sigma_Y^2)$, respectively. We shall test the null hypothesis $H_0: \mu_X - \mu_Y = 0$ against the alternative hypothesis $H_1: \mu_X - \mu_Y < 0$ using random samples of sizes $n = 9$ and $m = 11$ observations of X and Y. respectively.

(a) Define the test statistic and a critical region that has an $\alpha = 0.01$ significance level. Sketch a figure illustrating this critical region.

(b) Given the $n = 9$ observations of X,

$$0.9 \quad 1.1 \quad 0.1 \quad 0.7 \quad 0.4 \quad 0.9 \quad 0.8 \quad 1.0 \quad 0.4$$

and the $m = 11$ observations of Y,

$$1.5 \quad 0.9 \quad 1.6 \quad 0.5 \quad 1.4 \quad 1.9 \quad 1.0 \quad 1.2 \quad 1.3 \quad 1.6 \quad 2.1$$

calculate the value of the test statistic and clearly state your conclusion. Locate the value of the test statistic on your figure.

8.3-8 In Exercise 6.5-4, we let X and Y equal the blood volumes in milliliters for males who are paraplegics participating in vigorous physical activities and males who are able bodied participating in normal activities. We assumed that X was $N(\mu_X, \sigma_X^2)$ and Y was $N(\mu_Y, \sigma_Y^2)$. Using $n = 7$ observations of X

$$1612 \quad 1352 \quad 1456 \quad 1222 \quad 1560 \quad 1456 \quad 1924$$

and $m = 10$ observations of Y

$$\begin{array}{ccccc} 1082 & 1300 & 1092 & 1040 & 910 \\ 1248 & 1092 & 1040 & 1092 & 1288 \end{array}$$

you were asked to find a 95% confidence interval for $\mu_X - \mu_Y$.

(a) To construct the confidence interval, you would like to assume that the variances are equal. Is this a valid assumption? Test the hypothesis that the variances are equal against a two-sided alternative hypothesis. Let $\alpha = 0.05$.

(b) Test the hypothesis that the means are equal against a two-sided alternative hypothesis. Let $\alpha = 0.05$.

(c) Did 0 belong to the 95% confidence interval in Exercise 6.5-4? Did you reject the null hypothesis in this exercise? Are your answers to these questions compatible?

8.3-9 An office furniture manufacturer installed a new adhesive application process. To compare the new process to the old process, random samples were selected from the two processes, and "pull tests" were performed to determine the number of pounds of pressure that were required to pull apart the glued parts (destructive testing). Let X and Y denote the pounds of pressure needed for the new and old processes, respectively.

(a) Based on $n = m = 24$ observations, define the test statistic and critical region for testing $H_0: \mu_X - \mu_Y = 0$ against $H_1: \mu_X - \mu_Y > 0$. Let $\alpha = 0.05$. State assumptions.

(b) Use the following $n = 24$ observations of X

$$\begin{array}{cccccccc} 1250 & 1210 & 990 & 1310 & 1320 & 1200 & 1290 & 1360 \\ 1120 & 1360 & 1310 & 1110 & 1320 & 980 & 950 & 1430 \\ 960 & 1050 & 1310 & 1240 & 1420 & 1170 & 1470 & 1060 \end{array}$$

and the following $m = 24$ observations of Y

$$\begin{array}{cccccccc} 1180 & 1360 & 1310 & 1190 & 920 & 1060 & 1440 & 1010 \\ 1310 & 980 & 1310 & 1030 & 960 & 800 & 1280 & 1080 \\ 930 & 1050 & 1010 & 1310 & 940 & 860 & 1450 & 1070 \end{array}$$

to calculate the value of the test statistic and clearly state your conclusion.

(c) What is the approximate p-value of this test?

(d) Construct two box plots on the same graph. Do the box plots support your conclusion?

8.3-10 Let X and Y denote the tarsus lengths of male and female grackles, respectively. Assume that X is $N(\mu_X, \sigma_X^2)$ and Y is $N(\mu_Y, \sigma_Y^2)$. Given that $n = 25$, $\bar{x} = 33.80$, $s_x^2 = 4.88$, $m = 29$, $\bar{y} = 31.66$, and $s_y^2 = 5.81$, test

(a) $H_0: \sigma_X^2/\sigma_Y^2 = 1$ against a two-sided alternative with $\alpha = 0.02$.

(b) $H_0: \mu_X = \mu_Y$ against $H_1: \mu_X > \mu_Y$ with $\alpha = 0.01$.

8.3-11 Weight checks are done on the scales for an automatic bagger for water softener pellets. Each bagger has two scales, say south and north, that operate alternately to fill 80# bags of pellets. Let X and Y equal the weights of the bags of pellets from these two scales. Assume that the distributions of X and Y are $N(\mu_X, \sigma^2)$ and $N(\mu_Y, \sigma^2)$, respectively. Ten bags were selected randomly from each scale and weighed by hand on a scale with a "high" degree of accuracy, yielding the following weights.

X:	80.51	80.46	80.75	80.50	80.36
	80.32	80.36	80.78	80.26	80.34
Y:	80.51	80.28	80.40	80.35	80.38
	80.28	80.27	80.16	80.59	80.56

(a) Test $H_0: \mu_X = \mu_Y$ against $H_1: \mu_X \neq \mu_Y$. Assume that $\sigma_X^2 = \sigma_Y^2$. Give limits for the p-value of the test and state your conclusion.
(b) Test the assumption of equal variances against a two-sided alternative hypothesis.
(c) Draw box-and-whisker diagrams on the same graph. Does this figure confirm your answers?

8.3-12 Two different saws are used to cut columns for a gazebo that has a diameter of 16 feet. The company was interested in comparing the two saws. They suspect that one saw is cutting columns shorter than the other saw. Let the lengths of the columns cut by the two saws be observations of random variables X and Y that are $N(\mu_X, \sigma_X^2)$ and $N(\mu_Y, \sigma_Y^2)$, respectively. We shall use the following observations of X and Y.

X:	8.02	8.10	8.04	8.04	8.00	8.11	8.07	8.02	8.04
Y:	8.04	8.04	8.10	8.06	8.08	8.10	8.07	8.08	8.06

(a) Test $H_0: \mu_X = \mu_Y$ against $H_1: \mu_X < \mu_Y$. Assume that $\sigma_X^2 = \sigma_Y^2$. Give limits for the p-value of the test and state your conclusion.
(b) Test the assumption of equal variances against a two-sided alternative hypothesis.
(c) Draw box-and-whisker diagrams on the same graph. Does this figure confirm your answers?

8.3-13 When a stream is turbid, it is not completely clear due to suspended solids in the water. The higher the turbidity, the less clear the water. A stream was studied on 26 days, half during dry weather, say observations of X, and the other half immediately after a significant rainfall, say observations of Y. Assume that the distributions of X and Y are $N(\mu_X, \sigma^2)$ and $N(\mu_Y, \sigma^2)$, respectively. The following turbidities were recorded in units of NTUs (Nephelometric Turbidity Units).

X:	2.9	14.9	1.0	12.6	9.4	7.6	3.6
	3.1	2.7	4.8	3.4	7.1	7.2	
Y:	7.8	4.2	2.4	12.9	17.3	10.4	5.9
	4.9	5.1	8.4	10.8	23.4	9.7	

(a) Test the null hypothesis $H_0: \mu_X = \mu_Y$ against $H_1: \mu_X < \mu_Y$. Give limits for the p-value and state your conclusion.
(b) Draw box-and-whisker diagrams on the same graph. Does this figure confirm your answer?

8.3-14 Plants convert CO_2 in the atmosphere, along with water and energy from sunlight, into the energy they need for growth and reproduction. Experiments were performed with normal air atmospheric conditions and those with enriched CO_2 concentrations to determine the effect on plant growth. The plants were given the same amount of water and light for a four week period. The following table gives the plant growths in grams.

Normal Air	4.67	4.21	2.18	3.91	4.09	5.24	2.94	4.71
	4.04	5.79	3.80	4.38				
Enriched Air	5.04	4.52	6.18	7.01	4.36	1.81	6.22	5.70

Determine whether CO_2-enriched atmosphere increases plant growth based on these data.

8.3-15 Let X equal the fill weight in April and Y the fill weight in June for an 8-pound box of bleach. We shall test the null hypothesis H_0: $\mu_X - \mu_Y = 0$ against the alternative hypothesis H_1: $\mu_X - \mu_Y > 0$ given that $n = 90$ observations of X yielded $\bar{x} = 8.10$, $s_x = 0.117$ and $m = 110$ observations of Y yielded $\bar{x} = 8.07$ and $s_y = 0.054$.

    (a) What is your conclusion if $\alpha = 0.05$?
        HINT: Do the variances seem to be equal?
    (b) What is the approximate p-value of this test?

8.3-16 Let X and Y denote the lengths of male and female green Lynx spiders, respectively. Assume that the distributions of X and Y are $N(\mu_X, \sigma_X^2)$ and $N(\mu_Y, \sigma_Y^2)$. In Example 8.3-4, we concluded that $\sigma_Y^2 > \sigma_X^2$. Thus, use the modification of Z to test the hypothesis H_0: $\mu_X - \mu_Y = 0$ against the alternative hypothesis H_1: $\mu_X - \mu_Y < 0$.

    (a) Define the test statistic and a critical region that has a significance level of $\alpha = 0.025$.
    (b) Using the sample characteristics given in Example 8.3-4, calculate the value of the test statistic and state your conclusion.
    (c) Draw two box-and-whisker diagrams on the same figure. Does this confirm both the conclusion of this exercise and that of Example 8.3-4?

8.3-17 To measure air pollution in a home, let X and Y equal the amount of suspended particulate matter (in $\mu g/m^3$) measured during a 24-hour period in a home in which there is no smoker and a home in which there is a smoker, respectively. (Assume that the distributions of the independent random variables X and Y are $N(\mu_X, \sigma_X^2)$ and $N(\mu_Y, \sigma_Y^2)$, respectively.) We shall test the null hypothesis H_0: $\sigma_X^2/\sigma_Y^2 = 1$ against the one-sided alternative hypothesis H_1: $\sigma_X^2/\sigma_Y^2 > 1$. If a random sample of size $n = 9$ yielded $\bar{x} = 93$ and $s_x = 12.9$ while a random sample of size $m = 11$ yielded $\bar{y} = 132$ and $s_y = 7.1$, define a critical region and give your conclusion if $\alpha = 0.05$. Now test H_0: $\mu_X = \mu_Y$ against H_1: $\mu_X < \mu_Y$ if $\alpha = 0.05$.

8.3-18 Let X and Y equal the times in days required for maturation of Guardiola seeds from narrow-leaved and broad-leaved parents, respectively. Assume that X is $N(\mu_X, \sigma_X^2)$ and Y is $N(\mu_Y, \sigma_Y^2)$, and X and Y are independent. Test the hypothesis H_0: $\sigma_X^2/\sigma_Y^2 = 1$ against the alternative hypothesis H_1: $\sigma_X^2/\sigma_Y^2 > 1$ if a sample size $n = 13$ yielded $\bar{x} = 18.97, s_x^2 = 9.88$ and a sample of size $m = 9$ yielded $\bar{y} = 23.20, s_y^2 = 4.08$. Let $\alpha = 0.05$.

8.3-19 A random sample of $n = 7$ brand X light bulbs yielded $\bar{x} = 891$ hours and $s_x^2 = 9201$. A random sample of $m = 10$ brand Y light bulbs yielded $\bar{y} = 592$ hours and $s_y^2 = 4856$. Use these data to test H_0: $\sigma_X^2/\sigma_Y^2 = 1$ against H_1: $\sigma_X^2/\sigma_Y^2 > 1$. Let $\alpha = 0.05$ and state assumptions.

8.4* THE WILCOXON TESTS

As has been mentioned in the previous sections, sometimes it is clear that the normal assumptions are not met and other procedures, sometimes referred to as **nonparametric** or **distribution-free** methods should be considered. For illustration, suppose some hypothesis, say H_0: $m = m_0$ against H_1: $m \neq m_0$, is made about the unknown median, m, of a continuous-type distribution. We could, from the data, construct a $100(1 - \alpha)$ percent confidence interval for m; and if m_0 is not in that interval, we would reject H_0 at the α significance level.

We could also use the sign test. Let X be a continuous-type random variable. Let m denote the median of X. To test the hypothesis H_0: $m = m_0$ against an appropriate alternative hypothesis, a **sign test** can be used. That is, if $X_1, X_2, \ldots, X_n$ denote the observations of a random sample from this distribution and if we let Y equal the number of negative differences among $X_1 - m_0, X_2 - m_0, \ldots, X_n - m_0$, then Y has the binomial distribution $b(n, 1/2)$ under H_0 and is the test statistic for the sign test. If Y is too large or too small, we reject $H_0: m = m_0$.

EXAMPLE 8.4-1

Let X denote the length of time in seconds between two calls entering a college switchboard. Let m be the unique median of this continuous-type distribution. We test the null hypothesis $H_0: m = 6.2$ against the alternative hypothesis $H_1: m < 6.2$. If Y is the number of lengths of time in a random sample of size 20 that are less than 6.2, the critical region $C = \{y : y \geq 14\}$ has a significance level of $\alpha = 0.0577$ using Table II. A random sample of size 20 yielded the following data:

6.8	5.7	6.9	5.3	4.1	9.8	1.7	7.0
2.1	19.0	18.9	16.9	10.4	44.1	2.9	2.4
4.8	18.9	4.8	7.9				

Since $y = 9$, the null hypothesis is not rejected. ◄

The sign test can also be used to test the hypothesis that two continuous-type random variables X and Y are such that $p = P(X > Y) = 1/2$. To test the hypothesis $H_0: p = 1/2$ against an appropriate alternative hypothesis, consider the independent pairs $(X_1, Y_1), (X_2, Y_2), \ldots, (X_n, Y_n)$. Let W denote the number of pairs for which $X_i - Y_i > 0$. When H_0 is true, W is $b(n, 1/2)$, and the test can be based upon the statistic W. It is important to recognize that X_i and Y_i do not need to be independent and are frequently not independent in practice. If they were independent, then there are many ways in which $X_1, X_2, \ldots, X_n$ and $Y_1, Y_2, \ldots, Y_n$ can be paired up. If, however, X is the length of the right foot of a person and Y the length of the corresponding left foot, there is a natural pairing. Here $H_0: p = P(X > Y) = 1/2$ suggests that either foot of a particular individual is equally likely to be longer.

One major objection to the sign test is that it does not take into account the magnitude of the differences $X_1 - m_0, \ldots, X_n - m_0$. We now discuss a **test of Wilcoxon** that does take into account the magnitude of the differences $|X_i - m_0|, i = 1, 2, \ldots, n$. However, in addition to the assumption that the random variable X is of the continuous type, we must also assume that the p.d.f. of X is symmetric about the median in order to find the distribution of this new statistic. Because of the continuity assumption, we assume, in the following discussion, that no two observations are equal and that no observation is equal to the median.

We are interested in testing the hypothesis $H_0: m = m_0$, where m_0 is some given constant. With our random sample $X_1, X_2, \ldots, X_n$, we rank the absolute values $|X_1 - m_0|, |X_2 - m_0|, \ldots, |X_n - m_0|$ in ascending order according to magnitude. That is, for $i = 1, 2, \ldots, n$, we let R_i denote the rank of $|X_i - m_0|$ among $|X_1 - m_0|, |X_2 - m_0|, \ldots, |X_n - m_0|$. Note that $R_1, R_2, \ldots, R_n$ is a permutation of the first n positive integers, $1, 2, \ldots, n$. Now with each R_i we associate the sign of the difference $X_i - m_0$; that is, if $X_i - m_0 > 0$, we use R_i, but if $X_i - m_0 < 0$, we use $-R_i$. The Wilcoxon statistic W is the sum of these n signed ranks.

EXAMPLE 8.4-2   Consider a sunfish example in which the lengths of $n = 10$ fish are

$$x_i : 5.0 \ 3.9 \ 5.2 \ 5.5 \ 2.8 \ 6.1 \ 6.4 \ 2.6 \ 1.7 \ 4.3$$

We shall test $H_0: m = 3.7$ against the alternative hypothesis $H_1: m > 3.7$. Thus we have

$x_i - m_0$:	1.3,	0.2,	1.5,	1.8,	-0.9,	2.4,	2.7,	-1.1,	-2.0,	0.6
$\lvert x_i - m_0 \rvert$:	1.3,	0.2,	1.5,	1.8,	0.9,	2.4,	2.7,	1.1,	2.0,	0.6
Ranks:	5,	1,	6,	7,	3,	9,	10,	4,	8,	2
Signed Ranks:	5,	1,	6,	7,	-3,	9,	10,	-4,	-8,	2

Therefore, the Wilcoxon statistic is equal to

$$W = 5 + 1 + 6 + 7 - 3 + 9 + 10 - 4 - 8 + 2 = 25.$$

Incidentally, the positive answer seems reasonable because the number of the 10 lengths that are less than 3.7 is 3, which is the statistic used in the sign test. ◀

If the hypothesis $H_0: m = m_0$ is true, about one-half of the differences would be negative and thus about one-half of the signs would be negative. Thus it seems that the hypothesis $H_0: m = m_0$ is supported if the observed value of W is close to zero. If the alternative hypothesis is $H_1: m > m_0$, we would reject H_0 if the observed $W = w$ is too large, since, in this case, the larger deviations $\lvert X_i - m_0 \rvert$ would usually be associated with observations for which $x_i - m_0 > 0$. That is, the critical region would be of the form $\{w : w \geq c_1\}$. If the alternative hypothesis is $H_1: m < m_0$, the critical region would be of the form $\{w : w \leq c_2\}$. Of course, the critical region would be of the form $\{w : w \leq c_3 \text{ or } w \geq c_4\}$ for a two-sided alternative hypothesis $H_1: m \neq m_0$. In order to find the values of c_1, c_2, c_3, c_4 that yield desired significance levels, it is necessary to determine the distribution of W under H_0. We now consider certain characteristics of this distribution.

When $H_0: m = m_0$ is true,

$$P(X_i < m_0) = P(X_i > m_0) = \frac{1}{2}, \qquad i = 1, 2, \ldots, n.$$

Hence the probability is $1/2$ that a negative sign is associated with the rank R_i of $\lvert X_i - m_0 \rvert$. Moreover, the assignments of these n signs are independent because $X_1, X_2, \ldots, X_n$ are mutually independent. In addition, W is a sum that contains the integers $1, 2, \ldots, n$, each integer with a positive or negative sign. Since the underlying distribution is symmetric, it seems intuitively obvious that W has the same distribution as the random variable

$$V = \sum_{i=1}^{n} V_i,$$

where $V_1, V_2, \ldots, V_n$ are independent and

$$P(V_i = i) = P(V_i = -i) = \frac{1}{2}, \qquad i = 1, 2, \ldots, n.$$

That is, V is a sum that contains the integers $1, 2, \ldots, n$, and these integers receive their algebraic signs by independent assignments.

Since W and V have the same distribution, their means and variances are equal, and we can easily find those of V. Now the mean of V_i is

$$E(V_i) = -i\left(\frac{1}{2}\right) + i\left(\frac{1}{2}\right) = 0$$

and thus

$$E(W) = E(V) = \sum_{i=1}^{n} E(V_i) = 0.$$

The variance of V_i is

$$\text{Var}(V_i) = E(V_i^2) = (-i)^2\left(\frac{1}{2}\right) + (i)^2\left(\frac{1}{2}\right) = i^2.$$

Thus

$$\text{Var}(W) = \text{Var}(V) = \sum_{i=1}^{n} \text{Var}(V_i) = \sum_{i=1}^{n} i^2 = \frac{n(n+1)(2n+1)}{6}.$$

We shall not try to find the distribution of W in general, since that p.m.f. does not have a convenient expression. However, we demonstrate how we could find the distribution of W (or V) with enough patience and computer support. Recall that the moment-generating function of V_i is

$$M_i(t) = e^{t(-i)}\left(\frac{1}{2}\right) + e^{t(+i)}\left(\frac{1}{2}\right) = \frac{e^{-it} + e^{it}}{2}, \qquad i = 1, 2, \ldots, n.$$

Let $n = 2$; so the moment-generating function of $V_1 + V_2$ is

$$M(t) = E[e^{t(V_1+V_2)}].$$

From the independence of V_1 and V_2, we obtain

$$M(t) = E(e^{tV_1})E(e^{tV_2})$$

$$= \left(\frac{e^{-t} + e^{t}}{2}\right)\left(\frac{e^{-2t} + e^{2t}}{2}\right)$$

$$= \frac{e^{-3t} + e^{-t} + e^{t} + e^{3t}}{4}.$$

This means that each of the points $-3, -1, 1, 3$ in the support of $V_1 + V_2$ has probability $1/4$.

Next let $n = 3$, so the moment-generating function of $V_1 + V_2 + V_3$ is

$$M(t) = E[e^{t(V_1+V_2+V_3)}]$$

$$= E[e^{t(V_1+V_2)}]E(e^{tV_3})$$

$$= \left(\frac{e^{-3t} + e^{-t} + e^t + e^{3t}}{4}\right)\left(\frac{e^{-3t} + e^{3t}}{2}\right)$$

$$= \frac{e^{-6t} + e^{-4t} + e^{-2t} + 2e^0 + e^{-2t} + e^{-4t} + e^{-6t}}{8}.$$

Thus the points $-6, -4, -2, 0, 2, 4, 6$ in the support of $V_1 + V_2 + V_3$ have the respective probabilities $1/8, 1/8, 1/8, 2/8, 1/8, 1/8, 1/8$. Obviously, this procedure can be continued for $n = 4, 5, 6, \ldots$, but it is rather tedious. Fortunately, however, even though $V_1, V_2, \ldots, V_n$ are not identically distributed random variables, the sum V of them still has an approximate normal distribution. To obtain this normal approximation for V (or W), a more general form of the central limit theorem, due to Liapounov, can be used that allows us to say that the standardized random variable

$$Z = \frac{W - 0}{\sqrt{n(n+1)(2n+1)/6}}$$

is approximately $N(0, 1)$ when H_0 is true. We accept this without proof, and thus we can approximate probabilities such as $P(W \geq c; H_0) \approx P(Z \geq z_\alpha; H_0)$ when the sample size n is sufficiently large using this normal distribution. However, the next example illustrates this approximation.

EXAMPLE 8.4-3   The moment-generating function of W or of V is given by

$$M(t) = \prod_{i=1}^{n} \frac{e^{-it} + e^{it}}{2}.$$

Using a computer algebra system (CAS) such as *Maple*, it is possible to expand $M(t)$ and find the coefficients of e^{kt} which is equal to $P(W = k)$. In Figure 8.4-1 we have drawn a probability histogram for the distribution of W along with the approximating $N[0, n(n+1)(2n+1)/6]$ p.d.f. for $n = 4$ (a poor approximation) and for $n = 10$. It is important to note that the widths of the rectangles in the probability histogram are equal to 2, so the "half-unit correction for continuity" now becomes a unit correction.   ◀

EXAMPLE 8.4-4   Let m be the median of a symmetric distribution of the continuous type. To test the hypothesis $H_0: m = 160$ against the alternative hypothesis $H_1: m > 160$, we take a random sample of size $n = 16$. For an approximate significance level of $\alpha = 0.05$, H_0 is rejected if the computed $W = w$ is such that

$$z = \frac{w}{\sqrt{16(17)(33)/6}} \geq 1.645,$$

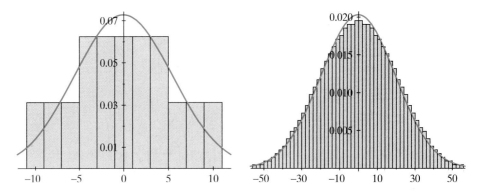

Figure 8.4-1: The Wilcoxon distribution for $n = 4$ and $n = 10$

or

$$w \geq 1.645\sqrt{\frac{16(17)(33)}{6}} = 63.626.$$

Say the observed values of a random sample are 176.9, 158.3, 152.1, 158.8, 172.4, 169.8, 159.7, 162.7, 156.6, 174.5, 184.4, 165.2, 147.8, 177.8, 160.1, 160.5. In Table 8.4-1 the magnitudes of the differences $|x_i - 160|$ have been ordered and ranked. Those differences $x_i - 160$ that were negative have been underlined and the ranks are under the ordered values. For this set of data

$$w = 1 - 2 + 3 - 4 - 5 + 6 + \cdots + 16 = 60.$$

Table 8.4-1: Ordered Absolute Differences from 160							
0.1	0.3	0.5	1.2	1.7	2.7	3.4	5.2
1	2	3	4	5	6	7	8
7.9	9.8	12.2	12.4	14.5	16.9	17.8	24.4
9	10	11	12	13	14	15	16

Since $60 < 63.626$, H_0 is not rejected at the 0.05 significance level. It is interesting to note that H_0 would have been rejected at $\alpha = 0.10$, since the approximate p-value is, making a unit correction for continuity,

$$p\text{-value} = P(W \geq 60)$$

$$= P\left(\frac{W - 0}{\sqrt{(16)(17)(33)/6}} \geq \frac{59 - 0}{\sqrt{(16)(17)(33)/6}}\right)$$

$$\approx P(Z \geq 1.525) = 0.0636.$$

(Using *Maple*, the p-value is equal to $4{,}251/65{,}536 = 0.0649$.) This would indicate to us that the data are too few to reject H_0, but if the pattern continues, we shall most certainly reject with a larger sample size.   ◀

Although theoretically we could ignore the possibilities that $x_i = m_0$, for some i, and that $|x_i - m_0| = |x_j - m_0|$ for some $i \neq j$, these situations do occur in applications. Usually, in practice, if $x_i = m_0$ for some i, that observation is deleted and the test is performed with a reduced sample size. If the absolute values of the differences from m_0 of two or more observations are equal, each observation is assigned the average of the corresponding ranks. The change this causes in the distribution of W is not very great, and thus we continue using the same normal approximation.

We now give an example that has some tied observations. This example also illustrates that the Wilcoxon rank sum test can be used to test hypotheses about the median of paired data.

EXAMPLE 8.4-5   We consider some paired data for percentage of body fat measured at the beginning and the end of a semester. Let m equal the median of the differences, $x - y$. We shall use the Wilcoxon statistic to test the null hypothesis H_0: $m = 0$ against the alternative hypothesis H_1: $m > 0$ with the differences given below. Since there are $n = 25$ nonzero differences, we reject H_0 if

$$z = \frac{w - 0}{\sqrt{(25)(26)(51)/6}} \geq 1.645$$

or, equivalently, if

$$w \geq 1.645\sqrt{\frac{(25)(26)(51)}{6}} = 122.27$$

at an approximate $\alpha = 0.05$ significance level. The 26 differences are

1.8	−3.1	0.1	1.1	0.6	−5.1	9.2	0.2	0.4
0.0	1.9	−0.4	−1.5	1.4	−1.0	2.2	0.8	−0.4
2.0	−5.8	−3.4	−2.3	3.0	2.7	0.2	3.2	

In Table 8.4-2 we list the ordered nonzero absolute values, underlining those that were originally negative. The rank is under each observation. Note that in the case of ties, the average of the ranks of the tied measurements is given.

Table 8.4-2: Ordered Absolute Values, Changes in Percentage of Body Fat												
0.1	0.2	0.2	0.4	0.4	0.4	0.6	0.8	1.0	1.1	1.4	1.5	1.8
1	2.5	2.5	5	5	5	7	8	9	10	11	12	13
1.9	2.0	2.2	2.3	2.7	3.0	3.1	3.2	3.4	5.1	5.8	9.2	
14	15	16	17	18	19	20	21	22	23	24	25	

The value of the Wilcoxon statistic is

$$w = 1 + 2.5 + 2.5 + 5 - 5 - 5 + \cdots + 25 = 51.$$

Since $51 < 122.27$, we fail to reject the null hypothesis. The approximate p-value of this test is

$$p\text{-value} = P(W \geq 51) = P(W \geq 50)$$

$$\approx P\left(Z \geq \frac{50 - 0}{\sqrt{(25)(26)(51)/6}}\right)$$

$$= P(Z \geq 0.673) = 0.2505. \quad \blacktriangleleft$$

There is another method due to Wilcoxon for testing the equality of two distributions of the continuous type that uses the magnitudes of the observations. For this test it is assumed that the populations have similar shapes. Order the combined sample of $X_1, X_2, \ldots, X_{n_1}$, and $Y_1, Y_2, \ldots, Y_{n_2}$ in increasing order of magnitude. Assign to the ordered values the ranks $1, 2, 3, \ldots, n_1 + n_2$. In the case of ties, assign the average of the ranks associated with the tied values. Let W equal the sum of the ranks of $Y_1, Y_2, \ldots, Y_{n_2}$. If the distribution of Y is shifted to the right of that of X, the values of Y would tend to be larger than the values of X and W would usually be larger than expected when $F(z) = G(z)$. If m_X and m_Y are the respective medians, the critical region for testing $H_0: m_X = m_Y$ against $H_1: m_X < m_Y$ would be of the form $w \geq c$. Similarly, if the alternative hypothesis is $m_X > m_Y$, the critical region would be of the form $w \leq c$.

We shall not derive the distribution of W. However, if n_1 and n_2 are both greater than 7, a normal approximation can be used. With $F(z) = G(z)$, the mean and variance of W are

$$\mu_W = \frac{n_2(n_1 + n_2 + 1)}{2}$$

and

$$\text{Var}(W) = \frac{n_1 n_2(n_1 + n_2 + 1)}{12},$$

and the statistic

$$Z = \frac{W - n_2(n_1 + n_2 + 1)/2}{\sqrt{n_1 n_2(n_1 + n_2 + 1)/12}}$$

is approximately $N(0, 1)$.

EXAMPLE 8.4-6   The weights of $n_1 = 8$ and $n_2 = 8$ tins of cinnamon packaged by companies A and B, respectively, selected at random, yielded the following observations of X:

$$117.1 \quad 121.3 \quad 127.8 \quad 121.9 \quad 117.4 \quad 124.5 \quad 119.5 \quad 115.1,$$

and the following observations of Y:

$$123.5 \quad 125.3 \quad 126.5 \quad 127.9 \quad 122.1 \quad 125.6 \quad 129.8 \quad 117.2.$$

The critical region for testing $H_0: m_X = m_Y$ against $H_1: m_X < m_Y$ is of the form $w \geq c$. Since $n_1 = n_2 = 8$, at an approximate $\alpha = 0.05$ significance level, H_0 is rejected if

$$z = \frac{w - 8(8 + 8 + 1)/2}{\sqrt{[(8)(8)(8 + 8 + 1)]/12}} \geq 1.645;$$

that is, if

$$w \geq 1.645 \sqrt{\frac{(8)(8)(17)}{12}} + 4(17) = 83.66.$$

From the data we see that the computed W is

$$w = 3 + 8 + 9 + 11 + 12 + 13 + 15 + 16 = 87 > 83.66.$$

Thus H_0 is rejected. The p-value of this test is, making a half-unit correction for continuity,

$$p\text{-value} = P(W \geq 87)$$

$$= P\left(\frac{W - 68}{\sqrt{90.667}} \geq \frac{86.5 - 68}{\sqrt{90.667}}\right)$$

$$\approx P(Z \geq 1.943) = 0.0260. \qquad \blacktriangleleft$$

Remark It should be mentioned that, shortly after Wilcoxon proposed his two-sample test, Mann and Whitney suggested a test based on the estimate of the probability $P(X < Y)$. Namely, they let U equal the number of times that $X_i < Y_j, i = 1, 2, \ldots, n_1$ and $j = 1, 2, \ldots, n_2$. Using the data in Example 8.4-6, we find that the computed U is $u = 51$ among all $n_1 n_2 = (8)(8) = 64$ pairs of (X, Y). Thus the estimate of $P(X < Y)$ is 51/64 or, in general, $u/n_1 n_2$. At the time of the Mann-Whitney suggestion, it was noted that U was just a linear function of Wilcoxon's W and hence really provided the same test. That relationship is

$$U = W - \frac{n_2(n_2 + 1)}{2},$$

which in our special case is

$$51 = 87 - \frac{8(9)}{2} = 87 - 36.$$

Thus we often read about the test of Mann, Whitney, and Wilcoxon. From this observation, this test could be thought of as one testing $H_0: P(X < Y) = 1/2$ against the alternative $H_1: P(X < Y) > 1/2$ with critical region of the form $w \geq c$.

It should be noted here that the two-sample Wilcoxon test is much less sensitive to extreme values than is the Student's t test based on $\overline{X} - \overline{Y}$. Therefore, if there is considerable skewness or contamination, these proposed distribution-free tests are much safer. In particular, that of Wilcoxon is quite good and does not lose too much in case the distributions are close to normal ones. It is important to note that the one-sample Wilcoxon requires symmetry of the underlying distribution, but the two-sample Wilcoxon does not and thus can be used for skewed distributions.

From theoretical developments beyond the scope of this text, the two Wilcoxon tests are strong competitors to the usual one- and two-sample tests based upon normal assumptions and these tests should be considered if those assumptions are questioned. ∎

EXERCISES

8.4-1 It is claimed that the median weight m of certain loads of candy is 40,000 pounds.

(a) Use the given 13 observations and the Wilcoxon statistic to test, at an approximate significance level of $\alpha = 0.05$, the null hypothesis H_0: $m = 40,000$ against the one-sided alternative hypothesis H_1: $m < 40,000$. Those 13 observations are

41,195	39,485	41,229	36,840	38,050	40,890	38,345
34,930	39,245	31,031	40,780	38,050	30,906	

(b) What is the approximate p-value of this test?
(c) Use the sign test to test the same hypothesis.
(d) Calculate the p-value and compare it to the p-value using the Wilcoxon test.

8.4-2 A course in economics was taught to two groups of students, one in a classroom situation and the other by TV. There were 24 students in each group. These students were first paired according to cumulative grade point averages and background in economics, and then assigned to the courses by a flip of a coin (this was repeated 24 times). At the end of the course each class was given the same final examination. Use the Wilcoxon test to test that the two methods of teaching are equally effective against a two-sided alternative. The differences in final scores for each pair of students, the TV student's score having been subtracted from the corresponding classroom student's scores, were as follows:

14	−4	−6	−2	−1	18
6	12	8	−4	13	7
2	6	21	7	−2	11
−3	−14	−2	17	−4	−5

8.4-3 Let X equal the weight (in grams) of a grape flavored jolly rancher. Denote the median of X by m. We shall test H_0: $m = 5.900$ against H_1: $m > 5.900$. A random sample of size $n = 25$ yielded the following ordered data.

5.625	5.665	5.697	5.837	5.863	5.870	5.878	5.884	5.908
5.967	6.019	6.020	6.029	6.032	6.037	6.045	6.049	
6.050	6.079	6.116	6.159	6.186	6.199	6.307	6.387	

(a) Use the sign test.
(b) Use the Wilcoxon test statistic.
(c) Use a t test.
(d) Write a short comparison of these three tests.

8.4-4 The outcomes on $n = 10$ simulations of a Cauchy random variable were −1.9415, 0.5901, −5.9848, −0.0790, −0.7757, −1.0962, 9.3820, −74.0216, −3.0678, 3.8545. For the Cauchy distribution, the mean does not exist; but, for this one, the median is believed to equal zero. Use the Wilcoxon test and these data to test H_0: $m = 0$ against the alternative hypothesis H_1: $m \neq 0$. Let $\alpha = 0.05$, approximately.

8.4-5 Let x equal a student's GPA in the fall semester and y the same student's GPA in the spring semester. Let m equal the median of the differences, $x - y$. We shall test the null hypothesis H_0: $m = 0$ against an appropriate alternative hypothesis that you select based on your past experience. Use a Wilcoxon test and the given 15 observations of paired data to test this hypothesis.

x	y	x	y
2.88	3.22	3.98	3.76
3.67	3.49	4.00	3.96
2.76	2.54	3.39	3.52
2.34	2.17	2.59	2.36
2.46	2.53	2.78	2.62
3.20	2.98	2.85	3.06
3.17	2.98	3.25	3.16
2.90	2.84		

8.4-6 Let m equal the median of the posttest grip strengths in the right arms of male freshmen in Health Dynamics. We shall use observations on $n = 15$ such students to test the null hypothesis H_0: $m = 50$ against the alternative hypothesis H_1: $m > 50$.

(a) Using the Wilcoxon statistic, define a critical region that has an approximate significance level of $\alpha = 0.05$.

(b) Given the observed values,

$$58 \quad 52.5 \quad 46 \quad 57.5 \quad 52 \quad 45.5 \quad 65.5 \quad 71$$
$$57 \quad 54 \quad 48 \quad 58 \quad 35.5 \quad 44 \quad 53$$

what is your conclusion?

(c) What is the p-value of this test?

8.4-7 Let X equal the weight in pounds of a "1-pound" bag of carrots. Let m equal the median weight of a population of these bags. Test the null hypothesis H_0: $m = 1.14$ against the alternative hypothesis H_1: $m > 1.14$.

(a) With a sample of size $n = 14$, use the Wilcoxon statistic to define a critical region. Use $\alpha = 0.10$, approximately.

(b) What is your conclusion if the observed weights were

$$1.12 \quad 1.13 \quad 1.19 \quad 1.25 \quad 1.06 \quad 1.31 \quad 1.12$$
$$1.23 \quad 1.29 \quad 1.17 \quad 1.20 \quad 1.11 \quad 1.18 \quad 1.23$$

(c) What is the p-value of your test?

8.4-8 A pharmaceutical company is interested in testing the effect of humidity on the weight of pills that are sold in aluminum packaging. Let X and Y denote the respective weights of pills and their packaging when the packaging is good and when it is defective after the pill has spent 1 week in a chamber containing 100% humidity and heated to 30°C.

(a) Use the Wilcoxon test to test H_0: $m_X = m_Y$ against H_0: $m_X - m_Y < 0$ using the following random samples of $n_1 = 12$ observations of X and $n_2 = 12$ observations of Y. What is the p-value?

X:	0.7565	0.7720	0.7776	0.7750	0.7494	0.7615
	0.7741	0.7701	0.7712	0.7719	0.7546	0.7719

Y:	0.7870	0.7750	0.7720	0.7876	0.7795	0.7972
	0.7815	0.7811	0.7731	0.7613	0.7816	0.7851

(b) Construct and interpret a q-q plot of these data. HINT: This is a q-q plot of the empirical distribution of X against that of Y.

8.4-9 Let us compare the failure times of a certain type of light bulb produced by two different manufacturers, X and Y, by testing 10 bulbs selected at random from each of the outputs. The data, in hundreds of hours used before failure, are

X:	5.6	4.6	6.8	4.9	6.1	5.3	4.5	5.8	5.4	4.7
Y:	7.2	8.1	5.1	7.3	6.9	7.8	5.9	6.7	6.5	7.1

(a) Use the Wilcoxon test to test the equality of medians of the two processes at the approximate 5% significance level. What is the p-value?

(b) Construct and interpret a q-q plot of these data. HINT: This is a q-q plot of the empirical distribution of X against that of Y.

8.4-10 Let X and Y denote the heights of blue spruce trees, measured in centimeters, in two large fields. We shall compare these heights by measuring 12 trees selected at random from each of the fields. Use the statistic W, the sum of the ranks of the observations of Y in the combined sample, to test at $\alpha = 0.05$, approximately, the hypothesis H_0: $m_X = m_Y$, against the alternative hypothesis H_1: $m_X < m_Y$, based on $n_1 = 12$ observations of X,

90.4	77.2	75.9	83.2	84.0	90.2
87.6	67.4	77.6	69.3	83.3	72.7

and $n_2 = 12$ observations of Y,

92.7	78.9	82.5	88.6	95.0	94.4
73.1	88.3	90.4	86.5	84.7	87.5

8.4-11 Let X and Y equal the sizes of grocery orders from a southside and a north-side foodstore of the same chain, respectively. We shall test the null hypothesis H_0: $m_X = m_Y$ against a two-sided alternative using the following observations that have been ordered.

X:	5.13	8.22	11.81	13.77	15.36
	23.71	31.39	34.65	40.17	75.58
Y:	4.42	6.47	7.12	10.50	12.12
	12.57	21.29	33.14	62.84	72.05

(a) Use the Wilcoxon test when $\alpha = 0.05$. What is the p-value of this two-sided test?

(b) Construct a q-q plot and interpret it. HINT: This is a q-q plot of the empirical distribution of X against that of Y.

8.4-12 A charter bus line has 48-passenger and 38-passenger buses. Let m_{48} and m_{38} denote the median number of miles traveled per day by the respective buses. With $\alpha = 0.05$, use the Wilcoxon statistic to test H_0: $m_{48} = m_{38}$ against the one-sided alternative H_1: $m_{48} > m_{38}$ using the following data, which give the numbers of miles traveled per day for respective random samples of sizes 9 and 11.

48-passenger buses:	331	308	300	414	253	
	323	452	396	104		
38-passenger buses:	248	393	260	355	279	184
	386	450	432	196	197	

8.4-13 A company manufactures and packages soap powder in 6-pound boxes. The quality assurance department was interested in comparing the fill weights of packages from the East and West lines. Taking random samples from the two lines they obtained the following weights.

East line (X):	6.06	6.04	6.11	6.06	6.06
	6.07	6.06	6.08	6.05	6.09
West line (Y):	6.08	6.03	6.04	6.07	6.11
	6.08	6.08	6.10	6.06	6.04

Let m_X and m_Y denote the median weights for the East and West lines, respectively. Test $H_0: m_X = m_Y$ against a two-sided alternative hypothesis using the Wilcoxon test with $\alpha = 0.05$, approximately. Find the p-value of this two-sided test.

(a) The Wilcoxon test with $\alpha = 0.05$, approximately. Find the p-value of this two-sided test.

(b) Construct and interpret a q-q plot of these data.

8.4-14 Let X and Y equal the CPU time in days per month for two different computers on a college campus. Let m_X and m_Y equal the medians of X and Y. Using $n_1 = n_2 = 14$ observations of each of X and Y, test the null hypothesis $H_0: m_X = m_Y$ against the alternative hypothesis $H_1: m_X < m_Y$. Use a normal approximation and the two-sample Wilcoxon test. Let $\alpha = 0.05$, approximately.

X:	7.1	8.5	9.9	13.8	2.2	2.4	2.3
	1.0	3.9	4.6	5.3	5.0	3.9	6.6

Y:	7.9	15.4	16.2	9.6	4.9	8.9	9.4
	15.2	12.1	4.3	5.0	2.4	2.3	5.4

8.4-15 With $\alpha = 0.05$, use the Wilcoxon statistic to test $H_0: m_X = m_Y$ against a two-sided alternative using the following observations of X and Y that have been ordered for your convenience.

X:	−2.3864	−2.2171	−1.9148	−1.9097	−1.4883
	−1.2007	−1.1077	−0.3601	0.4325	1.0598
	1.3035	1.5241	1.7133	1.7656	2.4912

Y:	−1.7613	−0.9391	−0.7437	−0.5530	−0.2469
	0.0647	0.2031	0.3219	0.3579	0.6431
	0.6557	0.6724	0.6762	0.9041	1.3571

8.4-16 Let X equal the playing time in minutes for a $33\frac{1}{3}$ rpm long-playing album and let Y equal the playing time for a compact disk. We shall test the null hypothesis $H_0: m_X = m_Y$ against the alternative hypothesis $H_1: m_X < m_Y$. Give the p-value and your conclusion for the Wilcoxon statistic. The observed data are

X:	40.83	43.18	35.72	38.68	37.17		
	39.75	24.76	34.58	33.98			

Y:	42.82	64.42	56.92	39.92	72.38	47.26	64.58
	38.20	72.75	39.09	39.07	33.70	62.02	

8.5   CHI-SQUARE GOODNESS OF FIT TESTS

We now consider applications of the very important chi-square statistic, first proposed by Karl Pearson in 1900. As the reader will see, it is a very adaptable test statistic and can be used for many different types of tests. In particular, one application of it allows us to test the appropriateness of different probabilistic models and, in this sense, is a competitor of the Kolmogorov-Smirnov test (Section 8.10*).

So that the reader can get some idea as to why Pearson first proposed his chi-square statistic, we begin with the binomial case. That is, let Y_1 be $b(n, p_1)$, where $0 < p_1 < 1$. According to the central limit theorem,

$$Z = \frac{Y_1 - np_1}{\sqrt{np_1(1 - p_1)}}$$

has a distribution that is approximately $N(0, 1)$ for large n, particularly when $np_1 \geq 5$ and $n(1 - p_1) \geq 5$. Thus it is not surprising that $Q_1 = Z^2$ is approximately $\chi^2(1)$. If we let $Y_2 = n - Y_1$ and $p_2 = 1 - p_1$, we see that Q_1 may be written as

$$Q_1 = \frac{(Y_1 - np_1)^2}{np_1(1 - p_1)} = \frac{(Y_1 - np_1)^2}{np_1} + \frac{(Y_1 - np_1)^2}{n(1 - p_1)}.$$

Since

$$(Y_1 - np_1)^2 = (n - Y_1 - n[1 - p_1])^2 = (Y_2 - np_2)^2,$$

we have

$$Q_1 = \frac{(Y_1 - np_1)^2}{np_1} + \frac{(Y_2 - np_2)^2}{np_2}.$$

Let us now carefully consider each term in this last expression for Q_1. Of course, Y_1 is the number of "successes," and np_1 is the expected number of "successes"; that is, $E(Y_1) = np_1$. Likewise, Y_2 and np_2 are, respectively, the number and the expected number of "failures." So each numerator consists of the square of a difference of the observed number and expected number. Note that Q_1 can be written as

$$Q_1 = \sum_{i=1}^{2} \frac{(Y_i - np_i)^2}{np_i}, \tag{8.5-1}$$

and we have seen intuitively that it has an approximate chi-square distribution with one degree of freedom. In a sense, Q_1 measure the "closeness" of the observed numbers to the corresponding expected numbers. For example, if the observed values of Y_1 and Y_2 equal their expected values, then the computed Q_1 is equal to $q_1 = 0$; but if they differ much from them, then the computed $Q_1 = q_1$ is relatively large.

To generalize, we let an experiment have k (instead of only two) mutually exclusive and exhaustive outcomes, say $A_1, A_2, \ldots, A_k$. Let $p_i = P(A_i)$ and thus $\sum_{i=1}^{k} p_i = 1$. The experiment is repeated n independent times, and we let Y_i represent the number of times the experiment results in A_i, $i = 1, 2, \ldots, k$. This joint distribution of $Y_1, Y_2, \ldots, Y_{k-1}$ is a straightforward generalization of the binomial distribution; this generalization is as follows.

In considering the joint p.m.f.,

$$f(y_1, y_2, \ldots, y_{k-1}) = P(Y_1 = y_1, Y_2 = y_2, \ldots, Y_{k-1} = y_{k-1}),$$

where $y_1, y_2, \ldots, y_{k-1}$ are nonnegative integers such that $y_1 + y_2 + \cdots + y_{k-1} \leq n$. Note that we do not need to consider Y_k since once the other $k - 1$ random variables are observed to equal $y_1, y_2, \ldots, y_{k-1}$, respectively, we know that

$$Y_k = n - y_1 - y_2 - \cdots - y_{k-1} = y_k, \quad \text{say.}$$

The probability of each particular arrangement of y_1 A_1s, y_2 A_2s, $\ldots$, y_k A_ks is, from the independence of the trials,

$$p_1^{y_1}\, p_2^{y_2}\, \cdots\, p_k^{y_k}.$$

The number of such arrangements is the multinomial coefficient

$$\binom{n}{y_1, y_2, \ldots, y_k} = \frac{n!}{y_1!\, y_2! \cdots y_k!}.$$

Hence the product of these two expressions gives the joint p.m.f. of $Y_1, Y_2, \ldots, Y_{k-1}$, namely,

$$f(y_1, y_2, \ldots, y_{k-1}) = \frac{n!}{y_1!\, y_2! \cdots y_k!}\, p_1^{y_1}\, p_2^{y_2} \cdots p_k^{y_k},$$

recalling that $y_k = n - y_1 - y_2 - \cdots - y_{k-1}$.

Pearson then constructed an expression similar to Q_1 (Equation 8.5-1), which involves Y_1 and $Y_2 = n - Y_1$, that we denote by Q_{k-1}, which involves $Y_1, Y_2, \ldots, Y_{k-1}$, and $Y_k = n - Y_1 - Y_2 - \cdots - Y_{k-1}$, namely,

$$Q_{k-1} = \sum_{i=1}^{k} \frac{(Y_i - np_i)^2}{np_i}.$$

He argued that Q_{k-1} has an approximate chi-square distribution with $k - 1$ degrees of freedom in much the same way we argued that Q_1 is approximately $\chi^2(1)$. We accept this fact as the proof is beyond the level of this text.

Some writers suggest that n should be large enough so that $np_i \geq 5$, $i = 1, 2, \ldots, k$, to be certain that the approximating distribution is adequate. This is probably good advice for the beginner to follow, although we have seen the approximation work very well when $np_i \geq 1$, $i = 1, 2, \ldots, k$. The important thing to guard against is allowing some particular np_i to become so small that the corresponding term in Q_{k-1}, namely $(Y_i - np_i)^2/np_i$, tends to dominate the others because of its small denominator. In any case it is important to realize that Q_{k-1} has only an approximate chi-square distribution.

We shall now show how we can use the fact that Q_{k-1} is approximately $\chi^2(k-1)$ to test hypotheses about probabilities of various outcomes. Let an experiment have k mutually exclusive and exhaustive outcomes, $A_1, A_2, \ldots, A_k$. We would like to test whether $p_i = P(A_i)$ is equal to a known number p_{i0}, $i = 1, 2, \ldots, k$. That is, we shall test the hypothesis

$$H_0: \ p_i = p_{i0}, \qquad i = 1, 2, \ldots, k.$$

In order to test such a hypothesis, we shall take a sample of size n, that is repeat the experiment n independent times. We tend to favor H_0 if the observed number of times that A_i occurred, say y_i, and the number of times A_i was expected to occur if H_0 were true, namely np_{i0}, are approximately equal. That is, if

$$q_{k-1} = \sum_{i=1}^{k} \frac{(y_i - np_{i0})^2}{np_{i0}}$$

is "small," we tend to favor H_0. Since the distribution of Q_{k-1} is approximately $\chi^2(k-1)$, we shall reject H_0 if $q_{k-1} \geq \chi_\alpha^2(k-1)$, where α is the desired significance level of the test.

EXAMPLE 8.5-1   If persons are asked to record a string of random digits, such as

$$3\ 7\ 2\ 4\ 1\ 9\ 7\ 2\ 1\ 5\ 0\ 8\ldots,$$

we usually find that they are reluctant to record the same or even the two closest numbers in adjacent positions. And yet, in true random digit generation, the probability of the next digit being the same as the preceding one is $p_{10} = 1/10$, the probability of the next being only one away from the preceding (assuming that 0 is one away from 9) is $p_{20} = 2/10$, and the probability of all other possibilities is $p_{30} = 7/10$. We shall test one person's concept of a random sequence by asking him to record a string of 51 digits that seems to represent a random generation. Thus we shall test

$$H_0\colon\ p_1 = p_{10} = \frac{1}{10}, \quad p_2 = p_{20} = \frac{2}{10}, \quad p_3 = p_{30} = \frac{7}{10}.$$

The critical region for an $\alpha = 0.05$ significance level is $q_2 \geq \chi_{0.05}^2(2) = 5.991$. The sequence of digits was as follows:

$$
\begin{array}{cccccccccccc}
5 & 8 & 3 & 1 & 9 & 4 & 6 & 7 & 9 & 2 & 6 & 3 & 0 \\
8 & 7 & 5 & 1 & 3 & 6 & 2 & 1 & 9 & 5 & 4 & 8 & 0 \\
3 & 7 & 1 & 4 & 6 & 0 & 4 & 3 & 8 & 2 & 7 & 3 & 9 \\
8 & 5 & 6 & 1 & 8 & 7 & 0 & 3 & 5 & 2 & 5 & 2
\end{array}
$$

We went through this listing and observed how many times the next digit was the same as or was one away from the preceding one.

	Frequency	Expected Number
Same	0	$50(1/10) = 5$
One away	8	$50(2/10) = 10$
Other	42	$50(7/10) = 35$
Total	50	50

The computed chi-square statistic is

$$\frac{(0-5)^2}{5} + \frac{(8-10)^2}{10} + \frac{(42-35)^2}{35} = 6.8 > 5.991 = \chi_{0.05}^2(2).$$

Thus we would say that this string of 51 digits does not seem to be random. ◀

One major disadvantage in the use of the chi-square test is that it is a many-sided test. That is, the alternative hypothesis is very general, and it would be difficult to restrict alternatives to situations such as, with $k = 3$, $H_1\colon p_1 > p_{10}$,

$p_2 > p_{20}$, $p_3 < p_{30}$. As a matter of fact, some statisticians would probably test H_0 against this particular alternative H_1 by using a linear function of Y_1, Y_2, and Y_3. However, this sort of discussion is beyond the scope of this book because it involves knowing more about the distributions of linear functions of the dependent random variables Y_1, Y_2, and Y_3. In any case, the student who truly recognizes that this chi-square statistic tests H_0: $p_i = p_{i0}$, $i = 1, 2, \ldots, k$, against all alternatives can usually appreciate the fact that it is more difficult to reject H_0 at a given significance level α using the chi-square statistic than it would be if some appropriate "one-sided" test statistic were available.

Many experiments yield a set of data, say $x_1, x_2, \ldots, x_n$, and the experimenter is often interested in determining whether these data can be treated as the observed values of a random sample $X_1, X_2, \ldots, X_n$ from a given distribution. That is, would this proposed distribution be a reasonable probabilistic model for these sample items? To see how the chi-square test can help us answer questions of this sort, consider a very simple example.

EXAMPLE 8.5-2   Let X denote the number of heads that occur when four coins are tossed at random. Under the assumptions that the four coins are independent and the probability of heads on each coin is $1/2$, X is $b(4, 1/2)$. One hundred repetitions of this experiment resulted in 0, 1, 2, 3, and 4 heads being observed on 7, 18, 40, 31, and 4 trials, respectively. Do these results support the assumptions? That is, is $b(4, 1/2)$ a reasonable model for the distribution of X? To answer this, we begin by letting $A_1 = \{0\}$, $A_2 = \{1\}$, $A_3 = \{2\}$, $A_4 = \{3\}$, $A_5 = \{4\}$. If $p_{i0} = P(X \in A_i)$ when X is $b(4, 1/2)$, then

$$p_{10} = p_{50} = \binom{4}{0}\left(\frac{1}{2}\right)^4 = \frac{1}{16} = 0.0625,$$

$$p_{20} = p_{40} = \binom{4}{1}\left(\frac{1}{2}\right)^4 = \frac{4}{16} = 0.25,$$

$$p_{30} = \binom{4}{2}\left(\frac{1}{2}\right)^4 = \frac{6}{16} = 0.375.$$

At an approximate $\alpha = 0.05$ significance level, the null hypothesis

$$H_0: \ p_i = p_{i0}, \qquad i = 1, 2, \ldots, 5,$$

is rejected if the observed value of Q_4 is greater than $\chi^2_{0.05}(4) = 9.488$. If we use the 100 repetitions of this experiment that resulted in the observed values of Y_1, Y_2, $\ldots$, Y_5 of $y_1 = 7$, $y_2 = 18$, $y_3 = 40$, $y_4 = 31$, and $y_5 = 4$, the computed value of Q_4 is

$$q_4 = \frac{(7 - 6.25)^2}{6.25} + \frac{(18 - 25)^2}{25} + \frac{(40 - 37.5)^2}{37.5} + \frac{(31 - 25)^2}{25} + \frac{(4 - 6.25)^2}{6.25}$$

$$= 4.47.$$

Since $4.47 < 9.488$, the hypothesis is not rejected. That is, the data support the hypothesis that $b(4, 1/2)$ is a reasonable probabilistic model for X. Recall that the mean of a chi-square random variable is its number of degrees of freedom. In this example the mean is 4 and the observed value of Q_4 is 4.47, just a little greater than the mean.   ◄

Thus far all the hypotheses H_0 tested with the chi-square statistic Q_{k-1} have been simple ones (i.e., completely specified, namely, in H_0: $p_i = p_{i0}$, $i = 1, 2, \ldots, k$, each p_{i0} has been known). This is not always the case, and it frequently happens that $p_{10}, p_{20}, \ldots, p_{k0}$ are functions of one or more unknown parameters. For example, suppose that the hypothesized model for X in Example 8.5-2 was H_0: X is $b(4, p), 0 < p < 1$. Then

$$p_{i0} = P(X \in A_i) = \frac{4!}{(i-1)!(5-i)!} p^{i-1}(1-p)^{5-i}, \qquad i = 1, 2, \ldots, 5,$$

which is a function of the unknown parameter p. Of course, if H_0: $p_i = p_{i0}$, $i = 1, 2, \ldots, 5$, is true, for large n,

$$Q_4 = \sum_{i=1}^{5} \frac{(Y_i - np_{i0})^2}{np_{i0}}$$

still has an approximate chi-square distribution with four degrees of freedom. The difficulty is that when $Y_1, Y_2, \ldots, Y_5$ are observed to be equal to $y_1, y_2, \ldots, y_5$, Q_4 cannot be computed, since $p_{10}, p_{20}, \ldots, p_{50}$ (and hence Q_4) are functions of the unknown parameter p.

One way out of the difficulty would be to estimate p from the data and then carry out the computations using this estimate. It is interesting to note the following. Say the estimation of p is carried out by minimizing Q_4 with respect to p yielding $\tilde{p}$. This $\tilde{p}$ is sometimes called a **minimum chi-square estimator** of p. If then this $\tilde{p}$ is used in Q_4, the statistic Q_4 still has an approximate chi-square distribution but with only $4 - 1 = 3$ degrees of freedom. That is, the number of degrees of freedom of the approximating chi-square distribution is reduced by one for each parameter estimated by the minimum chi-square technique. We accept this result without proof (as it is a rather difficult one). Although we have considered this when p_{i0}, $i = 1, 2, \ldots, k$, is a function of only one parameter, it holds when there is more than one unknown parameter, say d. Hence, in a more general situation, the test would be completed by computing Q_{k-1} using Y_i and the estimated p_{i0}, $i = 1, 2, \ldots, k$, to obtain q_{k-1} (i.e., q_{k-1} is the minimized chi-square). This value q_{k-1} would then be compared to a critical value $\chi_\alpha^2(k-1-d)$. In our special case, the computed (minimized) chi-square q_4 would be compared to $\chi_\alpha^2(3)$.

There is still one trouble with all of this: It is usually very difficult to find minimum chi-square estimators. Hence most statisticians usually use some reasonable method (maximum likelihood is satisfactory) of estimating the parameters. They then compute q_{k-1}, recognizing that it is somewhat larger than the minimized chi-square, and compare it with $\chi_\alpha^2(k-1-d)$. Note that this provides a slightly larger probability of rejecting H_0 than would the scheme in which the minimized chi-square were used because this computed q_{k-1} is larger than the minimum q_{k-1}.

EXAMPLE 8.5-3   Let X denote the number of alpha particles emitted by barium-133 in 1/10 of a second. The following 50 observations of X were taken with a Geiger counter in a fixed position:

$$\begin{array}{cccccccccc}
7 & 4 & 3 & 6 & 4 & 4 & 5 & 3 & 5 & 3 \\
5 & 5 & 3 & 2 & 5 & 4 & 3 & 3 & 7 & 6 \\
6 & 4 & 3 & 11 & 9 & 6 & 7 & 4 & 5 & 4 \\
7 & 3 & 2 & 8 & 6 & 7 & 4 & 1 & 9 & 8 \\
4 & 8 & 9 & 3 & 9 & 7 & 7 & 9 & 3 & 10
\end{array}$$

The experimenter is interested in determining whether X has a Poisson distribution. To test H_0: X is Poisson, we first estimate the mean of X, say λ, with the sample mean, $\bar{x} = 5.4$, of these 50 observations. We then partition the set of outcomes for this experiment into the sets $A_1 = \{0, 1, 2, 3\}$, $A_2 = \{4\}$, $A_3 = \{5\}$, $A_4 = \{6\}$, $A_5 = \{7\}$, and $A_6 = \{8, 9, 10, \ldots\}$. We combined $\{0, 1, 2, 3\}$ into one set A_1 and $\{8, 9, 10, \ldots\}$ into another A_6 so that the expected number of outcomes for each set is at least five when H_0 is true. In Table 8.5-1 the data are grouped, and the estimated probabilities specified by the hypothesis that X has a Poisson distribution with an estimated $\widehat{\lambda} = \bar{x} = 5.4$ are given. Since one parameter was estimated, Q_{6-1} has an approximate chi-square distribution with $r = 5 - 1 = 4$ degrees of freedom. Since

$$q_5 = \frac{[13 - 50(0.213)]^2}{50(0.213)} + \cdots + \frac{[10 - 50(0.178)]^2}{50(0.178)}$$

$$= 2.763 < 9.488 = \chi^2_{0.05}(4),$$

H_0 is not rejected at the 5% significance level. That is, with only these data, we are quite willing to accept the model that X has a Poisson distribution.   ◀

Table 8.5-1: Grouped Geiger Counter Data

	Outcome					
	A_1	A_2	A_3	A_4	A_5	A_6
Frequency	13	9	6	5	7	10
Probability	0.213	0.160	0.173	0.156	0.120	0.178
Expected ($50p_i$)	10.65	8.00	8.65	7.80	6.00	8.90

Let us now consider the problem of testing a model for the distribution of a random variable W of the continuous type. That is, if $F(w)$ is the distribution function of W, we wish to test

$$H_0: \ F(w) = F_0(w),$$

where $F_0(w)$ is some known distribution function of the continuous type. In Section 8.10* we will use a Kolmogorov-Smirnov test for H_0. Also recall that we have considered problems of this type using q-q plots. In order to use the chi-square statistic, we must partition the set of possible values of W into k sets. One way this can be done is as follows. Partition the interval $[0, 1]$ into k sets with the points $b_0, b_1, b_2, \ldots, b_k$, where

$$0 = b_0 < b_1 < b_2 < \cdots < b_k = 1.$$

Let $a_i = F_0^{-1}(b_i)$, $i = 1, 2, \ldots, k-1$; $A_1 = (-\infty, a_1]$, $A_i = (a_{i-1}, a_i]$ for $i = 2, 3, \ldots, k-1$, and $A_k = (a_{k-1}, \infty)$; $p_i = P(W \in A_i)$, $i = 1, 2, \ldots, k$. Let Y_i denote the number of times the observed value of W belongs to A_i, $i = 1, 2, \ldots, k$, in n independent repetitions of the experiment. Then $Y_1, Y_2, \ldots, Y_k$ have a multinomial distribution with parameters n, $p_1, p_2, \ldots, p_{k-1}$. Also let $p_{i0} = P(W \in A_i)$ when the distribution function of W is $F_0(w)$. The hypothesis that we actually test is a modification of H_0, namely

$$H_0': \ p_i = p_{i0}, \qquad i = 1, 2, \ldots, k.$$

This hypothesis is rejected if the observed value of the chi-square statistic

$$Q_{k-1} = \sum_{i=1}^{k} \frac{(Y_i - np_{i0})^2}{np_{i0}}$$

is at least as great as $\chi_\alpha^2(k-1)$. If the hypothesis $H_0': p_i = p_{i0}$, $i = 1, 2, \ldots, k$, is not rejected, we do not reject the hypothesis $H_0: F(w) = F_0(w)$.

EXAMPLE 8.5-4

Example 3.2-3 gives 105 observations of the times between calls to 911. Also given is a histogram of these data with the exponential p.d.f. with $\theta = 20$ superimposed. We shall now use a chi-square goodness of fit test to see whether or not this is an appropriate model for these data. That is, if X is equal to the time between calls to 911, we shall test the null hypothesis that the distribution of X is exponential with a mean of $\theta = 20$. Table 8.5-2 groups the data into 9 classes, gives the probabilities and expected values of these classes, and calculates the value of the chi-square statistic. The value of the chi-square goodness of fit statistic is

$$q_8 = \frac{(41 - 38.0520)^2}{38.0520} + \frac{(22 - 24.2655)^2}{24.2655} + \cdots + \frac{(2 - 2.8665)^2}{2.8665} = 4.6861.$$

The p-value associated with this test is 0.7905 which means that it is an extremely good fit.

   Note that we assumed that we knew $\theta = 20$. We could also have run this test letting $\theta = \bar{x}$, remembering that we then lose one degree of freedom. For this example, the outcome would be about the same.   ◀

Table 8.5-2: Summary of Times Between Calls to 911			
Class	**Frequency**	**Probability**	**Expected**
$A_1 = [0, 9]$	41	0.362	38.0520
$A_2 = (9, 18]$	22	0.231	24.2655
$A_3 = (18, 27]$	11	0.147	15.4665
$A_4 = (27, 36]$	10	0.094	9.8700
$A_5 = (36, 45]$	9	0.060	6.2895
$A_6 = (45, 54]$	5	0.038	4.0110
$A_7 = (54, 63]$	2	0.024	2.5620
$A_8 = (63, 72]$	3	0.016	1.6275
$A_9 = (72, \infty)$	2	0.027	2.8665

It is also true, in dealing with models of random variables of the continuous type, that we must frequently estimate unknown parameters. For example, let H_0 be that W is $N(\mu, \sigma^2)$, where μ and σ^2 are unknown. With a random sample $W_1, W_2, \ldots, W_n$, we first can estimate μ and σ^2, possibly with $\overline{w}$ and s_w^2. We partition the space $\{w: -\infty < w < \infty\}$ into k mutually disjoint sets $A_1, A_2, \ldots, A_k$. We then use the estimates of μ and σ^2, say $\overline{w}$ and $s^2 = s_w^2$, to estimate

$$p_{i0} = \int_{A_i} \frac{1}{s\sqrt{2\pi}} \exp\left[-\frac{(w - \overline{w})^2}{2s^2} \right] dw,$$

$i = 1, 2, \ldots, k$. Using the observed frequencies $y_1, y_2, \ldots, y_k$ of $A_1, A_2, \ldots, A_k$, respectively, from the observed random sample $w_1, w_2, \ldots, w_n$, and $\widehat{p}_{10}$, $\widehat{p}_{20}, \ldots, \widehat{p}_{k0}$ estimated with $\overline{w}$ and $s^2 = s_w^2$, we compare the computed

$$q_{k-1} = \sum_{i=1}^{k} \frac{(y_i - n\widehat{p}_{i0})^2}{n\widehat{p}_{i0}}$$

to $\chi_\alpha^2(k-1-2)$. This value q_{k-1} will again be somewhat larger than that which would be found using minimum chi-square estimation, and certain caution should be observed. Several exercises illustrate the procedure in which one or more parameters must be estimated. Finally, it should be noted that the methods given in this section frequently are classified under the more general title of goodness of fit tests. In particular, then, the tests in this section would be **chi-square goodness of fit tests**.

EXERCISES

8.5-1 A 1-pound bag of candy-coated chocolate-covered peanuts contained 224 pieces of candy colored brown, orange, green, and yellow. Test the null hypothesis that the machine filling these bags treats the four colors of candy equally likely; that is, test

$$H_0: \ p_B = p_O = p_G = p_Y = \frac{1}{4}.$$

The observed values were 42 brown, 64 orange, 53 green, and 65 yellow. You may select the significance level or give an approximate p-value.

8.5-2 A particular brand of candy-coated chocolate comes in five different colors that we shall denote as $A_1 = \{brown\}$, $A_2 = \{yellow\}$, $A_3 = \{orange\}$, $A_4 = \{green\}$, and $A_5 = \{coffee\}$. Let p_i equal the probability that the color of a piece of candy selected at random belongs to A_i, $i = 1, 2, \ldots, 5$. Test the null hypothesis

$$H_0: \ p_1 = 0.4, \ p_2 = 0.2, \ p_3 = 0.2, \ p_4 = 0.1, \ p_5 = 0.1$$

using a random sample of $n = 580$ pieces of candy whose colors yielded the respective frequencies 224, 119, 130, 48, and 59. You may select the significance level or give an approximate p-value.

8.5-3 In the Michigan Daily Lottery, each weekday a three-digit integer is generated one digit at a time. Let p_i denote the probability of generating digit i, $i = 0, 1, \ldots, 9$. Use the following 50 digits to test $H_0: p_0 = p_1 = \cdots = p_9 = 1/10$. Let $\alpha = 0.05$.

1	6	9	9	3	8	5	0	6	7
4	7	5	9	4	6	5	6	4	4
4	8	0	9	3	2	1	5	4	5
7	3	2	1	4	6	7	1	3	4
4	8	8	6	1	6	1	2	8	8

8.5-4 In a large bin of crocus bulbs it is claimed that 1/4 will produce yellow crocuses, 1/4 will produce white crocuses, and 1/2 will produce purple crocuses. If 40 bulbs produced 6 yellow, 7 white, and 27 purple crocuses, would the claim be rejected at the $\alpha = 0.05$ significance level?

8.5-5 In a biology laboratory the mating of two red-eyed fruit flies yielded $n = 432$ offspring, among which 254 were red-eyed, 69 were brown-eyed, 87 were scarlet-eyed, and 22 were white-eyed. Use these data to test, with $\alpha = 0.05$, the hypothesis that the ratio among the offspring would be 9:3:3:1, respectively.

8.5-6 In a biology laboratory students test the Mendelian theory of inheritance using corn. The Mendelian theory of inheritance claims that frequencies of the four categories smooth and yellow, wrinkled and yellow, smooth and purple, and wrinkled and purple will occur in the ratio 9:3:3:1. If a student counted 124, 30, 43, and 11, respectively, for these four categories, would these data support the Mendelian theory? Let $\alpha = 0.05$.

8.5-7 Let X equal the number of female children in a three-child family. We shall use a chi-square goodness of fit statistic to test the null hypothesis that the distribution of X is $b(3, 0.5)$.

   (a) Define the test statistic and critical region using an $\alpha = 0.05$ significance level.
   (b) Among students who were taking statistics, 52 came from families with 3 children. For these families, $x = 0, 1, 2$, and 3 for 5, 17, 24, and 6 families, respectively. Calculate the value of the test statistic and state your conclusion, considering how the sample was selected.

8.5-8 It has been claimed that, when spinning a penny minted in 1999 or earlier, the probability of observing heads in $p = 0.30$. Three students got together and they would each spin a penny and record the number of heads out of the three spins, say X. They repeated this $n = 200$ times, observing 0, 1, 2, and 3 heads 57, 95, 38, and 10 times, respectively. Use these data to test the hypotheses that X is $b(3, 0.30)$. Give limits for the p-value of this test. In addition, out of the 600 spins, calculate the number of heads occurring and then a 95% confidence interval for p.

8.5-9 In Example 2.6-3, it is claimed that X has a Poisson distribution and 100 observations of X are given in Table 2.6-1. Use a chi-square goodness of fit statistic to test whether this is true. Since $\bar{x} = 5.59$, let the estimate of λ be 5.6 so that you can use the Poisson probability table in the Appendix. Let $A_1 = \{x \leq 1\}$ and $A_{11} = \{x \geq 11\}$. Use an $\alpha = 0.05$ significance level.

8.5-10 In Exercise 2.6-12, we are asked whether the data look like observations of a Poisson random variable with mean $\lambda = 3$. Use a chi-square goodness of fit test to answer this question. Use $\alpha = 0.05$.

8.5-11 While testing a used tape for bad records, a computer operator counted the number of flaws per 100 feet of tape. Let X equal this random variable. Test the null hypothesis that X has a Poisson distribution with a mean of $\lambda = 2.4$ given that 40 observations of X yielded 5 zeros, 7 ones, 12 two, 9 threes, 5 fours, 1 five, and 1 six. Let $\alpha = 0.05$.
HINT: Combine five and six into one set; that is, the last set would be all x values ≥ 5.

8.5-12 Let X equal the distance between bad records on a used computer tape. Test the hypothesis that the distribution of X is exponential using the following 90 observations of X and 10 classes of equal probability. Use $\overline{x} = 42.4$ as an estimate of θ. Let $\alpha = 0.05$.

30	79	38	47	22	52	36	36	7	57
3	22	30	14	8	32	15	21	12	12
6	67	6	7	35	78	28	74	5	9
37	1	3	3	44	160	50	27	61	15
39	44	130	18	6	1	32	116	23	12
58	101	68	53	58	21	21	7	79	41
80	33	71	81	17	10	13	49	21	56
107	21	17	64	14	36	26	1	54	207
64	238	25	51	82	8	2	3	43	87

8.5-13 A sample of 100 2.2K ohms resistors was tested, yielding the following measurements:

2.17	2.18	2.17	2.19	2.23	2.18	2.16	2.23	2.29	2.20
2.24	2.13	2.18	2.22	2.21	2.22	2.23	2.18	2.23	2.25
2.24	2.20	2.21	2.20	2.25	2.15	2.20	2.25	2.16	2.19
2.20	2.18	2.18	2.19	2.22	2.19	2.20	2.19	2.22	2.21
2.22	2.18	2.18	2.28	2.19	2.23	2.21	2.19	2.21	2.21
2.19	2.18	2.21	2.17	2.20	2.18	2.18	2.21	2.22	2.18
2.22	2.23	2.19	2.23	2.18	2.19	2.18	2.21	2.18	2.15
2.20	2.23	2.20	2.20	2.23	2.20	2.19	2.22	2.17	2.20
2.20	2.17	2.19	2.19	2.25	2.19	2.19	2.20	2.20	2.19
2.21	2.14	2.24	2.21	2.19	2.23	2.18	2.22	2.23	2.19

   (a) Group these data into 9 classes of equal length with the class boundaries for the first class being 2.125–2.145.
   (b) Use a chi-square goodness of fit statistic to test whether these are observations of a normally distributed random variable, after grouping the first two classes and the last two classes to avoid small expected numbers. Estimate μ and σ with the mean and standard deviation of the grouped data. Let $\alpha = 0.05$.

8.5-14 Let X equal the amount of butterfat (in pounds) produced by 90 cows during a 305-day milk production period following their first calf. Test the hypothesis that the distribution of X is $N(\mu, \sigma^2)$ using $k = 10$ classes of equal probability. You may use $\overline{x} = 511.633$ and $s_x = 87.576$ as estimates of μ and σ, respectively.

486	537	513	583	453	510	570	500	458	555
618	327	350	643	500	497	421	505	637	599
392	574	492	635	460	696	593	422	499	524
539	339	472	427	532	470	417	437	388	481
537	489	418	434	466	464	544	475	608	444
573	611	586	613	645	540	494	532	691	478
513	583	457	612	628	516	452	501	453	643
541	439	627	619	617	394	607	502	395	470
531	526	496	561	491	380	345	274	672	509

8.6 CONTINGENCY TABLES

In this section we demonstrate the flexibility of the chi-square test. We first look at a method for testing whether two or more multinomial distributions are equal, sometimes called a test for homogeneity. Then we consider a test for independence of attributes of classification. Both of these lead to a similar test statistic.

Suppose that each of two independent experiments can end in one of the k mutually exclusive and exhaustive events $A_1, A_2, \ldots, A_k$. Let

$$p_{ij} = P(A_i), \qquad i = 1, 2, \ldots, k, \qquad j = 1, 2.$$

That is, $p_{11}, p_{21}, \ldots, p_{k1}$ are the probabilities of the events in the first experiment and $p_{12}, p_{22}, \ldots, p_{k2}$ are those associated with the second experiment. Let the experiments be repeated n_1 and n_2 independent times, respectively. Also let $Y_{11}, Y_{21}, \ldots, Y_{k1}$ be the frequencies of $A_1, A_2, \ldots, A_k$ associated with the n_1 independent trials of the first experiment. Similarly, let $Y_{12}, Y_{22}, \ldots, Y_{k2}$ be the respective frequencies associated with the n_2 trials of the second experiment. Of course, $\sum_{i=1}^{k} Y_{ij} = n_j$, $j = 1, 2$. From the sampling distribution theory corresponding to the basic chi-square test, we know that each of

$$\sum_{i=1}^{k} \frac{(Y_{ij} - n_j p_{ij})^2}{n_j p_{ij}}, \qquad j = 1, 2,$$

has an approximate chi-square distribution with $k-1$ degrees of freedom. Since the two experiments are independent (and thus the two chi-square statistics are independent), the sum

$$\sum_{j=1}^{2} \sum_{i=1}^{k} \frac{(Y_{ij} - n_j p_{ij})^2}{n_j p_{ij}}$$

is approximately chi-square with $k - 1 + k - 1 = 2k - 2$ degrees of freedom.

Usually, the p_{ij}, $i = 1, 2, \ldots, k$, $j = 1, 2$, are unknown, and frequently we wish to test the hypothesis

$$H_0: p_{11} = p_{12}, \ p_{21} = p_{22}, \ldots, \ p_{k1} = p_{k2};$$

that is, the hypothesis that the corresponding probabilities associated with the two independent experiments are equal. Under H_0, we can estimate the unknown

$$p_{i1} = p_{i2}, \qquad i = 1, 2, \ldots, k,$$

by using the relative frequency $(Y_{i1} + Y_{i2})/(n_1 + n_2)$, $i = 1, 2, \ldots, k$. That is, if H_0 is true, we can say that the two experiments are actually parts of a large one in which $Y_{i1} + Y_{i2}$ is the frequency of the event A_i, $i = 1, 2, \ldots, k$. Note that we only have to estimate the $k - 1$ probabilities $p_{i1} = p_{i2}$ using

$$\frac{Y_{i1} + Y_{i2}}{n_1 + n_2}, \qquad i = 1, 2, \ldots, k - 1,$$

since the sum of the k probabilities must equal one. That is, the estimator of $p_{k1} = p_{k2}$ is

$$1 - \frac{Y_{11} + Y_{12}}{n_1 + n_2} - \cdots - \frac{Y_{k-1,1} + Y_{k-1,2}}{n_1 + n_2} = \frac{Y_{k1} + Y_{k2}}{n_1 + n_2}.$$

Substituting these estimators, we have that

$$Q = \sum_{j=1}^{2} \sum_{i=1}^{k} \frac{[Y_{ij} - n_j(Y_{i1} + Y_{i2})/(n_1 + n_2)]^2}{n_j(Y_{i1} + Y_{i2})/(n_1 + n_2)}$$

has an approximate chi-square distribution with $2k - 2 - (k - 1) = k - 1$ degrees of freedom. Here $k - 1$ is subtracted from $2k - 2$ because that is the number of estimated parameters. The critical region for testing H_0 is of the form

$$q \geq \chi_\alpha^2(k-1).$$

EXAMPLE 8.6-1   To test two methods of instruction, 50 students are selected at random from each of two groups. At the end of the instruction period each student is assigned a grade (A, B, C, D, or F) by an evaluating team. The data are recorded as follows:

	Grade					
	A	B	C	D	F	Totals
Group I	8	13	16	10	3	50
Group II	4	9	14	16	7	50

Accordingly, if the hypothesis H_0 that the corresponding probabilities are equal is true, the respective estimates of the probabilities are

$$\frac{8+4}{100} = 0.12, \ 0.22, \ 0.30, \ 0.26, \ \frac{3+7}{100} = 0.10.$$

Thus the estimates of $n_1 p_{i1} = n_2 p_{i2}$ are 6, 11, 15, 13, 5, respectively. Hence the computed value of Q is

$$q = \frac{(8-6)^2}{6} + \frac{(13-11)^2}{11} + \frac{(16-15)^2}{15} + \frac{(10-13)^2}{13} + \frac{(3-5)^2}{5}$$

$$+ \frac{(4-6)^2}{6} + \frac{(9-11)^2}{11} + \frac{(14-15)^2}{15} + \frac{(16-13)^2}{13} + \frac{(7-5)^2}{5}$$

$$= \frac{4}{6} + \frac{4}{11} + \frac{1}{15} + \frac{9}{13} + \frac{4}{5} + \frac{4}{6} + \frac{4}{11} + \frac{1}{15} + \frac{9}{13} + \frac{4}{5} = 5.18.$$

Now, under H_0, Q has an approximate chi-square distribution with $k - 1 = 4$ degrees of freedom so the $\alpha = 0.05$ critical region is $q \geq 9.488 = \chi^2_{0.05}(4)$. Here $q = 5.18 < 9.488$, and hence H_0 is not rejected at the 5% significance level. Furthermore, the p-value for $q = 5.18$ is 0.268, which is greater than most significance levels. Thus with these data, we cannot say there is a difference between the two methods of instruction. ◀

It is fairly obvious how this procedure can be extended to testing the equality of h independent multinomial distributions. That is, let

$$p_{ij} = P(A_i), \qquad i = 1, 2, \ldots, k, \qquad j = 1, 2, \ldots, h,$$

and test

$$H_0: \ p_{i1} = p_{i2} = \cdots = p_{ih} = p_i, \qquad i = 1, 2, \ldots, k.$$

Repeat the jth experiment n_j independent times and let $Y_{1j}, Y_{2j}, \ldots, Y_{kj}$ denote the frequencies of the respective events $A_1, A_2, \ldots, A_k$. Now

$$Q = \sum_{j=1}^{h} \sum_{i=1}^{k} \frac{(Y_{ij} - n_j p_{ij})^2}{n_j p_{ij}}$$

has an approximate chi-square distribution with $h(k - 1)$ degrees of freedom. Under H_0, we must estimate $k - 1$ probabilities using

$$\widehat{p}_i = \frac{\displaystyle\sum_{j=1}^{h} Y_{ij}}{\displaystyle\sum_{j=1}^{h} n_j}, \qquad j = 1, 2, \ldots, k - 1,$$

because the estimate of p_k follows from $\widehat{p}_k = 1 - \widehat{p}_1 - \widehat{p}_2 - \cdots - \widehat{p}_{k-1}$. We use these estimates to obtain

$$Q = \sum_{j=1}^{h} \sum_{i=1}^{k} \frac{(Y_{ij} - n_j \widehat{p}_i)^2}{n_j \widehat{p}_i},$$

which has an approximate chi-square distribution with its degrees of freedom given by $h(k - 1) - (k - 1) = (h - 1)(k - 1)$.

Let us see how we can use the above procedures to test the equality of two or more independent distributions that are not necessarily multinomial. Suppose first that we are given random variables U and V with distribution functions $F(u)$ and $G(v)$, respectively. It is sometimes of interest to test the hypothesis $H_0: F(x) = G(x)$ for all x. We have previously considered tests of $\mu_U = \mu_V$, $\sigma_U^2 = \sigma_V^2$, and the two-sample Wilcoxon test. Now we shall only assume that the distributions are independent and of the continuous types.

We are interested in testing the hypothesis H_0: $F(x) = G(x)$ for all x. This hypothesis will be replaced by another one. Partition the real line into k mutually disjoint sets $A_1, A_2, \ldots, A_k$. Let

$$p_{i1} = P(U \in A_i), \qquad i = 1, 2, \ldots, k,$$

and

$$p_{i2} = P(V \in A_i), \qquad i = 1, 2, \ldots, k.$$

We observe that if $F(x) = G(x)$ for all x, then $p_{i1} = p_{i2}$, $i = 1, 2, \ldots, k$. We replace the hypothesis H_0: $F(x) = G(x)$ with the less restrictive hypothesis H_0': $p_{i1} = p_{i2}$, $i = 1, 2, \ldots, k$. That is, we are now essentially interested in testing the equality of two multinomial distributions.

Let n_1 and n_2 denote the number of independent observations of U and V, respectively. For $i = 1, 2, \ldots, k$, let Y_{ij} denote the number of these observations of U and V, $j = 1, 2$, respectively, that fall into a set A_i. At this point, we proceed to make the test of H_0' as described earlier. Of course, if H_0' is rejected at the (approximate) significance level α, then H_0 is rejected with the same probability. However, if H_0' is true, H_0 is not necessarily true. Thus if H_0' is accepted, it is probably better to say that we do not reject H_0 than to say H_0 is accepted.

In applications, the question of how to select $A_1, A_2, \ldots, A_k$ is frequently raised. Obviously, there is not a single choice for k nor the dividing marks of the partition. But it is interesting to observe that the combined sample can be used in this selection without upsetting the approximate distribution of Q. For example, suppose that $n_1 = n_2 = 20$. We could easily select the dividing marks of the partition so that $k = 4$ and one fourth of the combined sample falls in each of the four sets.

EXAMPLE 8.6-2   Select, at random, 20 cars of each of two comparable major-brand models. All 40 cars are submitted to accelerated life testing; that is, they are driven many miles over very poor roads in a short time and their failure times are recorded (in weeks):

Brand U:	25	31	20	42	39	19	35	36	44	26
	38	31	29	41	43	36	28	31	25	38
Brand V:	28	17	33	25	31	21	16	19	31	27
	23	19	25	22	29	32	24	20	34	26

If we use 23.5, 28.5, and 34.5 as dividing marks, we note that exactly one fourth of the 40 cars fall in each of the resulting four sets. Thus the data can be summarized as follows:

	A_1	A_2	A_3	A_4	Totals
Brand U	2	4	4	10	20
Brand V	8	6	6	0	20

The estimate of each p_i is $10/40 = 1/4$, which multiplied by $n_j = 20$ gives 5. Hence the computed Q is

$$q = \frac{(2-5)^2}{5} + \frac{(4-5)^2}{5} + \frac{(4-5)^2}{5} + \frac{(10-5)^2}{5} + \frac{(8-5)^2}{5}$$
$$+ \frac{(6-5)^2}{5} + \frac{(6-5)^2}{5} + \frac{(0-5)^2}{5}$$
$$= \frac{72}{5} = 14.4 > 7.815 = \chi^2_{0.05}(3).$$

Also the p-value is 0.0028. Hence it seems that the two brands of cars have different distributions for the length of life under accelerated life testing.   ◀

Again it should be clear how this can be extended to more than two distributions, and this extension will be illustrated in the exercises.

Now let us suppose that a random experiment results in an outcome that can be classified by two different attributes, such as height and weight. Assume that the first attribute is assigned to one and only one of k mutually exclusive and exhaustive events, say $A_1, A_2, \ldots, A_k$, and the second attribute falls in one and only one of h mutually exclusive and exhaustive events, say $B_1, B_2, \ldots, B_h$. Let the probability of $A_i \cap B_j$ be defined by

$$p_{ij} = P(A_i \cap B_j), \qquad i = 1, 2, \ldots, k, \qquad j = 1, 2, \ldots, h.$$

The random experiment is to be repeated n independent times, and Y_{ij} will denote the frequency of the event $A_i \cap B_j$. Since there are kh such events as $A_i \cap B_j$, the random variable

$$Q_{kh-1} = \sum_{j=1}^{h} \sum_{i=1}^{k} \frac{(Y_{ij} - np_{ij})^2}{np_{ij}}$$

has an approximate chi-square distribution with $kh - 1$ degrees of freedom, provided n is large.

Suppose that we wish to test the hypothesis of the independence of the A and B attributes, namely,

$$H_0: P(A_i \cap B_j) = P(A_i)P(B_j), \qquad i = 1, 2, \ldots, k, \qquad j = 1, 2, \ldots, h.$$

Let us denote $P(A_i)$ by $p_{i\cdot}$, and $P(B_j)$ by $p_{\cdot j}$; that is,

$$p_{i\cdot} = \sum_{j=1}^{h} p_{ij} = P(A_i), \qquad \text{and} \qquad p_{\cdot j} = \sum_{i=1}^{k} p_{ij} = P(B_j),$$

and, of course,

$$1 = \sum_{j=1}^{h}\sum_{i=1}^{k} p_{ij} = \sum_{j=1}^{h} p_{\cdot j} = \sum_{i=1}^{k} p_{i\cdot}.$$

Then the hypothesis can be formulated as

$$H_0: \quad p_{ij} = p_{i\cdot}p_{\cdot j}, \qquad i = 1, 2, \ldots, k, \qquad j = 1, 2, \ldots, h.$$

To test H_0, we can use Q_{kh-1} with p_{ij} replaced by $p_{i\cdot}p_{\cdot j}$. But if $p_{i\cdot}$, $i = 1, 2, \ldots, k$, and $p_{\cdot j}$, $j = 1, 2, \ldots, h$, are unknown, as they usually are in the applications, we cannot compute Q_{kh-1} once the frequencies are observed. In such a case we estimate these unknown parameters by

$$\widehat{p}_{i\cdot} = \frac{y_{i\cdot}}{n}, \qquad \text{where} \quad y_{i\cdot} = \sum_{j=1}^{h} y_{ij}$$

is the observed frequency of A_i, $i = 1, 2, \ldots, k$; and

$$\widehat{p}_{\cdot j} = \frac{y_{\cdot j}}{n}, \qquad \text{where} \quad y_{\cdot j} = \sum_{i=1}^{k} y_{ij}$$

is the observed frequency of B_j, $j = 1, 2, \ldots, h$. Since $\sum_{i=1}^{k} p_{i\cdot} = \sum_{j=1}^{h} p_{\cdot j} = 1$, we actually estimate only $k - 1 + h - 1 = k + h - 2$ parameters. So if these estimates are used in Q_{kh-1}, with $p_{ij} = p_{i\cdot}p_{\cdot j}$, then, according to the rule stated earlier, the random variable

$$Q = \sum_{j=1}^{h}\sum_{i=1}^{k} \frac{[Y_{ij} - n(Y_{i\cdot}/n)(Y_{\cdot j}/n)]^2}{n(Y_{i\cdot}/n)(Y_{\cdot j}/n)}$$

has an approximate chi-square distribution with $kh - 1 - (k + h - 2) = (k-1)(h-1)$ degrees of freedom, provided that H_0 is true. The hypothesis H_0 is rejected if the computed value of this statistic exceeds $\chi_\alpha^2[(k-1)(h-1)]$.

EXAMPLE 8.6-3   A random sample of 400 undergraduate students at the University of Iowa was taken. The students in the sample were classified according to the college in which they were enrolled and according to their gender. These results are recorded in Table 8.6-1; this table is called a $k \times h$ **contingency table**, where here $k = 2$ and $h = 5$. (Do not be concerned about the numbers in parentheses at this point.) Incidentally, these numbers do actually reflect the composition of the undergraduate colleges at Iowa, but they were modified a little to make the computations easier in this first example.

	College					
Gender	Business	Engineering	Liberal Arts	Nursing	Pharmacy	Totals
Male	21 (16.625)	16 (9.5)	145 (152)	2 (7.125)	6 (4.75)	190
Female	14 (18.375)	4 (10.5)	175 (168)	13 (7.875)	4 (5.25)	210
Totals	35	20	320	15	10	400

Table 8.6-1: Undergraduates at the University of Iowa

We desire to test the null hypothesis H_0: $p_{ij} = p_{i\cdot} p_{\cdot j}$, $i = 1, 2$ and $j = 1, 2, 3, 4, 5$, that the college in which a student enrolls is independent of the gender of that student. Under H_0, estimates of the probabilities are

$$\widehat{p}_{1\cdot} = \frac{190}{400} = 0.475 \qquad \text{and} \qquad \widehat{p}_{2\cdot} = \frac{210}{400} = 0.525$$

and

$$\widehat{p}_{\cdot 1} = \frac{35}{400} = 0.0875, \quad \widehat{p}_{\cdot 2} = 0.05, \quad \widehat{p}_{\cdot 3} = 0.8, \quad \widehat{p}_{\cdot 4} = 0.0375, \quad \widehat{p}_{\cdot 5} = 0.025.$$

The expected numbers $n(y_{i\cdot}/n)(y_{\cdot j}/n)$ are computed as follows:

$$400(0.475)(0.0875) = 16.625,$$
$$400(0.525)(0.0875) = 18.375,$$
$$400(0.475)(0.05) = 9.5,$$

and so on. These are the values recorded in the parentheses in Table 8.6-1. The computed chi-square statistic is

$$q = \frac{(21 - 16.625)^2}{16.625} + \frac{(14 - 18.375)^2}{18.375} + \cdots + \frac{(4 - 5.25)^2}{5.25}$$
$$= 1.15 + 1.04 + 4.45 + 4.02 + 0.32 + 0.29 + 3.69$$
$$+ 3.34 + 0.33 + 0.30 = 18.93.$$

Since the number of degrees of freedom equals $(k - 1)(h - 1) = 4$, this $q = 18.93 > 13.28 = \chi_{0.01}^2(4)$, and we reject H_0 at the $\alpha = 0.01$ significance level. Moreover, since the first two terms of q come from the business college, the next two from engineering, and so on, it is clear that the enrollments in engineering and nursing are more highly dependent on gender than in the other colleges because they have contributed the most to the value of the chi-square statistic. It is also interesting to note that one expected number is less than five, namely 4.75. However, as the associated term in q does not contribute an unusual amount to the chi-square value, it does not concern us. ◀

It is fairly obvious how to extend the testing procedure above to more than two attributes. For example, if the third attribute falls in one and only one of m mutually exclusive and exhaustive events, say $C_1, C_2, \ldots, C_m$, then we test the independence of the three attributes by using

$$Q = \sum_{r=1}^{m} \sum_{j=1}^{h} \sum_{i=1}^{k} \frac{[Y_{ijr} - n(Y_{i..}/n)(Y_{.j.}/n)(Y_{..r}/n)]^2}{n(Y_{i..}/n)(Y_{.j.}/n)(Y_{..r}/n)},$$

where Y_{ijr}, $Y_{i..}$, $Y_{.j.}$, and $Y_{..r}$ are the respective observed frequencies of the events $A_i \cap B_j \cap C_r$, A_i, B_j, and C_r in n independent trials of the experiment. If n is large and if the three attributes are independent, then Q has an approximate chi-square distribution with $khm - 1 - (k-1) - (h-1) - (m-1) = khm - k - h - m + 2$ degrees of freedom.

Rather than explore this extension further, it is more instructive to note some interesting uses of contingency tables.

EXAMPLE 8.6-4 Say we observed 30 values $x_1, x_2, \ldots, x_{30}$ that are claimed to be the values of a random sample. That is, the corresponding random variables $X_1, X_2, \ldots, X_{30}$ were supposed to be mutually independent and to have the same distribution. Say, however, by looking at the 30 values we detect an upward trend that indicates there might have been some dependence and/or the random variables did not actually have the same distribution. One simple way to test if they could be thought of as being observed values of a random sample is the following. Mark each x high (H) or low (L) depending on whether it is above or below the sample median. Then divide the x values into three groups: $x_1, \ldots, x_{10}$; $x_{11}, \ldots, x_{20}$; $x_{21}, \ldots, x_{30}$. Certainly if the observations are those of a random sample we would expect five H's and five L's in each group. That is, the attribute classified as H or L should be independent of the group number. The summary of these data provides a 3×2 contingency table. For example, say the 30 values are

5.6	8.2	7.8	4.8	5.5	8.1	6.7	7.7	9.3	6.9
8.2	10.1	7.5	6.9	11.1	9.2	8.7	10.3	10.7	10.0
9.2	11.6	10.3	11.7	9.9	10.6	10.0	11.4	10.9	11.1

The median can be taken to be the average of the two middle observations in magnitude, namely, 9.2 and 9.3. Marking each item H or L after comparing it with this median, we obtain the following 3×2 contingency table.

Group	L	H	Totals
1	9	1	10
2	5	5	10
3	1	9	10
Totals	15	15	30

Here each $n(y_{i\cdot}/n)(y_{\cdot j}/n) = 30(10/30)(15/30) = 5$ so that the computed value of Q is

$$q = \frac{(9-5)^2}{5} + \frac{(1-5)^2}{5} + \frac{(5-5)^2}{5} + \frac{(5-5)^2}{5} + \frac{(1-5)^2}{5} + \frac{(9-5)^2}{5}$$

$$= 12.8 > 5.991 = \chi^2_{0.05}(2),$$

since here $(k-1)(h-1) = 2$ degrees of freedom. (The p-value is 0.0017.) Hence we reject the conjecture that these 30 values could be the items of a random sample. Obviously, modifications could be made to this scheme: dividing the sample into more (or less) than three groups and rating items differently, such as low (L), middle (M), and high (H). ◀

It cannot be emphasized enough that the chi-square statistic can be used fairly effectively in almost any situation in which there should be independence. For illustration, suppose that we have a group of workers who have essentially the same qualifications (training, experience, etc.). Many believe that salary and gender of the workers should be independent attributes; yet there have been several claims in special cases that there is a dependence–or discrimination–in calculations associated with such a problem.

EXAMPLE 8.6-5   Two groups of workers have the same qualifications for a particular type of work. Their experience in salaries is summarized by the following 2×5 contingency table, in which the upper bound of each salary range is not included in that listing.

Salary (Thousands of Dollars)						
Group	12–14	14–16	16–18	18–20	20–	Totals
1	6	11	16	14	13	60
2	5	9	8	6	2	30
Totals	11	20	24	20	15	90

To test if the group assignment and the salaries seem to be independent with these data at the $\alpha = 0.05$ significance level, we compute

$$q = \frac{[6 - 90(60/90)(11/90)]^2}{90(60/90)(11/90)} + \cdots + \frac{[2 - 90(30/90)(15/90)]^2}{90(30/90)(15/90)}$$

$$= 4.752 < 9.488 = \chi^2_{0.05}(4).$$

Also, the p-value is 0.313. Hence with these limited data, group assignment and salaries seem to be independent. ◀

Before turning to the exercises, note that we could have thought of the last two examples in this section as testing the equality of two or more multinomial distributions. In Example 8.6-4 the three groups define three binomial distributions; and in Example 8.6-5 the two groups define two multinomial distributions. What would have happened if we had used the computations outlined earlier in this section? It is interesting to note that we obtain exactly the same value of chi-square and in each case the number of degrees of freedom is equal to $(k-1)(h-1)$. Hence it makes no difference whether we think of it as a test of independence or a test of the equality of several multinomial distributions. Our advice is to use the terminology that seems most natural for the particular situations.

EXERCISES

8.6-1 We wish to test to see if two groups of nurses distribute their time in six different categories about the same way. That is, the hypothesis under consideration is H_0: $p_{i1} = p_{i2}$, $i = 1, 2, \ldots, 6$. To test this, nurses are observed at random throughout several days, each observation resulting in a mark in one of the six categories. The summary is given by the following frequency table:

| | Category | | | | | | |
	1	2	3	4	5	6	Totals
Group I	95	36	71	21	45	32	300
Group II	53	26	43	18	32	28	200

Use a chi-square test with $\alpha = 0.05$.

8.6-2 Suppose that a third group of nurses was observed along with groups I and II of Exercise 8.6-1, resulting in the respective frequencies 130, 75, 136, 33, 61, and 65. Test H_0: $p_{i1} = p_{i2} = p_{i3}$, $i = 1, 2, \ldots, 6$, at the $\alpha = 0.025$ significance level.

8.6-3 Each of two comparable classes of 15 students responded to two different methods of instructions with the following scores on a standardized test.

Class U: 91 42 39 62 55 82 67 44
          51 77 61 52 76 41 59

Class V: 80 71 55 67 61 93 49 78
          57 88 79 81 63 51 75

Use a chi-square test with $\alpha = 0.05$ to test the equality of the distributions of test scores by dividing the combined sample into three equal parts (low, middle, high).

8.6-4 Suppose that a third class (W) of 15 students was observed along with classes U and V of Exercise 8.6-3, resulting in scores of

91 73 67 83 59 98 87 69
78 80 65 94 82 74 85

Again use a chi-square test with $\alpha = 0.05$ to test the equality of the three distributions by dividing the combined sample into three equal parts.

8.6-5 In a contingency table, 1015 individuals are classified by gender and by whether they favor, oppose, or have no opinion on a complete ban on smoking in public places. Test the null hypothesis that gender and opinion on smoking in public places are independent. Give the approximate p-value of this test.

Smoking in Public Places				
Gender	Favor	Oppose	No Opinion	Totals
Male	262	231	10	503
Female	302	205	5	512
Totals	564	436	15	1015

8.6-6 A random survey of 100 students asked each student to select the most preferred form of recreational activity from 5 choices. Test whether the choice is independent of the gender of the respondent. Approximate the p-value of the test. Would we reject at $\alpha = 0.05$?

Recreational Choice						
Gender	Basketball	Baseball Softball	Swimming	Jogging Running	Tennis	Totals
Male	21	5	9	12	13	60
Female	9	3	1	15	12	40
Totals	30	8	10	27	25	100

8.6-7 A random sample of 100 students were classified by gender and by the "instrument" that they "played." Test whether the selection of instrument is independent of the gender of the respondent. Approximate the p-value of this test.

Instrument						
Gender	Piano	Woodwind	Brass	String	Vocal	Totals
Male	4	11	15	6	9	45
Female	7	18	6	6	18	55
Totals	11	29	21	12	27	100

8.6-8 A student who uses the college recreational facilities was interested in whether there is a difference between the facilities used by men and women. Test the null hypothesis that facility and gender are independent attributes using the following data. Use $\alpha = 0.05$.

Facility		
Gender	Racquetball Court	Track	Totals
Male	51	30	81
Female	43	48	91
Totals	94	78	172

8.6-9 A survey of high school girls classified them by two attributes of classification: whether or not they participated in sports and whether or not they had (an) older brother(s). Test the null hypothesis that these two attributes of classification are independent.

	Participated in Sports		
Older Brother(s)	**Yes**	**No**	**Totals**
Yes	12	8	20
No	13	27	40
Totals	25	35	60

Approximate the p-value of this test. Do we reject the null hypothesis if $\alpha = 0.05$?

8.6-10 A random sample of 50 women who were tested for cholesterol were classified according to age and cholesterol level and grouped in the following contingency table.

	Cholesterol Level			
Age	**<180**	**180–210**	**>210**	**Totals**
< 50	5	11	9	25
≥ 50	4	3	18	25
Totals	9	14	27	50

Test the null hypothesis H_0: age and cholesterol level are independent attributes of classification. What is your conclusion if $\alpha = 0.01$?

8.6-11 Although high school grades and testing scores, such as SAT or ACT, can be used to predict first-year college grade-point average (GPA), many educators claim that a more important factor influencing that GPA is the living conditions of students. In particular, it is claimed that the roommate of the student will have a great influence on his or her grades. To test this, suppose we selected at random 200 students and classified each according to the following two attributes:

(a) Ranking of the student's roommate from 1 to 5, from a person who was difficult to live with and discouraged scholarship to one who was congenial but encouraged scholarship.

(b) The student's first-year GPA.

Say this gives the following 5 × 4 contingency table.

	Grade-Point Average				
Rank of Roommate	**Under 2.00**	**2.00–2.69**	**2.70–3.19**	**3.20–4.00**	**Totals**
1	8	9	10	4	31
2	5	11	15	11	42
3	6	7	20	14	47
4	3	5	22	23	53
5	1	3	11	12	27
Totals	23	35	78	64	200

Compute the chi-square statistic used to test the independence of the two attributes and compare it to the critical value associated with $\alpha = 0.05$.

8.6-12 (*How Americans Watch TV: A Nation of Grazers.*) A random sample of 395 people were classified by their age and by whether or not they change channels during programs when watching television. Use the following data to test the null hypothesis H_0: Changing channels and age are independent attributes.

(a) Give a significance level, α, for which H_0 is rejected. Why?

(b) Give a value of the significance level, α, for which H_0 is not rejected. Why?

	Age				
Change?	18–24	25–34	35–49	50–64	Totals
Yes	60	54	46	41	201
No	40	44	53	57	194
Totals	100	98	99	98	395

8.6-13 A study was conducted to determine the media credibility for reporting news. Those surveyed were asked to give their age, gender, education, and the most credible medium. Test whether

(a) media credibility and age are independent.

(b) media credibility and gender are independent.

(c) media credibility and education are independent.

(d) Give the approximate p-value for each test.

	Most Credible Medium			
Age	Newspaper	Television	Radio	Totals
Under 35	30	68	10	108
35–54	61	79	20	160
Over 54	98	43	21	162
Totals	189	190	51	430

	Most Credible Medium			
Gender	Newspaper	Television	Radio	Totals
Male	92	108	19	219
Female	97	81	32	210
Totals	189	189	51	429

	Most Credible Medium			
Education	Newspaper	Television	Radio	Totals
Grade School	45	22	6	73
High School	94	115	30	239
College	49	52	13	114
Totals	188	189	49	426

8.6-14 In a psychology experiment 140 students were divided into majors emphasizing left-hemisphere brain skills (e.g., philosophy, physics, and mathematics), and into majors emphasizing right-hemisphere skills (e.g., art, music, theater, and dance). They were also classified into one of three groups on the basis of hand posture (right noninverted, left inverted, left noninverted).

	LH	RH
RN	89	29
LI	5	4
LN	5	8

Do these data show sufficient evidence to reject the claim that the choice of college major is independent of hand posture? Let $\alpha = 0.025$.

8.6-15 A random sample of $n = 1362$ persons were classified according to the respondent's education level and whether the respondent was Protestant, Catholic, or Jewish. Use these data to test at an $\alpha = 0.05$ significance level the hypothesis that these attributes of classification are independent.

Education Level	Protestant	Catholic	Jewish
Less than high school	359	140	5
High school or junior college	462	200	17
Bachelor's degree	88	39	2
Graduate degree	37	10	3

8.6-16 In the Michigan Daily Lottery one three-digit integer is generated 6 days a week. Twenty weeks of lottery numbers have been classified by the day of the week and the magnitude of the number. At the 10% significance level, test whether these attributes of classification are independent.

Days	Magnitude of Number		Totals
	000–499	500–999	
Monday and Tuesday	22	18	40
Wednesday and Thursday	19	21	40
Friday and Saturday	13	27	40
Totals	54	66	120

8.7 ONE-FACTOR ANALYSIS OF VARIANCE

Frequently, experimenters want to compare more than two treatments: yields of several different corn hybrids, results due to three or more teaching techniques, or miles per gallon obtained from many different types of compact cars. Sometimes the different treatment distributions of the resulting observations are due to changing the level of a certain factor (e.g., different doses of a given drug). Thus the consideration of the equality of the different means of the various distributions comes under the analysis of a **one-factor experiment**.

In Section 8.3 we discussed how to compare two normal distributions. More generally, let us consider m normal distributions with unknown means $\mu_1, \mu_2, \ldots, \mu_m$, respectively, and an unknown but common variance σ^2. One inference that we wish to consider is a test of the equality of the m means, namely $H_0: \mu_1 = \mu_2 = \cdots = \mu_m = \mu$, μ unspecified, against all possible alternative hypotheses H_1. In order to test this hypothesis we shall take independent random samples from these distributions. Let $X_{i1}, X_{i2}, \ldots, X_{in_i}$ represent a random sample of size n_i from the normal distribution $N(\mu_i, \sigma^2)$, $i = 1, 2, \ldots, m$. In Table 8.7-1 we have indicated these random samples along with the row means (sample means) where, with $n = n_1 + n_2 + \cdots + n_m$,

$$\overline{X}_{..} = \frac{1}{n} \sum_{i=1}^{m} \sum_{j=1}^{n_i} X_{ij} \quad \text{and} \quad \overline{X}_{i.} = \frac{1}{n_i} \sum_{j=1}^{n_i} X_{ij}, \quad i = 1, 2, \ldots, m.$$

Table 8.7-1: One-factor Random Samples

					Means
X_1:	X_{11}	X_{12}	$\cdots$	X_{1n_1}	$\overline{X}_{1.}$
X_2:	X_{21}	X_{22}	$\cdots$	X_{2n_2}	$\overline{X}_{2.}$
$\vdots$	$\vdots$	$\vdots$	$\vdots$	$\vdots$	$\vdots$
X_m:	X_{m1}	X_{m2}	$\cdots$	X_{mn_m}	$\overline{X}_{m.}$
Grand Mean:					$\overline{X}_{..}$

The dot in the notation for the means, $\overline{X}_{..}$ and $\overline{X}_{i.}$, indicates the index over which the average is taken. Here $\overline{X}_{..}$ is an average taken over both indices while $\overline{X}_{i.}$ is just taken over the index j.

To determine a critical region for a test of H_0, we shall first partition the sum of squares associated with the variance of the combined samples into two parts. This sum of squares is given by

$$\text{SS(TO)} = \sum_{i=1}^{m} \sum_{j=1}^{n_i} (X_{ij} - \overline{X}_{..})^2$$

$$= \sum_{i=1}^{m} \sum_{j=1}^{n_i} (X_{ij} - \overline{X}_{i.} + \overline{X}_{i.} - \overline{X}_{..})^2$$

$$= \sum_{i=1}^{m} \sum_{j=1}^{n_i} (X_{ij} - \overline{X}_{i.})^2 + \sum_{i=1}^{m} \sum_{j=1}^{n_i} (\overline{X}_{i.} - \overline{X}_{..})^2$$

$$+ 2 \sum_{i=1}^{m} \sum_{j=1}^{n_i} (X_{ij} - \overline{X}_{i.})(\overline{X}_{i.} - \overline{X}_{..}).$$

The last term of the righthand member of this identity may be written as

$$2\sum_{i=1}^{m}\left[(\overline{X}_{i\cdot}-\overline{X}_{\cdot\cdot})\sum_{j=1}^{n_i}(X_{ij}-\overline{X}_{i\cdot})\right]=2\sum_{i=1}^{m}(\overline{X}_{i\cdot}-\overline{X}_{\cdot\cdot})(n_i\overline{X}_{i\cdot}-n_i\overline{X}_{i\cdot})=0,$$

and the preceding term may be written as

$$\sum_{i=1}^{m}\sum_{j=1}^{n_i}(\overline{X}_{i\cdot}-\overline{X}_{\cdot\cdot})^2=\sum_{i=1}^{m}n_i(\overline{X}_{i\cdot}-\overline{X}_{\cdot\cdot})^2.$$

Thus

$$\text{SS(TO)}=\sum_{i=1}^{m}\sum_{j=1}^{n_i}(X_{ij}-\overline{X}_{i\cdot})^2+\sum_{i=1}^{m}n_i(\overline{X}_{i\cdot}-\overline{X}_{\cdot\cdot})^2.$$

For notation let

$$\text{SS(TO)}=\sum_{i=1}^{m}\sum_{j=1}^{n_i}(X_{ij}-\overline{X}_{\cdot\cdot})^2,\text{ the total sum of squares.}$$

$$\text{SS(E)}=\sum_{i=1}^{m}\sum_{j=1}^{n_i}(X_{ij}-\overline{X}_{i\cdot})^2,\quad\begin{array}{l}\text{the sum of squares within treatments,}\\\text{groups, or classes, often called the error}\\\text{sum of squares.}\end{array}$$

$$\text{SS(T)}=\sum_{i=1}^{m}n_i(\overline{X}_{i\cdot}-\overline{X}_{\cdot\cdot})^2,\quad\begin{array}{l}\text{the sum of squares among the different}\\\text{treatments, groups, or classes, often called}\\\text{the between treatment sum of squares.}\end{array}$$

Thus

$$\text{SS(TO)}=\text{SS(E)}+\text{SS(T)}.$$

When H_0 is true, we may regard $X_{ij}, i = 1, 2, \ldots, m, j = 1, 2, \ldots, n_i$, as a random sample of size $n = n_1+n_2+\cdots+n_m$ from the normal distribution $N(\mu, \sigma^2)$. Then $\text{SS(TO)}/(n - 1)$ is an unbiased estimator of σ^2 because $\text{SS(TO)}/\sigma^2$ is $\chi^2(n-1)$ so that $E[\text{SS(TO)}/\sigma^2] = n - 1$ and $E[\text{SS(TO)}/(n - 1)] = \sigma^2$. An unbiased estimator of σ^2 based only on the sample from the ith distribution is

$$W_i = \frac{\displaystyle\sum_{j=1}^{n_i}(X_{ij}-\overline{X}_{i\cdot})^2}{n_i-1}\qquad\text{for }i=1,2,\ldots,m,$$

because $(n_i - 1)W_i/\sigma^2$ is $\chi^2(n_i-1)$. Thus

$$E\left[\frac{(n_i - 1)W_i}{\sigma^2}\right] = n_i - 1,$$

and so

$$E(W_i) = \sigma^2, \qquad i = 1, 2, \ldots, m.$$

It follows that the sum of m of these independent chi-square random variables, namely

$$\sum_{i=1}^{m} \frac{(n_i - 1)W_i}{\sigma^2} = \frac{SS(E)}{\sigma^2},$$

is also chi-square with $(n_1 - 1) + (n_2 - 1) + \cdots + (n_m - 1) = n - m$ degrees of freedom. Hence $SS(E)/(n - m)$ is an unbiased estimator of σ^2. We now have that

$$\frac{SS(TO)}{\sigma^2} = \frac{SS(E)}{\sigma^2} + \frac{SS(T)}{\sigma^2},$$

where

$$\frac{SS(TO)}{\sigma^2} \text{ is } \chi^2(n-1) \text{ and } \frac{SS(E)}{\sigma^2} \text{ is } \chi^2(n-m).$$

Because $SS(T) \geq 0$, there is a theorem (see following remark) that states that $SS(E)$ and $SS(T)$ are independent and the distribution of $SS(T)/\sigma^2$ is $\chi^2(m-1)$.

Remark The sums of squares, $SS(T)$, $SS(E)$, and $SS(TO)$, are examples of **quadratic forms** in the variables X_{ij}, $i = 1, 2, \ldots, m$, $j = 1, 2, \ldots, n_i$. That is, each term in these sums of squares is of second degree in X_{ij}. Furthermore the coefficients of the variables are real numbers, so these sums of squares are called **real quadratic forms**. The following theorem, stated without proof, is used in this chapter. [For a proof, see Hogg, McKean, and Craig, *Introduction to Mathematical Statistics*, 6th ed. (Prentice Hall, 2005).] ∎

Theorem 8.7-1

Let $Q = Q_1 + Q_2 + \cdots + Q_k$, where $Q, Q_1, \ldots, Q_k$ are $k+1$ real quadratic forms in n mutually independent random variables normally distributed with the same variance σ^2. Let $Q/\sigma^2, Q_1/\sigma^2, \ldots, Q_{k-1}/\sigma^2$ have chi-square distributions with $r, r_1, \ldots, r_{k-1}$ degrees of freedom, respectively. If Q_k is nonnegative, then

(a) $Q_1, \ldots, Q_k$ are mutually independent, and hence,
(b) Q_k/σ^2 has a chi-square distribution with $r - (r_1 + \cdots + r_{k-1}) = r_k$ degrees of freedom.

Since under H_0, $SS(T)/\sigma^2$ is $\chi^2(m-1)$, we have $E[SS(T)/\sigma^2] = m-1$ and hence $E[SS(T)/(m-1)] = \sigma^2$. Now the estimator of σ^2, which is based on SS(E), namely $SS(E)/(n-m)$, is always unbiased whether H_0 is true or false. However, if the means $\mu_1, \mu_2, \ldots, \mu_m$ are not equal, the expected value of the estimator based on SS(T) will be greater than σ^2. To make this last statement clear, we have

$$E[SS(T)] = E\left[\sum_{i=1}^{m} n_i(\overline{X}_{i\cdot} - \overline{X}_{\cdot\cdot})^2\right] = E\left[\sum_{i=1}^{m} n_i\overline{X}_{i\cdot}^2 - n\overline{X}_{\cdot\cdot}^2\right]$$

$$= \sum_{i=1}^{m} n_i\{\text{Var}(\overline{X}_{i\cdot}) + [E(\overline{X}_{i\cdot})]^2\} - n\{\text{Var}(\overline{X}_{\cdot\cdot}) + [E(\overline{X}_{\cdot\cdot})]^2\}$$

$$= \sum_{i=1}^{m} n_i\left\{\frac{\sigma^2}{n_i} + \mu_i^2\right\} - n\left\{\frac{\sigma^2}{n} + \overline{\mu}^2\right\}$$

$$= (m-1)\sigma^2 + \sum_{i=1}^{m} n_i(\mu_i - \overline{\mu})^2,$$

where $\overline{\mu} = (1/n)\sum_{i=1}^{m} n_i\mu_i$. If $\mu_1 = \mu_2 = \cdots = \mu_m = \mu$,

$$E\left(\frac{SS(T)}{m-1}\right) = \sigma^2.$$

If the means are not all equal,

$$E\left[\frac{SS(T)}{m-1}\right] = \sigma^2 + \sum_{i=1}^{m} n_i\frac{(\mu_i - \overline{\mu})^2}{m-1} > \sigma^2.$$

Exercise 8.7-4 also illustrates the fact that the estimator using SS(T) is usually greater than that using SS(E) when H_0 is false.

We can base our test of H_0 on the ratio of $SS(T)/(m-1)$ and $SS(E)/(n-m)$, both of which are unbiased estimators of σ^2, provided that $H_0: \mu_1 = \mu_2 = \cdots = \mu_m$ is true so that, under H_0, the ratio would assume values near one. However, in the case that the means $\mu_1, \mu_2, \ldots, \mu_m$ begin to differ, this ratio tends to become large, since $E[SS(T)/(m-1)]$ gets larger. Under H_0, the ratio

$$\frac{SS(T)/(m-1)}{SS(E)/(n-m)} = \frac{[SS(T)/\sigma^2]/(m-1)}{[SS(E)/\sigma^2]/(n-m)} = F$$

has an F distribution with $m-1$ and $n-m$ degrees of freedom because $SS(T)/\sigma^2$ and $SS(E)/\sigma^2$ are independent chi-square variables. We would reject H_0 if the observed value of F is too large because this would indicate that we have a relatively large SS(T), suggesting that the means are unequal. Thus the critical region is of the form $F \geq F_\alpha(m-1, n-m)$.

	Table 8.7-2: Analysis-of-Variance Table			
Source	**Sum of Squares (SS)**	**Degrees of Freedom**	**Mean Square (MS)**	**F Ratio**
Treatment	SS(T)	$m - 1$	$\mathrm{MS(T)} = \dfrac{\mathrm{SS(T)}}{m-1}$	$\dfrac{\mathrm{MS(T)}}{\mathrm{MS(E)}}$
Error	SS(E)	$n - m$	$\mathrm{MS(E)} = \dfrac{\mathrm{SS(E)}}{n-m}$	
Total	SS(TO)	$n - 1$		

The information for tests of the equality of several means is often summarized in an **analysis-of-variance table** or **ANOVA** table like that given in Table 8.7-2, where the mean square (MS) is the sum of squares (SS) divided by its degrees of freedom.

EXAMPLE 8.7-1 Let X_1, X_2, X_3, X_4 be independent random variables that have normal distributions $N(\mu_i, \sigma^2)$, $i = 1, 2, 3, 4$. We shall test

$$H_0: \mu_1 = \mu_2 = \mu_3 = \mu_4 = \mu$$

against all alternatives based on a random sample of size $n_i = 3$ from each of the four distributions. A critical region of size $\alpha = 0.05$ is given by

$$F = \frac{\mathrm{SS(T)}/(4-1)}{\mathrm{SS(E)}/(12-4)} \geq 4.07 = F_{0.05}(3, 8).$$

The observed data are given in Table 8.7-3. (Clearly, these data are not observations from normal distributions. They were selected to illustrate the calculations.)

For these data, the calculated SS(TO), SS(E), and SS(T) are

$$\mathrm{SS(TO)} = (13 - 11)^2 + (8 - 11)^2 + \cdots + (15 - 11)^2 + (10 - 11)^2 = 80;$$

$$\mathrm{SS(E)} = (13 - 10)^2 + (8 - 10)^2 + \cdots + (15 - 12)^2 + (10 - 12)^2 = 50;$$

$$\mathrm{SS(T)} = 3[(10 - 11)^2 + (13 - 11)^2 + (9 - 11)^2 + (12 - 11)^2] = 30.$$

Note that since SS(TO) = SS(E) + SS(T), only two of the three values need to be calculated from the data directly. Here the computed value of F is

$$\frac{30/3}{50/8} = 1.6 < 4.07,$$

and H_0 is not rejected. The p-value is the probability, under H_0, of observing an F that is at least as large as this observed F. It is often given by computer programs.

Table 8.7-3: Illustrative Data

	Observations			$\overline{X}_i.$
X_1:	13	8	9	10
X_2:	15	11	13	13
X_3:	8	12	7	9
X_4:	11	15	10	12
$\overline{X}..$				11

The information for this example is summarized in the ANOVA table, Table 8.7-4. Again we note that (here and elsewhere) the F statistic is the ratio of two appropriate mean squares. ◀

Table 8.7-4: ANOVA Table for Illustrative Data

Source	Sum of Squares (SS)	Degrees of Freedom	Mean Square (MS)	F Ratio	p-value
Treatment	30	3	30/3	1.6	0.264
Error	50	8	50/8		
Total	80	11			

Formulas that sometimes simplify the calculations of SS(TO), SS(T), and SS(E) are

$$\text{SS(TO)} = \sum_{i=1}^{m} \sum_{j=1}^{n_i} X_{ij}^2 - \frac{1}{n}\left[\sum_{i=1}^{m}\sum_{j=1}^{n_i} X_{ij}\right]^2 ,$$

$$\text{SS(T)} = \sum_{i=1}^{m} \frac{1}{n_i}\left[\sum_{j=1}^{n_i} X_{ij}\right]^2 - \frac{1}{n}\left[\sum_{i=1}^{m}\sum_{j=1}^{n_i} X_{ij}\right]^2 ,$$

and

$$\text{SS(E)} = \text{SS(TO)} - \text{SS(T)}.$$

It is interesting to note that in these formulas each square is divided by the number of observations in the sum being squared: X_{ij}^2 by one, $(\sum_{j=1}^{n_i} X_{ij})^2$ by n_i, and $(\sum_{i=1}^{m}\sum_{j=1}^{n_i} X_{ij})^2$ by n. These formulas are used in Example 8.7-2. Although these formulas are useful, you are encouraged to use appropriate statistical packages on a computer to aid you with these calculations.

If the sample sizes are all at least equal to 7, insight can be gained by plotting on the same figure box-and-whisker diagrams for each of the samples. This is also illustrated in Example 8.7-2.

EXAMPLE 8.7-2

A window that is manufactured for an automobile has five studs for attaching it. A company that manufactures these windows performs "pull out tests" to determine the force needed to pull a stud out of the window. Let X_i, $i = 1, 2, 3, 4, 5$, equal the force required at position i and assume that the distribution of X_i is $N(\mu_i, \sigma^2)$. We shall test the null hypothesis $H_0: \mu_1 = \mu_2 = \mu_3 = \mu_4 = \mu_5$ using 7 independent observations at each position. At an $\alpha = 0.01$ significance level, H_0 is rejected if the computed

$$F = \frac{\text{SS(T)}/(5 - 1)}{\text{SS(E)}/(35 - 5)} \geq 4.02 = F_{0.01}(4, 30).$$

The observed data along with certain sums are given in Table 8.7-5. For these data,

$$\text{SS(TO)} = 556{,}174 - \frac{1}{35}(4334)^2 = 19{,}500.97$$

$$\text{SS(T)} = \frac{1}{7}(645^2 + 721^2 + 970^2 + 1017^2 + 981^2)$$

$$- \frac{1}{35}(4334)^2 = 16{,}672.11$$

$$\text{SS(E)} = 19{,}500.97 - 16{,}672.11 = 2828.86.$$

Table 8.7-5: Pull-out Test Data

Observations							$\sum_{j=1}^{7} x_{ij}$	$\sum_{j=1}^{7} x_{ij}^2$
X_1: 92	90	87	105	86	83	102	645	59,847
X_2: 100	108	98	110	114	97	94	721	74,609
X_3: 143	149	138	136	139	120	145	970	134,936
X_4: 147	144	160	149	152	131	134	1017	148,367
X_5: 142	155	119	134	133	146	152	981	138,415
Totals							4334	556,174

Since the computed F is

$$F = \frac{16{,}672.11/4}{2828.86/30} = 44.20,$$

the null hypothesis is clearly rejected. This information is summarized in Table 8.7-6.

Source	Sum of Squares (SS)	Degrees of Freedom	Mean Square (MS)	F
Treatment	16,672.11	4	4,168.03	44.20
Error	2,828.86	30	94.30	
Total	19,500.97	34		

Table 8.7-6: ANOVA Table for Pull-out Tests

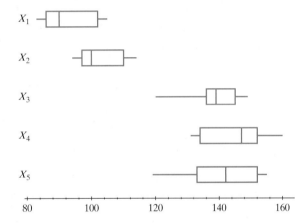

Figure 8.7-1: Box plots for pull out tests

But why is H_0 rejected? The box-and-whisker diagrams shown in Figure 8.7-1 help to answer this question. It looks like the forces required to pull out studs in positions 1 and 2 are similar and those in positions 3, 4, and 5 are quite similar but different from positions 1 and 2. (See Exercise 8.7-12.) An examination of the window would confirm that this is the case.   ◄

As with the two sample t test, the F test works quite well even if the underlying distributions are non-normal, unless they are highly skewed or the variances are quite different. In these latter cases, we might need to transform the observations to make the data more symmetric with about the same variances or to use certain nonparametric methods that are beyond the scope of this course.

EXERCISES

8.7-1 Let μ_1, μ_2, μ_3 be, respectively, the means of three normal distributions with a common but unknown variance σ^2. In order to test, at the $\alpha = 0.05$ significance level, the hypothesis H_0: $\mu_1 = \mu_2 = \mu_3$ against all possible alternative hypotheses, we take a random sample of size 4 from each of these distributions. Determine whether we accept or reject H_0 if the observed values from these three distributions are, respectively,

$$
\begin{array}{llllll}
X_1: & 5 & 9 & 6 & 8 \\
X_2: & 11 & 13 & 10 & 12 \\
X_3: & 10 & 6 & 9 & 9
\end{array}
$$

8.7-2 Let μ_i be the average yield in bushels per acre of variety i of corn, $i = 1, 2, 3, 4$. In order to test, at the 5% significance level, the hypothesis $H_0: \mu_1 = \mu_2 = \mu_3 = \mu_4$, four test plots for each of the four varieties of corn are planted. Determine whether we accept or reject H_0 if the yield in bushels per acre of the four varieties of corn are, respectively,

$$
\begin{array}{lllll}
X_1: & 68.82 & 76.99 & 74.30 & 78.73 \\
X_2: & 86.84 & 75.69 & 77.87 & 76.18 \\
X_3: & 90.16 & 78.84 & 80.65 & 83.58 \\
X_4: & 61.58 & 73.51 & 74.57 & 70.75
\end{array}
$$

8.7-3 Four groups of three pigs each were fed individually four different feeds for a specified length of time to test the hypothesis $H_0: \mu_1 = \mu_2 = \mu_3 = \mu_4$, where μ_i is the mean weight gain for each of the feeds, $i = 1, 2, 3, 4$. Determine whether the null hypothesis is accepted or rejected at a 5% significance level if the observed weight gains are, respectively,

$$
\begin{array}{llll}
X_1: & 194.11 & 182.80 & 187.43 \\
X_2: & 216.06 & 203.50 & 216.88 \\
X_3: & 178.10 & 189.20 & 181.33 \\
X_4: & 197.11 & 202.68 & 209.18
\end{array}
$$

8.7-4 For the following set of data show that the computed $SS(E)/(n - m) = 1$ and $SS(T)/(m - 1) = 75$. This suggests that the unbiased estimate of σ^2 based on $SS(T)$ is usually greater than σ^2 when the true means are unequal.

$$
\begin{array}{llll}
X_1: & 4 & 5 & 6 \\
X_2: & 9 & 10 & 11 \\
X_3: & 14 & 15 & 16
\end{array}
$$

8.7-5 In a tulip display area, several varieties of tulips have long stems. The lengths of the flower stems were measured for four varieties of tulips, the colors being white, pink, red trimmed with yellow, and dark red. We shall call them varieties 1, 2, 3, and 4. Let μ_i, $i = 1, 2, 3, 4$, denote the mean of the lengths of the stems for variety i. Test the hypothesis $H_0: \mu_1 = \mu_2 = \mu_3 = \mu_4$ against all alternatives, making reasonable assumptions. Use a 5% significance level. Clearly specify the critical region for this test.

$$
\begin{array}{llllll}
X_1: & 13.75 & 13.00 & 14.25 & 12.00 & 12.25 \\
X_2: & 13.75 & 14.50 & 12.75 & 14.50 & 13.25 \\
X_3: & 12.75 & 12.50 & 12.00 & 14.00 & 13.00 \\
X_4: & 16.75 & 14.25 & 14.50 & 15.50 & 13.75
\end{array}
$$

8.7-6 Hogg and Ledolter [see R. V. Hogg and J. Ledolter, *Applied Statistics for Engineers and Physical Scientists*, 2nd ed. (New York: Macmillan, 1992)] report that a civil engineer wishes to compare the strengths of three different types of beams, one (A) made of steel and two (B and C) made of different and more expensive alloys. A certain deflection (in units of 0.001 inch) was measured for each beam when submitted to a given force; thus a small deflection would indicate a beam of great strength. The order statistics for the three samples of sizes $n_1 = 8$, $n_2 = 6$, and $n_3 = 6$ are the following:

$$
\begin{array}{lllllllll}
A: & 79 & 82 & 83 & 84 & 85 & 86 & 86 & 87 \\
B: & 74 & 75 & 76 & 77 & 78 & 82 \\
C: & 77 & 78 & 79 & 79 & 79 & 82
\end{array}
$$

(a) Use these data and the F test to test the equality of the three means. Use $\alpha = 0.05$.

(b) Construct box-and-whisker diagrams for each set of data on the same figure and give an interpretation.

8.7-7 Montgomery considers the strengths of a synthetic fiber that is possibly affected by the percentage of cotton in the fiber. Five levels of this percentage are considered with five observations taken at each level.

Percentage of Cotton	Tensile Strength in Pounds per Square Inch				
15	7	7	15	11	9
20	12	17	12	18	18
25	14	18	18	19	19
30	19	25	22	19	23
35	7	10	11	15	11

Use the F test, with $\alpha = 0.05$, to see if there are differences in the breaking strengths due to the percentages of cotton used. [D. C. Montgomery, *Design and Analysis of Experiments*, 2nd ed. (New York: Wiley, 1984), p. 51.]

8.7-8 Let X_1, X_2, X_3, X_4 equal the cholesterol level of a woman under the age of 50, a man under 50, a woman 50 or older, and a man 50 or older, respectively. Assume that the distribution of X_i is $N(\mu_i, \sigma^2)$, $i = 1, 2, 3, 4$. We shall test the null hypothesis H_0: $\mu_1 = \mu_2 = \mu_3 = \mu_4$ using 7 observations of each X_i.

(a) Give a critical region for an $\alpha = 0.05$ significance level.

(b) Construct an ANOVA table and state your conclusion using the following data.

$$
\begin{array}{lllllllll}
X_1: & 221 & 213 & 202 & 183 & 185 & 197 & 162 \\
X_2: & 271 & 192 & 189 & 209 & 227 & 236 & 142 \\
X_3: & 262 & 193 & 224 & 201 & 161 & 178 & 265 \\
X_4: & 192 & 253 & 248 & 278 & 232 & 267 & 289
\end{array}
$$

(c) Give bounds on the p-value for this test.

(d) Construct box-and-whisker diagrams for each set of data on the same figure and give an interpretation.

8.7-9 Let X_i, $i = 1, 2, 3, 4$ equal the distance that a golf ball travels when hit from a tee, where i denotes the index of the ith manufacturer. Assume that the distribution of X_i is $N(\mu_i, \sigma^2)$, $i = 1, 2, 3, 4$ when hit by a certain golfer. We shall test the null hypothesis H_0: $\mu_1 = \mu_2 = \mu_3 = \mu_4$ using 3 observations of each random variable.

(a) Give a critical region for an $\alpha = 0.05$ significance level.

(b) Construct an ANOVA table and state your conclusion using the following data.

$$
\begin{array}{llll}
X_1: & 240 & 221 & 265 \\
X_2: & 286 & 256 & 272 \\
X_3: & 259 & 245 & 232 \\
X_4: & 239 & 215 & 223
\end{array}
$$

(c) What would your conclusion be if $\alpha = 0.025$?

(d) What is the approximate p-value of this test?

8.7-10 Different sizes of nails are packaged in "1-pound" boxes. Let X_i equal the weight of a box with nail size $(4i)C$, $i = 1, 2, 3, 4, 5$, where $4C, 8C, 12C, 16C$, and $20C$ are the sizes of the sinkers from smallest to largest. Assume that the distribution of X_i is $N(\mu_i, \sigma^2)$. To test the null hypothesis that the mean weights of "1-pound" boxes are all equal for different sizes of nails, we shall use random samples of size 7, weighing the nails to the nearest hundredth of a pound.

(a) Give a critical region for an $\alpha = 0.05$ significance level.

(b) Construct an ANOVA table and state your conclusion using the following data.

X_1:	1.03	1.04	1.07	1.03	1.08	1.06	1.07
X_2:	1.03	1.10	1.08	1.05	1.06	1.06	1.05
X_3:	1.03	1.08	1.06	1.02	1.04	1.04	1.07
X_4:	1.10	1.10	1.09	1.09	1.06	1.05	1.08
X_5:	1.04	1.06	1.07	1.06	1.05	1.07	1.05

(c) Construct box-and-whisker diagrams for each set of data on the same figure and give an interpretation.

8.7-11 The driver of a diesel-powered automobile decided to test the quality of three types of diesel fuel sold in the area based on mpg. Test the null hypothesis that the three means are equal using the following data. Make the usual assumptions and take $\alpha = 0.05$.

Brand A:	38.7	39.2	40.1	38.9	
Brand B:	41.9	42.3	41.3		
Brand C:	40.8	41.2	39.5	38.9	40.3

8.7-12 Based on the box-and-whisker diagrams in Figure 8.7-1, it looks like the means of X_1 and X_2 could be equal and also that the means of X_3, X_4, and X_5 could be equal but different from the first two.

(a) Using the data in Example 8.7-2, test H_0: $\mu_1 = \mu_2$ against a two-sided alternative hypothesis using a t test and an F test. Let $\alpha = 0.05$. Do F and t tests give the same result?

(b) Using the data in Example 8.7-2, test H_0: $\mu_3 = \mu_4 = \mu_5$. Let $\alpha = 0.05$.

8.7-13 For an aerosol product, there are three weights: the tare weight (container weight), the concentrate weight, and the propellant weight. Let X_1, X_2, X_3 denote the propellant weights on three different days. Assume that each of these independent random variables has a normal distribution with common variance and respective means, μ_1, μ_2, μ_3. We shall test the null hypothesis H_0: $\mu_1 = \mu_2 = \mu_3$ using 9 observations of each of the random variables.

(a) Give a critical region for an $\alpha = 0.01$ significance level.

(b) Construct an ANOVA table and state your conclusion using the following data.

X_1:	43.06	43.32	42.63	42.86	43.05
	42.87	42.94	42.80	42.36	
X_2:	42.33	42.81	42.13	42.41	42.39
	42.10	42.42	41.42	42.52	
X_3:	42.83	42.57	42.96	43.16	42.25
	42.24	42.20	41.97	42.61	

(c) Construct box-and-whisker diagrams for each set of data on the same figure and give an interpretation.

8.7-14 A particular process puts a coating on a piece of glass so that it is sensitive to touch. Randomly throughout the day, pieces of glass are selected from the production line and the resistance is measured at 12 different locations on the glass. On each of three different days, December 6, December 7, and December 22, the following data give the means of the 12 measurements on each of 11 pieces of glass.

December 6:	175.05	177.44	181.94	176.51	182.12	164.34
	163.20	168.12	171.26	171.92	167.87	
December 7:	175.93	176.62	171.39	173.90	178.34	172.90
	174.67	174.27	177.16	184.13	167.21	
December 22:	167.27	161.48	161.86	173.83	170.75	172.90
	173.27	170.82	170.93	173.89	177.68	

(a) Use these data to test whether the means on all three days are equal.

(b) Use box-and-whisker diagrams to confirm your answer.

8.8* TWO-FACTOR ANALYSIS OF VARIANCE

The test of the equality of several means, considered in Section 8.7 is an example of a statistical inference method called the analysis of variance (ANOVA). This method derives its name from the fact that the quadratic form $\text{SS(TO)} = (n-1)S^2$, the total sum of squares about the combined sample mean, is decomposed into component parts and analyzed. In this section, other problems in the analysis of variance will be investigated; here we restrict our considerations to the two-factor case, but the reader can see how it can be extended to three-factor and other cases.

Consider a situation in which it is desirable to investigate the effects of two factors that influence an outcome of an experiment. For example, a teaching method (lecture, discussion, computer-assisted, television, etc.) and the size of a class might influence a student's score on a standard test; or the type of car and the grade of gasoline used might change the number of miles per gallon. In this latter example, if the number of miles per gallon is not affected by the grade of gasoline, we would no doubt use the least expensive grade.

The first analysis of variance model that we discuss is referred to as a **two-way classification with one observation per cell**. In particular, assume that there are two factors (attributes), one of which has a levels and the other b levels. There are thus $n = ab$ possible combinations, each of which determines a cell. Let us think of these cells as being arranged in a rows and b columns. Here we take one observation per cell, and we denote the observation in the ith row and jth column by X_{ij}. Further assume that X_{ij} is $N(\mu_{ij}, \sigma^2)$, $i = 1, 2, \ldots, a$, and $j = 1, 2, \ldots, b$; and the $n = ab$ random variables are independent. [The assumptions of normality and homogeneous (same) variances can be somewhat relaxed in applications with little change in the significance levels of the resulting tests.] We shall assume that the means μ_{ij} are composed of a row effect, a column effect, and an overall effect in some additive way, namely $\mu_{ij} = \mu + \alpha_i + \beta_j$, where $\sum_{i=1}^{a} \alpha_i = 0$ and $\sum_{j=1}^{b} \beta_j = 0$. The parameter α_i represents the ith row effect, and the parameter β_j represents the jth column effect.

Remark   There is no loss in generality in assuming that

$$\sum_{i=1}^{a} \alpha_i = \sum_{j=1}^{b} \beta_j = 0.$$

To see this, let $\mu_{ij} = \mu' + \alpha'_i + \beta'_j$. Write

$$\overline{\alpha}' = \left(\frac{1}{a}\right)\sum_{i=1}^{a} \alpha'_i \quad \text{and} \quad \overline{\beta}' = \left(\frac{1}{b}\right)\sum_{j=1}^{b} \beta'_j.$$

We have

$$\mu_{ij} = (\mu' + \overline{\alpha}' + \overline{\beta}') + (\alpha'_i - \overline{\alpha}') + (\beta'_j - \overline{\beta}') = \mu + \alpha_i + \beta_j,$$

where $\sum_{i=1}^{a} \alpha_i = 0$ and $\sum_{j=1}^{b} \beta_j = 0$. The reader is asked to find μ, α_i, and β_j for one display of μ_{ij} in Exercise 8.8-2. ∎

To test the hypothesis that there is no row effect, we would test H_A: $\alpha_1 = \alpha_2 = \cdots = \alpha_a = 0$, since $\sum_{i=1}^{a} \alpha_i = 0$. Similarly, to test that there is no column effect, we would test H_B: $\beta_1 = \beta_2 = \cdots = \beta_b = 0$, since $\sum_{j=1}^{b} \beta_j = 0$. To test these hypotheses, we shall again partition the total sum of squares into several component parts. Letting

$$\overline{X}_{i\cdot} = \frac{1}{b} \sum_{j=1}^{b} X_{ij}, \ \overline{X}_{\cdot j} = \frac{1}{a} \sum_{i=1}^{a} X_{ij}, \ \overline{X}_{\cdot\cdot} = \frac{1}{ab} \sum_{i=1}^{a} \sum_{j=1}^{b} X_{ij},$$

we have

$$SS(TO) = \sum_{i=1}^{a} \sum_{j=1}^{b} (X_{ij} - \overline{X}_{\cdot\cdot})^2$$

$$= \sum_{i=1}^{a} \sum_{j=1}^{b} [(\overline{X}_{i\cdot} - \overline{X}_{\cdot\cdot}) + (\overline{X}_{\cdot j} - \overline{X}_{\cdot\cdot}) + (X_{ij} - \overline{X}_{i\cdot} - \overline{X}_{\cdot j} + \overline{X}_{\cdot\cdot})]^2$$

$$= b \sum_{i=1}^{a} (\overline{X}_{i\cdot} - \overline{X}_{\cdot\cdot})^2 + a \sum_{j=1}^{b} (\overline{X}_{\cdot j} - \overline{X}_{\cdot\cdot})^2$$

$$+ \sum_{i=1}^{a} \sum_{j=1}^{b} (X_{ij} - \overline{X}_{i\cdot} - \overline{X}_{\cdot j} + \overline{X}_{\cdot\cdot})^2$$

$$= SS(A) + SS(B) + SS(E),$$

where $SS(A)$ is the sum of squares among levels of factor A or among rows, $SS(B)$ is the sum of squares among levels of factor B or among columns, and $SS(E)$ is the error or residual sum of squares. In Exercise 8.8-4 the reader is asked to show that the three "cross-product" terms in the square of the trinomial sum to zero. The distribution of the error sum of squares does not depend on the mean μ_{ij}, provided that the additive model is correct, and hence its distribution is the same whether H_A or H_B is true or not; thus $SS(E)$ acts as a "measuring stick" as did $SS(E)$ in Section 8.7. This can be seen more clearly by writing

$$SS(E) = \sum_{i=1}^{a} \sum_{j=1}^{b} (X_{ij} - \overline{X}_{i\cdot} - \overline{X}_{\cdot j} + \overline{X}_{\cdot\cdot})^2$$

$$= \sum_{i=1}^{a} \sum_{j=1}^{b} [X_{ij} - (\overline{X}_{i\cdot} - \overline{X}_{\cdot\cdot}) - (\overline{X}_{\cdot j} - \overline{X}_{\cdot\cdot}) - \overline{X}_{\cdot\cdot}]^2$$

and noting the similarity of the summand in the righthand member with

$$X_{ij} - \mu_{ij} = X_{ij} - \alpha_i - \beta_j - \mu.$$

We now show that $SS(A)/\sigma^2$, $SS(B)/\sigma^2$, and $SS(E)/\sigma^2$ are independent chi-square variables, provided that both H_A and H_B are true, that is, when all the means μ_{ij} have a common value μ. To do this, we first note that $SS(TO)/\sigma^2$ is $\chi^2(ab-1)$. In addition, from Section 8.7, we see that expressions such as $SS(A)/\sigma^2$ and $SS(B)/\sigma^2$ are chi-square variables, namely $\chi^2(a-1)$ and $\chi^2(b-1)$, by replacing the n_i of Section 8.7 by a and b, respectively. Obviously, $SS(E) \geq 0$, and hence by Theorem 8.7-1 we have that $SS(A)/\sigma^2$, $SS(B)/\sigma^2$, and $SS(E)/\sigma^2$ are independent chi-square variables with $a - 1$, $b - 1$, and $ab - 1 - (a - 1) - (b - 1) = (a - 1)(b - 1)$ degrees of freedom, respectively.

To test the hypothesis $H_A: \alpha_1 = \alpha_2 = \cdots = \alpha_a = 0$, we shall use the row sum of squares $SS(A)$ and the residual sum of squares $SS(E)$. When H_A is true, $SS(A)/\sigma^2$ and $SS(E)/\sigma^2$ are independent chi-square variables with $a - 1$ and $(a - 1)(b - 1)$ degrees of freedom, respectively. Thus $SS(A)/(a - 1)$ and $SS(E)/[(a - 1)(b - 1)]$ are both unbiased estimators of σ^2 when H_A is true. However, $E[SS(A)/(a - 1)] > \sigma^2$ when H_A is not true, and hence we would reject H_A when

$$F_A = \frac{SS(A)/[\sigma^2(a - 1)]}{SS(E)/[\sigma^2(a - 1)(b - 1)]} = \frac{SS(A)/(a - 1)}{SS(E)/[(a - 1)(b - 1)]}$$

is "too large." Since F_A has an F distribution with $a - 1$ and $(a - 1)(b - 1)$ degrees of freedom when H_A is true, H_A is rejected if the observed value of $F_A \geq F_\alpha[a-1, (a-1)(b-1)]$.

Similarly, the test of the hypothesis $H_B: \beta_1 = \beta_2 = \cdots = \beta_b = 0$ against all alternatives can be based on

$$F_B = \frac{SS(B)/[\sigma^2(b - 1)]}{SS(E)/[\sigma^2(a - 1)(b - 1)]} = \frac{SS(B)/(b - 1)}{SS(E)/[(a - 1)(b - 1)]}$$

which has an F distribution with $b - 1$ and $(a - 1)(b - 1)$ degrees of freedom, provided that H_B is true.

In Table 8.8-1 we give the ANOVA table that summarizes the information needed for these tests of hypotheses. The formulas for F_A and F_B show that each of them is a ratio of two mean squares.

EXAMPLE 8.8-1 Each of three cars is driven with each of four different brands of gasoline. The number of miles per gallon driven for each of the $ab = (3)(4) = 12$ different combinations is recorded in Table 8.8-2.

We would like to test whether we can expect the same mileage for each of these four brands of gasoline. In our notation, we test the hypothesis

$$H_B: \beta_1 = \beta_2 = \beta_3 = \beta_4 = 0$$

	Table 8.8-1: Two-way ANOVA Table, One Observations Per Cell			
Source	**Sum of Squares (SS)**	**Degrees of Freedom**	**Mean Square (MS)**	**F**
Factor A (row)	SS(A)	$a - 1$	$MS(A) = \dfrac{SS(A)}{a - 1}$	$\dfrac{MS(A)}{MS(E)}$
Factor B (column)	SS(B)	$b - 1$	$MS(B) = \dfrac{SS(B)}{b - 1}$	$\dfrac{MS(B)}{MS(E)}$
Error	SS(E)	$(a - 1)(b - 1)$	$MS(E) = \dfrac{SS(E)}{(a - 1)(b - 1)}$	
Total	SS(TO)	$ab - 1$		

	Table 8.8-2: Gas Mileage Data				
		Gasoline			
Car	**1**	**2**	**3**	**4**	$\overline{X}_{i\cdot}$
1	16	18	21	21	19
2	14	15	18	17	16
3	15	15	18	16	16
$\overline{X}_{\cdot j}$	15	16	19	18	17

against all alternatives. At a 1% significance level we shall reject H_B if the computed F, namely

$$\frac{SS(B)/(4 - 1)}{SS(E)/[(3 - 1)(4 - 1)]} \geq 9.78 = F_{0.01}(3, 6).$$

We have

$$SS(B) = 3[(15 - 17)^2 + (16 - 17)^2 + (19 - 17)^2 + (18 - 17)^2] = 30;$$

$$SS(E) = (16 - 19 - 15 + 17)^2 + (14 - 16 - 15 + 17)^2 + \cdots$$
$$+ (16 - 16 - 18 + 17)^2 = 4.$$

Hence the computed F is

$$\frac{30/3}{4/6} = 15 > 9.78,$$

and the hypothesis H_B is rejected. That is, the gasolines seem to give different performances (at least with these three cars).

The information for this example is summarized in Table 8.8-3.   ◄

In a two-way classification problem, particular combinations of the two factors might interact differently from that expected from the additive model.

	Sum of Squares (SS)	Degrees of Freedom	Mean Square (MS)	F	p-value
Table 8.8-3: ANOVA Table for Gas Mileage Data					
Source					
Row (A)	24	2	12	18	0.003
Column (B)	30	3	10	15	0.003
Error	4	6	2/3		
Total	58	11			

For example, in Example 8.8-1 gasoline 3 seemed to be the best gasoline and car 1 the best car; however, it sometimes happens that the two best do not "mix" well and the joint performance is poor. That is, there might be a strange interaction between this combination of car and gasoline and accordingly the joint performance is not as good as expected. Sometimes it can happen that we get good results from a combination of some of the poorer levels of each factor. This is called interaction, and it frequently occurs in practice (e.g., in chemistry). In order to test for possible interaction, we shall consider a two-way classification problem in which $c > 1$ independent observations are taken per cell.

Assume that $X_{ijk}, i = 1, 2, \ldots, a; j = 1, 2, \ldots, b;$ and $k = 1, 2, \ldots, c$, are $n = abc$ random variables that are mutually independent and have normal distributions with a common, but unknown, variance σ^2. The mean of each $X_{ijk}, k = 1, 2, \ldots, c$, is $\mu_{ij} = \mu + \alpha_i + \beta_j + \gamma_{ij}$, where $\sum_{i=1}^{a} \alpha_i = 0, \sum_{j=1}^{b} \beta_j = 0$, $\sum_{i=1}^{a} \gamma_{ij} = 0$, and $\sum_{j=1}^{b} \gamma_{ij} = 0$. The parameter γ_{ij} is called the **interaction** associated with cell (i, j). That is, the interaction between the ith level of one classification and the jth level of the other classification is γ_{ij}. The reader is asked to determine μ, α_i, β_j, and γ_{ij} for some given μ_{ij} in Exercise 8.8-6.

To test the hypotheses that (a) the row effects are equal to zero, (b) the column effects are equal to zero, and (c) there is no interaction, we shall again partition the total sum of squares into several component parts. Letting

$$\overline{X}_{ij\cdot} = \frac{1}{c} \sum_{k=1}^{c} X_{ijk},$$

$$\overline{X}_{i\cdot\cdot} = \frac{1}{bc} \sum_{j=1}^{b} \sum_{k=1}^{c} X_{ijk},$$

$$\overline{X}_{\cdot j\cdot} = \frac{1}{ac} \sum_{i=1}^{a} \sum_{k=1}^{c} X_{ijk},$$

$$\overline{X}_{\cdots} = \frac{1}{abc} \sum_{i=1}^{a} \sum_{j=1}^{b} \sum_{k=1}^{c} X_{ijk},$$

we have

$$SS(TO) = \sum_{i=1}^{a}\sum_{j=1}^{b}\sum_{k=1}^{c}(X_{ijk} - \overline{X}_{...})^2$$

$$= bc\sum_{i=1}^{a}(\overline{X}_{i..} - \overline{X}_{...})^2 + ac\sum_{j=1}^{b}(\overline{X}_{.j.} - \overline{X}_{...})^2$$

$$+ c\sum_{i=1}^{a}\sum_{j=1}^{b}(\overline{X}_{ij.} - \overline{X}_{i..} - \overline{X}_{.j.} + \overline{X}_{...})^2 + \sum_{i=1}^{a}\sum_{j=1}^{b}\sum_{k=1}^{c}(X_{ijk} - \overline{X}_{ij.})^2$$

$$= SS(A) + SS(B) + SS(AB) + SS(E),$$

where SS(A) is the row sum of squares or the sum of squares among levels of factor A, SS(B) is the column sum of squares or the sum of squares among levels of factor B, SS(AB) is the interaction sum of squares, and SS(E) is the error sum of squares. Again we can show that the cross-product terms sum to zero.

To consider the joint distribution of SS(A), SS(B), SS(AB), and SS(E), let us assume that all the means equal the same value μ. Of course, we know that $SS(TO)/\sigma^2$ is $\chi^2(abc-1)$. Also, by letting the n_i of Section 8.7 equal bc and ac, respectively, we know that $SS(A)/\sigma^2$ and $SS(B)/\sigma^2$ are $\chi^2(a-1)$ and $\chi^2(b-1)$. Moreover,

$$\frac{\sum_{k=1}^{c}(X_{ijk} - \overline{X}_{ij.})^2}{\sigma^2}$$

is $\chi^2(c-1)$; hence $SS(E)/\sigma^2$ is the sum of ab independent chi-square variables such as this and thus is $\chi^2[ab(c-1)]$. Of course SS(AB) ≥ 0; so, according to Theorem 8.7-1, we have that $SS(A)/\sigma^2$, $SS(B)/\sigma^2$, $SS(AB)/\sigma^2$, and $SS(E)/\sigma^2$ are mutually independent chi-square variables with $a-1, b-1, (a-1)(b-1)$, and $ab(c-1)$ degrees of freedom, respectively.

To test the hypotheses concerning row, column, and interaction effects, we form F statistics in which the numerators are affected by deviations from the respective hypotheses whereas the denominator is a function of SS(E), whose distribution depends only on the value of σ^2 and not on the values of the cell means. Hence SS(E) acts as our measuring stick here.

The statistic for testing the hypothesis

$$H_{AB}: \gamma_{ij} = 0, \ i = 1, 2, \ldots, a, \ j = 1, 2, \ldots, b,$$

against all alternatives is

$$F_{AB} = \frac{c \sum_{i=1}^{a} \sum_{j=1}^{b} (\overline{X}_{ij\cdot} - \overline{X}_{i\cdot\cdot} - \overline{X}_{\cdot j\cdot} + \overline{X}_{\cdots})^2 / [\sigma^2 (a-1)(b-1)]}{\sum_{i=1}^{a} \sum_{j=1}^{b} \sum_{k=1}^{c} (X_{ijk} - \overline{X}_{ij\cdot})^2 / [\sigma^2 ab(c-1)]}$$

$$= \frac{SS(AB)/[(a-1)(b-1)]}{SS(E)/[ab(c-1)]},$$

which has an F distribution with $(a-1)(b-1)$ and $ab(c-1)$ degrees of freedom when H_{AB} is true. If the computed $F_{AB} \geq F_\alpha[(a-1)(b-1), ab(c-1)]$, we reject H_{AB} and say there is a difference among the means, since there seems to be interaction. Most statisticians do *not* proceed to test row and column effects if H_{AB} is rejected.

The statistic for testing the hypothesis

$$H_A : \alpha_1 = \alpha_2 = \cdots = \alpha_a = 0$$

against all alternatives is

$$F_A = \frac{bc \sum_{i=1}^{a} (\overline{X}_{i\cdot\cdot} - \overline{X}_{\cdots})^2 / [\sigma^2 (a-1)]}{\sum_{i=1}^{a} \sum_{j=1}^{b} \sum_{k=1}^{c} (X_{ijk} - \overline{X}_{ij\cdot})^2 / [\sigma^2 ab(c-1)]} = \frac{SS(A)/(a-1)}{SS(E)/[ab(c-1)]},$$

which has an F distribution with $a - 1$ and $ab(c - 1)$ degrees of freedom when H_A is true. The statistic for testing the hypothesis

$$H_B : \beta_1 = \beta_2 = \cdots = \beta_b = 0$$

against all alternatives is

$$F_B = \frac{ac \sum_{j=1}^{b} (\overline{X}_{\cdot j\cdot} - \overline{X}_{\cdots})^2 / [\sigma^2 (b-1)]}{\sum_{i=1}^{a} \sum_{j=1}^{b} \sum_{k=1}^{c} (X_{ijk} - \overline{X}_{ij\cdot})^2 / [\sigma^2 ab(c-1)]} = \frac{SS(B)/(b-1)}{SS(E)/[ab(c-1)]},$$

which has an F distribution with $b - 1$ and $ab(c - 1)$ degrees of freedom when H_B is true. Each of these hypotheses is rejected if the observed value of F is greater than a given constant that is selected to yield the desired significance level.

Table 8.8-4: Two-way ANOVA Table, c Observations Per Cell

Source	Sum of Squares (SS)	Degrees of Freedom	Mean Square (MS)	F
Factor A (row)	SS(A)	$a-1$	$MS(A) = \dfrac{SS(A)}{a-1}$	$\dfrac{MS(A)}{MS(E)}$
Factor B (column)	SS(B)	$b-1$	$MS(B) = \dfrac{SS(B)}{b-1}$	$\dfrac{MS(B)}{MS(E)}$
Factor AB (interaction)	SS(AB)	$(a-1)(b-1)$	$MS(AB) = \dfrac{SS(AB)}{(a-1)(b-1)}$	$\dfrac{MS(AB)}{MS(E)}$
Error	SS(E)	$ab(c-1)$	$MS(E) = \dfrac{SS(E)}{ab(c-1)}$	
Total	SS(TO)	$abc-1$		

Table 8.8-5: Illustration of Arrays for Numbers of Digits

| Type of Array | Number of Digits | | |
	1	2	3
Down	5	25	259
	3	69	567
	8	37	130
Across	$5 + 3 + 8 =$	$25 + 69 + 37 =$	$259 + 567 + 130 =$

In Table 8.8-4 we give the ANOVA table that summarizes the information needed for these tests of hypotheses.

EXAMPLE 8.8-2   Consider the following experiment. One hundred eight people were randomly divided into six groups with 18 people in each group. Each person was given sets of three numbers to add. The three numbers were either in a "down array" or an "across array," representing the two levels of factor A. The levels of factor B are determined by the number of digits in the numbers to be added: one-digit, two-digit, or three-digit numbers. Table 8.8-5 illustrates this with a sample problem for each cell; note, however, that an individual person only works problems of one of these types. Each person was placed in one of the six groups and was told to work as many problems as possible in 90 seconds. The measurement that was recorded was the average number of problems worked correctly in two trials.

Table 8.8-6: Cell, Row, and Column Means When Adding Numbers				
Type of Array	**Number of Digits**			
	1	**2**	**3**	**Row Means**
Down	23.806	10.694	6.278	13.593
Across	26.056	6.750	3.944	12.250
Column means	24.931	8.722	5.111	

Whenever this many subjects are used, a computer becomes an invaluable tool. A computer program provided the summary in Table 8.8-6 of the sample means of the rows, the columns, and the six cells. Each cell mean is the average for 18 people.

Simply considering these means, we can see clearly that there is a column effect. It is not surprising that it is easier to add one-digit than three-digit numbers.

The most interesting feature of these results is that they show the possibility of interaction. The largest cell mean occurs when adding one-digit numbers in an across array. Note, however, that for two- and three-digit numbers the down arrays have larger means than the across arrays.

Table 8.8-7: ANOVA Table When Adding Numbers					
Source	**Sum of Squares**	**Degrees of Freedom**	**Mean Square**	**F**	**p-value**
Factor A (array)	48.678	1	48.669	2.885	0.0925
Factor B (number of digits)	8022.73	2	4011.363	237.778	< 0.0001
Interaction	185.92	2	92.961	5.510	0.0053
Error	1720.76	102	16.870		
Total	9978.08	107			

The computer provided the ANOVA table given in Table 8.8-7. The number of degrees of freedom for SS(E) is not in our F table. However, the right column, obtained from the computer printout, provides the p-value of each test, namely the probability of obtaining an F as large as or larger than the calculated F ratio. Note, for example, that to test for interaction $F = 5.51$ and the p-value is 0.006. Thus the hypothesis of no interaction would be rejected at the $\alpha = 0.05$ or $\alpha = 0.01$ significance level but not with $\alpha = 0.001$. ◄

8.8-1 For the data given in Example 8.8-1, test the hypothesis H_A: $\alpha_1 = \alpha_2 = \alpha_3 = 0$ against all alternatives at the 5% significance level.

8.8-2 With $a = 3$ and $b = 4$, find μ, α_i, and β_j if $\mu_{ij}, i = 1, 2, 3$ and $j = 1, 2, 3, 4$, are given by

$$
\begin{array}{cccc}
6 & 3 & 7 & 8 \\
10 & 7 & 11 & 12 \\
8 & 5 & 9 & 10
\end{array}
$$

Note that in an "additive" model such as this one, one row (column) can be determined by adding a constant value to each of the elements of another row (column).

8.8-3 We wish to compare compressive strengths of concrete corresponding to $a = 3$ different drying methods (treatments). Concrete is mixed in batches that are just large enough to produce three cylinders. Although care is taken to achieve uniformity, we expect some variability among the $b = 5$ batches used to obtain the following compressive strengths. (There is little reason to suspect interaction and hence only one observation is taken in each cell.)

Treatment	Batch				
	B_1	B_2	B_3	B_4	B_5
A_1	52	47	44	51	42
A_2	60	55	49	52	43
A_3	56	48	45	44	38

(a) Use the 5% significance level and test H_A: $\alpha_1 = \alpha_2 = \alpha_3 = 0$ against all alternatives.

(b) Use the 5% significance level and test H_B: $\beta_1 = \beta_2 = \beta_3 = \beta_4 = \beta_5 = 0$ against all alternatives. [See R. V. Hogg and J. Ledolter, *Applied Statistics for Engineers and Physical Scientists*, 2nd ed., (New York: Macmillan, 1992).]

8.8-4 Show that the cross-product terms formed from $(X_{ij} - \overline{X}_{i\cdot} - \overline{X}_{\cdot j} + \overline{X}_{\cdot\cdot}), (\overline{X}_{i\cdot} - \overline{X}_{\cdot\cdot})$, and $(\overline{X}_{\cdot j} - \overline{X}_{\cdot\cdot})$, sum to zero, $i = 1, 2, \ldots a$ and $j = 1, 2, \ldots, b$.
HINT: For example, write

$$
\sum_{i=1}^{a} \sum_{j=1}^{b} (\overline{X}_{\cdot j} - \overline{X}_{\cdot\cdot})(X_{ij} - \overline{X}_{i\cdot} - \overline{X}_{\cdot j} + \overline{X}_{\cdot\cdot})
$$

$$
= \sum_{j=1}^{b} (\overline{X}_{\cdot j} - \overline{X}_{\cdot\cdot}) \sum_{i=1}^{a} [(X_{ij} - \overline{X}_{\cdot j}) - (\overline{X}_{i\cdot} - \overline{X}_{\cdot\cdot})]
$$

and sum each term in the inner summation as grouped here to get zero.

8.8-5 A psychology student was interested in testing how food consumption by rats would be affected by a particular drug. She used two levels of one attribute, namely drug and placebo, and four levels of a second attribute, namely male (M), castrated (C), female (F), and ovariectomized (O). For each cell she observed five rats. The amount of food consumed in grams per 24 hours is listed in the table. Test the hypotheses using a 5% significance level for each.

(a) $H_{AB}: \gamma_{ij} = 0, i = 1, 2, j = 1, 2, 3, 4.$
(b) $H_A: \alpha_1 = \alpha_2 = 0.$
(c) $H_B: \beta_1 = \beta_2 = \beta_3 = \beta_4 = 0.$
(d) How could you modify this model so that there are three attributes of classification, each with two levels?

	M	C	F	O
Drug	22.56	16.54	18.58	18.20
	25.02	24.64	15.44	14.56
	23.66	24.62	16.12	15.54
	17.22	19.06	16.88	16.82
	22.58	20.12	17.58	14.56
Placebo	25.64	22.50	17.82	19.74
	28.84	24.48	15.76	17.48
	26.00	25.52	12.96	16.46
	26.02	24.76	15.00	16.44
	23.24	20.62	19.54	15.70

8.8-6 With $a = 3$ and $b = 4$, find μ, α_i, β_j, and γ_{ij}, if $\mu_{ij}, i = 1, 2, 3$ and $j = 1, 2, 3, 4$, are given by

$$
\begin{array}{cccc}
6 & 7 & 7 & 12 \\
10 & 3 & 11 & 8 \\
8 & 5 & 9 & 10
\end{array}
$$

Note the difference between the layout here and that in Exercise 8.8-2. Does the interaction help explain these differences?

8.8-7 In order to test whether four brands of gasoline give equal performance in terms of mileage, each of three cars was driven with each of the four brands of gasoline. Then each of the $(3)(4) = 12$ possible combinations was repeated four times. The number of miles per gallon for each of the four repetitions in each cell is recorded in the table. Test each of the hypotheses H_{AB}, H_A, H_B at the 5% significance level.

Car	Brand of Gasoline							
	1		**2**		**3**		**4**	
1	21.0	14.9	16.3	20.0	15.8	19.4	17.8	17.3
	16.2	18.8	15.2	21.6	14.5	14.8	18.2	20.4
2	20.6	19.5	15.5	16.8	16.6	13.7	18.1	17.1
	20.8	18.9	17.4	19.4	18.2	16.1	21.5	19.1
3	14.2	13.1	17.4	18.1	15.2	16.7	16.3	16.4
	16.8	17.4	16.4	16.9	17.7	18.1	17.9	18.8

8.8-8 The data in the following table could represent the outcomes of some experiment, but in actuality they were generated on a computer. Recall that $\mu_{ij} = \mu + \alpha_i + \beta_j + \gamma_{ij}$ is the mean of cell (i, j) and in the simulation, we took $\mu = 15$, $\alpha_1 = -3$, $\alpha_2 = 2$, $\alpha_3 = 1$, $\beta_1 = -1.5$, $\beta_2 = 1.5$, $\gamma_{11} = 4$, $\gamma_{12} = -4$, $\gamma_{21} = -2$, $\gamma_{22} = 2$, $\gamma_{31} = -2$, $\gamma_{32} = 2$, and $\sigma^2 = 9$. Use the data and the two-factor analysis of variance model with four repetitions per cell to test the hypotheses H_{AB}: no interaction, H_A: no row effect, and H_B: no column effect, each at the $\alpha = 0.05$ significance level.

	B_1	B_2
A_1	14.552	13.980
	17.024	5.777
	10.132	7.638
	16.979	12.089
A_2	11.378	20.825
	12.578	17.333
	17.709	20.329
	11.037	25.277
A_3	12.345	22.223
	11.044	21.844
	9.458	16.176
	16.156	17.946

8.8-9 Hogg and Ledolter report on an engineer in a textile mill who studies the effects of temperature and time in a process involving dye on the brightness of a synthetic fabric (brightness is measured on a 50-point scale). Three observations were taken at each combination of temperature and time.

Time (cycles)	Temperature		
	350°F	375°F	400°F
40	38, 32, 30	37, 35, 40	36, 39, 43
50	40, 45, 36	39, 42, 46	39, 48, 47

Construct the ANOVA table and conduct tests for the interaction first and then, if appropriate, the main effects. Use $\alpha = 0.05$ for each test. [R. V. Hogg and J. Ledolter, *Applied Statistics for Engineers and Physical Scientists*, 2nd ed. (New York: Macmillan, 1992).]

8.8-10 There is another way of looking at Exercise 8.7-8, namely as a two-factor analysis of variance problem with the levels of gender being female and male, the levels of age being less than 50 and at least 50, and the measurement for each subject being their cholesterol level. The data would then be set up as follows:

(a) Test H_{AB}: $\gamma_{ij} = 0$, $i = 1, 2$; $j = 1, 2$ (no interaction).
(b) Test H_A: $\alpha_1 = \alpha_2 = 0$ (no row effect).
(c) Test H_B: $\beta_1 = \beta_2 = 0$ (no column effect).

Use a 5% significance level for each test.

	Age	
Gender	< 50	≥ 50
Female	221	262
	213	193
	202	224
	183	201
	185	161
	197	178
	162	265
Male	271	192
	192	253
	189	248
	209	278
	227	232
	236	267
	142	289

8.9* TESTS CONCERNING REGRESSION AND CORRELATION

In Section 6.11 we considered the estimation of the parameters of a very simple regression curve, namely a straight line. We can use confidence intervals for the parameters to test hypotheses about them. For illustration, with the same model as that in Section 6.11, we could test the hypothesis H_0: $\beta = \beta_0$ by using a t random variable that was used for a confidence interval with β replaced by β_0, namely

$$T_1 = \frac{\widehat{\beta} - \beta_0}{\sqrt{\dfrac{n\widehat{\sigma^2}}{(n-2)\displaystyle\sum_{i=1}^{n}(x_i - \overline{x})^2}}}.$$

The null hypothesis along with three possible alternative hypotheses are given in Table 8.9-1; these tests are equivalent to stating that we reject H_0 if β_0 is not in certain confidence intervals. For example, the first test is equivalent to rejecting H_0 if β_0 is not in the one-sided confidence interval with lower bound

$$\widehat{\beta} - t_\alpha(n-2)\sqrt{\frac{n\widehat{\sigma^2}}{(n-2)\displaystyle\sum_{i=1}^{n}(x_i - \overline{x})^2}}.$$

Often we let $\beta_0 = 0$ and test the hypothesis H_0: $\beta = 0$. That is, we test the null hypothesis that the slope is equal to zero.

Table 8.9-1: Tests About the Slope of the Regression Line		
H_0	H_1	**Critical Region**
$\beta = \beta_0$	$\beta > \beta_0$	$t_1 \geq t_\alpha(n-2)$
$\beta = \beta_0$	$\beta < \beta_0$	$t_1 \leq -t_\alpha(n-2)$
$\beta = \beta_0$	$\beta \neq \beta_0$	$|t_1| \geq t_{\alpha/2}(n-2)$

EXAMPLE 8.9-1 Let x equal a student's preliminary test score in a psychology course and y the same student's score on the final examination. With $n = 10$ students, we shall test $H_0: \beta = 0$ against $H_1: \beta \neq 0$. At the 0.01 significance level, the critical region is $|t_1| \geq t_{0.005}(8) = 3.355$. Using the data in Example 6.11-1, the observed value of T_1 is

$$t_1 = \frac{0.742 - 0}{\sqrt{10(21.7709)/8(756.1)}} = \frac{0.742}{0.1897} = 3.911.$$

Thus we reject H_0. ◀

We consider tests about the correlation coefficient ρ of a bivariate normal distribution. Let X and Y have a bivariate normal distribution. We know that if the correlation coefficient ρ is zero, then X and Y are independent random variables. Furthermore, the value of ρ gives a measure of the linear relationship between X and Y. We now give methods for using the sample correlation coefficient to test the hypothesis $H_0: \rho = 0$ and also to form a confidence interval for ρ.

Let $(X_1, Y_1), (X_2, Y_2), \ldots, (X_n, Y_n)$ denote a random sample from a bivariate normal distribution with parameters μ_X, μ_Y, σ_X^2, σ_Y^2, and ρ. That is, the n pairs of (X, Y) are independent, and each pair has the same bivariate normal distribution. The **sample correlation coefficient** is

$$R = \frac{\dfrac{1}{n-1}\displaystyle\sum_{i=1}^{n}(X_i - \overline{X})(Y_i - \overline{Y})}{\sqrt{\dfrac{1}{n-1}\displaystyle\sum_{i=1}^{n}(X_i - \overline{X})^2}\sqrt{\dfrac{1}{n-1}\displaystyle\sum_{i=1}^{n}(Y_i - \overline{Y})^2}} = \frac{S_{XY}}{S_X S_Y}.$$

We note that

$$R\,\frac{S_Y}{S_X} = \frac{S_{XY}}{S_X^2} = \frac{\dfrac{1}{n-1}\displaystyle\sum_{i=1}^{n}(X_i - \overline{X})(Y_i - \overline{Y})}{\dfrac{1}{n-1}\displaystyle\sum_{i=1}^{n}(X_i - \overline{X})^2}$$

is exactly the solution that we obtained for $\widehat{\beta}$ in Section 6.11 when the X values were fixed at $X_1 = x_1, X_2 = x_2, \ldots, X_n = x_n$. Let us consider these values fixed temporarily so that we are considering conditional distributions, given $X_1 = x_1, \ldots, X_n = x_n$. Moreover, if H_0: $\rho = 0$ is true, then the distributions of $Y_1, Y_2, \ldots, Y_n$ are independent of $x_1, x_2, \ldots, x_n$ and thus $\beta = \rho \sigma_Y / \sigma_X = 0$. Under these conditions. the conditional distribution of

$$\widehat{\beta} = \frac{\displaystyle\sum_{i=1}^{n} (x_i - \overline{x})(Y_i - \overline{Y})}{\displaystyle\sum_{i=1}^{n} (x_i - \overline{x})^2}$$

is $N[0, \sigma_Y^2/(n-1)s_x^2]$ when $s_x^2 > 0$. Moreover, recall from Section 6.11 that the conditional distribution, given $X_1 = x_1, \ldots, X_n = x_n$, of (see Exercise 8.9-8)

$$\frac{\sum_{i=1}^{n} [Y_i - \overline{Y} - (S_{xY}/s_x^2)(x_i - \overline{x})]^2}{\sigma_Y^2} = \frac{(n-1)S_Y^2(1 - R^2)}{\sigma_Y^2}$$

is $\chi^2(n-2)$ and is independent of $\widehat{\beta}$. Thus, when $\rho = 0$, the conditional distribution of

$$T = \frac{(RS_Y/s_x)/(\sigma_Y/\sqrt{n-1}\,s_x)}{\sqrt{[(n-1)\,S_Y^2\,(1 - R^2)/\sigma_Y^2][1/(n-2)]}} = \frac{R\sqrt{n-2}}{\sqrt{1 - R^2}}$$

is t with $n - 2$ degrees of freedom. However, since the conditional distribution of T, given $X_1 = x_1, \ldots, X_n = x_n$, does not depend on $x_1, x_2, \ldots, x_n$, the unconditional distribution of T must be t with $n - 2$ degrees of freedom and T and $(X_1, X_2, \ldots, X_n)$ are independent when $\rho = 0$.

Remark   It is interesting to note that in the discussion about the distribution of T, nothing was said about the distribution of $X_1, X_2, \ldots, X_n$. This means that if X and Y are independent and Y has a normal distribution, then T has a t distribution whatever the distribution of X. Obviously, the roles of X and Y can be reversed in all of this development. In particular, if X and Y are independent, then T and $Y_1, Y_2, \ldots, Y_n$ are also independent.   ∎

Now T can be used to test H_0: $\rho = 0$. If the alternative hypothesis is H_1: $\rho > 0$, we would use the critical region defined by observed $T \geq t_\alpha(n-2)$, since large T implies large R. Obvious modifications would be made for the alternative hypotheses H_1: $\rho < 0$ and H_1: $\rho \neq 0$, the latter leading to a two-sided test.

Using the p.d.f. $h(t)$ of T, we can find the distribution function and p.d.f. of R when $-1 < r < 1$, provided that $\rho = 0$:

$$G(r) = P(R \le r) = P\left(T \le \frac{r\sqrt{n-2}}{\sqrt{1-r^2}}\right)$$

$$= \int_{-\infty}^{r\sqrt{n-2}/\sqrt{1-r^2}} h(t)\, dt$$

$$= \int_{-\infty}^{r\sqrt{n-2}/\sqrt{1-r^2}} \frac{\Gamma[(n-1)/2]}{\Gamma(1/2)\,\Gamma[(n-2)/2]} \frac{1}{\sqrt{n-2}} \left(1 + \frac{t^2}{n-2}\right)^{-(n-1)/2} dt.$$

The derivative of $G(r)$, with respect to r, is (see Appendix A.4)

$$g(r) = h\left(\frac{r\sqrt{n-2}}{\sqrt{1-r^2}}\right) \frac{d(r\sqrt{n-2}/\sqrt{1-r^2})}{dr},$$

which equals

$$g(r) = \frac{\Gamma[(n-1)/2]}{\Gamma(1/2)\,\Gamma[(n-2)/2]} (1-r^2)^{(n-4)/2}, \qquad -1 < r < 1.$$

Thus, for example, to test the hypothesis $H_0: \rho = 0$ against the alternative hypothesis $H_1: \rho \ne 0$ at a significance level α, select either a constant $r_{\alpha/2}(n-2)$ or a constant $t_{\alpha/2}(n-2)$ so that

$$\alpha = P(|R| \ge r_{\alpha/2}(n-2); H_0) = P(|T| \ge t_{\alpha/2}(n-2); H_0),$$

depending on the availability of R or T tables.

It is interesting to graph the p.d.f. of R. Note in particular that if $n = 4$, $g(r) = 1/2$, $-1 < r < 1$, and if $n = 6$, $g(r) = (3/4)(1 - r^2)$, $-1 < r < 1$. The graphs of the p.d.f. of R when $n = 8$ and when $n = 14$ are given in Figure 8.9-1. Recall that this is the p.d.f. of R when $\rho = 0$. As n increases R is more likely to equal values close to 0.

Table XI in the Appendix lists selected values of the distribution function of R when $\rho = 0$. For example, if $n = 8$, the number of degrees of freedom is 6 and $P(R \le 0.7887) = 0.99$. Also, if $\alpha = 0.10$, then $r_{\alpha/2}(6) = r_{0.05}(6) = 0.6215$. See Figure 8.9-1(a).

It is also possible to obtain an approximate test of size α by using the fact that

$$W = \frac{1}{2} \ln \frac{1+R}{1-R}$$

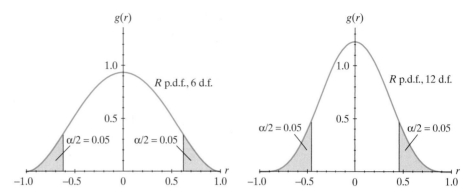

Figure 8.9-1: R p.d.f.s when $n = 8$ and $n = 14$

has an approximate normal distribution with mean $(1/2)\ln[(1+\rho)/(1-\rho)]$ and variance $1/(n-3)$. We accept this statement without proof (see Exercise 8.9-10). Thus a test of H_0: $\rho = \rho_0$ can be based on the statistic

$$Z = \frac{\dfrac{1}{2}\ln\dfrac{1+R}{1-R} - \dfrac{1}{2}\ln\dfrac{1+\rho_0}{1-\rho_0}}{\sqrt{\dfrac{1}{n-3}}},$$

which has a distribution that is approximately $N(0, 1)$.

EXAMPLE 8.9-2   We would like to test the hypothesis H_0: $\rho = 0$ against H_1: $\rho \neq 0$ at an $\alpha = 0.05$ significance level. A random sample of size 18 from a bivariate normal distribution yielded a sample correlation coefficient of $r = 0.35$. Using Table XI in the Appendix, since $0.35 < 0.4683$, H_0 is accepted (not rejected) at an $\alpha = 0.05$ significance level. Using the t distribution, we would reject H_0 if $|t| \geq 2.120 = t_{0.025}(16)$. Since

$$t = \frac{0.35\sqrt{16}}{\sqrt{1-(0.35)^2}} = 1.495,$$

H_0 is not rejected. If we had used the normal approximation for Z, H_0 would be rejected if $|z| \geq 1.96$. Since

$$z = \frac{(1/2)\ln[(1+0.35)/(1-0.35)] - 0}{\sqrt{1/(18-3)}} = 1.415,$$

H_0 is not rejected.   ◀

To develop an approximate $100(1 - \alpha)\%$ confidence interval for ρ, we use the normal approximation for the distribution of Z. Thus we select a constant

$c = z_{\alpha/2}$ from Table V in the Appendix so that

$$P\left(-c \le \frac{(1/2)\ln[(1+R)/(1-R)] - (1/2)\ln[(1+\rho)/(1-\rho)]}{\sqrt{1/(n-3)}} \le c\right) \approx 1 - \alpha.$$

After several steps, this becomes

$$P\left(\frac{1 + R - (1-R)\exp(2c/\sqrt{n-3})}{1 + R + (1-R)\exp(2c/\sqrt{n-3})} \le \rho \le\right.$$

$$\left.\frac{1 + R - (1-R)\exp(-2c/\sqrt{n-3})}{1 + R + (1-R)\exp(-2c/\sqrt{n-3})}\right) \approx 1 - \alpha.$$

EXAMPLE 8.9-3   Suppose that a random sample of size 12 from a bivariate normal distribution yielded a correlation coefficient of $r = 0.6$. An approximate 95% confidence interval for ρ would be

$$\left[\frac{1 + 0.6 - (1 - 0.6)\exp\left(\frac{2(1.96)}{3}\right)}{1 + 0.6 + (1 - 0.6)\exp\left(\frac{2(1.96)}{3}\right)}, \frac{1 + 0.6 - (1 - 0.6)\exp\left(\frac{-2(1.96)}{3}\right)}{1 + 0.6 + (1 - 0.6)\exp\left(\frac{-2(1.96)}{3}\right)}\right]$$

$$= [0.040, 0.873].$$

If the sample size had been $n = 39$ and $r = 0.6$, the approximate 95% confidence interval would have been $[0.351, 0.770]$.   ◄

EXERCISES

8.9-1 For the data given in Exercise 6.11-1, test $H_0: \beta = 0$ against $H_1: \beta > 0$ at the $\alpha = 0.025$ significance level using a t test.

8.9-2 For the data given in Exercise 6.11-2, test $H_0: \beta = 0$ against $H_1: \beta > 0$ at the $\alpha = 0.025$ significance level using a t test.

8.9-3 For the data given in Exercise 6.11-3, test $H_0: \beta = 0$ against $H_1: \beta > 0$ at the $\alpha = 0.025$ significance level using a t test.

8.9-4 For the candy data given in Exercise 6.11-11, with the usual assumptions, test $H_0: \beta = 0$ against $H_1: \beta > 0$ at an $\alpha = 0.01$ significance level.

8.9-5 A random sample of size $n = 27$ from a bivariate normal distribution yielded a sample correlation coefficient of $r = -0.45$. Would the hypothesis $H_0: \rho = 0$ be rejected in favor of $H_1: \rho \ne 0$ at an $\alpha = 0.05$ significance level?

8.9-6 When bowling it is often possible to score well on the first game and then bowl poorly on the second game, or vice versa. The following six pairs of numbers give the scores of the first and second games bowled by the same person on six consecutive Tuesday evenings. Assume a bivariate normal distribution and use these scores to test the hypothesis $H_0: \rho = 0$ against $H_1: \rho \ne 0$ at $\alpha = 0.10$.

$$\begin{array}{lllllll}
\text{Game 1:} & 170 & 190 & 200 & 183 & 187 & 178 \\
\text{Game 2:} & 197 & 178 & 150 & 176 & 205 & 153
\end{array}$$

8.9-7 A random sample of size 28 from a bivariate normal distribution yielded a sample correlation coefficient of $r = 0.65$. Find an approximate 90% confidence interval for ρ.

8.9-8 By squaring the binomial expression $[(Y_i - \overline{Y}) - (S_{xY}/s_x^2)(x_i - \overline{x})]$, show that

$$\sum_{i=1}^{n} [(Y_i - \overline{Y}) - (S_{xY}/s_x^2)(x_i - \overline{x})]^2$$

$$= \sum_{i=1}^{n} (Y_i - \overline{Y})^2 - 2\left(\frac{S_{xY}}{s_x^2}\right) \sum_{i=1}^{n} (x_i - \overline{x})(Y_i - \overline{Y}) + \frac{S_{xY}^2}{s_x^4} \sum_{i=1}^{n} (x_i - \overline{x})^2$$

equals $(n-1)S_Y^2(1 - R^2)$, where $X_1 = x_1, X_2 = x_2, \ldots, X_n = x_n$.
HINT: Replace $S_{xY} = \sum_{i=1}^{n} (x_i - \overline{x})(Y_i - \overline{Y})/(n-1)$ by $Rs_x S_Y$.

8.9-9 To help determine whether gallinules selected their mate on the basis of weight, 14 pairs of gallinules were captured and weighed. Test the null hypothesis $H_0 : \rho = 0$ against a two-sided alternative at an $\alpha = 0.01$ significance level. Given that the male and female weights for the $n = 14$ pairs of birds yielded a sample correlation coefficient of $r = -0.252$, would H_0 be rejected?

8.9-10 It is true, when sampling from a bivariate normal distribution, that the correlation coefficient R has an approximate normal distribution $N[\rho, (1 - \rho^2)^2/n]$ if the sample size n is large. Since, for large n, R is close to ρ, use two terms of the Taylor's expansion of $u(R)$ about ρ and determine that function $u(R)$ such that it has a variance (essentially) free of ρ. (The solution of this exercise explains why the transformation $(1/2) \ln[(1 + R)/(1 - R)]$ was suggested.)

8.9-11 When $\rho = 0$, show that

   (a) the points of inflection for the graph of the p.d.f. of R are at $r = \pm 1/\sqrt{n-5}$ for $n \geq 7$,
   (b) $E(R) = 0$,
   (c) $\text{Var}(R) = 1/(n-1), n \geq 3$. HINT: Note that $E(R^2) = E[1 - (1 - R^2)]$.

8.9-12 In a college health fitness program, let X equal the weight in kilograms of a female freshman at the beginning of the program and let Y equal her weight change during the semester. We shall use the following data for $n = 16$ observations of (x, y) to test the null hypothesis $H_0 : \rho = 0$ against a two-sided alternative hypothesis.

$$\begin{array}{llll}
(61.4, -3.2) & (62.9, 1.4) & (58.7, 1.3) & (49.3, 0.6) \\
(71.3, 0.2) & (81.5, -2.2) & (60.8, 0.9) & (50.2, 0.2) \\
(60.3, 2.0) & (54.6, 0.3) & (51.1, 3.7) & (53.3, 0.2) \\
(81.0, -0.5) & (67.6, -0.8) & (71.4, -0.1) & (72.1, -0.1)
\end{array}$$

   (a) What is the conclusion if $\alpha = 0.10$?
   (b) What is the conclusion if $\alpha = 0.05$?

8.9-13 Let X and Y have a bivariate normal distribution with correlation coefficient ρ. To test $H_0 : \rho = 0$ against $H_1 : \rho \neq 0$, a random sample of n pairs of observations is selected. Suppose that the sample correlation coefficient is $r = 0.68$. Using a significance level of $\alpha = 0.05$, find the smallest value of the sample size n so that H_0 is rejected.

8.9-14 In Exercise 6.11-5 data are given for horse power, time it takes a car to go from 0 to 60, and the weight in pounds of a car for 14 cars. Those data are repeated here.

Horse Power	0–60	Weight	Horse Power	0–60	Weight
230	8.1	3516	282	6.2	3627
225	7.8	3690	300	6.4	3892
375	4.7	2976	220	7.7	3377
322	6.6	4215	250	7.0	3625
190	8.4	3761	315	5.3	3230
150	8.4	2940	200	6.2	2657
178	7.2	2818	300	5.5	3518

(a) Let ρ be the correlation coefficient of horse power and weight. Test $H_0: \rho = 0$ against $H_1: \rho \neq 0$.

(b) Let ρ be the correlation coefficient of horse power and "0–60." Test $H_0: \rho = 0$ against $H_1: \rho < 0$.

(c) Let ρ be the correlation coefficient of weight and "0–60." Test $H_0: \rho = 0$ against $H_1: \rho \neq 0$.

8.9-15 In Exercise 6.11-15 the ACT composite score and the first year college gpa were given for 18 students. Those data are repeated here.

ACT	GPA	ACT	GPA
23	2.40	25	3.67
29	3.10	23	3.03
15	2.51	30	3.07
23	2.98	16	2.28
28	3.93	24	2.45
27	1.91	26	3.83
23	1.69	27	3.78
26	2.88	21	2.57
20	2.60	19	3.42

Let ρ be the correlation coefficient of these two variables. Test the null hypothesis $H_0: \rho = 0$ against the alternative hypothesis $H_1: \rho > 0$.

8.10* KOLMOGOROV-SMIRNOV GOODNESS OF FIT TEST

In this section we discuss a test that considers the goodness of fit between a hypothesized distribution function and an empirical distribution function. The empirical distribution function is given here in terms of the order statistics. Let $y_1 < y_2 < \cdots < y_n$ be the observed values of the order statistics of a random sample $x_1, x_2, \ldots, x_n$ of size n. When no two observations are equal, the empirical distribution function is defined by

$$F_n(x) = \begin{cases} 0, & x < y_1, \\ k/n, & y_k \leq x < y_{k+1}, \quad k = 1, 2, \ldots, n-1, \\ 1, & y_n \leq x. \end{cases}$$

In this case the empirical distribution function has a jump of magnitude $1/n$ occurring at each observation. If n_k observations are equal to x_k, a jump of magnitude n_k/n occurs at x_k. In any case, $F_n(x)$ is simply the fraction of sample observations that are less than or equal to x.

Suppose that a random sample of size n is taken from a distribution of the continuous type that has the distribution function $F(x)$. How can we measure the "closeness" of $F(x)$ and the empirical distribution function $F_n(x)$? How does the sample size affect this closeness? We give some theoretical results to help answer these questions and then give a test for goodness of fit.

Let $X_1, X_2, \ldots, X_n$ denote a random sample of size n from a distribution of the continuous type with the distribution function $F(x)$. Consider a fixed value of x. Then $W = F_n(x)$, the value of the empirical distribution function at x, can be thought of as a random variable that takes on the values $0, 1/n, 2/n, \ldots, 1$. Now $nW = k$ if, and only if, exactly k observations are less than or equal to x (say success) and $n - k$ observations are greater than x. The probability that an observation is less than or equal to x is given by $F(x)$. That is, the probability of success is $F(x)$. Because of the independence of the random variables $X_1, X_2, \ldots, X_n$, the probability of k successes is given by the binomial distribution, namely

$$P(nW = k) = P\left(W = \frac{k}{n}\right)$$

$$= \binom{n}{k}[F(x)]^k[1 - F(x)]^{n-k}, \qquad k = 0, 1, 2, \ldots, n.$$

Since nW has a binomial distribution with $p = F(x)$, the mean and variance of nW are given by

$$E(nW) = nF(x) \quad \text{and} \quad \text{Var}(nW) = n[F(x)][1 - F(x)].$$

Hence the mean and the variance of $W = F_n(x)$ are

$$E[F_n(x)] = E(W) = F(x)$$

and

$$\text{Var}[F_n(x)] = \text{Var}(W) = \frac{F(x)[1 - F(x)]}{n}.$$

Since the variance of $F_n(x)$ gets nearer to zero as n becomes large, $F_n(x)$ and its mean $F(x)$ tend to be closer with large n. As a matter of fact, there is a theorem by Glivenko, the proof of which is beyond the level of this book, which states that with probability one, $F_n(x)$ converges to $F(x)$ uniformly in x as $n \to \infty$.

Because of the convergence of the empirical distribution function to the theoretical distribution function, it makes sense to construct a goodness of fit test based on the closeness of the empirical and a hypothesized distribution function, say $F_n(x)$ and $F_0(x)$, respectively. We shall use the Kolmogorov-Smirnov statistic defined by

$$D_n = \sup_x [\,|F_n(x) - F_0(x)|\,].$$

That is, D_n is the least upper bound of all pointwise differences $|F_n(x) - F_0(x)|$.

The exact distribution of the statistic D_n can be derived. We shall not derive this distribution but do give some values of the distribution function of D_n, namely $P(D_n \leq d)$, in Table VIII in the Appendix that will be used for goodness of fit tests. We would like to point out that the distribution of D_n does not depend on the particular function $F_0(x)$ of the continuous type. [This is essentially due to the fact that $Y = F_0(X)$ has a uniform distribution $U(0, 1)$.] Thus D_n can be thought of as a distribution-free statistic.

We are interested in using the Kolmogorov-Smirnov statistic D_n to test the hypothesis H_0: $F(x) = F_0(x)$ against all alternatives, H_1: $F(x) \neq F_0(x)$, where $F_0(x)$ is some specified distribution function. Intuitively, we accept H_0 if the empirical distribution function $F_n(x)$ is sufficiently close to $F_0(x)$, that is, if the value of D_n is sufficiently small. The hypothesis H_0 is rejected if the observed value of D_n is greater than the critical value selected from Table VIII in the Appendix, where this critical value depends upon the desired significance level and sample size.

The use of the Kolmogorov-Smirnov statistic is illustrated by two examples.

EXAMPLE 8.10-1 We shall test the hypothesis H_0: $F(x) = F_0(x)$ against H_1: $F(x) \neq F_0(x)$, where

$$F_0(x) = \begin{cases} 0, & x < 0, \\ x, & 0 \leq x < 1, \\ 1, & 1 \leq x. \end{cases}$$

That is, the null hypothesis is that X is $U(0, 1)$. If the test is based on a sample of size $n = 10$ and if $a = 0.10$, the critical region is $C = \{d_{10} : d_{10} \geq 0.37\}$, where d_{10} is the observed value of the Kolmogorov-Smirnov statistic D_{10} Suppose that the observed values of the random sample are 0.62, 0.36, 0.23, 0.76, 0.65, 0.09, 0.55, 0.26, 0.38, and 0.24. In Figure 8.10-1 we have plotted the empirical and hypothesized distribution functions for $0 \leq x \leq 1$. We see that $d_{10} = F_{10}(0.65) - F_0(0.65) = 0.25$ and hence H_0 is not rejected. Note that the more carefully you plot the empirical and theoretical distribution functions, the more you can minimize the number of points at which you should calculate the possible value of the Kolmogorov-Smirnov goodness of fit statistic. ◀

EXAMPLE 8.10-2 When observing a Poisson process with a mean rate of arrivals $\lambda = 1/\theta$, the random variable W, which denotes the waiting time until the αth arrival, has a gamma distribution. The p.d.f. of W is

$$f(w) = \frac{w^{\alpha-1} e^{-w/\theta}}{\Gamma(\alpha)\theta^\alpha}, \qquad 0 \leq w < \infty.$$

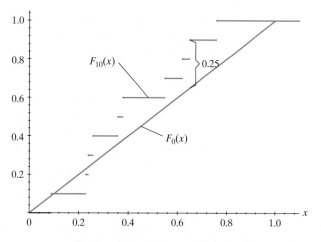

Figure 8.10-1: H_0: X is $U(0, 1)$

A Geiger counter was set up to record the waiting time W in seconds to observe $\alpha = 100$ alpha particle emissions of barium 133. It is claimed that the number of counts per second has a Poisson distribution with $\lambda = 14.7$ and hence $\theta = 0.068$. We shall test the hypothesis

$$H_0:\ F(w) = \int_{-\infty}^{w} f(t)\, dt,$$

where $f(t)$ is the gamma p.d.f. with $\theta = 0.068$ and $\alpha = 100$. Based on 25 observations, H_0 is rejected if $d_{25} \geq 0.24$ for $\alpha = 0.10$. For 25 observations, the empirical and theoretical distribution functions are depicted in Figure 8.10-2. For these data (Exercise 3.4-6) $d_{25} = 0.123$ and hence H_0 is not rejected.   ◀

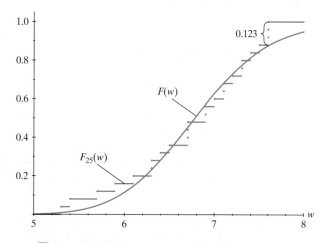

Figure 8.10-2: H_0: X has a gamma distribution

You will note that we have been assuming that $F(x)$ is a continuous function. That is, we have only considered random variables of the continuous type. This procedure may also be applied in the discrete case. However, in the discrete case, the true significance level will be at most α. That is, the resulting test will be conservative.

Another application of the Kolmogorov-Smirnov statistic is in forming a confidence band for an unknown distribution function $F(x)$. To form a confidence band based on a sample of size n, select a number d such that

$$P(D_n \geq d) = \alpha.$$

Then

$$
\begin{aligned}
1 - \alpha &= P[\sup_x |F_n(x) - F(x)| \leq d] \\
&= P[|F_n(x) - F(x)| \leq d \text{ for all } x] \\
&= P[F_n(x) - d \leq F(x) \leq F_n(x) + d \text{ for all } x].
\end{aligned}
$$

Let

$$
F_L(x) = \begin{cases} 0, & F_n(x) - d \leq 0, \\ F_n(x) - d, & F_n(x) - d > 0, \end{cases}
$$

and

$$
F_U(x) = \begin{cases} F_n(x) + d, & F_n(x) + d < 1, \\ 1, & F_n(x) + d \geq 0. \end{cases}
$$

The two-step functions $F_L(x)$ and $F_U(x)$ yield a $100(1 - \alpha)\%$ confidence band for the unknown distribution function $F(x)$.

EXAMPLE 8.10-3   A random sample of size $n = 15$ from an unknown distribution yielded the sample values 3.88, 3.97, 4.03, 2.49, 3.18, 3.08, 2.91, 3.43, 2.41, 1.57, 3.78, 3.25, 1.29, 2.57, and 3.40. Now

$$P(D_{15} \geq 0.30) = 0.10.$$

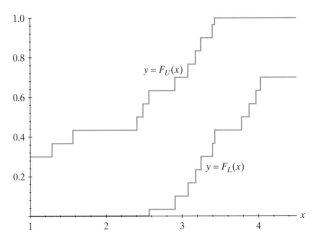

Figure 8.10-3: A 90% confidence band for an unknown distribution function

A 90% confidence band for the unknown distribution function $F(x)$ is depicted in Figure 8.10-3. ◀

EXERCISES

8.10-1 Five observations of X, which have been ordered, are 0.40, 0.51, 0.53, 0.62, 0.74. Use the Kolmogorov-Smirnov statistic to test the hypothesis that X is $U(0, 1)$. Let $\alpha = 0.20$.

8.10-2 Select 10 sets of 10 random numbers from Table IX in the Appendix. For each set of 10 random numbers, calculate the value of the Kolmogorov-Smirnov statistic d_{10} with $F_0(x) = x, 0 \leq x \leq 1$. Is it true that about 20% of the observations of D_{10} are greater than 0.32? (Compare your results with those of other students, provided that they selected different random numbers.)

8.10-3 A doctor of obstetrics used an ultrasound examination between the 16th and 25th weeks of pregnancy to measure in millimeters the widest diameter of the fetal head. Let X equal this diameter. Use the following 10 observations of X to test the hypothesis that the distribution of X is $N(60, 100)$.

$$56 \quad 65 \quad 47 \quad 57 \quad 62 \quad 48 \quad 68 \quad 75 \quad 79 \quad 49$$

8.10-4 In Exercise 8.4-4 the outcomes of 10 simulations of a Cauchy random variable are given.

(a) Sketch the empirical distribution function and superimpose the Cauchy distribution function.

(b) Use the Kolmogorov-Smirnov statistic to test whether the 10 observations represent a random sample from a Cauchy distribution. Let $\alpha = 0.10$.

8.10-5 In Exercise 3.1-14 diameters (in mm) of 30 individual grains of soil were given as follows.

1.24	1.36	1.28	1.31	1.35	1.20	1.39	1.35	1.41	1.31
1.28	1.26	1.37	1.49	1.32	1.40	1.33	1.28	1.25	1.39
1.38	1.34	1.40	1.27	1.33	1.36	1.43	1.33	1.29	1.34

(a) Calculate the sample mean and the sample variance for these data.

(b) Test the null hypothesis that these are observations of a normally distributed random variable using a Kolmogorov-Smirnov goodness of fit statistic.

8.10-6 Construct a 90% confidence band for the unknown distribution function $F(x)$ using the following 15 observations of X:

20.2	85.4	59.9	72.7	88.0	33.7	87.1	99.5	93.8
18.4	60.6	98.9	90.9	86.9	74.2			

8.10-7 While testing a computer tape for bad records, the computer operator counted the number of flaws per 100 feet. Let X equal this number and test the hypothesis that the distribution of X is Poisson with a mean of $\lambda = 2.4$ using the following 40 observations of X. Let $\alpha = 0.10$, approximately.

x	Frequency
0	5
1	7
2	12
3	9
4	5
5	1
6	1
	40

8.10-8 The computer operator in Exercise 8.10-7 also measured the distances between flaws. These were reported in Exercise 8.5-12 in which a chi-square goodness of fit test was used. Now use a Kolmogorov-Smirnov goodness of fit statistic to test the null hypothesis that we are observing observations of an exponential random variable. Are your conclusions consistent with each other?

30	79	38	47	22	52	36	36	7	57
3	22	30	14	8	32	15	21	12	12
6	67	6	7	35	78	28	74	5	9
37	1	3	3	44	160	50	27	61	15
39	44	130	18	6	1	32	116	23	12
58	101	68	53	58	21	21	7	79	41
80	33	71	81	17	10	13	49	21	56
107	21	17	64	14	36	26	1	54	207
64	238	25	51	82	8	2	3	43	87

8.10-9 In Exercise 8.11-14, 70 observations of a random variable X are listed that give the force required to pull a stud out of a window.

   (a) Test the null hypothesis that the distribution of X is $N(117.5, 252)$. Use the Kolmogorov-Smirnov statistic with a significance level of $\alpha = 0.20$. What is your conclusion?

   (b) Carefully analyzing the graphs of the empirical and theoretical distribution functions or by constructing a histogram, do you agree with the conclusion in part (a)? Why?

   (c) Construct and interpret a q-q plot for these data.

8.10-10 Let X equal the number of chocolate chips in a chocolate-chip cookie. Use the Kolmogorov-Smirnov goodness of fit statistic and the following data to test the hypothesis that the distribution of X is Poisson with a mean of $\lambda = 5.6$. Let $\alpha = 0.10$, approximately.

x	Frequency
0	0
1	0
2	2
3	8
4	7
5	13
6	13
7	10
8	4
9	4
10	1
	62

8.10-11 In Example 3.2-3 105 times between calls to 911 are listed. Also given is a graph of the empirical distribution function with an exponential distribution function superimposed. Does this graph and the Kolmogorov-Smirnov goodness of fit statistic support the suggestion that these are observations of an exponential random variable? Why?

8.10-12 In Exercise 6.10-15 the weights in grams of "14-gram" snack packages of peanuts were given as follows:

| 13.9 | 14.4 | 14.6 | 14.7 | 14.7 | 15.2 | 15.2 | 15.2 |
| 15.3 | 15.4 | 15.4 | 15.5 | 15.6 | 15.6 | 15.9 | 16.4 |

Test the null hypothesis that these weights are observations of a normally distributed random variable with mean $\mu = 15.3$ and standard deviation $\sigma = 0.6$. Use a 10% significance level.

8.10-13 *Parade Magazine*, March 29, 1992, reported the results of a study by Runzheimer International who gathered data from around the world on the cost in U.S. dollars per gallon for the least-expensive gasoline available. They compared each cost in December 1990 (before the Gulf War) with the cost in December 1991. Let X equal the difference, the 1991 cost minus the 1990 cost. Assume that the following $n = 18$ differences represent a random sample of observations of X.

-0.28	-0.40	0.06	0.04	-0.29	0.38
-0.08	-0.02	-0.22	-0.28	0.11	0.44
-0.17	-0.39	-0.36	-0.18	-0.39	-0.31

Test whether the times are observations of a shifted (two parameter) exponential distribution. Estimate δ and θ with their unbiased estimators. That is, letting y_1 equal the minimum value of the X observations, unbiased estimates of θ and δ are

$$\widehat{\theta} = [n/(n-1)][\overline{x} - y_1] \quad \text{and} \quad \widehat{\delta} = y_1 - \widehat{\theta}/n.$$

8.11*   RUN TEST AND TEST FOR RANDOMNESS

Under the assumption that the random variables X and Y are of the continuous type and have distribution functions $F(x)$ and $G(y)$, respectively, we describe another test of the hypothesis H_0: $F(z) = G(z)$. This new test can also be used to test for randomness. For these tests we need the concept of runs, which we now define.

Suppose that we have n_1 observations of the random variable X and n_2 observations of the random variable Y. The combination of two sets of independent observations into one collection of $n_1 + n_2$ observations, placed in ascending order of magnitude, might yield an arrangement

$$\underline{y\ y\ y}\ \underline{x\ x}\ \underline{y}\ \underline{x}\ \underline{y}\ \underline{x\ x}\ \underline{y\ y}$$

where x denotes an observation of X and y an observation of Y in the ordered arrangement. We have underlined groups of successive values of X and Y. Each underlined group is called a **run**. Thus we have a run of three values of Y, followed by a run of two values of X, followed by a run of one value of Y, and so on. In this example there are seven runs.

We give two more examples to show what might be indicated by the number of runs. If the five x's and seven y's had the orderings

$$\underline{x\ x\ x\ x}\ \underline{y}\ \underline{x}\ \underline{y\ y\ y\ y\ y\ y}$$

we might suspect that $F(z) \geq G(z)$. Note that there are four runs in this ordering. The ordered arrangement

$$\underline{y\ y\ y}\ \underline{x\ x}\ \underline{y}\ \underline{x\ x\ x}\ \underline{y\ y\ y}$$

might suggest that the medians of the two distributions are equal but that the spread of the Y distribution is greater than the spread of the X distribution, for example, that $\sigma_Y > \sigma_X$. These examples suggest that the hypothesis $F(z) = G(z)$ should be rejected if the number of runs is too small, where a small number of runs could be caused by differences in the location or in the spread of the two distributions.

Let the random variable R equal the number of runs in the combined ordered sample of n_1 observations of X and n_2 observations of Y. We shall find the distribution of R when $F(z) = G(z)$ and then describe a test of the hypothesis H_0: $F(z) = G(z)$.

Under H_0, all permutations of the n_1 observations of X and n_2 observations of Y have equal probabilities. We can select the n_1 positions for the n_1 values of X in

$$\binom{n_1 + n_2}{n_1}$$

ways, the probability of each arrangement being

$$\frac{1}{\binom{n_1 + n_2}{n_1}}.$$

To find $P(R = r)$, we must determine the number of permutations that yield r runs.

First suppose that $r = 2k$, where k is a positive integer. In this case the n_1 ordered values of X and the n_2 ordered values of Y must each be separated into k runs. We can form k runs of the n_1 values of X by inserting $k - 1$ dividers into the $n_1 - 1$ spaces between the values of X, with no more than one divider per space. This can be done in

$$\binom{n_1 - 1}{k - 1}$$

ways (see Exercise 8.11-4). Similarly, k runs of the n_2 values of Y can be formed in

$$\binom{n_2 - 1}{k - 1}$$

ways. These two sets of runs can be placed together to form $r = 2k$ runs, of which

$$\binom{n_1 - 1}{k - 1}\binom{n_2 - 1}{k - 1}$$

begin with a run of x's and

$$\binom{n_2 - 1}{k - 1}\binom{n_1 - 1}{k - 1}$$

begin with a run of y's. Thus

$$P(R = 2k) = \frac{2\binom{n_1 - 1}{k - 1}\binom{n_2 - 1}{k - 1}}{\binom{n_1 + n_2}{n_2}}, \qquad (8.11\text{-}1)$$

where $2k$ is an element of the space of R.

When $r = 2k + 1$, it is possible to have $k + 1$ runs of the ordered values of X and k runs of the ordered values of Y or k runs of Xs and $k + 1$ runs of Ys. We can form $k + 1$ runs of the n_1 values of X by inserting k dividers into the $n_1 - 1$ spaces between the values of X with no more than one divider per space in

$$\binom{n_1 - 1}{k}$$

ways. Similarly, k runs of n_2 values of Y can be done in

$$\binom{n_2 - 1}{k - 1}$$

ways. These two sets of runs can be placed together to form $2k + 1$ runs in

$$\binom{n_1 - 1}{k}\binom{n_2 - 1}{k - 1}$$

ways. In addition, $k + 1$ runs of the n_2 values of Y and k runs of the n_1 values of X can be placed together to form

$$\binom{n_2 - 1}{k}\binom{n_1 - 1}{k - 1}$$

sets of $2k + 1$ runs. Hence

$$P(R = 2k + 1) = \frac{\binom{n_1 - 1}{k}\binom{n_2 - 1}{k - 1} + \binom{n_1 - 1}{k - 1}\binom{n_2 - 1}{k}}{\binom{n_1 + n_2}{n_2}}, \qquad (8.11\text{-}2)$$

for $2k + 1$ in the space of R.

A test based on the number of runs can be used for testing the hypothesis $H_0 \colon F(z) = G(z)$. The hypothesis is rejected if the observed number of runs r is too small. That is, the critical region is of the form $r \leq c$, where the constant c is determined by using the p.d.f. of R to yield the desired significance level. The **run test** is sensitive to both differences in location and differences in spread of the two distributions.

EXAMPLE 8.11-1 Let X and Y equal the percentages of body fat for freshman women and men, respectively, with distribution functions $F(x)$ and $G(y)$. We shall use the run test to test the hypothesis H_0: $F(z) = G(z)$ against the alternative hypothesis H_1: $F(z) < G(z)$. (That is, the alternative hypothesis is that the X distribution is to the right of the Y distribution.) Ten observations of both X and Y that have been ordered are

$$
\begin{array}{llllll}
X: & 16.6 & 16.7 & 18.5 & 19.2 & 21.5 \\
   & 22.4 & 22.6 & 23.2 & 24.2 & 26.3
\end{array}
$$

$$
\begin{array}{llllll}
Y: & 9.4 & 9.7 & 11.3 & 11.8 & 13.3 \\
   & 15.6 & 16.1 & 16.5 & 18.2 & 21.7
\end{array}
$$

The critical region is of the form $r \le c$. To determine the value of c, we use Formulas 8.11-1 and 8.11-2 with $n_1 = n_2 = 10$. Table I in the Appendix is very useful for evaluating these probabilities. We have

$$
P(R = 2) = \frac{2}{84{,}756}; \quad P(R = 3) = \frac{18}{84{,}756};
$$

$$
P(R = 4) = \frac{162}{84{,}756}; \quad P(R = 5) = \frac{648}{84{,}756};
$$

$$
P(R = 6) = \frac{2592}{84{,}756}; \quad P(R = 7) = \frac{6048}{84{,}756}.
$$

The sum of these six probabilities is $9470/184{,}756 = 0.051$, so we can take for our critical region $C = \{r : r \le 7\}$ with a significance level of $\alpha = 0.051$. To determine the number of runs, we order the combined samples and underline adjacent x and y values.

$$
\begin{array}{llllllllll}
\underline{9.4} & \underline{9.7} & \underline{11.3} & \underline{11.8} & \underline{13.3} & \underline{15.6} & \underline{16.1} & \underline{16.5} & \underline{16.6} & \underline{16.7} \\
\underline{18.2} & \underline{18.5} & \underline{19.2} & \underline{21.5} & \underline{21.7} & \underline{22.4} & \underline{22.6} & \underline{23.2} & \underline{24.2} & \underline{26.3}
\end{array}
$$

We see that the number of runs is $r = 6$ so we reject the null hypothesis. Note that the p-value of this test is

$$
p\text{-value} = P(R \le 6) = \frac{3422}{84{,}756} = 0.0185.
$$

When n_1 and n_2 are large, say each is at least equal to 10, R can be approximated with a normally distributed random variable. That is, it can be shown that

$$
\mu_R = E(R) = \frac{2n_1 n_2}{n_1 + n_2} + 1,
$$

$$
\text{Var}(R) = \frac{(\mu_R - 1)(\mu_R - 2)}{n_1 + n_2 - 1} = \frac{2n_1 n_2 (2n_1 n_2 - n_1 - n_2)}{(n_1 + n_2)^2 (n_1 + n_2 - 1)},
$$

and

$$Z = \frac{R - \mu_R}{\sqrt{\mathrm{Var}(R)}}$$

is approximately $N(0, 1)$. The critical region for testing the null hypothesis H_0: $F(z) = G(z)$ is of the form $z \leq -z_\alpha$, where α is the desired significance level.

EXAMPLE 8.11-2   We use the normal approximation to calculate the significance level and the p-value for Example 8.11-1. With $n_1 = n_2 = 10$,

$$\mu_R = \frac{2(10)(10)}{10 + 10} + 1 = 11; \qquad \sigma_R^2 = \frac{(11 - 1)(11 - 2)}{19} = \frac{90}{19}.$$

With the critical region $C = \{r : \ r \leq 7\}$, the approximate significance level, using a half-unit correction for continuity, is

$$\alpha = P(R \leq 7) = P\left(\frac{R - 11}{\sqrt{90/19}} \leq \frac{7.5 - 11}{\sqrt{90/19}}\right)$$

$$\approx P(Z \leq -1.608) = 0.0539.$$

Note that this value compares very favorably with $\alpha = 0.051$ given in Example 8.11-1. Since $r = 6$, the approximate p-value, using a normal approximation, is

$$p\text{-value} = P(R \leq 6)$$

$$\approx P\left(Z \leq \frac{6.5 - 11}{\sqrt{90/19}}\right) = P(Z \leq -2.068) = 0.0193,$$

which is close to the p-value given in Example 8.11-1.   ◀

Applications of the run test include tests for randomness. Analysis of runs can also be useful in quality-control studies. To illustrate these applications, let $x_1, x_2, \ldots, x_k$ be the observed values of a random variable X, where the subscripts now designate the order in which the outcomes were observed and the observations are not arranged in the order of magnitude. It is possible, in a quality-control situation, that the observations are made systematically every hour, for example. Assume that k is even. The median divides the k numbers into a lower and an upper half. Replace each observation by L if it falls below the median and by U if it falls above the median. Then, for example, a sequence such as

$$U \, U \, U \, L \, U \, L \, L \, L$$

might suggest a trend toward decreasing values of X. If trend is the alternative hypothesis to randomness, the critical region would be of the form $r \leq c$. On the other hand, if we have a sequence such as

$$U \, L \, U \, L \, U \, L \, U \, L,$$

we would suspect a cyclic effect and would reject the hypothesis of randomness if r were too large. To test both for trend and cyclic effect, the critical region for testing the hypothesis of randomness is of the form $r \leq c_1$ or $r \geq c_2$.

If the sample size k is odd, the number of observations in the "upper half" and "lower half" will differ by one. In this case we will always put the extra observation in the upper group and, of course, $n_2 = n_1 + 1$. If the median is equal to a value that is tied with other values, we will again put the tied values in the upper group and then perform the test in which n_1 and n_2 are not equal to each other.

EXAMPLE 8.11-3   We shall use a sample of size $k = 14$ to test for both trend and cyclic effect. To determine the critical region for rejecting the hypothesis of randomness, we use the p.m.f. of R with $n_1 = n_2 = 7$. Since

$$P(R = 2) = P(R = 14) = \frac{2}{3432},$$

$$P(R = 3) = P(R = 13) = \frac{12}{3432},$$

$$P(R = 4) = P(R = 12) = \frac{72}{3432},$$

the critical region $\{r : r \leq 4$ or $r \geq 12\}$ would yield a test at a significance level of $\alpha = 172/3432 = 0.05$. The 14 observations are

$$
\begin{array}{ccccccc}
81.4 & 76.3 & 85.6 & 76.4 & 88.4 & 80.2 & 85.6 \\
84.6 & 78.3 & 82.8 & 88.1 & 85.4 & 87.7 & 86.6
\end{array}
$$

The median of these outcomes is $(84.6 + 85.4)/2 = 85.0$. Replacing each outcome with L if it falls below 85.0 and U if it falls above 85.0 yields the sequence

$$\underline{L\,L}\,\underline{U}\,\underline{L}\,\underline{U}\,\underline{L}\,\underline{U}\,\underline{L\,L\,L}\,\underline{U\,U\,U\,U}.$$

Since $r = 8$, the hypothesis of randomness is not rejected.   ◄

EXERCISES

8.11-1 Let the total lengths of the male and female trident lynx spiders be denoted by X and Y, respectively, with corresponding distribution functions $F(x)$ and $G(y)$. Measurement of the lengths, in millimeters, of eight male and eight female spiders yielded the following observations of X:

$$
\begin{array}{cccccccc}
5.40 & 5.55 & 6.00 & 5.00 & 5.70 & 5.20 & 5.45 & 4.95
\end{array}
$$

and of Y:

$$
\begin{array}{cccccccc}
6.20 & 6.25 & 5.75 & 5.85 & 6.55 & 6.05 & 5.50 & 6.65
\end{array}
$$

(a) Use these data to test the hypothesis H_0: $F(z) = G(z)$ against the alternative H_1: $F(z) > G(z)$. Let $\alpha = 0.10$, approximately, and use a run test.

(b) Use the two-sample Wilcoxon test to test H_0: $m_X = m_Y$ against the alternative H_1: $m_X < m_Y$ with $\alpha = 0.10$, approximately.

8.11-2 Let X and Y denote the times in hours per week that students in two different schools watch television. Let $F(x)$ and $G(y)$ denote the respective distribution functions. To test the hypothesis H_0: $F(z) = G(z)$, a random sample of eight students was selected from each school. Test this hypothesis using the following observations:

X:	16.75	19.25	22.00	20.50	22.50	15.50	17.25	20.75
Y:	24.75	21.50	19.75	17.50	22.75	23.50	13.00	19.00

8.11-3 A parade consists of 6 bands and 12 floats. How many different parade line-ups are possible if the parade begins and ends with a band and there is at least 1 float between each pair of bands?

8.11-4 Given six values of x, list the $\binom{5}{3} = 10$ ways in which three dividers can be inserted between the six values of x, no more than one divider per space.

8.11-5 Use the run test to test H_0: $F(z) = G(z)$ using the observations of X and Y that are given in Exercise 8.4-15.

(a) Show that if $C = \{r : r \leq 11\}$, then the significance level is $\alpha = 0.0473$, approximately.

(b) Calculate the value of r and state your conclusion.

(c) What is the p-value of this test?

8.11-6 Use the run test to test H_0: $F(z) = G(z)$ using the following observations of X and Y that have been ordered for your convenience. Let $\alpha = 0.05$, approximately.

X:	−2.0482	−1.5748	−0.8797	−0.7170	−0.4907
	−0.2051	0.1651	0.2893	0.3186	0.3550
	0.4056	0.6975	0.7113	0.7377	1.7356

Y:	−1.2311	−1.0228	−0.8836	−0.6684	−0.6157
	−0.5755	−0.1019	−0.0297	0.3781	0.7400
	0.8479	1.0901	1.1397	1.1748	1.2921

NOTE: These data were simulated on a computer using the standard normal distribution for X and a $U(-1.5, 1.5)$ for Y.

8.11-7 Use the following sample of size $k = 14$ to test the hypothesis of randomness against the alternative hypothesis of a trend effect at an $\alpha = 0.025$ significance level.

12.4	14.2	11.7	14.0	12.7	15.7	12.8
14.1	17.9	18.4	17.5	20.2	20.8	20.3

8.11-8 Use the following sample of size $k = 16$ to test the hypothesis of randomness against the alternative hypothesis of a cyclic effect:

12.4	31.8	22.2	24.5	17.9	24.6	15.7	27.3
22.7	26.0	14.5	22.0	21.8	31.9	11.5	28.3

What is your conclusion if **(a)** $\alpha = 0.0317$? **(b)** $\alpha = 0.10$?

8.11-9 A powdered soap manufacturer checks periodically throughout a day the weights of soap in the company's 6-pound boxes. At each of 22 times, four boxes are selected at random, and the average of the weights of the soap in the boxes is recorded. Use the following 22 average weights to test the hypothesis of randomness against the alternative hypothesis of a trend at an approximate significance level of $\alpha = 0.025$.

6.050	6.038	6.003	6.015	6.025	6.063	6.033	6.010
5.995	6.020	6.060	6.060	6.065	6.050	6.043	6.040
6.045	6.065	6.055	6.060	6.060	6.070		

8.11-10 Each hour a manufacturer of mints selects four mints at random from the production line and finds the average weight in grams. For 1 week the following average weights were observed. Use these weights to test the hypothesis of randomness against the alternative hypothesis of a cyclic effect.

```
21.2   21.7   21.3   21.4   21.8   21.9   21.6   21.7   21.3   21.9
21.3   22.0   21.3   21.5   21.6   21.3   21.6   21.9   21.3   21.6
21.9   22.0   21.9   21.4   21.4   21.3   21.7   21.6   21.5   21.7
21.3   21.7   21.0   21.3   21.3   21.6   20.9   21.4
```

8.11-11 It is claimed that X is $U(0, 1)$. A random sample of 14 observations of X yielded the following data:

$$
\begin{array}{ccccccc}
0.15 & 0.67 & 0.05 & 0.47 & 0.29 & 0.23 & 0.10 \\
0.01 & 0.96 & 0.92 & 0.51 & 0.73 & 0.91 & 0.82
\end{array}
$$

   (a) Test H_0: $m = 0.5$ against the two-sided alternative hypothesis H_1: $m \neq 0.5$ using the sign test with significance level $\alpha = 0.0574$.

   (b) Use a run test with these observations of X to test for both trend and cyclic effect at the approximate significance level $\alpha = 0.05$.

8.11-12 On an introductory statistics test, 27 form A and 29 form B tests were used. The order in which the tests were returned is listed below.

```
A   B   B   A   B   A   B   A   A   B   B   A   B   B
A   B   B   A   A   A   B   A   B   A   A   A   B   B
B   A   B   A   B   A   B   A   A   A   B   A   B   B
A   B   A   A   B   A   B   B   A   B   A   B   B   B
```

   (a) Use a run test to test the null hypothesis that this is a random sequence against the alternative hypothesis that there is a cyclic effect. Use the normal approximation and an $\alpha = 0.01$ significance level.

   (b) Based on your conclusion, what advice would you give to the professor who gave this test?

8.11-13 In Exercise 10.1-4 the United States birth rates are given for 1960–2002. In order, they are

```
23.7   23.3   22.4   21.7   21.1   19.4   18.4   17.8   17.6
17.9   18.4   17.2   15.6   14.8   14.8   14.6   14.6   15.1
15.0   15.6   15.9   15.8   15.9   15.5   15.5   15.8   15.6
15.7   15.9   16.3   16.7   16.2   15.8   15.4   15.0   14.6
14.4   14.2   14.3   14.2   14.4   14.1   13.9
```

Use the run test to test whether this is a random sequence against the alternative that there is a trend effect. Give the p-value of your test.

8.11-14 A window is manufactured that will be inserted into a car. For attaching the window, five studs are located in the frame in locations A, B, C, D, and E. Periodically, a window is selected randomly from the production line and a standard "stud pullout test" is performed that measures the force required to pull a stud out of the window. For each window, the tests are always performed in the order A, B, C, D, and E. Seventy observations that resulted from 14 windows were

```
140   159   138   102   84   126   147   126   103   92
149   155   135   120   94   149   143   109   101   86
144   154   120   105   97   149   151   140   103   99
157   140   120    96   87   146   137   120    93   89
149   154   139   100   84   148   142   130    98   81
112   135   109    84   87   112   135   109    84   87
126   118   135    90   77   125   126   131    78   75
```

For these data, the median is equal to 120. Use the run test to test whether this is a random sequence against the alternative hypothesis of a cyclic effect. Give the approximate p-value of the test. Interpret your conclusion. (See Exercise 8.11-14.)

CHAPTER

9

THEORY OF STATISTICAL TESTS

9.1* POWER OF A STATISTICAL TEST

In Chapter 8, we gave several tests of fairly common statistical hypotheses in such a way that we described the significance level α and the p-values of each. Of course, these tests were based on good (sufficient) statistics of the parameters, when the latter exist. In this section, we consider the probability of making the other type of error: accepting the null hypothesis H_0 when the alternative hypothesis H_1 is true. This consideration leads to ways to find most powerful tests of the null hypothesis H_0 against the alternative hypothesis H_1.

The first example introduces a new concept using a test about p, the probability of success. The sample size is kept small so that Appendix Table II can be used for finding probabilities. The application is one that you can actually perform.

EXAMPLE 9.1-1 Assume that a person, when given a name tag, puts it on either the right or left side. Let p equal the probability that the name tag is placed on the right side. We shall test the null hypothesis H_0: $p = 1/2$ against the composite alternative hypothesis H_1: $p < 1/2$. (Included with the null hypothesis are those values of p that are greater than 1/2. That is, we could think of H_0 as H_0: $p \geq 1/2$.) We shall give name tags to a random sample of $n = 20$ people, denoting the placements of their name tags with Bernoulli random variables, $X_1, X_2, \ldots, X_{20}$, where $X_i = 1$ if a person places

591

the name tag on the right and $X_i = 0$ if a person places the name tag on the left. We can then use for our test statistic $Y = \sum_{i=1}^{20} X_i$, which has a binomial distribution, $b(20, p)$. Say the critical region is defined by $C = \{y : y \leq 6\}$ or, equivalently, by $\{(x_1, x_2, \ldots, x_{20}) : \sum_{i=1}^{20} x_i \leq 6\}$. Since Y is $b(20, 1/2)$ if $p = 1/2$, the significance level of the corresponding test is

$$\alpha = P\left(Y \leq 6; \ p = \frac{1}{2}\right) = \sum_{y=0}^{6} \binom{20}{y}\left(\frac{1}{2}\right)^{20} = 0.0577,$$

using Table II in the Appendix. Of course, the probability β of the Type II error has different values with different values of p selected from the composite alternative hypothesis $H_1: p < 1/2$. For illustration, with $p = 1/4$,

$$\beta = P\left(7 \leq Y \leq 20; \ p = \frac{1}{4}\right) = \sum_{y=7}^{20} \binom{20}{y}\left(\frac{1}{4}\right)^{y}\left(\frac{3}{4}\right)^{20-y} = 0.2142,$$

whereas with $p = 1/10$,

$$\beta = P\left(7 \leq Y \leq 20; \ p = \frac{1}{10}\right) = \sum_{y=7}^{20} \binom{20}{y}\left(\frac{1}{10}\right)^{y}\left(\frac{9}{10}\right)^{20-y} = 0.0024.$$

Instead of considering the probability β of accepting H_0 when H_1 is true, we could compute the probability K of rejecting H_0 when H_1 is true. After all, β and $K = 1 - \beta$ provide the same information. Since K is a function of p, we denote this explicitly by writing $K(p)$. The probability

$$K(p) = \sum_{y=0}^{6} \binom{20}{y} p^y (1 - p)^{20-y}, \qquad 0 < p \leq \frac{1}{2},$$

is called the **power function of the test**. Of course, $\alpha = K(1/2) = 0.0577$, $1 - K(1/4) = 0.2142$ and $1 - K(1/10) = 0.0024$. The value of the power function at a specified p is called the **power** of the test at that point. For illustration, $K(1/4) = 0.7858$ and $K(1/10) = 0.9976$ are the powers at $p = 1/4$ and $p = 1/10$, respectively. An acceptable power function is one that assumes small values when H_0 is true and larger values when p differs much from $p = 1/2$. See Figure 9.1-1 for a graph of this power function.

In Example 9.1-1 we introduced the new concept of the power function of a test. We now show how the sample size can be selected so as to create a test with appropriate power.

EXAMPLE 9.1-2 Let $X_1, X_2, \ldots, X_n$ be a random sample of size n from the normal distribution $N(\mu, 100)$, which we can suppose is a possible distribution of scores of students in a statistics course that uses a new method of teaching (e.g., computer-related materials). We wish to decide between $H_0: \mu = 60$ (*no change* hypothesis

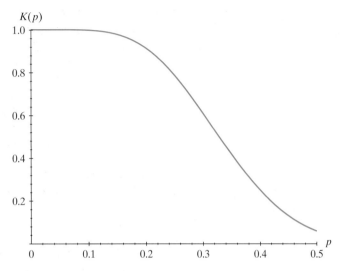

Figure 9.1-1: Power function: $K(p) = P(Y \leq 6; p)$, where Y is $b(20, p)$

because, let us say, this was the mean by the previous method of teaching) and the research worker's hypothesis H_1: $\mu > 60$. Let us consider a sample of size $n = 25$. Of course, the sample mean $\overline{X}$ is the maximum likelihood estimator of μ, and thus it seems reasonable to base our decision on this statistic. Initially, we use the rule to reject H_0 and accept H_1 if and only if $\overline{x} \geq 62$. What are the consequences of this test? These are summarized in the power function of the test.

We first find the probability of rejecting H_0: $\mu = 60$ for various values of $\mu \geq 60$. The probability of rejecting H_0 is given by

$$K(\mu) = P(\overline{X} \geq 62; \mu)$$

because this test calls for the rejection of H_0: $\mu = 60$ when $\overline{x} \geq 62$. When the new process has the general mean μ, we know that $\overline{X}$ has the normal distribution $N(\mu, 100/25 = 4)$. Accordingly,

$$K(\mu) = P\left(\frac{\overline{X} - \mu}{2} \geq \frac{62 - \mu}{2}; \mu\right)$$

$$= 1 - \Phi\left(\frac{62 - \mu}{2}\right), \qquad 60 \leq \mu,$$

is the probability of rejecting H_0: $\mu = 60$ using this particular test. Several values of $K(\mu)$ are given in Table 9.1-1. Figure 9.1-2 depicts the graph of the function $K(\mu)$.

The probability $K(\mu)$ of rejecting H_0: $\mu = 60$ is called the power function of the test. At the value μ_1 of the parameter, $K(\mu_1)$ is the power at μ_1. The power at $\mu = 60$ is $K(60) = 0.1587$, and this is the probability of rejecting H_0:

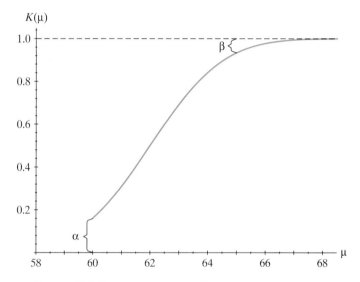

μ	$K(\mu)$
60	0.1587
61	0.3085
62	0.5000
63	0.6915
64	0.8413
65	0.9332
66	0.9772

Table 9.1-1: Values of the Power Function

Figure 9.1-2: Power function: $K(\mu) = 1 - \Phi([62 - \mu]/2)$

$\mu = 60$ when H_0 is true. That is, $K(60) = 0.1587 = \alpha$ is the probability of the Type I error and is called the significance level of the test.

The power at $\mu = 65$ is $K(65) = 0.9332$, and this is the probability of making the correct decision (namely, rejecting H_0: $\mu = 60$ when $\mu = 65$). Hence we are pleased that here it is large. When $\mu = 65$, $1 - K(65) = 0.0668$ is the probability of not rejecting H_0: $\mu = 60$ when $\mu = 65$; that is, it is the probability of a Type II error and is denoted by $\beta = 0.0668$. These α and β values are displayed on Figure 9.1-2. Clearly, the probability $\beta = 1 - K(\mu_1)$ of the Type II error depends on which value, say μ_1, is taken in the alternative hypothesis H_1: $\mu > 60$. Thus while $\beta = 0.0668$ when $\mu = 65$, β is equal to $1 - K(63) = 0.3085$ when $\mu = 63$.

Frequently, statisticians like to have the significance level α smaller than 0.1587, say around 0.05 or less, because it is a probability of an error, namely the Type I error. Hence if we would like $\alpha = 0.05$, then we can no longer use, with $n = 25$, the critical region $\overline{x} \geq 62$, but $\overline{x} \geq c$, where c is selected so that

$$K(60) = P(\overline{X} \geq c; \mu = 60) = 0.05.$$

However, when $\mu = 60$, $\overline{X}$ is $N(60, 4)$ and

$$K(60) = P\left(\frac{\overline{X} - 60}{2} \geq \frac{c - 60}{2}; \mu = 60\right)$$

$$= 1 - \Phi\left(\frac{c - 60}{2}\right) = 0.05.$$

From Table Va in the Appendix we have that

$$\frac{c - 60}{2} = 1.645 = z_{0.05} \qquad \text{and} \qquad c = 60 + 3.29 = 63.29.$$

Although this change reduces α from 0.1587 to 0.05, it increases β from 0.0668 at $\mu = 65$ to

$$\beta = 1 - P(\overline{X} \geq 63.29; \mu = 65)$$

$$= 1 - P\left(\frac{\overline{X} - 65}{2} \geq \frac{63.29 - 65}{2}; \mu = 65\right)$$

$$= \Phi(-0.855) = 0.1963.$$

In general, without changing the sample size or the type of test of the hypothesis, a decrease in α causes an increase in β, and a decrease in β causes an increase in α. Both probabilities α and β of the two types of errors can be decreased only by increasing the sample size or, in some way, constructing a better test of the hypothesis.

For example, if $n = 100$ and we desire a test with significance level $\alpha = 0.05$, then

$$\alpha = P(\overline{X} \geq c; \mu = 60) = 0.05$$

means, since $\overline{X}$ is $N(\mu, 100/100 = 1)$,

$$P\left(\frac{\overline{X} - 60}{1} \geq \frac{c - 60}{1}; \mu = 60\right) = 0.05$$

and $c - 60 = 1.645$. Thus $c = 61.645$. The power function is

$$K(\mu) = P(\overline{X} \geq 61.645; \mu)$$

$$= P\left(\frac{\overline{X} - \mu}{1} \geq \frac{61.645 - \mu}{1}; \mu\right) = 1 - \Phi(61.645 - \mu).$$

In particular, this means that β at $\mu = 65$ is

$$\beta = 1 - K(\mu) = \Phi(61.645 - 65) = \Phi(-3.355) \approx 0;$$

so, with $n = 100$, both α and β have decreased from their respective original values of 0.1587 and 0.0668 when $n = 25$.

Rather than guess at the value of n, an ideal power function determines the sample size. Let us use a critical region of the form $\overline{x} \geq c$. Further, suppose that we want $\alpha = 0.025$ and, when $\mu = 65$, $\beta = 0.05$. Thus, since $\overline{X}$ is $N(\mu, 100/n)$,

$$0.025 = P(\overline{X} \geq c; \mu = 60) = 1 - \Phi\left(\frac{c - 60}{10/\sqrt{n}}\right)$$

and

$$0.05 = 1 - P(\overline{X} \geq c; \mu = 65) = \Phi\left(\frac{c - 65}{10/\sqrt{n}}\right).$$

That is,

$$\frac{c - 60}{10/\sqrt{n}} = 1.96 \qquad \text{and} \qquad \frac{c - 65}{10/\sqrt{n}} = -1.645.$$

Solving these equations simultaneously for c and $10/\sqrt{n}$, we obtain

$$c = 60 + 1.96 \, \frac{5}{3.605} = 62.718;$$

$$\frac{10}{\sqrt{n}} = \frac{5}{3.605}.$$

Thus

$$\sqrt{n} = 7.21 \qquad \text{and} \qquad n = 51.98.$$

Since n must be an integer, we would use $n = 52$ and obtain $\alpha = 0.025$ and $\beta = 0.05$, approximately.   ◀

The next example is an extension of Example 9.1-1.

EXAMPLE 9.1-3   To test H_0: $p = 1/2$ against H_1: $p < 1/2$, we take a random sample of Bernoulli trials, $X_1, X_2, \ldots, X_n$, and use for our test statistic $Y = \sum_{i=1}^{n} X_i$, which has a binomial distribution $b(n, p)$. Let the critical region be defined by $C = \{y : y \leq c\}$. The power function for this test is defined by $K(p) = P(Y \leq c; p)$. We shall find the values of n and c so that $K(1/2) = 0.05$ and $K(1/4) = 0.90$, approximately. That is, we would like the significance level to be $\alpha = K(1/2) = 0.05$ and the power at $p = 1/4$ to equal 0.90. We proceed as follows:

$$0.05 = P\left(Y \leq c; \; p = \frac{1}{2}\right) = P\left(\frac{Y - n/2}{\sqrt{n(1/2)(1/2)}} \leq \frac{c - n/2}{\sqrt{n(1/2)(1/2)}}\right)$$

implies that

$$(c - n/2)/\sqrt{n/4} \approx -1.645;$$

and

$$0.90 = P\left(Y \leq c; \; p = \frac{1}{4}\right) = P\left(\frac{Y - n/4}{\sqrt{n(1/4)(3/4)}} \leq \frac{c - n/4}{\sqrt{n(1/4)(3/4)}}\right)$$

implies that

$$(c - n/4)/\sqrt{3n/16} \approx 1.282.$$

Therefore,

$$\frac{n}{4} \approx 1.645\sqrt{\frac{n}{4}} + 1.282\sqrt{\frac{3n}{16}} \qquad \text{and} \qquad \sqrt{n} \approx 4(1.378) = 5.512.$$

Thus n is about equal to 31, and from either of the first two approximate equalities we find that c is about equal to 10.9. Using $n = 31$ and $c = 10.9$ means that $K(1/2) = 0.05$ and $K(1/4) = 0.90$ are only approximate. In fact, since Y must be an integer, we could let $c = 10.5$. Then, with $n = 31$,

$$\alpha = K\left(\frac{1}{2}\right) = P\left(Y \leq 10.5; \; p = \frac{1}{2}\right) \approx 0.0362;$$

$$K\left(\frac{1}{4}\right) = P\left(Y \leq 10.5; \; p = \frac{1}{4}\right) \approx 0.8729.$$

Or, we could let $c = 11.5$ and $n = 31$, in which case

$$\alpha = K\left(\frac{1}{2}\right) = P\left(Y \leq 11.5; \; p = \frac{1}{2}\right) \approx 0.0558;$$

$$K\left(\frac{1}{4}\right) = P\left(Y \leq 11.5; \; p = \frac{1}{4}\right) \approx 0.9235. \qquad \blacktriangleleft$$

EXERCISES

9.1-1 A certain size bag is designed to hold 25 pounds of potatoes. A farmer fills such bags in the field. Assume that the weight X of potatoes in a bag is $N(\mu, 9)$. We shall test the null hypothesis $H_0: \mu = 25$ against the alternative hypothesis $H_1: \mu < 25$. Let X_1, X_2, X_3, X_4 be a random sample of size 4 from this distribution, and let the critical region C for this test be defined by $\bar{x} \leq 22.5$, where $\bar{x}$ is the observed value of $\bar{X}$.

(a) What is the power function $K(\mu)$ of this test? In particular, what is the significance level $\alpha = K(25)$ for your test?

(b) If the random sample of four bags of potatoes yielded the values $x_1 = 21.24$, $x_2 = 24.81$, $x_3 = 23.62$, $x_4 = 26.82$, would you accept or reject H_0 using your test?

(c) What is the p-value associated with the $\overline{x}$ in part (b)?

9.1-2 Let X equal the number of milliliters in a bottle that has a label volume of 350 milliliters. Assume that the distribution of X is $N(\mu, 4)$. To test the null hypothesis H_0: $\mu = 355$ against the alternative hypothesis H_1: $\mu < 355$, let the critical region be defined by $C = \{\overline{x}: \overline{x} \leq 354.05\}$, where $\overline{x}$ is the sample mean of the contents of a random sample of $n = 12$ bottles.

(a) Find the power function $K(\mu)$ for this test.

(b) What is the (approximate) significance level of this test?

(c) Find the values of $K(354.05)$ and $K(353.1)$ and sketch the graph of the power function.

(d) Use the following 12 observations to state your conclusion of this test.

$$350 \quad 353 \quad 354 \quad 356 \quad 353 \quad 352$$
$$354 \quad 355 \quad 357 \quad 353 \quad 354 \quad 355$$

(e) What is the approximate p-value of this test?

9.1-3 Assume that SAT mathematics scores of students who attend small liberal arts colleges are $N(\mu, 8100)$. We shall test H_0: $\mu = 530$ against the alternative hypothesis H_1: $\mu < 530$. Given a random sample of size $n = 36$ SAT mathematics scores, let the critical region be defined by $C = \{\overline{x}: \overline{x} \leq 510.77\}$, where $\overline{x}$ is the observed mean of the sample.

(a) Find the power function, $K(\mu)$, for this test.

(b) What is the value of the significance level of this test?

(c) What is the value of $K(510.77)$?

(d) Sketch the graph of the power function.

(e) What is the p-value associated with (i) $\overline{x} = 507.35$; (ii) $\overline{x} = 497.45$?

9.1-4 Let X be $N(\mu, 100)$. To test H_0: $\mu = 80$ against H_1: $\mu > 80$, let the critical region be defined by $C = \{(x_1, x_2, \ldots, x_{25}): \overline{x} \geq 83\}$, where $\overline{x}$ is the sample mean of a random sample of size $n = 25$ from this distribution.

(a) What is the power function $K(\mu)$ for this test?

(b) What is the significance level of this test?

(c) What are the values of $K(80)$, $K(83)$, and $K(86)$?

(d) Sketch the graph of the power function.

(e) What is the p-value corresponding to $\overline{x} = 83.41$?

9.1-5 Let X equal the yield of alfalfa in tons per acre per year. Assume that X is $N(1.5, 0.09)$. It is hoped that new fertilizer will increase the average yield. We shall test the null hypothesis H_0: $\mu = 1.5$ against the alternative hypothesis H_1: $\mu > 1.5$. Assume that the variance continues to equal $\sigma^2 = 0.09$ with the new fertilizer. Using $\overline{X}$, the mean of a random sample of size n, as the test statistic, reject H_0 if $\overline{x} \geq c$. Find n and c so that the power function $K(\mu) = P(\overline{X} \geq c : \mu)$ is such that $\alpha = K(1.5) = 0.05$ and $K(1.7) = 0.95$.

9.1-6 Let X equal the number of pounds of butterfat produced by a Holstein cow during the 305-day milking period following the birth of a calf. Assume that the distribution of X is $N(\mu, 140^2)$. To test the null hypothesis H_0: $\mu = 715$ against the alternative hypothesis H_1: $\mu < 715$, let the critical region be defined by $C = \{\overline{x}: \overline{x} \leq 668.94\}$, where $\overline{x}$ is the sample mean of $n = 25$ butterfat weights from 25 cows selected at random.

(a) Find the power function $K(\mu)$ for this test.

(b) What is the significance level of this test?

(c) What are the values of $K(668.94)$ and $K(622.88)$?

(d) Sketch a graph of the power function.

(e) What is your conclusion using the following 25 observations of X?

425	710	661	664	732	714	934	761	744
653	725	657	421	573	535	602	537	405
874	791	721	849	567	468	975		

(f) What is the approximate p-value of this test?

9.1-7 In Exercise 9.1-6, let $C = \{\bar{x} : \bar{x} \leq c\}$ be the critical region. Find values for n and c so that the significance level of this test is $\alpha = 0.05$ and the power at $\mu = 650$ is 0.90.

9.1-8 Let X have a Bernoulli distribution with p.m.f.

$$f(x; p) = p^x (1-p)^{1-x}, \qquad x = 0, 1, \qquad 0 \leq p \leq 1.$$

We would like to test the null hypothesis H_0: $p \leq 0.4$ against the alternative hypothesis H_1: $p > 0.4$. For the test statistic use $Y = \sum_{i=1}^{n} X_i$, where $X_1, X_2, \ldots, X_n$ is a random sample of size n from this Bernoulli distribution. Let the critical region be of the form $C = \{y : y \geq c\}$.

(a) Let $n = 100$. On the same set of axes, sketch the graphs of the power functions corresponding to the three critical regions, $C_1 = \{y : y \geq 40\}$, $C_2 = \{y : y \geq 50\}$, and $C_3 = \{y : y \geq 60\}$. Use the normal approximation to compute the probabilities.

(b) Let $C = \{y : y \geq 0.45n\}$. On the same set of axes, sketch the graphs of the power functions corresponding to the three samples of sizes 10, 100, and 1000.

9.1-9 Let p denote the probability that, for a particular tennis player, the first serve is good. Since $p = 0.40$, this player decided to take lessons in order to increase p. When the lessons are completed, the hypothesis H_0: $p = 0.40$ will be tested against H_1: $p > 0.40$ based on $n = 25$ trials. Let y equal the number of first serves that are good, and let the critical region be defined by $C = \{y : y \geq 14\}$.

(a) Find the power function $K(p)$ for this test.

(b) What is the value of the significance level, $\alpha = K(0.40)$? Use Appendix Table II.

(c) Evaluate $K(p)$ at $p = 0.45, 0.50, 0.60, 0.70, 0.80$, and 0.90. Use Table II.

(d) Sketch the graph of the power function.

(e) If $y = 15$ following the lessons, would H_0 be rejected?

(f) What is the p-value associated with $y = 15$?

9.1-10 Let $X_1, X_2, \ldots, X_8$ be a random sample of size $n = 8$ from a Poisson distribution with mean λ. Reject the simple null hypothesis H_0: $\lambda = 0.5$ and accept H_1: $\lambda > 0.5$ if the observed sum $\sum_{i=1}^{8} x_i \geq 8$.

(a) Compute the significance level α of the test.

(b) Find the power function $K(\lambda)$ of the test as a sum of Poisson probabilities.

(c) Using Table III in the Appendix, determine $K(0.75)$, $K(1)$, and $K(1.25)$.

9.1-11 Let p equal the fraction defective of a certain manufactured item. To test H_0: $p = 1/26$ against H_1: $p > 1/26$, we inspect n items selected at random and let Y be the number of defective items in this sample. We reject H_0 if the observed $y \geq c$. Find n and c so that $\alpha = K(1/26) = 0.05$ and $K(1/10) = 0.90$ approximately, where $K(p) = P(Y \geq c; p)$.

HINT: Use either the normal or Poisson approximation to help solve this exercise.

9.1-12 Let X_1, X_2, X_3 be a random sample of size $n = 3$ from an exponential distribution with mean $\theta > 0$. Reject the simple null hypothesis H_0: $\theta = 2$ and accept the composite alternative hypothesis H_1: $\theta < 2$ if the observed sum $\sum_{i=1}^{3} x_i \leq 2$.

(a) What is the power function $K(\theta)$ written as an integral?
(b) Using integration by parts, define the power function as a summation.
(c) With the help of Table III in the Appendix, determine $\alpha = K(2)$, $K(1)$, $K(1/2)$, and $K(1/4)$.

BEST CRITICAL REGIONS

In this section we consider the properties a satisfactory test (or critical region) should possess. To introduce our investigation, we begin with a nonstatistical example.

EXAMPLE 9.2-1

Say that you have α dollars with which to buy books. Suppose further that you are not interested in the books themselves but only in filling as much of your bookshelves as possible. How do you decide which books to buy? Does the following approach seem reasonable? First of all, take all the available free books. Then start choosing those books for which the cost of filling an inch of bookshelf is smallest. That is, choose those books for which the ratio c/w is a minimum, where w is the width of the book in inches and c is the cost of the book. Continue choosing books this way until you have spent the α dollars. ◀

To see how Example 9.2-1 provides the background for selecting a good critical region of size α, let us consider a test of the simple hypothesis H_0: $\theta = \theta_0$ against a simple alternative hypothesis H_1: $\theta = \theta_1$. In this discussion we assume that the random variables $X_1, X_2, \ldots, X_n$ under consideration have a joint p.m.f. of the discrete type, which we here denote by $L(\theta; x_1, x_2, \ldots, x_n)$. That is,

$$P(X_1 = x_1, X_2 = x_2, \ldots, X_n = x_n) = L(\theta; x_1, x_2, \ldots, x_n).$$

A critical region C of size α is a set of points $(x_1, x_2, \ldots, x_n)$ with the probability of α when $\theta = \theta_0$. For a good test, this set C of points should have a large probability when $\theta = \theta_1$ because, under H_1: $\theta = \theta_1$, we wish to reject H_0: $\theta = \theta_0$. Accordingly, the first point we would place in the critical region C is the one with the smallest ratio

$$\frac{L(\theta_0; x_1, x_2, \ldots, x_n)}{L(\theta_1; x_1, x_2, \ldots, x_n)}.$$

That is, the "cost" in terms of probability under H_0: $\theta = \theta_0$ is small compared to the probability that we can "buy" if $\theta = \theta_1$. The next point to add to C would be the one with the next smallest ratio. We would continue to add points to C in this manner until the probability of C, under H_0: $\theta = \theta_0$, equals α. In this way, we have achieved, for the given significance level α, the region C with the largest probability when H_1: $\theta = \theta_1$ is true. We now formalize this discussion by defining a best critical region and proving the well-known Neyman-Pearson lemma.

Definition 9.2-1   Consider the test of the simple null hypothesis $H_0: \theta = \theta_0$ against the simple alternative hypothesis $H_1: \theta = \theta_1$. Let C be a critical region of size α; that is, $\alpha = P(C; \theta_0)$. Then C is a *best critical region of size α* if, for every other critical region D of size $\alpha = P(D; \theta_0)$, we have that

$$P(C; \theta_1) \geq P(D; \theta_1).$$

That is, when $H_1: \theta = \theta_1$ is true, the probability of rejecting $H_0: \theta = \theta_0$ using the critical region C is at least as great as the corresponding probability using any other critical region D of size α.

Thus a best critical region of size α is the critical region that has the greatest power among all critical regions of size α. The Neyman-Pearson lemma gives sufficient conditions for a best critical region of size α.

Theorem 9.2-1   (**Neyman-Pearson Lemma**) Let $X_1, X_2, \ldots, X_n$ be a random sample of size n from a distribution with p.d.f. or p.m.f. $f(x; \theta)$, where θ_0 and θ_1 are two possible values of θ. Denote the joint p.d.f. or p.m.f. of $X_1, X_2, \ldots, X_n$ by the likelihood function

$$L(\theta) = L(\theta; x_1, x_2, \ldots, x_n) = f(x_1; \theta) f(x_2; \theta) \ldots f(x_n; \theta).$$

If there exist a positive constant k and a subset C of the sample space such that

(a) $P[(X_1, X_2, \ldots, X_n) \in C; \theta_0] = \alpha,$

(b) $\dfrac{L(\theta_0)}{L(\theta_1)} \leq k$ for $(x_1, x_2, \ldots, x_n) \in C,$

(c) $\dfrac{L(\theta_0)}{L(\theta_1)} \geq k$ for $(x_1, x_2, \ldots, x_n) \in C',$

then C is a best critical region of size α for testing the simple null hypothesis $H_0: \theta = \theta_0$ against the simple alternative hypothesis $H_1: \theta = \theta_1$.

Proof: We prove the theorem when the random variables are of the continuous type; for discrete-type random variables, replace the integral signs by summation signs. To simplify the exposition, we shall use the following notation:

$$\int_B L(\theta) = \int \cdots \int_B L(\theta; x_1, x_2, \ldots, x_n) \, dx_1 \, dx_2 \cdots dx_n.$$

Assume that there exists another critical region of size α, say D, such that, in this new notation,

$$\alpha = \int_C L(\theta_0) = \int_D L(\theta_0).$$

So we have

$$0 = \int_C L(\theta_0) - \int_D L(\theta_0)$$

$$= \int_{C \cap D'} L(\theta_0) + \int_{C \cap D} L(\theta_0) - \int_{C \cap D} L(\theta_0) - \int_{C' \cap D} L(\theta_0)$$

and hence

$$0 = \int_{C \cap D'} L(\theta_0) - \int_{C' \cap D} L(\theta_0).$$

By hypothesis (b), $kL(\theta_1) \geq L(\theta_0)$ at each point in C and therefore in $C \cap D'$; thus

$$k \int_{C \cap D'} L(\theta_1) \geq \int_{C \cap D'} L(\theta_0).$$

By hypothesis (c), $kL(\theta_1) \leq L(\theta_0)$ at each point in C', and therefore in $C' \cap D$; thus we obtain

$$k \int_{C' \cap D} L(\theta_1) \leq \int_{C' \cap D} L(\theta_0).$$

Therefore,

$$0 = \int_{C \cap D'} L(\theta_0) - \int_{C' \cap D} L(\theta_0) \leq (k) \left\{ \int_{C \cap D'} L(\theta_1) - \int_{C' \cap D} L(\theta_1) \right\}.$$

That is.

$$0 \leq (k) \left\{ \int_{C \cap D'} L(\theta_1) + \int_{C \cap D} L(\theta_1) - \int_{C \cap D} L(\theta_1) - \int_{C' \cap D} L(\theta_1) \right\}$$

or, equivalently,

$$0 \leq (k) \left\{ \int_C L(\theta_1) - \int_D L(\theta_1) \right\}.$$

Thus

$$\int_C L(\theta_1) \geq \int_D L(\theta_1);$$

that is, $P(C; \theta_1) \geq P(D; \theta_1)$. Since that is true for every critical region D of size α, C is a best critical region of size α. ◯

For a realistic application of the Neyman-Pearson lemma, consider the following, in which the test is based on a random sample from a normal distribution.

EXAMPLE 9.2-2   Let $X_1, X_2, \ldots, X_n$ be a random sample from the normal distribution $N(\mu, 36)$. We shall find the best critical region for testing the simple hypothesis H_0: $\mu = 50$ against the simple alternative hypothesis H_1: $\mu = 55$. Using the ratio of the likelihood functions, namely $L(50)/L(55)$, we shall find those points in the sample space for which this ratio is less than or equal to some constant k. That is, we shall solve the following inequality:

$$\frac{L(50)}{L(55)} = \frac{(72\pi)^{-n/2} \exp\left[-\left(\dfrac{1}{72}\right) \sum_{i=1}^{n} (x_i - 50)^2\right]}{(72\pi)^{-n/2} \exp\left[-\left(\dfrac{1}{72}\right) \sum_{i=1}^{n} (x_i - 55)^2\right]}$$

$$= \exp\left[-\left(\frac{1}{72}\right)\left(10 \sum_{i=1}^{n} x_i + n50^2 - n55^2\right)\right] \le k.$$

If we take the natural logarithm of each member of the inequality, we find that

$$-10 \sum_{i=1}^{n} x_i - n50^2 + n55^2 \le (72)\ln k.$$

Thus

$$\frac{1}{n} \sum_{i=1}^{n} x_i \ge -\frac{1}{10n} [n50^2 - n55^2 + (72)\ln k]$$

or, equivalently,

$$\overline{x} \ge c,$$

where $c = -(1/10n)[n50^2 - n55^2 + (72)\ln k]$. Thus $L(50)/L(55) \le k$ is equivalent to $\overline{x} \ge c$. A best critical region is, according to the Neyman-Pearson lemma,

$$C = \{(x_1, x_2, \ldots, x_n): \overline{x} \ge c\},$$

where c is selected so that the size of the critical region is α. Say $n = 16$ and $c = 53$. Since $\overline{X}$ is $N(50, 36/16)$ under H_0, we have

$$\alpha = P(\overline{X} \ge 53; \mu = 50)$$

$$= P\left(\frac{\overline{X} - 50}{6/4} \ge \frac{3}{6/4}; \mu = 50\right) = 1 - \Phi(2) = 0.0228. \qquad \blacktriangleleft$$

This last example illustrates what is often true, namely, that the inequality

$$L(\theta_0)/L(\theta_1) \le k$$

can be expressed in terms of a function $u(x_1, x_2, \ldots, x_n)$, say

$$u(x_1, \ldots, x_n) \le c_1$$

or

$$u(x_1, \ldots, x_n) \ge c_2,$$

where c_1 or c_2 is selected so that the size of the critical region is α. Thus the test can be based on the statistic $u(x_1, \ldots, x_n)$. Also, for illustration, if we want α to be a given value, say 0.05, we could then choose our c_1 or c_2. In Example 9.2-2, with $\alpha = 0.05$ we want

$$0.05 = P(\overline{X} \ge c; \ \mu = 50)$$

$$= P\left(\frac{\overline{X} - 50}{6/4} \ge \frac{c - 50}{6/4}; \ \mu = 50 \right) = 1 - \Phi\left(\frac{c - 50}{6/4} \right).$$

Hence it must be true that $(c - 50)/(3/2) = 1.645$, or, equivalently,

$$c = 50 + \frac{3}{2}(1.645) \approx 52.47.$$

EXAMPLE 9.2-3 Let $X_1, X_2, \ldots, X_n$ denote a random sample of size n from a Poisson distribution with mean λ. A best critical region for testing H_0: $\lambda = 2$ against H_1: $\lambda = 5$ is given by

$$\frac{L(2)}{L(5)} = \frac{2^{\Sigma x_i} e^{-2n}}{x_1! x_2! \ldots x_n!} \frac{x_1! x_2! \ldots x_n!}{5^{\Sigma x_i} e^{-5n}} \le k.$$

This inequality is equivalent to

$$\left(\frac{2}{5} \right)^{\Sigma x_i} e^{3n} \le k \qquad \text{and} \qquad (\Sigma x_i) \ln\left(\frac{2}{5} \right) + 3n \le \ln k.$$

Since $\ln(2/5) < 0$, this is the same as

$$\sum_{i=1}^{n} x_i \ge \frac{\ln k - 3n}{\ln(2/5)} = c.$$

If $n = 4$ and $c = 13$, then

$$\alpha = P\left(\sum_{i=1}^{4} X_i \geq 13; \; \lambda = 2\right) = 1 - 0.936 = 0.064,$$

from Table III in the Appendix, since $\sum_{i=1}^{4} X_i$ has a Poisson distribution with mean 8 when $\lambda = 2$. ◀

When $H_0: \theta = \theta_0$ and $H_1: \theta = \theta_1$ are both simple hypotheses, a critical region of size α is a best critical region if the probability of rejecting H_0 when H_1 is true is a maximum when compared with all other critical regions of size α. The test using the best critical region is called a **most powerful test** because it has the greatest value of the power function at $\theta = \theta_1$ when compared with that of other tests of significance level α. If H_1 is a composite hypothesis, the power of a test depends on each simple alternative in H_1.

Definition 9.2-2   A test, defined by a critical region C of size α, is a *uniformly most powerful test* if it is a most powerful test against each simple alternative in H_1. The critical region C is called a *uniformly most powerful critical region of size α*.

Let us reconsider Example 9.2-2 when the alternative hypothesis is composite.

EXAMPLE 9.2-4   Let $X_1, \ldots, X_n$ be a random sample from $N(\mu, 36)$. We have seen that when testing $H_0: \mu = 50$ against $H_1: \mu = 55$, a best critical region C is defined by $C = \{(x_1, x_2, \ldots, x_n): \overline{x} \geq c\}$, where c is selected so that the significance level is α. Now consider testing $H_0: \mu = 50$ against the one-sided composite alternative hypothesis $H_1: \mu > 50$. For each simple hypothesis in H_1, say $\mu = \mu_1$, the quotient of the likelihood functions is

$$\frac{L(50)}{L(\mu_1)} = \frac{(72\pi)^{-n/2} \exp\left[-\left(\dfrac{1}{72}\right) \sum_{i=1}^{n} (x_i - 50)^2\right]}{(72\pi)^{-n/2} \exp\left[-\left(\dfrac{1}{72}\right) \sum_{i=1}^{n} (x_i - \mu_1)^2\right]}$$

$$= \exp\left[-\frac{1}{72}\left\{2(\mu_1 - 50)\sum_{i=1}^{n} x_i + n(50^2 - \mu_1^2)\right\}\right].$$

Now $L(50)/L(\mu_1) \leq k$ if and only if

$$\overline{x} \geq \frac{(-72)\ln(k)}{2n(\mu_1 - 50)} + \frac{50 + \mu_1}{2} = c.$$

Thus the best critical region of size α for testing $H_0: \mu = 50$ against $H_1: \mu = \mu_1$, where $\mu_1 > 50$, is given by $C = \{(x_1, x_2, \ldots, x_n): \overline{x} \geq c\}$, where c is selected such that $P(\overline{X} \geq c; H_0: \mu = 50) = \alpha$. Note that the same value of c can be used for each $\mu_1 > 50$, but (of course) k does not remain the same. Since the critical region C defines a test that is most powerful against each simple alternative $\mu_1 > 50$, this is a uniformly most powerful test, and C is a uniformly most powerful critical region of size α. Again if $\alpha = 0.05$, then $c \approx 52.47$. ◀

EXAMPLE 9.2-5

Let Y have the binomial distribution $b(n, p)$. To find a uniformly most powerful test of the simple null hypothesis $H_0: p = p_0$ against the one-sided alternative hypothesis $H_1: p > p_0$, consider, with $p_1 > p_0$,

$$\frac{L(p_0)}{L(p_1)} = \frac{\binom{n}{y} p_0^y (1 - p_0)^{n-y}}{\binom{n}{y} p_1^y (1 - p_1)^{n-y}} \leq k.$$

This is equivalent to

$$\left[\frac{p_0(1 - p_1)}{p_1(1 - p_0)} \right]^y \left[\frac{1 - p_0}{1 - p_1} \right]^n \leq k$$

and

$$y \ln \left[\frac{p_0(1 - p_1)}{p_1(1 - p_0)} \right] \leq \ln k - n \ln \left[\frac{1 - p_0}{1 - p_1} \right].$$

Since $p_0 < p_1$, $p_0(1 - p_1) < p_1(1 - p_0)$, then $\ln[p_0(1 - p_1)/p_1(1 - p_0)] < 0$. Thus we have

$$\frac{y}{n} \geq \frac{\ln k - n \ln[(1 - p_0)/(1 - p_1)]}{n \ln[p_0(1 - p_1)/p_1(1 - p_0)]} = c$$

for each $p_1 > p_0$.

It is interesting to note that if the alternative hypothesis is the one-sided H_1: $p < p_0$, then a uniformly most powerful test is of the form $(y/n) \leq c$. Thus we see that the tests of $H_0: p = p_0$ against the one-sided alternatives given in Table 8.1-1 are uniformly most powerful. ◀

Exercise 9.2-5 will demonstrate that uniformly most powerful tests do not always exist; in particular, they do not usually exist when the composite alternative hypothesis is two-sided.

Remark* We close this section with one easy but important observation. If a sufficient statistic $Y = u(X_1, X_2, \ldots, X_n)$ exists for θ, then, by the factorization theorem,

$$\frac{L(\theta_0)}{L(\theta_1)} = \frac{\phi[u(x_1, x_2, \dots, x_n); \theta_0] h(x_1, x_2, \dots, x_n)}{\phi[u(x_1, x_2, \dots, x_n); \theta_1] h(x_1, x_2, \dots, x_n)}$$

$$= \frac{\phi[u(x_1, x_2, \dots, x_n); \theta_0)}{\phi[u(x_1, x_2, \dots, x_n); \theta_1]}.$$

Thus $L(\theta_0)/L(\theta_1) \leq k$ provides a critical region that is a function of the observations $x_1, x_2, \dots x_n$ only through the observed value of the sufficient statistic $y = u(x_1, x_2, \dots, x_n)$. Hence, best critical and uniformly most powerful critical regions are based upon sufficient statistics when they exist. ∎

EXERCISES

9.2-1 Let $X_1, X_2, \dots, X_n$ be a random sample from a normal distribution $N(\mu, 64)$.

    (a) Show that $C = \{(x_1, x_2, \dots, x_n): \bar{x} \leq c\}$ is a best critical region for testing H_0: $\mu = 80$ against H_1: $\mu = 76$.

    (b) Find n and c so that $\alpha = 0.05$ and $\beta = 0.05$, approximately.

9.2-2 Let $X_1, X_2, \dots, X_n$ be a random sample from $N(0, \sigma^2)$.

    (a) Show that $C = \{(x_1, x_2, \dots, x_n): \sum_{i=1}^{n} x_i^2 \geq c\}$ is a best critical region for testing H_0: $\sigma^2 = 4$ against H_1: $\sigma^2 = 16$.

    (b) If $n = 15$, find the value of c so that $\alpha = 0.05$.
HINT: Recall that $\sum_{i=1}^{n} X_i^2/\sigma^2$ is $\chi^2(n)$.

    (c) If $n = 15$ and c is the value found in part (b), find the approximate value of $\beta = P(\sum_{i=1}^{n} X_i^2 < c; \sigma^2 = 16)$.

9.2-3 Let X have an exponential distribution with a mean of θ; that is, the p.d.f. of X is $f(x; \theta) = (1/\theta)e^{-x/\theta}, 0 < x < \infty$. Let $X_1, X_2, \dots, X_n$ be a random sample from this distribution.

    (a) Show that a best critical region for testing H_0: $\theta = 3$ against H_1: $\theta = 5$ can be based on the statistic $\sum_{i=1}^{n} X_i$.

    (b) If $n = 12$, use the fact that $(2/\theta)\sum_{i=1}^{12} X_i$ is $\chi^2(24)$ to find a best critical region of size $\alpha = 0.10$.

    (c) If $n = 12$, find a best critical region of size $\alpha = 0.10$ for testing H_0: $\theta = 3$ against H_1: $\theta = 7$.

    (d) If H_1: $\theta > 3$, is the common region found in parts (b) and (c) a uniformly most powerful critical region of size $\alpha = 0.10$?

9.2-4 Let $X_1, X_2, \dots, X_n$ be a random sample of Bernoulli trials $b(1, p)$.

    (a) Show that a best critical region for testing H_0 $p = 0.9$ against H_1: $p = 0.8$ can be based on the statistic $Y = \sum_{i=1}^{n} X_i$, which is $b(n, p)$.

    (b) If $C = \{(x_1, x_2, \dots, x_n): \sum_{i=1}^{n} x_i \leq n(0.85)\}$ and $Y = \sum_{i=1}^{n} X_i$, find the value of n such that $\alpha = 0.10 = P[Y \leq n(0.85); p = 0.9]$, approximately.
HINT: Use the normal approximation for the binomial distribution.

    (c) What is the approximate value of $\beta = P[Y > n(0.85); p = 0.8]$ for the test given in part (b)?

    (d) Is the test of part (b) a uniformly most powerful test when the alternative hypothesis is H_1: $p < 0.9$?

9.2-5 Let $X_1, X_2, \dots, X_n$ be a random sample from the normal distribution $N(\mu, 36)$.

    (a) Show that a uniformly most powerful critical region for testing H_0: $\mu = 50$ against H_1: $\mu < 50$ is given by $C_2 = \{\bar{x}: \bar{x} \leq c\}$.

(b) With this result and that of Example 9.2-4, argue that a uniformly most powerful test for testing H_0: $\mu = 50$ against H_1: $\mu \neq 50$ does not exist.

9.2-6 Let $X_1, X_2, \ldots, X_n$ be a random sample from the normal distribution $N(\mu, 9)$. To test the hypothesis H_0: $\mu = 80$ against H_1: $\mu \neq 80$, consider the following three critical regions: $C_1 = \{\bar{x}: \bar{x} \geq c_1\}$, $C_2 = \{\bar{x}: \bar{x} \leq c_2\}$, and $C_3 = \{\bar{x}: |\bar{x} - 80| \geq c_3\}$.

(a) If $n = 16$, find the values of c_1, c_2, c_3 such that the size of each critical region is 0.05. That is, find c_1, c_2, c_3 such that

$$0.05 = P(\bar{X} \in C_1; \mu = 80) = P(\bar{X} \in C_2; \mu = 80)$$
$$= P(\bar{X} \in C_3; \mu = 80).$$

(b) Sketch, on the same graph paper, the power functions for these three critical regions.

9.2-7 Let $X_1, X_2, \ldots, X_{10}$ be a random sample of size 10 from a Poisson distribution with mean μ.

(a) Show that a uniformly most powerful critical region for testing H_0: $\mu = 0.5$ against H_1: $\mu > 0.5$ can be defined using the statistic $\sum_{i=1}^{10} X_i$.

(b) What is a uniformly most powerful critical region of size $\alpha = 0.068$? Recall that $\sum_{i=1}^{10} X_i$ has a Poisson distribution with mean 10μ.

(c) Sketch the power function of this test.

9.3* LIKELIHOOD RATIO TESTS

In this section we consider a general test-construction method that is applicable when both the null and alternative hypotheses, say H_0 and H_1, are composite. We continue to assume that the functional form of the p.d.f. is known but that it depends on an unknown parameter or unknown parameters. That is, we assume that the p.d.f. of X is $f(x; \theta)$, where θ represents one or more unknown parameters. We let Ω denote the total parameter space, that is, the set of all possible values of the parameter θ given by either H_0 or H_1. These hypotheses will be stated as follows:

$$H_0: \theta \in \omega, \qquad H_1: \theta \in \omega',$$

where ω is a subset of Ω and ω' is the complement of ω with respect to Ω. The test will be constructed by using a ratio of likelihood functions that have been maximized in ω and Ω, respectively. In a sense, this is a natural generalization of the ratio appearing in the Neyman-Pearson lemma when the two hypotheses were simple.

Definition 9.3-1

The *likelihood ratio* is the quotient

$$\lambda = \frac{L(\hat{\omega})}{L(\hat{\Omega})},$$

where $L(\hat{\omega})$ is the maximum of the likelihood function with respect to θ when $\theta \in \omega$, and $L(\hat{\Omega})$ is the maximum of the likelihood function with respect to θ when $\theta \in \Omega$.

Because λ is the quotient of nonnegative functions, $\lambda \geq 0$. In addition, since $\omega \subset \Omega$, we have that $L(\widehat{\omega}) \leq L(\widehat{\Omega})$ and hence $\lambda \leq 1$. Thus $0 \leq \lambda \leq 1$. If the maximum of L in ω is much smaller than that in Ω, it would seem that the data $x_1, x_2, \ldots, x_n$ do not support the hypothesis $H_0: \theta \in \omega$. That is, a small value of the ratio $\lambda = L(\widehat{\omega})/L(\widehat{\Omega})$ would lead to the rejection of H_0. On the other hand, a value of the ratio λ that is close to one would support the null hypothesis H_0. This leads us to the following definition.

Definition 9.3-2	To test $H_0: \theta \in \omega$ against $H_1: \theta \in \omega'$, the *critical region for the likelihood ratio test* is the set of points in the sample space for which

$$\lambda = \frac{L(\widehat{\omega})}{L(\widehat{\Omega})} \leq k,$$

where $0 < k < 1$ and k is selected so that the test has a desired significance level α.

The following example illustrates these definitions.

EXAMPLE 9.3-1	Assume that the weight X in ounces of a "10-pound" bag of sugar is $N(\mu, 5)$. We shall test the hypothesis $H_0: \mu = 162$ against the alternative hypothesis $H_1: \mu \neq 162$. Thus $\Omega = \{\mu : -\infty < \mu < \infty\}$ and $\omega = \{162\}$. To find the likelihood ratio, we need $L(\widehat{\omega})$ and $L(\widehat{\Omega})$. When H_0 is true, μ can take on only one value, namely $\mu = 162$. Thus $L(\widehat{\omega}) = L(162)$. To find $L(\widehat{\Omega})$, we must find the value of μ that maximizes $L(\mu)$. We recall that $\widehat{\mu} = \overline{x}$ is the maximum likelihood estimate of μ. Thus $L(\widehat{\Omega}) = L(\overline{x})$ and the likelihood ratio $\lambda = L(\widehat{\omega})/L(\widehat{\Omega})$ is given by

$$\lambda = \frac{(10\pi)^{-n/2} \exp\left[-\left(\dfrac{1}{10}\right) \displaystyle\sum_{i=1}^{n} (x_i - 162)^2\right]}{(10\pi)^{-n/2} \exp\left[-\left(\dfrac{1}{10}\right) \displaystyle\sum_{i=1}^{n} (x_i - \overline{x})^2\right]}$$

$$= \frac{\exp\left[-\left(\dfrac{1}{10}\right) \displaystyle\sum_{i=1}^{n} (x_i - \overline{x})^2 - \left(\dfrac{n}{10}\right)(\overline{x} - 162)^2\right]}{\exp\left[-\left(\dfrac{1}{10}\right) \displaystyle\sum_{i=1}^{n} (x_i - \overline{x})^2\right]}$$

$$= \exp\left[-\frac{n}{10}(\overline{x} - 162)^2\right].$$

A value of $\overline{x}$ close to 162 would tend to support H_0 and in that case λ is close to 1. On the other hand, an $\overline{x}$ that differs from 162 by too much would tend to support H_1. See Figure 9.3-1 for the graph of this likelihood ratio when $n = 5$.

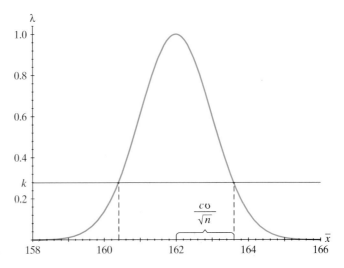

Figure 9.3-1: The likelihood ratio for testing $H_0: \mu = 162$

A likelihood ratio critical region is given by $\lambda \le k$, where k is selected so that the significance level of the test is α. Using this criterion and simplifying the inequality as we do using the Neyman-Pearson lemma, we have that $\lambda \le k$ is equivalent to each of the following inequalities:

$$-\left(\frac{n}{10}\right)(\bar{x} - 162)^2 \le \ln k,$$

$$(\bar{x} - 162)^2 \ge -\left(\frac{10}{n}\right)\ln k,$$

$$\frac{|\bar{x} - 162|}{\sigma/\sqrt{n}} \ge \frac{\sqrt{-(10/n)\ln k}}{\sigma/\sqrt{n}} = c.$$

Since $Z = (\overline{X} - 162)/(\sigma/\sqrt{n})$ is $N(0, 1)$ when $H_0: \mu = 162$ is true, let $c = z_{\alpha/2}$. Thus the critical region is

$$C = \left\{\bar{x}: \frac{|\bar{x} - 162|}{\sigma/\sqrt{n}} \ge z_{\alpha/2}\right\}$$

and, for illustration, if $\alpha = 0.05$, we have that $z_{0.025} = 1.96$. ◄

As illustrated in Example 9.3-1, the inequality $\lambda \le k$ can often be expressed in terms of a statistic whose distribution is known. Also, note that although the likelihood ratio test is an intuitive test, it leads to the same critical region as that given by the Neyman-Pearson lemma when H_0 and H_1 are both simple hypotheses.

Suppose now that the random sample $X_1, X_2, \ldots, X_n$ arises from the normal population $N(\mu, \sigma^2)$. where both μ and σ^2 are unknown. Let us consider the likelihood ratio test of the null hypothesis $H_0: \mu = \mu_0$ against the two-sided alternative hypothesis $H_1: \mu \neq \mu_0$. For this test

$$\omega = \{(\mu, \sigma^2): \mu = \mu_0,\ 0 < \sigma^2 < \infty\}$$

and

$$\Omega = \{(\mu, \sigma^2): -\infty < \mu < \infty,\ 0 < \sigma^2 < \infty\}.$$

If $(\mu, \sigma^2) \in \Omega$, the observed maximum likelihood estimates are $\widehat{\mu} = \bar{x}$ and $\widehat{\sigma^2} = (1/n) \sum_{i=1}^{n}(x_i - \bar{x})^2$. Thus

$$L(\widehat{\Omega}) = \left[\frac{1}{2\pi \left(\dfrac{1}{n}\right) \sum_{i=1}^{n}(x_i - \bar{x})^2} \right]^{n/2} \exp\left[-\frac{\sum_{i=1}^{n}(x_i - \bar{x})^2}{\left(\dfrac{2}{n}\right) \sum_{i=1}^{n}(x_i - \bar{x})^2} \right]$$

$$= \left[\frac{ne^{-1}}{2\pi \sum_{i=1}^{n}(x_i - \bar{x})^2} \right]^{n/2}.$$

Similarly, if $(\mu, \sigma^2) \in \omega$, the observed maximum likelihood estimates are $\widehat{\mu} = \mu_0$ and $\widehat{\sigma^2} = (1/n) \sum_{i=1}^{n}(x_i - \mu_0)^2$. Thus

$$L(\widehat{\omega}) = \left[\frac{1}{2\pi \left(\dfrac{1}{n}\right) \sum_{i=1}^{n}(x_i - \mu_0)^2} \right]^{n/2} \exp\left[-\frac{\sum_{i=1}^{n}(x_i - \mu_0)^2}{\left(\dfrac{2}{n}\right) \sum_{i=1}^{n}(x_i - \mu_0)^2} \right]$$

$$= \left[\frac{ne^{-1}}{2\pi \sum_{i=1}^{n}(x_i - \mu_0)^2} \right]^{n/2}.$$

The likelihood ratio $\lambda = L(\widehat{\omega})/L(\widehat{\Omega})$ for this test is

$$\lambda = \frac{\left[\dfrac{ne^{-1}}{2\pi \sum_{i=1}^{n}(x_i - \mu_0)^2} \right]^{n/2}}{\left[\dfrac{ne^{-1}}{2\pi \sum_{i=1}^{n}(x_i - \bar{x})^2} \right]^{n/2}} = \left[\frac{\sum_{i=1}^{n}(x_i - \bar{x})^2}{\sum_{i=1}^{n}(x_i - \mu_0)^2} \right]^{n/2}.$$

However, note that

$$\sum_{i=1}^{n}(x_i - \mu_0)^2 = \sum_{i=1}^{n}(x_i - \bar{x} + \bar{x} - \mu_0)^2 = \sum_{i=1}^{n}(x_i - \bar{x})^2 + n(\bar{x} - \mu_0)^2.$$

If this substitution is made in the denominator of λ, we have

$$\lambda = \left[\frac{\displaystyle\sum_{i=1}^{n}(x_i - \bar{x})^2}{\displaystyle\sum_{i=1}^{n}(x_i - \bar{x})^2 + n(\bar{x} - \mu_0)^2}\right]^{n/2}$$

$$= \left[\frac{1}{1 + \dfrac{n(\bar{x} - \mu_0)^2}{\dfrac{n}{\displaystyle\sum_{i=1}^{n}(x_i - \bar{x})^2}}}\right]^{n/2}.$$

Note that λ is close to one when $\bar{x}$ is close to μ_0 and it is small when $\bar{x}$ and μ_0 differ by a great deal. The likelihood ratio test, given by the inequality $\lambda \leq k$, is the same as

$$\frac{1}{1 + \dfrac{n(\bar{x} - \mu_0)^2}{\dfrac{n}{\displaystyle\sum_{i=1}^{n}(x_i - \bar{x})^2}}} \leq k^{2/n}$$

or, equivalently,

$$\frac{n(\bar{x} - \mu_0)^2}{\dfrac{\displaystyle\sum_{i=1}^{n}(x_i - \bar{x})^2}{n}} \geq (n - 1)(k^{-2/n} - 1).$$

When H_0 is true, $\sqrt{n}(\bar{X} - \mu_0)/\sigma$ is $N(0, 1)$, and $\sum_{i=1}^{n}(X_i - \bar{X})^2/\sigma^2$ has an independent chi-square distribution $\chi^2(n-1)$. Hence, under H_0,

$$T = \frac{\sqrt{n}(\bar{X} - \mu_0)/\sigma}{\sqrt{\displaystyle\sum_{i=1}^{n}(X_i - \bar{X})^2/[\sigma^2(n - 1)]}} = \frac{\sqrt{n}(\bar{X} - \mu_0)}{\sqrt{\displaystyle\sum_{i=1}^{n}(X_i - \bar{X})^2/(n - 1)}} = \frac{\bar{X} - \mu_0}{S/\sqrt{n}}$$

has a t distribution with $r = n - 1$ degrees of freedom. In accordance with the likelihood ratio test criterion, H_0 is rejected if the observed

$$T^2 \geq (n - 1)(k^{-2/n} - 1).$$

That is, we reject H_0: $\mu = \mu_0$ and accept H_1: $\mu \neq \mu_0$ if the observed $|T| \geq t_{\alpha/2}(n-1)$.

The reader should observe that this test is exactly the same as that listed in Table 8.2-2 for testing H_0: $\mu = \mu_0$ against H_1: $\mu \neq \mu_0$. That is, the test listed there is a likelihood ratio test. As a matter of fact, all six of the tests given in Tables 8.2-2 and 8.3-1 are likelihood ratio tests. Thus the examples and exercises associated with those tables are illustrations of the use of likelihood ratio tests.

The final development of this section concerns a test about the variance of a normal population. Let $X_1, X_2, \ldots, X_n$ be a random sample from $N(\mu, \sigma^2)$, where μ and σ^2 are unknown. We wish to test H_0: $\sigma^2 = \sigma_0^2$ against H_1: $\sigma^2 \neq \sigma_0^2$. For this, we have

$$\omega = \{(\mu, \sigma^2): -\infty < \mu < \infty, \ \sigma^2 = \sigma_0^2\}$$

and

$$\Omega = \{(\mu, \sigma^2): -\infty < \mu < \infty, \ 0 < \sigma^2 < \infty\}.$$

As in the test concerning the mean, we obtain

$$L(\widehat{\Omega}) = \left[\frac{ne^{-1}}{2\pi \sum\limits_{i=1}^{n} (x_i - \bar{x})^2} \right]^{n/2}.$$

If $(\mu, \sigma^2) \in \omega$, then $\widehat{\mu} = \bar{x}$ and $\widehat{\sigma^2} = \sigma_0^2$; thus

$$L(\widehat{\omega}) = \left(\frac{1}{2\pi \sigma_0^2} \right)^{n/2} \exp\left[-\frac{\sum_{i=1}^{n} (x_i - \bar{x})^2}{2\sigma_0^2} \right].$$

Accordingly, the likelihood ratio test $\lambda = L(\widehat{\omega})/L(\widehat{\Omega})$ is

$$\lambda = \left(\frac{w}{n} \right)^{n/2} \exp\left(-\frac{w}{2} + \frac{n}{2} \right) \leq k,$$

where $w = \sum_{i=1}^{n}(x_i - \bar{x})^2/\sigma_0^2$. Solving this inequality for w, we obtain a solution of the form $w \leq c_1$ or $w \geq c_2$, where the constants c_1 and c_2 are appropriate functions of the constants k and n so as to achieve the desired significance level α. Since $W = \sum_{i=1}^{n}(X_i - \bar{X})^2/\sigma_0^2$ is $\chi^2(n-1)$ if H_0: $\sigma^2 = \sigma_0^2$ is true, most statisticians modify this test slightly by taking $c_1 = \chi^2_{1-\alpha/2}(n-1)$ and $c_2 = \chi^2_{\alpha/2}(n-1)$. That is, this test and the other tests listed in Table 8.2-3 are either likelihood ratio tests or slight modifications of likelihood ratio tests. As a matter of fact, most tests involving normal assumptions are likelihood ratio tests or modifications of them; these include those tests involving regression and analysis of variance.

Remark*   Note that likelihood ratio tests are based on sufficient statistics when they exist as was also true of best critical and uniformly most powerful critical regions.   ∎

EXERCISES

9.3-1 In Example 9.3-1, if $n = 20$ and $\bar{x} = 161.1$, is H_0 accepted at a significance level of size α

    (a) If $\alpha = 0.10$?
    (b) If $\alpha = 0.05$?
    (c) What is the p-value of this test?

9.3-2 Assume that the weight X in ounces of a "10-ounce" box of corn flakes is $N(\mu, 0.03)$. Let $X_1, X_2, \ldots, X_n$ be a random sample from this distribution.

    (a) To test the hypothesis H_0: $\mu \geq 10.35$ against the alternative hypothesis H_1: $\mu < 10.35$, what is the critical region of size $\alpha = 0.05$ specified by the likelihood ratio test criterion?
        HINT: If $\mu \geq 10.35$ and $\bar{x} < 10.35$, note that $\hat{\mu} = 10.35$.
    (b) If a random sample of n = 50 boxes yielded a sample mean of $\bar{x} = 10.31$, is H_0 rejected?
        HINT: Find the critical value z_α when H_0 is true by taking $\mu = 10.35$, which is the extreme value in $\mu \geq 10.35$.
    (c) What is the p-value of this test?

9.3-3 Let $X_1, X_2, \ldots, X_n$ be a random sample from the normal distribution $N(\mu, 100)$.

    (a) To test H_0: $\mu = 230$ against H_1: $\mu > 230$, what is the critical region specified by the likelihood ratio test criterion?
    (b) Is this test uniformly most powerful?
    (c) If a random sample of $n = 16$ yielded $\bar{x} = 232.6$, is H_0 accepted at a significance level of $\alpha = 0.10$?
    (d) What is the p-value of this test?

9.3-4 Let $X_1, X_2, \ldots, X_n$ be a random sample from the normal distribution $N(\mu, 225)$.

    (a) To test H_0: $\mu = 59$ against H_1: $\mu \neq 59$, what is the critical region of size $\alpha = 0.05$ specified by the likelihood ratio test criterion?
    (b) If a sample of size $n = 100$ yielded $\bar{x} = 56.13$, is H_0 accepted?
    (c) What is the p-value of this test? Note that H_1 is a two-sided alternative.

9.3-5 It is desired to test the hypothesis H_0: $\mu = 30$ against the alternative hypothesis H_1: $\mu \neq 30$, where μ is the mean of a normal distribution and σ^2 is unknown. If a random sample of size $n = 9$ has $\bar{x} = 32.8$ and $s = 4$, is H_0 accepted at an $\alpha = 0.05$ significance level? What is the approximate p-value of this test?

9.3-6 To test H_0: $\mu = 335$ against H_1: $\mu < 335$, under normal assumptions, a random sample of size 17 yielded $\bar{x} = 324.8$ and $s = 40$. Is H_0 accepted at an $\alpha = 0.10$ significance level?

9.3-7 Let X have a normal distribution, in which μ and σ^2 are both unknown. It is desired to test $H_0: \mu = 1.80$ against $H_1: \mu > 1.80$ at an $\alpha = 0.10$ significance level. If a random sample of size $n = 121$ yielded $\bar{x} = 1.84$ and $s = 0.20$, is H_0 accepted or rejected? What is the p-value of this test?

9.3-8 Let $X_1, X_2, \ldots, X_n$ be a random sample from an exponential distribution with mean θ. Show that the likelihood ratio test of $H_0: \theta = \theta_0$ against $H_1: \theta \neq \theta_0$ has a critical region of the form $\sum_{i=1}^{n} x_i \leq c_1$ or $\sum_{i=1}^{n} x_i \geq c_2$. How would you modify this so that chi-square tables can be used easily?

9.3-9 Let independent random samples of sizes n and m be taken respectively from two normal distributions with unknown means μ_X and μ_Y and unknown variances σ_X^2 and σ_Y^2.

   (a) Show that the likelihood ratio for testing $H_0: \mu_X = \mu_Y$ against $H_1: \mu_X \neq \mu_Y$ when $\sigma_X^2 = \sigma_Y^2$ is a function of the usual two-sample t statistic.

   (b) Show that the likelihood ratio for testing $H_0: \sigma_X^2 = \sigma_Y^2$ against $H_1: \sigma_X^2 \neq \sigma_Y^2$ is a function of the usual two-sample F statistic.

   The tests described in parts (a) and (b) are listed (the second is modified slightly) in Tables 8.3-1 and 8.3-2, respectively.

CHAPTER 10

QUALITY IMPROVEMENT THROUGH STATISTICAL METHODS

10.1 TIME SEQUENCES

Thus far we have treated $x_1, x_2, \ldots, x_n$ as if they were observations from some random experiment. That is, each one of these observations has the same distribution with mean μ and variance σ^2 and we thus might use $\bar{x}$ and s^2 as estimates of μ and σ^2. In practice, however, we often find that $x_1, x_2, \ldots, x_n$ are observed in **time sequence**, there is frequently change in the distribution as we continue to observe the x values. Thus we do not want to use $\bar{x}$ and s^2 as estimates of those characteristics. To see what is actually going on in these cases, we should plot the data in the order that it is collected. In a **time sequence**, observations are recorded in the order in which they were collected as ordered pairs where the usual x coordinate denotes the time t (in order of collection, which could be days, weeks, years, etc.) and the usual y coordinate records the observation x. If measurements are taken sequentially and plotted in this time sequence, trends, cycles, or major changes in process (like government interventions) are often observed. For illustration, consider the three time sequences in Figures 10.1-1(a), 10.1-1(b), 10.1-1(c).

Figure 10.1-1(a) might be a recording of sales of a company for the last 12 years. Despite occasional drops in sales from one year to the next, there has been an upward trend. Suppose, however, that only the last two points had been recorded for the stockholders. Since sales went down, they might

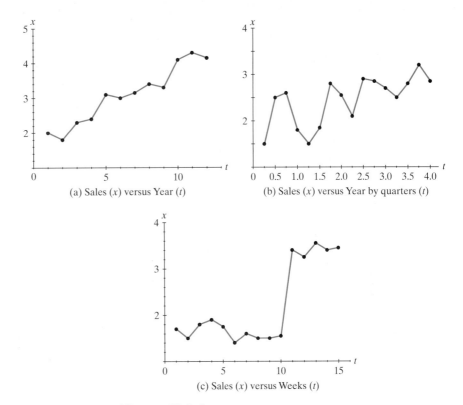

Figure 10.1-1: Sales of company over time

believe that there are some difficulties and that changes should be made. If they look at the entire sequence, however, they clearly see the long-term improvement and only variability about that upward trend accounts for the last drop. On the other hand, if several successive drops appear, then some action might be needed.

The points in Figure 10.1-1(b) might also represent sales, but now plotted by quarters. Clearly, in this business under consideration, there is a cyclic effect with increases from the first to second quarters and from second to third but decreases from the third to the fourth and the fourth to the first quarter of the next year. Again we note that there seems to be an overall upward trend.

Figure 10.1-1(c) could be the plot of some measure of success (possibly sales) of the company. Clearly, something happened after the tenth week. Perhaps management did not like the past level of performance and instituted a new program (possibly one based on statistical quality control). On the other hand, there might be some outside intervention (like war) that increased the sales of this particular product.

We have listed three possible scenarios associated with these three time sequences. There might be others; but, in any case, we clearly see that plotting the points in a time sequence often helps explore the variability of the outcomes associated with a particular system. That is, time is often an important factor in dealing with variability, possibly when purchasing stock. The following is a simple example based on one student's experience.

EXAMPLE 10.1-1 A student monitored some stock that she purchased in March for $5 per share. Every two weeks she recorded the value of a share yielding the following paired data, (t, x), with t equal to the time in weeks after purchase and x the value of a share at that time.

$$(0, 5.00) \quad (2, 5.00) \quad (4, 5.25) \quad (6, 5.25) \quad (8, 5.50)$$
$$(10, 5.50) \quad (12, 5.25) \quad (14, 5.75) \quad (16, 6.25) \quad (18, 6.50)$$
$$(20, 6.25) \quad (22, 7.25) \quad (24, 7.50)$$

A time sequence of these data is given in Figure 10.1-2.   ◀

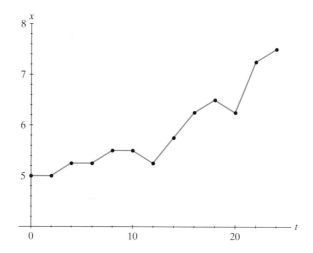

Figure 10.1-2: Value of a share of stock plotted biweekly

Sometimes, however, time is not an important factor in the response variable and yet time sequence plots still help us understand the concept of variation. Suppose that a process or individual performs at some average level, sometimes doing better and sometimes worse than that average. For example, suppose a certain worker's output is plotted by the month in Figure 10.1-3.

Let us say the manager of the company decides to do the following. On good months (like the second month) the worker is given a "pat on the back" (maybe a bonus or even being named "worker of the month"). What happens? Usually the performance goes down the next month. Recall that the worker is averaging below the performance of that high month; that is, she was performing above average when she was made worker of the month. So there is a tendency to do worse the month following a very good performance. (Statisticians sometimes call this "regression toward the mean.")

This explains what is often called the "sophomore jinx" in sports. A player might have an unusually good first year, actually one far above his average. So he tends to do worse the next year. Fans will say it's the sophomore jinx;

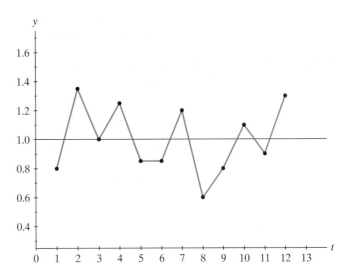

Figure 10.1-3: Average output of worker (y) versus month (t)

he probably got over confident and didn't work as hard. Although there might be some truth to this, it could also be that he worked just as hard as before but still went down because his first year was over his average. So, often the "rookie of the year" will do worse the second year because not many players win that award with an average year. That is, he was performing far above his average when he won the award.

Back to the worker described by Figure 10.1-3. On bad months (like the eighth month), the manager could call in the worker and have a serious talk with her or, in some way, penalize her. What happens? Usually, she does better the next month (again regression toward the mean). So the boss gets the idea that it is better to get tough because the workers seem to do better after these little talks. Of course, like the "pat on the back" technique, this does not really work because the workers keep varying about their average. They usually do better after a very poor performance and worse after an excellent performance.

If the manager does not like the the average level of performance, then he or she must do something to change the system. Most of the time, the individual workers can do very little to change the process; major decisions for improvements must be made by the manager or even in the board room. After these decisions, the workers must be given "road maps directing them to use these improvements." Most workers (as well as students) want to do better and would like to take pride in their performance. Managers (as well as teachers) should provide for the opportunity to improve the self esteem of those working for them.

The following example could help some of you in analyzing performances in events that are repeated over time.

EXAMPLE 10.1-2 Consider the following 20 quiz scores of a student in a two-semester calculus course. Each quiz was graded on a 20-point scale and they are given in the order in which they were taken.

18	20	16	12	15	15	10	13	14	9
12	10	8	11	12	11	10	11	8	6

We could calculate the average of all 20 quiz scores which is 12.05. We could calculate the average of the first 10 and the second 10. A graphic display of what has happened to this student over the course of the two semesters is shown in Figure 10.1-4. Since the time history is indicated in a time sequence, it is clear that the student was not making a great effort throughout the year in this calculus class. If interested, students can correct trends like this, but they must make changes in their behavior to do so. By continuing the past pattern nothing will change and the downward trend will continue. What improvements could be made to change this? Each student must answer this in his or her own way. ◄

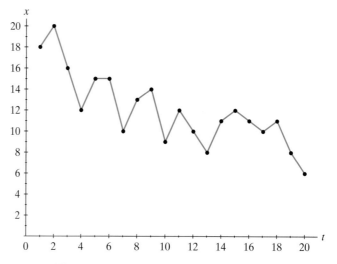

Figure 10.1-4: Calculus scores over time

The next example illustrates how, in a production process, tracking something over time can be beneficial.

EXAMPLE 10.1-3 Containers of a prewash concentrate are selected regularly from the production line and the concentrate is weighed. The target weight is 330.5–336.5 grams. We now construct a time-sequence plot (see Figure 10.1-5) for the following 60 weights that were observed in the order left to right, line by line.

337.9	338.1	337.7	338.3	337.3	338.1	337.3	338.0	337.9	338.9
337.6	338.2	338.1	338.6	338.0	338.4	334.4	334.3	334.5	333.9
335.1	334.0	335.6	334.1	335.0	334.8	335.2	335.2	335.2	334.0
334.5	334.6	335.4	335.9	335.6	335.8	334.5	335.4	335.0	335.5
335.1	335.0	334.1	334.6	334.4	334.7	333.8	335.0	334.0	334.3
334.0	333.7	334.7	333.5	334.8	333.9	335.7	335.1	335.1	334.6

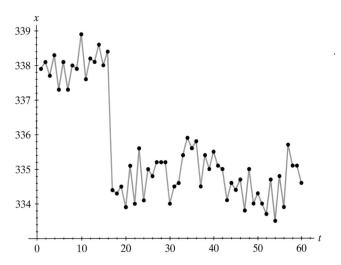

Figure 10.1-5: Prewash concentrate fill weights

It is pretty clear that during the beginning of production, the fill weights were slightly above the target weight. This is good news for the customer as long as the containers do not overflow. However, if the trend had continued, this would have resulted in a loss of profit for the company. After adjustments were made, the process seemed to come under control, with the weights reflecting the normal variability for a process like this. ◀

EXERCISES

10.1-1 The following data give the percentage of solids in a product. The specifications call for the percentage to be between 60.0% and 72.0%. The data were observed in the order left to right, line by line.

69.8	71.3	65.6	66.3	70.1	71.9	69.6	71.9	71.1	71.7
71.9	69.8	66.8	68.3	64.4	64.2	65.1	63.7	66.2	61.9
66.1	62.9	66.9	67.3	63.3	63.4	67.2	67.4	65.5	66.2
67.5	67.3	66.9	66.5	65.5	63.9	64.6	62.3	66.2	67.2
66.0	69.8	69.7	71.0	69.8	66.0	70.3	65.5	67.0	66.8
67.6	68.6	66.5	66.2	70.4	68.1	64.3	65.2	68.0	65.1

(a) Construct a time sequence of these percentages.
(b) Interpret your output.

10.1-2 Twenty-four "3-pound" bags of apples were filled and weighed automatically in the following time order by scale number 5, yielding the following weights.

3.26	3.62	3.39	3.12	3.53	3.30	3.10	3.26
3.19	3.22	3.14	3.39	3.31	3.21	3.49	3.41
3.02	3.17	3.20	3.12	3.42	3.36	3.21	3.26

(a) Construct and interpret a time sequence of these weights.
(b) Compare scale number 5 with scale number 6 in the next exercise.

10.1-3 Twenty-four "3-pound" bags of apples were filled and weighed automatically in the following time order by scale number 6, yielding the following weights.

3.22	2.97	3.00	3.08	3.04	3.09	3.04	3.21
3.00	3.05	3.11	3.00	3.10	3.13	3.08	3.22
3.12	3.04	3.03	3.02	3.10	3.09	3.18	3.00

(a) Construct and interpret a time sequence of these weights.
(b) Compare scale number 6 with scale number 5 in the last exercise.

10.1-4 The year, x, and the birth rate, y (number of births per 1000 population) for 1960–2002 were as follows.

(1960, 23.7)	(1961, 23.3)	(1962, 22.4)	(1963, 21.7)
(1964, 21.1)	(1965, 19.4)	(1966, 18.4)	(1967, 17.8)
(1968, 17.6)	(1969, 17.9)	(1970, 18.4)	(1971, 17.2)
(1972, 15.6)	(1973, 14.8)	(1974, 14.8)	(1975, 14.6)
(1976, 14.6)	(1977, 15.1)	(1978, 15.0)	(1979, 15.6)
(1980, 15.9)	(1981, 15.8)	(1982, 15.9)	(1983, 15.5)
(1984, 15.5)	(1985, 15.8)	(1986, 15.6)	(1987, 15.7)
(1988, 15.9)	(1989, 16.3)	(1990, 16.7)	(1991, 16.2)
(1992, 15.8)	(1993, 15.4)	(1994, 15.0)	(1995, 14.6)
(1996, 14.4)	(1997, 14.2)	(1998, 14.3)	(1999, 14.2)
(2000, 14.4)	(2001, 14.1)	(2002, 13.9)	

(a) Construct a time sequence of these birth rates.
(b) Interpret your output.

10.1-5 Dish drop viscosity has a specification of 120–230 cps (centipoise). One hundred observations were selected in the following time order and measured, yielding the following viscosities (read across, line by line).

158	147	158	159	169	151	166	151	143	169
153	174	151	164	185	168	140	180	176	154
160	187	145	164	158	169	153	149	144	157
156	183	157	140	162	158	160	180	154	160
164	168	154	158	164	159	153	170	158	170
150	161	169	166	154	157	138	155	134	165
161	172	156	145	153	143	152	152	156	163
179	157	135	172	143	154	165	145	152	145
171	189	144	154	147	187	147	159	167	151
153	168	148	188	152	165	155	140	157	176

(a) Construct a time sequence of these viscosities.
(b) If there seems to be no significant time factor, construct a histogram of these viscosities.
(c) Use the two graphs to interpret these measurements.

10.1-6 A manufacturer of car windows has studs in the windows for attaching them to the car. A standard "stud pull-out test" measures the force required to pull each of the five studs out of a window. Because of the shape of the window, the two studs on one side are attached in one way and the three studs on the other side in a different way. The target mean is 123. Seventy observations, going around each of 14 windows in a clockwise direction, were as follows.

140	159	138	102	84	126	147	126	103	92
149	155	135	120	94	149	143	109	101	86
144	154	120	105	97	149	151	140	103	99
157	140	120	96	87	146	137	120	93	89
149	154	139	100	84	148	142	130	98	81
112	135	109	84	87	112	135	109	84	87
118	135	90	77	125	126	131	78	75	126

(a) Construct a time sequence of these forces.
(b) Construct a histogram of these forces.
(c) Give an interpretation.

10.1-7 The amount of fluoride in a certain brand of toothpaste has a specification of 0.85–1.10 mg/g. One hundred tubes of toothpaste were selected randomly from the production line and the amounts of fluoride were measured, yielding the following data:

0.98	0.92	0.89	0.90	0.94	0.99	0.86	0.85	1.06	1.01
1.03	0.85	0.95	0.90	1.03	0.87	1.02	0.88	0.92	0.88
0.88	0.90	0.98	0.96	0.98	0.93	0.98	0.92	1.00	0.95
0.88	0.90	1.01	0.98	0.85	0.91	0.95	1.01	0.88	0.89
0.99	0.95	0.90	0.88	0.92	0.89	0.90	0.95	0.93	0.96
0.93	0.91	0.92	0.86	0.87	0.91	0.89	0.93	0.93	0.95
0.92	0.88	0.87	0.98	0.98	0.91	0.93	1.00	0.90	0.93
0.89	0.97	0.98	0.91	0.88	0.89	1.00	0.93	0.92	0.97
0.97	0.91	0.85	0.92	0.87	0.86	0.91	0.92	0.95	0.97
0.88	1.05	0.91	0.89	0.92	0.94	0.90	1.00	0.90	0.93

  (a) Construct a time sequence of these measurements.
  (b) Construct a histogram of these measurements.
  (c) Construct a box-plot of these measurements.
  (d) Give an interpretation.

10.1-8 Use the data in Exercise 3.1-15 for this exercise.

  (a) Construct and interpret a time sequence of the durations of eruptions of Old Faithful Geyser. Note the starting times when making your interpretation.
  (b) Construct and interpret a time sequence of the times between eruptions of Old Faithful Geyser. Note the starting times when making your interpretation.

10.1-9 A college computer center monitored the number of off campus calls into their computer facility. The following data give the number of calls per hour, beginning at midnight and continuing for three days.

69	38	10	6	6	7	20	37	65	58	83	87
81	82	76	107	111	88	115	117	116	111	110	101
78	26	18	9	8	6	22	40	53	54	67	59
89	67	86	117	98	114	108	108	124	34	121	114
87	53	15	8	7	7	17	38	72	56	62	86
75	68	80	74	116	112	128	131	125	119	128	118

Construct and interpret a time sequence of these numbers of calls.

10.1-10 Continuing with Exercise 10.1-9, the college has 40 ports for receiving calls to their computer. When a call is received, the ports are always checked in order from the first until an open port is found. If it gets to the fortieth (last) port and all are busy, the caller receives a busy signal. The numbers of users on the 40 ports during a 24-hour period were as follows, in order from left to right.

74	69	54	79	59	61	73	77	67	71
72	56	56	72	66	55	62	43	42	46
39	33	36	36	44	35	41	32	30	28
30	26	24	15	17	17	15	16	19	13

The numbers of minutes used on each of the 40 ports were the following.

1200	1180	1140	1119	1077	1107	1059	1022	995	997
982	994	1009	911	878	849	793	763	767	740
760	753	740	683	612	616	610	532	565	579
449	454	375	346	355	330	339	306	284	288

Construct and interpret time sequences for these two sets of data.

STATISTICAL QUALITY CONTROL

Statistical methods can be used in many scientific fields, such as medical research, engineering, chemistry, and psychology. Often it is necessary to compare two ways of doing something, say the old way and a possible new way. We collect data on each, quite possibly in a laboratory situation, and try to decide if the new is actually better than the old. Needless to say, it would be terrible to change the new way at great expense only to find out that it is really not any better than the old. That is, suppose the lab results indicate, by some statistical method, that the new is seemingly better than the old. Can we actually extrapolate those outcomes in the lab to the situations in the real world? Clearly, statisticians can not make these decisions, but they should be made by some professional who knows that specialty very well. The statistical analysis might provide helpful guidelines, but we still need the expert to make the final decision.

However, even before investigating possible changes in any process, it is first extremely important to determine exactly what the process in question is doing at the present time. Often people in charge of an organization do not understand the capabilities of many of its processes. Simply measuring what is going on often leads to improvements. In many cases measurement is easy, such as determining the diameter of a bolt, but sometimes it is extremely difficult, as in evaluating good teaching or many other service activities. But if at all possible, we encourage those involved to begin to "listen" to their processes; that is, they should measure what is going on in their organization. These measurements alone often are the beginning of desirable improvements. While most of our remarks in this chapter concern measurements made in manufacturing, service industries frequently find them just as useful.

At one time, some manufacturing plants would make parts to be used in the construction of some piece of equipment. Say a particular line in the plant making a certain part might produce several hundreds of them each day. These items would then be sent on to an inspection cage, where they would be checked for goodness, often several days or even weeks later. Occasionally, the inspectors would discover many defectives among the items made two weeks ago, say. There was little that could be done at that point except scrap or rework the defective parts, both expensive outcomes.

In the 1920s, W. A. Shewhart, who was working for AT&T Bell Labs, recognized that this was an undesirable situation and suggested that, with some suitable frequency, a sample of these parts should be taken as they were being made. If the sample indicated the items were satisfactory, the manufacturing process would continue. But if these sampled parts were not satisfactory, corrections should be made then so that things were satisfactory. This idea led to what are commonly called *Shewhart control charts* that are the basis of what was called *statistical quality control* in those early days; today it is often referred to as *statistical process control*.

These control charts consist of calculated values of a statistic, say $\bar{x}$, plotted in time sequence. That is, in making products, every so often (each hour, each day, each week depending upon how many items are being produced) a sample of size n of them is taken, and they are measured resulting in the observations $x_1, x_2, \ldots, x_n$. The average $\bar{x}$ and the standard deviation s are computed. This is done k times, and the k values of $\bar{x}$ and s are averaged resulting in $\bar{\bar{x}}$ and $\bar{s}$, respectively; usually k is equal to some number between 10 and 30.

If the true mean μ and standard deviation σ of the process were known, then the central limit theorem states that almost all of the $\bar{x}$ values would plot between $\mu - 3\sigma/\sqrt{n}$ and $\mu + 3\sigma/\sqrt{n}$, unless the system has actually changed. However, we know neither μ nor σ, and thus μ is estimated by $\bar{\bar{x}}$ and $3\sigma/\sqrt{n}$ by $A_3\bar{s}$, where $\bar{\bar{x}}$ and $\bar{s}$ are the means of the k observations of $\bar{x}$ and s. A_3 is a factor depending upon n that can be found in books on statistical quality control. A few values of A_3 (and some other constants that will be used later) are given in Table 10.2-1 for typical values of n.

Table 10.2-1: Some Constants Used with Control Charts						
n	A_3	B_3	B_4	A_2	D_3	D_4
4	1.63	0	2.27	0.73	0	2.28
5	1.43	0	2.09	0.58	0	2.11
6	1.29	0.03	1.97	0.48	0	2.00
8	1.10	0.185	1.815	0.37	0.14	1.86
10	0.98	0.28	1.72	0.31	0.22	1.78
20	0.68	0.51	1.49	0.18	0.41	1.59

The estimates of $\mu \pm 3\sigma/\sqrt{n}$ are called the upper control limit (UCL), $\bar{\bar{x}} + A_3\bar{s}$, and the lower control limit (LCL), $\bar{\bar{x}} - A_3\bar{s}$, and $\bar{\bar{x}}$ provides the estimate of the centerline. A typical plot is given in Figure 10.2-1. Here, in the 13th sampling period, $\bar{x}$ is outside the control limits, indicating the process has changed and some investigation and action is needed to correct this change, which seems like a shift upward in the process.

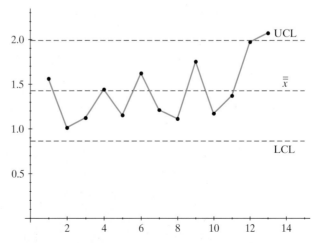

Figure 10.2-1: Typical control chart

It should be noted that there is a control chart for the s values, too. From sampling distribution theory, values of B_3 and B_4 have been determined and are given in Table 10.2-1 so that we know that almost all s values should be between $B_3\bar{s}$ and $B_4\bar{s}$ if there is no change in the underlying distribution. So again, if an individual s value is outside these control limits, some action should be taken as it seems as if there has been a change in the variation of the underlying distribution.

Often, when these charts are first constructed after $k = 10$ to 30 sampling periods, many points fall outside the control limits. A team consisting of workers, the foreman, the supervisor, an engineer, and even a statistician should try to find the reasons why this has occurred, and the situation should be corrected. After this is done and the points plot within the control limits, the process is "in statistical control." However, being in statistical control is not a guarantee of satisfaction with the products. Since $A_3\bar{s}$ is an estimate of $3\sigma/\sqrt{n}$, then $\sqrt{n}A_3\bar{s}$ is an estimate of 3σ, and with an underlying distribution close to a normal one, almost all items would be between $\bar{\bar{x}} \pm \sqrt{n}A_3\bar{s}$. If these limits are too wide, then corrections must be made again.

If the variation is under control (i.e., $\bar{x}$ and s within their control limits), we say the variations seen in $\bar{x}$ and s are due to common causes. If products made under such a system with these existing common causes are satisfactory, then the production continues. If either $\bar{x}$ or s, however, is outside the control limits, that is an indication that some special causes are at work, which must be corrected. That is, a team should investigate why and some action should be taken.

EXAMPLE 10.2-1

A company produces a storage console. Twice a day nine critical characteristics are tested on five consoles that are selected randomly from the production line. One of these characteristics is the time it takes the lower storage component door to open completely. Table 10.2-2 lists the opening times in seconds for the consoles that were tested during one week. Also included in the table are the sample means, sample standard deviations, and the ranges.

The upper control limit (UCL) and the lower control limit (LCL) for $\bar{x}$ are found using A_3 from Table 10.2-1 with $n = 5$ as follows:

$$\text{UCL} = \bar{\bar{x}} + A_3\bar{s} = 1.174 + 1.43(0.20) = 1.460$$

and

$$\text{LCL} = \bar{\bar{x}} - A_3\bar{s} = 1.174 - 1.43(0.20) = 0.888.$$

These control limits and the sample means are plotted on the $\bar{x}$ chart in Figure 10.2-2. There should be some concern about the fifth sampling period; thus there should be an investigation to determine why that $\bar{x}$ is below the LCL.

Group	x_1	x_2	x_3	x_4	x_5	$\overline{x}$	s	R
1	1.2	1.8	1.7	1.3	1.4	1.480	0.259	0.60
2	1.5	1.2	1.0	1.0	1.8	1.300	0.346	0.80
3	0.9	1.6	1.0	1.0	1.0	1.100	0.283	0.70
4	1.3	0.9	0.9	1.2	1.2	1.060	0.182	0.40
5	0.7	0.8	0.9	0.6	0.8	0.760	0.114	0.30
6	1.2	0.9	1.1	1.0	1.0	1.040	0.104	0.30
7	1.1	0.9	1.1	1.0	1.4	1.100	0.187	0.50
8	1.4	0.9	0.9	1.1	1.0	1.060	0.207	0.50
9	1.3	1.4	1.1	1.5	1.6	1.380	0.192	0.50
10	1.6	1.5	1.4	1.3	1.5	1.460	0.114	0.30

Table 10.2-2: Console Opening Times

$$\overline{\overline{x}} = 1.174 \qquad \overline{s} = 0.200 \qquad \overline{R} = 0.49$$

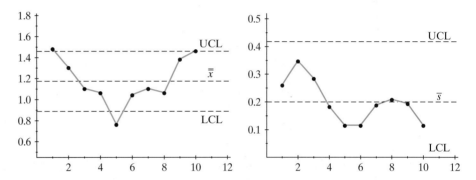

Figure 10.2-2: The $\overline{x}$ chart and s chart

The UCL and LCL for s are found using B_3 and B_4 in Table 10.2-1 and $n = 5$ as follows:

$$\text{UCL} = B_4\overline{s} = 2.09(0.200) = 0.418$$

and

$$\text{LCL} = B_3\overline{s} = 0(0.200) = 0.$$

These control limits and the sample standard deviations are plotted on the s chart in Figure 10.2-2.

Almost all of the observations should lie between $\overline{\overline{x}} \pm \sqrt{n}\, A_3\overline{s}$, namely,

$$1.174 + \sqrt{5}\,(1.43)(0.20) = 1.814$$

and

$$1.174 - \sqrt{5}\,(1.43)(0.20) = 0.535.$$

This is illustrated in Figure 10.2-3, in which all 50 observations do fall within these control limits.   ◀

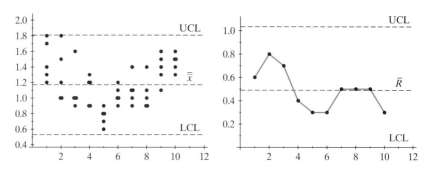

Figure 10.2-3: Plot of 50 console opening times and R chart

In most books on statistical quality control, there is an alternate way of constructing the limits on an $\bar{x}$ chart. For each sample, compute the range, R, which is the absolute value of the difference of the extremes of the sample. This computation is much easier than that for calculating s. After k samples are taken, compute the average of these R values, obtaining $\bar{R}$ as well as $\bar{\bar{x}}$. The statistic $A_2 \bar{R}$ serves as an estimate of $3\sigma/\sqrt{n}$, where A_2 is found in Table 10.2-1. Thus the estimates of $\mu \pm 3\sigma/\sqrt{n}$, namely $\bar{\bar{x}} \pm A_2 \bar{R}$, can be used as the UCL and LCL of an $\bar{x}$ chart.

In addition, $\sqrt{n} A_2 \bar{R}$ is an estimate of 3σ; so with an underlying distribution that is close to a normal one, we find that almost all observations are within the limits $\bar{\bar{x}} \pm \sqrt{n} A_2 \bar{R}$.

Moreover, an R chart can be constructed with centerline $\bar{R}$ and control limits equal to $D_3 \bar{R}$ and $D_4 \bar{R}$, where D_3 and D_4 are given in Table 10.2-1 and were determined so that almost all R values should be between the control limits if there is no change in the underlying distribution. Thus an R falling outside those limits would indicate a change in the spread of the underlying distribution, and some corrective action should be considered.

The use of R, rather than s, is illustrated in the next example.

EXAMPLE 10.2-2   Using the data in Example 10.2-1, we compute UCL and LCL for an $\bar{x}$ chart using $\bar{\bar{x}} \pm A_2 \bar{R}$ as follows:

$$\text{UCL} = \bar{\bar{x}} + A_2 \bar{R} = 1.174 + 0.58(0.49) = 1.458$$

and

$$\text{LCL} = \bar{\bar{x}} - A_2 \bar{R} = 1.174 - 0.58(0.49) = 0.890.$$

Note that these are very close to the limits that were found for the $\bar{x}$ chart in the $\bar{x}$ chart in Figure 10.2-2 using $\bar{\bar{x}} \pm A_3 \bar{s}$. In addition, almost all of the observations should lie within the limits $\bar{\bar{x}} + \sqrt{n} A_2 \bar{R}$. These limits are

$$\text{UCL} = 1.174 + \sqrt{5}(0.58)(0.49) = 1.809$$

and

$$\text{LCL} = 1.174 - \sqrt{5}\,(0.58)(0.49) = 0.539.$$

Note that these are almost the same as the limits found in Example 10.2-1 and plotted in Figure 10.2-3.

An R chart can be constructed with centerline $\overline{R} = 0.49$ and control limits given by

$$\text{UCL} = D_4 \overline{R} = 2.11(0.49) = 1.034$$

and

$$\text{LCL} = D_3 \overline{R} = 0(0.49) = 0.$$

Figure 10.2-3 illustrates this control chart for the range, and we see its pattern is similar to that of the s chart in Figure 10.2-2.   ◄

There are two other Shewhart control charts, the p- and c charts. The central limit theorem, which provided a justification for the three sigma limits in the $\overline{x}$ chart, also justifies the control limits in the p chart. Suppose the number of defectives among n items that are selected randomly, say D, has a binomial distribution $b(n, p)$. The limits $p \pm 3\sqrt{p(1 - p)/n}$ should include almost all of the D/n values. However, p must be approximated by observing k values of D, say $D_1, D_2, \ldots .D_k$, and computing what is called $\overline{p}$ in the statistical quality control literature, namely,

$$\overline{p} = \frac{D_1 + D_2 + \cdots + D_k}{kn}.$$

Thus the LCL and UCL for the fraction defective, D/n, are given by

$$\text{LCL} = \overline{p} - 3\sqrt{\overline{p}(1 - \overline{p})/n}$$

and

$$\text{UCL} = \overline{p} + 3\sqrt{\overline{p}(1 - \overline{p})/n}.$$

If the process is in control, almost all D/n values are between LCL and UCL. This may not be satisfactory and improvements might be needed to decrease $\overline{p}$. If satisfactory, however, let the process continue under these common causes of variation until a point, D/n, outside the control limits would indicate that some special cause has changed the variation. Incidentally, if D/n is below the LCL, this might very well indicate that some type of change for the better has been made, and we want to find out why. In general, outlying statistics can often suggest that good (as well as bad) breakthroughs have been made.

The following example gives the results of a simple experiment that you can easily duplicate.

EXAMPLE 10.2-3 Let D_i equal the number of yellow candies in a 1.69-ounce bag. Because the number of pieces of candy varies slightly from bag to bag, we shall use an average value for n when we construct the control limits. Table 10.2-3 lists, for 20 packages, the number of pieces of candy in the package, the number of yellow ones, and the proportion of yellow ones.

Table 10.2-3: Data on Yellow Candies

Package	n_i	D_i	$\bar{p}_i$	Package	n_i	D_i	$\bar{p}_i$
1	56	8	0.14	11	57	10	0.18
2	55	13	0.24	12	59	8	0.14
3	58	12	0.21	13	54	10	0.19
4	56	13	0.23	14	55	11	0.20
5	57	14	0.25	15	56	12	0.21
6	54	5	0.09	16	57	11	0.19
7	56	14	0.25	17	54	6	0.11
8	57	15	0.26	18	58	7	0.12
9	54	11	0.20	19	58	12	0.21
10	55	13	0.24	20	58	14	0.24

For these data,

$$\sum_{i=1}^{20} n_i = 1124 \quad \text{and} \quad \sum_{i=1}^{20} D_i = 219.$$

It follows that

$$\bar{p} = \frac{219}{1124} = 0.195 \quad \text{and} \quad \bar{n} = \frac{1124}{20} \approx 56.$$

Thus the UCL and LCL are given by

$$\text{LCL} = \bar{p} - 3\sqrt{\bar{p}(1 - \bar{p})/56} = 0.195 - 3\sqrt{0.195(0.805)/56} = 0.036$$

and

$$\text{UCL} = \bar{p} + 3\sqrt{\bar{p}(1 - \bar{p})/56} = 0.195 + 3\sqrt{0.195(0.805)/56} = 0.354.$$

The control chart for p is depicted in Figure 10.2-4. (For your information the "true" value for p is 0.20.) ◀

Consider the following explanation of the c chart. Suppose the number of flaws, say C, on some product has a Poisson distribution with parameter λ. If λ is sufficiently large, as in Example 5.5-5, we considered approximating

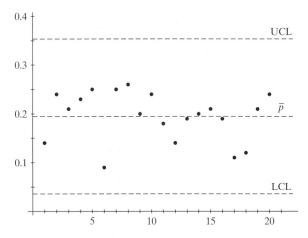

Figure 10.2-4: The p chart

the discrete Poisson distribution with the continuous $N(\lambda, \lambda)$ distribution. Thus the interval from $\lambda - 3\sqrt{\lambda}$ to $\lambda + 3\sqrt{\lambda}$ contains virtually all of the C values. Since λ is unknown, however, it must be approximated by $\bar{c}$, the average of the k values, $c_1, c_2, \ldots .c_k$. Hence the two control limits for C are computed as

$$\text{LCL} = \bar{c} - 3\sqrt{\bar{c}} \qquad \text{and} \qquad \text{UCL} = \bar{c} + 3\sqrt{\bar{c}}.$$

The remarks made about the $\bar{x}$- and $\bar{p}$ charts apply to the c chart as well, but we must remember that each c value is the number of flaws on one manufactured item, not an average $\bar{x}$ or a fraction defective D/n.

STATISTICAL COMMENTS   In observing time-sequence plots, as with control charts, do not read too much into a short sequence of points. For example, after a pep talk by the coach of a women's golf team, three successive decreasing scores might indicate that the coach had some influence. But even if the process has not changed, the probability of three decreasing scores (ruling out ties) is 1/6, since there are 3! = 6 equally likely ways of arranging three points. Often business people are worse offenders because they might think two increasing sales points indicate improvement. We know that with no change in the system, the probability of that is 1/2. Now maybe 4 points that successively increase after an intervention would be cause for claiming improvement because the probability of such an event is 1/4! = 1/24 if no change has actually been made. This probability is small enough to justify a celebration. ■

EXERCISES

10.2-1 It is important to control the viscosity of liquid dishwasher soap so that it flows out of the container but does not run out too rapidly. Thus samples are taken randomly throughout the day and the viscosity is measured. Use the following 20 sets of 5 observations for this exercise.

Observations					$\bar{x}$	s	R
158	147	158	159	169	158.20	7.79	22
151	166	151	143	169	156.00	11.05	26
153	174	151	164	185	165.40	14.33	34
168	140	180	176	154	163.60	16.52	40
160	187	145	164	158	162.80	15.29	42
169	153	149	144	157	154.40	9.48	25
156	183	157	140	162	159.60	15.47	43
158	160	180	154	160	162.40	10.14	26
164	168	154	158	164	161.60	5.55	14
159	153	170	158	170	162.00	7.65	17
150	161	169	166	154	160.00	7.97	19
157	138	155	134	165	149.80	13.22	31
161	172	156	145	153	157.40	10.01	27
143	152	152	156	163	153.20	7.26	20
179	157	135	172	143	157.20	18.63	44
154	165	145	152	145	152.20	8.23	20
171	189	144	154	147	161.00	18.83	45
187	147	159	167	151	162.20	15.85	40
153	168	148	188	152	161.80	16.50	40
165	155	140	157	176	158.60	13.28	36

(a) Calculate the values of $\bar{\bar{x}}$, $\bar{s}$, and $\bar{R}$.
(b) Construct an $\bar{x}$ chart using the value of A_3 and $\bar{s}$.
(c) Construct an s chart.
(d) Construct an $\bar{x}$ chart using A_2 and $\bar{R}$.
(e) Construct an R chart.
(f) Do the charts indicate that viscosity is in statistical control?

10.2-2 It is necessary to control the percentage of solids in a product, so samples are taken randomly throughout the day and the percentage of solids is measured. Use the following 20 sets of 5 observations for this exercise.

Observations					$\bar{x}$	s	R
69.8	71.3	65.6	66.3	70.1	68.62	2.51	5.7
71.9	69.6	71.9	71.1	71.7	71.24	0.97	2.3
71.9	69.8	66.8	68.3	64.4	68.24	2.86	7.5
64.2	65.1	63.7	66.2	61.9	64.22	1.61	4.3
66.1	62.9	66.9	67.3	63.3	65.30	2.06	4.4
63.4	67.2	67.4	65.5	66.2	65.94	1.61	4.0
67.5	67.3	66.9	66.5	65.5	66.74	0.79	2.0
63.9	64.6	62.3	66.2	67.2	64.84	1.92	4.9
66.0	69.8	69.7	71.0	69.8	69.26	1.90	5.0
66.0	70.3	65.5	67.0	66.8	67.12	1.88	4.8
67.6	68.6	66.5	66.2	70.4	67.86	1.71	4.2
68.1	64.3	65.2	68.0	65.1	66.14	1.78	3.8
64.5	66.6	65.2	69.3	62.0	65.52	2.69	7.3
67.1	68.3	64.0	64.9	68.2	66.50	1.96	4.3
67.1	63.8	71.4	67.5	63.7	66.70	3.17	7.7
60.7	63.5	62.9	67.0	69.6	64.74	3.53	8.9
71.0	68.6	68.1	67.4	71.7	69.36	1.88	4.3
69.5	61.5	63.7	66.3	68.6	65.92	3.34	8.0
66.7	75.2	79.0	75.3	79.2	75.08	5.07	12.5
77.3	67.2	69.3	67.9	65.6	69.46	4.58	11.7

(a) Calculate the values of $\bar{\bar{x}}$, $\bar{s}$, and $\bar{R}$.
(b) Construct an $\bar{x}$ chart using the value of A_3 and $\bar{s}$.
(c) Construct an s chart.
(d) Construct an $\bar{x}$ chart using A_2 and $\bar{R}$.
(e) Construct an R chart.
(f) Do the charts indicate that the percentage of solids in this product is in statistical control?

10.2-3 It is important to control the net weight of a packaged item; thus items are selected randomly throughout the day from the production line and their weights are recorded. Use the following 20 sets of 5 weights (in grams) for this exercise. (Note that a weight recorded here is the actual weight minus 330.)

Observations					$\bar{x}$	s	R
7.97	8.10	7.73	8.26	7.30	7.872	0.3740	0.96
8.11	7.26	7.99	7.88	8.88	8.024	0.5800	1.62
7.60	8.23	8.07	8.51	8.05	8.092	0.3309	0.91
8.44	4.35	4.33	4.48	3.89	5.098	1.8815	4.55
5.11	4.05	5.62	4.13	5.01	4.784	0.6750	1.57
4.79	5.25	5.19	5.23	3.97	4.886	0.5458	1.28
4.47	4.58	5.35	5.86	5.61	5.174	0.6205	1.39
5.82	4.51	5.38	5.01	5.54	5.252	0.5077	1.31
5.06	4.98	4.13	4.58	4.35	4.620	0.3993	0.93
4.74	3.77	5.05	4.03	4.29	4.376	0.5199	1.28
4.05	3.71	4.73	3.51	4.76	4.152	0.5748	1.25
3.94	5.72	5.07	5.09	4.61	4.886	0.6599	1.78
4.63	3.79	4.69	5.13	4.66	4.580	0.4867	1.34
4.30	4.07	4.39	4.63	4.47	4.372	0.2079	0.56
4.05	4.14	4.01	3.95	4.05	4.040	0.0693	0.19
4.20	4.50	5.32	4.42	5.24	4.736	0.5094	1.12
4.54	5.23	4.32	4.66	3.86	4.522	0.4999	1.37
5.02	4.10	5.08	4.94	5.18	4.864	0.4360	1.08
4.80	4.73	4.82	4.69	4.27	4.662	0.2253	0.55
4.55	4.76	4.45	4.85	4.02	4.526	0.3249	0.83

(a) Calculate the values of $\bar{\bar{x}}$, $\bar{s}$, and $\bar{R}$.
(b) Construct an $\bar{x}$ chart using the value of A_3 and $\bar{s}$.
(c) Construct an s chart.
(d) Construct an $\bar{x}$ chart using A_2 and $\bar{R}$.
(e) Construct an R chart.
(f) Do the charts indicate that these fill weights are in statistical control?

10.2-4 The following data give fill weights (in grams) of "four ounce bags" of candy as packed by a vertical form, fill, and seal machine.

118.8	116.6	116.2	113.8	116.9	118.1	116.2	115.6	113.5	119.2
115.2	117.0	113.6	115.1	113.5	117.0	113.6	115.5	113.5	117.0
119.0	117.5	119.2	118.1	121.2	117.2	113.5	119.7	118.4	116.6
119.0	113.6	118.8	118.1	117.7	116.0	117.1	119.1	116.8	118.7
115.9	116.8	118.7	115.9	116.8	117.2	115.5	120.0	117.6	117.8
117.0	116.2	119.9	118.9	118.2	118.8	118.9	116.5	116.8	116.6

119.3	116.2	116.4	114.5	114.8	121.1	118.3	113.6	116.7	118.0
118.3	117.3	115.2	114.6	114.2	117.2	115.8	113.6	121.6	120.8
117.2	117.0	115.4	118.2	119.4	117.5	113.5	119.1	116.3	118.1
119.7	118.5	119.6	119.2	116.7	117.3	115.6	115.0	117.5	114.1
118.4	118.3	117.4	117.2	116.3	116.5	116.7	117.8	118.8	118.0
115.4	113.5	118.2	116.5	117.4	113.6	113.5	119.0	116.7	117.0
117.5	117.9	117.2	118.7	117.6	118.2	120.5	116.4	114.9	116.8
116.7	116.5	117.8	116.4	116.9	118.4	116.8	120.8	118.2	119.3
118.9	118.0	120.1	114.9	119.5	119.0	115.5	115.8	115.8	115.4
120.8	118.3	115.9	117.7	114.8	116.6	118.1	114.0	113.4	113.4
116.0	121.8	119.3	117.7	118.3	118.4	116.8	118.0	117.8	116.1
119.1	115.9	118.8	117.9	118.5					

(a) Taking the data in consecutive sets of 5 (row by row, 118.8, 116.6, 116.2, 113.8, 116.9, etc.), construct $\bar{x}$-R and $\bar{x}$-s charts.

(b) Describe what these charts show. Is the machine performing satisfactorily?

10.2-5 To give some indication of how the values in Table 10.2-1 are calculated, values of A_3 are found in this exercise. Let $X_1, X_2, \ldots, X_n$ be a random sample of size n from the normal distribution $N(\mu, \sigma^2)$. Let S^2 equal the sample variance of this random sample.

(a) Show that $E[S^2] = \sigma^2$ using the fact that $Y = (n-1)S^2/\sigma^2$ has a distribution that is $\chi^2(n-1)$.

(b) Using the $\chi^2(n-1)$ p.d.f., find the value of $E(\sqrt{Y})$.

(c) Show that

$$E\left[\frac{\sqrt{n-1}\,\Gamma\left(\dfrac{n-1}{2}\right)}{\sqrt{2}\,\Gamma\left(\dfrac{n}{2}\right)}\,S\right] = \sigma.$$

(d) Verify that

$$\frac{3}{\sqrt{n}}\left[\frac{\sqrt{n-1}\,\Gamma\left(\dfrac{n-1}{2}\right)}{\sqrt{2}\,\Gamma\left(\dfrac{n}{2}\right)}\right] = A_3,$$

found in Table 10.2-1 for $n = 5$ and $n = 6$. Thus $A_3\bar{s}$ approximates $3\sigma/\sqrt{n}$.

10.2-6 A company has been producing bolts that are about $\bar{p} = 0.02$ defective, and this is satisfactory. To monitor the quality of this process, 100 bolts are selected at random each hour and the number of defective bolts counted. With $\bar{p} = 0.02$, compute the UCL and LCL of the $\bar{p}$ chart. Then suppose that, over the next 24 hours, the following numbers of defective bolts are observed:

$$4\ 1\ 1\ 0\ 5\ 2\ 1\ 3\ 4\ 3\ 1\ 0\ 0\ 4\ 1\ 1\ 6\ 2\ 0\ 0\ 2\ 8\ 7\ 5$$

Would any action have been required during this time?

10.2-7 In the past, $n = 50$ fuses are tested each hour and $\bar{p} = 0.03$ have been defective. Calculate the UCL and LCL. After an a production error, say the true p shifts to $p = 0.05$.

(a) What is the probability that the next observation exceeds the UCL?

(b) What is the probability that at least one of the next five observations exceeds the UCL?

HINT: Assume independence and compute the probability that none of the next five exceeds the UCL.

10.2-8 In a woolen mill, 100-yard pieces are inspected. In the last 20 observations the following numbers of flaws were found:

$$2\ 4\ 0\ 1\ 0\ 3\ 4\ 1\ 1\ 2\ 4\ 0\ 0\ 1\ 0\ 3\ 2\ 3\ 5\ 0$$

(a) Compute the control limits of the c chart.
(b) Is the process in statistical control?

10.2-9 In 50-foot tin strips, we find that the number of blemishes average about $\bar{c} = 1.4$. Calculate the control limits. Say the process has gone out of control and this average has increased to 3.

(a) What is the probability that the next observation will exceed the UCL?
(b) What is the probability that at least 1 of the next 10 observations will exceed the UCL?

10.3* GENERAL FACTORIAL AND 2^k FACTORIAL DESIGNS

In Section 8.8* we studied two-factor experiments in which the A factor is performed at a levels and the B factor has b levels. Without replications, we need ab level combinations and, with c replications with each of these combinations, we need a total of abc experiments.

Let us consider a situation with three factors, say A, B, and C with a, b, and c levels, respectively. Here there are a total of abc level combinations; and if at each of these combinations we have d replications, there is need for $abcd$ experiments. Once these experiments are run, in some random order, and the data collected, there are computer programs available to calculate the entries in the ANOVA table, as in Table 10.3-1.

Table 10.3-1: ANOVA Table				
Source	**SS**	**d.f.**	**MS**	**F**
A	SS(A)	$a-1$	MS(A)	MS(A)/MS(E)
B	SS(B)	$b-1$	MS(B)	MS(B)/MS(E)
C	SS(C)	$c-1$	MS(C)	MS(C)/MS(E)
AB	SS(AB)	$(a-1)(b-1)$	MS(AB)	MS(AB)/MS(E)
AC	SS(AC)	$(a-1)(c-1)$	MS(AC)	MS(AC)/MS(E)
BC	SS(BC)	$(b-1)(c-1)$	MS(BC)	MS(BC)/MS(E)
ABC	SS(ABC)	$(a-1)(b-1)(c-1)$	MS(ABC)	MS(ABC)/MS(E)
Error	SS(E)	$abc(d-1)$	MS(E)	
Total	SS(TO)	$abcd-1$		

The main effects (A, B, and C) and the two-factor interactions (AB, AC, and BC) have the same interpretations as in the two-factor ANOVA. The three-factor interaction represents that part of the model for the means μ_{ijh}; $i = 1, 2, \ldots, a$; $j = 1, 2, \ldots, b$; $h = 1, 2, \ldots, c$ that cannot be explained by a model including only the main effects and two-factor interactions. In particular, if for each fixed h, the "plane" created by μ_{ijh} is "parallel" to the "plane" created by every other fixed h, then the three-factor interaction is equal to zero. Usually, higher-order interactions tend to be small.

In the testing sequence, we test the three-factor interaction first by checking to see whether or not

$$MS(ABC)/MS(E) \geq F_\alpha[(a-1)(b-1)(c-1), abc(d-1)].$$

If this inequality holds, the ABC interaction is significant at the α level. We would then not continue testing the two-factor interactions and the main effects with those F values, but analyze the data otherwise. For example, we could, for each fixed h, look at a two-factor ANOVA for factors A and B. Of course, if the inequality does not hold, we next check the two-factor interactions with the appropriate F values. If these are not significant, we check the main effects, A, B, and C.

Factorial experiments with three or more factors require many experiments, particularly if each factor has several levels. Often, in the health, social, and physical sciences, experimenters want to consider several factors (maybe as many as 10 or 20 or hundreds), and they cannot afford to run that many experiments. This is particularly true with preliminary or screening investigations in which they want to detect the factors that seem most important. In these cases they often consider factorial experiments such that each of k factors is run only at two levels, frequently without replication. We consider only this situation, although the reader should recognize that there are many variations of this. In particular, there are methods for investigating only *fractions of these* 2^k *designs*. The reader interested in more information should refer to a good book on design of experiments, such as that by Box, Hunter, and Hunter (see references). Many statisticians in industry believe these statistical methods are the most useful in improving product and process designs. Hence it is clearly an extremely important topic as many industries are greatly concerned about the quality of their products.

In factorial experiments in which each of the k factors is considered at only two levels, those levels are selected at some reasonable low and high values. That is, with the help of someone in the field, the typical range of each factor is considered. For illustration, if we are considering baking temperatures in the range of $300°$ to $375°$, a representative low is selected, say $320°$, and a representative high is selected, say $355°$. There is no formula for these selections, and someone familiar with the experiment would help make these selections. Often it happens that only two different types of material are considered, say fabric, and one is called low and the other high.

Thus we select a low and high for each factor; these will be coded as -1 and $+1$ or, more simply, as $-$ and $+$, respectively. We give three 2^k designs, for $k = 2, 3$, and 4, in standard order in Tables 10.3-2, 10.3-3, and 10.3-4, respectively. From these three tables, we can easily note what is meant by standard order. The A column starts with a minus sign and then the sign alternates. The B column begins with two minus signs, then the signs alternate in blocks of two. The C column has 4 minus signs and then 4 plus signs and so on. The D column starts with 8 minus signs and then 8 plus signs. It is easy to extend this idea to 2^k designs, where $k \geq 5$. For illustration, under the E column in a 2^5 design, we have 16 minus signs followed by 16 plus signs, which account for the 32 experiments.

Table 10.3-2: 2^2 Design			
	2^2 Design		
Run	A	B	Observation
1	−	−	X_1
2	+	−	X_2
3	−	+	X_3
4	+	+	X_4

Table 10.3-3: 2^3 Design				
	2^3 Design			
Run	A	B	C	Observation
1	−	−	−	X_1
2	+	−	−	X_2
3	−	+	−	X_3
4	+	+	−	X_4
5	−	−	+	X_5
6	+	−	+	X_6
7	−	+	+	X_7
8	+	+	+	X_8

Table 10.3-4: 2^4 Design					
	2^4 Design				
Run	A	B	C	D	Observation
1	−	−	−	−	X_1
2	+	−	−	−	X_2
3	−	+	−	−	X_3
4	+	+	−	−	X_4
5	−	−	+	−	X_5
6	+	−	+	−	X_6
7	−	+	+	−	X_7
8	+	+	+	−	X_8
9	−	−	−	+	X_9
10	+	−	−	+	X_{10}
11	−	+	−	+	X_{11}
12	+	+	−	+	X_{12}
13	−	−	+	+	X_{13}
14	+	−	+	+	X_{14}
15	−	+	+	+	X_{15}
16	+	+	+	+	X_{16}

Table 10.3-5: 2^3 Design Decomposition								

	2^3 Design							
Run	**A**	**B**	**C**	**AB**	**AC**	**BC**	**ABC**	**Observation**
1	$-$	$-$	$-$	$+$	$+$	$+$	$-$	X_1
2	$+$	$-$	$-$	$-$	$-$	$+$	$+$	X_2
3	$-$	$+$	$-$	$-$	$+$	$-$	$+$	X_3
4	$+$	$+$	$-$	$+$	$-$	$-$	$-$	X_4
5	$-$	$-$	$+$	$+$	$-$	$-$	$+$	X_5
6	$+$	$-$	$+$	$-$	$+$	$-$	$-$	X_6
7	$-$	$+$	$+$	$-$	$-$	$+$	$-$	X_7
8	$+$	$+$	$+$	$+$	$+$	$+$	$+$	X_8

To be absolutely certain what these runs mean, consider run number 12 in Table 10.3-4: A is set at its high level, B at its high, C at its low, and D at its high level. The value X_{12} is the random observation resulting from this one combination of these four settings. It must be emphasized, however, that the runs are not necessarily performed in the order $1, 2, 3, \ldots, 2^k$; but they should be performed in a random order, if at all possible. That is, in a 2^3 design, we might perform the experiment in the order: 3, 2, 8, 6, 5, 1, 4, 7 if this, in fact, was a random selection of a permutation of the first eight positive integers.

Once all 2^k experiments have been run, it is possible to consider the total sum of squares

$$\sum_{i=1}^{2^k} (X_i - \overline{X})^2$$

and decompose it very easily into $2^k - 1$ parts, which represent the respective measurements (estimators) of the k main effects, $\binom{k}{2}$ two-factor interactions, $\binom{k}{3}$ three-factor interactions, and so on until we have the one k-factor interaction. We illustrate this decomposition with the 2^3 design in Table 10.3-5. Note the column AB is found by formally multiplying the elements of column A by the corresponding ones in B. Likewise, AC is found by multiplying column A and column C, and so on until column ABC is the product of the corresponding elements of columns A, B, C. We then construct seven linear forms using these seven columns of signs with the corresponding observations. The resulting measures (estimates) of the main effects (A, B, C), the two-factor interactions (AB, AC, BC), and the three-factor interaction (ABC) are then found by dividing these linear forms by $2^k = 2^3 = 8$ (some statisticians divide by $2^{k-1} = 2^{3-1} = 4$). These are denoted by

$$[A] = (-X_1 + X_2 - X_3 + X_4 - X_5 + X_6 - X_7 + X_8)/8,$$
$$[B] = (-X_1 - X_2 + X_3 + X_4 - X_5 - X_6 + X_7 + X_8)/8,$$

$$[C] = (-X_1 - X_2 - X_3 - X_4 + X_5 + X_6 + X_7 + X_8)/8,$$
$$[AB] = (+X_1 - X_2 - X_3 + X_4 + X_5 - X_6 - X_7 + X_8)/8,$$
$$[AC] = (+X_1 - X_2 + X_3 - X_4 - X_5 + X_6 - X_7 + X_8)/8,$$
$$[BC] = (+X_1 + X_2 - X_3 - X_4 - X_5 - X_6 + X_7 + X_8)/8,$$
$$[ABC] = (-X_1 + X_2 + X_3 - X_4 + X_5 - X_6 - X_7 + X_8)/8.$$

With assumptions of normality, mutual independence, and common variance σ^2, under the overall null hypothesis of the equality of all the means, each of these measures has a normal distribution with mean zero and variance $\sigma^2/8$ (in general $\sigma^2/2^k$). This implies that the square of each measure divided by $\sigma^2/8$ is $\chi^2(1)$. Moreover, it can be shown (see Exercise 10.3-2) that

$$\sum_{i=1}^{8} (X_i - \overline{X})^2$$

$$= 8([A]^2 + [B]^2 + [C]^2 + [AB]^2 + [AC]^2 + [BC]^2 + [ABC]^2).$$

So by Theorem 8.7-1, the terms on the righthand side, divided by σ^2, are mutually independent random variables, each being $\chi^2(1)$. While it requires a little more theory, it follows that the linear forms [A], [B], [C], [AB], [AC], [BC], and [ABC] are mutually independent $N(0, \sigma^2/8)$ random variables.

Since we have assumed that we have not run any replications, how can we obtain an estimate of σ^2 to see if any of the main effects or interactions are significant? To help us, we fall back on the use of a q-q plot because, under the overall null hypothesis, those seven measures are mutually independently normally distributed with the same mean and variance. Thus a q-q plot of the normal percentiles against the corresponding ordered values of the measures should be about on a straight line if, in fact, the null hypothesis is true. If one of these points is "out of line," we might believe that the overall null hypothesis is not true and that the effect associated with it is significant. It is possible that two or three points might be out of line, then all corresponding effects (main or interaction) should be investigated. Clearly, this is not a formal test, but it has been extremely successful in practice.

As an illustration, we use the data from an experiment designed to evaluate the effects of laundering on a certain fire-retardant treatment for fabrics. These data, somewhat modified were taken from *Experimental Statistics, National Bureau of Standards Handbook 91* by Mary G. Natrella (Washington, D.C.: U.S. Government Printing Office, 1963). Factor A is the type of fabric (sateen or monk's cloth), factor B corresponds to two different fire-retardant treatments, and factor C describes the laundering conditions (no laundering, after one laundering). The observation are inches burned, measured on a standard size fabric after a flame test. They are, in standard order.

$$x_1 = 41.0, \quad x_2 = 30.5, \quad x_3 = 47.5, \quad x_4 = 27.0,$$
$$x_5 = 39.5, \quad x_6 = 26.5, \quad x_7 = 48.0, \quad x_8 = 27.5.$$

Thus the measures of the effects are

$$[A] = (-41.0 + 30.5 - 47.5 + 27.0 - 39.5 + 26.5 - 48.0 + 27.5)/8 = -8.06,$$
$$[B] = (-41.0 - 30.5 + 47.5 + 27.0 - 39.5 - 26.5 + 48.0 + 27.5)/8 = 1.56,$$
$$[C] = (-41.0 - 30.5 - 47.5 - 27.0 + 39.5 + 26.5 + 48.0 + 27.5)/8 = 0.56,$$
$$[AB] = (+41.0 - 30.5 - 47.5 + 27.0 + 39.5 - 26.5 - 48.0 + 27.5)/8 = -2.19,$$
$$[AC] = (+41.0 - 30.5 + 47.5 - 27.0 - 39.5 + 26.5 - 48.0 + 27.5)/8 = -0.31,$$
$$[BC] = (+41.0 + 30.5 - 47.5 - 27.0 - 39.5 - 26.5 + 48.0 + 27.5)/8 = 0.81,$$
$$[ABC] = (-41.0 + 30.5 + 47.5 - 27.0 + 39.5 - 26.5 - 48.0 + 27.5)/8 = 0.31.$$

In Table 10.3-6 we order these seven measures, determine their percentiles, and find the corresponding percentiles of the standard normal distribution.

Table 10.3-6: Seven Measures Ordered			
Identity of Effect	**Ordered Effect**	**Percentile**	**Percentile from $N(0, 1)$**
[A]	−8.06	12.5	−1.15
[AB]	−2.19	25.0	−0.67
[AC]	−0.31	37.5	−0.32
[ABC]	0.31	50.0	0.00
[C]	0.56	62.5	0.32
[BC]	0.81	75.0	0.67
[B]	1.56	87.5	1.15

The q-q plot is given in Figure 10.3-1. Each point has been identified with its effect. A straight line fits six of those points reasonably well, but the point

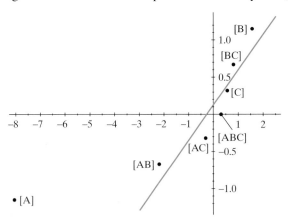

Figure 10.3-1: A q-q plot of normal percentiles versus estimated effects

associated with $[A] = -8.06$ is far from this straight line. Hence the main effect of factor A (the type of fabric) seems to be significant. It is interesting to note that the laundering factor, C, does not seem to be a significant factor.

EXERCISES

10.3-1 Write out a 2^2 design, displaying the A, B, and AB columns for the four runs.

    (a) If X_1, X_2, X_3, X_4 are the four observations for the respective runs in standard order, write out the three linear forms, $[A]$, $[B]$, and $[AB]$, that measure the two main effects and the interaction. These linear forms should include the divisor $2^2 = 4$.

    (b) Show that $\sum_{i=1}^{4} (X_i - \overline{X})^2 = 4([A]^2 + [B]^2 + [AB]^2)$.

    (c) Under the null hypothesis that all the means are equal and with the usual assumptions (normality, mutually independent, and common variance), what can you say about the distributions of the expressions in (b) after each is divided by σ^2?

10.3-2 Show that in a 2^3 design, it is true that

$$\sum_{i=1}^{8} (X_i - \overline{X})^2 = 8([A]^2 + [B]^2 + [C]^2 + [AB]^2 + [AC]^2 + [BC]^2 + [ABC]^2).$$

HINT: Since both the right and the left members of this equation are symmetric in the variables $X_1, X_2, \ldots, X_8$, it is only necessary to show that the corresponding coefficients of $X_1 X_i$, $i = 1, 2, \ldots, 8$, are the same in each member of the equation. Of course, recall $\overline{X} = (X_1 + X_2 + \cdots + X_8)/8$.

10.3-3 Show that the unbiased estimator of the variance σ^2 from a sample of size $n = 2$ is one-half of the square of the difference of the two observations. Thus, if a 2^k design is replicated, say with X_{i1} and X_{i2}, $i = 1, 2, \ldots, 2^k$, show that the estimate of the common σ^2 is

$$\frac{1}{2^{k+1}} \sum_{i=1}^{2^k} (X_{i1} - X_{i2})^2 = MS(E).$$

Under the usual assumptions, this implies that each of $2^k[A]^2/MS(E)$, $2^k[B]^2/MS(E)$, $2^k[AB]^2/MS(E)$, and so on has an $F(1, 2^k)$ distribution under the null hypothesis. This, of course, would provide tests for the significance of the various effects, including interactions.

10.3-4 Hogg and Ledolter note that percent yields from a certain chemical reaction for changing temperature (factor A), reaction time (factor B), and concentration (factor C) are $x_1 = 79.7$, $x_2 = 74.3$, $x_3 = 76.7$, $x_4 = 70.0$, $x_5 = 84.0$, $x_6 = 81.3$, $x_7 = 87.3$, $x_8 = 73.7$ in standard order with a 2^3 design.

    (a) Estimate the main effects, the three two-factor interactions, and the three-factor interaction.

    (b) Construct an appropriate q-q plot to see if any of these effects seem to be significantly larger than the others. [R. V. Hogg and J. Ledolter, *Applied Statistics for Engineers and Physical Scientists*, 2nd ed. (New York: Macmillan, 1992).]

10.3-5 Box, Hunter, and Hunter studied the effects of catalyst charge (10 pounds $= -1$, 20 pounds $= +1$), temperature ($220°C = -1$, $240°C = +1$), pressure (50 psi $= -1$, 80 psi $= +1$), and concentration (10% $= -1$, 12% $= +1$) on percent conversion (X) of a certain chemical. The results of a 2^4 design, in standard order, are

$$x_1 = 71, \quad x_2 = 61, \quad x_3 = 90, \quad x_4 = 82, \quad x_5 = 68, \quad x_6 = 61,$$
$$x_7 = 87, \quad x_8 = 80, \quad x_9 = 61, \quad x_{10} = 50, \quad x_{11} = 89,$$
$$x_{12} = 83, \quad x_{13} = 59, \quad x_{14} = 51, \quad x_{15} = 85, \quad x_{16} = 78.$$

(a) Estimate the main effects and the two-, three-, and four-factor interactions.

(b) Construct an appropriate q-q plot and assess the significance of the various effects.

10.4 MORE ON DESIGN OF EXPERIMENTS

Many observations that are taken occur without active participation of an experimenter. The analysis of such data represents a passive use of statistics that is often extremely helpful in making future decisions. However, in trying to create better products or services, there are frequently a number of factors (independent variables or "parameters") that we can control, and we want to set these at levels which will produce the best products.

For example, in making a cake mix, a company must decide how much sugar, flour, vanilla, salt, butter, egg mix, and milk to use, and then also recommend what time and temperature should be used in baking the mixture. Each of these variables (from amount of sugar to temperature) is called a factor, and the experimenters must decide which level is best for each factor. If we can list nine or so factors for making a cake, think of how many factors would be associated with a more complicated product.

Much can be learned about a process by considering the most influential factors and changing them according to a specified plan to try to find the best level of each. How does a cake mix company know that a certain mixture should be baked at 350° for 35 minutes? Why not use 325° for 40 minutes? Or 375° for 30 minutes? The company must experiment with these levels of time and temperature, as well as with those of the other factors, to determine the optimal. Only by experimenting can the company determine the best level for each factor.

Such investigations involve collecting and analyzing data from carefully designed experiments. Since a poorly designed experiment may not shed light on a particular question that we want answered, it is important to plan an experiment before starting to perform it. Trying to rescue a poorly designed experiment after it has been run and the data have been collected usually leads to less-than-satisfactory results. Thus it is important to design the experiment ahead of time so that the validity of the experimental results is assured. The *design of experiments* is a very important area of statistics that teaches us how to perform experiments to maximize our knowledge of the situation under consideration with a relatively few number of experimental runs.

As an example of such experimentation, let us return to the time and temperature needed to bake the best cake. Suppose that this is a completely new cake mix and the company has no idea about the right time and temperature, except that they will probably be in the ranges of 25–50 minutes and 250°–400°, respectively. Suppose that we have some measure of the "goodness" of a cake, possibly scores of professional tasters who grade each cake on a 100-point scale.

Unknown to anyone, suppose that the real goodness response to the cakes made at different times and temperatures is given in Figure 10.4-1. That is,

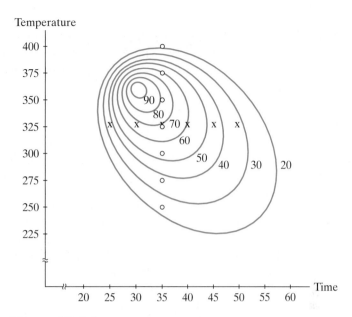

Figure 10.4-1: Goodness response to cake-baking experiment

this response surface is depicted by contour curves which are like "isotherm curves." For example, a cake baked at 300° for 35 minutes would get a score of 52 from the tasters. (Although there would be variation among tasters, let us say they average 52 for such a cake.) Clearly, if we knew this response surface ahead of time, we should bake this cake mix at about 360° for about 31 minutes to obtain a score above 90. However, we do *not* know this surface. It is our job to find out something about it so that we can determine the optimal time and temperature at which to bake this cake. We do this by baking the cake mixture at different times and temperatures.

One procedure, called "changing one factor at a time" is sometimes used; but, as we see in Figure 10.4-1, that scheme often misses the optimal levels of the factors. In it, we set each of the factors except the first at some middle value. In our case, we only have two factors; so we set the second at 325°, say. Now let us run the experiment 6 times: at 25, 30, 35, 40, 45, and 50 minutes. The respective scores are *about* 35, 64, 69, 60, 46, and 32 by interpolating between contour curves at the *x* points. That is, with the temperature set at 325°, the maximum score occurs at 35 minutes. Thus we now set the time at 35 minutes and perform six more experiments with respective temperatures of 250°, 275°, 300°, 350°, 375°, and 400°, recalling that 35 minutes and 325° has already produced a score of about 69. The seven scores, with 69 inserted in the correct position, are about 23, 37, 53, 69, 79, 61, and 20. Thus it looks as if the maximum occurs at 350° for 35 minutes; also, the maximum value is 79. Clearly, we have missed the true optimum levels of these two factors by using the "changing one factor at a time" scheme.

Wouldn't it have been much better to run at first four experiments changing the levels of each factor? That is, take two middle values of each factor: say 35 and 40 minutes, 300° and 325°. Then run four experiments at (35, 300°), (35, 325°), (40, 300°), and (40, 325°), obtaining scores of about 53, 69, 52, and 60, respectively. These values are plotted in Figure 10.4-2.

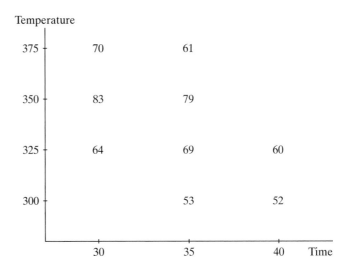

Figure 10.4-2: Results of cake-baking experiments

It seems fairly clear that the surface is increasing as we move to the left and upward. Thus we might run three more experiments at (30, 325°), (30, 350°), and (35, 350°) obtaining scores of about 64, 83, and 79, respectively; these are also noted in Figure 10.4-2. Clearly, we must continue upward and we perform two more experiments at (30, 375°) and (35, 375°) obtaining 70 and 61. At this point, it seems as if we should investigate points in the neighborhood of 30 minutes and temperatures about 350° to 375°, which will just about give us an optimal solution. But note how changing both factors in experimenting is better than changing one at a time. This is even more true as the number of factors increases.

In the preceding discussion, in which we first ran four experiments, with two factors at two levels each, is often called a 2^2 factorial experiment. If we had three factors involved and ran each at two levels, that would give us $2 \times 2 \times 2 = 8$ experimental runs, which is called a 2^3 factorial experiment. In general, with k factors at 2 levels each, we have a 2^k factorial experiment. Note that with $k = 10$ factors and only two levels each, we would need $2^{10} = 1024$ experimental runs. Often this is too many, and in advanced courses on design of experiments we learn how to run only a fraction of this number without losing too much information. (See also Section 10.3*.)

Let us say that we now wish to compare two different methods of doing something (say teaching statistics). In essence we have only one factor, teaching methods, and we have only two levels: Method I (M_1) and Method 2 (M_2). If we had 50 students who were participating in this study, it would be desirable to place 25 in each group. The first group uses M_1 and the second M_2. Clearly, if the investigators favor M_2 over M_1, it would not be fair to let them place the better students in the second group. The assignments should be made at random so that each combination of 50 students taken 25 at a time has the same chance of being the first group. One way of doing this is placing the 50 names on equal size slips of paper and then drawing 25 from a bowl at random.

Suppose that we knew a little more about these students and were able to pair them up by mathematical ability, say. Student S_1 is about the same as S_2, S_3 about the same as S_4, and so on. Then let us select one from each pair at random (like with the flip of a coin) for the first group. We would say that the pair S_1, S_2 is a block; pair S_3, S_4 is a block; and so on. This procedure of assigning students to the respective groups is called **blocking**.

To illustrate blocking, suppose that we wish to test the wearing ability of two different types of soles for tennis shoes. Say we have 30 grade-school children who would participate. One way to select 15 for each group would be at random, and the first group would wear shoes with the first sole and the second group the second sole. After one month, say, we would compare the amount of wear for each group. A better way would be to block as follows: Let each student use a pair of shoes that has both types of soles, one on the right foot and the other on the left, as we recognize that each foot must take essentially the same number of steps (to be safe, we might let 15 students, selected at random, wear one type of sole on the right foot and the other 15 wear the other type on the right foot). In this experiment each student is a block. Sometimes in experiments, siblings or twins or two rats from the same litter could be used as a block.

Actually, a lot of good design of experiments is just plain common sense, but sometimes it is helpful for statisticians to note a few things about careful designs for those interested in running important experiments and studies. Our simple recommendation is to block what you can and randomize what you cannot block. However, if a person is involved in a complicated experiment or study, he or she should study advanced methods in the design of experiments or see a friendly statistician.

A block design can be used with more than two levels of one factor. Two illustrations of a *randomized block design* are given.

1. Suppose we wish to test three methods of teaching statistics, say M_1, M_2, and M_3; and we have selected 60 students with whom to experiment. We could divide these sixty into three groups at random to be used with the respective methods, M_1, M_2, and M_3. This would be good randomized design, but it might happen "by the luck of the draw" that one of the methods had better math students than the other two and hence would seemingly have an advantage. Possibly a better way would be to put these students into 4 blocks, consisting of 15 students each. The first block would be the best math students as measured by the highest 15 ACT math scores. The second block would consist of the 15 students with the next highest ACT math scores. This is continued until we have 4 blocks, each with 15 students. After doing this, we would select 5 students at random from each

block to use M_1, 5 students to use M_2, and 5 students to use M_3. This seems fairer for now each method has the same number of students with essentially the same math ability as measured by the ACT math exam. We now observe the scores on some evaluating test at the end of the semester and compare the differences.

2. Suppose we wish to test four different feeds, say F_1, F_2, F_3, F_4, to be used on newly weaned pigs. It is recognized that there might be difference in pigs from different litters, some sows producing better pigs than others. Hence we block as follows. Select 4 pigs (after ruling out the runts) from each of several litters. The four from one litter are assigned at random to each of the four feeds. This is done for each litter; that is, we have blocked on sows' litters. So the experiment seems perfectly fair as each feed has one pig from each litter. At the end of the experiment we compare the differences of the weights gained with each feed.

EXERCISES

These and similar problems could be assigned as projects if time permits.

10.4-1 Design an experiment to find out how the distance to the basket and the angle from the perpendicular to the backboard affects your shooting percentage in basketball. Make certain you randomize the order of the positions from which you shoot to avoid "warm up" and "fatigue" effects. Display your results. (Essentially, you are estimating a response surface and you might draw contours.)

10.4-2 Suppose that we want to test the effects of two factors in making popcorn. One factor is the type of popper with two levels: oil-based or air-based. The other factor, type of popcorn, has three levels: gourmet, national brand, generic brand. Clearly, one response measurement would be the amount of popcorn that is obtained from $\frac{1}{2}$ cup of popcorn. But you could also try some sort of taste measurement, the amount of popcorn obtained from 10 cents' worth of each of popcorn. or some combination of taste and cost. What are your conclusions?

10.4-3 Suppose that we plan to compare two detergents, say A and B. We will wash equally dirty white clothes with each detergent, also changing two other factors: hot or cold water, soft or hard water. Thus, we have a three-factor experiment in which each factor has two levels; a 2^3 factorial experiment is under consideration and, without replication, we needed 8 runs. Replications might be desirable using 16 or 24 runs. What is the response variable?

10.4-4 We want to investigate the effects of five factors on the growth of pinto beans. They are

- Soaking fluid: water or regular beer
- Salinity: no salt or salt
- Acidity: no vinegar or vinegar
- Temperature: refrigerator or room
- Soaking time: 2 hours or 4 hours

Use 3 tablespoons of soaking fluid. For salt add $\frac{1}{4}$ teaspoon to soaking liquid. For vinegar, add 1 teaspoon to the soaking fluid. Note that this is a 2^5 factorial experiment and, without replications, 32 runs are needed to cover all possible combinations.

10.4-5 Thirty-six overweight undergraduate college males agree to participate in a study of the effectiveness of four reducing plans, say R_1, R_2, R_3, R_4. Before starting these plans, the ideal weight of each participant is determined and from that how much he is overweight. These amounts of overweight ranged roughly from 15 to 80 pounds. Explain how you might run a randomized block design with three blocks, in which you block on pounds overweight.

EPILOGUE

Clearly there is much more to applied and theoretical statistics than can be studied in one book. As mentioned in the Prologue, there is a huge demand for statistical scientists in many fields to make sense out of the increasing volumes of data. Statistics is needed to turn quality data into useful information upon which decisions can be made. The striking thing about any data set is that there is variation; all the data points simply do not lie on the pattern. It is the statistician's job to find that pattern and describe the variation about it. Done properly, this clearly helps the decision maker significantly.

One observations about variation should be noted before any major adjustments or decisions are made. Frequently persons in charge jump at conclusions too quickly; that is, major decisions are often made after too few observations. For illustration, we know that if X_1, X_2, X_3 are independent and identically distributed continuous-type observations, then

$$P(X_1 < X_2 < X_3) = \frac{1}{3!} = \frac{1}{6}.$$

Yet if this occurred and these observations were taken on sales, say, and plotted in time sequence, the fact that these three points were "going up" might suggest to management that "the company is on a roll." In some cases, only two increasing points might cause this reaction. If there is no change in the system, two or three increasing points have respective probabilities of 1/2 and 1/6 and those probabilities do not warrant that sort of reaction. If, on the other hand, we observe four or five such points, with respective probabilities of 1/24 and 1/120, then an appropriate reaction is in order.

An interesting question to ask is why statisticians treat $1/20 = 0.05$ as the value at which the probability of an event is considered small and often suggests some type of action. Possibly it was because Ronald A. Fisher suggested that one out of twenty seemed small enough. Obviously the value of 0.05 is not written in stone, but statisticians seem to look for differences of two or three standard deviations as a guide for action. Since many estimators have approximate normal distributions, such differences do have small probabilities.

For illustration, suppose a candidate believes that he or she has at least 50% of the votes. Yet in a poll of $n = 400$ only 160 favor that candidate; thus the standardized value

$$\frac{160 - 400(1/2)}{\sqrt{400(1/2)(1/2)}} = -4$$

suggests that the candidate does not in fact have 50% of the votes. Depending upon the financial situation, the candidate must change the approach and/or work harder, or possibly even consider dropping from the race. However, we note that this simple statistical analysis can be a guide in the decision process.

Another example is provided by W. A. Shewhart's efforts at Bell Telephone Laboratories in the 1920s in his quality improvement through statistical methods. He conceived of the idea of sampling occasionally during the manufacturing process rather than waiting until the items arrived at an inspection center at the end of the line. To monitor the need of possible adjustments, certain statistics, like $\overline{X}$ and S, were taken. If one of these had an *unusual* value, corrective action was taken before too many more defective items were produced. If T is one of those statistics, an unusual value of T was one that was outside the *control limits* $\mu_T \pm 3\sigma_T$, where the mean μ_T and the standard deviation σ_T of T would often be estimated from past data coming from a satisfactory (*in control*) process.

While the three sigma limits have proved to be an effective guide, many companies are now considering six sigma limits. The Six Sigma program was first started by Motorola, and it has its statistical underpinnings in the following. There are specifications in manufacturing items that are usually set by the engineers–or possibly by customers. These are usually given in terms of a target value and upper and lower specification limits, USL and LSL, respectively. The mean μ of the values resulting from the process is hopefully close to the target value. The "specs" are six standard deviations, 6σ, away from the target, where σ is the standard deviation associated with the process. However, the mean μ is often dynamic, and these Six Sigma companies try to keep it at least 4.5σ from the closest spec. If the values of the items are distributed normally and nearest spec is 4.5σ from μ, there are only 3.4 defectives per million. This is the goal of the Six Sigma companies; and they use many of Deming's ideas, some of which are outlined in the Centerpiece, in their attempts to achieve this very worthwhile goal.

While understanding of quality improvement ideas and basic statistical methods, like those in Chapter 10, is extremely important to the Six Sigma programs, possibly the major factor is the attitude of the CEOs and other important administrators. There were many total quality management programs in the 1980s; but they were not as successful as Six Sigma, for now each CEO is demanding "Show me the money!" That is, companies hire a Six Sigma expert to come in for four one-week periods, about one month apart, for a fee of about $15,000 per person. If they have 20 participants, this cost is $300,000 plus the four weeks of time "lost" to the work process by each trainee. That is a great deal of money. However, each of these individuals (sometimes a pair) has a project associated with some process that has not been very efficient. They work on these projects using statistical methods during the "off months" and report to the Six Sigma expert during the training weeks to get advice. Often these projects, if successful, will save millions of dollars for a company, and the expenses are well worth the benefits. As a matter of fact, if a participant's project saves the company at least one million dollars, he or she earns a "six sigma Black Belt." So the CEO does see the money saved, and thus the bottom line looks extremely good to him or her.

While the reader has now studied enough statistics to appreciate the importance of understanding variation, there are many more useful statistical techniques to be studied if the reader is so inclined. For example, there are courses in regression and time series in which we learn how to predict future observations. Or a study of design of experiment can help an investigator select the most efficient levels of the various factors. After all, if we have 10 factors and we run each at only two levels, we have created $2^{10} = 1024$ runs. Can we perform only a fraction of these runs without losing too much information? Additional study of multivariate analysis can lead to interesting problems in classification. Say a doctor takes several measurements on a patient and then classifies the patent's disease as one of many possible diseases. There are errors of misclassification and the statisticians can help reduce the probabilities of those errors. Doctors–and statisticians–can made mistakes and second or third opinions should be asked for if there is some doubt.

As mentioned in the Prologue, the computer has opened the door to a wide variety of new statistical techniques; and researchers, computer scientists, and statisticians are working together to reduce huge amounts of data into nuggets of quality information on which important decisions can be made. Statistics is an exciting field that finds many useful applications in the social, health, and physical sciences. The authors have found statistics to be a great profession; we hope a few of you find it that way too. In any case, we hope that statistical thinking will make you more aware of the need of understanding variation, which can be a great influence on your daily life.

REVIEW OF SELECTED MATHEMATICAL TECHNIQUES

A.1 ALGEBRA OF SETS

The totality of objects under consideration is called the **universal set** and is denoted by S. Each object in S is called an **element** of S. If a set A is a collection of elements that are also in S, then A is said to be a **subset** of S. In applications of probability, S will usually denote the **sample space**. An **event** A will be a collection of possible outcomes of the experiment and will be a subset of S. We say that event A *has occurred* if the outcome of the experiment is an element of A. The set or event A may be described by listing all of its elements or by defining the properties that its elements must satisfy.

EXAMPLE A.1-1

A four-sided die, called a tetrahedron, has four faces that are equilateral triangles. These faces are numbered 1, 2, 3, 4. When the tetrahedron is rolled, the outcome of the experiment is the number of the face that is down. If this tetrahedron is rolled twice and we keep track of the first roll and the second roll, then the sample space is that displayed in Figure A.1-1.

Let A be the event that the second roll is a 1 or a 2. That is,

$$A = \{(x, y): \ y = 1 \text{ or } y = 2\}.$$

Let

$$B = \{(x, y): \ x + y = 6\} = \{(2, 4), (3, 3), (4, 2)\},$$

651

and let

$$C = \{(x, y): \ x + y \geq 7\} = \{(4, 3), (3, 4), (4, 4)\}.$$

Events A, B, and C are shown in Figure A.1-1.

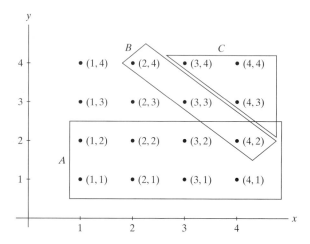

Figure A.1-1: Sample space for two rolls of a four-sided die

When a is an element in A, we write $a \in A$. When a is not an element in A, we write $a \notin A$. So, in Example A.1-1, we have $(3, 1) \in A$ and $(1, 3) \notin A$. If every element of a set A is also an element in a set B, then A is a **subset** of B. We write $A \subset B$. In probability, if event B occurs whenever event A occurs, then $A \subset B$. The two sets A and B are equal (i.e., $A = B$), if $A \subset B$ and $B \subset A$. Note that it is always true that $A \subset A$ and $A \subset S$, where S is the universal set. We denote the subset that contains no elements by $\emptyset$. This set is called the **null** or **empty** set. For all sets A, $\emptyset \subset A$.

The set of elements in either A or B or possibly both A and B is called the **union** of A and B and is denoted $A \cup B$. The set of elements in both A and B is called the **intersection** of A and B and is denoted $A \cap B$. The **complement** of a set A is the set of elements in the universal set S that are not in the set A and is denoted A'. In probability, if A and B are two events, the event that at least one of the two events has occurred is denoted by $A \cup B$ and the event that both events have occurred is denoted by $A \cap B$. The event that A has not occurred is denoted by A', and the event that A has not occurred but B has occurred is denoted by $A' \cap B$. If $A \cap B = \emptyset$, we say that A and B are **mutually exclusive**. In Example A.1-1, $B \cup C = \{(x, y): \ x + y \geq 6\}$, $A \cap B = \{(4, 2)\}$, and $A \cap C = \emptyset$. Note that A and C are mutually exclusive. Also, $C' = \{(x, y): \ x + y \leq 6\}$.

The operations of union and intersection may be extended to more than two sets. Let $A_1, A_2, \ldots, A_n$ be a finite collection of sets. Then the **union**

$$A_1 \cup A_2 \cup \cdots \cup A_n = \bigcup_{k=1}^{n} A_k$$

is the set of elements that belong to at least one A_k, $k = 1, 2, \ldots, n$. The **intersection**

$$A_1 \cap A_2 \cap \cdots \cap A_n = \bigcap_{k=1}^{n} A_k$$

is the set of all elements that belong to every A_k, $k = 1, 2, \ldots, n$. Similarly, let $A_1, A_2, \ldots, A_n, \ldots$ be a countable collection of sets. Then x belongs to the **union**

$$A_1 \cup A_2 \cup A_3 \cup \cdots = \bigcup_{k=1}^{\infty} A_k$$

if x belongs to at least one A_k, $k = 1, 2, 3, \ldots$. Also x belongs to the **intersection**

$$A_1 \cap A_2 \cap A_3 \cap \cdots = \bigcap_{k=1}^{\infty} A_k$$

if x belongs to every A_k, $k = 1, 2, 3, \ldots$.

EXAMPLE A.1-2   Let

$$A_k = \left\{ x \colon \frac{10}{k+1} \leq x \leq 10 \right\}, \qquad k = 1, 2, 3, \ldots.$$

Then

$$\bigcup_{k=1}^{8} A_k = \left\{ x \colon \frac{10}{9} \leq x \leq 10 \right\};$$

$$\bigcup_{k=1}^{\infty} A_k = \{ x \colon 0 < x \leq 10 \}.$$

Note that the number zero is not in this latter union, since it is not in one of the sets $A_1, A_2, A_3, \ldots$. Also,

$$\bigcap_{k=1}^{8} A_k = \{ x \colon 5 \leq x \leq 10 \} = A_1;$$

$$\bigcap_{k=1}^{\infty} A_k = \{ x \colon 5 \leq x \leq 10 \} = A_1,$$

since $A_1 \subset A_k$, $k = 1, 2, 3, \ldots$.   ◄

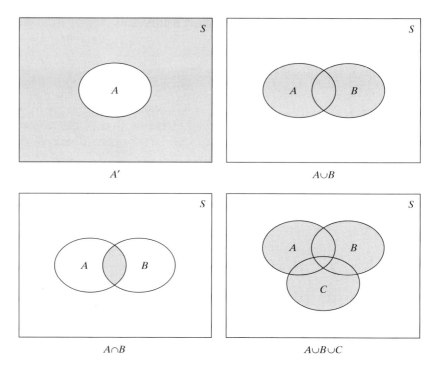

Figure A.1-2: Algebra of sets

A convenient way to illustrate operations on sets is with a **Venn** diagram. In Figure A.1-2 the universal set S is represented by the rectangle and its interior and the subsets of S by the points enclosed by the circles. The sets under consideration are the shaded regions.

Set operations satisfy several properties. For example, if A, B, and C are subsets of S, we have the following:

Commutative Laws

$$A \cup B = B \cup A$$
$$A \cap B = B \cap A$$

Associative Laws

$$(A \cup B) \cup C = A \cup (B \cup C)$$
$$(A \cap B) \cap C = A \cap (B \cap C)$$

Distributive Laws

$$A \cap (B \cup C) = (A \cap B) \cup (A \cap C)$$
$$A \cup (B \cap C) = (A \cup B) \cap (A \cup C)$$

De Morgan's Laws

$$(A \cup B)' = A' \cap B'$$

$$(A \cap B)' = A' \cup B'$$

A Venn diagram will be used to justify the first of De Morgan's laws. In Figure A.1-3(a), $A \cup B$ is represented by horizontal lines, and thus $(A \cup B)'$ is the region represented by vertical lines. In Figure A.1-3(b), A' is indicated with horizontal lines, and B' is indicated with vertical lines. An element belongs to $A' \cap B'$ if it belongs to both A' and B'. Thus the crosshatched region represents $A' \cap B'$. Clearly, this crosshatched region is the same as that shaded with vertical lines in Figure A.1-3(a).

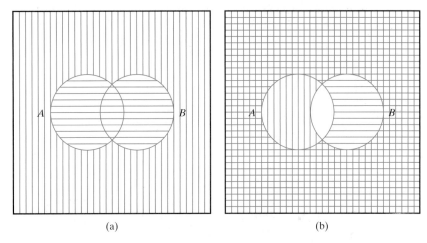

(a)   (b)

Figure A.1-3: Venn diagrams illustrating DeMorgan's laws

A.2 **MATHEMATICAL TOOLS FOR THE HYPER-GEOMETRIC DISTRIBUTION**

Let X have a hypergeometric distribution. That is, the p.m.f. of X is

$$f(x) = \frac{\binom{N_1}{x}\binom{N_2}{n-x}}{\binom{N_1+N_2}{n}}$$

$$= \frac{\binom{N_1}{x}\binom{N_2}{n-x}}{\binom{N}{n}}, \qquad x \leq n, \ x \leq N_1, \ n-x \leq N_2.$$

To show that $\sum_{x=0}^{n} f(x) = 1$ and to find the mean and variance of X, we use the following theorem.

Theorem A.2-1

$$\binom{N}{n} = \sum_{x=0}^{n} \binom{N_1}{x}\binom{N_2}{n-x},$$

where $N = N_1 + N_2$ and it is understood that $\binom{k}{j} = 0$ if $j > k$.

Proof: Because $N = N_1 + N_2$, we have the identity

$$(1 + y)^N \equiv (1 + y)^{N_1}(1 + y)^{N_2}. \tag{A.2-1}$$

We will expand each of these binomials; and, since the polynomials on each side are identically equal, the coefficients of y^n on each side of Equation A.2-1 must be equal. Using the binomial expansion, we find that the expansion of the left side of Equation A.2-1 is

$$(1 + y)^N = \sum_{k=0}^{N} \binom{N}{k} y^k$$

$$= \binom{N}{0} + \binom{N}{1}y + \cdots + \binom{N}{n}y^n + \cdots + \binom{N}{N}y^N.$$

The right side of Equation A.2-1 becomes

$$(1 + y)^{N_1}(1 + y)^{N_2} = \left[\binom{N_1}{0} + \binom{N_1}{1}y + \cdots + \binom{N_1}{n}y^n + \cdots + \binom{N_1}{N_1}y^{N_1}\right]$$

$$\times \left[\binom{N_2}{0} + \binom{N_2}{1}y + \cdots + \binom{N_2}{n}y^n + \cdots + \binom{N_2}{N_2}y^{N_2}\right].$$

The coefficient of y^n in this product is

$$\binom{N_1}{0}\binom{N_2}{n} + \binom{N_1}{1}\binom{N_2}{n-1} + \cdots + \binom{N_1}{n}\binom{N_2}{0} = \sum_{x=0}^{n} \binom{N_1}{x}\binom{N_2}{n-x}$$

and this sum must be equal to $\binom{N}{n}$, the coefficient of y^n on the left side of Equation A.2-1.   ○

Using Theorem A.2-1, we find that it follows that if X has a hypergeometric distribution with p.m.f. $f(x)$, then

$$\sum_{x=0}^{n} f(x) = \sum_{x=0}^{n} \frac{\binom{N_1}{x}\binom{N_2}{n-x}}{\binom{N}{n}} = 1.$$

To find the mean and variance of a hypergeometric random variable, it is useful to note that, with $n > 0$,

$$\binom{N}{n} = \frac{N!}{n!\,(N-n)!} = \frac{N}{n} \cdot \frac{(N-1)!}{(n-1)!\,(N-n)!} = \frac{N}{n}\binom{N-1}{n-1}.$$

The mean of a hypergeometric random variable X is

$$\mu = \sum_{x=0}^{n} x f(x)$$

$$= \frac{\displaystyle\sum_{x=1}^{n} x \cdot \frac{N_1!}{x!\,(N_1-x)!} \cdot \frac{N_2!}{(n-x)!\,(N_2-n+x)!}}{\displaystyle\binom{N}{n}}$$

$$= \frac{N_1 \displaystyle\sum_{x=1}^{n} \frac{(N_1-1)!}{(x-1)!\,(N_1-x)!} \cdot \frac{N_2!}{(n-x)!\,(N_2-n+x)!}}{\displaystyle\binom{N}{n}}.$$

If we now make the change of variables $k = x - 1$ in the summation, and replace

$$\binom{N}{n} \qquad \text{with} \qquad \binom{N}{n}\binom{N-1}{n-1}$$

in the denominator, this becomes

$$\mu = \frac{N_1}{\displaystyle\binom{N}{n}} \frac{\displaystyle\sum_{k=0}^{n-1} \frac{(N_1-1)!}{k!\,(N_1-1-k)!} \cdot \frac{N_2!}{(n-k-1)!\,(N_2-n+k+1)!}}{\displaystyle\binom{N-1}{n-1}}$$

$$= n\left(\frac{N_1}{N}\right) \frac{\displaystyle\sum_{k=0}^{n-1} \binom{N_1-1}{k}\binom{N_2}{n-1-k}}{\displaystyle\binom{N-1}{n-1}} = n\left(\frac{N_1}{N}\right)$$

because the summation in the expression for μ is equal to $\binom{N-1}{n-1}$ from Theorem A.2-1.

Note that

$$\text{Var}(X) = \sigma^2 = E[(X - \mu)^2]$$
$$= E[X^2] - \mu^2$$
$$= E[X(X - 1)] + E(X) - \mu^2.$$

So to find the variance of X, we first find $E[X(X - 1)]$:

$$E[X(X - 1)] = \sum_{x=0}^{n} x(x - 1) f(x)$$

$$= \frac{\displaystyle\sum_{x=2}^{n} x(x - 1) \frac{N_1!}{x!(N_1 - x)!} \cdot \frac{N_2!}{(n - x)!(N_2 - n + x)!}}{\dbinom{N}{n}}$$

$$= N_1(N_1 - 1) \frac{\displaystyle\sum_{x=2}^{n} \frac{(N_1 - 2)!}{(x - 2)!(N_1 - x)!} \cdot \frac{N_2!}{(n - x)!(N_2 - n + x)!}}{\dbinom{N}{n}}.$$

In the summation, let $k = x - 2$, and in the denominator, note that

$$\binom{N}{n} = \frac{N!}{n!(N - n)!} = \frac{N(N - 1)}{n(n - 1)} \binom{N - 2}{n - 2}.$$

Thus

$$E[X(X - 1)] = \frac{N_1(N_1 - 1)}{\dfrac{N(N - 1)}{n(n - 1)}} \sum_{k=0}^{n-2} \frac{\dbinom{N_1 - 2}{k}\dbinom{N_2}{n - 2 - k}}{\dbinom{N - 2}{n - 2}}$$

$$= \frac{N_1(N_1 - 1)(n)(n - 1)}{N(N - 1)}$$

from Theorem A.2-1. Thus the variance of a hypergeometric random variable is, after some algebra,

$$\sigma^2 = \frac{N_1(N_1 - 1)(n)(n - 1)}{N(N - 1)} + \frac{nN_1}{N} - \left(\frac{nN_1}{N}\right)^2$$

$$= n\left(\frac{N_1}{N}\right)\left(\frac{N_2}{N}\right)\left(\frac{N - n}{N - 1}\right).$$

A.3 **LIMITS**

We refer the reader to the many fine books on calculus for the definition of a limit and the other concepts used in that subject. Here we simply remind you of some of the techniques we find most useful in probability and statistics.

Early in a calculus course the existence of the following limit is discussed and it is denoted by the letter e:

$$e = \lim_{t \to 0} (1 + t)^{1/t} = \lim_{n \to \infty} \left(1 + \frac{1}{n}\right)^n.$$

Of course, e is an irrational number, which to six significant figures equals 2.71828.

Often it is rather easy to see the value of certain limits. For example, with $-1 < r < 1$, the sum of the geometric progression allows us to write

$$\lim_{n \to \infty} (1 + r + r^2 + \cdots + r^{n-1}) = \lim_{n \to \infty} \left(\frac{1 - r^n}{1 - r}\right) = \frac{1}{1 - r}.$$

That is, the limit of the ratio $(1 - r^n)/(1 - r)$ is not difficult to find because $\lim_{n \to \infty} r^n = 0$ when $-1 < r < 1$.

However, it is not that easy to determine the limit of every ratio; for example, consider

$$\lim_{b \to \infty} (be^{-b}) = \lim_{b \to \infty} \left(\frac{b}{e^b}\right).$$

Since both the numerator and the denominator of the latter ratio are unbounded, we can use **L'Hôpital's rule**, taking the limit of the ratio of the derivative of the numerator and the derivative of the denominator. We have

$$\lim_{b \to \infty} \left(\frac{b}{e^b}\right) = \lim_{b \to \infty} \left(\frac{1}{e^b}\right) = 0.$$

This result can be used in the evaluation of the integral

$$\int_0^\infty xe^{-x}\, dx = \lim_{b \to \infty} \int_0^b xe^{-x}\, dx$$

$$= \lim_{b \to \infty} [-xe^{-x} - e^{-x}]_0^b$$

$$= \lim_{b \to \infty} [1 - be^{-b} - e^{-b}] = 1.$$

Note that

$$D_x[-xe^{-x} - e^{-x}] = xe^{-x} - e^{-x} + e^{-x} = xe^{-x};$$

that is, $-xe^{-x} - e^{-x}$ is the antiderivative of xe^{-x}.

Another limit of importance is

$$\lim_{n\to\infty}\left(1+\frac{b}{n}\right)^n = \lim_{n\to\infty} e^{n\,\ln(1+b/n)},$$

where b is a constant.

Since the exponential function is continuous, the limit can be taken to the exponent. That is,

$$\lim_{n\to\infty}\exp[n\,\ln(1+b/n)] = \exp[\lim_{n\to\infty} n\,\ln(1+b/n)].$$

The limit in the exponent is equal to

$$\lim_{n\to\infty}\frac{\ln(1+b/n)}{1/n} = \lim_{n\to\infty}\frac{\dfrac{-b/n^2}{1+b/n}}{-1/n^2} = \lim_{n\to\infty}\frac{b}{1+b/n} = b$$

by L'Hôpital's rule. Since this limit is equal to b, the original limit is

$$\lim_{n\to\infty}\left(1+\frac{b}{n}\right)^n = e^b.$$

Applications of this limit in probability occur with $b = -1$ yielding

$$\lim_{n\to\infty}\left(1-\frac{1}{n}\right)^n = e^{-1}.$$

A.4 INFINITE SERIES

A function $f(x)$ possessing derivatives of all orders at $x = b$ can be expanded in the following **Taylor's series**:

$$f(x) = f(b) + \frac{f'(b)}{1!}(x-b) + \frac{f''(b)}{2!}(x-b)^2 + \frac{f'''(b)}{3!}(x-b)^3 + \cdots.$$

If $b = 0$, we obtain the special case that is often called **Maclaurin's series**;

$$f(x) = f(0) + \frac{f'(0)}{1!}x + \frac{f''(0)}{2!}x^2 + \frac{f'''(0)}{3!}x^3 + \cdots.$$

For example, if $f(x) = e^x$ so that all derivatives of $f(x) = e^x$ are $f^{(r)}(x) = e^x$, then $f^{(r)}(0) = 1$, for $r = 1, 2, 3, \ldots$. Thus the Maclaurin's series expansion of $f(x) = e^x$ is

$$e^x = 1 + \frac{x}{1!} + \frac{x^2}{2!} + \frac{x^3}{3!} + \frac{x^4}{4!} + \cdots.$$

The **ratio test**

$$\lim_{n \to \infty} \left| \frac{x^n/n!}{x^{n-1}/(n-1)!} \right| = \lim_{n \to \infty} \left| \frac{x}{n} \right| = 0$$

shows that this series expansion of e^x converges for all real values of x.
Note, for examples, that

$$e = 1 + \frac{1}{1!} + \frac{1}{2!} + \frac{1}{3!} + \cdots$$

and

$$e^{-1} = 1 - \frac{1}{1!} + \frac{1}{2!} - \frac{1}{3!} + \cdots + \frac{(-1)^n}{n!} + \cdots .$$

For another example, consider

$$h(w) = (1 - w)^{-r},$$

where r is a positive integer. Here

$$h'(w) = r(1 - w)^{-(r+1)}$$

$$h''(w) = (r)(r + 1)(1 - w)^{-(r+2)}$$

$$h'''(w) = (r)(r + 1)(r + 2)(1 - w)^{-(r+3)}$$

$$\vdots$$

In general, $h^{(k)}(0) = (r)(r + 1) \cdots (r + k - 1) = (r + k - 1)!/(r - 1)!$. Thus

$$(1 - w)^{-r} = 1 + \frac{(r + 1 - 1)!}{(r - 1)! \, 1!} w + \frac{(r + 2 - 1)!}{(r - 1)! \, 2!} w^2 + \cdots + \frac{(r + k - 1)!}{(r - 1)! \, k!} w^k + \cdots$$

$$= \sum_{k=0}^{\infty} \binom{r + k - 1}{r - 1} w^k.$$

This is often called the negative binomial series. Using the ratio test,

$$\lim_{n \to \infty} \left| \frac{w^n (r + n - 1)!/[(r - 1)! \, n!]}{w^{n-1}(r + n - 2)!/[(r - 1)! \, (n - 1)!]} \right| = \lim_{n \to \infty} \left| \frac{w(r + n - 1)}{n} \right| = |w|.$$

Thus the series converges when $|w| < 1$ or $-1 < w < 1$.

A negative binomial random variable receives its name from this negative binomial series. Before showing that relationship, we note that, for $-1 < w < 1$,

$$h(w) = \sum_{k=0}^{\infty} \binom{r+k-1}{r-1} w^k = (1-w)^{-r}$$

$$h'(w) = \sum_{k=1}^{\infty} \binom{r+k-1}{r-1} k w^{k-1} = r(1-w)^{-r-1}$$

$$h''(w) = \sum_{k=2}^{\infty} \binom{r+k-1}{r-1} k(k-1) w^{k-2} = r(r+1)(1-w)^{-r-2}.$$

The p.m.f. of a negative binomial random variable X is

$$g(x) = \binom{x-1}{r-1} p^r q^{x-r}, \qquad x = r, r+1, r+2, \ldots.$$

In the series expansion for $h(w) = (1-w)^{-r}$, let $x = k + r$. Then

$$\sum_{x=r}^{\infty} \binom{x-1}{r-1} w^{x-r} = (1-w)^{-r}.$$

Letting $w = q$ in this equation, we see that

$$\sum_{x=r}^{\infty} g(x) = \sum_{x=r}^{\infty} \binom{x-1}{r-1} p^r q^{x-r} = p^r (1-q)^{-r} = 1.$$

That is, $g(x)$ does satisfy the properties of a p.m.f.

To find the mean of X, we first find

$$E(X - r) = \sum_{x=r}^{\infty} (x-r) \binom{x-1}{r-1} p^r q^{x-r} = \sum_{x=r+1}^{\infty} (x-r) \binom{x-1}{r-1} p^r q^{x-r}.$$

Letting $k = x - r$ in this latter summation and using the expansion of $h'(w)$ gives us

$$E(X - r) = \sum_{k=1}^{\infty} (k) \binom{r+k-1}{r-1} p^r q^k$$

$$= p^r q \sum_{k=1}^{\infty} (k) \binom{r+k-1}{r-1} k q^{k-1}$$

$$= p^r q r (1-q)^{-r-1} = r\left(\frac{q}{p}\right).$$

Thus

$$E(X) = r + r\left(\frac{q}{p}\right)$$

$$= r\left(1 + \frac{q}{p}\right) = r\left(\frac{1}{p}\right).$$

Similarly, you can show, using $h''(w)$, that

$$E[(X - r)(X - r - 1)] = \left(\frac{q^2}{p^2}\right)(r)(r + 1)$$

and thus

$$\text{Var}(X) = \text{Var}(X - r) = \left(\frac{q^2}{p^2}\right)(r)(r + 1) + r\left(\frac{q}{p}\right) - r^2\left(\frac{q^2}{p^2}\right) = r\left(\frac{q}{p^2}\right).$$

A very special case of the negative binomial series occurs when $r = 1$, obtaining the well known geometric series

$$(1 - w)^{-1} = 1 + w + w^2 + w^3 + \cdots$$

provided that $-1 < w < 1$.

The geometric series gives its name to the geometric probability distribution. Perhaps you recall the geometric series being written, for $-1 < r < 1$,

$$g(r) = \sum_{k=0}^{\infty} ar^k = \frac{a}{1 - r}.$$

To find the mean and the variance of a geometric random variable X, simply let $r = 1$ in the formulas for the mean and the variance of a negative binomial random variable. However, if you want to find the mean and variance directly, you can use

$$g'(r) = \sum_{k=1}^{\infty} akr^{k-1} = \frac{a}{(1 - r)^2}$$

and

$$g''(r) = \sum_{k=2}^{\infty} ak(k - 1)r^{k-2} = \frac{2a}{(1 - r)^3}$$

to find $E(X)$ and $E[X(X - 1)]$, respectively.

In applications associated with the geometric random variable, it is also useful to recall that the nth partial sum of a geometric series is

$$s_n = \sum_{k=0}^{n-1} ar^k = \frac{a(1 - r^n)}{1 - r}.$$

A bonus in this section is the following logarithmic series that produces a useful tool in daily life. Consider

$$f(x) = \ln(1 + x)$$
$$f'(x) = (1 + x)^{-1}$$
$$f''(x) = (-1)(1 + x)^{-2}$$
$$f'''(x) = (-1)(-2)(1 + x)^{-3}$$
$$\vdots \ .$$

Thus $f^{(r)}(0) = (-1)^{r-1}(r - 1)!$ and

$$\ln(1 + x) = \frac{0!}{1!}x - \frac{1!}{2!}x^2 + \frac{2!}{3!}x^3 - \frac{3!}{4!}x^4 + \cdots$$

$$= x - \frac{x^2}{2} + \frac{x^3}{3} - \frac{x^4}{4} + \cdots,$$

which converges for $-1 < x < 1$.

Now consider the following question:"How long does it take money to double in value if the interest rate is i?" Assuming the compounding is on an annual basis and that you begin with \$1, after one year you have \$$(1 + i)$ and after two years the number of dollars you have is

$$(1 + i) + i(1 + i) = (1 + i)^2.$$

Continuing this process, the equation that we have to solve is

$$(1 + i)^n = 2,$$

the solution of which is

$$n = \frac{\ln 2}{\ln(1 + i)}.$$

To approximate the value of n, recall that $\ln 2 \approx 0.693$ and use the series expansion of $f(x) = \ln(1 + x)$ to obtain

$$n \approx \frac{0.693}{i - \frac{i^2}{2} + \frac{i^3}{3} - \cdots}.$$

Due to the alternating series in the denominator, the denominator is a little less than i. Frequently brokers increase the numerator a little (say to 0.72) and simply divide by i, obtaining the "well known Rule of 72," namely

$$n \approx \frac{72}{100i}.$$

For example, if $i = 0.08$, then $n \approx 72/8 = 9$ provides an excellent approximation (the answer is about 9.006). Many persons find that the Rule of 72 is extremely useful when dealing with money matters.

A.5 INTEGRATION   Say $F'(t) = f(t)$, $a \leq t \leq b$, then

$$\int_a^b f(t)\, dt = F(b) - F(a).$$

Thus if $u(x)$ is such that $u'(x)$ exists and $a \leq u(x)$, then

$$\int_a^{u(x)} f(t)\, dt = F[u(x)] - F(a).$$

Taking derivatives of this latter equation, we obtain

$$D_x \left[\int_a^{u(x)} f(t)\, dt \right] = F'[u(x)]u'(x) = f[u(x)]u'(x).$$

For example, with $0 < v$,

$$D_v \left[2 \int_0^{\sqrt{v}} \frac{1}{\sqrt{2\pi}} e^{-z^2/2}\, dz \right] = \left(\frac{2}{\sqrt{2\pi}} e^{-v/2} \right) \frac{1}{2\sqrt{v}} = \frac{v^{(1/2)-1} e^{-v/2}}{\sqrt{\pi}\, 2^{1/2}}.$$

This is needed in proving that if Z is $N(0, 1)$, then Z^2 is $\chi^2(1)$.

The preceding example could be worked by first changing variables in the integral. That is, first using the fact that

$$\int_a^b f(x)\, dx = \int_{u(a)}^{u(b)} f[w(y)]\, w'(y)\, dy,$$

where the monotone increasing (decreasing) function $x = w(y)$ has derivative $w'(y)$ and inverse function $y = u(x)$. In that example, $a = 0$, $b = \sqrt{v}$, $z = \sqrt{t}$, $z' = 1/2\sqrt{t}$, and $t = z^2$ so that

$$2 \int_0^{\sqrt{v}} \frac{1}{\sqrt{2\pi}} e^{-z^2/2}\, dz = 2 \int_0^v \frac{1}{\sqrt{2\pi}} e^{-t/2} \left(\frac{1}{2\sqrt{t}} \right) dt.$$

The derivative of the latter, by one form of the fundamental theorem of calculus, is

$$2\frac{1}{\sqrt{2\pi}}e^{-v/2}\left(\frac{1}{2\sqrt{v}}\right) = \frac{v^{(1/2)-1}e^{-v/2}}{\sqrt{\pi}\,2^{1/2}}.$$

Integration by parts is frequently needed. It is based upon the derivative of the product of two functions of x, say $u(x)$ and $v(x)$, namely

$$D_x[u(x)v(x)] = u(x)v'(x) + v(x)u'(x).$$

Thus

$$[u(x)v(x)]_a^b = \int_a^b u(x)v'(x)\,dx + \int_a^b v(x)u'(x)\,dx$$

or, equivalently,

$$\int_a^b u(x)v'(x)\,dx = [u(x)v(x)]_a^b - \int_a^b v(x)u'(x)\,dx.$$

For example, by letting $u(x) = x$ and $v'(x) = e^{-x}$, we have that

$$\int_0^b xe^{-1}\,dx = \left[-xe^{-x}\right]_0^b - \int_0^b (1)(-e^{-x})\,dx$$

$$= -be^{-b} + \left[-e^{-x}\right]_0^b = -be^{-b} - e^{-b} + 1$$

because $u'(x) = 1$ and $v(x) = -e^{-x}$.

With some thought about the product rule of differentiation, it is not always necessary to assign $u(x)$ and $v'(x)$, however. For illustration, an integral like

$$\int_0^b x^3e^{-x}\,dx$$

would require integration by parts three times, the first of which would assign $u(x) = x^3$ and $v'(x) = e^{-x}$. But note that

$$D_x(-x^3e^{-x}) = x^3e^{-x} - 3x^2e^{-x}.$$

That is, $-x^3e^{-x}$ is "almost" the antiderivative of x^3e^{-x} except for the undesirable term, $-3x^2e^{-x}$. Clearly,

$$D_x(-x^3e^{-x} - 3x^2e^{-x}) = x^3e^{-x} - 3x^2e^{-x} + 3x^2e^{-x} - 6xe^{-x} = x^3e^{-x} - 6xe^{-x}.$$

So we eliminated that undesirable term $-3x^2e^{-x}$, but got another one, namely $-6xe^{-x}$. However,

$$D_x(-x^3e^{-x} - 3x^2e^{-x} - 6xe^{-x}) = x^3e^{-x} - 6e^{-x}$$

and finally

$$D_x(-x^3e^{-x} - 3x^2e^{-x} - 6xe^{-x} - 6e^{-x}) = x^3e^{-x}.$$

That is,

$$-x^3e^{-x} - 3x^2e^{-x} - 6xe^{-x} - 6e^{-x}$$

is the antiderivative of x^3e^{-x} and can be written down without ever assigning u and v.

As practice in this technique, consider

$$\int_0^{\pi/2} x^2 \cos x \, dx = \left[x^2 \sin x + 2x \cos x - 2 \sin x \right]_0^{\pi/2}.$$

Now $x^2 \sin x$ is our first guess because we obtain $x^2 \cos x$ when we differentiate the $\sin x$ factor. But we get the undesirable term $2x \sin x$. That is why we add $2x \cos x$ as the derivative of the $\cos x$ is $-\sin x$ and $-2x \sin x$ eliminates $2x \sin x$. But the second term of the derivative of $2x \cos x$ is $2 \cos x$, which we get rid of by taking the derivative of the next term, $-2 \sin x$.

Possibly the best advice is to take the derivative of the righthand member, here

$$x^2 \sin x + 2x \cos x - 2 \sin x,$$

and note how the terms cancel, leaving only $x^2 \cos x$. Then practice on integrals like

$$\int x^4 e^{-x} \, dx, \quad \int x^3 \sin x \, dx, \quad \int x^5 e^x \, dx.$$

A.6 MULTIVARIATE CALCULUS

We really only make some suggestions about functions of two variables, say

$$z = f(x, y).$$

But these remarks can be extended to more than two variables. The two first *partial derivatives* with respect to x and y, denoted by $\dfrac{\partial z}{\partial x}$ and $\dfrac{\partial z}{\partial y}$, can be found in the usual manner of differentiating by treating the "other" variable as a constant. For illustration,

$$\frac{\partial(x^2y + \sin x)}{\partial x} = 2xy + \cos x$$

and

$$\frac{\partial(e^{xy^2})}{\partial y} = (e^{xy^2})(2xy).$$

The second partial derivatives are simply first partial derivatives of the first partial derivatives. If $z = e^{xy^2}$, then

$$\frac{\partial}{\partial x}\left(\frac{\partial z}{\partial y}\right) = \frac{\partial}{\partial x}(2xye^{xy^2}) = 2xye^{xy^2}(y^2) + 2ye^{xy^2}.$$

For notation we use

$$\frac{\partial}{\partial x}\left(\frac{\partial z}{\partial x}\right) = \frac{\partial^2 z}{\partial x^2}, \qquad \frac{\partial}{\partial x}\left(\frac{\partial z}{\partial y}\right) = \frac{\partial^2 z}{\partial x \partial y},$$

$$\frac{\partial}{\partial y}\left(\frac{\partial z}{\partial x}\right) = \frac{\partial^2 z}{\partial y \partial x}, \qquad \frac{\partial}{\partial y}\left(\frac{\partial z}{\partial y}\right) = \frac{\partial^2 z}{\partial y^2},$$

In general,

$$\frac{\partial^2 z}{\partial x \partial y} = \frac{\partial^2 z}{\partial y \partial x},$$

provided the partial derivatives involved are continuous functions.

As you might guess, at a relative maximum or minimum of $z = f(x, y)$, we have

$$\frac{\partial z}{\partial x} = 0 \qquad \text{and} \qquad \frac{\partial z}{\partial y} = 0,$$

provided that the derivatives exist. To assure us that we have a maximum or minimum, we need

$$\left(\frac{\partial^2 z}{\partial x \partial y}\right)^2 - \left(\frac{\partial^2 z}{\partial x^2}\right)\left(\frac{\partial^2 z}{\partial y^2}\right) < 0.$$

Moreover, it is a relative minimum if $\dfrac{\partial^2 z}{\partial x^2} > 0$ and a relative maximum if $\dfrac{\partial^2 z}{\partial x^2} < 0$.

A major problem in statistics, called least squares, is to find a and b to minimize

$$K(a, b) = \sum_{i=1}^{n} (y_i - a - bx_i)^2.$$

Thus the solution of the two equations

$$\frac{\partial K}{\partial a} = \sum_{i=1}^{n} 2(y_i - a - bx_i)(-1) = 0$$

$$\frac{\partial K}{\partial b} = \sum_{i=1}^{n} 2(y_i - a - bx_i)(-x_i) = 0$$

would possibly give us a point (a, b) that minimizes $K(a, b)$. Taking second partial derivatives,

$$\frac{\partial^2 K}{\partial a^2} = \sum_{i=1}^{n} 2(-1)(-1) = 2n > 0$$

$$\frac{\partial^2 K}{\partial b^2} = \sum_{i=1}^{n} 2(-x_i)(-x_i) = 2\sum_{i=1}^{n} x_i^2 > 0$$

$$\frac{\partial^2 K}{\partial a \partial b} = \sum_{i=1}^{n} 2(-1)(-x_i) = 2\sum_{i=1}^{n} x_i,$$

we note that

$$\left(2\sum_{i=1}^{n} x_i\right)^2 - (2n)\left(2\sum_{i=1}^{n} x_i^2\right) < 0$$

because $(\sum_{i=1}^{n} x_i)^2 < n \sum_{i=1}^{n} x_i^2$ provided all x_i are not equal. Noting that $\frac{\partial^2 z}{\partial x^2} > 0$, we see that the solution of the two equations, $\frac{\partial K}{\partial a} = 0$ and $\frac{\partial K}{\partial b} = 0$, provides the only minimizing solution.

The value of the *double integral*

$$\iint_A f(x, y)\, dx\, dy$$

can usually be evaluated by an iterated process; that is, by evaluating two successive single integrals. For example, say $A = \{(x, y): 0 \le x \le 1, 0 \le y \le x\}$ as given in Figure A.6-1.

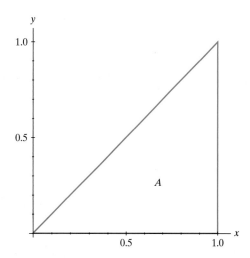

Figure A.6-1: $A = \{(x, y) : 0 \le x \le 1, 0 \le y \le x\}$

Then

$$\iint_A (x + x^3 y^2)\, dx\, dy = \int_0^1 \left[\int_0^x (x + x^3 y^2)\, dy \right] dx$$

$$= \int_0^1 \left[xy + \frac{x^3 y^3}{3} \right]_0^x dx$$

$$= \int_0^1 \left(x^2 + \frac{x^6}{3} \right) dx = \left[\frac{x^3}{3} + \frac{x^7}{3 \cdot 7} \right]_0^1$$

$$= \frac{1}{3} + \frac{1}{21} = \frac{8}{21}.$$

When placing the limits on the iterated integral, note that for each fixed x between zero and one, y is restricted to the interval zero to x. Also in the inner integral on y, x is treated as a constant.

In evaluating this double integral we could have restricted y to the interval zero to one, then x would be between y and one. That is, we would have evaluated the iterated integral

$$\int_0^1 \left[\int_y^1 (x + x^3 y^2)\, dx \right] dy = \int_0^1 \left[\frac{x^2}{2} + \frac{x^4 y^2}{4} \right]_y^1 dy$$

$$= \int_0^1 \left[\frac{1}{2} + \frac{y^2}{4} - \frac{y^2}{2} - \frac{y^6}{4} \right] dy$$

$$= \left[\frac{y}{2} - \frac{y^3}{3 \cdot 4} - \frac{y^7}{7 \cdot 4} \right]_0^1$$

$$= \frac{1}{2} - \frac{1}{12} - \frac{1}{28} = \frac{8}{21}.$$

Finally, we can *change variables* in a double integral

$$\iint_A f(x, y)\, dx\, dy.$$

If $f(x, y)$ is a joint p.d.f. of random variables X and Y of the continuous type, then the double integral represents $P[(X, Y) \in A]$. Consider only one-to-one transformations, say $z = u_1(x, y)$ and $w = u_2(x, y)$ with inverse transformation given by $x = v_1(z, w)$ and $y = v_2(z, w)$. The determinant of order two

$$J = \begin{vmatrix} \dfrac{\partial x}{\partial z} & \dfrac{\partial x}{\partial w} \\[2ex] \dfrac{\partial y}{\partial z} & \dfrac{\partial y}{\partial w} \end{vmatrix}$$

is called the *Jacobian* of the inverse transformation. Moreover, say the region A maps onto B in the (z, w) space. Since we are usually dealing with probabilities in this book, we fixed the *sign* of the integral so that it is positive (by using the absolute value of the Jacobian). Then it is true that

$$\iint\limits_{A} f(x, y) \, dx \, dy = \iint\limits_{B} f[v_1(z, w), v_2(z, w)] \, |J| \, dz \, dw.$$

For illustration, let

$$f(x, y) = \frac{1}{2\pi} e^{-(x^2 + y^2)/2}, \qquad -\infty < x < \infty, \quad -\infty < y < \infty,$$

which is the joint p.d.f. of two independent normal variables, each with mean zero and variance one. Say $A = \{(x, y): 0 \le x^2 + y^2 \le 1\}$ and consider

$$P(A) = \iint\limits_{A} f(x, y) \, dx \, dy.$$

This is impossible to deal with directly in the x, y variables. However, consider the inverse transformation to polar coordinates, namely

$$x = r \cos \theta, \qquad y = r \sin \theta$$

with Jacobian

$$J = \begin{vmatrix} \cos \theta & -r \sin \theta \\ \sin \theta & r \cos \theta \end{vmatrix} = r(\cos^2 \theta + \sin^2 \theta) = r.$$

Since A maps onto $B = \{(r, \theta): 0 \le r \le 1, \ 0 \le \theta < 2\pi\}$, we have that

$$P(A) = \int_0^{2\pi} \left(\int_0^1 \frac{1}{2\pi} e^{-r^2/2} r \, dr \right) d\theta$$

$$= \int_0^{2\pi} \left[-\frac{1}{2\pi} e^{-r^2/2} \right]_0^1 d\theta$$

$$= \int_0^{2\pi} \frac{1}{2\pi} (1 - e^{-1/2}) \, d\theta$$

$$= \frac{1}{2\pi} (1 - e^{-1/2}) \, 2\pi = 1 - e^{-1/2}.$$

APPENDIX B

REFERENCES

Aspin, A. A., "Tables for Use in Comparisons Whose Accuracy Involves Two Variances, Separately Estimated," *Biometrika*, **36** (1949), pp. 290–296.

Barnett, A., "How Numbers Can Trick You,", *Technology Review*, 1994.

Basu, D., "On Statistics Independent of a Complete Sufficient Statistic," Sankhya, 15, 377 (1955).

Bernstein, P. L., *Against the Gods: The Remarkable Story of Risk*, John Wiley & Sons, Inc. 1996.

Berry, D. A., *Statistics: A Bayesian Perspective*, Duxbury Press, An Imprint of Wadsworth Publishing Co., 1996.

Box, G. E. P., W. G. Hunter, and J. S. Hunter, *Statistics for Experimenters*, John Wiley & Sons, Inc., New York, 1978.

Box, G. E. P., and M. E. Muller, "A Note on the Generation of Random Normal Deviates," *Ann. Math. Statist.*," **29** (1958), p. 610.

Crisman, R., "Shortest Confidence Interval for the Standard Deviation of a Normal Distribution," *J. Undergrad. Math.*, **7**, 2 (1975), p. 57.

Guenther, William C., "Shortest Confidence Intervals," *Amer. Statist.*, **23**, 1(1969), p. 22.

Hogg, R. V., "Testing the Equality of Means of Rectangular Populations," *Ann. Math. Statist.*, **24** (1953), p. 691.

Hogg, R. V., and J. Ledolter, *Applied Statistics for Engineers and Physical Scientists*, 2nd ed., Macmillan Publishing Company, New York, 1992.

Hogg, R. V., J. W. McKean, and A. T. Craig, *Introduction to Mathematical Statistics*, 6th ed., Prentice Hall, NJ, 2005.

Hogg, R. V., and A. T. Craig, "On the Decomposition of Certain Chi-Square Variables," *Ann. Math. Statist.*, **29** (1958), p. 608.

Johnson, V. E. and J. H. Albert, *Ordinal Data Modeling*, Springer-Verlag, New York, 1999.

Karian, Z. A., and E. A. Tanis, *Probability & Statistics: Explorations with MAPLE*, 2nd ed., Prentice Hall, NJ, 1999.

Keating, Jerome P., and Scott, David W., "Ask Dr. STATS," *Stats, The Magazine for Students of Statistics*, **25**, Spring, 1999, pp. 16–22.

Kinney, J., "Mathematica As an Aid in Teaching Probability and Statistics," *Proceedings of the Statistical Computing Section*, (1998), American Statistical Association, pp. 25–32.

Montgomery, D. C., *Design and Analysis of Experiments*, 2nd ed., John Wiley & Sons, Inc., New York, 1984.

Natrella, M. G., *Experimental Statistics, National Bureau of Standards Handbook 91*, U.S. Government Printing Office, Washington DC, 1963.

Pearson, K., "On the Criterion That a Given System of Deviations from the Probable in the Case of a Correlated System of Variables Is Such That It Can Be Reasonably Supposed to Have Arisen from Random Sampling," *Phil. Mag.*, Series 5, **50** (1900), p. 157.

Quain, J. R., "Going Mainstream," *PC Magazine*, February, 1994.

Rafter, J. A., M. L. Abell, J. P. Braselton, *Statistics with Maple*, Academic Press, An imprint of Elsevier Science (USA), 2003.

Raspe, Rudolph Erich, *The Surprising Adventures of Baron Munchausen*, IndyPublish.com, 2001.

Snee, R. D., L. B. Hare, and J. R. Trout, *Experiments in Industry*, American Society of Quality Conrol, Milwaukee, Wis., 1985.

Stigler, S. M., *The History of Statistics: The Measurement of Uncertainty Before 1900*, Harvard University Press, 1986.

Tanis, E. A., "Maple Integrated Into the Instruction of Probability and Statistics," *Proceedings of the Statistical Computing Section*, (1998), American Statistical Association, pp. 19–24.

Tate, R. F., and G. W. Klett, "Optimum Confidence Intervals for the Variance of a Normal Distribution," *J. Am. Statist. Assoc.*, **54** (1959), p. 674.

Tsutakawa, R. K., G. L. Shoop, and C. J. Marienfeld, "Empirical Bayes estimation of cancer mortality rates," *Statistics in Medicine*, **4**, (1985), pp. 201–212, 1985.

Tukey, John W., *Exploratory Data Analysis*, Addison-Wesley Publishing Company, Reading, Mass., 1977.

Velleman, P. F., D. C. Hoaglin, *Applications, Basics, and Computing of Exploratory Data Analysis*, Duxbury Press, Boston, 1981.

Wilcoxon, F., "Individual Comparisons by Ranking Methods," *Biometrics Bull.*, **1** (1945), p. 80.

APPENDIX C

TABLES

Table I: Binomial Coefficients

$$\binom{n}{r} = \frac{n!}{r!(n-r)!} = \binom{n}{n-r}$$

n	$\binom{n}{0}$	$\binom{n}{1}$	$\binom{n}{2}$	$\binom{n}{3}$	$\binom{n}{4}$	$\binom{n}{5}$	$\binom{n}{6}$	$\binom{n}{7}$	$\binom{n}{8}$	$\binom{n}{9}$	$\binom{n}{10}$	$\binom{n}{11}$	$\binom{n}{12}$	$\binom{n}{13}$
0	1													
1	1	1												
2	1	2	1											
3	1	3	3	1										
4	1	4	6	4	1									
5	1	5	10	10	5	1								
6	1	6	15	20	15	6	1							
7	1	7	21	35	35	21	7	1						
8	1	8	28	56	70	56	28	8	1					
9	1	9	36	84	126	126	84	36	9	1				
10	1	10	45	120	210	252	210	120	45	10	1			
11	1	11	55	165	330	462	462	330	165	55	11	1		
12	1	12	66	220	495	792	924	792	495	220	66	12	1	
13	1	13	78	286	715	1,287	1,716	1,716	1,287	715	286	78	13	1
14	1	14	91	364	1,001	2,002	3,003	3,432	3,003	2,002	1,001	364	91	14
15	1	15	105	455	1,365	3,003	5,005	6,435	6,435	5,005	3,003	1,365	455	105
16	1	16	120	560	1,820	4,368	8,008	11,440	12,870	11,440	8,008	4,368	1,820	560
17	1	17	136	680	2,380	6,188	12,376	19,448	24,310	24,310	19,448	12,376	6,188	2,380
18	1	18	153	816	3,060	8,568	18,564	31,824	43,758	48,620	43,758	31,824	18,564	8,568
19	1	19	171	969	3,876	11,628	27,132	50,388	75,582	92,378	92,378	75,582	50,388	27,132
20	1	20	190	1,140	4,845	15,504	38,760	77,520	125,970	167,960	184,756	167,960	125,970	77,520
21	1	21	210	1,330	5,985	20,349	54,264	116,280	203,490	293,930	352,716	352,716	293,930	203,490
22	1	22	231	1,540	7,315	26,334	74,613	170,544	319,770	497,420	646,646	705,432	646,646	497,420
23	1	23	253	1,771	8,855	33,649	100,947	245,157	490,314	817,190	1,144,066	1,352,078	1,352,078	1,144,066
24	1	24	276	2,024	10,626	42,504	134,596	346,104	735,471	1,307,504	1,961,256	2,496,144	2,704,156	2,496,144
25	1	25	300	2,300	12,650	53,130	177,100	480,700	1,081,575	2,042,975	3,268,760	4,457,400	5,200,300	5,200,300
26	1	26	325	2,600	14,950	65,780	230,230	657,800	1,562,275	3,124,550	5,311,735	7,726,160	9,657,700	10,400,600

For $r > 13$ you may use the identity $\binom{n}{r} = \binom{n}{n-r}$.

Table II: The Binomial Distribution

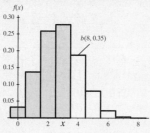

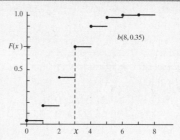

$$F(x) = P(X \le x) = \sum_{k=0}^{x} \frac{n!}{k!(n-k)!} p^k (1-p)^{n-k}$$

							p				
n	x	0.05	0.10	0.15	0.20	0.25	0.30	0.35	0.40	0.45	0.50
2	0	0.9025	0.8100	0.7225	0.6400	0.5625	0.4900	0.4225	0.3600	0.3025	0.2500
	1	0.9975	0.9900	0.9775	0.9600	0.9375	0.9100	0.8775	0.8400	0.7975	0.7500
	2	1.0000	1.0000	1.0000	1.0000	1.0000	1.0000	1.0000	1.0000	1.0000	1.0000
3	0	0.8574	0.7290	0.6141	0.5120	0.4219	0.3430	0.2746	0.2160	0.1664	0.1250
	1	0.9928	0.9720	0.9392	0.8960	0.8438	0.7840	0.7182	0.6480	0.5748	0.5000
	2	0.9999	0.9990	0.9966	0.9920	0.9844	0.9730	0.9571	0.9360	0.9089	0.8750
	3	1.0000	1.0000	1.0000	1.0000	1.0000	1.0000	1.0000	1.0000	1.0000	1.0000
4	0	0.8145	0.6561	0.5220	0.4096	0.3164	0.2401	0.1785	0.1296	0.0915	0.0625
	1	0.9860	0.9477	0.8905	0.8192	0.7383	0.6517	0.5630	0.4752	0.3910	0.3125
	2	0.9995	0.9963	0.9880	0.9728	0.9492	0.9163	0.8735	0.8208	0.7585	0.6875
	3	1.0000	0.9999	0.9995	0.9984	0.9961	0.9919	0.9850	0.9744	0.9590	0.9375
	4	1.0000	1.0000	1.0000	1.0000	1.0000	1.0000	1.0000	1.0000	1.0000	1.0000
5	0	0.7738	0.5905	0.4437	0.3277	0.2373	0.1681	0.1160	0.0778	0.0503	0.0312
	1	0.9774	0.9185	0.8352	0.7373	0.6328	0.5282	0.4284	0.3370	0.2562	0.1875
	2	0.9988	0.9914	0.9734	0.9421	0.8965	0.8369	0.7648	0.6826	0.5931	0.5000
	3	1.0000	0.9995	0.9978	0.9933	0.9844	0.9692	0.9460	0.9130	0.8688	0.8125
	4	1.0000	1.0000	0.9999	0.9997	0.9990	0.9976	0.9947	0.9898	0.9815	0.9688
	5	1.0000	1.0000	1.0000	1.0000	1.0000	1.0000	1.0000	1.0000	1.0000	1.0000
6	0	0.7351	0.5314	0.3771	0.2621	0.1780	0.1176	0.0754	0.0467	0.0277	0.0156
	1	0.9672	0.8857	0.7765	0.6553	0.5339	0.4202	0.3191	0.2333	0.1636	0.1094
	2	0.9978	0.9842	0.9527	0.9011	0.8306	0.7443	0.6471	0.5443	0.4415	0.3438
	3	0.9999	0.9987	0.9941	0.9830	0.9624	0.9295	0.8826	0.8208	0.7447	0.6562
	4	1.0000	0.9999	0.9996	0.9984	0.9954	0.9891	0.9777	0.9590	0.9308	0.8906
	5	1.0000	1.0000	1.0000	0.9999	0.9998	0.9993	0.9982	0.9959	0.9917	0.9844
	6	1.0000	1.0000	1.0000	1.0000	1.0000	1.0000	1.0000	1.0000	1.0000	1.0000
7	0	0.6983	0.4783	0.3206	0.2097	0.1335	0.0824	0.0490	0.0280	0.0152	0.0078
	1	0.9556	0.8503	0.7166	0.5767	0.4449	0.3294	0.2338	0.1586	0.1024	0.0625
	2	0.9962	0.9743	0.9262	0.8520	0.7564	0.6471	0.5323	0.4199	0.3164	0.2266
	3	0.9998	0.9973	0.9879	0.9667	0.9294	0.8740	0.8002	0.7102	0.6083	0.5000
	4	1.0000	0.9998	0.9988	0.9953	0.9871	0.9712	0.9444	0.9037	0.8471	0.7734
	5	1.0000	1.0000	0.9999	0.9996	0.9987	0.9962	0.9910	0.9812	0.9643	0.9375

Table II: *continued*

n	x	0.05	0.10	0.15	0.20	0.25	0.30	0.35	0.40	0.45	0.50
							p				
	6	1.0000	1.0000	1.0000	1.0000	0.9999	0.9998	0.9994	0.9984	0.9963	0.9922
	7	1.0000	1.0000	1.0000	1.0000	1.0000	1.0000	1.0000	1.0000	1.0000	1.0000
8	0	0.6634	0.4305	0.2725	0.1678	0.1001	0.0576	0.0319	0.0168	0.0084	0.0039
	1	0.9428	0.8131	0.6572	0.5033	0.3671	0.2553	0.1691	0.1064	0.0632	0.0352
	2	0.9942	0.9619	0.8948	0.7969	0.6785	0.5518	0.4278	0.3154	0.2201	0.1445
	3	0.9996	0.9950	0.9786	0.9437	0.8862	0.8059	0.7064	0.5941	0.4770	0.3633
	4	1.0000	0.9996	0.9971	0.9896	0.9727	0.9420	0.8939	0.8263	0.7396	0.6367
	5	1.0000	1.0000	0.9998	0.9988	0.9958	0.9887	0.9747	0.9502	0.9115	0.8555
	6	1.0000	1.0000	1.0000	0.9999	0.9996	0.9987	0.9964	0.9915	0.9819	0.9648
	7	1.0000	1.0000	1.0000	1.0000	1.0000	0.9999	0.9998	0.9993	0.9983	0.9961
	8	1.0000	1.0000	1.0000	1.0000	1.0000	1.0000	1.0000	1.0000	1.0000	1.0000
9	0	0.6302	0.3874	0.2316	0.1342	0.0751	0.0404	0.0207	0.0101	0.0046	0.0020
	1	0.9288	0.7748	0.5995	0.4362	0.3003	0.1960	0.1211	0.0705	0.0385	0.0195
	2	0.9916	0.9470	0.8591	0.7382	0.6007	0.4628	0.3373	0.2318	0.1495	0.0898
	3	0.9994	0.9917	0.9661	0.9144	0.8343	0.7297	0.6089	0.4826	0.3614	0.2539
	4	1.0000	0.9991	0.9944	0.9804	0.9511	0.9012	0.8283	0.7334	0.6214	0.5000
	5	1.0000	0.9999	0.9994	0.9969	0.9900	0.9747	0.9464	0.9006	0.8342	0.7461
	6	1.0000	1.0000	1.0000	0.9997	0.9987	0.9957	0.9888	0.9750	0.9502	0.9102
	7	1.0000	1.0000	1.0000	1.0000	0.9999	0.9996	0.9986	0.9962	0.9909	0.9805
	8	1.0000	1.0000	1.0000	1.0000	1.0000	1.0000	0.9999	0.9997	0.9992	0.9980
	9	1.0000	1.0000	1.0000	1.0000	1.0000	1.0000	1.0000	1.0000	1.0000	1.0000
10	0	0.5987	0.3487	0.1969	0.1074	0.0563	0.0282	0.0135	0.0060	0.0025	0.0010
	1	0.9139	0.7361	0.5443	0.3758	0.2440	0.1493	0.0860	0.0464	0.0233	0.0107
	2	0.9885	0.9298	0.8202	0.6778	0.5256	0.3828	0.2616	0.1673	0.0996	0.0547
	3	0.9990	0.9872	0.9500	0.8791	0.7759	0.6496	0.5138	0.3823	0.2660	0.1719
	4	0.9999	0.9984	0.9901	0.9672	0.9219	0.8497	0.7515	0.6331	0.5044	0.3770
	5	1.0000	0.9999	0.9986	0.9936	0.9803	0.9527	0.9051	0.8338	0.7384	0.6230
	6	1.0000	1.0000	0.9999	0.9991	0.9965	0.9894	0.9740	0.9452	0.8980	0.8281
	7	1.0000	1.0000	1.0000	0.9999	0.9996	0.9984	0.9952	0.9877	0.9726	0.9453
	8	1.0000	1.0000	1.0000	1.0000	1.0000	0.9999	0.9995	0.9983	0.9955	0.9893
	9	1.0000	1.0000	1.0000	1.0000	1.0000	1.0000	1.0000	0.9999	0.9997	0.9990
	10	1.0000	1.0000	1.0000	1.0000	1.0000	1.0000	1.0000	1.0000	1.0000	1.0000
11	0	0.5688	0.3138	0.1673	0.0859	0.0422	0.0198	0.0088	0.0036	0.0014	0.0005
	1	0.8981	0.6974	0.4922	0.3221	0.1971	0.1130	0.0606	0.0302	0.0139	0.0059
	2	0.9848	0.9104	0.7788	0.6174	0.4552	0.3127	0.2001	0.1189	0.0652	0.0327
	3	0.9984	0.9815	0.9306	0.8389	0.7133	0.5696	0.4256	0.2963	0.1911	0.1133
	4	0.9999	0.9972	0.9841	0.9496	0.8854	0.7897	0.6683	0.5328	0.3971	0.2744
	5	1.0000	0.9997	0.9973	0.9883	0.9657	0.9218	0.8513	0.7535	0.6331	0.5000
	6	1.0000	1.0000	0.9997	0.9980	0.9924	0.9784	0.9499	0.9006	0.8262	0.7256
	7	1.0000	1.0000	1.0000	0.9998	0.9988	0.9957	0.9878	0.9707	0.9390	0.8867
	8	1.0000	1.0000	1.0000	1.0000	0.9999	0.9994	0.9980	0.9941	0.9852	0.9673
	9	1.0000	1.0000	1.0000	1.0000	1.0000	1.0000	0.9998	0.9993	0.9978	0.9941

						p					
n	x	0.05	0.10	0.15	0.20	0.25	0.30	0.35	0.40	0.45	0.50
	10	1.0000	1.0000	1.0000	1.0000	1.0000	1.0000	1.0000	1.0000	0.9998	0.9995
	11	1.0000	1.0000	1.0000	1.0000	1.0000	1.0000	1.0000	1.0000	1.0000	1.0000
12	0	0.5404	0.2824	0.1422	0.0687	0.0317	0.0138	0.0057	0.0022	0.0008	0.0002
	1	0.8816	0.6590	0.4435	0.2749	0.1584	0.0850	0.0424	0.0196	0.0083	0.0032
	2	0.9804	0.8891	0.7358	0.5583	0.3907	0.2528	0.1513	0.0834	0.0421	0.0193
	3	0.9978	0.9744	0.9078	0.7946	0.6488	0.4925	0.3467	0.2253	0.1345	0.0730
	4	0.9998	0.9957	0.9761	0.9274	0.8424	0.7237	0.5833	0.4382	0.3044	0.1938
	5	1.0000	0.9995	0.9954	0.9806	0.9456	0.8822	0.7873	0.6652	0.5269	0.3872
	6	1.0000	0.9999	0.9993	0.9961	0.9857	0.9614	0.9154	0.8418	0.7393	0.6128
	7	1.0000	1.0000	0.9999	0.9994	0.9972	0.9905	0.9745	0.9427	0.8883	0.8062
	8	1.0000	1.0000	1.0000	0.9999	0.9996	0.9983	0.9944	0.9847	0.9644	0.9270
	9	1.0000	1.0000	1.0000	1.0000	1.0000	0.9998	0.9992	0.9972	0.9921	0.9807
	10	1.0000	1.0000	1.0000	1.0000	1.0000	1.0000	0.9999	0.9997	0.9989	0.9968
	11	1.0000	1.0000	1.0000	1.0000	1.0000	1.0000	1.0000	1.0000	0.9999	0.9998
	12	1.0000	1.0000	1.0000	1.0000	1.0000	1.0000	1.0000	1.0000	1.0000	1.0000
13	0	0.5133	0.2542	0.1209	0.0550	0.0238	0.0097	0.0037	0.0013	0.0004	0.0001
	1	0.8646	0.6213	0.3983	0.2336	0.1267	0.0637	0.0296	0.0126	0.0049	0.0017
	2	0.9755	0.8661	0.6920	0.5017	0.3326	0.2025	0.1132	0.0579	0.0269	0.0112
	3	0.9969	0.9658	0.8820	0.7473	0.5843	0.4206	0.2783	0.1686	0.0929	0.0461
	4	0.9997	0.9935	0.9658	0.9009	0.7940	0.6543	0.5005	0.3530	0.2279	0.1334
	5	1.0000	0.9991	0.9924	0.9700	0.9198	0.8346	0.7159	0.5744	0.4268	0.2905
	6	1.0000	0.9999	0.9987	0.9930	0.9757	0.9376	0.8705	0.7712	0.6437	0.5000
	7	1.0000	1.0000	0.9998	0.9988	0.9944	0.9818	0.9538	0.9023	0.8212	0.7095
	8	1.0000	1.0000	1.0000	0.9998	0.9990	0.9960	0.9874	0.9679	0.9302	0.8666
	9	1.0000	1.0000	1.0000	1.0000	0.9999	0.9993	0.9975	0.9922	0.9797	0.9539
	10	1.0000	1.0000	1.0000	1.0000	1.0000	0.9999	0.9997	0.9987	0.9959	0.9888
	11	1.0000	1.0000	1.0000	1.0000	1.0000	1.0000	1.0000	0.9999	0.9995	0.9983
	12	1.0000	1.0000	1.0000	1.0000	1.0000	1.0000	1.0000	1.0000	1.0000	0.9999
	13	1.0000	1.0000	1.0000	1.0000	1.0000	1.0000	1.0000	1.0000	1.0000	1.0000
14	0	0.4877	0.2288	0.1028	0.0440	0.0178	0.0068	0.0024	0.0008	0.0002	0.0001
	1	0.8470	0.5846	0.3567	0.1979	0.1010	0.0475	0.0205	0.0081	0.0029	0.0009
	2	0.9699	0.8416	0.6479	0.4481	0.2811	0.1608	0.0839	0.0398	0.0170	0.0065
	3	0.9958	0.9559	0.8535	0.6982	0.5213	0.3552	0.2205	0.1243	0.0632	0.0287
	4	0.9996	0.9908	0.9533	0.8702	0.7415	0.5842	0.4227	0.2793	0.1672	0.0898
	5	1.0000	0.9985	0.9885	0.9561	0.8883	0.7805	0.6405	0.4859	0.3373	0.2120
	6	1.0000	0.9998	0.9978	0.9884	0.9617	0.9067	0.8164	0.6925	0.5461	0.3953
	7	1.0000	1.0000	0.9997	0.9976	0.9897	0.9685	0.9247	0.8499	0.7414	0.6047
	8	1.0000	1.0000	1.0000	0.9996	0.9978	0.9917	0.9757	0.9417	0.8811	0.7880
	9	1.0000	1.0000	1.0000	1.0000	0.9997	0.9983	0.9940	0.9825	0.9574	0.9102
	10	1.0000	1.0000	1.0000	1.0000	1.0000	0.9998	0.9989	0.9961	0.9886	0.9713

| Table II: *continued* | | | | | | | | | | |

		p									
n	*x*	0.05	0.10	0.15	0.20	0.25	0.30	0.35	0.40	0.45	0.50
	11	1.0000	1.0000	1.0000	1.0000	1.0000	1.0000	0.9999	0.9994	0.9978	0.9935
	12	1.0000	1.0000	1.0000	1.0000	1.0000	1.0000	1.0000	0.9999	0.9997	0.9991
	13	1.0000	1.0000	1.0000	1.0000	1.0000	1.0000	1.0000	1.0000	1.0000	0.9999
	14	1.0000	1.0000	1.0000	1.0000	1.0000	1.0000	1.0000	1.0000	1.0000	1.0000
15	0	0.4633	0.2059	0.0874	0.0352	0.0134	0.0047	0.0016	0.0005	0.0001	0.0000
	1	0.8290	0.5490	0.3186	0.1671	0.0802	0.0353	0.0142	0.0052	0.0017	0.0005
	2	0.9638	0.8159	0.6042	0.3980	0.2361	0.1268	0.0617	0.0271	0.0107	0.0037
	3	0.9945	0.9444	0.8227	0.6482	0.4613	0.2969	0.1727	0.0905	0.0424	0.0176
	4	0.9994	0.9873	0.9383	0.8358	0.6865	0.5155	0.3519	0.2173	0.1204	0.0592
	5	0.9999	0.9978	0.9832	0.9389	0.8516	0.7216	0.5643	0.4032	0.2608	0.1509
	6	1.0000	0.9997	0.9964	0.9819	0.9434	0.8689	0.7548	0.6098	0.4522	0.3036
	7	1.0000	1.0000	0.9994	0.9958	0.9827	0.9500	0.8868	0.7869	0.6535	0.5000
	8	1.0000	1.0000	0.9999	0.9992	0.9958	0.9848	0.9578	0.9050	0.8182	0.6964
	9	1.0000	1.0000	1.0000	0.9999	0.9992	0.9963	0.9876	0.9662	0.9231	0.8491
	10	1.0000	1.0000	1.0000	1.0000	0.9999	0.9993	0.9972	0.9907	0.9745	0.9408
	11	1.0000	1.0000	1.0000	1.0000	1.0000	0.9999	0.9995	0.9981	0.9937	0.9824
	12	1.0000	1.0000	1.0000	1.0000	1.0000	1.0000	0.9999	0.9987	0.9989	0.9963
	13	1.0000	1.0000	1.0000	1.0000	1.0000	1.0000	1.0000	1.0000	0.9999	0.9995
	14	1.0000	1.0000	1.0000	1.0000	1.0000	1.0000	1.0000	1.0000	1.0000	1.0000
	15	1.0000	1.0000	1.0000	1.0000	1.0000	1.0000	1.0000	1.0000	1.0000	1.0000
16	0	0.4401	0.1853	0.0743	0.0281	0.0100	0.0033	0.0010	0.0003	0.0001	0.0000
	1	0.8108	0.5147	0.2839	0.1407	0.0635	0.0261	0.0098	0.0033	0.0010	0.0003
	2	0.9571	0.7892	0.5614	0.3518	0.1971	0.0994	0.0451	0.0183	0.0066	0.0021
	3	0.9930	0.9316	0.7899	0.5981	0.4050	0.2459	0.1339	0.0651	0.0281	0.0106
	4	0.9991	0.9830	0.9209	0.7982	0.6302	0.4499	0.2892	0.1666	0.0853	0.0384
	5	0.9999	0.9967	0.9765	0.9183	0.8103	0.6598	0.4900	0.3288	0.1976	0.1051
	6	1.0000	0.9995	0.9944	0.9733	0.9204	0.8247	0.6881	0.5272	0.3660	0.2272
	7	1.0000	0.9999	0.9989	0.9930	0.9729	0.9256	0.8406	0.7161	0.5629	0.4018
	8	1.0000	1.0000	0.9998	0.9985	0.9925	0.9743	0.9329	0.8577	0.7441	0.5982
	9	1.0000	1.0000	1.0000	0.9998	0.9984	0.9929	0.9771	0.9417	0.8759	0.7728
	10	1.0000	1.0000	1.0000	1.0000	0.9997	0.9984	0.9938	0.9809	0.9514	0.8949
	11	1.0000	1.0000	1.0000	1.0000	1.0000	0.9997	0.9987	0.9951	0.9851	0.9616
	12	1.0000	1.0000	1.0000	1.0000	1.0000	1.0000	0.9998	0.9991	0.9965	0.9894
	13	1.0000	1.0000	1.0000	1.0000	1.0000	1.0000	1.0000	0.9999	0.9994	0.9979
	14	1.0000	1.0000	1.0000	1.0000	1.0000	1.0000	1.0000	1.0000	0.9999	0.9997
	15	1.0000	1.0000	1.0000	1.0000	1.0000	1.0000	1.0000	1.0000	1.0000	1.0000
	16	1.0000	1.0000	1.0000	1.0000	1.0000	1.0000	1.0000	1.0000	1.0000	1.0000
20	0	0.3585	0.1216	0.0388	0.0115	0.0032	0.0008	0.0002	0.0000	0.0000	0.0000
	1	0.7358	0.3917	0.1756	0.0692	0.0243	0.0076	0.0021	0.0005	0.0001	0.0000
	2	0.9245	0.6769	0.4049	0.2061	0.0913	0.0355	0.0121	0.0036	0.0009	0.0002
	3	0.9841	0.8670	0.6477	0.4114	0.2252	0.1071	0.0444	0.0160	0.0049	0.0013
	4	0.9974	0.9568	0.8298	0.6296	0.4148	0.2375	0.1182	0.0510	0.0189	0.0059

Table II: *continued*

n	x	0.05	0.10	0.15	0.20	0.25	0.30	0.35	0.40	0.45	0.50
	5	0.9997	0.9887	0.9327	0.8042	0.6172	0.4164	0.2454	0.1256	0.0553	0.0207
	6	1.0000	0.9976	0.9781	0.9133	0.7858	0.6080	0.4166	0.2500	0.1299	0.0577
	7	1.0000	0.9996	0.9941	0.9679	0.8982	0.7723	0.6010	0.4159	0.2520	0.1316
	8	1.0000	0.9999	0.9987	0.9900	0.9591	0.8867	0.7624	0.5956	0.4143	0.2517
	9	1.0000	1.0000	0.9998	0.9974	0.9861	0.9520	0.8782	0.7553	0.5914	0.4119
	10	1.0000	1.0000	1.0000	0.9994	0.9961	0.9829	0.9468	0.8725	0.7507	0.5881
	11	1.0000	1.0000	1.0000	0.9999	0.9991	0.9949	0.9804	0.9435	0.8692	0.7483
	12	1.0000	1.0000	1.0000	1.0000	0.9998	0.9987	0.9940	0.9790	0.9420	0.8684
	13	1.0000	1.0000	1.0000	1.0000	1.0000	0.9997	0.9985	0.9935	0.9786	0.9423
	14	1.0000	1.0000	1.0000	1.0000	1.0000	1.0000	0.9997	0.9984	0.9936	0.9793
	15	1.0000	1.0000	1.0000	1.0000	1.0000	1.0000	1.0000	0.9997	0.9985	0.9941
	16	1.0000	1.0000	1.0000	1.0000	1.0000	1.0000	1.0000	1.0000	0.9997	0.9987
	17	1.0000	1.0000	1.0000	1.0000	1.0000	1.0000	1.0000	1.0000	1.0000	0.9998
	18	1.0000	1.0000	1.0000	1.0000	1.0000	1.0000	1.0000	1.0000	1.0000	1.0000
	19	1.0000	1.0000	1.0000	1.0000	1.0000	1.0000	1.0000	1.0000	1.0000	1.0000
	20	1.0000	1.0000	1.0000	1.0000	1.0000	1.0000	1.0000	1.0000	1.0000	1.0000
25	0	0.2774	0.0718	0.0172	0.0038	0.0008	0.0001	0.0000	0.0000	0.0000	0.0000
	1	0.6424	0.2712	0.0931	0.0274	0.0070	0.0016	0.0003	0.0001	0.0000	0.0000
	2	0.8729	0.5371	0.2537	0.0982	0.0321	0.0090	0.0021	0.0004	0.0001	0.0000
	3	0.9659	0.7636	0.4711	0.2340	0.0962	0.0332	0.0097	0.0024	0.0005	0.0001
	4	0.9928	0.9020	0.6821	0.4207	0.2137	0.0905	0.0320	0.0095	0.0023	0.0005
	5	0.9988	0.9666	0.8385	0.6167	0.3783	0.1935	0.0826	0.0294	0.0086	0.0020
	6	0.9998	0.9905	0.9305	0.7800	0.5611	0.3407	0.1734	0.0736	0.0258	0.0073
	7	1.0000	0.9977	0.9745	0.8909	0.7265	0.5118	0.3061	0.1536	0.0639	0.0216
	8	1.0000	0.9995	0.9920	0.9532	0.8506	0.6769	0.4668	0.2735	0.1340	0.0539
	9	1.0000	0.9999	0.9979	0.9827	0.9287	0.8106	0.6303	0.4246	0.2424	0.1148
	10	1.0000	1.0000	0.9995	0.9944	0.9703	0.9022	0.7712	0.5858	0.3843	0.2122
	11	1.0000	1.0000	0.9999	0.9985	0.9893	0.9558	0.8746	0.7323	0.5426	0.3450
	12	1.0000	1.0000	1.0000	0.9996	0.9966	0.9825	0.9396	0.8462	0.6937	0.5000
	13	1.0000	1.0000	1.0000	0.9999	0.9991	0.9940	0.9745	0.9222	0.8173	0.6550
	14	1.0000	1.0000	1,0000	1.0000	0.9998	0.9982	0.9907	0.9656	0.9040	0.7878
	15	1.0000	1.0000	1.0000	1.0000	1.0000	0.9995	0.9971	0.9868	0.9560	0.8852
	16	1.0000	1.0000	1.0000	1.0000	1.0000	0.9999	0.9992	0.9957	0.9826	0.9461
	17	1.0000	1.0000	1.0000	1.0000	1.0000	1.0000	0.9998	0.9988	0.9942	0.9784
	18	1.0000	1.0000	1.0000	1.0000	1.0000	1.0000	1.0000	0.9997	0.9984	0.9927
	19	1.0000	1.0000	1.0000	1.0000	1.0000	1.0000	1.0000	0.9999	0.9996	0.9980
	20	1.0000	1.0000	1.0000	1.0000	1.0000	1.0000	1.0000	1.0000	0.9999	0.9995
	21	1.0000	1.0000	1.0000	1.0000	1.0000	1.0000	1.0000	1.0000	1.0000	0.9999
	22	1.0000	1.0000	1.0000	1.0000	1.0000	1.0000	1.0000	1.0000	1.0000	1.0000
	23	1.0000	1.0000	1.0000	1.0000	1.0000	1.0000	1.0000	1.0000	1.0000	1.0000
	24	1.0000	1.0000	1.0000	1.0000	1.0000	1.0000	1.0000	1.0000	1.0000	1.0000
	25	1.0000	1.0000	1.0000	1.0000	1.0000	1.0000	1.0000	1.0000	1.0000	1.0000

Table III: The Poisson Distribution

$$F(x) = P(X \le x) = \sum_{k=0}^{x} \frac{\lambda^k e^{-\lambda}}{k!}$$

x	$\lambda = E(X)$									
	0.1	0.2	0.3	0.4	0.5	0.6	0.7	0.8	0.9	1.0
0	0.905	0.819	0.741	0.670	0.607	0.549	0.497	0.449	0.407	0.368
1	0.995	0.982	0.963	0.938	0.910	0.878	0.844	0.809	0.772	0.736
2	1.000	0.999	0.996	0.992	0.986	0.977	0.966	0.953	0.937	0.920
3	1.000	1.000	1.000	0.999	0.998	0.997	0.994	0.991	0.987	0.981
4	1.000	1.000	1.000	1.000	1.000	1.000	0.999	0.999	0.998	0.996
5	1.000	1.000	1.000	1.000	1.000	1.000	1.000	1.000	1.000	0.999
6	1.000	1.000	1.000	1.000	1.000	1.000	1.000	1.000	1.000	1.000

x	1.1	1.2	1.3	1.4	1.5	1.6	1.7	1.8	1.9	2.0
0	0.333	0.301	0.273	0.247	0.223	0.202	0.183	0.165	0.150	0.135
1	0.699	0.663	0.627	0.592	0.558	0.525	0.493	0.463	0.434	0.406
2	0.900	0.879	0.857	0.833	0.809	0.783	0.757	0.731	0.704	0.677
3	0.974	0.966	0.957	0.946	0.934	0.921	0.907	0.891	0.875	0.857
4	0.995	0.992	0.989	0.986	0.981	0.976	0.970	0.964	0.956	0.947
5	0.999	0.998	0.998	0.997	0.996	0.994	0.992	0.990	0.987	0.983
6	1.000	1.000	1.000	0.999	0.999	0.999	0.998	0.997	0.997	0.995
7	1.000	1.000	1.000	1.000	1.000	1.000	1.000	0.999	0.999	0.999
8	1.000	1.000	1.000	1.000	1.000	1.000	1.000	1.000	1.000	1.000

x	2.2	2.4	2.6	2.8	3.0	3.2	3.4	3.6	3.8	4.0
0	0.111	0.091	0.074	0.061	0.050	0.041	0.033	0.027	0.022	0.018
1	0.355	0.308	0.267	0.231	0.199	0.171	0.147	0.126	0.107	0.092
2	0.623	0.570	0.518	0.469	0.423	0.380	0.340	0.303	0.269	0.238
3	0.819	0.779	0.736	0.692	0.647	0.603	0.558	0.515	0.473	0.433
4	0.928	0.904	0.877	0.848	0.815	0.781	0.744	0.706	0.668	0.629
5	0.975	0.964	0.951	0.935	0.916	0.895	0.871	0.844	0.816	0.785
6	0.993	0.988	0.983	0.976	0.966	0.955	0.942	0.927	0.909	0.889
7	0.998	0.997	0.995	0.992	0.988	0.983	0.977	0.969	0.960	0.949
8	1.000	0.999	0.999	0.998	0.996	0.994	0.992	0.988	0.984	0.979
9	1.000	1.000	1.000	0.999	0.999	0.998	0.997	0.996	0.994	0.992
10	1.000	1.000	1.000	1.000	1.000	1.000	0.999	0.999	0.998	0.997
11	1.000	1.000	1.000	1.000	1.000	1.000	1.000	1.000	0.999	0.999
12	1.000	1.000	1.000	1.000	1.000	1.000	1.000	1.000	1.000	1.000

Table III: *continued*

x	4.2	4.4	4.6	4.8	5.0	5.2	5.4	5.6	5.8	6.0
0	0.015	0.012	0.010	0.008	0.007	0.006	0.005	0.004	0.003	0.002
1	0.078	0.066	0.056	0.048	0.040	0.034	0.029	0.024	0.021	0.017
2	0.210	0.185	0.163	0.143	0.125	0.109	0.095	0.082	0.072	0.062
3	0.395	0.359	0.326	0.294	0.265	0.238	0.213	0.191	0.170	0.151
4	0.590	0.551	0.513	0.476	0.440	0.406	0.373	0.342	0.313	0.285
5	0.753	0.720	0.686	0.651	0.616	0.581	0.546	0.512	0.478	0.446
6	0.867	0.844	0.818	0.791	0.762	0.732	0.702	0.670	0.638	0.606
7	0.936	0.921	0.905	0.887	0.867	0.845	0.822	0.797	0.771	0.744
8	0.972	0.964	0.955	0.944	0.932	0.918	0.903	0.886	0.867	0.847
9	0.989	0.985	0.980	0.975	0.968	0.960	0.951	0.941	0.929	0.916
10	0.996	0.994	0.992	0.990	0.986	0.982	0.977	0.972	0.965	0.957
11	0.999	0.998	0.997	0.996	0.995	0.993	0.990	0.988	0.984	0.980
12	1.000	0.999	0.999	0.999	0.998	0.997	0.996	0.995	0.993	0.991
13	1.000	1.000	1.000	1.000	0.999	0.999	0.999	0.998	0.997	0.996
14	1.000	1.000	1.000	1.000	1.000	1.000	0.999	0.999	0.999	0.999
15	1.000	1.000	1.000	1.000	1.000	1.000	1.000	1.000	1.000	0.999
16	1.000	1.000	1.000	1.000	1.000	1.000	1.000	1.000	1.000	1.000

x	6.5	7.0	7.5	8.0	8.5	9.0	9.5	10.0	10.5	11.0
0	0.002	0.001	0.001	0.000	0.000	0.000	0.000	0.000	0.000	0.000
1	0.011	0.007	0.005	0.003	0.002	0.001	0.001	0.000	0.000	0.000
2	0.043	0.030	0.020	0.014	0.009	0.006	0.004	0.003	0.002	0.001
3	0.112	0.082	0.059	0.042	0.030	0.021	0.015	0.010	0.007	0.005
4	0.224	0.173	0.132	0.100	0.074	0.055	0.040	0.029	0.021	0.015
5	0.369	0.301	0.241	0.191	0.150	0.116	0.089	0.067	0.050	0.038
6	0.527	0.450	0.378	0.313	0.256	0.207	0.165	0.130	0.102	0.079
7	0.673	0.599	0.525	0.453	0.386	0.324	0.269	0.220	0.179	0.143
8	0.792	0.729	0.662	0.593	0.523	0.456	0.392	0.333	0.279	0.232
9	0.877	0.830	0.776	0.717	0.653	0.587	0.522	0.458	0.397	0.341
10	0.933	0.901	0.862	0.816	0.763	0.706	0.645	0.583	0.521	0.460
11	0.966	0.947	0.921	0.888	0.849	0.803	0.752	0.697	0.639	0.579
12	0.984	0.973	0.957	0.936	0.909	0.876	0.836	0.792	0.742	0.689
13	0.993	0.987	0.978	0.966	0.949	0.926	0.898	0.864	0.825	0.781
14	0.997	0.994	0.990	0.983	0.973	0.959	0.940	0.917	0.888	0.854
15	0.999	0.998	0.995	0.992	0.986	0.978	0.967	0.951	0.932	0.907
16	1.000	0.999	0.998	0.996	0.993	0.989	0.982	0.973	0.960	0.944
17	1.000	1.000	0.999	0.998	0.997	0.995	0.991	0.986	0.978	0.968
18	1.000	1.000	1.000	0.999	0.999	0.998	0.096	0.993	0.988	0.982
19	1.000	1.000	1.000	1.000	0.999	0.999	0.998	0.997	0.994	0.991
20	1.000	1.000	1.000	1.000	1.000	1.000	0.999	0.998	0.997	0.995
21	1.000	1.000	1.000	1.000	1.000	1.000	1.000	0.999	0.999	0.998
22	1.000	1.000	1.000	1.000	1.000	1.000	1.000	1.000	0.999	0.999
23	1.000	1.000	1.000	1.000	1.000	1.000	1.000	1.000	1.000	1.000

					Table III: *continued*					
x	11.5	12.0	12.5	13.0	13.5	14.0	14.5	15.0	15.5	16.0
0	0.000	0.000	0.000	0.000	0.000	0.000	0.000	0.000	0.000	0.000
1	0.000	0.000	0.000	0.000	0.000	0.000	0.000	0.000	0.000	0.000
2	0.001	0.001	0.000	0.000	0.000	0.000	0.000	0.000	0.000	0.000
3	0.003	0.002	0.002	0.001	0.001	0.000	0.000	0.000	0.000	0.000
4	0.011	0.008	0.005	0.004	0.003	0.002	0.001	0.001	0.001	0.000
5	0.028	0.020	0.015	0.011	0.008	0.006	0.004	0.003	0.002	0.001
6	0.060	0.046	0.035	0.026	0.019	0.014	0.010	0.008	0.006	0.004
7	0.114	0.090	0.070	0.054	0.041	0.032	0.024	0.018	0.013	0.010
8	0.191	0.155	0.125	0.100	0.079	0.062	0.048	0.037	0.029	0.022
9	0.289	0.242	0.201	0.166	0.135	0.109	0.088	0.070	0.055	0.043
10	0.402	0.347	0.297	0.252	0.211	0.176	0.145	0.118	0.096	0.077
11	0.520	0.462	0.406	0.353	0.304	0.260	0.220	0.185	0.154	0.127
12	0.633	0.576	0.519	0.463	0.409	0.358	0.311	0.268	0.228	0.193
13	0.733	0.682	0.629	0.573	0.518	0.464	0.413	0.363	0.317	0.275
14	0.815	0.772	0.725	0.675	0.623	0.570	0.518	0.466	0.415	0.368
15	0.878	0.844	0.806	0.764	0.718	0.669	0.619	0.568	0.517	0.467
16	0.924	0.899	0.869	0.835	0.798	0.756	0.711	0.664	0.615	0.566
17	0.954	0.937	0.916	0.890	0.861	0.827	0.790	0.749	0.705	0.659
18	0.974	0.963	0.948	0.930	0.908	0.883	0.853	0.819	0.782	0.742
19	0.986	0.979	0.969	0.957	0.942	0.923	0.901	0.875	0.846	0.812
20	0.992	0.988	0.983	0.975	0.965	0.952	0.936	0.917	0.894	0.868
21	0.996	0.994	0.991	0.986	0.980	0.971	0.960	0.947	0.930	0.911
22	0.999	0.997	0.995	0.992	0.989	0.983	0.976	0.967	0.956	0.942
23	0.999	0.999	0.998	0.996	0.994	0.991	0.986	0.981	0.973	0.963
24	1.000	0.999	0.999	0.998	0.997	0.995	0.992	0.989	0.984	0.978
25	1.000	1.000	0.999	0.999	0.998	0.997	0.996	0.994	0.991	0.987
26	1.000	1.000	1.000	1.000	0.999	0.999	0.998	0.997	0.995	0.993
27	1.000	1.000	1.000	1.000	1.000	0.999	0.999	0.998	0.997	0.996
28	1.000	1.000	1.000	1.000	1.000	1.000	0.999	0.999	0.999	0.998
29	1.000	1.000	1.000	1.000	1.000	1.000	1.000	1.000	0.999	0.999
30	1.000	1.000	1.000	1.000	1.000	1.000	1.000	1.000	1.000	0.999
31	1.000	1.000	1.000	1.000	1.000	1.000	1.000	1.000	1.000	1.000
32	1.000	1.000	1.000	1.000	1.000	1.000	1.000	1.000	1.000	1.000
33	1.000	1.000	1.000	1.000	1.000	1.000	1.000	1.000	1.000	1.000
34	1.000	1.000	1.000	1.000	1.000	1.000	1.000	1.000	1.000	1.000
35	1.000	1.000	1.000	1.000	1.000	1.000	1.000	1.000	1.000	1.000

Table IV: The Chi-Square Distribution

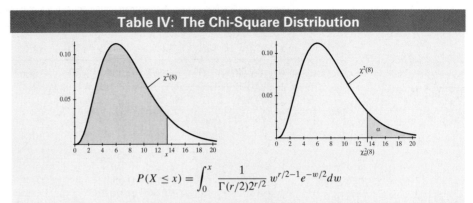

$$P(X \le x) = \int_0^x \frac{1}{\Gamma(r/2)2^{r/2}}\, w^{r/2-1} e^{-w/2}\, dw$$

				$P(X \le x)$				
	0.010	0.025	0.050	0.100	0.900	0.950	0.975	0.990
r	$\chi^2_{0.99}(r)$	$\chi^2_{0.975}(r)$	$\chi^2_{0.95}(r)$	$\chi^2_{0.90}(r)$	$\chi^2_{0.10}(r)$	$\chi^2_{0.05}(r)$	$\chi^2_{0.025}(r)$	$\chi^2_{0.01}(r)$
1	0.000	0.001	0.004	0.016	2.706	3.841	5.024	6.635
2	0.020	0.051	0.103	0.211	4.605	5.991	7.378	9.210
3	0.115	0.216	0.352	0.584	6.251	7.815	9.348	11.34
4	0.297	0.484	0.711	1.064	7.779	9.488	11.14	13.28
5	0.554	0.831	1.145	1.610	9.236	11.07	12.83	15.09
6	0.872	1.237	1.635	2.204	10.64	12.59	14.45	16.81
7	1.239	1.690	2.167	2.833	12.02	14.07	16.01	18.48
8	1.646	2.180	2.733	3.490	13.36	15.51	17.54	20.09
9	2.088	2.700	3.325	4.168	14.68	16.92	19.02	21.67
10	2.558	3.247	3.940	4.865	15.99	18.31	20.48	23.21
11	3.053	3.816	4.575	5.578	17.28	19.68	21.92	24.72
12	3.571	4.404	5.226	6.304	18.55	21.03	23.34	26.22
13	4.107	5.009	5.892	7.042	19.81	22.36	24.74	27.69
14	4.660	5.629	6.571	7.790	21.06	23.68	26.12	29.14
15	5.229	6.262	7.261	8.547	22.31	25.00	27.49	30.58
16	5.812	6.908	7.962	9.312	23.54	26.30	28.84	32.00
17	6.408	7.564	8.672	10.08	24.77	27.59	30.19	33.41
18	7.015	8.231	9.390	10.86	25.99	28.87	31.53	34.80
19	7.633	8.907	10.12	11.65	27.20	30.14	32.85	36.19
20	8.260	9.591	10.85	12.44	28.41	31.41	34.17	37.57
21	8.897	10.28	11.59	13.24	29.62	32.67	35.48	38.93
22	9.542	10.98	12.34	14.04	30.81	33.92	36.78	40.29
23	10.20	11.69	13.09	14.85	32.01	35.17	38.08	41.64
24	10.86	12.40	13.85	15.66	33.20	36.42	39.36	42.98
25	11.52	13.12	14.61	16.47	34.38	37.65	40.65	44.31
26	12.20	13.84	15.38	17.29	35.56	38.88	41.92	45.64
27	12.88	14.57	16.15	18.11	36.74	40.11	43.19	46.96
28	13.56	15.31	16.93	18.94	37.92	41.34	44.46	48.28
29	14.26	16.05	17.71	19.77	39.09	42.56	45.72	49.59
30	14.95	16.79	18.49	20.60	40.26	43.77	46.98	50.89
40	22.16	24.43	26.51	29.05	51.80	55.76	59.34	63.69
50	29.71	32.36	34.76	37.69	63.17	67.50	71.42	76.15
60	37.48	40.48	43.19	46.46	74.40	79.08	83.30	88.38
70	45.44	48.76	51.74	55.33	85.53	90.53	95.02	100.4
80	53.34	57.15	60.39	64.28	96.58	101.9	106.6	112.3

This table is abridged and adapted from Table III in *Biometrika Tables for Statisticians*, edited by E.S.Pearson and H.O.Hartley.

Table Va: The Normal Distribution

$$P(Z \le z) = \Phi(z) = \int_{-\infty}^{z} \frac{1}{\sqrt{2\pi}} e^{-w^2/2} \, dw$$

$$\Phi(-z) = 1 - \Phi(z)$$

z	0.00	0.01	0.02	0.03	0.04	0.05	0.06	0.07	0.08	0.09
0.0	0.5000	0.5040	0.5080	0.5120	0.5160	0.5199	0.5239	0.5279	0.5319	0.5359
0.1	0.5398	0.5438	0.5478	0.5517	0.5557	0.5596	0.5636	0.5675	0.5714	0.5753
0.2	0.5793	0.5832	0.5871	0.5910	0.5948	0.5987	0.6026	0.6064	0.6103	0.6141
0.3	0.6179	0.6217	0.6255	0.6293	0.6331	0.6368	0.6406	0.6443	0.6480	0.6517
0.4	0.6554	0.6591	0.6628	0.6664	0.6700	0.6736	0.6772	0.6808	0.6844	0.6879
0.5	0.6915	0.6950	0.6985	0.7019	0.7054	0.7088	0.7123	0.7157	0.7190	0.7224
0.6	0.7257	0.7291	0.7324	0.7357	0.7389	0.7422	0.7454	0.7486	0.7517	0.7549
0.7	0.7580	0.7611	0.7642	0.7673	0.7703	0.7734	0.7764	0.7794	0.7823	0.7852
0.8	0.7881	0.7910	0.7939	0.7967	0.7995	0.8023	0.8051	0.8078	0.8106	0.8133
0.9	0.8159	0.8186	0.8212	0.8238	0.8264	0.8289	0.8315	0.8340	0.8365	0.8389
1.0	0.8413	0.8438	0.8461	0.8485	0.8508	0.8531	0.8554	0.8577	0.8599	0.8621
1.1	0.8643	0.8665	0.8686	0.8708	0.8729	0.8749	0.8770	0.8790	0.8810	0.8830
1.2	0.8849	0.8869	0.8888	0.8907	0.8925	0.8944	0.8962	0.8980	0.8997	0.9015
1.3	0.9032	0.9049	0.9066	0.9082	0.9099	0.9115	0.9131	0.9147	0.9162	0.9177
1.4	0.9192	0.9207	0.9222	0.9236	0.9251	0.9265	0.9279	0.9292	0.9306	0.9319
1.5	0.9332	0.9545	0.9357	0.9370	0.9382	0.9394	0.9406	0.9418	0.9429	0.9441
1.6	0.9452	0.9463	0.9474	0.9484	0.9495	0.9505	0.9515	0.9525	0.9535	0.9545
1.7	0.9554	0.9564	0.9573	0.9582	0.9591	0.9599	0.9608	0.9616	0.9625	0.9633
1.8	0.9641	0.9649	0.9656	0.9664	0.9671	0.9678	0.9686	0.9693	0.9699	0.9706
1.9	0.9713	0.9719	0.9726	0.9732	0.9738	0.9744	0.9750	0.9756	0.9761	0.9767
2.0	0.9772	0.9778	0.9783	0.9788	0.9793	0.9798	0.9803	0.9808	0.9812	0.9817
2.1	0.9821	0.9826	0.9830	0.9834	0.9838	0.9842	0.9846	0.9850	0.9854	0.9857
2.2	0.9861	0.9864	0.9868	0.9871	0.9875	0.9878	0.9881	0.9884	0.9887	0.9890
2.3	0.9893	0.9896	0.9898	0.9901	0.9904	0.9906	0.9909	0.9911	0.9913	0.9916
2.4	0.9918	0.9920	0.9922	0.9925	0.9927	0.9929	0.9931	0.9932	0.9934	0.9936
2.5	0.9938	0.9940	0.9941	0.9943	0.9945	0.9946	0.9948	0.9949	0.9951	0.9952
2.6	0.9953	0.9955	0.9956	0.9957	0.9959	0.9960	0.9961	0.9962	0.9963	0.9964
2.7	0.9965	0.9966	0.9967	0.9968	0.9969	0.9970	0.9971	0.9972	0.9973	0.9974
2.8	0.9974	0.9975	0.9976	0.9977	0.9977	0.9978	0.9979	0.9979	0.9980	0.9981
2.9	0.9981	0.9982	0.9982	0.9983	0.9984	0.9984	0.9985	0.9985	0.9986	0.9986
3.0	0.9987	0.9987	0.9987	0.9988	0.9988	0.9989	0.9989	0.9989	0.9990	0.9990

α	0.400	0.300	0.200	0.100	0.050	0.025	0.020	0.010	0.005	0.001
z_α	0.253	0.524	0.842	1.282	1.645	1.960	2.054	2.326	2.576	3.090
$z_{\alpha/2}$	0.842	1.036	1.282	1.645	1.960	2.240	2.326	2.576	2.807	3.291

Table Vb: The Normal Distribution

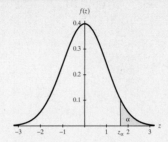

$$P(Z > z_\alpha) = \alpha$$

$$P(Z > z) = 1 - \Phi(z) = \Phi(-z)$$

z_α	0.00	0.01	0.02	0.03	0.04	0.05	0.06	0.07	0.08	0.09
0.0	0.5000	0.4960	0.4920	0.4880	0.4840	0.4801	0.4761	0.4721	0.4681	0.4641
0.1	0.4602	0.4562	0.4522	0.4483	0.4443	0.4404	0.4364	0.4325	0.4286	0.4247
0.2	0.4207	0.4168	0.4129	0.4090	0.4052	0.4013	0.3974	0.3936	0.3897	0.3859
0.3	0.3821	0.3783	0.3745	0.3707	0.3669	0.3632	0.3594	0.3557	0.3520	0.3483
0.4	0.3446	0.3409	0.3372	0.3336	0.3300	0.3264	0.3228	0.3192	0.3156	0.3121
0.5	0.3085	0.3050	0.3015	0.2981	0.2946	0.2912	0.2877	0.2843	0.2810	0.2776
0.6	0.2743	0.2709	0.2676	0.2643	0.2611	0.2578	0.2546	0.2514	0.2483	0.2451
0.7	0.2420	0.2389	0.2358	0.2327	0.2296	0.2266	0.2236	0.2206	0.2177	0.2148
0.8	0.2119	0.2090	0.2061	0.2033	0.2005	0.1977	0.1949	0.1922	0.1894	0.1867
0.9	0.1841	0.1814	0.1788	0.1762	0.1736	0.1711	0.1685	0.1660	0.1635	0.1611
1.0	0.1587	0.1562	0.1539	0.1515	0.1492	0.1469	0.1446	0.1423	0.1401	0.1379
1.1	0.1357	0.1335	0.1314	0.1292	0.1271	0.1251	0.1230	0.1210	0.1190	0.1170
1.2	0.1151	0.1131	0.1112	0.1093	0.1075	0.1056	0.1038	0.1020	0.1003	0.0985
1.3	0.0968	0.0951	0.0934	0.0918	0.0901	0.0885	0.0869	0.0853	0.0838	0.0823
1.4	0.0808	0.0793	0.0778	0.0764	0.0749	0.0735	0.0721	0.0708	0.0694	0.0681
1.5	0.0668	0.0655	0.0643	0.0630	0.0618	0.0606	0.0594	0.0582	0.0571	0.0559
1.6	0.0548	0.0537	0.0526	0.0516	0.0505	0.0495	0.0485	0.0475	0.0465	0.0455
1.7	0.0446	0.0436	0.0427	0.0418	0.0409	0.0401	0.0392	0.0384	0.0375	0.0367
1.8	0.0359	0.0351	0.0344	0.0336	0.0329	0.0322	0.0314	0.0307	0.0301	0.0294
1.9	0.0287	0.0281	0.0274	0.0268	0.0262	0.0256	0.0250	0.0244	0.0239	0.0233
2.0	0.0228	0.0222	0.0217	0.0212	0.0207	0.0202	0.0197	0.0192	0.0188	0.0183
2.1	0.0179	0.0174	0.0170	0.0166	0.0162	0.0158	0.0154	0.0150	0.0146	0.0143
2.2	0.0139	0.0136	0.0132	0.0129	0.0125	0.0122	0.0119	0.0116	0.0113	0.0110
2.3	0.0107	0.0104	0.0102	0.0099	0.0096	0.0094	0.0091	0.0089	0.0087	0.0084
2.4	0.0082	0.0080	0.0078	0.0075	0.0073	0.0071	0.0069	0.0068	0.0066	0.0064
2.5	0.0062	0.0060	0.0059	0.0057	0.0055	0.0054	0.0052	0.0051	0.0049	0.0048
2.6	0.0047	0.0045	0.0044	0.0043	0.0041	0.0040	0.0039	0.0038	0.0037	0.0036
2.7	0.0035	0.0034	0.0033	0.0032	0.0031	0.0030	0.0029	0.0028	0.0027	0.0026
2.8	0.0026	0.0025	0.0024	0.0023	0.0023	0.0022	0.0021	0.0021	0.0020	0.0019
2.9	0.0019	0.0018	0.0018	0.0017	0.0016	0.0016	0.0015	0.0015	0.0014	0.0014
3.0	0.0013	0.0013	0.0013	0.0012	0.0012	0.0011	0.0011	0.0011	0.0010	0.0010
3.1	0.0010	0.0009	0.0009	0.0009	0.0008	0.0008	0.0008	0.0008	0.0007	0.0007
3.2	0.0007	0.0007	0.0006	0.0006	0.0006	0.0006	0.0006	0.0005	0.0005	0.0005
3.3	0.0005	0.0005	0.0005	0.0004	0.0004	0.0004	0.0004	0.0004	0.0004	0.0003
3.4	0.0003	0.0003	0.0003	0.0003	0.0003	0.0003	0.0003	0.0003	0.0003	0.0002

Table VI: The *t* Distribution

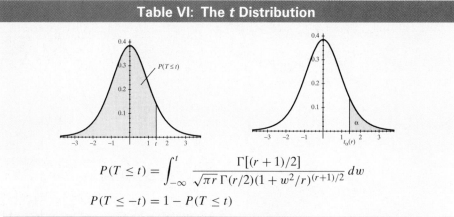

$$P(T \le t) = \int_{-\infty}^{t} \frac{\Gamma[(r+1)/2]}{\sqrt{\pi r}\,\Gamma(r/2)(1+w^2/r)^{(r+1)/2}}\, dw$$

$$P(T \le -t) = 1 - P(T \le t)$$

				$P(T \le t)$			
	0.60	0.75	0.90	0.95	0.975	0.99	0.995
r	$t_{0.40}(r)$	$t_{0.25}(r)$	$t_{0.10}(r)$	$t_{0.05}(r)$	$t_{0.025}(r)$	$t_{0.01}(r)$	$t_{0.005}(r)$
1	0.325	1.000	3.078	6.314	12.706	31.821	63.657
2	0.289	0.816	1.886	2.920	4.303	6.965	9.925
3	0.277	0.765	1.638	2.353	3.182	4.541	5.841
4	0.271	0.741	1.533	2.132	2.776	3.747	4.604
5	0.267	0.727	1.476	2.015	2.571	3.365	4.032
6	0.265	0.718	1.440	1.943	2.447	3.143	3.707
7	0.263	0.711	1.415	1.895	2.365	2.998	3.499
8	0.262	0.706	1.397	1.860	2.306	2.896	3.355
9	0.261	0.703	1.383	1.833	2.262	2.821	3.250
10	0.260	0.700	1.372	1.812	2.228	2.764	3.169
11	0.260	0.697	1.363	1.796	2.201	2.718	3.106
12	0.259	0.695	1.356	1.782	2.179	2.681	3.055
13	0.259	0.694	1.350	1.771	2.160	2.650	3.012
14	0.258	0.692	1.345	1.761	2.145	2.624	2.997
15	0.258	0.691	1.341	1.753	2.131	2.602	2.947
16	0.258	0.690	1.337	1.746	2.120	2.583	2.921
17	0.257	0.689	1.333	1.740	2.110	2.567	2.898
18	0.257	0.688	1.330	1.734	2.101	2.552	2.878
19	0.257	0.688	1.328	1.729	2.093	2.539	2.861
20	0.257	0.687	1.325	1.725	2.086	2.528	2.845
21	0.257	0.686	1.323	1.721	2.080	2.518	2.831
22	0.256	0.686	1.321	1.717	2.074	2.508	2.819
23	0.256	0.685	1.319	1.714	2.069	2.500	2.807
24	0.256	0.685	1.318	1.711	2.064	2.492	2.797
25	0.256	0.684	1.316	1.708	2.060	2.485	2.787
26	0.256	0.684	1.315	1.706	2.056	2.479	2.779
27	0.256	0.684	1.314	1.703	2.052	2.473	2.771
28	0.256	0.683	1.313	1.701	2.048	2.467	2.763
29	0.256	0.683	1.311	1.699	2.045	2.462	2.756
30	0.256	0.683	1.310	1.697	2.042	2.457	2.750
∞	0.253	0.674	1.282	1.645	1.960	2.326	2.576

This table is taken from Table III of Fisher and Yates: *Statistical Tables for Biological, Agricultrual, and Medical Research*, published by Longman Group Ltd., London (previously published by Oliver and Boyd, Edinburgh).

Table VII: The F Distribution

$$P(F \leq f) = \int_0^f \frac{\Gamma[(r_1 + r_2)/2](r_1/r_2)^{r_1/2} w^{r_1/2-1}}{\Gamma(r_1/2)\Gamma(r_2/2)(1 + r_1 w/r_2)^{(r_1+r_2)/2}} \, dw$$

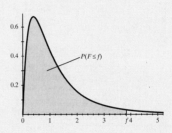

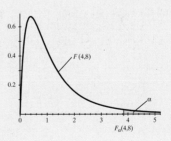

Table VII: *continued*

$$P(F \leq f) = \int_0^f \frac{\Gamma[(r_1 + r_2)/2](r_1/r_2)^{r_1/2} w^{r_1/2-1}}{\Gamma(r_1/2)\Gamma(r_2/2)(1 + r_1 w/r_2)^{(r_1+r_2)/2}}\, dw$$

α	$P(F \leq f)$	Den. d.f. r_2	1	2	3	4	5	6	7	8	9	10
								Numerator Degrees of Freedom, r_1				
0.05	0.95	1	161.4	199.5	215.7	224.6	230.2	234.0	236.8	238.9	240.5	241.9
0.025	0.975		647.79	799.50	864.16	899.58	921.85	937.11	948.22	956.66	963.28	968.63
0.01	0.99		4052	4999.5	5403	5625	5764	5859	5928	5981	6022	6056
0.05	0.95	2	18.51	19.00	19.16	19.25	19.30	19.33	19.35	19.37	19.38	19.40
0.025	0.975		38.51	39.00	39.17	39.25	39.30	39.33	39.36	39.37	39.39	39.40
0.01	0.99		98.50	99.00	99.17	99.25	99.30	99.33	99.36	99.37	99.39	99.40
0.05	0.95	3	10.13	9.55	9.28	9.12	9.01	8.94	8.89	8.85	8.81	8.79
0.025	0.975		17.44	16.04	15.44	15.10	14.88	14.73	14.62	14.54	14.47	14.42
0.01	0.99		34.12	30.82	29.46	28.71	28.24	27.91	27.67	27.49	27.35	27.23
0.05	0.95	4	7.71	6.94	6.59	6.39	6.26	6.16	6.09	6.04	6.00	5.96
0.025	0.975		12.22	10.65	9.98	9.60	9.36	9.20	9.07	8.98	8.90	8.84
0.01	0.99		21.20	18.00	16.69	15.98	15.52	15.21	14.98	14.80	14.66	14.55
0.05	0.95	5	6.61	5.79	5.41	5.19	5.05	4.95	4.88	4.82	4.77	4.74
0.025	0.975		10.01	8.43	7.76	7.39	7.15	6.98	6.85	6.76	6.68	6.62
0.01	0.99		16.26	13.27	12.06	11.39	10.97	10.67	10.46	10.29	10.16	10.05
0.05	0.95	6	5.99	5.14	4.76	4.53	4.39	4.28	4.21	4.15	4.10	4.06
0.025	0.975		8.81	7.26	6.60	6.23	5.99	5.82	5.70	5.60	5.52	5.46
0.01	0.99		13.75	10.92	9.78	9.15	8.75	8.47	8.26	8.10	7.98	7.87
0.05	0.95	7	5.59	4.74	4.35	4.12	3.97	3.87	3.79	3.73	3.68	3.64
0.025	0.975		8.07	6.54	5.89	5.52	5.29	5.12	4.99	4.90	4.82	4.76
0.01	0.99		12.25	9.55	8.45	7.85	7.46	7.19	6.99	6.84	6.72	6.62
0.05	0.95	8	5.32	4.46	4.07	3.84	3.69	3.58	3.50	3.44	3.39	3.35
0.025	0.975		7.57	6.06	5.42	5.05	4.82	4.65	4.53	4.43	4.36	4.30
0.01	0.99		11.26	8.65	7.59	7.01	6.63	6.37	6.18	6.03	5.91	5.81
0.05	0.95	9	5.12	4.26	3.86	3.63	3.48	3.37	3.29	3.23	3.18	3.14
0.025	0.975		7.21	5.71	5.08	4.72	4.48	4.32	4.20	4.10	4.03	3.96
0.01	0.99		10.56	8.02	6.99	6.42	6.06	5.80	5.61	5.47	5.35	5.26
0.05	0.95	10	4.96	4.10	3.71	3.48	3.33	3.22	3.14	3.07	3.02	2.98
0.025	0.975		6.94	5.46	4.83	4.47	4.24	4.07	3.95	3.85	3.78	3.72
0.01	0.99		10.04	7.56	6.55	5.99	5.64	5.39	5.20	5.06	4.94	4.85

Table VII: *continued*

$$P(F \le f) = \int_0^f \frac{\Gamma[(r_1 + r_2)/2](r_1/r_2)^{r_1/2}w^{r_1/2-1}}{\Gamma(r_1/2)\Gamma(r_2/2)(1 + r_1 w/r_2)^{(r_1+r_2)/2}}\, dw$$

α	$P(F \le f)$	Den. d.f. r_2	Numerator Degrees of Freedom, r_1									
			1	2	3	4	5	6	7	8	9	10
0.05	0.95	12	4.75	3.89	3.49	3.26	3.11	3.00	2.91	2.85	2.80	2.75
0.025	0.975		6.55	5.10	4.47	4.12	3.89	3.73	3.61	3.51	3.44	3.37
0.01	0.99		9.33	6.93	5.95	5.41	5.06	4.82	4.64	4.50	4.39	4.30
0.05	0.95	15	4.54	3.68	3.29	3.06	2.90	2.79	2.71	2.64	2.59	2.54
0.025	0.975		6.20	4.77	4.15	3.80	3.58	3.41	3.29	3.20	3.12	3.06
0.01	0.99		8.68	6.36	5.42	4.89	4.56	4.32	4.14	4.00	3.89	3.80
0.05	0.95	20	4.35	3.49	3.10	2.87	2.71	2.60	2.51	2.45	2.39	2.35
0.025	0.975		5.87	4.46	3.86	3.51	3.29	3.13	3.01	2.91	2.84	2.77
0.01	0.99		8.10	5.85	4.94	4.43	4.10	3.87	3.70	3.56	3.46	3.37
0.05	0.95	24	4.26	3.40	3.01	2.78	2.62	2.51	2.42	2.36	2.30	2.25
0.025	0.975		5.72	4.32	3.72	3.38	3.15	2.99	2.87	2.78	2.70	2.64
0.01	0.99		7.82	5.61	4.72	4.22	3.90	3.67	3.50	3.36	3.26	3.17
0.05	0.95	30	4.17	3.32	2.92	2.69	2.53	2.42	2.33	2.27	2.21	2.16
0.025	0.975		5.57	4.18	3.59	3.25	3.03	2.87	2.75	2.65	2.57	2.51
0.01	0.99		7.56	5.39	4.51	4.02	3.70	3.47	3.30	3.17	3.07	2.98
0.05	0.95	40	4.08	3.23	2.84	2.61	2.45	2.34	2.25	2.18	2.12	2.08
0.025	0.975		5.42	4.05	3.46	3.13	2.90	2.74	2.62	2.53	2.45	2.39
0.01	0.99		7.31	5.18	4.31	3.83	3.51	3.29	3.12	2.99	2.89	2.80
0.05	0.95	60	4.00	3.15	2.76	2.53	2.37	2.25	2.17	2.10	2.04	1.99
0.025	0.975		5.29	3.93	3.34	3.01	2.79	2.63	2.51	2.41	2.33	2.27
0.01	0.99		7.08	4.98	4.13	3.65	3.34	3.12	2.95	2.82	2.72	2.63
0.05	0.95	120	3.92	3.07	2.68	2.45	2.29	2.17	2.09	2.02	1.96	1.91
0.025	0.975		5.15	3.80	3.23	2.89	2.67	2.52	2.39	2.30	2.22	2.16
0.01	0.99		6.85	4.79	3.95	3.48	3.17	2.96	2.79	2.66	2.56	2.47
0.05	0.95	∞	3.84	3.00	2.60	2.37	2.21	2.10	2.01	1.94	1.88	1.83
0.025	0.975		5.02	3.69	3.12	2.79	2.57	2.41	2.29	2.19	2.11	2.05
0.01	0.99		6.63	4.61	3.78	3.32	3.02	2.80	2.64	2.51	2.41	2.32

Table VII: continued

$$P(F \leq f) = \int_0^f \frac{\Gamma[(r_1 + r_2)/2](r_1/r_2)^{r_1/2} w^{r_1/2-1}}{\Gamma(r_1/2)\Gamma(r_2/2)(1 + r_1 w/r_2)^{(r_1+r_2)/2}}\, dw$$

Den. d.f. r_2	$P(F \leq f)$	α	12	15	20	24	30	40	60	120	∞
1	0.95	0.05	243.9	245.9	248.0	249.1	250.1	251.1	252.2	253.3	254.3
	0.975	0.025	976.71	984.87	993.10	997.25	1001.4	1005.6	1009.8	1014.0	1018.3
	0.99	0.01	6106	6157	6209	6235	6261	6287	6313	6339	6366
2	0.95	0.05	19.41	19.43	19.45	19.45	19.46	19.47	19.48	19.49	19.50
	0.975	0.025	39.42	39.43	39.45	39.46	39.47	39.47	39.48	39.49	39.50
	0.99	0.01	99.42	99.43	99.45	99.46	99.47	99.47	99.48	99.49	99.50
3	0.95	0.05	8.74	8.70	8.66	8.64	8.62	8.59	8.57	8.55	8.53
	0.975	0.025	14.34	14.25	14.17	14.12	14.08	14.04	13.99	13.95	13.90
	0.99	0.01	27.05	26.87	26.69	26.60	26.50	26.41	26.32	26.22	26.13
4	0.95	0.05	5.91	5.86	5.80	5.77	5.75	5.72	5.69	5.66	5.63
	0.975	0.025	8.75	8.66	8.56	8.51	8.46	8.41	8.36	8.31	8.26
	0.99	0.01	14.37	14.20	14.02	13.93	13.84	13.75	13.65	13.56	13.46
5	0.95	0.05	4.68	4.62	4.56	4.53	4.50	4.46	4.43	4.40	4.36
	0.975	0.025	6.52	6.43	6.33	6.28	6.23	6.18	6.12	6.07	6.02
	0.99	0.01	9.89	9.72	9.55	9.47	9.38	9.29	9.20	9.11	9.02
6	0.95	0.05	4.00	3.94	3.87	3.84	3.81	3.77	3.74	3.70	3.67
	0.975	0.025	5.37	5.27	5.17	5.12	5.07	5.01	4.96	4.90	4.85
	0.99	0.01	7.72	7.56	7.40	7.31	7.23	7.14	7.06	6.97	6.88
7	0.95	0.05	3.57	3.51	3.41	3.41	3.38	3.34	3.30	3.27	3.23
	0.975	0.025	4.67	4.57	4.47	4.42	4.36	4.31	4.25	4.20	4.14
	0.99	0.01	6.47	6.31	6.16	6.07	5.99	5.91	5.82	5.74	5.65
8	0.95	0.05	3.28	3.22	3.15	3.12	3.08	3.04	3.01	2.97	2.93
	0.975	0.025	4.20	4.10	4.00	3.95	3.89	3.84	3.78	3.73	3.67
	0.99	0.01	5.67	5.52	5.36	5.28	5.20	5.12	5.03	4.95	4.86
9	0.95	0.05	3.07	3.01	2.94	2.90	2.86	2.83	2.79	2.75	2.71
	0.975	0.025	3.87	3.77	3.67	3.61	3.56	3.51	3.45	3.39	3.33
	0.99	0.01	5.11	4.96	4.81	4.73	4.65	4.57	4.48	4.40	4.31

Numerator Degrees of Freedom, r_1

Table VII: continued

$$P(F \le f) = \int_0^f \frac{\Gamma[(r_1+r_2)/2](r_1/r_2)^{r_1/2}w^{r_1/2-1}}{\Gamma(r_1/2)\Gamma(r_2/2)(1+r_1w/r_2)^{(r_1+r_2)/2}}\, dw$$

Den. d.f. r_2	$P(F \le f)$	α	Numerator Degrees of Freedom, r_1								
			12	15	20	24	30	40	60	120	∞
10	0.95	0.05	2.91	2.85	2.77	2.74	2.70	2.66	2.62	2.58	2.54
	0.975	0.025	3.62	3.52	3.42	3.37	3.31	3.26	3.20	3.14	3.08
	0.99	0.01	4.71	4.56	4.41	4.33	4.25	4.17	4.08	4.00	3.91
12	0.95	0.05	2.69	2.62	2.54	2.51	2.47	2.43	2.38	2.34	2.30
	0.975	0.025	3.28	3.18	3.07	3.02	2.96	2.91	2.85	2.79	2.72
	0.99	0.01	4.16	4.01	3.86	3.78	3.70	3.62	3.54	3.45	3.36
15	0.95	0.05	2.48	2.40	2.33	2.29	2.25	2.20	2.16	2.11	2.07
	0.975	0.025	2.96	2.86	2.76	2.70	2.64	2.59	2.52	2.46	2.40
	0.99	0.01	3.67	3.52	3.37	3.29	3.21	3.13	3.05	2.96	2.87
20	0.95	0.05	2.28	2.20	2.12	2.08	2.04	1.99	1.95	1.90	1.84
	0.975	0.025	2.68	2.57	2.46	2.41	2.35	2.29	2.22	2.16	2.09
	0.99	0.01	3.23	3.09	2.94	2.86	2.78	2.69	2.61	2.52	2.42
24	0.95	0.05	2.18	2.11	2.03	1.98	1.94	1.89	1.84	1.79	1.73
	0.975	0.025	2.54	2.44	2.33	2.27	2.21	2.15	2.08	2.01	1.94
	0.99	0.01	3.03	2.89	2.74	2.66	2.58	2.49	2.40	2.31	2.21
30	0.95	0.05	2.09	2.01	1.93	1.89	1.84	1.79	1.74	1.68	1.62
	0.975	0.025	2.41	2.31	2.20	2.14	2.07	2.01	1.94	1.87	1.79
	0.99	0.01	2.84	2.70	2.55	2.47	2.39	2.30	2.21	2.11	2.01
40	0.95	0.05	2.00	1.92	1.84	1.79	1.74	1.69	1.64	1.58	1.51
	0.975	0.025	2.29	2.18	2.07	2.01	1.94	1.88	1.80	1.72	1.64
	0.99	0.01	2.66	2.52	2.37	2.29	2.20	2.11	2.02	1.92	1.80
60	0.95	0.05	1.92	1.84	1.75	1.70	1.65	1.59	1.53	1.47	1.39
	0.975	0.025	2.17	2.06	1.94	1.88	1.82	1.74	1.67	1.58	1.48
	0.99	0.01	2.50	2.35	2.20	2.12	2.03	1.94	1.84	1.73	1.60
120	0.95	0.05	1.83	1.75	1.66	1.61	1.55	1.50	1.43	1.35	1.25
	0.975	0.025	2.05	1.95	1.82	1.76	1.69	1.61	1.53	1.43	1.31
	0.99	0.01	2.34	2.19	2.03	1.95	1.86	1.76	1.66	1.53	1.38
∞	0.95	0.05	1.75	1.67	1.57	1.52	1.46	1.39	1.32	1.22	1.00
	0.975	0.025	1.94	1.83	1.71	1.64	1.57	1.48	1.39	1.27	1.00
	0.99	0.01	2.18	2.04	1.88	1.79	1.70	1.59	1.47	1.32	1.00

Table VIII: Kolmogorov-Smirnov Acceptance Limits

$$D_n = \sup_x[\,|F_n(x) - F_0(x)|\,]$$
$$\alpha = 1 - P(D_n \le d)$$

n	α			
	0.20	0.10	0.05	0.01
1	0.90	0.95	0.98	0.99
2	0.68	0.78	0.84	0.93
3	0.56	0.64	0.71	0.83
4	0.49	0.56	0.62	0.73
5	0.45	0.51	0.56	0.67
6	0.41	0.47	0.52	0.62
7	0.38	0.44	0.49	0.58
8	0.36	0.41	0.46	0.54
9	0.34	0.39	0.43	0.51
10	0.32	0.37	0.41	0.49
11	0.31	0.35	0.39	0.47
12	0.30	0.34	0.38	0.45
13	0.28	0.32	0.36	0.43
14	0.27	0.31	0.35	0.42
15	0.27	0.30	0.34	0.40
16	0.26	0.30	0.33	0.39
17	0.25	0.29	0.32	0.38
18	0.24	0.28	0.31	0.37
19	0.24	0.27	0.30	0.36
20	0.23	0.26	0.29	0.35
25	0.21	0.24	0.26	0.32
30	0.19	0.22	0.24	0.29
35	0.18	0.21	0.23	0.27
40	0.17	0.19	0.21	0.25
45	0.16	0.18	0.20	0.24
Large n	$\dfrac{1.07}{\sqrt{n}}$	$\dfrac{1.22}{\sqrt{n}}$	$\dfrac{1.36}{\sqrt{n}}$	$\dfrac{1.63}{\sqrt{n}}$

Table IX: Random Numbers on the Interval $(0, 1)$

3407	1440	6960	8675	5649	5793	1514
5044	9859	4658	7779	7986	0520	6697
0045	4999	4930	7408	7551	3124	0527
7536	1448	7843	4801	3147	3071	4749
7653	4231	1233	4409	0609	6448	2900
6157	1144	4779	0951	3757	9562	2354
6593	8668	4871	0946	3155	3941	9662
3187	7434	0315	4418	1569	1101	0043
4780	1071	6814	2733	7968	8541	1003
9414	6170	2581	1398	2429	4763	9192
1948	2360	7244	9682	5418	0596	4971
1843	0914	9705	7861	6861	7865	7293
4944	8903	0460	0188	0530	7790	9118
3882	3195	8287	3298	9532	9066	8225
6596	9009	2055	4081	4842	7852	5915
4793	2503	2906	6807	2028	1075	7175
2112	0232	5334	1443	7306	6418	9639
0743	1083	8071	9779	5973	1141	4393
8856	5352	3384	8891	9189	1680	3192
8027	4975	2346	5786	0693	5615	2047
3134	1688	4071	3766	0570	2142	3492
0633	9002	1305	2256	5956	9256	8979
8771	6069	1598	4275	6017	5946	8189
2672	1304	2186	8279	2430	4896	3698
3136	1916	8886	8617	9312	5070	2720
6490	7491	6562	5355	3794	3555	7510
8628	0501	4618	3364	6709	1289	0543
9270	0504	5018	7013	4423	2147	4089
5723	3807	4997	4699	2231	3193	8130
6228	8874	7271	2621	5746	6333	0345
7645	3379	8376	3030	0351	8290	3640
6842	5836	6203	6171	2698	4086	5469
6126	7792	9337	7773	7286	4236	1788
4956	0215	3468	8038	6144	9753	3131
1327	4736	6229	8965	7215	6458	3937
9188	1516	5279	5433	2254	5768	8718
0271	9627	9442	9217	4656	7603	8826
2127	1847	1331	5122	8332	8195	3322
2102	9201	2911	7318	7670	6079	2676
1706	6011	5280	5552	5180	4630	4747
7501	7635	2301	0889	6955	8113	4364
5705	1900	7144	8707	9065	8163	9846
3234	2599	3295	9160	8441	0085	9317
5641	4935	7971	8917	1978	5649	5799
2127	1868	3664	9376	1984	6315	8396

Table X: Divisors for the Confidence Interval for σ of Minimum Length

Let $X_1, X_2, \ldots, X_n$ be a random sample of size n from $N(\mu, \sigma^2)$; let $\bar{x} = (1/n) \sum_{i=1}^{n} x_i$ and $s^2 = (1/[n-1]) \sum_{i=1}^{n} (x_i - \bar{x})^2$. A $(100\gamma)\%$ confidence interval for σ which has minimum length is given by $[\sqrt{(n-1)s^2/b}, \sqrt{(n-1)s^2/a}]$. In the table, a is the upper number, b the lower number; $r = n - 1$, the number of degrees of freedom; and $\gamma = 1 - \alpha$, the confidence level.

r	$\gamma = 1 - \alpha$ 0.90	0.95	0.99	r	$\gamma = 1 - \alpha$ 0.90	0.95	0.99
2	0.206	0.101	0.020	16	8.774	7.604	5.649
	12.521	15.111	20.865		29.233	32.072	38.097
3	0.565	0.345	0.114	17	9.505	8.282	6.226
	13.153	15.589	20.973		30.480	33.362	39.469
4	1.020	0.692	0.294	18	10.242	8.969	6.814
	14.18	16.573	21.838		31.721	34.647	40.835
5	1.535	1.109	0.546	19	10.986	9.663	7.413
	15.350	17.743	22.985		32.959	35.927	42.195
6	2.093	1.578	0.837	20	11.736	10.365	8.021
	16.581	18.996	24.262		34.192	37.202	43.550
7	2.683	2.085	1.214	21	12.492	11.073	8.638
	17.839	20.286	25.602		35.420	38.472	44.899
8	3.298	2.623	1.611	22	13.253	11.788	9.264
	19.110	21.595	26.975		36.646	39.738	46.243
9	3.934	3.187	2.039	23	14.019	12.509	9.898
	20.385	22.912	28.364		37.867	41.000	47.586
10	4.588	3.773	2.496	24	14.790	13.236	10.539
	21.660	24.230	29.760		39.084	42.257	48.914
11	5.257	4.377	2.976	25	15.565	13.968	11.186
	22.933	25.547	31.158		40.299	43.510	50.243
12	5.940	4.997	3.477	26	16.344	14.704	11.841
	24.202	26.862	32.554		41.509	44.760	51.566
13	6.634	5.631	3.997	27	17.127	15.446	12.501
	25.467	28.172	33.947		42.717	46.006	52.886
14	7.338	6.278	4.533	28	17.914	16.192	13.168
	26.727	29.477	35.336		43.922	47.248	54.200
15	8.052	6.936	5.084	29	18.705	16.942	13.840
	27.982	30.777	36.719		45.123	48.487	55.511

Table XI: Distribution Function of the Correlation Coefficient R, $\rho = 0$

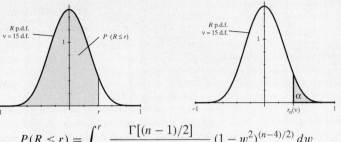

$$P(R \leq r) = \int_{-1}^{r} \frac{\Gamma[(n-1)/2]}{\Gamma(1/2)\Gamma[(n-2)/2]} (1-w^2)^{(n-4)/2)} \, dw$$

$\nu = n - 2$ degrees of freedom	$P(R \leq r)$			
	0.95	·0.975	0.99	0.995
	$r_{0.05}(\nu)$	$r_{0.25}(\nu)$	$r_{0.01}(\nu)$	$r_{0.005}(\nu)$
1	0.9877	0.9969	0.9995	0.9999
2	0.9000	0.9500	0.9800	0.9900
3	0.8053	0.8783	0.9343	0.9587
4	0.7292	0.8113	0.8822	0.9172
5	0.6694	0.7544	0.8329	0.8745
6	0.6215	0.7067	0.7887	0.8343
7	0.5822	0.6664	0.7497	0.7977
8	0.5493	0.6319	0.7154	0.7646
9	0.5214	0.6020	0.6850	0.7348
10	0.4972	0.5759	0.6581	0.7079
11	0.4761	0.5529	0.6338	0.6835
12	0.4575	0.5323	0.6120	0.6613
13	0.4408	0.5139	0.5922	0.6411
14	0.4258	0.4973	0.5742	0.6226
15	0.4123	0.4821	0.5577	0.6054
16	0.4000	0.4683	0.5425	0.5897
17	0.3887	0.4555	0.5285	0.5750
18	0.3783	0.4437	0.5154	0.5614
19	0.3687	0.4328	0.5033	0.5487
20	0.3597	0.4226	0.4920	0.5367
25	0.3232	0.3808	0.4450	0.4869
30	0.2959	0.3494	0.4092	0.4487
35	0.2746	0.3246	0.3809	0.4182
40	0.2572	0.3044	0.3578	0.3931
45	0.2428	0.2875	0.3383	0.3721
50	0.2306	0.2732	0.3218	0.3541
60	0.2108	0.2500	0.2948	0.3248
70	0.1954	0.2318	0.2736	0.3017
80	0.1829	0.2172	0.2565	0.2829
90	0.1725	0.2049	0.2422	0.2673
100	0.1638	0.1946	0.2300	0.2540

Table XII: Discrete Distributions

Probability Distribution and Parameter Values	Probability Mass Function	Moment-Generating Function	Mean $E(X)$	Variance $Var(X)$	Examples
Bernoulli $0 < p < 1$ $q = 1 - p$	$p^x q^{1-x}, \quad x = 0, 1$	$q + pe^t$	p	pq	Experiment with two possible outcomes, say success and failure, $p = P(\text{success})$
Binomial $n = 1, 2, 3, \ldots$ $0 < p < 1$	$\binom{n}{x} p^x q^{n-x}$, $x = 0, 1, \ldots, n$	$(q + pe^t)^n$	np	npq	Number of successes in a sequence of n Bernoulli trials, $p = P(\text{success})$
Geometric $0 < p < 1$ $q = 1 - p$	$q^{x-1} p$, $x = 1, 2, \ldots$	$\dfrac{pe^t}{1 - qe^t}$	$\dfrac{1}{p}$	$\dfrac{q}{p^2}$	The number of trials to obtain the first success in a sequence of Bernoulli trials
Hypergeometric $x \le n, x \le N_1$ $n - x \le N_2$ $N = N_1 + N_2$ $N_1 > 0, \; N_2 > 0$	$\dfrac{\binom{N_1}{x}\binom{N_2}{n-x}}{\binom{N}{n}}$		$n\left(\dfrac{N_1}{N}\right)$	$n\left(\dfrac{N_1}{N}\right)\left(\dfrac{N_2}{N}\right)\left(\dfrac{N-n}{N-1}\right)$	Selecting r objects at random without replacement from a set composed of two types of objects
Negative Binomial $r = 1, 2, 3, \ldots$ $0 < p < 1$	$\binom{x-1}{r-1} p^r q^{x-r}$, $x = r, r+1, \ldots$	$\dfrac{(pe^t)^r}{(1 - qe^t)^r}$	$\dfrac{r}{p}$	$\dfrac{rq}{p^2}$	The number of trials to obtain the rth success in a sequence of Bernoulli trials
Poisson $0 < \lambda$	$\dfrac{\lambda^x e^{-\lambda}}{x!}$, $x = 0, 1, \ldots$	$e^{\lambda(e^t - 1)}$	λ	λ	Number of events occurring in a unit interval, events are occurring randomly at a mean rate of λ per unit interval
Uniform $m > 0$	$\dfrac{1}{m}$, $x = 1, 2, \ldots, m$		$\dfrac{m+1}{2}$	$\dfrac{m^2 - 1}{12}$	Select an integer randomly from $1, 2, \ldots, m$

Table XIII: Continuous Distributions

Probability Distribution and Parameter Values	Probability Density Function	Moment-Generating Function	Mean $E(X)$	Variance $\mathrm{Var}(X)$	Examples
Beta $0 < \alpha$ $0 < \beta$	$\dfrac{\Gamma(\alpha+\beta)}{\Gamma(\alpha)\Gamma(\beta)}\, x^{\alpha-1}(1-x)^{\beta-1},$ $0 < x < 1$		$\dfrac{\alpha}{\alpha+\beta}$	$\dfrac{\alpha\beta}{(\alpha+\beta+1)(\alpha+\beta)^2}$	$X = X_1/(X_1+X_2),$ where X_1 and X_2 have independent gamma distributions with same θ
Chi-square $r = 1, 2, \ldots$	$\dfrac{x^{r/2-1}e^{-x/2}}{\Gamma(r/2)2^{r/2}},$ $0 < x < \infty$	$\dfrac{1}{(1-2t)^{r/2}},\ t < \dfrac{1}{2}$	r	$2r$	Gamma distribution, $\theta = 2$, $\alpha = r/2$; sum of squares of r independent $N(0,1)$ random variables
Exponential $0 < \theta$	$\dfrac{1}{\theta}e^{-x/\theta},\ 0 \le x < \infty$	$\dfrac{1}{1-\theta t},\ t < \dfrac{1}{\theta}$	θ	θ^2	Waiting time to first arrival when observing a Poisson process with a mean rate of arrivals equal to $\lambda = 1/\theta$
Gamma $0 < \alpha$ $0 < \theta$	$\dfrac{x^{\alpha-1}e^{-x/\theta}}{\Gamma(\alpha)\theta^\alpha},$ $0 < x < \infty$	$\dfrac{1}{(1-\theta t)^\alpha},\ t < \dfrac{1}{\theta}$	$\alpha\theta$	$\alpha\theta^2$	Waiting time to αth arrival when observing a Poisson process with a mean rate of arrivals equal to $1/\theta$
Normal $-\infty < \mu < \infty$ $0 < \sigma$	$\dfrac{e^{-(x-\mu)^2/2\sigma^2}}{\sigma\sqrt{2\pi}},$ $-\infty < x < \infty$	$e^{\mu t+\sigma^2 t^2/2}$	μ	σ^2	Errors in measurements; heights of children; breaking strengths
Uniform $-\infty < a < b < \infty$	$\dfrac{1}{b-a},\ a \le x \le b$	$\dfrac{e^{tb}-e^{ta}}{t(b-a)},\ t \ne 0$ $1,\qquad\qquad t = 0$	$\dfrac{a+b}{2}$	$\dfrac{(b-a)^2}{12}$	Select a point at random from the interval $[a, b]$

Table XIV: Tests and Confidence Intervals

Distribution	θ: The parameter of interest	W: The variable used to test $H_0: \theta = \theta_0$	Two-sided $1-\alpha$ Confidence Interval for θ	Comments
$N(\mu, \sigma^2)$ or n large σ^2 known	μ	$\dfrac{\overline{X} - \theta_0}{\sigma/\sqrt{n}}$	$\overline{x} \pm z_{\alpha/2}\dfrac{\sigma}{\sqrt{n}}$	W is $N(0,1)$; $P(W \geq z_{\alpha/2}) = \alpha/2$
$N(\mu, \sigma^2)$ σ^2 unknown	μ	$\dfrac{\overline{X} - \theta_0}{S/\sqrt{n}}$	$\overline{x} \pm t_{\alpha/2}(n-1)\dfrac{s}{\sqrt{n}}$	W has a t distribution with $n-1$ degrees of freedom; $P[W \geq t_{\alpha/2}(n-1)] = \alpha/2$
Any distribution with known variance, σ^2	μ	$\dfrac{\overline{X} - \theta_0}{\sigma/\sqrt{n}}$	$\overline{x} \pm z_{\alpha/2}\dfrac{\sigma}{\sqrt{n}}$	W has an approximate $N(0,1)$ distribution for n sufficiently large
$N(\mu_X, \sigma_X^2)$ $N(\mu_Y, \sigma_Y^2)$ σ_X^2, σ_Y^2 known	$\mu_X - \mu_Y$	$\dfrac{\overline{X} - \overline{Y} - \theta_0}{\sqrt{\dfrac{\sigma_X^2}{n} + \dfrac{\sigma_Y^2}{m}}}$	$\overline{x} - \overline{y} \pm z_{\alpha/2}\sqrt{\dfrac{\sigma_X^2}{n} + \dfrac{\sigma_Y^2}{m}}$	W is $N(0,1)$
$N(\mu_X, \sigma_X^2)$ $N(\mu_Y, \sigma_Y^2)$ σ_X^2, σ_Y^2 unknown	$\mu_X - \mu_Y$	$\dfrac{\overline{X} - \overline{Y} - \theta_0}{\sqrt{\dfrac{S_X^2}{n} + \dfrac{S_Y^2}{m}}}$	$\overline{x} - \overline{y} \pm z_{\alpha/2}\sqrt{\dfrac{s_x^2}{n} + \dfrac{s_y^2}{m}}$	W is approximately $N(0,1)$ if sample sizes are large
$N(\mu_X, \sigma_X^2)$ $N(\mu_Y, \sigma_Y^2)$ $\sigma_X^2 = \sigma_Y^2$, unknown	$\mu_X - \mu_Y$	$\dfrac{\overline{X} - \overline{Y} - \theta_0}{\sqrt{\dfrac{(n-1)S_X^2 + (m-1)S_Y^2}{n+m-2}\left(\dfrac{1}{n} + \dfrac{1}{m}\right)}}$	$\overline{x} - \overline{y} \pm t_{\alpha/2}(n+m-2)s_p\sqrt{\dfrac{1}{n} + \dfrac{1}{m}}$ $s_p = \sqrt{\dfrac{(n-1)s_x^2 + (m-1)s_y^2}{n+m-2}}$	W has a t distribution with $r = n+m-2$ degrees of freedom
$D = X - Y$ is $N(\mu_X - \mu_Y, \sigma_D^2)$ X and Y dependent	$\mu_X - \mu_Y$	$\dfrac{\overline{D} - \theta_0}{S_D/\sqrt{n}}$	$\overline{d} \pm t_{\alpha/2}(n-1)\dfrac{s_d}{\sqrt{n}}$	W has a t distribution with $n-1$ degrees of freedom

Table XIV: *continued*

Distribution	θ: The parameter of interest	W: The variable used to test $H_0: \theta = \theta_0$	Two-sided $1 - \alpha$ Confidence Interval for θ	Comments
$N(\mu, \sigma^2)$ μ unknown	σ^2	$\dfrac{(n-1)S^2}{\theta_0}$	$\dfrac{(n-1)s^2}{\chi^2_{\alpha/2}(n-1)}, \dfrac{(n-1)s^2}{\chi^2_{1-\alpha/2}(n-1)}$	W is $\chi^2(n-1)$, $P[W \le \chi^2_{1-\alpha/2}(n-1)] = \alpha/2$, $P[W \ge \chi^2_{\alpha/2}(n-1)] = \alpha/2$
$N(\mu, \sigma^2)$ μ unknown	σ	$\dfrac{(n-1)S^2}{\theta_0^2}$	$\sqrt{\dfrac{(n-1)s^2}{b}}, \sqrt{\dfrac{(n-1)s^2}{a}}$	W is $\chi^2(n-1)$. Select a and b from Table X in the Appendix for interval of minimum length
$N(\mu_X, \sigma_X^2)$ $N(\mu_Y, \sigma_Y^2)$ μ_X, μ_Y unknown	$\dfrac{\sigma_X^2}{\sigma_Y^2}$	$\dfrac{S_Y^2}{S_X^2}\,\theta_0$	$\dfrac{s_x^2/s_y^2}{F_{\alpha/2}(n-1, m-1)}, F_{\alpha/2}(m-1, n-1)\dfrac{s_x^2}{s_y^2}$	W has an F distribution with $m-1$ and $n-1$ degrees of freedom
$b(n, p)$	p	$\dfrac{\dfrac{Y}{n} - \theta_0}{\sqrt{\left(\dfrac{Y}{n}\right)\left(1 - \dfrac{Y}{n}\right)\big/ n}}$	$\dfrac{y}{n} \pm z_{\alpha/2}\sqrt{\left(\dfrac{y}{n}\right)\left(1 - \dfrac{y}{n}\right)\big/ n}$	W is approximately $N(0, 1)$ for n sufficiently large
$b(n_1, p_1)$ $b(n_2, p_2)$	$p_1 - p_2$	$\dfrac{\dfrac{Y_1}{n_1} - \dfrac{Y_2}{n_2} - \theta_0}{\sqrt{\left(\dfrac{Y_1+Y_2}{n_1+n_2}\right)\left(1 - \dfrac{Y_1+Y_2}{n_1+n_2}\right)\left(\dfrac{1}{n_1} + \dfrac{1}{n_2}\right)}}$	$\dfrac{y_1}{n_1} - \dfrac{y_2}{n_2} \pm$ $z_{\alpha/2}\sqrt{\dfrac{y_1}{n_1}\left(1 - \dfrac{y_1}{n_1}\right)\big/ n_1 + \dfrac{y_2}{n_2}\left(1 - \dfrac{y_2}{n_2}\right)\big/ n_2}$	W is approximately $N(0, 1)$ when n_1 and n_2 are sufficiently large

APPENDIX D

ANSWERS TO
ODD-NUMBERED
EXERCISES

CHAPTER 1

1.1-1 (a) $S = \{1, 2, 3, \ldots, 36\}$;

(b) $S = \{w: \ 19 \le w \le 23\}$, where w is the weight in grams;

(c) $S = \{HHH, \ HHT, \ HTH, \ THH, \ HTT, \ THT, \ TTH, \ TTT\}$.

1.1-3 (a) Litter size: 1 2 3 4
Frequency: 5 10 4 1

(d) Yes, 2.

1.1-5 (a) Clutch size: 4 5 6 7 8 9 10
Frequency: 4 5 6 4 6 7 1

(c) No.

1.1-7 (a) Number: 0 1 2 3 4 5
Frequency: 3 12 33 37 13 2

1.1-11 (a) 0.261, 0.196;

(b) 0.327, 0.315;

(c) 0.270, 0.173;

(d) Example of Simpson's paradox.

1.1-13 (a) $0.889 > 0.856, \ \ 0.948 > 0.921, \ \ 0.914 > 0.855, \ \ 0.831 > 0.713,$
$0.858 > 0.767$;

(b) $0.867 < 0.891$;

(c) Example of Simpson's paradox.

1.2-1 (a) 12/52; **(b)** 2/52; **(c)** 16/52; **(d)** 1; **(e)** 0.

1.2-3 27/64.

1.2-5 (a) **(i)** 4/8, **(ii)** 2/8, **(iii)** 3/8, **(iv)** 4/8;

(b) **(i)** 1/8, **(ii)** 0, **(iii)** 2/8;

(c) **(i)** 5/8, **(ii)** 5/8, **(iii)** 5/8.

1.2-7   0.6.

1.2-9 **(a)** $3(1/3) - 3(1/3)^2 + (1/3)^3$;

(b) $P(A_1 \cup A_2 \cup A_3) = 1 - [1 - 3(1/3) + 3(1/3)^2 - (1/3)^3] = 1 - (1 - 1/3)^3$.

1.2-11 **(a)** $1 - (1/2)^{10}$;   **(b)** $1 - (1/2)^{20}$;   **(c)** $1 - (1/2)^{20}$;

(d) $1 - (1/2)^{10}$;   **(e)** $(1/2)^{10} - (1/2)^{20}$;   **(f)** $(1/2)^{20}$.

1.2-13 **(a)** $S = \{00, 0, 1, 2, 3, \ldots, 36\}$;

(b) $P(A) = 2/38$;

(c) $P(B) = 4/38$;

(d) $P(D) = 18/38$.

1.2-15   2/3.

1.3-1   4096.

1.3-3 **(a)** 6,760,000; **(b)** 17,576,000.

1.3-5 **(a)** 24; **(b)** 256.

1.3-7 **(a)** 0.0024; **(b)** 0.0012; **(c)** 0.0006; **(d)** 0.0004.

1.3-9 **(a)** 60; **(b)** 125.

1.3-11 **(a)** 2; **(b)** 8; **(c)** 20; **(d)** 40.

1.3-13 **(a)** 96; **(b)** 393,216; **(c)** 2,973,696.

1.3-15 **(a)** 362,880; **(b)** 84; **(c)** 512.

1.3-17 **(a)** 0.00024; **(b)** 0.00144; **(c)** 0.02113; **(d)** 0.04754; **(e)** 0.42257.

1.3-19   14,702,688.

1.3-21 **(a)** 560; **(b)** 3,688.

1.4-1 **(a)** 5000/1,000,000; **(b)** 78,515/1,000,000;

(c) 73,630/995,000; **(d)** 4,885/78,515.

1.4-3 **(a)** 5/35; **(b)** 26/35; **(c)** 5/19; **(d)** 9/23; **(e)** left.

1.4-5 **(a)** 3/10; **(b)** 3/5; **(c)** 3/7.

1.4-7 **(a)** $S = \{(R, R), (R, W), (W, R), (W, W)\}$; **(b)** 1/3.

1.4-9   1/5.

1.4-11 **(a)** 1/220; **(b)** 3/110; **(c)** 9/55.

1.4-13 **(b)** 8/36; **(c)** 5/11;

(e) $8/36 + 2[(5/36)(5/11) + (4/36)(4/10) + (3/36)(3/9)] = 0.49293$.

1.4-15 **(a)** 365^r; **(b)** $_{365}P_r$; **(c)** $1 - \,_{365}P_r/365^r$; **(d)** 23.

1.4-17 **(a)** 49/153; **(b)** 4/7.

1.4-19 **(f)** $1 - 1/e$.

1.5-1 (a) 0.14; **(b)** 0.76; **(c)** 0.86.

1.5-3 (a) 1/6; **(b)** 1/12; **(c)** 1/4; **(d)** 1/4; **(e)** 1/2.

1.5-5 Yes; $0.9 = 0.8 + 0.5 - (0.8)(0.5)$.

1.5-7 (a) 0.29; **(b)** 0.44.

1.5-9 (a) 0.36; **(b)** 0.49; **(c)** 0.01.

1.5-11 (a) No.

    (b) Only if $P(A) = 0$ or $P(B) = 1$.

1.5-13 $(2/3)^3(1/3)^2$; $(2/3)^3(1/3)^2$.

1.5-15 (a) $10(1/2)^6$;

    (b) $[(120)(45)/15{,}504][7/15]$;

    (c) neither model is very good.

1.5-17 (a) $1 - (11/12)^{12}$; **(b)** $1 - (11/12)^{11}$.

1.5-19 (b) $1 - 1/e$.

1.6-1 (a) 21/32; **(b)** 16/21.

1.6-3 (a) 11/24; **(b)** 2/11.

1.6-5 $60/95 = 0.632$.

1.6-7 $50/95 = 0.526$.

1.6-9 (a) $495/30{,}480 = 0.016$; **(b)** $29{,}985/30{,}480 = 0.984$.

CHAPTER 2

2.1-3 (a) 10; **(b)** 1/55; **(c)** 3; **(d)** 1/30; **(e)** $n(n + 1)/2$.

2.1-5 (b)

x	Frequency	Relative Frequency	$f(x)$
1	38	0.38	0.40
2	27	0.27	0.30
3	21	0.21	0.20
4	14	0.14	0.10

2.1-7 (a) $f(x) = \dfrac{13 - 2x}{36}$, $x = 1, 2, 3, 4, 5, 6$;

    (c) $g(0) = \dfrac{6}{36}, g(y) = \dfrac{12 - 2y}{36}$, $y = 1, 2, 3, 4, 5$.

2.1-11 0.416.

2.1-13 2/5.

2.1-15 (a) 0; **(b)** 0.228; **(c)** 0.413.

2.2-1 \$360.

2.2-3 (a) 3; **(b)** 7; **(c)** 4/3; **(d)** 7/3; **(e)** $(2n + 1)/3$.

2.2-5 $-91/300$ or -30.33 cents.

2.2-7 $E(X) = -17/216 = -0.0787$.

2.2-9 $E(X) = \sum_{x=1}^{n} \frac{1}{x}\left(\frac{1}{n}\right); \dfrac{137}{300} = 0.4567;$

$\dfrac{1}{100}\left[\dfrac{\ln(101) + 1 + \ln(100)}{2}\right] = 0.05110 \approx 0.05187.$

2.2-11 (a) $-1/19$; **(b)** $-1/37$.

2.2-13 $-\$0.01414$.

2.3-1 (a) 15; 50; **(b)** 5; 0; **(c)** 5/3; 5/9.

2.3-3 (a) 16; **(b)** 6; **(c)** 16.

2.3-5 (a) (ii) 127/64; **(iii)** $\sqrt{7359}/64$; **(iv)** 1.6635;
(b) (ii) 385/64; **(iii)** $\sqrt{7359}/64$; **(iv)** -1.6635.

2.3-7 (a) $25/8 = 3.125$, $55/64 = 0.8594$, $\sqrt{55}/8 = 0.9270$;
(b) $79/25 = 3.16$, $662/825 = 0.8024$, $\sqrt{21{,}846}/165 = 0.8958$.

2.3-11 (a) 41/20; 259/380;
(b) 938/117; 8,657/3,393;
(c) 226/33; 215/66;
(d) 251/100; 10,499/9,900.

2.3-13   7.

2.3-15   2.

2.3-17 (a) 394/7; **(b)** 5,452/97.

2.3-19 $\$1809.80$.

2.4-1 $f(x) = (7/18)^x(11/18)^{1-x}$, $x = 0, 1$; $\mu = 7/18$; $\sigma^2 = 77/324$.

2.4-3 (a) $(1/5)^2(4/5)^4 = 0.0164$;
(b) $\dfrac{6!}{2!4!}(1/5)^2(4/5)^4 = 0.2458$.

2.4-5 (a) 0.9825; **(b)** 0.0060; **(c)** 0.0115; **(d)** $17.5, 5.25, \sqrt{5.25}$.

2.4-7 (a) $b(2000, \pi/4)$; **(b)** 1570.796, 337.096, 18.360; **(c)** π;
(f) $V_n = \pi^{n/2}/\Gamma(n/2 + 1)$ is the volume of a ball of radius one in n-space.

2.4-9 (a) $b(9, 0.10)$; **(b)** 0.9, 0.81, 0.9; **(c) (i)** 0.1722, **(ii)** 0.2252.

2.4-11 (a) 0.8725; **(b)** 0.5841; **(c)** 0.1597.

2.4-13   0.1268.

2.4-15 (a) $\mu = 7.2, \sigma^2 = 1.44, \sigma = 1.2$;
(b) $\bar{x} = 7.32, s^2 = 1.4117, s = 1.1882$.

2.4-17 (a) 0.6513; **(b)** 0.7941.

2.4-19

x	$b(8, 0.2)$	$N_1 = 8, N_2 = 32$	$N_1 = 16, N_2 = 64$
0	0.1678	0.1368	0.1527
1	0.3355	0.3501	0.3429
2	0.2936	0.3299	0.3104
3	0.1468	0.1466	0.1473
4	0.0459	0.0327	0.0399
5	0.0092	0.0036	0.0063
6	0.0011	0.0002	0.0006
7	0.0001	0.0000	0.0000
8	0.0000	0.0000	0.0000

2.4-21   0.9841.

2.4-23   Take longer route.

2.5-1 **(a)** $b(1, 2/3)$; **(b)** $b(12, 0.75)$.

2.5-3 $\mu = 2, \sigma^2 = 4/5$,
$$f(x) = \begin{cases} 2/5, & x = 1, \\ 1/5, & x = 2, \\ 2/5, & x = 3. \end{cases}$$

2.5-5 $(4/5)^3(1/5)$.

2.5-7 **(a)** 0.4604; **(b)** 0.5580; **(c)** 0.0184.

2.5-9 **(a)** 0.0670; **(b)** 0.4823; **(c)** 0.1319.

2.5-11 **(a)** $f(x) = (1/2)^{x-1}, x = 2, 3, 4, \ldots$;
   (b) 3, 2, 1.414;
   (c) **(i)** 3/4, **(ii)** 1/8, **(iii)** 1/4.

2.5-13 **(a)** $f(x) = (x - 1)/2^x, \ x = 2, 3, \ldots$;
   (b) 4, 4, 2;
   (c) **(i)** 1/2, **(ii)** 5/16, **(iii)** 1/4.

2.5-15   25/3.

2.5-17 **(a)** 29.29; **(b)** about 8 years; **(c)** 49.77.

2.5-21 **(a)** $\mu = 1, \sigma^2 = 1$; **(b)** 19/30.

2.5-23   0.395.

2.6-1 **(a)** 0.693; **(b)** 0.762; **(c)** 0.433.

2.6-3   0.540.

2.6-5   0.558.

2.6-7   0.947.

2.6-9 **(a)** 0.217; **(b)** 0.971.

2.6-11 **(a)** $\bar{x} = 7.94, s^2 = 7.73$, yes.

2.6-13 **(a)** $\bar{x} = 4/3 = 1.333, s^2 = 88/69 = 1.275$, yes; **(c)** yes.

2.6-15 (b) $\bar{x} = 5.597$, $s^2 = 3.490$, possibly.

2.6-17 (a) 2.967; **(b)** 2.792; **(c)** 1.671.

2.6-19   If $n = 80$, $Ac = 3$, $Oc(0.02) \approx 0.92$, $Oc(0.08) \approx 0.12$.

2.6-21 (a) 0.564 using binomial, 0.560 using Poisson approximation.

(b) $ 598.56 using Poisson approximation, $ 614.14 using binomial.

2.6-23 (a) 1.941; **(b)** $n = 6$.

CHAPTER 3

3.1-1 (a) $\bar{x} = 1.1$;

(b) $s^2 = 0.0350$;

(c) $s = 0.1871$.

3.1-3 (a) Frequencies: [7, 3, 4, 5, 5, 4];

(c) $\bar{x} = 67.32$;

(d) $s = 35.53$.

3.1-5 (a) Frequencies: [3, 15, 8, 1, 0, 5, 8, 2];

(b) $\bar{x} = 17.30$; $s_x = 2.34$.

3.1-7 (a) Frequencies: [1, 1, 3, 4, 20, 23, 16, 10, 3, 0, 1];

(d) $\bar{x} = 4.01$;

(e) $s_x = 0.81$.

3.1-9 (a) $\bar{x} = 112.12$; $s = 231.3576$;

(d) Half of the observations are less than 48.

3.1-11 (a) Frequencies: [6, 18, 5, 21, 8, 3, 5];

(c) $\bar{x} = 407.48$; $s_x^2 = 1454.3767$; $s_x = 38.14$;

(d) $\bar{u} = 409.55$; $s_u^2 = 1484.1594$; $s_u = 38.52$.

3.1-13 (a) $\bar{x} = 56.29$; $s^2 = 56.21$;

(b) Frequencies: [1, 4, 6, 8, 19, 21, 20, 11, 5, 3].

3.1-15 (a) With class boundaries 90.5, 108.5, 126.5, ..., 306.5:

Frequencies: [8, 11, 4, 1, 0, 0, 0, 1, 2, 12, 12, 3];

(b) $\bar{x} = 201$;

(c) With class boundaries 47.5, 52.5, ..., 107.5:

Frequencies: [4, 4, 9, 4, 4, 0, 3, 9, 7, 5, 4, 1];

(d) $\bar{x} = 76.35$.

3.1-17 **(a)**

Stems	Leaves	Frequency	Depths
127	8	1	1
128	8	1	2
129	5 8 9	3	5
130	8	1	6
131	2 3 4 4 5 5 7	7	(7)
132	2 7 7 8	4	7
133	7 9	2	3
134	8	1	1

(Multiply numbers by 10^{-1}.)

(b) 131.3, 7.0, 2.575, 131.45, 131.47, 3.034.

3.1-19

Stems	Leaves	Frequency	Depths
$3f$	40 44 47 50	4	4
$3s$	60 61 63 64 64 65 66 68 70 70 70 71 74 74 75	15	19
$3\bullet$	80 80 80 80 80 90 90	7	26
$4*$	00 00 00 10 10 10 10 10 10 14 15 17	12	(12)
$4t$	20 20 20 20 20 28 30 30 30 30 30 30 32 34 38 38	17	28
$4f$	40 40 40	3	11
$4s$	60 62 64	3	8
$4\bullet$	80 80 80 82	4	5
$5*$	00	1	1

3.2-1 **(b)** $F(x) = \begin{cases} 0, & x < 0, \\ x(2-x), & 0 \le x < 1, \\ 1, & 1 \le x. \end{cases}$

(c) **(i)** 3/4, **(ii)** 1/2, **(iii)** 0, **(iv)** 1/16.

3.2-3 **(a)** **(i)** 3; **(ii)** $F(x) = x^4$, $0 \le x \le 1$;

(b) **(i)** 3/16; **(ii)** $F(x) = (1/8)x^{3/2}$, $0 \le x \le 4$;

(c) **(i)** 1/4; **(ii)** $F(x) = x^{1/4}$, $0 \le x \le 1$.

3.2-5 **(a)** 4/5, 2/75, $\sqrt{6}/15$;

(b) 12/5, 192/175; $8\sqrt{21}/35$;

(c) 1/5, 16/225, 4/15.

3.2-7 **(a)** $\pi/2$, $(\pi^2 - 8)/4$;

(c) $F(x) = (1 - \cos x)/2$, $0 \le x \le \pi$.

3.2-9 **(a)** d = 2; **(b)** $E(Y) = 2$; **(c)** $E(Y^2)$ is unbounded.

3.2-11 **(a)** $\mu = 0$, $\sigma^2 = 3/5$;

(b) $\mu = 0, \sigma^2 = 1/3$;

(c) $\mu = 0, \sigma^2 = 1/6$.

3.2-13 (a) $\mu = 0$; **(b)** $\sigma^2 = 1$.

3.2-15 $\qquad f(x) = \dfrac{e^{-x}}{(1 + e^{-x})^2} = \dfrac{e^{-x}}{(1 + e^{-x})^2} \dfrac{e^{2x}}{e^{2x}} = \dfrac{e^x}{(e^x + 1)^2} = f(-x).$

3.2-17 (a) $\pi_{0.5} = 0$; **(b)** $\pi_{0.25} = -0.50$; **(c)** $\pi_{0.90} = 0.80$.

3.2-19 (a) $F_n(x) = nx, \ 0 \le x \le 1/n$.

3.2-21 $\qquad \theta = 2.33$.

3.2-23 $\qquad \$ 740.74$.

3.2-25 (a) $1/e = 0.368$;

(b) $1/e^{2.375} = 0.093$.

3.3-3 (a) $f(x) = 1/10, \ 0 < x < 10$; **(b)** 0.2; **(c)** 0.6;

(d) $\mu = 5$; (e) $\sigma^2 = 25/3$.

3.3-5 (a) $G(w) = (w - a)/(b - a), \ a \le w \le b$;

(b) $U(a, b)$.

3.3-7 (b) $\bar{x} = 20.486, s^2 = 20.344$;

(c) $0.472 \approx 0.467$;

(d) $0.472 \approx 0.462 = 12/26$.

3.3-9 (a) $f(x) = (1/3)e^{-x/3}, \ 0 < x < \infty$; 3; 9;

(b) $f(x) = 3e^{-3x}, \ 0 < x < \infty$; 1/3; 1/9.

3.3-11 $\qquad P(X > x + y \mid X > x) = \dfrac{P(X > x + y)}{P(X > x)} = \dfrac{e^{-(x+y)/\theta}}{e^{-x/\theta}} = P(X > y).$

3.3-13 (a) $\lambda = 0.025$; **(b)** exponential distribution, $\theta = 40$;

(c) $\mu = 40, \sigma^2 = 1600$; **(d) (i)** $1 - e^{-1/2}$, **(ii)** e^{-1}, **(iii)** e^{-1}.

3.3-15 (a) 10.524; 9.320; yes.

3.3-17 (a) $F(x) = 1 - e^{-(x-\delta)/\theta}, \ \delta \le x < \infty$;

(b) $\theta + \delta$; θ^2.

3.3-19 (a) $-\theta \ln 0.75 = 0.2877 \, \theta$; **(b)** $(\ln 0.75 + 1)\theta = 0.7123 \, \theta$;

(c) $-\theta \ln 0.25 = 1.3863 \, \theta$; **(d)** $(-\ln 0.25 - 1) \theta = 0.3863 \, \theta$.

3.4-1 (a) $f(x) = \dfrac{1}{\Gamma(10)(3/2)^{10}} x^9 e^{-2x/3}, \ 0 \le x < \infty$;

(b) $M(t) = (1 - 3t/2)^{-10}, \ t < 2/3; \ \mu = 15; \ \sigma^2 = 45/2$.

3.4-5 $\qquad f(x) = \dfrac{1}{\Gamma(20)7^{20}} x^{19} e^{-x/7}, \ 0 \le x < \infty; \ \mu = 140; \ \sigma^2 = 980$.

3.4-7 (a) $\mu = 240/7 = 34.286, \bar{x} = 32.636$;

(b) $\sigma^2 = 28{,}800/49 = 587.755, s^2 = 548.338$;

(c) $0.605 \approx 0.591 = 13/22$.

3.4-9 (a) 0.025; **(b)** 0.05; **(c)** 0.94; **(d)** 8.672; **(e)** 30.19.

3.4-11 (a) 0.80; **(b)** $a = 11.69, b = 38.08$;

   (c) $\mu = 23, \sigma^2 = 46$; **(d)** 35.17, 13.09.

3.4-13 (a) $r - 2$; **(b)** $x = r - 2 \pm \sqrt{2r - 4}, r \geq 4$.

3.4-15   0.9444.

3.5-1 $g(y) = 2y, \ 0 < y < 1$.

3.5-3 $g(y) = (1/8)y^5 e^{-y^2/2}, \ 0 < y < \infty$.

3.5-7 (a) $F(r) = (x - 0.03)/0.04, \ 0.03 < r < 0.07; \ f(r) = 1/0.04, \ 0.03 < r < 0.07$;

   (b) The interest for each of n equal parts is R/n. The amount at the end of the year is $50,000(1 + R/n)^n$; the limit as $n \to \infty$ is $50,000 \, e^R$.

3.5-9 (b) Exponential;

   (c) $\exp(-e^{-2}) = 0.873$.

3.5-11 $M(t) = (1 - 2t)^{-1}, \ t < 1/2$; exponential distribution with $\theta = 2$.

3.5-13 (a) $\dfrac{1}{2} - \dfrac{\arctan 1}{\pi} = 0.25$;

   (b) $\dfrac{1}{2} - \dfrac{\arctan 5}{\pi} = 0.0628$;

   (c) $\dfrac{1}{2} - \dfrac{\arctan 10}{\pi} = 0.0317$.

3.6-1 (a) $e^{-(5/10)^2} = e^{-1/4} = 0.7788$.

3.6-3   Weibull with parameters α and $\beta/3^{1/\alpha}$.

3.6-5 (a) 0.5; **(b)** 0; **(c)** 0.25; **(d)** 0.75; **(e)** 0.625; **(f)** 0.75.

3.6-7 (b) $\mu = 31/24, \sigma^2 = 167/567$;

   (c) 15/64; 1/4; 0; 11/16.

3.6-9 (a)
$$F(x) = \begin{cases} 0, & x < 0, \\ x/2, & 0 \leq x < 1, \\ 1/2, & 1 \leq x < 2, \\ 4/6, & 2 \leq x < 4, \\ 5/6, & 4 \leq x < 6, \\ 1, & 6 \leq x. \end{cases}$$

   (b) $ 2.25.

3.6-11 $3 + 5 e^{-3/5} = 5.744$.

3.6-13   226.21, 1,486.92.

CHAPTER 4

4.1-1 (a) $f_1(x) = (2x + 5)/16$, $x = 1, 2$;

(b) $f_2(y) = (2y + 3)/32$, $y = 1, 2, 3, 4$;

(c) 3/32; **(d)** 9/32; **(e)** 3/16; **(f)** 1/4; **(g)** dependent.

4.1-3 (b) $f(x, y) = 1/16$, $x = 1, 2, 3, 4$; $y = x + 1, x + 2, x + 3, x + 4$;

(c) $f_1(x) = 1/4$, $x = 1, 2, 3, 4$;

(d) $f_2(y) = (4 - |y - 5|)/16$, $y = 2, 3, 4, 5, 6, 7, 8$;

(e) dependent because the space is not rectangular.

4.1-5   11/30.

4.1-7 (a) $f(x, y) = \dfrac{15!}{x!\, y!\, (15 - x - y)!} \left(\dfrac{6}{10}\right)^x \left(\dfrac{3}{10}\right)^y \left(\dfrac{1}{10}\right)^{15-x-y}$, $0 \le x + y \le 15$;

(b) no, because the space is not rectangular;

(c) 0.0735;

(d) X is $b(15, 0.6)$;

(e) 0.9095.

4.1-9 $f_1(x) = 2e^{-2x}$, $0 < x < \infty$; $f_2(y) = 2e^{-y}(1 - e^{-y})$, $0 < y < \infty$; no.

4.1-11 $f_1(x) = x/2$, $0 \le x \le 2$; $f_2(y) = 3y^2/8$, $0 \le y \le 2$; yes.

4.2-1 $\mu_X = 25/16$; $\mu_Y = 45/16$; $\sigma_X^2 = 63/256$; $\sigma_Y^2 = 295/256$;
$\mathrm{Cov}(X, Y) = -5/256$; $\rho = -\sqrt{2{,}065}/1{,}239 = -0.0367$; dependent.

4.2-3 (a) $f(x, y) = 1/16$, $x = 1, 2, 3, 4$; $y = x + 1, x + 2, x + 3, x + 4$;
$f_1(x) = 1/4$, $x = 1, 2, 3, 4$; $f_2(y) = (4 - |y - 5|)/16$, $y = 2, 3, \ldots, 8$;

(b) $\mu_X = 5/2$; $\mu_Y = 5$; $\sigma_X^2 = 5/4$; $\sigma_Y^2 = 5/2$; $\mathrm{Cov}(X, Y) = 5/4$; $\rho = \sqrt{2}/2$;

(c) $y = x + 5/2$.

4.2-5 $a = \mu_Y - \mu_X b$, $b = \mathrm{Cov}(X, Y)/\sigma_X^2$.

4.2-7 (a) No; **(b)** $\mathrm{Cov}(X, Y) = 0$, $\rho = 0$.

4.2-9 (a) $f_1(1) = 0.15$, $f_1(2) = 0.25$, $f_1(3) = 0.45$, $f_1(4) = 0.15$;
$f_2(1) = 0.35$, $f_2(2) = 0.65$;
$\mu_X = 2.60$; $\mu_Y = 1.65$; $\sigma_X^2 = 0.8400$; $\sigma_Y^2 = 0.2275$;

(b) $\mathrm{Cov}(X, Y) = -0.0900$; $\rho = -0.2059$;

(c) $E(C) = 32.45$.

4.2-11 (a) $f_1(x) = x + 1/2$, $0 < x < 1$, $f_2(y) = y + 1/2$, $0 < y < 1$;

(b) $\mu_X = 7/12$; $\mu_Y = 7/12$; $\sigma_X^2 = 11/144$; $\sigma_Y^2 = 11/144$;
$\mathrm{Cov}(X, Y) = -1/144$; $\rho = -1/11$;

(c) $y = 7/11 - x/11$.

4.3-1 (d) 9/14, 7/18, 5/9; **(e)** 20/7, 55/49.

4.3-3 (a) $f(x, y) = \dfrac{50!}{x!\, y!\, (50 - x - y)!}\, (0.02)^x (0.90)^y (0.08)^{50-x-y},\ 0 \le x + y \le 50$;

(b) Y is $b(50, 0.90)$;

(c) $b(47, 0.90/0.98)$;

(d) $2115/49$; **(e)** $\rho = -3/7$.

4.3-5 (a) $E(Y \mid x) = 2(2/3) - (2/3)x,\ x = 1, 2$; **(b)** yes.

4.3-7 $E(Y \mid x) = x + 5/2,\ x = 1, 2, 3, 4$; it is linear; yes.

4.3-9 (a) $f_1(x) = 1/8,\ x = 0, 1, \ldots, 7$;

(b) $h(y \mid x) = 1/3,\ y = x, x + 1, x + 2$, for $x = 0, 1, \ldots, 7$;

(c) $E(Y \mid x) = x + 1,\ x + 0, 1, \ldots, 7$;

(d) $\sigma_Y^2 = 2/3$;

(e)
$$f_2(y) = \begin{cases} 1/24, & y = 0, 9, \\ 2/24, & y = 1, 8, \\ 3/24, & y = 2, 3, 4, 5, 6, 7. \end{cases}$$

4.3-11 $E(Y) = x$ and $E(X) = .700$; thus $\$\,700$.

4.3-13 (b) $f_1(x) = 1/10,\ 0 \le x \le 10$;

(c) $h(y \mid x) = 1/4,\ 10 - x \le y \le 14 - x$ for $0 \le x \le 10$;

(d) $E(Y \mid x) = 12 - x$.

4.3-15 (a) $f(x, y) = 1/(2x^2),\ 0 < x < 2,\ 0 < y < x^2$;

(b) $f_2(y) = (2 - \sqrt{y})/(4\sqrt{y}),\ 0 < y < 4$;

(c) $E(X \mid y) = [2\sqrt{y}\, \ln(2/\sqrt{y})]/[2 - \sqrt{y}]$;

(d) $E(Y \mid x) = x^2/2$.

4.3-17 (a) $h(y \mid x) = e^{-x},\ 0 < y < e^x$ for $0 < x < 1$;

(b) $E(Y \mid x) = e^x/2,\ 0 < x < 1$;

(c) $f(x, y) = e^{-x},\ 0 < x < 1,\ 0 < y < e^x$;

(d) $f_2(y) = \begin{cases} 1 - 1/e, & 0 < y < 1, \\ 1/y - 1/e, & 1 \le y < e. \end{cases}$

4.4-1 $g(y_1, y_2) = (1/4)e^{-y_2/2},\ 0 < y_1 < y_2 < \infty$;
$g_1(y_1) = (1/2)e^{-y_1/2},\ 0 < y_1 < \infty$;
$g_2(y_2) = (y_2/4)e^{-y_2/2},\ 0 < y_2 < \infty$; no.

4.4-3 $\mu = \dfrac{r_2}{r_2 - 2},\ r_2 > 2$;

$\sigma^2 = \dfrac{2r_2^2(r_1 + r_2 - 2)}{r_1(r_2 - 2)^2(r_2 - 4)},\ r_2 > 4$.

4.4-9 840.

4.4-11 (a) 0.1792; **(b)** 0.1792.

4.4-13 (a) $G(y_1, y_2) = \displaystyle\int_0^{y_1} \int_u^{y_2} 2(1/1000^2) \exp[-(u+v)/1000]\, dv\, du$

$$= 2\exp[-(y_1 + y_2)/1000] - \exp[-y_1/500]$$
$$- 2\exp[-y_2/1000] + 1, \quad 0 < y_1 < y_2 < \infty;$$

(b) $2e^{-6/5} - e^{-12/5} \approx 0.5117.$

4.4-15   Each has mean 1/3 (8 hours). The respective standard deviations are 0.0943, 0.0673, and 0.0552. In case (3), the probability that a person sleeps between 7 and 9 hours is 0.546 and between 6 and 10 hours is 0.869. If we let $\alpha = 48$ and $\beta = 96$, then $\sigma = 0.0391$ and the probability a person sleeps between 7 and 9 hours is 0.7116.

4.5-1 (a) 0.0182; **(b)** 0.0337.

4.5-3 (a) 36/125; **(b)** 2/7.

4.5-5 (a) $\mu = 10, \sigma^2 = 20$;

(b) $(1 - 2t)^{-10/2}, \; t < 1/2$;

(c) $\mu = 10, \sigma^2 = 20$, equal.

4.5-7   2/5.

4.5-9 (a) 729/4096; **(b)** $\mu = 3/2$; $\sigma^2 = 3/40$.

4.5-11 (a) 0.0035; **(b)** 8; **(c)** $\mu = 4, \sigma^2 = 7/3$.

4.5-13 (c) Using *Maple*, $\mu = 13{,}315{,}424/3{,}011{,}805 = 4.4211$.

4.5-15 $1 - e^{-3/100} \approx 0.03$.

4.5-17 $G(y) = 1 - ([y/500]\,e^{-y/500} + e^{-y/500})^3, \; 0 < y < \infty$;

$P(Y < 300) = 1 - ([8/5]e^{-3/5})^3 \approx 0.323.$

4.5-19 $G(y) = y^{13}, \; 0 < y < 1$; $\; g(y) = 13y^{12}, \; 0 < y < 1$.

4.6-1

$$g(y) = \begin{cases} 1/36, & y = 2, \\ 4/36, & y = 3, \\ 10/36, & y = 4, \\ 12/36, & y = 5, \\ 9/36, & y = 6; \end{cases}$$

$\mu = 14/3, \sigma^2 = 10/9.$

4.6-3 (a)

$$g(y) = \begin{cases} 1/64, & y = 3, 12, \\ 3/64, & y = 4, 11, \\ 6/64, & y = 5, 10, \\ 10/64, & y = 6, 9, \\ 12/64, & y = 7, 8. \end{cases}$$

4.6-5 $\mu_Y = 1, \sigma_Y^2 = 61.$

4.6-7 (a) $M(t) = e^{7(e^t - 1)}$; **(b)** Poisson, $\lambda = 7$; **(c)** 0.800.

4.6-9 (a) $M(t) = (pe^t)^5/(1 - qe^t)^5, \; t < -\ln q$;

(b) negative binomial with parameters $r = 5$ and p.

4.6-11 (a) $M(t) = 1/(1 - 5t)^{21}$, $t < 1/5$;

   (b) gamma distribution, $\alpha = 21$, $\theta = 5$.

4.6-13 (a) $(1/24)(e^{2t} + 2e^{3t} + 3e^{4t} + 4e^{5t} + 4e^{6t} + 4e^{7t} + 3e^{8t} + 2e^{9t} + e^{10t})$;

   (b)
$$g(w) = \begin{cases} 1/24, & w = 2, 10, \\ 2/24, & w = 3, 9, \\ 3/24, & w = 4, 8, \\ 4/24, & w = 5, 6, 7. \end{cases}$$

4.6-15 (a) $g(w) = 1/8$, $w = 0, 1, 2, \ldots, 7$;

   (b) $h(w) = 1/16$, $w = 0, 1, 2, \ldots, 15$.

4.6-17 (a)
$$h_1(w_1) = \begin{cases} 1/36, & w_1 = 0, \\ 4/36, & w_1 = 1, \\ 10/36, & w_1 = 2, \\ 12/36, & w_1 = 3, \\ 9/36, & w_1 = 4. \end{cases}$$

   (b) $h_2(w) = h_1(w)$;

   (c)
$$h(w) = \begin{cases} 1/1296, & w = 0 \\ 8/1296, & w = 1, \\ 36/1296, & w = 2, \\ 104/1296, & w = 3, \\ 214/1296, & w = 4, \\ 312/1296, & w = 5, \\ 324/1296, & w = 6, \\ 216/1296, & w = 7, \\ 81/1296, & w = 8. \end{cases}$$

4.6-19 (b) $\mu_Y = 25/3$, $\sigma_Y^2 = 130/9$;

   (c)
$$P(Y = y) = \begin{cases} 96/1024, & y = 4, \\ 144/1024, & y = 5, \\ 150/1024, & y = 6, \\ 135/1024, & y = 7. \end{cases}$$

4.6-21 $Y - X + 25$ is $b(50, \ 1/2)$;

$$P(Y - X \geq 2) = \sum_{k=27}^{50} \binom{50}{k} \left(\frac{1}{2}\right)^{50} = 0.3359.$$

4.6-23 (a) $\sum_{k=3}^{10} \binom{10}{k}(1 - 2/e)^k (2/e)^{10-k} = 0.5167$;

   (b) 0.792.

4.6-25 $1 - 17/2e^3 = 0.5678$.

4.6-27 21,816.

4.7-1 (a) 0.84; **(b)** 0.082.

4.7-3 $k = 1.464$; 8/15.

4.7-5 (a) 0.25; **(b)** 0.85; **(c)** 0.925.

4.7-7 675.

CHAPTER 5

5.2-1 **(a)** 0.2784; **(b)** 0.7209; **(c)** 0.9616; **(d)** 0.0019;
(e) 0.9500; **(f)** 0.6826; **(g)** 0.9544; **(h)** 0.9974.

5.2-3 **(a)** 2.326; **(b)** −2.576; **(c)** 1.67; **(d)** −2.17.

5.2-5 **(a)** 1.96; **(b)** 1.96; **(c)** 1.645; **(d)** 1.645.

5.2-7 **(a)** 0.3849; **(b)** 0.5403; **(c)** 0.0603; **(d)** 0.0013;
(e) 0.6826; **(f)** 0.9544; **(g)** 0.9974; **(h)** 0.99.

5.2-9 **(a)** 0.6326; **(b)** 50.

5.2-11 **(a)** Gamma ($\alpha = 1/2$, $\theta = 8$); **(b)** Gamma ($\alpha = 1/2$, $\theta = 2\sigma^2$).

5.2-13 X could be normal because of the linearity of the q-q plot.

5.2-15 **(a)**

Stems	Leaves	Frequency	Depths
7●	938	1	1
8*	032 089	2	3
8t	222 268 383	3	6
8f	442 490 528 572	4	(4)
8s	674 734 786	3	9
8●	850 873 920	3	6
9*	069 150	2	3
9t	243	1	1

(Multiply numbers by 10^{-4}.)

(b) Normal quantiles: −1.645, −1.282, −1.036, −0.842, −0.674, −0.524, −0.385, −0.253, −0.126, 0, 0.126, 0.253, 0.385, 0.524, 0.674, 0.842, 1.036, 1.282, 1.645.

(c) Yes, because of the linearity of the q-q plot.

5.2-17   0.025.

5.2-19 **(a)** 0.0228; **(b)** 0.8904.

5.2-21 **(a)** The frequencies of 450, 451, 452, and so on are 2, 3, 3, 5, 6, 15, 19, 22, 11, 5, 5, 3, 0, 1; **(b)** yes.

5.2-23 **(a)** $\sigma = 0.043$; **(b)** $\mu = 12.116$.

5.2-25 **(a)** 0.0228; **(b)** 0.6844.

5.2-27 **(a)** $N(0, 1)$; **(b)** $N(-1, 1)$; **(c)** $N(2, 1)$.

5.3-1 **(a)** 0.4772; **(b)** 0.8561.

5.3-3 **(a)** 46.58, 2.56; **(b)** 0.8447.

5.3-5   0.925.

5.3-7 **(b)** 0.05466; 0.3102.

5.3-9 **(a)** 0.8962; **(b)** 0.8962; **(c)** 87.31; 229.58.

5.3-11   0.8997.

5.3-13 **(a)** 0.3085; **(b)** 0.2267.

5.3-15   25.

5.3-17 $E(Z) = 0; E(Z^2) = 1; E(1/\sqrt{U}) = \Gamma[(r-1)/2]/\{\}\Gamma[r/2]\sqrt{2}\};$
$E(1/U) = 1/(r-2).$

5.3-19 $0.8413 > 0.7734$, select X.

5.4-1   0.4772.

5.4-3   0.8185.

5.4-5 **(a)** $\chi^2(18)$; **(b)** 0.0756, 0.9974.

5.4-7   0.6247.

5.4-9   0.95.

5.4-11 **(a)** The frequencies are 5, 6, 12, 18, 31, 31, 20, 17, 6, 4; the fifth and sixth classes; **(c)** yes.

5.4-13 $ 444,338.13.

5.4-15   0.9522.

5.4-17 **(a)** $\displaystyle\int_0^{25} \frac{1}{\Gamma(13)2^{13}} y^{13-1} e^{-y/2} \, dy = 0.4810$ (using *Maple*);

(b) 0.4449 using normal approximation.

5.5-1 **(a)** 0.2878, 0.2881; **(b)** 0.4428, 0.4435; **(c)** 0.1550, 0.1554.

5.5-3   0.9258 using normal approximation, 0.9258 using binomial.

5.5-5   0.6915 using normal approximation, 0.7030 using binomial.

5.5-7   0.3085.

5.5-9   0.6247 using normal approximation, 0.6148 using Poisson.

5.5-11 **(a)** 0.5548; **(b)** 0.3823.

5.5-13   0.6813 using normal approximation, 0.6788 using binomial.

5.5-15 **(a)** 0.3802; **(b)** 0.7571.

5.5-17   0.4734 using normal approximation, 0.4749 using Poisson
approximation with $\lambda = 50$, 0.4769 using $b(5000, \ 0.01)$.

5.5-19   0.6455 using normal approximation, 0.6449 using Poisson.

5.5-21 **(a)** 0.8289 using normal approximation, 0.8294 using Poisson.

(b) 0.0261 using tables in book, 0.0218 using *Maple*.

5.6-1 **(a)** 0.6006; **(b)** 0.7888; **(c)** 0.8185; **(d)** 0.9371.

5.6-3 **(a)** 0.5746; **(b)** 0.7357.

5.6-5 **(a)** $N(86.4, 40.96)$; **(b)** 0.4192.

5.6-7 **(a)** 0.8248;

(b) $E(Y \mid x) = 457.1735 - 0.2655x$;

(c) $\text{Var}(Y \mid x) = 645.9375$;

(d) 0.8079.

5.6-9 $a(x) = x - 11, \ b(x) = x + 5.$

5.6-11 (a) 0.2857; **(b)** $\mu_{Y|x} = -0.2x + 4.7$;
  (c) $\sigma^2_{Y|x} = 8.0784$; **(d)** 0.4230.
5.7-1 (a) 0.9984; **(b)** 0.998.
5.7-3
$$M(t) = \left[1 - \frac{2t\sigma^2}{n-1}\right]^{-(n-1)/2} \rightarrow e^{\sigma^2 t}.$$

CHAPTER 6

6.1-1 (a) 394/7;
  (b) 5,452/97.
6.1-3 (a) $\bar{x} = 8.0489$, $\bar{y} = 8.0700$, $s_x = 0.0372$, $s_y = 0.0224$;
  (b) 8.00, 8.02, 8.04, 0.085, 8.11; 8.04, 0.05, 0.07, 0.09, 0.10.
6.1-5 (a) Baby carrots: 1.02, 1.03, 1.04, 1.04, 1.06;
    Regular size carrots: 1.00, 1.15, 1.21, 1.26, 1.43;
  (c) Regular size packages tend to be heavier.
6.1-7 (a)

Stems	Leaves	Frequency	Depths
127	8	1	1
128	8	1	2
129	5 8 9	3	5
130	8	1	6
131	2 3 4 4 5 5 7	7	(7)
132	2 7 7 8	4	7
133	7 9	2	3
134	8	1	1

  (Multiply numbers by 10^{-1}.)
  (b) 131.3, 7.0, 2.575, 131.45, 131.47, 3.034.
6.1-9 (a) 1, 6.75, 32, 90.75, 527;
  (b) 84, 216.75, 342.75.
6.1-11 (a)

Stems	Leaves	Frequency	Depths
1	8 9 9	3	3
2	0 3 5 5	4	7
3		0	7
4	4 5 5	3	10
5	3 5 8	3	13
6	5	1	14
7	3 5 7	3	14
8	0 4	2	11
9	3 8	2	9
10	4 8 8	3	7
11	2 5 6	3	4
12	8	1	1

(b) 18, 29.75, 69, 102.50, 128;

(c) no.

6.2-3 **(b)** $\bar{x} = 89/40 = 2.225$.

6.2-5 **(a)** $\hat{\theta} = \bar{X}/2$; **(b)** $\hat{\theta} = \bar{X}/3$; **(c)** $\hat{\theta}$ equals the sample median.

6.2-7 **(c) (i)** $\hat{\theta} = 0.5493, \tilde{\theta} = 0.5975$,

(ii) $\hat{\theta} = 2.2101, \tilde{\theta} = 2.4004$,

(iii) $\hat{\theta} = 0.9588, \tilde{\theta} = 0.8646$.

6.2-9 **(c)** $\bar{x} = 3.48$.

6.2-11 $1/9.5 = 20/190 = 0.1053$.

6.2-15 **(a)** $\tilde{\theta} = \bar{X}$; **(b)** yes; **(c)** 7.382; **(d)** 7.485.

6.2-17 **(a)** $\tilde{\lambda} = \bar{X}$;

(b) $\bar{x} = 1.17$;

(c) $s^2 = 1.24$, yes.

6.2-19 7.2.

6.3-1 **(a)** $f(x; p) = \exp\{x \ln(1 - p) + \ln[p/(1 - p)]\}$;

$K(x) = x;$ $\sum_{i=1}^{n} X_i$ is sufficient.

(b) $\bar{X}$.

6.3-3 **(a)** $\sum_{i=1}^{n} X_i^2$; **(b)** $\widehat{\sigma^2} = \left(\dfrac{1}{n}\right)\sum_{i=1}^{n} X_i^2$; **(c)** yes.

6.4-1 $[71.35, 76.25]$.

6.4-3 $[2.03, 2.15]$.

6.4-5 $[48.467, 72.266]$ or $[48.076, 72.657]$.

6.4-7 $[19.47, 22.33]$.

6.4-9 **(a)** 3.24; **(b)** $s^2 = 0.2372, s = 0.487$; **(c)** $[2.99, 3.50]$.

6.4-11 **(a)** 14.72, 1.381; **(b)** $[14.292, 15.148]$.

6.4-13 $\bar{x} = 1.3347, s = 0.0630, [1.3151, 1.3542]$.

6.4-15 $[22.74, 25.48]$.

6.4-17 $[0.757, \infty)$.

6.4-19 **(a)** 29.49, 3.41; **(b)** $[0, 31.259]$;

(c) yes, because of the linearity of the q-q plot of the data and the corresponding normal quantiles.

6.4-21 **(a)** $\bar{x} = 25.475, s = 2.4935$; **(b)** $[24.059, \infty)$.

6.5-1 $[-59.725, -43.275]$.

6.5-3 $[-5.845, 0.845]$.

6.5-5 $[-\infty, -1.828]$.

6.5-7 (a) Yes; **(b)** [11.5, 13.7]; **(d)** do not change.

6.5-9 (a) $\bar{d} = 0.450$; **(b)** [−0.494, 1.394].

6.5-11 (a) [−0.556, 1.450];

   (b) [0.367, 1.863];

   (c) no for men, yes for women.

6.5-13   [157.227, ∞).

6.6-1 (a) $s = 6.144$; **(b)** [4.406, 10.142] or [4.107, 9.521].

6.6-3 (a) $\bar{x} = 273.04$, $s^2 = 3{,}155.54$; **(b)** [2,079.43, 5,468.08];

   (c) [45.60, 73.95]; **(d)** [44.02, 71.56]; **(e)** yes.

6.6-5 $$\left[\frac{\sum_{i=1}^{n} (X_i - \mu)^2}{\chi_{\alpha/2}^2(n)} , \frac{\sum_{i=1}^{n} (X_i - \mu)^2}{\chi_{1-\alpha/2}^2(n)} \right].$$

6.6-7 (a) [130.268, 364.503]; **(b)** $\bar{x} = 204$, $s = 195.45$, yes.

6.6-9 (a) 0.5688; **(b)** [0, 1.905], yes.

6.6-11 (a) 0.4987; **(b)** [0, 1.835].

6.6-13 $g(a) = g(b)$ and $a = -b$; so $a = -t_{\alpha/2}(n-1) = -b$.

6.6-15 (a) [0.041, 0.599]; **(b)** it is questionable.

6.7-1 (a) 0.0374; **(b)** [0.0227, 0.0521]; **(c)** [0.0252, 0.0550]; **(d)** [0, 0.0497].

6.7-3 (a) 0.69; **(b)** [0.6613, 0.7187].

6.7-5 (a) 0.506; **(b)** [0.461, 0.551]; **(c)** not necessarily.

6.7-7   [0.074, 0.286].

6.7-9 (a) 0.680; **(b)** [0.646, 0.714].

6.7-11 (a) [0.445, 0.515]; **(b)** 91.04%.

6.7-13   [0.207, 0.253].

6.7-15 (a) 0.2115; **(b)** [0.155, 0.268].

6.7-17 (a) 0.2182; **(b)** [0.1756, 1].

6.8-1   117.

6.8-3 (a) 1083; **(b)** [6.047, 6.049]; **(c)** \$58,800; **(d)** 0.0145.

6.8-5 (a) 257; **(b)** yes.

6.8-7 (a) 1068; **(b)** 2401; **(c)** 752.

6.8-9   2305.

6.8-11   451.

6.8-13 (a) 38; **(b)** [0.621, 0.845].

6.8-15   601.

6.8-17   235.

6.9-1 (a) $y_1 = 5$, $y_2 = 6$, $y_3 = 7$, $y_4 = 9, \ldots, y_{10} = 78$;

(b) $\tilde{m} = 31.5, \tilde{\pi}_{0.80} = 72.6$;

(c) $\tilde{q}_1 = 6.75, \tilde{q}_3 = 68.75$.

6.9-3 **(a)** $g_3(y) = 10(1 - e^{-y/3})^2 e^{-y}, \ 0 < y < \infty$;

(b) $5(1 - e^{-5/3})^4 e^{-5/3} + (1 - e^{-5/3})^5 = 0.7599$;

(c) $e^{-5/3} = 0.1889$.

6.9-5 **(a)** 0.2553; **(b)** 0.7483.

6.9-7 **(a)** $g_1(y) = 19(e^{-y/\theta})^{18} \dfrac{1}{\theta} e^{-y/\theta}, \ 0 < y < \infty$;

(b) 1/20.

6.9-9 **(a)** $L(\theta) = \exp[-\sum_{i=1}^{10}(x_i - \theta)]$ is a maximum when θ is as large as possible. Thus $\hat{\theta} = Y_1$.

(b) $g_1(y_1) = 10e^{-10(y_1 - \theta)}, \ \theta \leq y_1 < \infty$.

(c) $[y_1 - (1/10)\ln(20), \ y_1]$.

6.10-1 **(a)** 0.7812; **(b)** 0.7844; **(c)** 0.4528.

6.10-3 **(a)** (6.31, 7.40); **(b)** (6.58, 7.22), 0.8204.

6.10-5   (2.72, 2.84), 93.46%.

6.10-7   (15.40, 17.05).

6.10-9 **(a)**

Stems	Leaves	Frequency	Depths
0	09 11 13 24 34 51 57 75 79 98	10	10
1	36 52 61 98	4	(4)
2	15 38 44 72 88	5	11
3	15 15 50	3	6
4	45	1	3
5		0	2
6	20	1	2
7	23	1	1

(b) $\tilde{\pi}_{0.25} = 54, \tilde{m} = 161, \tilde{\pi}_{0.75} = 301.5$;

(c) **(i)** (13, 98), 89.66%; **(ii)** (79, 244), 89.22%; **(iii)** (238, 455), 89.66%.

(d) [142.7, 267.1], 89.22%; [141.2, 268.6], 90%; they are quite different because data are skewed (exponential) rather than normal.

6.10-11 **(a)** $\tilde{\pi}_{0.50} = \tilde{m} = 0.92$;

(b) $(y_{41}, y_{60}) = (0.92, 0.93)$;

0.9426 using normal approximation, 0.9431 using binomial;

(c) $\tilde{\pi}_{0.25} = 0.89$;

(d) $(y_{17}, y_{34}) = (0.88, 0.90)$;

0.9504 using normal approximation, 0.9513 using binomial;

(e) $\widetilde{\pi}_{0.75} = 0.97$;

(f) $(y_{67}, y_{84}) = (0.95, 0.98)$;

0.9504 using normal approximation, 0.9513 using binomial.

6.10-13 $y_{11} = 370 < \pi_{0.25} < y_{25} = 405, 94.81\%$;

$y_{27} = 409 < \pi_{0.5} < y_{44} = 435, 95.86\%$;

$y_{46} = 437 < \pi_{0.75} < y_{60} = 472, 94.81\%$.

6.10-15 (a) 14.75; **(b)** 0.8995.

6.11-1 (a) $\widehat{y} = 86.8 + (842/829)(x - 74.5)$;

(c) $\widehat{\sigma^2} = 17.9998$.

6.11-3 (a) $\widehat{y} = 6.478 + 4.483x$.

6.11-5 (a) $\widehat{y} = 10.6 - 0.015x$;

(c) $\widehat{y} = 5.47 + 0.0004x$;

(e) horse power.

6.11-7   Solve for α: $-t_{\alpha/2}(n-2) \leq \dfrac{\widehat{\alpha} - \alpha}{\sqrt{\widehat{\sigma^2}/(n-1)}} \leq t_{\alpha/2}(n-2)$.

6.11-9 $[83.341, 90.259]$, $[0.478, 1.553]$, $[10.265, 82.578]$.

6.11-11 (a) $\widehat{y} = 15.695 + 0.620x$;

(c) $\widehat{\sigma^2} = 0.2595$;

(d) $[0.4416, 0.7984]$.

6.11-13 (a) $\widehat{y} = 1.896 + 0.538x$;

(c) $\widehat{\alpha} = 6.9667, \widehat{\beta} = 0.5385, \widehat{\sigma^2} = 0.0549$;

(d) $[6.8263, 7.1070]$, $[0.2081, 0.8688]$, $[0.0333, 0.1643]$.

6.11-15 (a) $\widehat{y} = 1.49 + 0.0597x$;

(c) $[2.5830, 3.2059]$, $[-0.0166, 0.1359]$, $[0.2156, 0.9002]$.

6.12-1 (a) $[75.283, 85.113]$, $[83.838, 90.777]$, $[89.107, 99.728]$;

(b) $[68.206, 92.190]$, $[75.833, 98.783]$, $[82.258, 106.577]$.

6.12-3 (a) $[62.793, 129.495]$, $[101.854, 135.266]$, $[119.483, 162.470]$;

(b) $[28.926, 163.361]$, $[57.856, 179.265]$, $[78.784, 203.169]$.

6.12-5 (a) $[4.897, 8.444]$, $[9.464, 12.068]$, $[12.718, 17.004]$;

(b) $[1.899, 11.442]$, $[6.149, 15.383]$, $[9.940, 19.782]$.

6.12-7 (a) $[19.669, 26.856]$, $[22.122, 27.441]$, $[24.048, 28.551]$,
$[25.191, 30.445]$, $[25.791, 32.882]$;

(b) $[15.530, 30.996]$, $[17.306, 32.256]$, $[18.915, 33.684]$,
$[20.351, 35.285]$, $[21.618, 37.055]$.

6.12-9 $\widehat{y} = 1.1037 + 2.0327x - 0.2974x^2 + 0.6204x^3$.

6.12-13 (a) $r = 0.143$;

(b) $\hat{y} = 37.68 + 0.83x$;

(d) no;

(e) $\hat{y} = 12.845 + 22.566x - 3.218x^2$;

(f) yes.

6.12-15 $\hat{y} = 1.735042 - 0.000377x + 0.000124x^2$, yes.

6.14-1 **(b)** σ^2/n; **(c)** 1/5.

6.14-3 **(a)** $2\theta^2/n$; **(b)** $N(\theta, 2\theta^2/n)$;

      (c) $\chi^2(n)$.

CHAPTER 7

7.1-3 $3800(10/38) = 1000$; 307.79.

7.1-5   2.67, 8, 16, 21.33, 64.

7.2-1 **(a)** $k(\theta \mid y) \propto \theta^{\alpha+y-1} e^{-\theta(n+1/\beta)}$.

      Thus the posterior p.d.f. of θ is gamma with parameters $\alpha + y$ and $1/(n + 1/\beta)$.

    (b) $w(y) = E(\theta \mid y) = (\alpha + y)/(n + 1/\beta)$.

    (c) $w(y) = \left(\dfrac{y}{n}\right)\left(\dfrac{n}{n + 1/\beta}\right) + (\alpha\beta)\left(\dfrac{1/\beta}{n + 1/\beta}\right)$.

7.2-3 **(a)** $E[\{w(Y) - \theta\}^2] = \{E[w(Y) - \theta]\}^2 + \mathrm{Var}[w(Y)]$
                    $= (74\theta^2 - 114\theta + 45)/500$;

    (b) $\theta = 0.569$ to $\theta = 0.872$.

7.3-5 **(c)** $2yz$ is $\chi^2(2n)$; $\left(\dfrac{2y}{\chi^2_{\alpha/2}(2n)}, \dfrac{2y}{\chi^2_{1-\alpha/2}(2n)}\right)$.

CHAPTER 8

8.1-1 $\alpha = 7/27$; $\beta = 7/27$.

8.1-3 **(a)** 0.3032 using $b(100, 0.08)$, 0.313 using Poisson approximation,
        0.2902 using normal approximation;

    (b) 0.1064 using $b(100, 0.04)$, 0.111 using Poisson approximation,
        0.1010 using normal approximation.

8.1-5 **(a)** $\alpha = 0.1056$; **(b)** $\beta = 0.3524$.

8.1-7 **(a)** $z = 2.266 > 1.645$, reject H_0;

    (b) $z = 2.266 < 2.326$, do not reject H_0;

    (c) p-value $= 0.0117$.

8.1-9 **(a)** $z = 1.752 > 1.645$, reject H_0;

    (b) $z = 1.752 < 1.96$, do not reject H_0;

    (c) p-value $= 0.0399$.

8.1-11 (a) $z = \dfrac{y/n - 0.40}{\sqrt{(0.40)(0.60)/n}} \geq 1.645$;

    (b) $z = 2.215 > 1.645$, reject H_0.

8.1-13 (a) H_0: $p = 0.40$, H_1: $p > 0.40$, $z \geq 2.326$;

    (b) $2.236 < 2.326$, do not reject H_0.

8.1-15 (a) H_0: $p = 0.037$, H_1: $p > 0.037$;

    (b) $z \geq 2.326$;

    (c) $z = 2.722 > 2.326$, reject H_0.

8.1-17 (a) $|z| \geq 1.960$;

    (b) $1.726 < 1.960$, do not reject H_0.

8.1-19   [0.007, 0.071], yes.

8.2-1 (a) $1.4 < 1.645$, do not reject H_0;

    (b) $1.4 > 1.282$, reject H_0.

    (c) p-value $= 0.0808$.

8.2-3 (a) $z = (\bar{x} - 170)/2$, $z \geq 1.645$;

    (b) $1.260 < 1.645$, do not reject H_0;

    (c) 0.1038.

8.2-5 (a) $t = (\bar{x} - 3{,}315)/(s/\sqrt{30}) \leq -1.699$;

    (b) $-1.414 > -1.699$, do not reject H_0;

    (c) $0.05 < p$-value < 0.10 or p-value ≈ 0.08.

8.2-7 (a) $t = (\bar{x} - 47)/(s/\sqrt{20}) \leq -1.729$;

    (b) $-1.789 < -1.729$, reject H_0;

    (c) $0.025 < p$-value < 0.05, p-value ≈ 0.045.

8.2-9 (a) $t \geq 2.764$;

    (b) $4.028 > 2.764$, reject H_0;

    (c) p-value < 0.005;

    (d) $\chi^2 \leq 3.940$;

    (e) $4.223 > 3.940$, do not reject H_0;

    (f) $0.05 < p$-value < 0.10.

8.2-11 (a) $\chi^2 = 24s^2/140^2$, $\chi^2 \leq 36.42$;

    (b) $29.18 < 36.42$, do not reject H_0;

    (c) [0, 725.432].

8.2-13 (a) $59.53 < 64.28$, reject H_0;

    (b) $0.025 < p$-value < 0.05.

8.2-15 (a) -4.60, p-value < 0.0001;

    (b) clearly reject H_0;

    (c) [0, 14.573].

8.2-17 $1.607 < 1.645$, do not reject H_0.

8.2-19 $1.477 < 1.833$, do not reject H_0.

8.2-21 $\bar{d} = 0.357$, $s_d = 0.561$, $t = 2.985$, p-value < 0.005.

8.3-1 **(a)** (a) $t \leq -1.734$; **(b)** $t = -2.221 < -1.734$, reject H_0.

8.3-3 **(a)** $2.759 < 4.20$, do not reject H_0 at $\alpha = 0.05$;

    (b) $|t| = 0.374 < 2.086$, do not reject H_0 at $\alpha = 0.05$.

8.3-5 **(a)** $t < -1.706$; **(b)** $-1.714 < -1.706$, reject H_0;

    (c) $0.025 < p$-value < 0.05; **(e)** $1.836 < 3.28$, do not reject.

8.3-7 **(a)** $t < -2.552$; **(b)** $t = -3.638 < -2.552$, reject H_0.

8.3-9 **(a)** $z > 1.645$; **(b)** $1.957 > 1.645$, reject H_0; **(c)** p-value $= 0.0252$.

8.3-11 **(a)** $t = 1.20$, $0.20 < p$-value < 0.50, p-value $= 0.246$, do not reject H_0;

    (b) $F = 1.624$, do not reject H_0.

8.3-13 **(a)** $t = -1.67$, $0.05 < p$-value < 0.10, p-value $= 0.054$, fail to reject H_0.

8.3-15 **(a)** $z = 2.245 > 1.645$, reject H_0; **(b)** p-value $= 0.0124$.

8.3-17 $F = 3.30 > 3.07$, reject $\sigma_X^2 = \sigma_Y^2$;
    Welch's test: $t = -8.119 < -t_{0.05}(12) = -1.782$.

8.3-19 $F = 1.895 < 3.37$, do not reject H_0.

8.4-1 **(a)** $-55 < -47.08$, reject H_0; **(b)** 0.0296;

    (c) $9 < 10$, do not reject H_0; **(d)** p-value $= 0.1334$.

8.4-3 **(a)** $y = 17$, p-value $= 0.0539$;

    (b) $w = 171$, p-value $= 0.0111$;

    (c) $t = 2.608$, p-value $= 0.0077$.

8.4-5 $w = 54$, $z = 1.533$, p-value $= 0.0661$, do not reject H_0.

8.4-7 **(a)** Let Y equal the number of weights less that 1.14 pounds.
    Then $C = \{y: \ y \leq 4\}$, $\alpha = 0.0898$.

    (b) $y = 5$, fail to reject H_0;

    (c) p-value $= 0.2120$;

    (d) $z = 2.072 > 1.282$, p-value $= 0.0207$, reject H_0.

8.4-9 **(a)** $w = 145 > 126$ or $z = 3.024 > 1.645$, reject H_0; p-value ≈ 0.0014.

8.4-11 **(a)** $C = \{w: \ w \leq 79 \ \text{or} \ w \geq 131\}$, $\alpha \approx 0.0539$, $w = 95$, do not reject H_0.

8.4-13 **(a)** $C = \{w: \ w \leq 79 \ \text{or} \ w \geq 131\}$, $\alpha \approx 0.0539$, $w = 107.5$,
    p-value $= 0.8798$, do not reject H_0.

8.4-15 $C = \{w: \ w \leq 184 \ \text{or} \ w \geq 280\}$, $\alpha \approx 0.0489$, $w = 241$, do not reject H_0.

8.5-1 $6.25 < 7.815$, do not reject if $\alpha = 0.05$; p-value ≈ 0.10.

8.5-3 $7.60 < 16.92$, do not reject H_0.

8.5-5 $3.65 < 7.815$, do not reject H_0.

8.5-7 **(a)** $q_3 \geq 7.815$;

    (b) $q_3 = 1.744 < 7.815$; do not reject H_0.

8.5-9 $6.113 < 16.92$, do not reject H_0.

8.5-11 $2.010 < 11.07 = \chi^2_{0.05}(5)$, do not reject null hypothesis.

8.5-13 (a) Frequencies are 2, 4, 21, 33, 20, 14, 4, 1, 1;

   (b) $q = 4.653 < 9.488$, do not reject H_0.

8.6-1 $3.23 < 11.07$, do not reject H_0.

8.6-3 $2.40 < 5.991$, do not reject H_0.

8.6-5 $5.975 < \chi^2_{0.05}(2) = 5.991$, do not reject the null hypothesis; however, p-value ≈ 0.05.

8.6-7 $8.449 < \chi^2_{0.05}(4) = 9.488$, do not reject the null hypothesis; $0.05 < p$-value < 0.10; p-value $= 0.076$.

8.6-9 $4.149 > \chi^2_{0.05}(1) = 3.841$, reject the null hypothesis; $0.025 < p$-value < 0.05; p-value ≈ 0.042.

8.6-11 $23.78 > 21.03$, reject hypothesis of independence.

8.6-13 (a) $39.591 > 9.488$, reject hypothesis of independence;

   (b) $7.117 > 5.991$, reject hypothesis of independence;

   (c) $11.398 > 9.488$, reject hypothesis of independence.

8.6-15 $9.925 < 12.59 = \chi^2_{0.05}(6)$, do not reject independence.

8.7-1 $7.875 > 4.26$, reject H_0.

8.7-3 $13.773 > 4.07$, reject H_0.

8.7-5 $5.171 > 3.25$, reject H_0.

8.7-7 $14.757 > 4.43$, reject H_0.

8.7-9 (a) $F \geq 4.07$;

   (b) $4.106 > 4.07$, reject H_0;

   (c) $4.106 < 5.42$, do not reject H_0;

   (d) $0.025 < p$-value < 0.05, p-value ≈ 0.05.

8.7-11 $10.224 > 4.26$, reject H_0.

8.7-13 (a) $F \geq 5.61$;

   (b) $6.337 > 5.61$; reject H_0.

8.8-1 $18.00 > 5.14$, reject H_A.

8.8-3 (a) $7.624 > 4.46$, reject H_A;

   (b) $15.538 > 3.84$, reject H_B.

8.8-5 (a) $1.723 < 2.90$, accept H_{AB};

   (b) $5.533 > 4.15$, reject H_A;

   (c) $28.645 > 2.90$, reject H_B.

8.8-7 (a) $1.727 < 2.37$, accept H_{AB};

   (b) $2.238 < 3.27$, do not reject H_A;

   (c) $2.063 < 2.87$, do not reject H_B.

8.8-9 (a) $0.111 < 3.89$, accept H_{AB};

(b) $9.692 > 4.75$, reject H_A;

(c) $2.606 < 3.89$, do not reject H_B.

8.9-1 $4.359 > 2.306$, reject H_0.

8.9-3 $2.148 < 2.160$, do not reject H_0.

8.9-5 $-0.45 < -0.3808$, reject H_0.

8.9-7 $[0.419, 0.802]$.

8.9-9 $|r| = 0.252 < 0.6613$, do not reject H_0.

8.9-13 $n = 9$.

8.9-15 $r = 0.383 < 0.4000 = r_{0.05}(16)$ so fail to reject H_0.

8.10-1 $d_5 = 0.40 < 0.45$, do not reject H_0.

8.10-3 $d_{10} = 0.1643$ at $x = 49$, do not reject H_0.

8.10-5 (a) $\bar{x} = 1.3347$, $s^2 = 0.00397$;

(b) $d_{30} = 0.0738$ at $x = 1.28$, do not reject H_0.

8.10-7 $d_{40} = 0.04559$, do not reject H_0.

8.10-9 (a) $d_{70} = 0.1295 > 1.07/\sqrt{70} = 0.1279$, reject H_0;

(b) yes, the histogram is bimodal.

8.10-11 $d_{105} = 0.07$ at $x = 6$; fit is very good.

8.10-13 $\widehat{\theta} = 0.28588$, $\widehat{\delta} = -0.41588$, $d_{18} = 0.0873$ at $x = -0.31$.

8.11-1 (a) $C = \{r: r \leq 6\}$, $\alpha = 43/429$, $r = 6$, reject H_0;

(b) $w = 95$, p-value ≈ 0.0027, clearly reject H_0.

8.11-3 $(6!)(12!)\binom{11}{4} = 113,810,780,160,000.$

8.11-5 (b) $r = 11$, reject H_0; (c) p-value $= 0.0457$.

8.11-7 $r = 4$, p-value $= 43/1,716 = 0.025$, reject the hypothesis of randomness. There appears to be a trend effect.

8.11-9 $r = 7$, p-value $= 1,331/58,786 = 0.0226$, reject the hypothesis of randomness.

8.11-11 (a) $C = \{y: y \leq 3 \text{ or } y \geq 11\}$, $y = 7$, do not reject H_0;

(b) $C = \{r: r \leq 4 \text{ or } r \geq 12\}$, $\alpha = 43/858 = 0.05$, $r = 4$,
reject the hypothesis of randomness. There appears to be a trend effect.

8.11-13 $C = \{r: r \leq 10\}$, $\alpha = 919/22,287 = 0.041$,
$r = 6$, p-value $= 1,133/1,002,915 = 0.0011$, reject H_0.

CHAPTER 9

9.1-1 (a) $K(\mu) = \Phi\left(\dfrac{22.5 - \mu}{3/2}\right)$; $\alpha = 0.0478$;

(b) $\bar{x} = 24.1225 > 22.5$, do not reject H_0;

(c) 0.2793.

9.1-3 (a) $K(\mu) = \Phi\left(\dfrac{510.77 - \mu}{15}\right)$;

(b) $\alpha = 0.10$;

(c) 0.5000;

(e) (i) 0.0655, **(ii)** 0.0150.

9.1-5 $n = 25, c = 1.6$.

9.1-7 $n = 40, c = 678.38$.

9.1-9 (a) $K(p) = \sum_{y=14}^{25} \binom{25}{y} p^y (1-p)^{25-y}, \ 0.40 \le p \le 1.0$;

(b) $\alpha = 0.0778$;

(c) 0.1827, 0.3450, 0.7323, 0.9558, 0.9985, 1.0000;

(e) yes;

(f) 0.0344.

9.1-11   With $n = 130, c = 8.5, \alpha \approx 0.055, \beta \approx 0.094$.

9.2-1 (a) $\dfrac{L(80)}{L(76)} = \exp\left[\dfrac{6}{128}\sum_{i=1}^{n} x_i - \dfrac{624n}{128}\right] \le k \ \text{ or } \ \overline{x} \le c$;

(b) $n = 43, c = 78$.

9.2-3 (a) $\dfrac{L(3)}{L(5)} \le k$ if and only if $\displaystyle\sum_{i=1}^{n} x_i \ge (-15/2)[\ln(k) - \ln(5/3)^n] = c$;

(b) $\overline{x} \ge 4.15$;

(c) $\overline{x} \ge 4.15$;

(d) yes.

9.2-5 (a) $\dfrac{L(50)}{L(\mu_1)} \le k$ if and only if $\overline{x} \le \dfrac{(-72)\ln(k)}{2n(\mu_1 - 50)} + \dfrac{50 + \mu_1}{2} = c$.

9.2-7 (a) $\dfrac{L(0.5)}{L(\mu)} \le k$ if and only if $\displaystyle\sum_{i=1}^{n} x_i \ge \dfrac{\ln(k) + n(0.05 - \mu)}{\ln(0.5/\mu)} = c$;

(b) $\displaystyle\sum_{i=1}^{10} x_i \ge 9$.

9.3-1 (a) $|-1.80| > 1.645$, reject H_0;

(b) $|-1.80| < 1.96$, do not reject H_0;

(c) p-value $= 0.0718$.

9.3-3 (a) $\overline{x} \ge 230 + 10z_\alpha/\sqrt{n}$ or $\dfrac{\overline{x} - 230}{10/\sqrt{n}} \ge z_\alpha$;

(b) yes; **(c)** $1.04 < 1.282$, do not reject H_0; **(d)** p-value $= 0.1492$.

9.3-5 (a) $|2.10| < 2.306$, do not reject H_0; $0.05 < p$-value < 0.10.

9.3-7 $2.20 > 1.282$, reject H_0; p-value $= 0.0139$.

CHAPTER 10

10.2-1 (a) 158.97, 12.1525, 30.55; **(f)** yes.

10.2-3 (a) 5.176(335.176), 0.5214, 1.294; **(f)** no.

10.2-5 (b) $E(\sqrt{Y}) = \dfrac{\sqrt{2}\,\Gamma\left(\dfrac{n}{2}\right)}{\Gamma\left(\dfrac{n-1}{2}\right)}$;

(c) $S = \dfrac{\sigma\sqrt{Y}}{\sqrt{n-1}}$ so $E(S) = \dfrac{\sqrt{2}\,\Gamma\left(\dfrac{n}{2}\right)}{\sqrt{n-1}\,\Gamma\left(\dfrac{n-1}{2}\right)}\,\sigma$.

10.2-7 $UCL = 0.1024$; $LCL = 0$;

(a) 0.0378; **(b)** 0.1752.

10.2-9 $UCL = 4.9496$; $LCL = 0$;

(a) 0.185; **(b)** 0.871.

10.3-1

	2^2 Design			
Run	**A**	**B**	**AB**	**Observations**
1	$-$	$-$	$+$	X_1
2	$+$	$-$	$-$	X_2
3	$-$	$+$	$-$	X_3
4	$+$	$+$	$+$	X_4

(a) $[A] = (-X_1 + X_2 - X_3 + X_4)/4$,
$[B] = (-X_1 - X_2 + X_3 + X_4)/4$,
$[AB] = (X_1 - X_2 - X_3 + X_4)/4$.

(b) It is sufficient to compare the coefficients on both sides of the equations of X_1^2, X_1X_2, X_1X_3, and X_1X_4, which are $3/4$, $-1/2$, $-1/2$, and $-1/2$, respectively.

(c) Each is $\chi^2(1)$.

10.3-3 $[A]$ is $N(0, \sigma^2/2)$ so $E[(X_2 - X_1)^2/4] = \sigma^2/2$ or $E[(X_2 - X_1)^2/2] = \sigma^2$.

10.3-5 (a) $[A] = -4$, $[B] = 12$, $[C] = -1.125$, $[D] = -2.75$, $[AB] = 0.5$, $[AC] = 0.375$, $[AD] = 0$, $[BC] = -0.625$, $[BD] = 2.25$, $[CD] = -0.125$, $[ABC] = -0.375$, $[ABD] = 0.25$, $[ACD] = -0.125$, $[BCD] = -0.375$, $[ABCD] = -0.125$.

(b) There is clearly a temperature (B) effect. There is also a catalyst charge (A) effect and probably a concentration (D) and a temperature-concentration (BD) effect.

INDEX

Robert Hogg/Elliot Tanis
Probability and Statistical Inference, 7th Edition, Data CD
0-13-146414-0
© **2005 Pearson Education, Inc.**
Pearson Prentice Hall
Pearson Education, Inc.
Upper Saddle River, NJ 07458
Pearson Prentice Hall™ is a trademark of Pearson Education, Inc.

YOU SHOULD CAREFULLY READ THE TERMS AND CONDITIONS BEFORE USING THE CD-ROM PACKAGE. USING THIS CD-ROM PACKAGE INDICATES YOUR ACCEPTANCE OF THESE TERMS AND CONDITIONS.

Pearson Education, Inc. provides this program and licenses its use. You assume responsibility for the selection of the program to achieve your intended results, and for the installation, use, and results obtained from the program. This license extends only to use of the program in the United States or countries in which the program is marketed by authorized distributors.

LICENSE GRANT

You hereby accept a nonexclusive, nontransferable, permanent license to install and use the program ON A SINGLE COMPUTER at any given time. You may copy the program solely for backup or archival purposes in support of your use of the program on the single computer. You may not modify, translate, disassemble, decompile, or reverse engineer the program, in whole or in part.

TERM

The License is effective until terminated. Pearson Education, Inc. reserves the right to terminate this License automatically if any provision of the License is violated. You may terminate the License at any time. To terminate this License, you must return the program, including documentation, along with a written warranty stating that all copies in your possession have been returned or destroyed.

LIMITED WARRANTY

THE PROGRAM IS PROVIDED "AS IS" WITHOUT WARRANTY OF ANY KIND, EITHER EXPRESSED OR IMPLIED, INCLUDING, BUT NOT LIMITED TO, THE IMPLIED WARRANTIES OR MERCHANTABILITY AND FITNESS FOR A PARTICULAR PURPOSE. THE ENTIRE RISK AS TO THE QUALITY AND PERFORMANCE OF THE PROGRAM IS WITH YOU. SHOULD THE PROGRAM PROVE DEFECTIVE, YOU (AND NOT PEARSON EDUCATION, INC. OR ANY AUTHORIZED DEALER) ASSUME THE ENTIRE COST OF ALL NECESSARY SERVICING, REPAIR, OR CORRECTION. NO ORAL OR WRITTEN INFORMATION OR ADVICE GIVEN BY PEARSON EDUCATION, INC., ITS DEALERS, DISTRIBUTORS, OR AGENTS SHALL CREATE A WARRANTY OR INCREASE THE SCOPE OF THIS WARRANTY. SOME STATES DO NOT ALLOW THE EXCLUSION OF IMPLIED WARRANTIES, SO THE ABOVE EXCLUSION MAY NOT APPLY TO YOU. THIS WARRANTY GIVES YOU SPECIFIC LEGAL RIGHTS AND YOU MAY ALSO HAVE OTHER LEGAL RIGHTS THAT VARY FROM STATE TO STATE.

Pearson Education, Inc. does not warrant that the functions contained in the program will meet your requirements or that the operation of the program will be uninterrupted or error-free. However, Pearson Education, Inc. warrants the CD-ROM(s) on which the program is furnished to be free from defects in material and workmanship under normal use for a period of ninety (90) days from the date of delivery to you as evidenced by a copy of your receipt. The program should not be relied on as the sole basis to solve a problem whose incorrect solution could result in injury to person or property. If the program is employed in such a manner, it is at the user's own risk and Pearson Education, Inc. explicitly disclaims all liability for such misuse.

LIMITATION OF REMEDIES

Pearson Education, Inc.'s entire liability and your exclusive remedy shall be:

1. the replacement of any CD-ROM not meeting Pearson Education, Inc.'s "LIMITED WARRANTY" and that is returned to Pearson Education, or
2. if Pearson Education is unable to deliver a replacement CD-ROM that is free of defects in materials or workmanship, you may terminate this agreement by returning the program.

IN NO EVENT WILL PEARSON EDUCATION, INC. BE LIABLE TO YOU FOR ANY DAMAGES, INCLUDING ANY LOST PROFITS, LOST SAVINGS, OR OTHER INCIDENTAL OR CONSEQUENTIAL DAMAGES ARISING OUT OF THE USE OR INABILITY TO USE SUCH PROGRAM EVEN IF PEARSON EDUCATION, INC. OR AN AUTHORIZED DISTRIBUTOR HAS BEEN ADVISED OF THE POSSIBILITY OF SUCH DAMAGES, OR FOR ANY CLAIM BY ANY OTHER PARTY.SOME STATES DO NOT ALLOW FOR THE LIMITATION OR EXCLUSION OF LIABILITY FOR INCIDENTAL OR CONSEQUENTIAL DAMAGES, SO THE ABOVE LIMITATION OR EXCLUSION MAY NOT APPLY TO YOU.

GENERAL

You may not sublicense, assign, or transfer the license of the program. Any attempt to sublicense, assign or transfer any of the rights, duties, or obligations hereunder is void. This Agreement will be governed by the laws of the State of New York. Should you have any questions concerning this Agreement, you may contact Pearson Education, Inc. by writing to:

ESM Media Development
Higher Education Division
Pearson Education, Inc.
One Lake Street
Upper Saddle River, NJ 07458

Should you have any questions concerning technical support, you may write to:

New Media Production
Higher Education Division
Pearson Education, Inc.
One Lake Street
Upper Saddle River, NJ 07458

YOU ACKNOWLEDGE THAT YOU HAVE READ THIS AGREEMENT, UNDERSTAND IT, AND AGREE TO BE BOUND BY ITS TERMS AND CONDITIONS. YOU FURTHER AGREE THAT IT IS THE COMPLETE AND EXCLUSIVE STATEMENT OF THE AGREEMENT BETWEEN US THAT SUPERSEDES ANY PROPOSAL OR PRIOR AGREEMENT, ORAL OR WRITTEN, AND ANY OTHER COMMUNICATIONS BETWEEN US RELATING TO THE SUBJECTMATTER OF THIS AGREEMENT.